PERIODIC TABLE OF THE ELEMENTS

Period \ Group	1	2	3	4	5	6	7	8	9	10	11	12	13	14	15	16	17	18
1	1 H hydrogen 1.0079 $1s^1$																	2 He helium 4.00 $1s^2$
2	3 Li lithium 6.94 $2s^1$	4 Be beryllium 9.01 $2s^2$											5 B boron 10.81 $2s^22p^1$	6 C carbon 12.01 $2s^22p^2$	7 N nitrogen 14.01 $2s^22p^3$	8 O oxygen 16.00 $2s^22p^4$	9 F fluorine 19.00 $2s^22p^5$	10 Ne neon 20.18 $2s^22p^6$
3	11 Na sodium 22.99 $3s^1$	12 Mg magnesium 24.31 $3s^2$											13 Al aluminium 26.98 $3s^23p^1$	14 Si silicon 28.09 $3s^23p^2$	15 P phosphorus 30.97 $3s^23p^3$	16 S sulfur 32.06 $3s^23p^4$	17 Cl chlorine 35.45 $3s^23p^5$	18 Ar argon 39.95 $3s^23p^6$
4	19 K potassium 39.10 $4s^1$	20 Ca calcium 40.08 $4s^2$	21 Sc scandium 44.96 $3d^14s^2$	22 Ti titanium 47.87 $3d^24s^2$	23 V vanadium 50.94 $3d^34s^2$	24 Cr chromium 52.00 $3d^54s^1$	25 Mn manganese 54.94 $3d^54s^2$	26 Fe iron 55.84 $3d^64s^2$	27 Co cobalt 58.93 $3d^74s^2$	28 Ni nickel 58.69 $3d^84s^2$	29 Cu copper 63.55 $3d^{10}4s^1$	30 Zn zinc 65.41 $3d^{10}4s^2$	31 Ga gallium 69.72 $4s^24p^1$	32 Ge germanium 72.64 $4s^24p^2$	33 As arsenic 74.92 $4s^24p^3$	34 Se selenium 78.96 $4s^24p^4$	35 Br bromine 79.90 $4s^24p^5$	36 Kr krypton 83.80 $4s^24p^6$
5	37 Rb rubidium 85.47 $5s^1$	38 Sr strontium 87.62 $5s^2$	39 Y yttrium 88.91 $4d^15s^2$	40 Zr zirconium 91.22 $4d^25s^2$	41 Nb niobium 92.91 $4d^45s^1$	42 Mo molybdenum 95.94 $4d^55s^1$	43 Tc technetium (98) $4d^55s^2$	44 Ru ruthenium 101.07 $4d^75s^1$	45 Rh rhodium 102.90 $4d^85s^1$	46 Pd palladium 106.42 $4d^{10}$	47 Ag silver 107.87 $4d^{10}5s^1$	48 Cd cadmium 112.41 $4d^{10}5s^2$	49 In indium 114.82 $5s^25p^1$	50 Sn tin 118.71 $5s^25p^2$	51 Sb antimony 121.76 $5s^25p^3$	52 Te tellurium 127.60 $5s^25p^4$	53 I iodine 126.90 $5s^25p^5$	54 Xe xenon 131.29 $5s^25p^6$
6	55 Cs caesium 132.91 $6s^1$	56 Ba barium 137.33 $6s^2$	57 La lanthanum 138.91 $5d^16s^2$	72 Hf hafnium 178.49 $5d^26s^2$	73 Ta tantalum 180.95 $5d^36s^2$	74 W tungsten 183.84 $5d^46s^2$	75 Re rhenium 186.21 $5d^56s^2$	76 Os osmium 190.23 $5d^66s^2$	77 Ir iridium 192.22 $5d^76s^2$	78 Pt platinum 195.08 $5d^96s^1$	79 Au gold 196.97 $5d^{10}6s^1$	80 Hg mercury 200.59 $5d^{10}6s^2$	81 Tl thallium 204.38 $6s^26p^1$	82 Pb lead 207.2 $6s^26p^2$	83 Bi bismuth 208.98 $6s^26p^3$	84 Po polonium (209) $6s^26p^4$	85 At astatine (210) $6s^26p^5$	86 Rn radon (222) $6s^26p^6$
7	87 Fr francium (223) $7s^1$	88 Ra radium (226) $7s^2$	89 Ac actinium (227) $6d^17s^2$	104 Rf rutherfordium (261) $6d^27s^2$	105 Db dubnium (262) $6d^37s^2$	106 Sg seaborgium (263) $6d^47s^2$	107 Bh bohrium (262) $6d^57s^2$	108 Hs hassium (265) $6d^67s^2$	109 Mt meitnerium (266) $6d^77s^2$	110 Ds darmstadtium (271) $6d^87s^2$	111 Rg roentgenium (272) $6d^97s^2$	112 ? copernicium ? $6d^{10}7s^2$	113	114	115	116	117	118

Period															
6	58 Ce cerium 140.12 $4f^15d^16s^2$	59 Pr praseodymium 140.91 $4f^36s^2$	60 Nd neodymium 144.24 $4f^46s^2$	61 Pm prometheum (145) $4f^56s^2$	62 Sm samarium 150.36 $4f^66s^2$	63 Eu europium 151.96 $4f^76s^2$	64 Gd gadolinium 157.25 $4f^75d^16s^2$	65 Tb terbium 158.93 $4f^96s^2$	66 Dy dysprosium 162.50 $4f^{10}6s^2$	67 Ho holmium 164.93 $4f^{11}6s^2$	68 Er erbium 167.26 $4f^{12}6s^2$	69 Tm thulium 168.93 $4f^{13}6s^2$	70 Yb ytterbium 173.04 $4f^{14}6s^2$	71 Lu lutetium 174.97 $5d^16s^2$	Lanthanoids (lanthanides)
7	90 Th thorium 232.04 $6d^27s^2$	91 Pa protactinium 231.04 $5f^26d^17s^2$	92 U uranium 238.03 $5f^36d^17s^2$	93 Np neptunium (237) $5f^46d^17s^2$	94 Pu plutonium (244) $5f^67s^2$	95 Am americium (243) $5f^77s^2$	96 Cm curium (247) $5f^76d^17s^2$	97 Bk berkelium (247) $5f^97s^2$	98 Cf californium (251) $5f^{10}7s^2$	99 Es einsteinium (252) $5f^{11}7s^2$	100 Fm fermium (257) $5f^{12}7s^2$	101 Md mendelevium (258) $5f^{13}7s^2$	102 No nobelium (259) $5f^{14}7s^2$	103 Lr lawrencium (262) $6d^17s^2$	Actinoids (actinides)

Molar masses (atomic weights) quoted to the number of significant figures given here can be regarded as typical of most naturally occurring samples

About the cover image: the chemical connection

The cover shows a cheetah (*Acinonyx jubatus*), which can run at speeds of up to 110 kilometres per hour. Such energy-consuming processes are ultimately powered by the high-energy molecule, adenosine triphosphate (ATP). During acceleration, the cheetah uses ATP at a rate of around 55 g per second. This corresponds to a rate of consumption around 40 times greater than your current rate of consumption of ATP as you sit reading this (which is a rate of around just 1.4 g per second).

The design of limbs is different in cheetahs as compared with humans; these differences give the cheetah the ability to run using significantly less energy than humans. Indeed, as an approximation, the ATP consumed when a human sprinter runs 100 m would be enough to power a cheetah for a far greater distance – somewhere between 500 m to 1 km.

The efficient anatomical design of the cheetah leg is reflected in interesting ways: the special prosthetic lower legs used by the South African amputee athlete Oscar Pistorius are known as 'cheetahs'.

We learn more about the concept of energy, and the energy of biochemical processes in Chapter 13. We encounter ATP, a critically important molecule in biological systems, many times throughout this book, beginning with the opening paragraphs of Chapter 1.

Chemistry for the **Biosciences**

The essential concepts

SECOND EDITION

Jonathan Crowe
Oxford, UK

Tony Bradshaw
Oxford Brookes University, Oxford, UK

OXFORD
UNIVERSITY PRESS

Great Clarendon Street, Oxford OX2 6DP

Oxford University Press is a department of the University of Oxford.
It furthers the University's objective of excellence in research, scholarship,
and education by publishing worldwide in

Oxford New York

Auckland Cape Town Dar es Salaam Hong Kong Karachi
Kuala Lumpur Madrid Melbourne Mexico City Nairobi
New Delhi Shanghai Taipei Toronto

With offices in

Argentina Austria Brazil Chile Czech Republic France Greece
Guatemala Hungary Italy Japan Poland Portugal Singapore
South Korea Switzerland Thailand Turkey Ukraine Vietnam

Oxford is a registered trade mark of Oxford University Press
in the UK and in certain other countries

Published in the United States
by Oxford University Press Inc., New York

First published 2006

Second edition published 2010

British Library Cataloguing in Publication Data
Data available

Library of Congress Cataloging in Publication Data
Data available

Typeset by MPS Limited, A Macmillan Company
Printed in Italy on acid-free paper by
L.E.G.O. S.p.A.

ISBN 978–0–19–957087–4

10 9 8 7 6 5 4 3 2 1

JWC: For Mum, Nick, and Christina, and the latest generation: Aimee, Samuel, Jacques, Kenai, and Sebastian.

TKB: For Mum, and the family who were always there; and for my newly discovered American family.

Physical quantities commonly used in chemistry

Physical quantity	Symbol	SI unit
Amount of substance, chemical amount	n	mol
Atomic mass		amu (atomic mass unit)
Molar mass	M	kg mol^{-1}
Molecular weight	M_r	
Atomic weight	A_r	
Concentration of B	[B]	mol L^{-1}
Wavelength	λ	metre, m

Base SI units

Physical quantity	Symbol	SI unit
Length	l	metre, m
Mass	m	kilogram, kg
Time	t	seconds, s
Thermodynamic temperature	T	kelvin, K

Other physical quantities

Physical quantity	SI unit
Area	m^2
Volume	m^3
Force	newton, N
Energy	joule, J
Pressure	pascal, Pa

Prefixes to form the names and symbols of the decimal multiples and submultiples of SI units

	Multiple	Prefix	Symbol
	10^{-15}	femto	f
	10^{-12}	pico	p
	10^{-9}	nano	n
	10^{-6}	micro	*m*
	10^{-3}	milli	m
	10^{-2}	centi	c
↑ Getting smaller	10^{-1}	deci	d
↓ Getting larger	10	deca	da
	10^2	hecto	h
	10^3	kilo	k
	10^6	mega	M
	10^9	giga	G

Contents at a glance

Contents

Acknowledgements

Firstly, we'd like to gratefully acknowledge those individuals at Oxford University Press who have helped make this second edition a reality. We thank Kirsty Reade and Denny Einav for their valuable support, all of which helped to see this project through to completion without either author (quite) losing their sanity; and our production editors, Emma Lonie and Chantal Peacock, for bringing their guidance, expertise and general positive spirits to the production process. We also thank Ian Kingston and Graham Bliss, our copy editor and proof reader respectively, from whose eagle eyes the book has doubtless benefited greatly.

We are also grateful for the input from a number of colleagues during the preparation of this second edition. We extend a huge thank you to Professor Nicholas Price, University of Glasgow, whose constructive feedback and attention to detail has without doubt enhanced the final text. Big thanks are also due to colleagues at Oxford Brookes University, Dr Caroline Griffiths, Dr Peter Grebenik, Dr Alwyn Griffiths, and Professor David Fell, who all contributed with helpful discussion and scrutiny of parts of the text. We also thank those reviewers who provided invaluable critical feedback both on the first edition and on draft chapters from this edition:

Nicholas Brewer, University of Dundee
Hazel Davey, Aberystwyth University
Mark Howard, University of Kent
James Mason, King's College London
Peter Scott, University of Sussex

We also give warm thanks to Reverend Dr Paul Monk, whose valuable contribution as co-author of the first edition continues to pervade this second one.

It goes without saying that any mistakes or errors that remain are solely our responsibility.

We also received valuable help and support in preparing and sourcing artwork for the book. We are indebted to Dr James Keeler, University of Cambridge, for preparing the orbital images that appear in Chapters 2, 3, and 8; and to Professor Peter Atkins, University of Oxford, for providing the electrostatic surface plots that appear in Chapters 3 and 4.

Finally, we give grateful thanks to our friends and colleagues who have offered encouragement, moral support, and much-needed nights in the pub when self-imposed solitary confinement with the laptop has started to take its toll.

Chemical structures were prepared using a mix of ChemDraw Standard 8.0, and Chem3D Ultra 9.0. Biological structures were generated using the open-source software PyMOL (DeLano, W. L. *The PyMOL Molecular Graphics System* (2002) on the World Wide Web at **http://www.pymol.org/**), using data from the Protein Data Bank (**http://www.rcsb.org/pbd/**).

JWC, Oxford
TKB, Oxford
August 2009

Welcome to the book

For the lecturer: about this book

A relatively small number of students follow a full chemistry degree programme. However, many more students outside the realms of full chemistry degrees need at least a qualitative understanding of the essentials of chemistry to really make sense of their own areas of interest.

Why is this? Chemical principles pervade much of the life sciences. Indeed, a biological scientist can only effectively probe the many questions surrounding biological systems that remain unanswered if they use **all** the tools at their disposal. Chemical tools are among the most powerful available to the biological scientist; it follows that chemical concepts should form a central part of any biosciences degree programme.

This book seeks to fill the gap between texts for honours chemistry students (or 'chemistry majors') and the huge range of US-originated freshman 'general chemistry' texts. It is principally intended for those students with **no more than a GCSE in science**[1], and so assumes very little in the way of prior knowledge.

It may be that your students need to know chemistry in somewhat more depth than that offered by this book. In essence, what we have written here is intended as a 'springboard' into the many excellent undergraduate chemistry texts that are already available. As such, it offers a bridge from GCSE to undergraduate level, giving an introduction to those essential chemical concepts that life sciences students typically need to know, with the intention that it will give students the confidence and motivation to progress to the 'mainstream' chemistry texts if they need to know more.

The book is written in a deliberately conversational style. Its focus is on getting students to grasp the essential concepts, not on exhaustive coverage of the field (which we believe, at this level, can only really lead to rote learning of facts). We cover some relatively challenging concepts (for example, the notion of sigma and pi orbitals) but do so in a qualitative way – in the case of sigma and pi orbitals, merely using them to help provide a conceptual understanding of why molecules have the shape that they do. Throughout the book, we are writing for the biosciences student, not the chemistry

1 While really aimed at those with GCSE science, it's quite possible that the book may also be a good refresher for those students with an AS or A level in chemistry. We'd hope that these students would be at least familiar with much of the content of this book, however.

lecturer. One external reviewer of the first edition noted that the essential concepts we present in Chapter 1 aren't the concepts that a chemist would come up with. This is undoubtedly true, but misses the point: we want this book to be relevant to biosciences students, so we picked concepts that we hope will make them think 'Ah, so chemistry is relevant to me after all, and I **can** grasp what's going on.'

As is common with any book of this type, we faced a fine balance between simplifying material enough for students to grasp the concepts readily, and simplifying to the point of inaccuracy. We hope that, on the whole, we've got the balance right. However, writing a book is an ongoing process of refinement; we would be glad to know where you think we didn't get things right, so we can refine them in the future.

You can send us your feedback via the book's Online Resource Centre at http://www.oxfordtextbooks.co.uk/orc/crowe2e/.

New to the second edition

The second edition has presented us with a welcome opportunity to enhance the text's coverage in a number of ways, both in terms of content and presentation.

Content

Key changes include

- The more extensive provision of biological examples throughout, particularly in later chapters
- Significantly enhanced coverage of reaction mechanisms, which now occupies the two closing chapters of this edition
- Expanded coverage of enzyme kinetics in Chapter 14
- Expanded treatment of acids, bases, and buffer solutions in Chapter 16
- Enhanced coverage of chemical analysis in Chapter 11, which now introduces centrifugation, and expands its coverage of spectroscopy (including UV–visible spectroscopy)

Presentation

- A new, two-colour design, enabling the enhanced presentation of figures
- A rationalized use of boxes, with shorter boxes being absorbed into the main text

- A rationalized use of margin comments, with supplementary information now appearing in footnotes
- A restructuring to bring the isomerism chapter closer to the two chapters on molecular shape and structure, and to position the two reaction mechanisms chapters at the end of the book, where they draw upon various concepts introduced in earlier chapters

Learning support

New Maths Tools provide an opportunity for more extensive coverage of several mathematical/numerical skills that are used within the book. These Maths Tools include:

- Rearranging equations (Chapter 12)
- Measuring the gradient of a curve (Chapter 14)
- Handling brackets (Chapter 15)
- The exponential and logarithmic functions (Chapter 16)
- From ratios to percentages (Chapter 16)

Teaching support

This book is accompanied by an **Online Resource Centre** at www.oxfordtextbooks.co.uk/orc/crowe2e/ which features:

Figures from the book, available to download

A time-saving resource to support your lecture preparation.

Test bank of multiple-choice questions

A bank of multiple-choice questions, with answers linked to the book. The test bank features an average of ten questions per chapter, giving you a pool of questions from which to develop your own customized question bank to use for formative or summative assessment purposes.

For the student: using this book

We've written this book to try to make learning the essentials of chemistry as painless as possible. You might be asking yourself 'Why should I bother with chemistry at all?'. We've devoted Chapter 1 to answering this question; take a look at that chapter before you start reading the rest of the book.

Learning support

We've included a number of features to make this book as effective a learning tool as possible. These features include the following.

Margin comments

You'll see two types of comment in the margin as you read through the book.

- **Cross-references to other parts of the book**, so you can see how different topics fit together to give the overall 'big picture' of chemistry.
- **Additional guidance** – things to remember, common misconceptions, and other notes to help you grasp the key concepts as quickly and easily as possible.

Footnotes

Footnotes provide additional information to supplement what is being said in the text, without adding unnecessary detail to the text itself.

Boxes

Each chapter features a number of boxes which supplement the main text. In some cases, these boxes provide a little more detail about a concept being covered which isn't central to your grasping the subject matter, but which you might find interesting. Most often, however, they show how the chemical concepts being introduced apply to biological systems – examples of where chemistry quite literally comes to life. If you're reading the main text and find yourself thinking 'So what?', try reading some of the boxes to see how chemistry really is important in biological systems.

Maths tools

Many students find the use of mathematics in biology to be hugely intimidating. However, mathematical and numerical tools are amongst some of the most useful tools at the biologist's disposal – helping us, for example, to analyze and make sense of data. In fact – and despite first impressions – maths can actually help to simplify the biological world, by helping us to model and describe processes and relationships that in a way that would be impossible otherwise. In recognition of the central role maths can play (alongside chemistry) in understanding biological systems, we have included a small number of Maths Tools, which give an overview of some of the mathematical tools that we use in the text. If you're confident in the use of maths then you won't need to look at these Tools; if you're less confident, however, you may want to take a look at them, just to refresh your memory, or develop your understanding. We include five Maths Tools in this edition:

- Rearranging equations (Chapter 12)
- Measuring the gradient of a curve (Chapter 14)
- Handling brackets (Chapter 15)
- The exponential and logarithmic functions (Chapter 16)
- From ratios to percentages (Chapter 16)

MATHS TOOL 12.1 **Rearranging equations**

When we rearrange an equation, we typically want to get an unknown term (the variable we're trying to calculate) by itself on one side of the equation, and any known terms (the variables we already know the value of) on the other side of the equation.

To rearrange an equation, we effectively move the terms around one at a time to separate the known and unknown terms. In order to move a term from one side of an equation to the other, we need to reverse whatever mathematical operation is being applied to the term in the original equation:

- To reverse an addition, we have to subtract
- To reverse a subtraction, we have to add
- To reverse a multiplication, we have to divide
- To reverse a division, we have to multiply.

EXAMPLE

Imagine we have the equation $a + b = c$, and our unknown term is a.

Strategy

We need to isolate a on one side of the equation. This means we need to move b to the other side of the equation. In the original equation, b is being added to a. So, to reverse this operation, we subtract b from both sides of the equation.

Solution

$$a + \cancel{b} - \cancel{b} = c - b$$
$$a = c - b$$

There are two key things to note here:

(1) We're performing the same action on *both* sides of the equation;

(2) The two bs on the left-hand side of the equation cancel each other out, isolating a as required.

EXAMPLE

Now imagine that our original equation is $\frac{a}{b} = c$. Again, our unknown term is a.

Strategy

To isolate a we need to remove b from the left-hand side of the equation. In the original equation a is being divided by b so, to reverse this operation, we have to multiply by b.

Solution

$$\frac{a}{\cancel{b}} \times \cancel{b} = c \times b$$
$$a = c \times b$$

Again, notice how the two b terms on the left-hand side of the equation cancel each other out, while b then appears on the right-hand side of the equation instead.

If we need to move more than one term to isolate a particular, unknown term, we just need to move one term at a time, by reversing the operation being applied to it in the original equation.

EXAMPLE

Imagine our original equation is $\frac{a}{b} - c = d$. This time our unknown term is b.

Strategy

This time, we need to get a, c, and d together, leaving b by itself. We could do this in the following three steps.

Solution

Move c to the right-hand side by adding it to both sides (remembering that adding will reverse the effect of c being subtracted from the left-hand side of the original equation):

$$\frac{a}{b} - c + c = d + c$$
$$\frac{a}{b} = d + c$$

Next, we multiply both sides of the equation by b, to extract b from the bottom of the fraction on the left-hand side of the equation:

$$\frac{a}{b} = d + c$$
$$\frac{a}{b} \times b = (d + c) \times b$$
$$a = (d + c) \times b$$

Notice how we write d and c together in brackets here. This is because we're multiplying both d and c by b, and this saves us writing $(d \times b) + (c \times b)$.

Finally, we need to isolate b. At this stage, b is the subject of a multiplication operation; to reverse this we divide by $(d + c)$:

$$a = (d + c) \times b$$
$$\frac{a}{(d + c)} = \frac{(d + c) \times b}{(d + c)}$$
$$\frac{a}{(d + c)} = b$$

Let us now return to our original equation.

$$\frac{\text{mass of sample}}{\text{molar mass}} = \text{number of moles present in sample}$$

Self-check questions

We recommend that you try doing the self-check questions as you work through each chapter. They are an ideal way of checking whether you've properly got to grips with the concepts being introduced. Answers to self-check questions are given at the end of the book so you can quickly check whether you're on the right lines. Full solutions to numerical questions are provided in the book's Online Resource Centre as described below.

Notice how there are two absorption maxima (λ_{max} values) at approximately 190 nm and 280 nm, which correspond with the characteristic λ_{max} values for a carbonyl group, C=O.

Self-check 11.6 A sample containing an unidentified amino acid produces a UV–visible spectrum with a prominent peak at a wavelength of around 200 nm. Which of the following amino acids is most likely to be in the sample? (Look back at Figure 7.2 to remind yourself of the structures of the different amino acid side chains.) (a) Aspartic acid; (b) Phenylalanine; (c) Serine.

Read Box 11.8 to see how UV–visible can not only be used to characterize the chemical composition of a given compound, but can also be used to characterize biological processes.

Figure annotations

Many of the figures include additional annotations to make it as clear as possible what the figure is showing. We also annotate many equations so you can see what they represent.

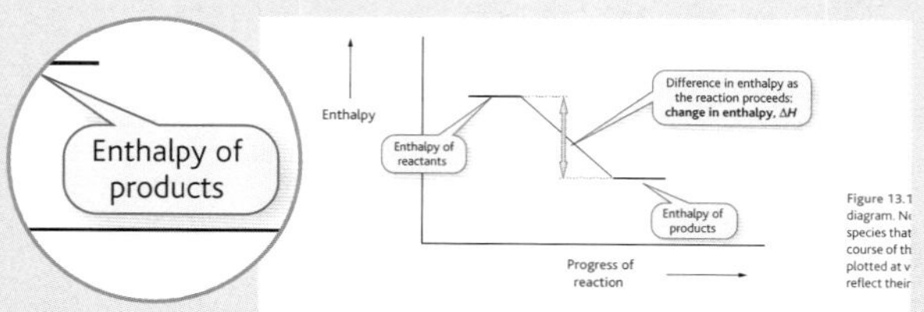

Key points

Key points appear throughout many chapters, and state the main take-home messages of the sections in which they appear. If you don't quite grasp the key points, then try re-reading the section.

• The mc
comp

- It identifies the different elements present in the compound.
- It tells us the number of atoms of each element in one molecule of the compound.

A molecule of glucose comprises six atoms of carbon, twelve atoms of hydrogen, and six atoms of oxygen. Its molecular formula is $C_6H_{12}O_6$. When writing a molecular formula, we list the composite elements in alphabetical order, rather than according to the number of atoms of each element that are present. (So, for example, we write $C_6H_{12}O_6$ rather than $C_6O_6H_{12}$.)

- *The molecular formula tells us the composition of one molecule of a covalent compound.*

Checklists of key concepts

Each chapter ends with a checklist of key concepts, which summarize the key learning points covered in the chapter. Use this checklist as a quick and easy way of revising the main points addressed, and as a prompt for going back to re-read any sections you're not sure about.

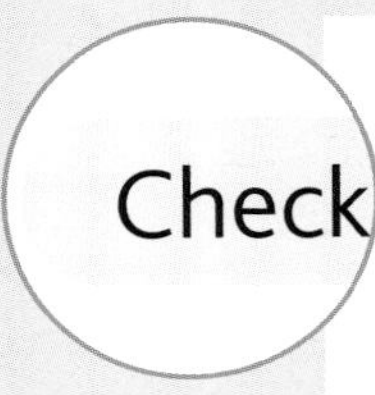

Check your understanding

To check that you've mastered the key concepts presented in this chapter, review the checklist of key concepts below, and attempt the multiple-choice questions available in the book's Online Resource Centre at http://www.oxfordtextbooks.co.uk/orc/crowe2e/.

Checklist of key concepts

Energy

- Energy is the capacity something has to do work
- The total energy that a substance has is its internal energy, *U*
- The internal energy has two components: work, *w*, and heat, *q*.
- Energy cannot be created or destroyed. It can only be transformed from one form to another
- Many different types of energy contribute to the internal energy–the main contributors are kinetic energy, potential energy, and chemical energy

- Energy is transferred betwe surroundings in the form of

Work and heat

- Work is any process that ca
- Heat is the transfer of ener
- Heat and temperature are r
- A substance's temperature a system with a high tempe energy; a system with a low

Online support

The Online Resource Centre to accompany this book features a number of additional materials to support your study of chemistry. Go to www.oxfordtextbooks.co.uk/orc/crowe2e/ to find out more.

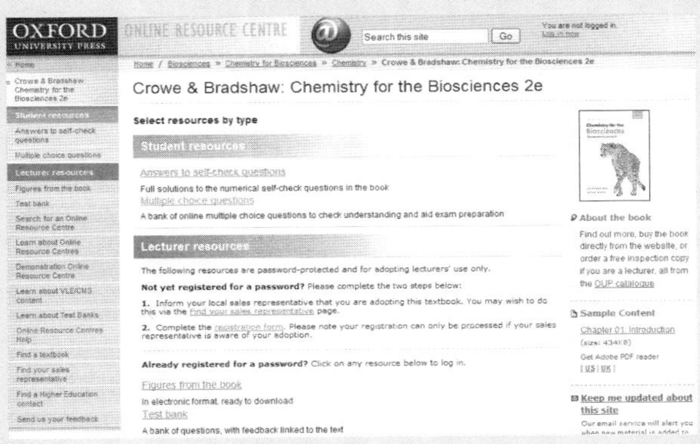

Online quizzing

To supplement the self-check questions provided throughout the text, the Online Resource Centre features a bank of over 150 additional questions in multiple-choice question format. Particularly useful as a way of checking what you do or don't understand prior to exams (to help focus your revision time), the multiple-choice questions are accompanied by feedback that refers back to the book so you can re-read topics that you're less confident about.

Full solutions to self-check questions

The answers to self-check questions provided at the end of this book show the final answer to a question, but do not show every step in a calculation (where a question requires a calculation to get to the answer). Instead, we provide full step-by-step solutions to calculations in the Online Resource Centre, grouped by chapter. Take a look at these solutions if you're not sure of the strategy required to complete a particular calculation.

1 Introduction: why biologists need chemistry

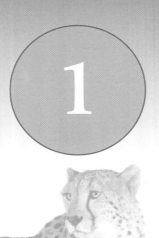

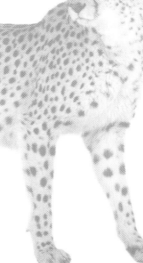

INTRODUCTION

Whatever else it may be, at the level of chemistry life is fantastically mundane: carbon, hydrogen, oxygen, and nitrogen, a little calcium, a dash of sulfur, a light dusting of other very ordinary elements–nothing you wouldn't find in an ordinary pharmacy–and that's all you need.

Bill Bryson, *A Short History of Nearly Everything* (2003)

Eukaryotic cells, including those that make up own bodies, are divided into subcellular compartments (or 'organelles'), each with its own distinctive biological function. One of these organelles is the mitochondrion, which is often referred to as the 'powerhouse' of the cell. Why is this? Mitochondria are responsible for synthesizing adenosine triphosphate (ATP), a substance that is used by the cell as a source of energy to power numerous processes that are necessary to make a cell function as it should.

Within the inner membrane of a mitochondrion are embedded multiple copies of a remarkable molecular machine, ATP synthase. ATP synthase comprises a series of different components which aggregate to form a multi-subunit complex. One part of this assembly is anchored within the mitochondrial membrane, while another part juts out into the mitochondrial matrix, the aqueous interior of the mitochondrion.

ATP synthase is a sophisticated and ingenious assembly, and operates as a true machine, complete with moving parts. When ATP synthase is in operation, the subunit embedded in the membrane physically rotates. This rotation is coupled to the process through which ATP is synthesized in the other subunit.

But what has ATP synthase, and its vital role in biology, to do with chemistry? We discover the answer when we start to consider the following questions:

- What is the structure of the ATP synthase, and how is its correct structure maintained?
- How does one part of the assembly remain anchored in the mitochondrial membrane? What stops it from simply sliding out, or the overall assembly inserting into the membrane the wrong way round?

- What drives the rotation of the membrane-bound subunit?
- Why is ATP a good energy source? How is it used by the cell to power biological processes?

Throughout this book, we will be exploring questions such as these to discover how chemistry lies at the heart of biological systems and the ways in which they operate. We learn:

- How interactions at the molecular level help to determine the three-dimensional structure of a complex such as ATP synthase;
- How the composition of a substance at the molecular level determines how it interacts with its surrounding environment–whether, for example, it binds tightly to the cell membrane or remains free in aqueous solution;
- How change–both structural and chemical–happens at the molecular level, and how the cell obtains energy to power such change.

At heart, every component of a biological system–the cell membranes that give cells their integrity, the chromosomes that store biological information and provide a means of transmitting this information from generation to generation, the blood that carries life-giving oxygen to every living tissue (and, indeed, oxygen itself)–is nothing more than a collection of atoms, the basic building blocks of chemistry. To echo Bill Bryson's sentiments above, when we consider life at the level of the atoms on which it is based, we could be forgiven for thinking that life itself is pretty mundane. The truly remarkable nature of the chemistry of life is summed up by the next sentence that Bryson writes; turn to the Epilogue on p. 657 to see what this sentence is–and see if you agree.

In this chapter, we consider some key concepts on which the remainder of the book is based. Before this, however, let us pause to consider why it's so useful for there to be interplay between different scientific disciplines–disciplines such as chemistry and biology.

1.1 Science: revealing our world

'Science' seeks to understand and explain our world, be that its physical composition (geology), chemical composition (chemistry), the way its composite matter interacts (physics), the organisms that inhabit it (zoology), or how they behave (psychology). We can only get a real idea of what life is all about by piecing together information from each discipline to give us the big picture, just as a forensic scientist pieces together different types of evidence to build a full picture of what happened at the crime scene.

For example, an ecologist may be studying the interactions within a particular group of organisms. Beyond merely observing how the organisms interact, the ecologist may use molecular biology to study patterns of interaction linked to reproduction; they may use chemistry to characterize pheromones, the hormones that are emitted by organisms to attract potential mates; and they may use mathematics and statistics to model how organisms interact, and so try to predict interactions within other groups of organisms.

It is quite normal for scientists of today to become specialists in rather narrow disciplines–molecular evolution; physical biochemistry; developmental neurobiology;

community ecology–yet this represents a stark shift from previous centuries, in which practising scientists would have interests ranging from chemistry to biology, or from earth science to physics. Take, for example, the seventeenth-century scientist Robert Hooke. Hooke, like other scientists of his time, was described as a 'natural philosopher'–individuals who asked wide-ranging questions about all aspects of life that surrounded them.

Hooke, for example, is particularly well known for initiating the field of microscopy (his most famous work being *Micrographia*, or *Some physiological descriptions of minute bodies made by magnifying glasses, with observations and inquiries thereupon*, which was published in 1665). Yet his interests spanned meteorology (he developed barometers and rain gauges), astronomy and physics, and mathematics; he even invented the spring-regulated watch. Natural philosophers asked all manner of questions about the world around them, not limiting themselves to questions only of 'biology' or 'chemistry', 'physics' or 'mathematics'. They simply used their enquiring minds to explore, regardless of the directions in which these explorations took them.

Today, science is arguably coming full circle, to the point that key questions about the world around us can't be probed without looking outside our own discipline and drawing inspiration from other areas of science. More than ever before, solving biological questions means using information that is provided by other related disciplines–using all the tools that are available across science as a whole to probe and question, and formulate new and exciting answers.

I'm a biologist: what has chemistry to do with me?

We don't have to look very far to realize that it's difficult, if not impossible, to separate biology from chemistry. After all, our body is a bag of chemicals. The proteins that form our hair, nails, and muscle fibres are chemicals; the minerals that are the basis of our bones and teeth are chemicals; even the food and drink we consume every day are chemicals. In fact, any and every object that we see around us is an example of chemistry in action: all are formed from a collection of millions of tiny atoms. Biology and chemistry explore the same thing–the world around us–but simply at different scales.

Chemistry explores life at the level of atoms and molecules: it is really all about understanding how atoms interact to form larger, more complicated substances, how these substances react with each other to form new substances, and how these substances behave–whether they are flexible or rigid; liquid or solid; stable or reactive. Biology then looks at how these substances behave when they are combined on a larger scale–the scale of cells, tissues, organisms, populations, or ecosystems.

Chemistry is encapsulated by a handful of essential concepts that are epitomized by the world around us, whether we realize it or not. Just grasping these concepts is sufficient for us to get to grips with many of the key chemical principles that underpin biology.

In the next section, we encounter the ten essential concepts that represent all the chemistry that a biologist needs to know, and we see how these ten concepts, which we refer back to throughout this book, can be illustrated with examples from everyday life. During the rest of this book we merely learn how these general concepts are reflected in some important *chemical* (rather than everyday) principles. If we understand the essential concepts, then we're well on our way to understanding how they apply in a chemical context.

1.2 The essential concepts

Let's now consider the ten essential concepts upon which this book is based, and get a snapshot of how each concept appears in this book:

1 Small, repetitive units join together to form larger, more complicated structures

How do we see this concept in everyday life?

Consider a house: a house is built from hundreds of bricks, which are stacked together in a repetitive way to give the house its shape. The bricks are small units which are brought together to form the house; the house is a larger, more complicated structure than each individual brick.

How do we see this concept in biology?

Biology operates on a range of scales, governed by the principle of small, repetitive units forming larger assemblies. Atoms form molecules; molecules form cellular components; cells form tissues; tissues form organisms; organisms form populations; populations form ecosystems, and so on.

Where do we learn more?

We learn about the composition of atoms and molecules in Chapters 2 and 3, and see how biological molecules are formed from their composite subunits in Chapter 7.

2 Only slight variety can lead to great diversity

How do we see this concept in everyday life?

Think about mixing paints on a palette: we may start with just a few colours–red, yellow, and blue, for example. By combining these colours in different ways, however, we can produce a wide variety of different hues: we can combine the red and yellow to produce orange; the red and blue to give purple; the yellow and blue to produce green. We can generate a diverse range of colours, despite having a limited variety of colours initially.

How do we see this concept in biology?

Just four nucleotides are responsible for the seemingly infinite range of DNA sequences that make each of us genetically unique[1]. And a pool of twenty amino acids forms the huge range of proteins that are central to life. Similarly, the function of many molecules in biology is conserved–that is, the same kinds of molecules are used to do the same kinds of things in a vast array of different organisms–yet the organisms themselves vary hugely in their biological attributes.

Where do we learn more?

We learn about the generation of nucleic acids and proteins from their limited pools of starting materials in Chapter 7. We learn how identical sets of atoms can be used to create a variety of molecules in Chapter 10.

3 Composition determines physical form; physical form dictates function

How do we see this concept in everyday life?

The substances from which objects around us are made (their **composition**) determine the objects' physical form and function. For example, clothes are made from substances such as cotton and polyester, which are soft to touch and very flexible. By contrast, furniture is made from substances such as woods, metals, and plastics,

1 With the exception of identical twins

which are hard and durable. A substance can only function as it should if its composition is appropriate: just as in the case of the proverbial chocolate teapot, clothes made of metal would be no more functional than a chair made of cotton (unless you happened to be a knight, in which case the metal-based armour would be entirely appropriate).

How do we see this concept in biology?

Proteins are comprised of multiple amino acids, joined together to form a continuous chain. However, the identity of the amino acid at each point in the chain must be correct if the protein is to function as it should. Sometimes, a change in just one amino acid is enough to disrupt the protein's function to the extent that the organism producing the protein carries a mutation.

Where do we learn more?

We learn about the molecular composition of biological molecules in Chapter 7, and explore the shape and structure of biological molecules, and how structure is intrinsically linked to function, in Chapter 9.

4 An object has a simple framework to give it form, and additional physical attributes to give it function

How do we see this concept in everyday life?

Consider a car: a car has a chassis–a simple framework to give the car its structure and strength. Many other components–bodywork, engine, wheels, etc.–are then added to the simple chassis to give the car its function. A car needs both the simple framework *and* the additional components to make it function as it should.

How do we see this concept in biology?

Many organisms have a broadly similar 'framework': a skeleton; internal organs such as the heart and lungs to carry out key biological processes; and skin to protect the organism from its environment. Yet, alongside these basic components, organisms possess an array of additional physical features, which vary from species to species–wings, legs, paws, claws, scales, hair, spines, horns–to make them better adapted to the particular environment in which they live.

Where do we learn more?

We learn about variety at the level of chemical compounds in Chapters 5 and 6.

5 Stability is preferable to instability

How do we see this concept in everyday life?

Consider a camping tent: the last thing we want is for the tent to blow down in the slightest gust of wind. To prevent this, and to give the tent more stability, we use tent pegs to secure the tent to the ground. Without the tent pegs, the tent is much more unstable, and a less desirable place in which to sleep!

How do we see this concept in biology?

All biological systems are subject to tight regulation and control which maintain stability, rather than letting things become unstable. The importance of such regulation becomes apparent when it breaks down and processes such as cell growth spiral out of control, leading to conditions such as cancer.

Where do we learn more?

We learn about the molecular interactions that confer stability at the molecular level in Chapter 4, and explore how the stability of different entities is affected by the energy they possess in Chapter 13.

6 Change involves movement

How do we see this concept in everyday life?

Think of any activity that involves 'change': changing a light bulb, or changing clothes, for example. Each of these activities requires movement to some degree: both movement by us (to effect the change), and movement of the object being affected by the change.

How do we see this concept in biology?

Many biological processes involve physical movement–for example, muscle contraction involves physical movement of the proteins forming the muscle fibre, and we see at the start of this chapter how the activity of the ATP synthase is linked to the physical rotation of one of its structural components. Even at the level of chemical change, movement is required–in this instance, the movement of electrons.

Where do we learn more?

We learn about the movement of electrons during the course of chemical reactions in Chapters 17 and 18, and discover how movement is central to the activity of biological macromolecules in Chapter 9.

7 Change involves energy

How do we see this concept in everyday life?

Consider heating up water in a kettle–*changing* its temperature from cold to hot. This process requires energy, which is provided by the kettle's heating element. Also, consider the activities mentioned in concept 6; each requires energy (to power our muscles to enable us to change the light bulb, for example) to bring about change.

How do we see this concept in biology?

Biological processes are all about change: breaking foods down into smaller molecules; building up large molecules from small subunits; or dividing cells into two daughter cells during growth and repair. At a fundamental level, such life processes require the provision of energy. As we see at the start of the chapter, one of the most important sources of energy is the energy carrier ATP, which transfers energy to processes that would otherwise be unable to proceed.

Where do we learn more?

We learn about chemical energy, and the energy associated with biological processes, in Chapter 13.

8 Some processes are more favourable than others; some processes need a helping hand to get them under way

How do we see this concept in everyday life?

Many events in the world around require a spark (quite literally) to get them going: we need the flame from a lighted match to light a fire (after which the fire will burn on its own); a car engine needs the spark from a spark plug to ignite the fuel within it. The readiness with which such processes happen also depends on the environment: a car engine will start more readily on a warm day than in the depths of winter: on a frosty morning, a car engine may need more of a helping hand–a longer turning of the ignition–to bring it to life.

How do we see this concept in biology?

Many biological processes would happen too slowly for life to be maintained if left to happen spontaneously. Instead, our cells rely on biological catalysts–enzymes–to give processes the helping hand they need to proceed.

Where do we learn more?
We explore the role of enzymes in the catalysis of biological processes in Chapter 14.

9 Change can often be reversed
How do we see this concept in everyday life?
Consider an ice-cube tray full of water. We can put the ice-cube tray in the freezer to produce ice cubes: the water changes from liquid to solid. We then take the tray out of the freezer and put the ice cubes in a drink. After a few minutes the water defrosts: the water changes back to liquid from solid. So the process of freezing–changing from liquid to solid–is one that can be reversed.

How do we see this concept in biology?
The storage and breakdown of sugars in our cells is a prime example of the reversibility of many biological processes. When glucose levels in the blood are high, the hormone insulin stimulates the conversion of glucose to glycogen, which is stored in the liver and muscle. When blood glucose levels drop, the hormone glucagon stimulates the liver to break down the stored glycogen back into glucose. Hence, by acting in concert, the two hormones allow blood sugar levels to be carefully regulated.

Where do we learn more?
We discuss the way that chemical reactions can happen in both forward and reverse directions in Chapter 15.

10 Change can happen at different rates
How do we see this concept in everyday life?
Consider the process of defrosting once again: if we take frozen food out of the freezer and leave it at room temperature, it defrosts over the course of a few hours. If we transfer the food from freezer to fridge (rather than to room temperature), the change happens more slowly.

How do we see this concept in biology?
During vigorous exercise, the body breaks down stored food reserves much more quickly than it does at rest. Indeed, when an animal hibernates–the antithesis of vigorous exercise–the biological processes happening within it slow down to a barely perceptible rate.

Where do we learn more?
We explore the field of kinetics–how fast reactions happen–in Chapter 14.
During the rest of this book we see how these concepts apply not just to our everyday life, but also at the scale of atoms and molecules: at the chemical scale.

1.3 The language of chemistry

Before we begin our exploration of chemistry, we need to consider briefly how we communicate chemical ideas and concepts.

We rely on language to communicate with those around us–to share ideas, views, and experiences. Chemistry also has its own 'language' associated with it so that we can share chemical ideas and concepts with others in a way that makes sense not only to us, but to our audience too.

There exists an international body that sets down the accepted conventions–the terms and notation–that are used throughout the world to describe chemical systems. This body is the **International Union of Pure and Applied Chemistry**, or **IUPAC**. (IUPAC is the chemical equivalent of FIFA, which sets down the regulations under which football (or 'soccer') is played internationally, so that teams all play by the same rules, regardless of their country of origin.)

Throughout this book we adopt the currently accepted IUPAC conventions for naming chemical compounds and describing chemical systems. These conventions come together to form what we call **chemical nomenclature**. For more details about chemical nomenclature as set down by IUPAC, visit their website at **http://www.iupac.org/**.

IUPAC conventions govern both the naming of chemical compounds and the use of units and symbols. We learn more about the naming of compounds in Chapters 5 and 6, when we explore chemical substances associated with the field of organic chemistry. However, the use of units and symbols is too important for getting to grips with chemistry for us to delay finding out more about them.

Units: making sense of numbers

We may think that we're surrounded every day by numbers. However, if we look a bit closer, it's not *numbers* that we're surrounded by most often, but **quantities**.

A quantity has two components: a **number** and a **unit**.

The number tells us the magnitude (size) of the thing that we're measuring; the unit tells us what it is that we are actually measuring.

A measurement is meaningless unless we include the units that we are measuring in. For example, suppose that we're visiting the doctor, and they want to know our weight. Imagine that we have a weight of 70 kg. There would be no point in us telling the doctor that we weigh '70'. This would tell the doctor nothing. For all they know, we could be measuring in kilograms, stones, ounces, or pounds. The number only makes sense if we include the unit of measurement: kg.

The same rule applies whatever we're measuring. If we are buying a new television, we measure the size of the screen in *inches*. If we buy a new car, we measure the size of the engine in *litres*. We don't ask for a '17 television' or a '1.6 car'; we ask for a 17 *inch* television, or a 1.6 *litre* car.

- ***As a general rule, a measurement of any kind is meaningless without its appropriate unit, telling us what is actually being measured.***

A word of warning: there are some instances in which a measurement does not have units. We say that they are **dimensionless**. However, these are the exceptions to the rule. We should assume that a measurement has units unless told otherwise.

There are often different ways of quoting what is essentially the same measurement, i.e. we can quote a measurement using different units. We might measure our height in metres, or feet and inches; we might measure our weight in kilograms, or stones and pounds.

To ensure consistency, measurements in chemistry are quoted using an internationally accepted system of units (just as the naming of chemical compounds is governed by IUPAC's nomenclature guidelines). We call the international system of units the **Système International d'Unités**, or **SI units**.

Look at the table on p. vi to see the SI units for those quantities that we encounter most often in this book.

Symbols

IUPAC not only governs the naming of compounds but also sets out standard symbols that are used throughout chemistry to represent common terms. Symbols are a valuable way of saving time (and space); they are a shorthand way of representing a particular term or expression.

The table on p. vi also includes symbols for a number of terms and expressions that we encounter in this book.

Checklist of key concepts

- To explore biological systems fully, we need to gather information from *different* scientific disciplines
- Chemistry is a central part of biology: everything in biology is made of chemicals
- The chemistry of biology is based on a handful of concepts, which we're all familiar with from everyday life
- We use particular terms and conventions to describe to others chemicals and chemical systems, just as we use particular words and phrases when talking on a day-to-day basis
- The **International Union of Pure and Applied Chemistry** (IUPAC) governs the terms and notations that are used to describe chemical systems
- We must use units to turn a number into a quantity
- The units that are assigned to chemical quantities are those specified in the **Système International d'Unités**

2 Atoms: the foundations of life

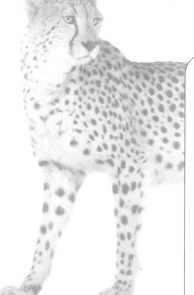

INTRODUCTION

Look around you now. You are probably surrounded by things that are made from a huge variety of materials. These materials may be artificial or may occur naturally–plastics, glass, ceramics, metals, wood, or stone. They can be brought together in different combinations to construct all the things you see around you, and indeed all the material objects that populate our lives: our homes, the cars we drive, even the books we read. However different they may seem to the observer, though, such objects, and indeed all matter, actually comprise a surprisingly small range of ingredients, the chemical elements.

In this chapter we look at the proton, neutron, and electron–the three fundamental particles from which all materials on the planet, and those organisms that inhabit it, are formed. We see how these particles come together to form atoms, and how different combinations of these particles generate the range of chemical elements which create the variety we see in the world around us.

2.1 The chemical elements

The quest to identify the range of substances from which all living things are made involved a number of great scientists over the course of hundreds of years. The Ancient Greeks were the first to postulate that all matter could be reduced down to the four so-called 'humours': Earth, Fire, Air, and Water. However, further study revealed a far more complicated picture. In time, it was established that everything around us is comprised of a large, but finite, number of different particles, the so-called **chemical elements**.

It has taken many years to isolate and characterize the different elements. To date, 112 elements have been formally identified; 90 of these are naturally occurring[1]. Additional elements have also been created in the lab, and we continue to synthesize new elements even today. The latest to be verified is the element copernicium (Cp), which was officially added to the periodic table in June 2009.

1 Some elements do not exist in a stable form, and so cannot persist in nature. Some sources say that there are 92 naturally occurring elements–but at least two of these (technetium and promethium) are too unstable to exist naturally.

Given the huge range of substances formed from them, 90 different elements may not sound like many. In fact, the reality of life is even more amazing: only a small proportion of the 90 naturally occurring elements are vital for life. Humans require only 25 elements for a healthy life–to create all the blood, bone, muscle, and other tissue from which our bodies are formed. These so-called 'essential elements' are shown in Table 2.1. A selection of just *eleven* elements are absolutely vital to all living things, including humans, as shown in Table 2.2. Another ten are required by most living organisms, and a further eight are vital to just some animals or plants.

While eleven elements may be absolutely vital, just four elements–hydrogen, carbon, oxygen, and nitrogen–are the real key players. These elements make up almost 97% of the total body mass, and constitute a staggering 99% of the total atoms from which the body is formed.

Every element has been given a name (for example, hydrogen, nitrogen, and oxygen) and, for convenience, has also been assigned a one- or two-letter shorthand symbol. Just as we have a shorthand notation for the nucleotide bases that make up DNA and RNA (G, C, A, T, and U), so the shorthand symbols for the chemical elements give us a way of communicating exactly what a particular substance or material contains at a glance. So oxygen is represented by the letter O, hydrogen by the letter H, and nitrogen by the letter N. We will employ these **chemical symbols** throughout this book.

The **periodic table** of the elements is the best place to look if we want to know the symbol for a particular element. A copy of the periodic table is reproduced on p. i. The periodic table was devised, in its present form, by the Russian Dmitri Mendeleev, in 1869. Mendeleev arranged the chemical elements sequentially in rows of eight. This choice of row length was not arbitrary: the layout of the table was such that elements fell into horizontal rows, called **periods**, and vertical columns, called **groups**. There are seven periods and eighteen groups in the periodic table. As shown in the periodic table on p. i, the periods are numbered 1 to 7 from the top of the table down; the groups are numbered 1 to 18 from left to right.

Table 2.1 The 25 essential elements for the human body.

Element	Proportion of total body mass (%)
Oxygen	61
Carbon	23
Hydrogen	10
Nitrogen	2.6
Calcium	1.4
Phosphorus	1.1
Sulfur	0.2
Potassium	0.2
Sodium	0.14
Chlorine	0.12
Magnesium	0.027
Silicon	0.026
Iron	0.006
Fluorine	0.0037
Zinc	0.0033
Copper	0.001
Manganese	0.0002
Tin	0.0002
Iodine	0.0002
Selenium	0.0002
Nickel	0.0002
Molybdenum	0.0001
Vanadium	0.0001
Chromium	0.00003
Cobalt	0.00002

Table 2.2 The eleven chemical elements that are essential for all biological systems.

Chemical symbol	Element name
H	Hydrogen
O	Oxygen
C	Carbon
N	Nitrogen
Na	Sodium
K	Potassium
Ca	Calcium
Mg	Magnesium
P	Phosphorus
S	Sulfur
Cl	Chlorine

Self-check 2.1 To which period and group does aluminium, Al, belong?

Importantly, Mendeleev noted that elements which fall in the same group (within the same vertical column on the periodic table) tend to exhibit similar physical and chemical properties, and that these properties follow a regular pattern of change as we move from group to group across a period.

This observation of gradual, incremental changes in physical and chemical properties from element to element as we move across a period is known as **periodicity**. This also manifests itself in the way that elements falling in the *same* group in the periodic table exhibit *similar* physical and chemical properties. We revisit the concept of periodicity again in later chapters.

An **element** is defined as a single, simple substance that cannot be split into any more separate substances by chemical means. For example, carbon is an element. We can't break carbon down into additional substances–it is 100% carbon. Water, on the other hand, is not an element–it can be split into its two component elements of hydrogen and oxygen. If we removed all the impurities from water, we would be left with 'pure' water–100% water with nothing else added. However, pure water is still not an element: it can be chemically split into hydrogen and oxygen, which conflicts with the definition of an element we have just seen.

The smallest unit in which an element can exist is the **atom**. An atom is defined as the smallest particle into which an element can be divided while still retaining the properties of the element. A piece of elemental carbon ('pure' carbon–for example, diamond) comprises a huge number of identical atoms, as illustrated in Figure 2.1.

- ***An element is a single substance that cannot be split into any more separate substances by chemical means.***
- ***An atom is the smallest particle into which an element can be divided, while still retaining the properties of the element.***

2.2 Atomic composition

If atoms are as small as we can get (in chemical terms), what determines the unique properties of different atoms? For example, what makes an atom of carbon different

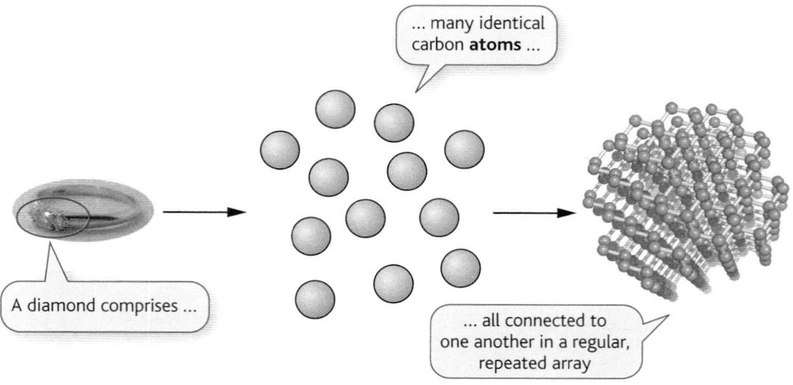

Figure 2.1 Diamond is 'pure' (elemental) carbon. It comprises millions of identical carbon atoms joined in a characteristic three-dimensional array.

to an atom of nitrogen? The answer lies with the three **subatomic particles**: protons, neutrons, and electrons[2].

Atoms of all the chemical elements contain protons, neutrons, and electrons. (This is with the exception of the simplest atom, hydrogen, which contains just one proton and one electron, but no neutron.) It is the *number* of each of these particles in an atom, however, that gives the atom its character and ensures each element is identifiably distinct. An atom of nitrogen contains 7 protons, 7 neutrons, and 7 electrons, while an atom of calcium contains 20 protons, 20 neutrons, and 20 electrons. Both elements contain protons, neutrons, and electrons, but it is because they contain *different* numbers of these particles that they have different chemical characteristics.

The definition of an atom given above stated that it was the smallest particle into which an element could be divided while still retaining the characteristics of that element. An atom of an element can be separated into its component subatomic particles, but these particles no longer behave like the element. If we separate a nitrogen atom into its composite electrons, protons, and neutrons, the separate particles no longer behave like nitrogen. It is this grouping, with a specific number of protons, electrons, and neutrons, that gives the atom its distinct properties.

The existence of atoms as real, physical entities was driven home by some remarkable work by D. M. Eigler and E. K. Schweizer at the IBM Research Division in 1990, as explained in Box 2.1.

Protons, electrons, and electrical charge

Electrical charge is a vital part of biology. For instance, our nervous system consists of a complicated network of nerve cells that use electrical charge to transmit nerve impulses throughout the body. When we touch something with a finger, an electrical charge is passed along a series of nerve cells from the finger to our brain, where the action of touching is turned into a sensation. Only then do we know we've 'touched' the object.

Electrical charges are also vital on the smaller scale of atoms rather than cells. Protons and electrons are both **charged**–a proton carries a single positive charge, while an electron carries a single negative charge. Neutrons carry no charge–they are electrically neutral, hence their name.

- *A proton carries a single positive charge.*
- *An electron carries a single negative charge.*
- *A neutron carries no charge.*

---

2 The prefix 'sub' literally means 'less than', so 'subatomic' means 'less than (or smaller than) an atom'. Particle physicists have identified other subatomic particles in addition to the proton, neutron, and electron, including several types of neutrino, and the positron. For our purposes, however, protons, neutrons, and electrons are the only subatomic particles we need to consider as we explore the chemistry of life.

BOX 2.1 At a push: art at the atomic scale

The early 1980s saw the invention of the scanning tunneling microscope (STM), which uses electrons to visualize the surface of materials at atomic resolution. In a unique application of STM, Don Eigler and Erhard Schweizer in 1990 used the tip of an STM to physically drag atoms across a surface–clear evidence, if you needed it that atoms are distinct physical entities.

Specifically, Eigler and Schweizer worked with nickel coated with a layer of xenon atoms, cooled to a distinctly chilly 4 K (–269 °C). They used the tip of the STM to pull atoms of xenon one by one across the surface of the nickel. In so doing, they organized the xenon atoms into distinct patterns, including the one illustrated below–a fitting tribute, perhaps, to the organization that supported Eigler's research.

Image © Don Eigler

Even though protons and electrons always bear an electrical charge, all atoms are **uncharged**. This fact has one important implication–in any uncharged atom, the number of protons must exactly equal the number of electrons. Look at Figure 2.2, which shows how the positive charges on the protons are 'cancelled out' by the equivalent number of negative charges on the electrons, leaving the atom uncharged overall.

Identifying the composition of an atom: atomic number and mass number

Earlier in the chapter, we saw how the different chemical elements have one- or two-letter chemical symbols which act as abbreviations for their full names. Similarly, each element has two numbers associated with it which enable us to identify its atomic composition at a glance. These two numbers are the **atomic number** and **mass number**.

The atomic number (sometimes called the **proton number**) tells us the number of **protons** within the atom. By implication, this number also indicates the number of electrons in the atom. (Remember: an atom is a neutral particle. For this to be the case, it *must* contain exactly the same number of protons and electrons.)

The mass number indicates the total number of protons and neutrons that the atom contains. To determine the number of neutrons in an atom, we merely subtract the atomic number (number of protons) from the mass number (number of protons + neutrons).

- ***The atomic number is equal to both the number of protons and the number of electrons in a neutral atom of a given element.***
- ***The mass number indicates the number of protons*** **plus** ***the number of neutrons that an atom of a given element contains.***

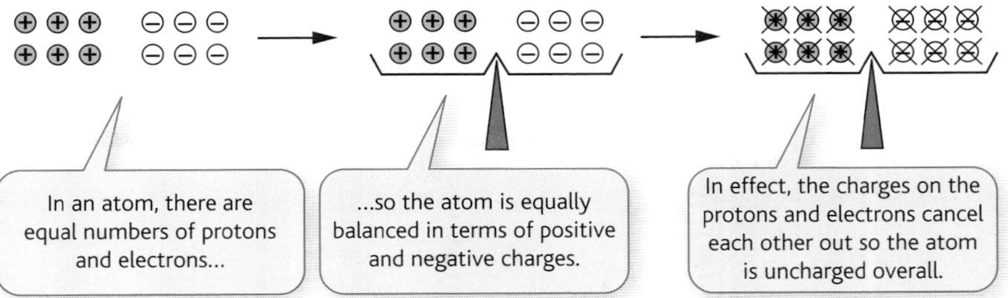

In an atom, there are equal numbers of protons and electrons...
...so the atom is equally balanced in terms of positive and negative charges.
In effect, the charges on the protons and electrons cancel each other out so the atom is uncharged overall.

Figure 2.2 The balancing of electrical charge in an atom. An atom contains an equal number of protons and electrons, so it has no overall electrical charge.

Let's look at an example:

EXAMPLE

The potassium atom has an atomic number of 19, and a mass number of 39.

Straight away, this information tells us an atom of potassium contains 19 protons and 19 electrons.

We also know the mass number = number of protons + number of neutrons in the atom. We can rewrite this relationship as follows:

number of neutrons = mass number − number of protons

Since the mass number = 39 and the number of protons in the potassium atom = 19

number of neutrons = 39 − 19 = 20

We learn more about the rearrangement of equations like this in Maths Tool 12.1 on p. 382.

Self-check 2.2 An atom of sulfur has the atomic number 16, and a mass number of 32. How many protons, electrons, and neutrons are there in one sulfur atom?

We have mentioned already how the periodic table lists all the known elements, along with their chemical symbols. Most versions of the periodic table also indicate the atomic number for each element. Look at the periodic table on p. i of this book, and notice where the atomic number for each element is given.

The formation of ions

The atoms of any element are uncharged because they contain an equal number of protons and electrons. If the exact balance between protons and electrons is disturbed, however, then we form a charged **ion**. If an atom of a given element *gains* one (or more) electrons so that there are now more electrons present than protons, the overall charge on the atom is negative, and a negative ion is formed. We call a negative ion an **anion**.

Likewise, if an atom *loses* one or more electrons, the positive charge on the protons outweighs the overall negative charge held by the smaller number of electrons, and the overall charge is positive. Consequently, a positive ion is formed. We call a positive ion a **cation**.

For example, look at Figure 2.3(a), which illustrates the formation of a cation from an atom of lithium. An atom of lithium has an atomic number of 3, so it comprises

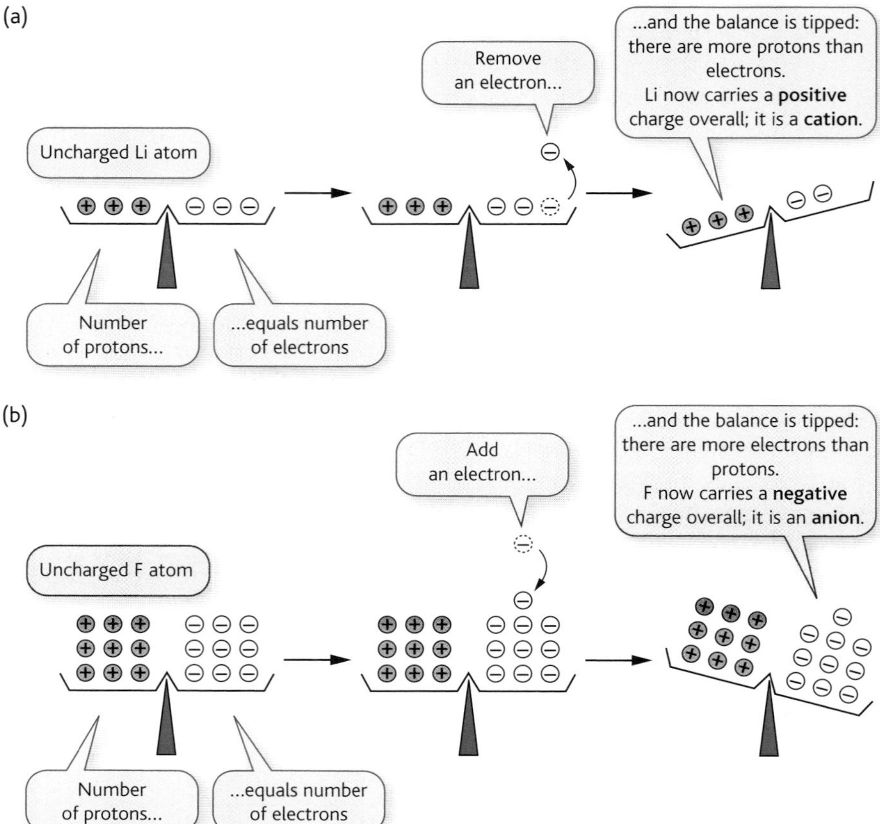

Figure 2.3 The formation of ions. (a) When lithium loses an electron it becomes a cation, and has a positive charge overall. (b) When fluorine gains an electron, it becomes an anion, and has a negative charge overall.

three protons and three electrons. Under certain conditions, however, the lithium atom loses one electron to form a positively charged lithium ion; it now has three protons and two electrons. By contrast, an atom of fluorine has an atomic number of 9, and so comprises nine protons and nine electrons. Figure 2.3(b) shows how an atom of fluorine can gain one electron to form a negatively charged fluoride ion[3]; it now has nine protons and ten electrons.

- *A cation carries an overall positive charge, and contains more protons than electrons.*
- *An anion carries an overall negative charge, and contains more electrons than protons.*

3 We recognize the change undergone by the fluorine atom by changing its name to fluoride. In fact, 'ide' at the end of a name means a negatively charged ion of an element. Other examples include bromine → bromide, oxygen → oxide, and sulfur → sulfide.

We indicate the charge on an ion with the following simple chemical notation: a superscript '+' signifies an overall positive charge (where there are more protons than electrons in the ion), and a superscript '–' signifies an overall negative charge on the ion (where there are more electrons than protons in the ion). Therefore we represent the sodium ion by writing Na^+, and represent the fluoride ion by writing F^-.

Atoms of different elements can gain or lose more than one electron to form ions carrying more than one positive or negative charge. If an atom has lost *two* electrons to form an ion with two more protons than electrons, the ion carries *two* positive charges. For example, a calcium atom can lose two electrons to form an ion with 20 protons and 18 electrons. This calcium cation carries two positive charges.

In addition to the superscript '+' or '–', the chemical symbol for an ion is written with a superscript number to indicate the total number of positive or negative charges on that ion. Therefore the calcium ion described above is written in chemical notation as Ca^{2+}–the superscript 2+ tells us that the ion is carrying *two* positive charges. However, we need to note that convention says a superscript number to indicate the ion's charge is only included if the number of charges is greater than one. If an ion carries just a single positive or negative charge, then no number is included with the '+' or '–' sign: for example, the sodium ion is written as Na^+, not Na^{1+}.

Self-check 2.3 An oxygen atom can gain two electrons to form a negatively charged ion. How many negative charges does the ion carry, and how is this represented in chemical notation?

Quite often, atoms form ions as part of their reaction with other atoms to form molecules, a process we examine more closely in Chapter 3. Indeed, it is the movement and sharing of electrons between atoms and molecules that lies at the heart of virtually all chemical processes–including those occurring in the cells of your body this minute. So the readiness with which an atom gains or loses electrons really dictates its reactivity; elements whose atoms gain or lose electrons readily tend to be very reactive, while those whose atoms are reluctant to gain or lose electrons will be chemically inert–they will be very unreactive.

We look at the reactivity of different elements–how readily they react–in Chapter 3.

Isotopes: varying the number of neutrons

An atom of a particular element always possesses the same number of protons and electrons. However, the same is not true concerning the number of neutrons.

Unlike protons and electrons, neutrons do not carry an electrical charge. Therefore the number of neutrons in an atom can vary without affecting the atom's overall charge. Indeed, many elements exist naturally as a mixture of atoms with identical numbers of electrons and protons, but with varying numbers of neutrons. Atoms that have the same number of protons and electrons, but a different number of neutrons, are called **isotopes**.

For example, oxygen exists naturally as a mixture of three different isotopes. The atoms of each isotope have eight protons and eight electrons. However, some atoms contain eight neutrons, some nine neutrons, and some ten. A difference of just one neutron is enough to make quite a difference in biology, as we discover in Box 2.2.

BOX 2.2 **The world of plants: a hotbed of discrimination**

We might think that isotopes of a single element are so very similar that they are virtually indistinguishable. Does a difference of one neutron really carry any significance? To plants, at least, the answer appears to be 'yes'.

Plants generate their own food supplies by photosynthesizing–they absorb carbon dioxide from the atmosphere, and combine it with water to generate the sugar glucose. This process occurs in the chloroplasts, specialized organelles within plant leaf cells, and is powered by energy absorbed by the plant from sunlight.

Atmospheric carbon dioxide (CO_2) contains two isotopes of carbon, ^{12}C and ^{13}C, in proportions 98.9% ^{12}C and 1.1% ^{13}C. (It also contains a tiny proportion of ^{14}C–just 0.000 000 000 1%.)

Research has found that plants discriminate *against* ^{13}C, and so preferentially metabolize $^{12}CO_2$ rather than $^{13}CO_2$. But how do plants discriminate between atoms of carbon that differ by only one neutron? The answer seems to come down to weight. The difference in one neutron is enough to make $^{13}CO_2$ heavier than $^{12}CO_2$. The extra weight causes $^{13}CO_2$ to diffuse more slowly than $^{12}CO_2$ within the chloroplasts; $^{12}CO_2$ gets to where it is needed more quickly, and so is preferentially metabolized.

Different types of plant discriminate against $^{13}CO_2$ to different degrees, however. Plants can be divided into three different groups, which are distinguished by the type of photosynthesis that occurs within them: C_3, C_4, and CAM. Research has found that C_3 plants discriminate against $^{13}CO_2$ to a greater degree than C_4 plants, and so C_3 plants absorb less $^{13}CO_2$ than C_4 plants. This difference in ^{13}C uptake can tell us some interesting things. For example, sucrose (table sugar) comes from either sugar beet (a C_3 plant) or sugar cane (a C_4 plant). If a sample of sucrose contains a relatively high amount of ^{13}C then we can deduce that it came from sugar cane and not sugar beet (because C_4 plants absorb ^{13}C (as $^{13}CO_2$) to a greater degree than C_3 plants).

Measuring ^{13}C levels in fossilized plants has even given us an insight into how plants evolved. ^{13}C levels in fossilized plants suddenly increased about 7 million years ago, suggesting that C_4 plants evolved at about that time.

Identifying different isotopes

An element's mass number indicates the total number of protons plus neutrons per atom. As isotopes of the same element contain different numbers of neutrons, each isotope must have a different mass number. In the case of oxygen, which has isotopes with eight, nine, and ten neutrons, the mass numbers of the three isotopes are 16, 17, and 18 respectively. (All the isotopes of oxygen contain eight electrons and eight protons.)

Where more than one isotope of an element exists, we often include the mass number as a superscript figure to the left of the element's chemical symbol to differentiate between isotopes of a single element. So, in the case of oxygen, the three isotopes are represented as ^{16}O, ^{17}O, and ^{18}O.

Relative abundances and atomic weight

In the natural world, we are rarely exposed to single isotopes of a given element. Instead, if we extract a sample of an element from a natural source it will invariably comprise a mix of different isotopes. For example, a sample of oxygen typically comprises 99.757% ^{16}O, just 0.038% ^{17}O, and 0.205% ^{18}O. The percentages are termed the **relative abundances** of the different isotopes. We see that the vast majority of naturally occurring oxygen is ^{16}O.

Similarly, magnesium occurs naturally as a mixture of three isotopes, with mass numbers of 24, 25, and 26. The relative abundance of each isotope is shown below, and illustrated graphically in Figure 2.4:

^{24}Mg 78.99%
^{25}Mg 10.00%
^{26}Mg 11.01%

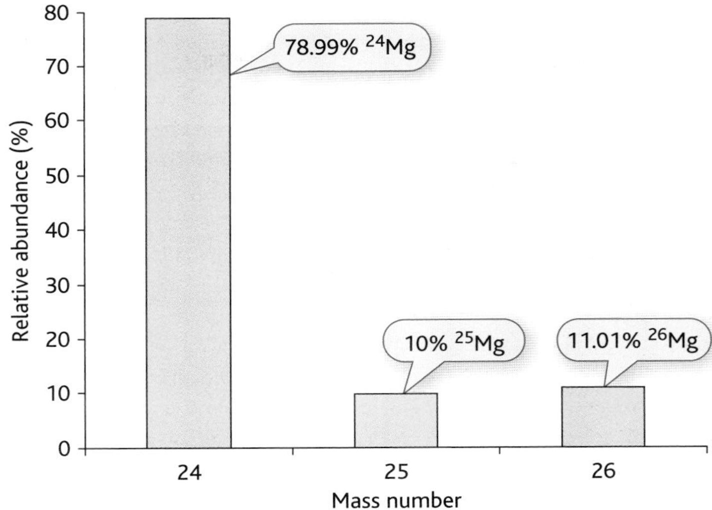

Figure 2.4 The relative abundance of the three isotopes of magnesium, Mg.

This means that if we take a sample of pure ('elemental') magnesium, 78.99% of the atoms in the sample have a mass number of 24, 10.00% have a mass number of 25, and 11.01% have a mass number of 26. In terms of 10 000 atoms, this is equivalent to saying that 7899 are ^{24}Mg, 1000 are ^{25}Mg, and 1101 are ^{26}Mg.

To find the mass number for the sample as a whole we calculate the **weighted** average of the mass numbers of each of the isotopes in the sample. The average mass number of all of the naturally occurring isotopes of an element is given by the **atomic weight, A_r**, of that element, also called the **relative atomic mass (RAM)**. The atomic weight is a dimensionless quantity–that is, it is quoted as a number *without units*.

EXAMPLE

For magnesium, the atomic weight is:

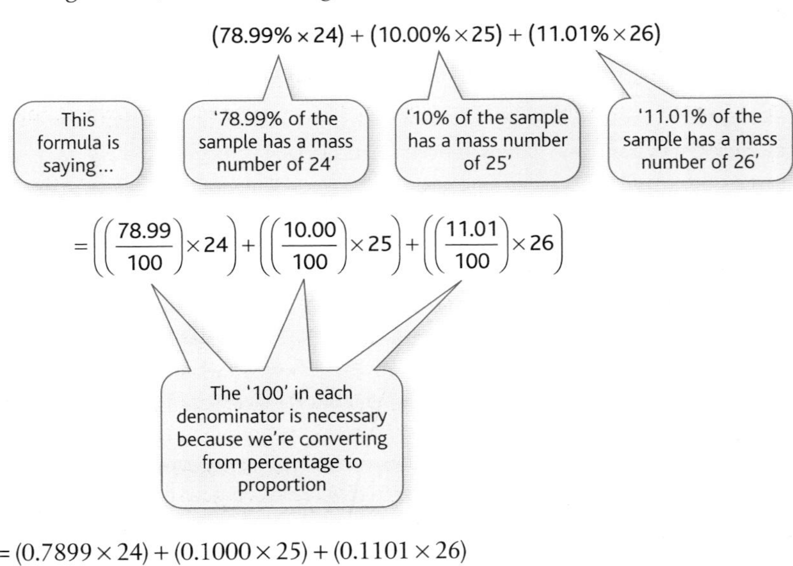

$= (0.7899 \times 24) + (0.1000 \times 25) + (0.1101 \times 26)$

$= 18.96 + 2.5 + 2.86$

$= 24.32$

Notice how, when we we add up the contribution of each isotope to the overall atomic weight, a majority of the sample (78.99%) has a mass number of 24, and so the atomic weight–the average mass number–is skewed towards a value of 24.

Self-check 2.4 Chlorine occurs naturally as a mixture of two isotopes, ^{35}Cl and ^{37}Cl, which have the following relative abundances:

^{35}Cl: 75.78% ^{37}Cl: 24.22%

What is the atomic weight of chlorine?

An element's atomic weight is often featured in the periodic table. Look at the periodic table on p. i, and notice where the atomic weight is quoted for each element.

As we noted on p. 13, hydrogen is a unique element: atoms of hydrogen contain no neutrons at all. However, hydrogen, too, exists as two isotopes–one with one proton, one electron, but no neutron (denoted ^{1}H), and one with one proton, one electron, and *one* neutron. This second isotope is given its own name, deuterium, and is represented by the chemical symbol ^{2}D. (Note that deuterium doesn't appear in the periodic table: it isn't a unique element because it has the same number of electrons and protons as hydrogen.) Only 0.015% of naturally occurring hydrogen exists as the isotope deuterium. But what would happen if this abundance was higher? Would this difference have a significant impact on biological systems? We discover the answer in Box 2.3.

BOX 2.3 Bring in the heavies...?

Deuterium was originally discovered as a natural isotope in water, where just 0.015% of water was found to be HDO, but just one molecule in 41 million was found to be D_2O rather than H_2O. (D_2O is given the name 'heavy water', to reflect the extra neutrons it contains.) But what effect does this extra neutron have on the properties of heavy water versus its lighter cousin? Are there any differences, and are they positive or negative, as far as biological systems are concerned?

Studies that examined the effect of exposing organisms to different levels of heavy water revealed some intriguing differences between D_2O and its lighter counterpart. High levels of D_2O (around 90%) proved quickly fatal to fish, flatworms, and the fruit fly *Drosophila*. At lower levels, around 25%, mice, rats, and dogs were found to be seemingly healthy for long periods, but were sterile (despite producing eggs and sperm).

High levels of D_2O are thought to disrupt cellular function in several ways: they impair haematopoiesis, the development and maturation of blood cells (leading to conditions such as anaemia, a reduced red blood cell count), and inhibit mitosis, the process through which cells duplicate their chromosomes, and then divide into two daughter cells. (In the latter case, D_2O is thought to affect the mitotic spindle, the cellular structure along which duplicated chromosomes align before being pulled apart into separate daughter cells.) It has also been observed to disturb nerve and muscle action.

But the story is not all bad. Recently, the intriguing possibility has been raised that drinking heavy water could have positive effects, particularly in combating the effects of ageing. One of the most prominent causes of ageing is thought to be free radical damage, whereby highly reactive chemical species damage biological molecules, including proteins and DNA, ultimately leading to the disruption of biochemical processes, and the kind of degradation associated with ageing. In proteins, free radicals are thought to have their damaging effect by targeting C–H bonds. So why might drinking heavy water help to combat this damage?

The answer lies in the strength of C–H bonds versus C–D bonds. The C–D bond breaks about ten times more slowly than the C–H bond, effectively making it far harder to break. And if its bonds were harder to break, a protein would be more resistant to the kind of damage wreaked by free radicals, potentially slowing down the ageing process as a result.

BOX 2.3 **Continued**

So the question is: can deuterium be administered at levels that could be therapeutically useful, without the toxic effects noted above being incurred? The answer appears to be 'yes': studies suggest that up to a fifth of the water in your body could be replaced with heavy water without you suffering ill effects as a result.

These studies open up intriguing possibilities for the incorporation of deuterium in your diet as a means of 'strengthening' biological molecules such as proteins and DNA, and slowing down the ageing process. Indeed, studies at the Institute for the Biology of Ageing are reported to have shown low levels of heavy water extending the lifespan of fruit flies by up to 30%.

So, perhaps one day, we'll be faced with the refreshing prospect of drinking ourselves to a longer life? A novel idea indeed.

We need to pause for a moment for a note on terminology: quite often, the term 'atomic weight' is used interchangeably with 'atomic mass'. However, such interchangeable use isn't strictly correct, and should be avoided. An element's **atomic mass** refers to the mass of a single atom of a *single isotope*, whereas atomic weight (or relative atomic mass) is the weighted mass of a sample of an element, which may comprise a mixture of *different* isotopes.

We revisit the atomic mass of elements in Chapter 11 when we look at measurements and quantities of chemical species in more detail.

In biological systems, we invariably encounter elements as a mixture of different isotopes (for this is how they exist in nature), so it is more appropriate for us to consider atomic weights than atomic masses.

- ***An element's atomic weight (or relative atomic mass) is a measure of the average mass of an atom of that element.***
- ***An element's atomic mass is the mass of a single atom of a single, specific isotope.***

Protons and chemical identity

In the preceding section we see how a particular element is characterized by the number of protons, electrons, and neutrons from which it is composed. An atom of a given element may vary in the number of **neutrons** it contains, giving rise to different isotopes of that element. And the element will exist as an ion if it gains or loses **electrons**.

The one constant feature–the characteristic of an element that dictates its true chemical identity–is the number of protons a given element contains.

Lithium has an atomic number of 3 and exists as a mixture of two isotopes, ^{6}Li and ^{7}Li. So a lithium atom comprises three electrons and three protons, and either three or four neutrons. We also see in Figure 2.3(a) how a lithium atom can also lose one electron to form a positively charged ion, Li^{+} (a cation). This lithium ion contains three protons, either three or four neutrons (depending on its isotope), and just two electrons.

The constant feature–the characteristic which dictates that the ion or atom is lithium–is the presence of three protons. Similarly, an atom or ion with two protons is always identified as the element helium, He, and an atom or ion with four protons is the element beryllium, Be.

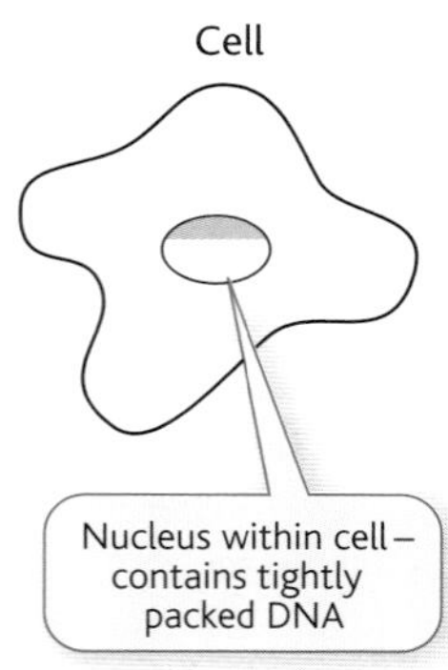

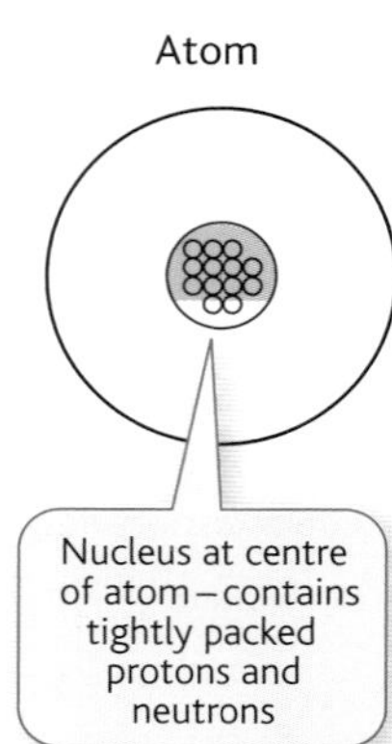

Figure 2.5 The nucleus occupies the centre of the atom, just as a cell's nucleus occupies (roughly) the centre of the cell.

2.3 Atomic structure

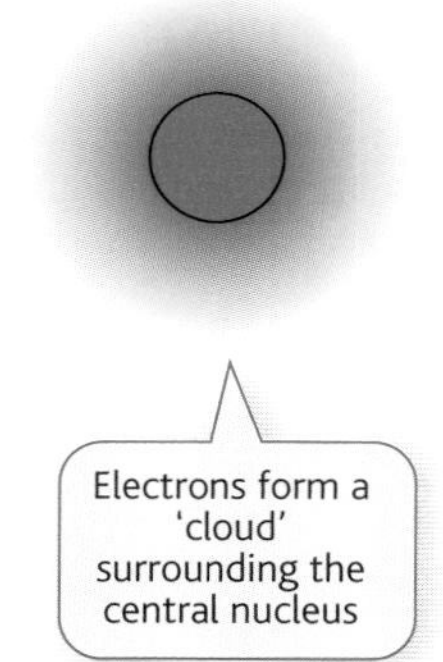

Figure 2.6 Electrons fill the majority of the space occupied by an atom. While the nucleus is a tightly packed structure, the electrons form a diffuse 'cloud'.

Protons, neutrons, and electrons do not simply 'clump' together haphazardly to form the atom. Instead, they are organized in a particular way. This structural organization underpins the behaviour of atoms, and the chemical properties they exhibit.

Just as a eukaryotic cell has a nucleus containing its DNA, so every atom has a nucleus at its centre, as illustrated in Figure 2.5. The nucleus contains the atom's protons and neutrons, packed together very tightly such that the nucleus occupies a mere fraction of the volume of the atom as a whole. An atom's nucleus has a diameter which is roughly ten thousand times smaller than that of the atom itself. To illustrate the relative difference in size here, if the nucleus were 1 mm wide–as long as this line:

-

the atom itself would be slightly longer than a double-decker bus.

Surrounding the nucleus are the atom's electrons, which fill the majority of the total space taken up by the atom. The electrons aren't stationary, but move continually around the nucleus to form an electron 'cloud', as shown in Figure 2.6[4]. However, the location of these electrons isn't random. Imagine a wasp trapped inside a jam jar. There is no restriction on where the wasp can fly within the jar, but it is unable to escape from it. Similarly, an atom's electrons are confined to moving within specific volumes of space we call **orbitals**–like the wasp, they can move freely within an orbital but are not free to move outside it.

Atomic orbitals

The orbitals in which electrons reside have different shapes, and are located at varying distances from the nucleus–some occupy space very close to the nucleus, while others lie further away. Figure 2.7 illustrates four different types of orbital–s, p, d, and f–with each type having a distinctive shape.

Orbitals are grouped into families we call **shells**, which are identified by a number: 1, 2, 3, 4, etc.

Each shell contains a unique combination of orbitals:

4 Electrons can often be portrayed as orbiting the atom's nucleus on fixed, circular paths, in much the same way the Earth orbits the Sun. Though this is a relatively easy model to visualize, it does not give a good impression of how electrons are really distributed in the atom.

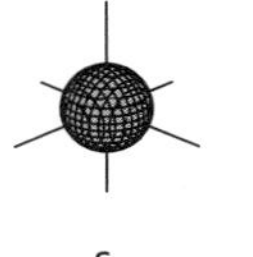

Figure 2.7 The four types of atomic orbital–s, p, d, and f. Each type of orbital has a characteristic shape.

- Shell 1 contains one s orbital.
- Shell 2 contains one s orbital and three p orbitals.
- Shell 3 contains one s orbital, three p orbitals and five d orbitals.
- Shell 4 contains one s orbital, three p orbitals, five d orbitals, and seven f orbitals.

A group of equivalent orbitals (for example, the three p orbitals in shell 3) are termed a **subshell**. We can see that shell 3 contains a total of three subshells, one containing the single s orbital, one containing the three p orbitals, and one containing the five d orbitals.

Look at Figure 2.8, which illustrates how orbitals arrange to form shells and subshells. Notice how different shells have different numbers and types of subshell.

Just as we locate a house within a town with two elements of its address (the name of its street and the number within that street), so we locate an atomic orbital by its shell number and the orbital type. We say an electron residing in one of the three p orbitals in shell 3 is a 3p electron, while an electron residing in the s orbital of shell 2 is a 2s electron. These residencies are illustrated in Figure 2.9.

- *There are four different types of orbital: s, p, d and f.*
- *Orbitals are grouped into shells.*
- *Equivalent orbitals occupying the same shell are grouped into subshells.*

Any orbital, regardless of whether it is an s, p, d, or f orbital, can hold a maximum of *two* electrons. Therefore, shell 1 can hold a total of two electrons because it comprises only one orbital, and shell 2 can hold a total of eight electrons (two in each of the single s orbital and the three p orbitals, giving eight altogether).

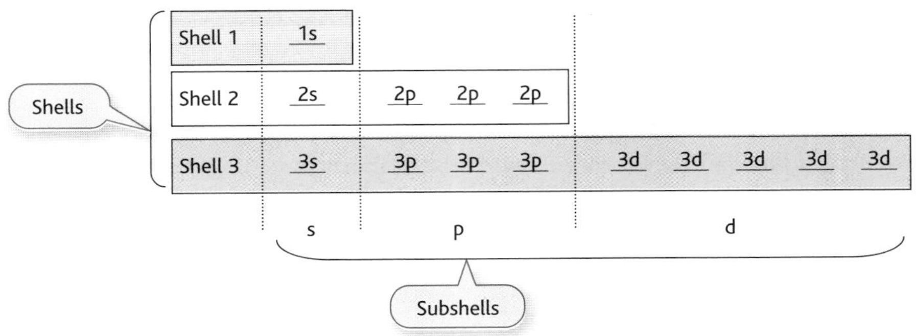

Figure 2.8 The grouping of atomic orbitals into shells and subshells. Notice how different shells comprise different numbers of subshells.

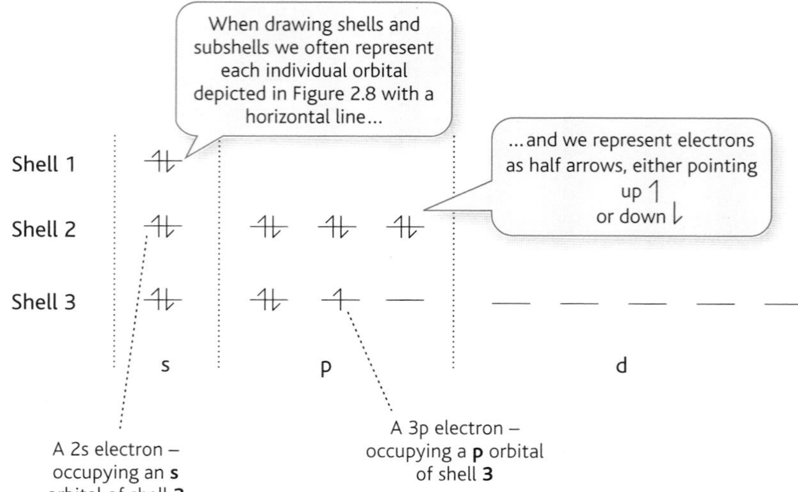

Figure 2.9 Determining an electron's 'address'. We identify the atomic orbital in which an electron is located by its shell number and orbital type. Each single-headed arrow represents an electron.

The restriction that an orbital can hold a maximum of two electrons is described by the **Pauli exclusion principle**, which has an important impact on the physical space that an atom occupies.

The majority of every living thing is actually empty space. An atom, with its tightly packed nucleus at the very centre, is over 99% emptiness. The reason that atoms do not condense to save this seemingly wasted volume is a consequence of the Pauli exclusion principle. Whether they like it or not, as their number increases, electrons are forced to occupy orbitals of progressively higher energy–orbitals that lie further and further away from the nucleus–rather than crowding round the nucleus. The tiny nucleus really belies the volume that an atom actually occupies.

Self-check 2.5 What is the maximum number of electrons that shell 3 can hold?

2.4 The energy of atoms

Orbitals and energy levels

Orbital energies differ according to their distance from the nucleus, with the orbital energy increasing as its separation from the nucleus increases.

Figure 2.10 illustrates the relative energies of those orbitals shown in Figure 2.8. Look at this figure and notice the following:

- Shells 1 to 4 possess progressively higher energies: shell 1 is of the lowest energy, shell 4 of the highest.
- s, p, and d orbitals within the same shell possess progressively higher energies: s orbitals are the lowest energy, d orbitals the highest.

The energy of an electron is equivalent to the energy of the orbital in which it is located–if an electron is in a low-energy orbital then it has low energy; if it is in a high-energy orbital then it has high energy. For example, an electron in a 3s orbital possesses more energy than an electron in a 2p orbital. However, an electron in a 3p orbital possesses more energy than an electron in a 3s orbital.

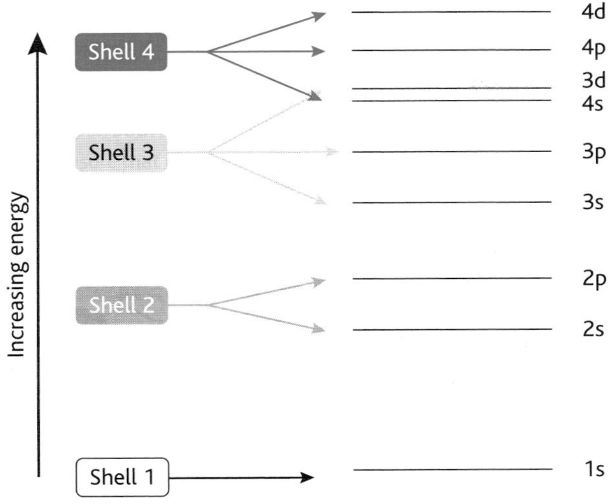

Figure 2.10 The relative energies of atomic orbitals. The vertical height at which an orbital is drawn here is proportional to its energy.

Filling up orbitals—the building-up principle

An electron will not enter an orbital at random. Rather, the orbitals are filled sequentially. Orbitals of lowest energy are filled first, and only later will higher-energy orbitals receive an electron. An analogous example is filling up a cup with marbles: the first marble falls to the bottom of the cup (lowest energy), and only as the number of marbles increases will the marbles reside elsewhere in the cup. As a good generalization, low-energy orbitals are positioned closer to the nucleus, and higher-energy orbitals extend further away, spatially.

We call the principle of electrons sequentially entering orbitals of progressively higher energy the **Aufbau**, or 'building-up', principle.

The *order* in which orbitals are filled can be summarized by the illustration in Figure 2.11.

Let's look at two examples:

A helium atom has two electrons, which enter the lowest-energy orbital: the s orbital in shell 1. We say that the helium atom has two 1s electrons.

An atom of lithium has three electrons; two of these enter the single s orbital in shell 1 (as per helium). This shell is now full, so the third electron moves to the shell of next highest energy, shell 2, and enters the lowest-energy orbital in that shell: again, an s orbital. So, lithium has two 1s electrons and one 2s electron.

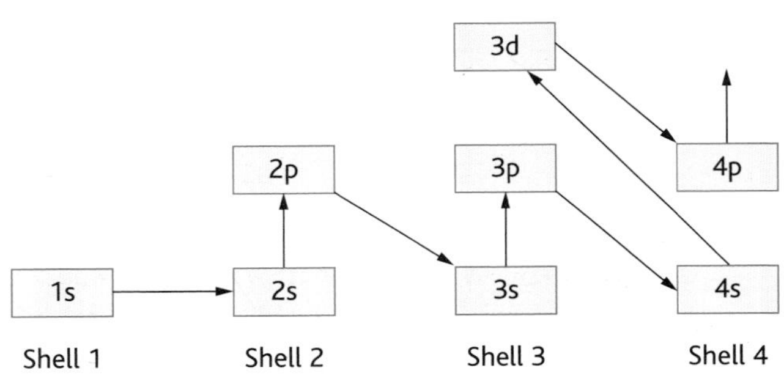

Figure 2.11 The order of filling of atomic orbitals. Notice how the 4s orbital fills before the 3d orbital. This is because the 4s orbital is of slightly lower energy than the 3d orbital, despite occupying a higher shell.

Notice that the two electrons in helium both occupy the 1s orbital, rather than one electron occupying the 1s orbital, and the other occupying the 2s orbital, the orbital of next-lowest energy. The two electrons both occupy the 1s orbital because, as a general rule, the lowest-energy shell must be completely filled before subsequent shells become occupied.

- ***Low-energy orbitals become filled with electrons before high-energy orbitals.***

Self-check 2.6 An atom of carbon has six electrons. Which orbitals are these electrons located in?

There are exceptions to this general rule, however. Look again at Figure 2.11. Notice how the 4s orbital fills before the 3d orbital. This is because the 4s orbital is of slightly lower energy than the 3d orbital (despite being part of a higher-numbered shell). Shell 4 is generally of higher energy than shell 3, but the 3d and 4s orbitals don't follow this general trend.

So why do orbitals fill in such an organized way? One of our essential concepts from Chapter 1 is that entities with lower energy are more stable than those with higher energy. An atom is most stable if its electrons have the lowest energy possible. To achieve this condition, electrons locate themselves as close to the nucleus as they can, entering low-energy orbitals closest to the nucleus before filling high-energy orbitals that are further away.

So, the order in which electrons fill orbitals can really be explained in terms of an atom trying to remain as stable as possible, by arranging its electrons to minimize their energy.

Electronic configurations

The way in which electrons are distributed within the orbitals of an atom is termed the atom's **electronic configuration**. An atom's electronic configuration is indicated by writing, from left to right, the subshells which contain electrons, in the order in which they become filled. The number of electrons in each subshell is then indicated by a superscript number to the right of the subshell. We add up these numbers to check if we have written the correct superscript numbers. They must equal the total number of electrons in the atom, which equals the atomic number.

So, for example, helium's electronic configuration is $1s^2$, while lithium's electronic configuration is:

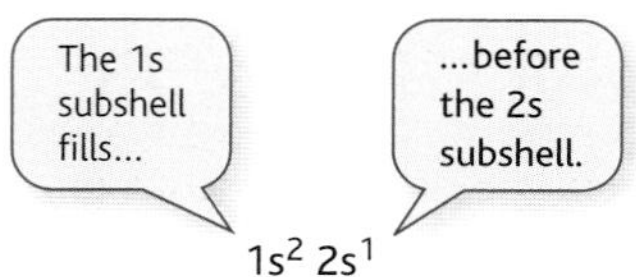
The 1s subshell fills...
...before the 2s subshell.
1s2 2s1

Self-check 2.7 What is the electronic configuration of a carbon atom?

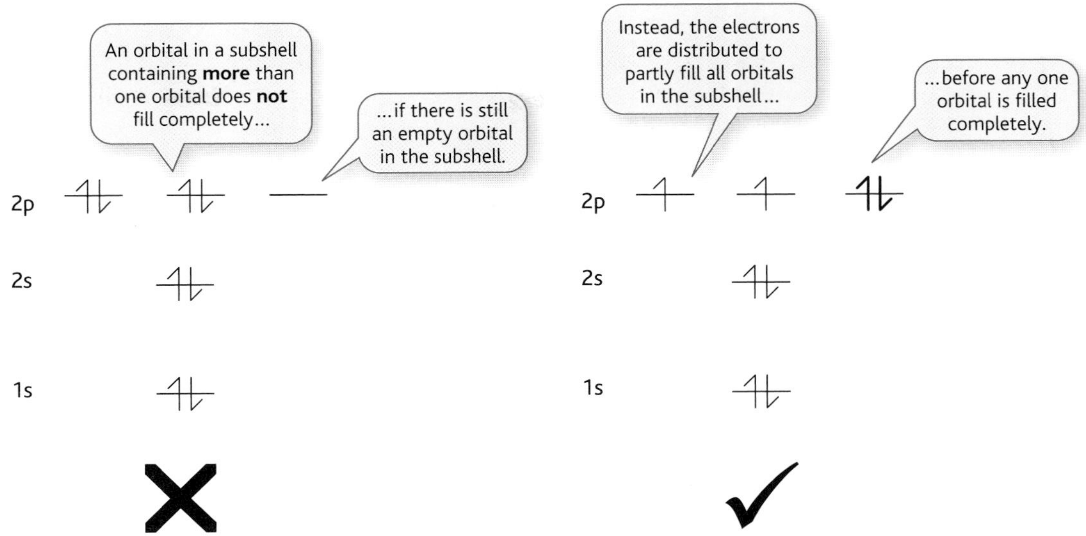

Figure 2.12 Filling degenerate orbitals. The subshells of degenerate orbitals all become partially filled before any one orbital becomes fully occupied.

The filling of subshells

Where a subshell contains more than one orbital (i.e. a subshell containing either p, d, or f orbitals), all the orbitals in that subshell have the *same* energy (we say they are **degenerate**). So, the three p orbitals in a p subshell have the same energy, and the five d orbitals in a d subshell have the same energy.

This doesn't mean that electrons enter these subshells in a random order, so that we might have two p orbitals that are completely full (holding two electrons each) whilst one orbital remains empty, as depicted in the left-hand diagram of Figure 2.12. Instead, an electron has a slightly lower energy if it enters an empty orbital rather than one already containing an electron.

Look at the right-hand diagram of Figure 2.12 and notice how each of the orbitals in a subshell accepts one electron to become partially filled before any of the orbitals accepts a second electron to become full. For example, one electron enters each of the three orbitals in a p subshell before any fills by accepting a second electron.

- ***Each degenerate orbital in a subshell must be occupied by one electron before any one of the orbitals becomes completely filled.***

Now look at Figure 2.13, which illustrates the electronic configurations of boron and carbon. Notice how the additional 2p electron in carbon occupies an empty 2p orbital, rather than occupying the 2p orbital that is already partially filled.

The concept of equivalent orbitals being filled with one electron each before a second electron enters is captured by **Hund's rule**.

Look at Figure 2.14, which simultaneously shows the Aufbau principle, the Pauli exclusion principle, and Hund's rule in action. This figure illustrates the order of filling of the atomic orbitals in the various shells and subshells in atoms of neon, sodium, magnesium, aluminium, silicon, and phosphorus. We start with neon (element 10), and then show the subsequent elements, each of which has more electrons.

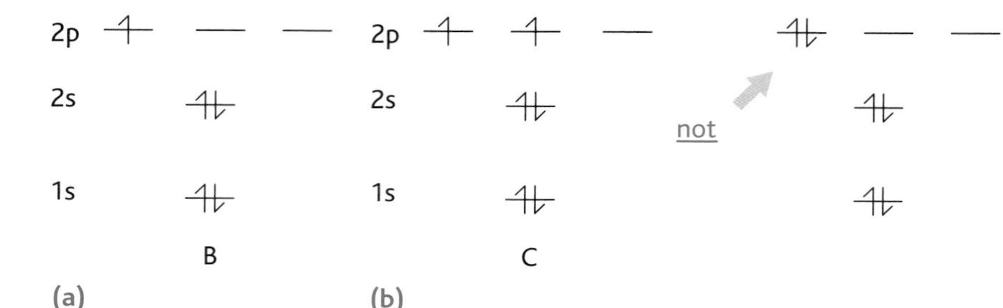

Figure 2.13 The electronic configurations of (a) boron and (b) carbon.

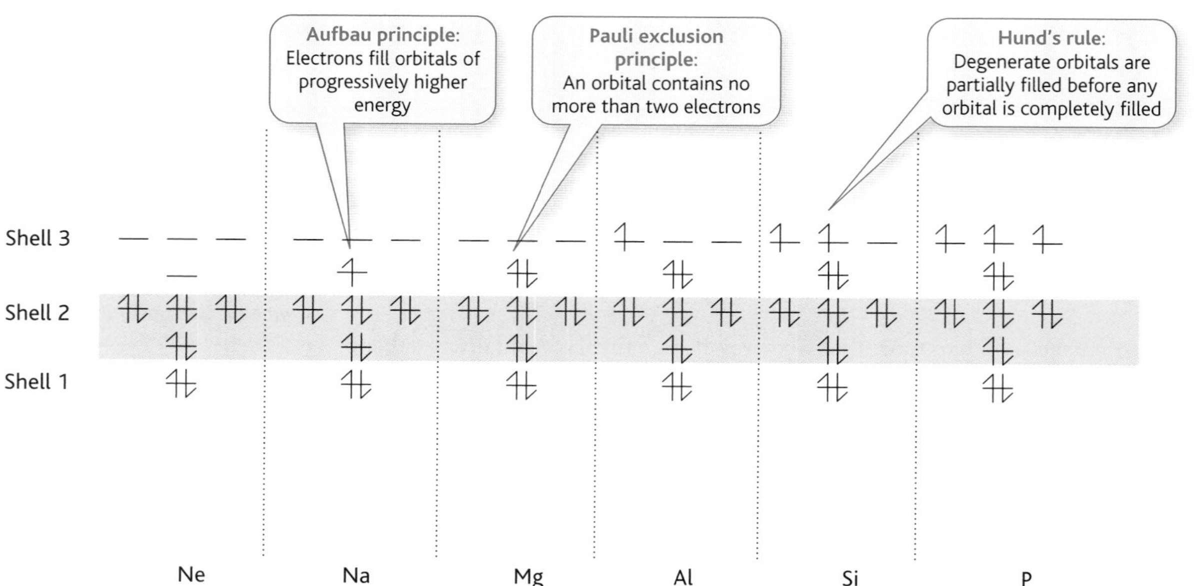

Figure 2.14 The Aufbau principle, Pauli exclusion principle, and Hund's rule in action.

In particular, look out for the following:

- As we increase the number of electrons, orbitals are filled at progressively higher energies. For example, neon has ten electrons, with the highest energy electrons being located in the three 2p orbitals, all of which are full. When there is one additional electron, as for the sodium atom, it enters the next-highest available orbital, the 3s orbital.

We are seeing the **Aufbau principle** in action.

- There are no more than two electrons in any atomic orbital in any of the atoms. In some cases, an orbital may contain just one electron (look at the 3p orbitals in aluminium, silicon, and phosphorus, for example).

We are seeing the **Pauli exclusion principle** in action.

- Equivalent orbitals in a subshell all partially fill (each occupied by a single electron) before any of them becomes totally filled with two electrons. Look at aluminium, silicon, and phosphorus. Notice how the additional electron in the silicon atom, compared to the aluminium atom, has entered the second 3p orbital, rather than joining the electron that is already located in the first 3p orbital. Likewise, the additional electron in the phosphorus atom has entered the third, empty, 3p orbital, rather than entering either of the partially filled 3p orbitals.

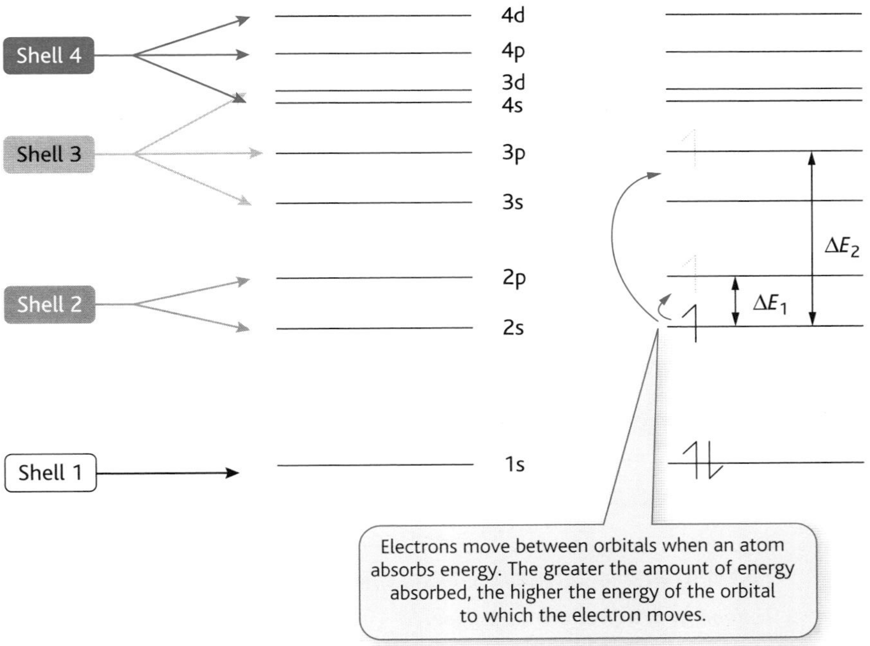
Shell 4
Shell 3
Shell 2
Shell 1
4d
4p
3d
4s
3p
3s
2p
2s
1s
ΔE2
ΔE1
Electrons move between orbitals when an atom absorbs energy. The greater the amount of energy absorbed, the higher the energy of the orbital to which the electron moves.

Figure 2.15 Electron excitation: moving between atomic orbitals. The vertical position of an atomic orbital in this figure represents its relative energy. Notice how atomic orbitals have fixed energies–they are specific distances apart. An electron can move between orbitals, but only if the atom absorbs energy that's exactly equivalent to the difference in energy between two orbitals (e.g. ΔE_1 or ΔE_2).

We are seeing **Hund's rule** in action.

Self-check 2.8 What is the arrangement of electrons in an atom of sulfur, the element with one more electron than phosphorus?

Moving between orbitals: electron excitation

Electrons can only possess specific energies and occupy defined orbitals, but they are not restricted to *always* staying in one particular orbital. If an electron in a low-energy orbital is given a sufficiently large boost of energy it is able to move to an orbital of higher energy. (Note that the higher-level orbital need not be completely empty, but must have at least one vacancy, because promotion to an orbital containing two electrons would yield an orbital possessing an 'illegal' number of three electrons.)

Look at Figure 2.15, which depicts a set of atomic orbitals. The vertical heights of the horizontal lines represent the relative energies of the different orbitals, which progressively increase as we move vertically from orbital to orbital.

Consider a ladder that has rungs set at different distances apart. The rungs may not be evenly spaced, but the distance between adjacent rungs is fixed. Similarly, while the difference in energy between two orbitals depends on their identities (the difference in energy between a 2s and a 2p orbital is less than the difference in energy between a 2p and a 3s orbital, for example), the difference in energy between a specific pair of orbitals is fixed. Look at the horizontal lines representing the 2s and 2p orbitals. The distance between these two lines represents the difference in energy between the two orbitals, represented in Figure 2.15 as ΔE_1[5]. This is a fixed difference in energy.

5 The symbol Δ means 'change in', and we use the symbol *E* to denote 'energy'. So Δ*E* means 'change in energy'. The subscript 1 and 2 (in ΔE_1 and ΔE_2) indicate which energy change we're talking about.

Similarly, the difference in energy between the 2s and 3p orbitals is a fixed amount, represented by ΔE_2.

Now focus on the single electron in the 2s orbital in shell 2. This electron has quite a low energy. Imagine that we give this electron a boost of energy equal to the value of ΔE_1. The electron now has an energy equivalent to that of the 2p orbital and is able to move to this higher-energy orbital. If the energy boost were larger (equal to ΔE_2, for example) then the electron would have sufficient energy to move to the 3p orbital.

This process of electrons 'jumping' to higher-level orbitals after absorbing 'packets' of energy is known as **excitation**.

Excitation is just a transient event; in other words, an electron only retains its extra energy briefly. Remember that atomic species with higher energy are less stable than those with lower energy. By absorbing more energy, and moving to a higher-energy orbital, an electron is less stable than when it is in its normal or **ground** state. So, once it has jumped to the higher-energy orbital, the electron releases the energy it gained and falls back to its original orbital to regain its previous stability. We say the electron **emits** energy.

- ***Electrons can jump between orbitals of different energy.***
- ***The energy of a given orbital is fixed.***

Excitation is central to a number of biological processes, including visual perception, as explained in Box 2.4.

BOX 2.4 What a sight! The power of excitation

The phenomenon of excitation lies behind a variety of vital processes in biology. In fact, you are relying on excitation at this very second, merely by reading the words on this page.

As we read (or look at anything) our eyes detect photons of light, which enter through the iris, travel through the ocular fluid (the aqueous medium that fills the eyeball), and strike the retina (a layer of photosensitive cells covering the inner, rear wall of our eyeball). The striking of the retina by a photon of light sends an electrical impulse to the brain via the optic nerve, and the brain transforms this impulse into our perception of 'seeing' something.

But how is the striking of the retina by a photon of light converted into an electrical impulse? The answer lies in excitation. The cells of our retina contain a light-sensitive pigment, rhodopsin. Rhodopsin is comprised of a protein to which is attached the molecule 11-*cis*-retinal:

CHO

When rhodopsin absorbs a photon of light, an electron in the 11-*cis*-retinal molecule undergoes excitation. (The energy possessed by the photon of light provides the 'boost' required to excite the electron.) This excitation induces a change in the shape of the 11-*cis*-retinal which, in turn, activates the rhodopsin. The activation of rhodopsin triggers a cascade of biochemical reactions, which ultimately result in the generation of the electrical impulse that is converted by our brains into vision.

Humans aren't the only organisms to depend on excitation. When plants photosynthesize (Box 2.2) they absorb energy from sunlight. Just as our eyes contain the pigment rhodopsin, plants contain the pigment chlorophyll. When chlorophyll absorbs light, an electron is excited in a manner analogous to an electron in 11-*cis*-retinal. In this instance, however, the excitation is coupled to the synthesis of sugar, rather than the sensation of vision.

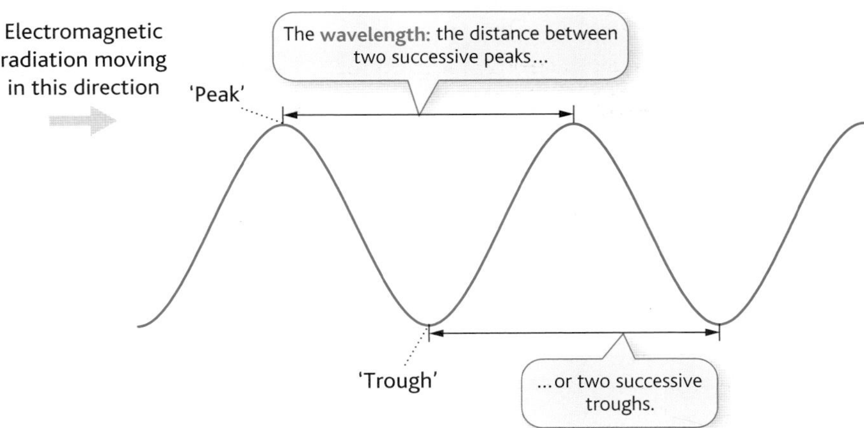
Electromagnetic radiation moving in this direction
'Peak'
The wavelength: the distance between two successive peaks...
'Trough'
...or two successive troughs.

Figure 2.16 The wavelength of electromagnetic radiation. The wavelength is equal to the distance between two successive peaks or two successive troughs.

The electromagnetic spectrum

We have just seen how electrons can absorb and emit energy. To be precise, electrons emit energy in the form of **electromagnetic radiation**. Electromagnetic radiation pervades all of our lives, in both positive and negative ways. A microwave oven cooks food by bombarding it with one form of electromagnetic radiation: microwaves. Conversely, doctors advise us to apply sunblock before sunbathing in intense sunlight to protect any exposed skin from the damaging effects of another form of electromagnetic radiation: ultraviolet (UV) radiation.

Electromagnetic radiation can be characterized by its **wavelength**: electromagnetic radiation moves as a wave of energy, as illustrated in Figure 2.16, and the wavelength is equal to the distance (usually expressed in metres) between peaks (or troughs) of successive waves.

The wavelength of electromagnetic radiation varies enormously. The range across which the wavelength of electromagnetic radiation varies is represented by the **electromagnetic spectrum**, and is illustrated in Figure 2.17. Cosmic rays, at one end of the spectrum, have a wavelength of 10^{-14} m, while radio waves, at the other end of the spectrum, have wavelengths in excess of 10^{0} m (or 1 m).

The energy that a particular type of electromagnetic radiation possesses is dictated by its wavelength. **High-energy** radiation such as cosmic radiation has **short** wavelengths; **low-energy** radiation such as radio waves has **long** wavelengths.

Self-check 2.9 Look at Figure 2.17. Which has the most energy: ultraviolet or infrared light? Why?

The region of the electromagnetic spectrum with which we are most familiar is that which encompasses electromagnetic radiation with wavelengths between 3.5×10^{-7} and 7×10^{-7} m. This is the **visible spectrum** and encompasses the wavelengths of electromagnetic radiation that we can detect with our eyes: **visible light**.

A rainbow is one of the most stunning ways of visualizing the visible spectrum. Visible light is a mixture of electromagnetic radiation with different wavelengths that span the visible spectrum. A rainbow forms when visible light is separated out into separate wavelengths as it passes through raindrops. The eye perceives these

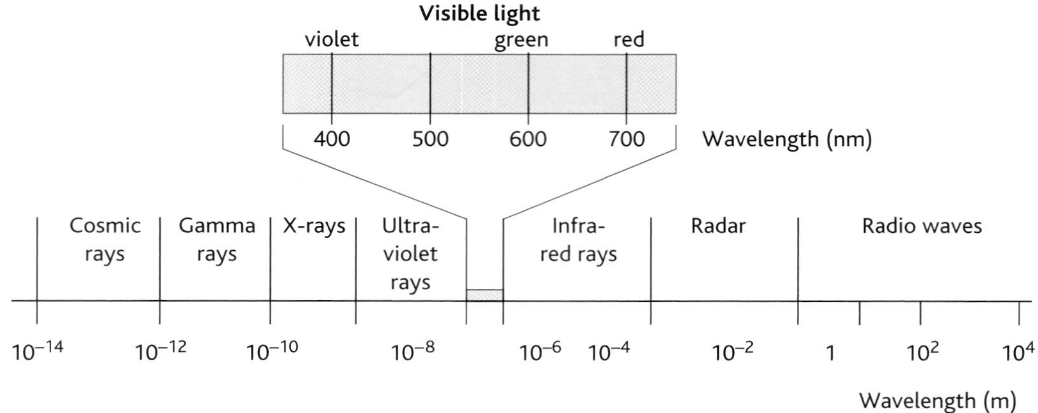

Figure 2.17 The electromagnetic spectrum. Different regions of the electromagnetic spectrum are assigned characteristic names; each region spans a characteristic range of wavelengths.

separate wavelengths of visible light as a specific colour, producing the distinct colours of the rainbow.

- ***High-energy radiation has a short wavelength.***
- ***Low-energy radiation has a long wavelength.***

Energy levels and quantization

The energy of an electron is dictated by the energy of the orbital in which it is located (p. 24). In consequence, electrons can only possess specific amounts of energy. Just as electrons aren't randomly located within the atom, but are restricted to specific volumes of space (the orbitals), so they can't have random energy levels–their energy levels must match those of the orbitals in which they are located. The concept of energy being restricted to specific levels is known as **quantization**, so an electron's energy is said to be **quantized**. Read Box 2.5 to find out how the phenomenon of quantization has been demonstrated experimentally.

BOX 2.5 How do we know that electrons are quantized?

To demonstrate the occurrence of quantization, we look at the effects of electrons being boosted to a higher energy level, and subsequently releasing that extra energy as they fall back to their original lower-level orbital.

Scientists studied hydrogen, a simple element to work with as it contains only one electron. They gave extra energy to the electron to excite it, and then monitored the energy emitted as the electron 'relaxed' back to its original orbital. The amount of energy associated with electromagnetic radiation is dictated by its wavelength (p. 31). Scientists found that the electron emitted electromagnetic radiation of specific wavelengths–i.e. only certain amounts of energy. However, these 'certain amounts' of energy spanned the electromagnetic spectrum, as illustrated in Figure 2.18.

This figure depicts the line emission spectrum of hydrogen. Each line represents electromagnetic radiation of a particular wavelength emitted by the electron 'relaxing' from a high-energy orbital to a lower-energy one. The shorter the wavelength at which the line appears, the higher the amount of energy released by the electron, and the greater the difference in energy of the two orbitals between which the electron has moved.

Let's go back to our example on p. 29 of an electron moving between a 2s and a 2p orbital to release an amount of energy equal to ΔE_1, and between a 2s and 3p orbital to release an amount of energy equal to ΔE_2. How would these energy emissions be represented on a line emission spectrum similar to that in Figure 2.18?

BOX 2.5 **Continued**

The difference in energy between the 2s and 3p orbital (ΔE_2) is greater than the difference in energy between the 2s and 2p orbital (ΔE_1). Therefore, an electron moving from a 3p orbital to a 2s orbital would release energy of a shorter wavelength (higher energy) than an electron moving from a 2p orbital to a 2s orbital (lower energy), and the line for the 3p to 2s movement would appear to the left of the line for the 2p to 2s movement, as illustrated in Figure 2.19.

The finding that lines on the line emission spectrum appeared only at fixed wavelengths–rather than the energies generating a continuous, inseparable 'blur'–confirmed that excited electrons were restricted to moving to orbitals of fixed higher energy, and hence to releasing fixed amounts of energy when they fell back from these higher-energy orbitals to their original ground state orbitals. However, the number of lines on the spectrum showed just how many different energy levels a single electron could occupy.

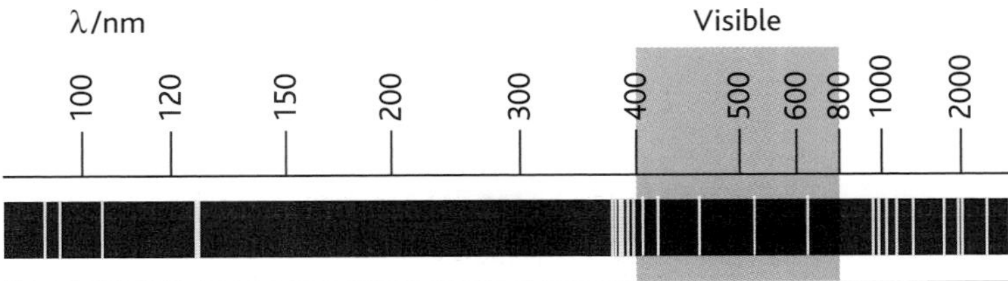

Figure 2.18 A representation of the line emission spectrum for hydrogen.

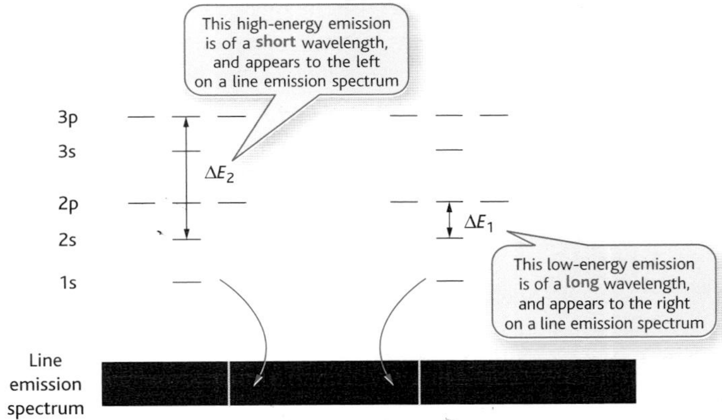

Figure 2.19 The correlation between the energy emitted by an atom during electron excitation and the line generated on a line emission spectrum. A low-energy emission generates a line to the right on a line emission spectrum; a high-energy emission generates a line to the left.

2.5 Valence shells and valence electrons

An atom's protons and neutrons are clustered together within the nucleus at the centre of the atom, whilst the electrons are located in orbitals, which expand out to fill the rest of the atom.

The shell which holds those electrons located furthest from the nucleus in a particular atom is called the **valence shell**. We can visualize the valence shell of an atom as representing the exterior boundary (the 'outside') of the atom. The volume of space occupied by the electrons in the valence shell effectively delineates the outer limits of the atom, as illustrated in Figure 2.20.

The identity of the valence shell varies from atom to atom. For example, looking at an oxygen atom (depicted in Figure 2.21), the valence shell is shell 2. Oxygen has eight electrons: two of these electrons fill the lowest-energy 1s orbital, two go on to fill the 2s orbital, and the remaining four partially fill the three 2p orbitals. In contrast, for a chlorine atom, which has 17 electrons, the valence shell is shell 3, with the outermost 3p orbitals being partially filled.

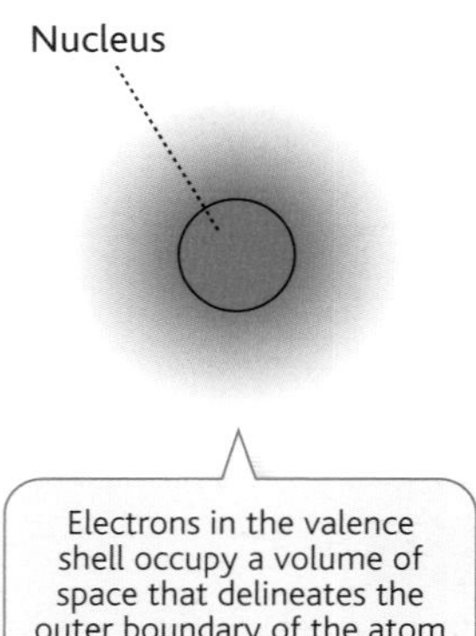
Nucleus
Electrons in the valence shell occupy a volume of space that delineates the outer boundary of the atom

Figure 2.20 An atom's valence shell delineates the outer limits of the atom.

Self-check 2.10 Which shell is the valence shell of calcium? What is the electronic configuration of the valence shell?

Electrons located in the valence shell are termed **valence electrons**. Critically, it is an atom's valence electrons which dictate the chemical reactivity of that atom. But why is this the case?

If we pick up an apple in each hand, and move them together until they are touching, the skins are the parts of the two apples which are in contact, not the cores at the centre of the apples. Similarly, Figure 2.22 shows how it is the valence electrons, at the exterior boundary of the atom, which come into contact with a neighbouring atom when two atoms come together, and not the electrons located further inside the atom.

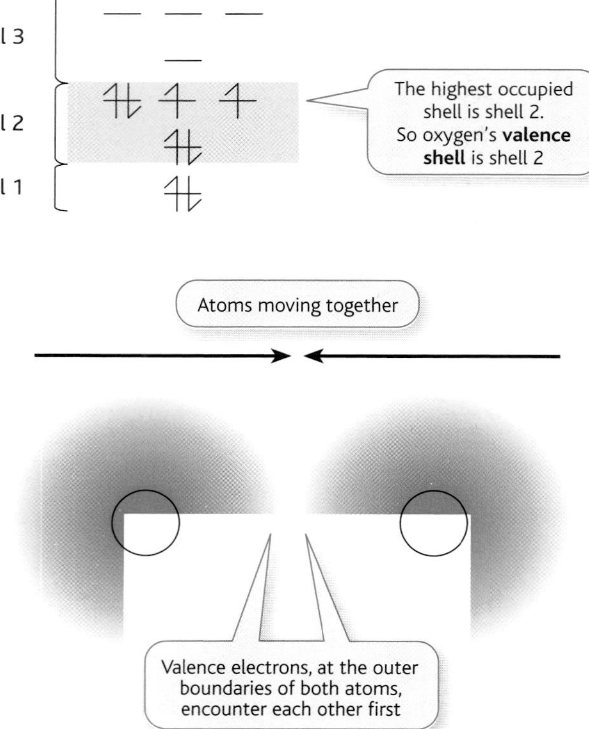
Shell 3
Shell 2
Shell 1
The highest occupied shell is shell 2. So oxygen's **valence shell** is shell 2
Atoms moving together
Valence electrons, at the outer boundaries of both atoms, encounter each other first

Figure 2.21 The electronic configuration of an oxygen atom. Oxygen's valence shell is shell 2.

Figure 2.22 The valence electrons of neighbouring atoms are those components of the atoms that interact when the two atoms approach one another. It is this interaction between valence electrons on neighbouring atoms that is the basis of chemical reactions.

Chemical reactions occur when valence electrons move between atoms (as we see in Chapter 17). So, chemical reactions are possible when contact occurs between atoms, so that the valence electrons of one atom come into sufficiently close proximity with another atom for movement between atoms to occur.

The **reactivity** of a particular atom is dependent upon the readiness with which its valence electrons will interact with a neighbouring atom. If an atom's valence electrons are reluctant to interact with a neighbouring atom, then no chemical reaction will occur. However, if an atom's valence electrons interact readily with another atom, then a chemical reaction will be much more likely. So, valence electrons really dictate the chemical properties of the different chemical elements.

The exact reactivity of the valence electrons varies greatly from element to element, as explained in the next chapter, when we discover how atoms interact with one another to form compounds.

We learn more about the involvement of valence electrons in chemical reactions in Chapter 17.

- ***The valence shell holds those electrons located furthest from the nucleus, called valence electrons.***
- ***Valence electrons lie at the heart of an element's chemical reactivity.***

Valence electrons and the underlying logic of the periodic table

We see in Section 2.1 how the periodic table, as it exists today, was devised by Dmitri Mendeleev, who grouped the chemical elements into rows of eight.

The number of electrons in an element increases by one as we move from left to right across a period. (For example, carbon has six electrons, nitrogen has seven, and oxygen has eight.) Elements from the same group, however, possess an *identical* number of valence electrons. Look at Figure 2.23, which shows the electronic configurations of lithium, sodium, and potassium, the first three elements of Group 1. Notice how all three elements have *one* electron in their outer (valence) shells. Elements from the same group exhibit similar chemical and physical properties as a consequence of the fact that they possess the *same* number of valence electrons, and so are likely to exhibit similar degrees of chemical reactivity.

It is quite remarkable that Mendeleev came up with a way of tabulating the chemical elements that not only provided such an enduring reference source, but—perhaps fortuitously—also encapsulated the chemical and physical trends displayed

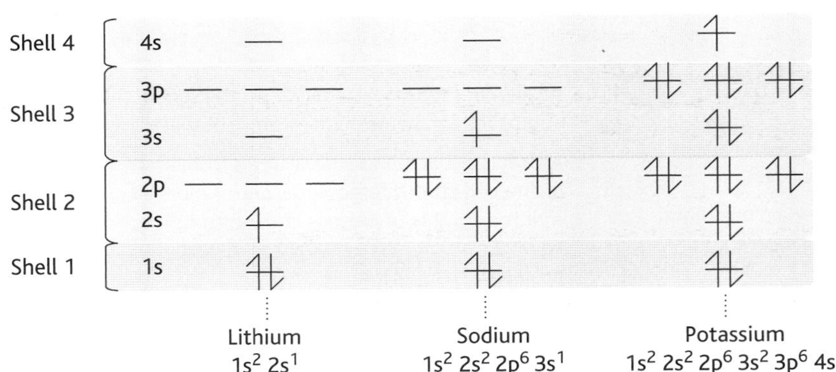

Figure 2.23 The electronic configurations of lithium, sodium, and potassium.

by the chemical elements with such elegant simplicity, with trends emerging as one migrates across the table, but similarities being captured by the vertical groups.

The variety of life: not so varied after all?

Let us end this chapter by reflecting on how biological systems exploit the chemical elements that are available. We might think that Nature would need to make use of the full range of chemical ingredients at its disposal in order to generate the huge variety that is exhibited by all living organisms. However, only a small proportion of the available elements are essential to life (as we see at the start of this chapter). If we look out for some of the most important biological elements in the periodic table (oxygen, nitrogen, carbon, sodium, for example) we see something striking: these elements fall within a small range of atomic numbers. It is an amazing biological feat: a difference of just a few protons, neutrons, and electrons provides sufficient chemical variety to generate the complexity and diversity of all living things.

Check your understanding

To check that you've mastered the key concepts presented in this chapter, review the checklist of key concepts below, and attempt the multiple-choice questions available in the book's Online Resource Centre at **http://www.oxfordtextbooks.co.uk/orc/crowe2e/**.

Checklist of key concepts

The chemical elements

- All matter is composed of a range of substances called the **chemical elements**
- Each chemical element has a name and a shorthand **chemical symbol**
- An **element** is defined as a single, simple substance that cannot be split into any more separate substances by chemical means
- An **atom** is the smallest unit in which an element can exist

The periodic table

- The **periodic table** displays all the known chemical elements in order of ascending atomic number
- The periodic table comprises a series of seven horizontal rows, called **periods**, and a series of 18 vertical columns, called **groups**
- **Periodicity** is the gradual change in chemical property from element to element as we move across a period, and the similarity of chemical and physical properties exhibited by elements within the same group

Atomic composition

- An atom comprises three subatomic particles: the proton, electron, and neutron
- A proton carries a single positive charge; an electron carries a single negative charge; a neutron is uncharged
- An atom is uncharged and so must comprise exactly the same number of protons and electrons
- An element's **atomic number** states both the number of protons and the number of electrons in an atom of that element
- An element's **mass number** states the sum of the number of protons and neutrons in an atom of that element
- An **ion** is a charged species which is formed when an uncharged atom gains or loses one or more electrons such that the number of protons and electrons is no longer equal. An **anion** is a negatively charged ion; a **cation** is a positively charged ion.
- **Isotopes** are atoms of the same element that contain different numbers of neutrons
- Isotopes of a single element occur in different **abundances**

- An element's atomic weight is a weighted average of the mass numbers of its naturally occurring isotopes
- An element's atomic mass is the mass of an atom of a single isotope

Atomic structure

- An atom has a defined structure: the protons and neutrons are confined to a tightly packed nucleus at the atom's centre; the electrons are restricted to being located in a series of atomic orbitals–specific volumes of space surrounding the nucleus
- There are four types of atomic orbital: s, p, d, and f
- An atom's orbitals are grouped into a series of shells, each of which comprises a series of subshells

Electronic configuration

- The order in which electrons fill the atomic orbitals is governed by three principles:
    - The Pauli exclusion principle dictates that an orbital can hold a maximum of two electrons
    - The Aufbau (or 'building-up') principle dictates that electrons sequentially enter orbitals of progressively higher energy
    - Hund's rule states that equivalent (degenerate) orbitals must each be occupied by one electron before any of the degenerate set is occupied by a second electron
- The distribution of electrons in an atom's orbitals is the atom's electronic configuration

The energy of atoms

- Orbitals generally possess progressively higher energy as their distance from the nucleus increases
- Electrons possess equivalent energy to the orbital which they occupy
- If an electron absorbs extra energy, it becomes excited and may move to a higher-energy orbital
- Excitation makes an atom more unstable; an excited electron emits its extra energy in the form of electromagnetic radiation and returns to its previous lower-energy orbital
- The energy of electromagnetic radiation is dictated by its wavelength: radiation with a short wavelength has high energy; radiation with a long wavelength has low energy
- The range across which the wavelength of electronic radiation varies is represented by the electromagnetic spectrum
- The visible spectrum incorporates electromagnetic radiation possessing wavelengths which we can detect with our eyes
- Orbitals, and the electrons which occupy them, possess restricted levels of energy–this phenomenon is termed quantization

Valence electrons

- The outermost shell of an atom is termed the valence shell, and the electrons occupying this shell are termed valence electrons
- Valence electrons are those components of an atom which interact with neighbouring atoms, and so dictate the atom's chemical reactivity

3 Compounds and chemical bonding: bringing atoms together

INTRODUCTION

The theme of small building blocks joining together to form large, more complicated structures recurs throughout biology, and occurs on many different scales. At the molecular scale, amino acids and nucleotides combine to form proteins and nucleic acids respectively. These and other substances combine in increasingly intricate ways at the cellular scale to form organelles such as the mitochondrion, the ribosome, and the nucleus. These organelles, in turn, come together to form the cell itself–the ultimate building block of every living organism.

In this chapter, we stay at the atomic scale to investigate how atoms interact with one another to form compounds. We ask what it is that holds atoms together, and why different atoms join together in different ways.

3.1 The formation of compounds

See Section 3.5 to find out what the notation '$CaCO_3$' is actually telling us.

Atoms aren't restricted to existing in total separation from one another. Many of the substances around us are formed from atoms of two or more elements which have joined together. For example, chalk is formed from the combination of calcium (Ca), carbon (C), and oxygen (O) to form calcium carbonate, $CaCO_3$.

Substances such as calcium carbonate that comprise *more* than one element are called **compounds**. Water (H_2O) and carbon dioxide (CO_2) are also compounds, consisting of hydrogen and oxygen, and carbon and oxygen respectively.

- *A compound is a substance that is composed of atoms of more than one element.*

The chemical bond: bridging the gap between atoms

Just as we use glue to hold two objects together, so atoms effectively need 'gluing' to one another to help them associate in a stable way. Chemical **bonds** are the glue that

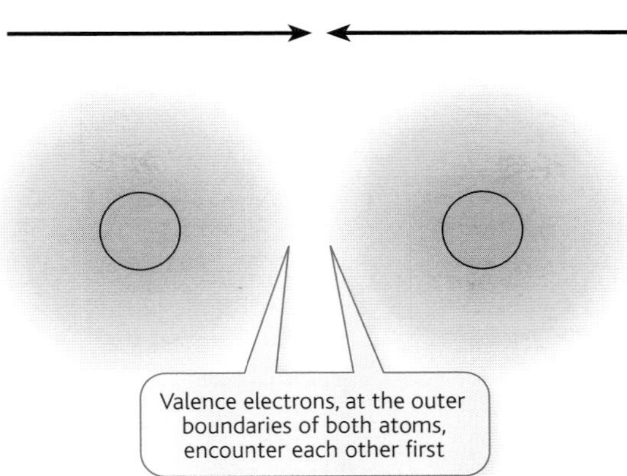
Valence electrons, at the outer boundaries of both atoms, encounter each other first

Figure 3.1 Valence electrons, which occupy the volume of the atom furthest from the nucleus, encounter each other first when two atoms come into close proximity.

holds atoms together. Just as we have different types of glue to fix together different types of material, so bonds form between different types of atom in different ways. But what gives rise to these bonds?

We find out more about the different ways in which chemical bonds form later in the chapter.

Valence electrons hold the key to chemical bonding. It is the rearrangement and redistribution of valence electrons between atoms which enable atoms to interact in a stable way–to 'bond' to one another. Valence electrons are ideally placed to drive chemical bonding. Figure 3.1 shows how it is the valence electrons, which occupy the volume of the atom furthest from its nucleus, that encounter each other first when two atoms come into close proximity. (This is a prerequisite for bonding to occur. You can't glue two objects together without them coming into contact; similarly, a bond cannot form between two atoms if the atoms aren't sufficiently close to one another.)

An atom's protons and neutrons, within its nucleus, play no part in the formation of compounds and aren't involved directly in chemical bonding. It is a job that is solely down to electrons.

Why should atoms bond with one another?

Atoms bond with one another in order to gain more stability (echoing one of our key themes from Chapter 1). Electrons in atoms distribute themselves within atomic orbitals in arrangements that have the lowest possible energy, to give atoms the greatest possible stability. (So, as we see in Chapter 2, there is a particular order in which the atom's orbitals are filled, starting with those that have the lowest energy, and progressing to those of higher energy.)

In a similar vein, electrons are rearranged into more stable configurations (so that atoms possess less energy) when atoms join together to form compounds than if the atoms remain separate from one another. Chemical bonding is a result of this rearrangement of electrons.

Which electron configuration is most stable?

The most stable atoms are those that have **full valence shells**. Only very few atoms naturally have full valence shells, however (see Box 3.1 on p. 43). Instead, atoms of different elements possess different numbers of electrons in their valence (outer) shells (as we see in Chapter 2). For example, sodium has one electron in its valence shell, while oxygen has six.

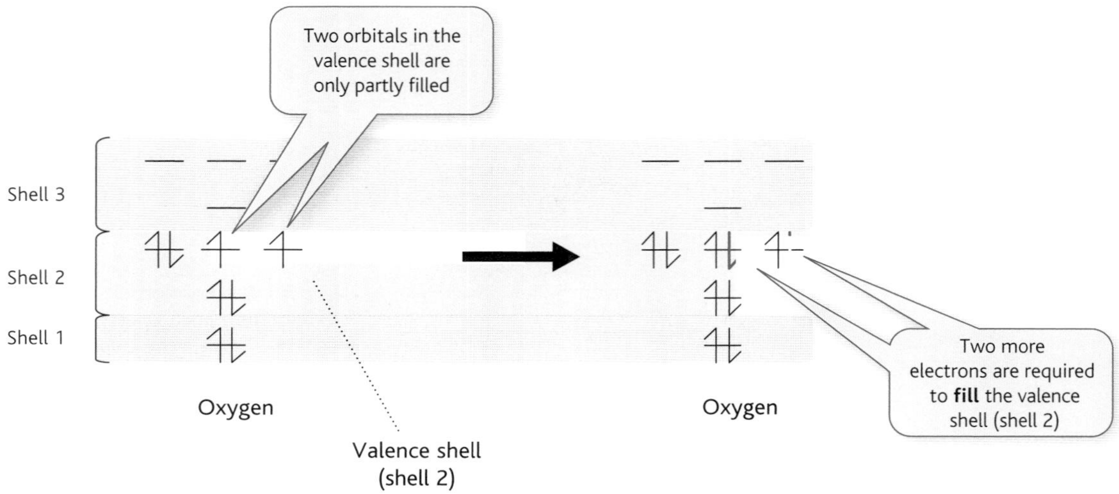

Figure 3.2 Oxygen's valence shell is shell 2. Oxygen requires two additional electrons to achieve a full valence shell.

Self-check 3.1 Which orbitals do the valence electrons of sodium and oxygen occupy?

These electronic configurations represent the most stable way of distributing those electrons already present in the atoms of these two elements. However, both of these atoms would be *more* stable if they had *full* valence shells.

When chemical bonds form between atoms to generate compounds, the atoms' valence electrons redistribute themselves between the different atoms so that all the atoms in the compound gain full valence shells. The resulting distribution of electrons in the compound is more stable than in the original constituent atoms. Indeed, we could consider the principal aim of chemical bond formation to be the generation of full valence shells.

There is one important point to note: when we refer to 'full valence shells' in this context, it is only the **s and p orbitals** of any given shell that must be filled. Look at Figure 3.2. Notice how oxygen has the electronic structure $1s^2\ 2s^2\ 2p^4$, and requires an additional two electrons to totally fill the second (valence) shell and achieve the electronic structure $1s^2\ 2s^2\ 2p^6$. Now look at Figure 3.3. By contrast with oxygen, the valence shell of sulfur is the third shell and sulfur has the electronic structure $1s^2\ 2s^2\ 2p^6\ 3s^2\ 3p^4$. Technically speaking, the third shell of a sulfur atom requires a further twelve electrons for it to be completely full: two to completely fill the three 3p orbitals, and another ten to fill the five 3d orbitals. However, we only need to consider the s and p orbitals, and can ignore the d orbitals. Look again at Figure 3.3 and notice how the valence shell of a sulfur atom needs just two electrons to be full (to complete the filling of the three 3p orbitals) rather than ten. A much more manageable task!

As we only have to consider the s and p orbitals, a valence shell only ever needs a total of *eight* electrons to be full (to be as stable as possible)[1]. This concept is captured

1 A full valence shell containing eight electrons is sometimes called a noble gas configuration. As we see in Box 3.1, the noble gas elements all have full valence shells.

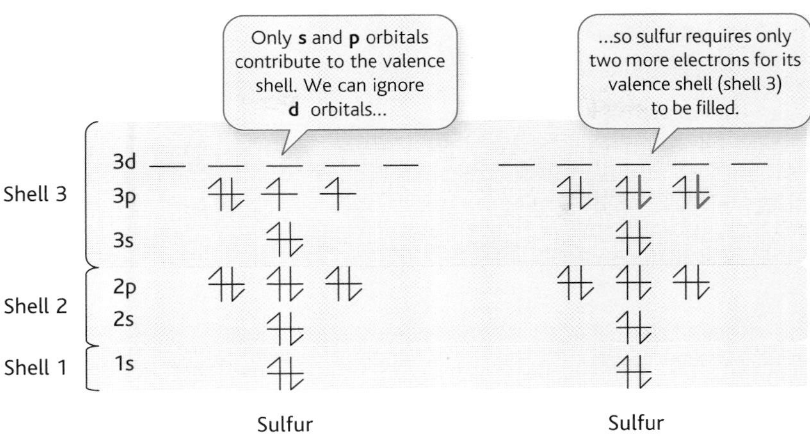

Figure 3.3 Sulfur's valence shell is shell 3. We ignore the five 3d orbitals in shell 3, and imagine that sulfur's valence shell comprises just one 3s and three 3p orbitals. So sulfur needs only two additional electrons to achieve a full valence shell.

by the **octet rule**, reflecting the fact that eight is the key number. The octet rule states that, to achieve maximum stability, an atom seeks to gain a full valence shell containing eight electrons.

- ***We consider a valence shell to be full once it contains eight electrons.***

A word of caution, however: the octet rule is not an absolute rule. Having said that we can consider an atom's valence shell to be full once it contains eight electrons (and therefore we only need to consider s and p orbitals, and can ignore d orbitals), there are some elements whose valence shells can accommodate more than eight electrons, primarily because their valence shells exploit available d orbitals too. We learn more about these expanded valence shells–which are surprisingly common in biological systems–in Section 3.7.

3.2 Valence shells and Lewis dot symbols

When we are investigating chemical reactions, we can often ignore the electrons that lie in inner shells, close to the nucleus, and focus solely on the valence electrons in the valence shell. (It is the movement of valence electrons that lies at the heart of many chemical reactions, including the formation of chemical bonds.)

We can visualize the valence shell in a simple way by representing it with a **Lewis dot symbol**. A Lewis dot symbol represents the valence electrons of an atom as a series of dots surrounding the chemical symbol for the element. Lewis dot symbols representing the valence shells of the atoms of several elements are shown in Figure 3.4. Dots are placed on four sides of the chemical symbols, depicting the four orbitals that comprise the valence shell (one s and three p orbitals). The exceptions are hydrogen and helium, whose valence shells comprise a single 1s orbital.

A single dot represents an unpaired electron in a partially filled orbital, while a pair of dots represents a pair of electrons in a fully occupied orbital. Note that Lewis dot symbols don't allow us to distinguish between the one s and three p orbitals that

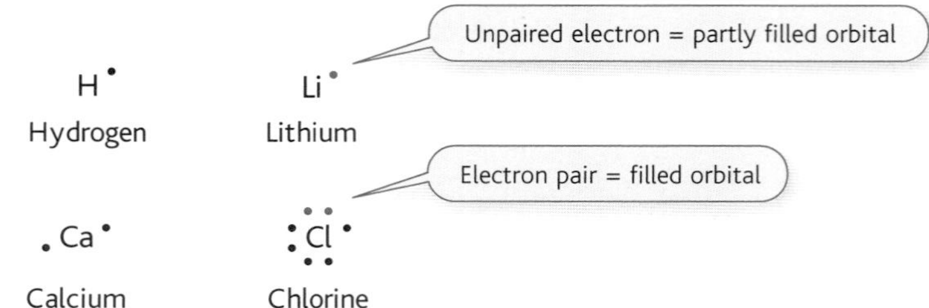

Figure 3.4 Lewis dot symbols representing the valence shells of the atoms of several different elements.

comprise the valence shell; the Lewis dot symbol is just a general representation of the valence electrons present, and doesn't specify the subshells that they occupy.

We see how several Lewis dot symbols can be combined to form a Lewis structure when we explore how atoms form molecules in Box 3.4.

Self-check 3.2 Represent the valence shells of the following two elements in the form of Lewis dot symbols:

(a) magnesium, Mg

(b) bromine, Br

Non-bonding pairs of electrons

An atom's valence shell may comprise singly occupied orbitals (containing one unpaired electron) or fully occupied orbitals (containing a pair of electrons). A pair of electrons in a fully occupied orbital is often called a **lone pair**.

When it comes to considering electrons in molecules, however, it is often more informative to use the term **non-bonding pair** instead of lone pair. Why is this? The aim of chemical bonding is to enable atoms to achieve full valence shells (Section 3.1). In practice, the filling of the valence shell is achieved by filling any partially occupied orbitals–adding one electron to those orbitals which contain an unpaired electron. Orbitals containing lone pairs of electrons are already full; they have no requirement for further electrons and do not usually participate directly in chemical bonding. Consequently, lone pairs on atoms that form part of a compound are those that are *not* shared between atoms as the result of chemical bonding–that is, they are non-bonding.

Look at Figure 3.5, which illustrates the valence shell of a nitrogen atom as a Lewis dot symbol. Notice how nitrogen has three unpaired electrons and one non-bonding pair of electrons.

Self-check 3.3 Represent the valence shell of oxygen in the form of a Lewis dot symbol. How many non-bonding pairs of electrons does an oxygen atom have?

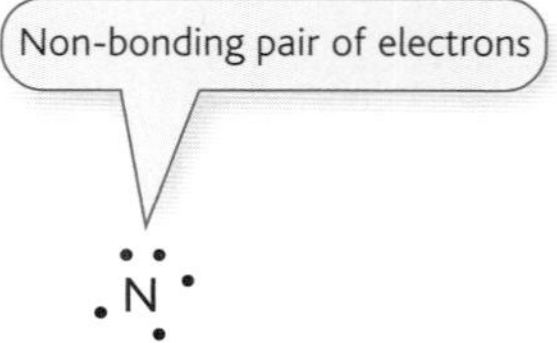

Figure 3.5 The valence shell of a nitrogen atom represented as a Lewis dot symbol.

It is important to note, however, that the term 'non-bonding' describes the nature of an electron pair only in a particular chemical context–that is, within a given molecule. When a molecule undergoes change as the result of a chemical reaction, any non-bonding pairs it contains may be involved in the reaction such that they contribute to a chemical bond in the molecule formed by the reaction. If this happens, they are no longer considered to be non-bonding after the reaction has occurred.

BOX 3.1 Antisocial behaviour: the noble elements

An increase in stability is the driving force behind atoms reacting with one another to form compounds. The achievement of a full valence shell represents the holy grail of stability, the state to which atoms aspire.

Pause for a moment to look at the electronic structure of neon, Ne. It has an atomic number of 10 and so has ten electrons per atom, with the electronic configuration $1s^2\ 2s^2\ 2p^6$ as illustrated in Figure 3.6. What do you notice about this arrangement of electrons?

Look at neon's valence shell (in the case of neon, this is shell 2) and notice that it contains eight electrons. Its valence shell is already full. Neon achieves the high level of stability to which most other elements aspire without having to interact with other atoms. As a result, neon is totally reluctant to interact with the atoms of other elements–it is chemically unreactive, or inert.

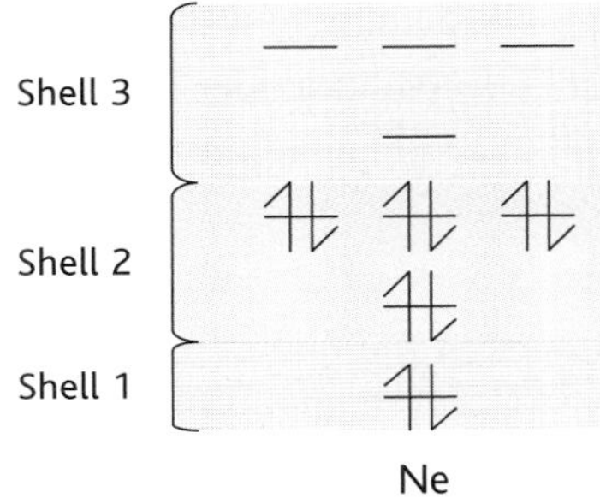

Figure 3.6 The electronic configuration of neon, Ne, is $1s^22s^22p^6$.

Neon shares its unwillingness to interact with neighbouring atoms with the other members of Group 18 in the periodic table (helium, argon, krypton, xenon, and radon), the atoms of which all have full valence shells. This group of elements is sometimes referred to as the noble gases. (As a further manifestation of the way these elements do not interact, all of these elements exist as gases under standard conditions.) Hence, when atoms of other elements redistribute their electrons when reacting with one another to gain full valence shells, they are often said to have achieved a noble gas structure, mirroring the electron arrangement that neon and the rest of its group achieve so effortlessly.

The lack of reactivity of the noble gases has one important consequence: they have no known role in biological systems. They can't cooperate with other atoms to form the many compounds that are so important in biology. Nor can they participate in the vital chemical reactions that give us energy, help cells to communicate, and so on. From a biological point of view, being sociable, and being willing to interact with your neighbours, is a must, leaving the noble gases as social outcasts when it comes to the chemistry of life.

Though non-bonding pairs typically do not contribute to chemical bonds within a compound, they do still have a crucial role in dictating the structure of many compounds, as described in Section 8.3.

We discover in Section 3.8 how dative covalent bonding is the exception to the rule that non-bonding pairs do not participate in chemical bonding. We also see in Chapter 17 how non-bonding pairs participate in chemical reactions.

- ***Non-bonding pairs of electrons do not normally participate in chemical bonding.***

3.3 Bond formation: redistributing valence electrons

The redistribution of valence electrons during chemical bond formation gives rise to two different types of compound which are distinguished by the way in which the electrons are redistributed, as illustrated in Figure 3.7:

- In a **covalent** compound one or more pairs of electrons are shared equally between the atoms, and a covalent bond is said to have formed.
- In an **ionic** compound one or more electrons are totally transferred from one atom to another, and an ionic bond is said to have formed.

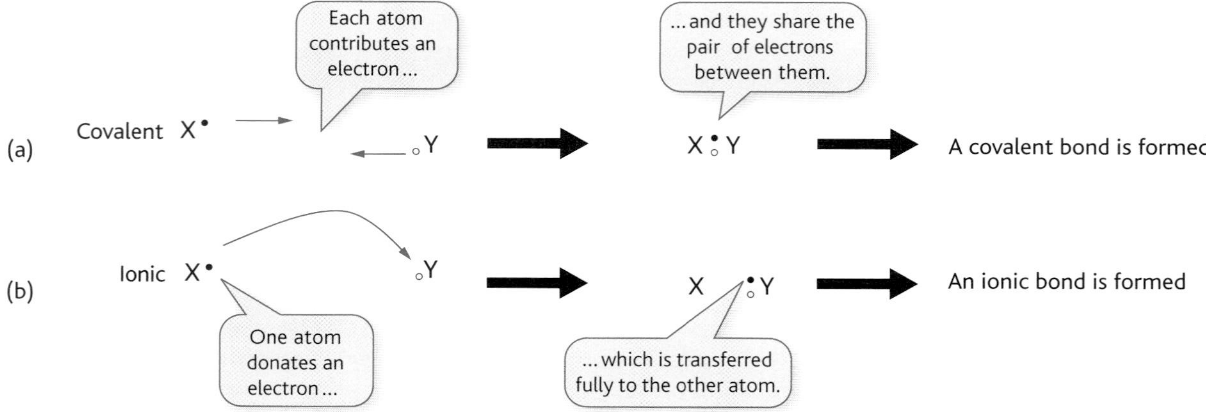

Figure 3.7 Covalent versus ionic compounds. (a) In a covalent compound one or more pairs of electrons are shared between two atoms. (b) In an ionic compound, one or more electrons are totally transferred from one atom to another.

We learn more about metals and non-metals in Box 3.2.

Whether an atom forms an ionic or covalent compound in association with other atoms depends on its chemical identity–which element it is. (The key factor that determines whether an element is most likely to undergo ionic covalent bonding is its electronegativity, as explained in Section 3.11.)

In general:

- two non-metals will react to form a covalent compound, held together by covalent bonds
- a non-metal will react with a metal to form an ionic compound, held together by ionic bonds.

BOX 3.2 Metals and non-metals

Metals and non-metals are characterized by their contrasting physical properties. If we are asked to describe a typical metal, we might think of a shiny substance that can be rolled into sheets and easily manipulated to hold different shapes–from the cylinders of food cans, to the smooth contours of car bodywork. Non-metals have almost directly opposing properties: they have a dull appearance, and are brittle. The contrasting physical properties of metals and non-metals are summarized in Table 3.1.

In general, metals fall on the left-hand side of the periodic table (see p. i), while the non-metals fall on the right-hand side. For example, the elements of Groups 1 and 2 (which include sodium, potassium, and calcium) are classed as metals, whereas the elements of Groups 16 and 17 (which include oxygen and chlorine) are classed as non-metals. Once again, hydrogen is an anomaly: despite being in Group 1, it is classed as a non-metal.

Table 3.1 Characteristic physical properties of metals and non-metals.

Characteristic	Metal	Non-metal
Appearance	Shiny	Dull
Flexibility	Very ductile	Brittle
Malleability	Malleable (can be hammered into sheets)	Shatters when hammered
Electrical properties	Good conductor of heat and electricity	Poor conductor; good insulator

Self-check 3.4 What kind of bonding would you expect to occur between:

(a) sodium and chlorine

(b) hydrogen and oxygen?

(You'll have a chance to check your prediction later in the chapter.)

Let's now look in more detail at these two types of bond, and what is happening to the electrons that are being redistributed.

3.4 The ionic bond: transferring electrons

Ionic bonds are formed when one or more electrons are *fully* transferred from one atom to another.

When this complete transfer of electrons from one atom to another occurs, the two atoms become **ions** (charged particles; see p. 15). This change from neutral to charged particle happens because the atoms no longer have an equal number of electrons and protons after the electrons have been transferred.

Let's take one of the simplest ionic compounds, sodium chloride, as an example. Sodium chloride is formed when one electron is transferred from a sodium atom to a chlorine atom, as illustrated in Figure 3.8. Notice how the sodium atom loses an electron to become a positively charged cation, Na^+: it goes from having 11 protons and 11 electrons to having 11 protons and 10 electrons. Likewise, the chlorine atom gains an electron to become a negatively charged anion, Cl^-: it goes from having 17 protons and 17 electrons to having 17 protons and 18 electrons.

It is the attraction between the oppositely charged cations and anions, from which an ionic compound is formed, that constitutes the 'bond' between the ions. This attraction between oppositely charged ions is called an **electrostatic interaction** and is the physical phenomenon through which ionic bonding is achieved and maintained. In effect, the electrostatic interaction between oppositely charged ions is the 'glue' which holds ionic compounds together.

Just as the opposing north and south poles of a magnet attract one another, in the case of sodium chloride the opposing positive and negative charges on the sodium and chloride ions attract one another so that the two ions remain tightly associated with one another to form the compound.

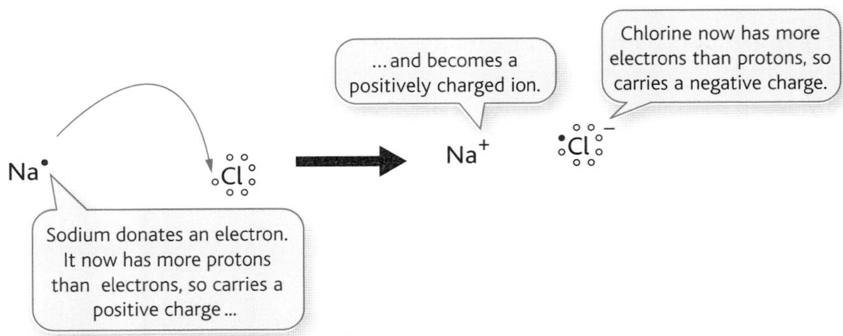

Na
Cl
Sodium donates an electron. It now has more protons than electrons, so carries a positive charge ...
... and becomes a positively charged ion.
Na+
Chlorine now has more electrons than protons, so carries a negative charge.
Cl−

Figure 3.8 The formation of the ionic compound sodium chloride, NaCl. Sodium chloride is formed when one electron is transferred from a sodium atom to a chlorine atom.

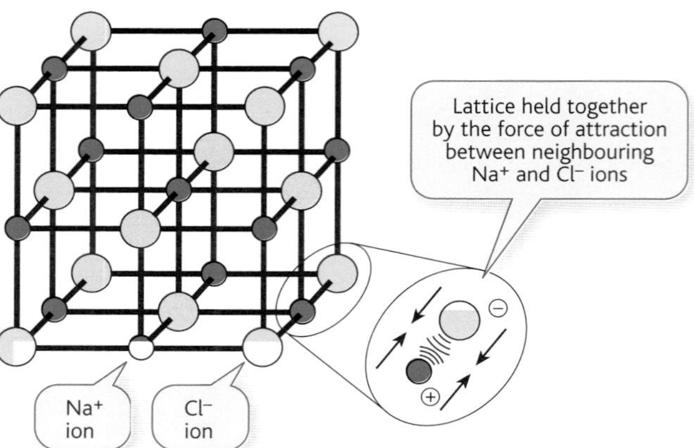

Figure 3.9 The structure of solid sodium chloride. Notice how sodium chloride exists as an extended lattice, in which the anions and cations are arranged in a regular, repeating pattern.

We compare the occurrence of ionic compounds as an extended repeating lattice of anions and cations to that of smaller, discrete arrangements of atoms seen in covalent compounds in Section 3.6.

Under normal conditions, ionic compounds exist as extended **lattices**–a network of cations and anions arranged in a regular, repeating pattern, all held together by electrostatic interactions. For example, Figure 3.9 shows how sodium chloride exists as a network of Na^+ and Cl^- atoms arranged in an organized, repetitive way.

- ***An ionic bond is the strong interaction between two oppositely charged ions.***
- ***An ionic bond is formed when an electron is completely transferred from one atom to another.***

Ionic bonding and full shells: how many electrons are transferred?

We see in Section 3.1 how the key outcome of chemical bonding is for the atoms involved in the chemical bond to achieve full valence shells. Atoms of different elements have to gain or lose different numbers of electrons to achieve full valence shells.

For example, look at Figure 3.10, which depicts the valence shells of sodium and chlorine[2].

A chlorine atom can **accept** one electron from another atom. A chlorine atom needs to gain one electron in order to achieve a full valence shell. By gaining one electron its electronic configuration changes from $1s^2\ 2s^2\ 2p^6\ 3s^2\ 3p^5$ to $1s^2\ 2s^2\ 2p^6\ 3s^2\ 3p^6$. In keeping with the octet rule, shell 3 now contains *eight* electrons and is full.

By contrast, a sodium atom can **donate** one electron to another atom. Sodium needs to lose one electron to gain a full valence shell: by losing one electron from the singly occupied 3s orbital, its electronic configuration changes from $1s^2\ 2s^2\ 2p^6\ 3s^1$ to $1s^2\ 2s^2\ 2p^6$. In this case, shell 2 satisfies the octet rule and is full.

But what determines whether an atom of a particular element accepts or donates electrons to achieve a full shell? As a general rule, the *smallest number of electrons possible* are redistributed in an atom in order to achieve a full valence shell. This may

2 Remember–you can discount the five d orbitals in shell 3; you only need to think about the s and p orbitals and how many electrons are needed to fill these.

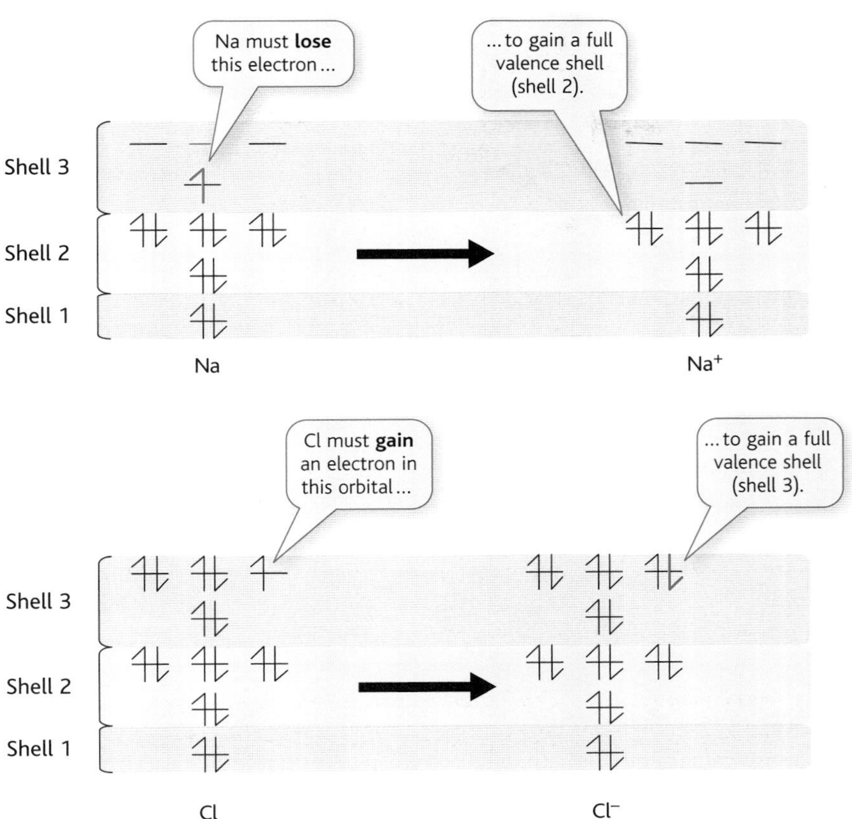

Figure 3.10 Changes in the electronic configurations of sodium and chlorine during the formation of sodium chloride. Sodium *loses* one electron from its valence shell, while chlorine *gains* one electron in its valence shell.

be achieved by emptying an existing valence shell so that the new full shell is the one next closest to the nucleus, rather than filling the existing valence shell.

Let's return to the electronic configuration of the sodium atom. We have just seen how the sodium atom loses one electron to achieve a full valence shell. In so doing, the valence shell changes from being shell 3 to shell 2. An alternative would be for the sodium atom to gain *seven* extra electrons to fill shell 3, to fill the 3s and three 3p orbitals, as illustrated in Figure 3.11.

Following the rule that the smallest possible number of electrons are redistributed (even if it means changing the identity of the valence shell), the sodium atom will lose one electron to achieve a full shell 2 rather than gain seven electrons to achieve a full shell 3. Likewise, chlorine will gain one electron to fill its valence shell rather than lose all seven electrons that are already there.

Q **Self-check 3.5** Does bromine gain or lose electrons to achieve a full valence shell? How many electrons must be gained or lost?

- ***The smallest number of electrons possible are redistributed—either accepted or donated—to achieve a full valence shell.***

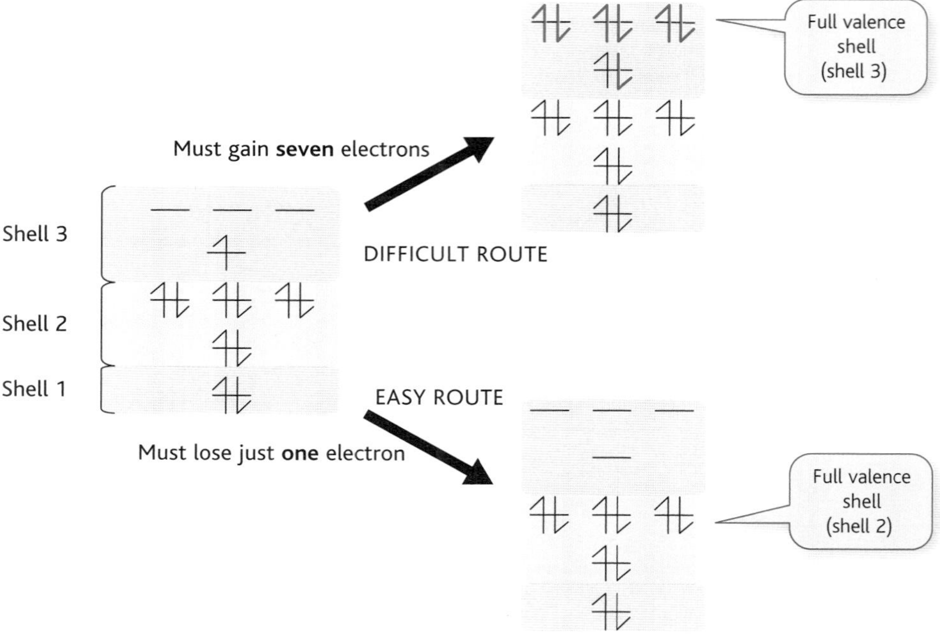

Figure 3.11 Two possible routes by which sodium can achieve a full valence shell. It could either gain *seven* electrons, or lose just *one* electron.

The transfer of multiple electrons

Some atoms gain or lose more than one electron to form ions that carry more than one positive or negative charge (p. 17). Look at Figure 3.12. Notice how a single magnesium atom can donate *two* electrons to one or more neighbouring atoms and form a magnesium ion, Mg^{2+}, which carries two positive charges (the ion has two more protons than electrons). (Magnesium, with an electronic configuration of $1s^2$ $2s^2$ $2p^6$ $3s^2$, must lose two electrons to gain a full valence shell.)

In contrast, a single oxygen atom can accept two electrons from a neighbouring atom to form the O^{2-} ion, which carries two negative charges. (An oxygen atom needs to gain two electrons to fill its valence shell.)

The donation of electrons by one atom must be coupled to the acceptance of electrons by another atom. Therefore, a magnesium atom cannot donate two electrons to become a Mg^{2+} ion unless there are neighbouring atoms in its vicinity that are able to accept the electrons that the magnesium atom is donating. An atom cannot simply release an electron into its surroundings; the electron must have a 'home' to go to.

The overall charge on an ionic compound is zero: an ionic compound is electrically neutral. For this to occur, there must be an equal number of positive and negative charges present within the compound as a whole. (This is analogous to the equal number of protons and electrons in an atom, which renders the atom neutral overall.)

For example, the ionic compound sodium chloride is neutral: the single positive charge on the sodium ion is balanced by the single negative charge on the chloride ion.

- ***Ionic compounds have an overall charge of zero: there must be an equal number of positive and negative charges within the compound overall.***

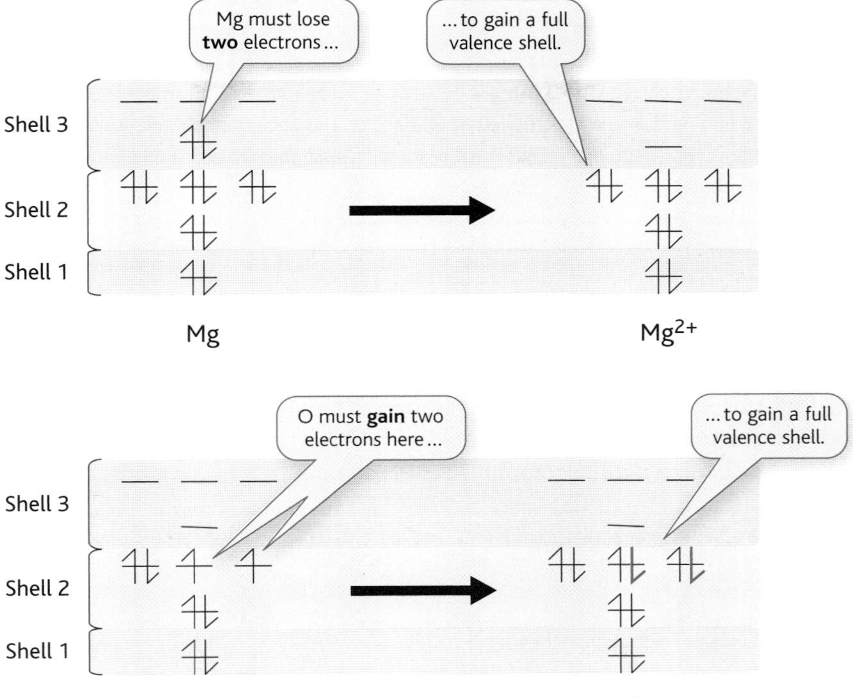

Figure 3.12 Changes in the electronic configurations of magnesium and oxygen during the formation of magnesium oxide. Magnesium loses *two* electrons from its valence shell, while oxygen gains *two* electrons in its valence shell.

An atom of one element can donate electrons to or accept electrons from more than one atom of another element in order for an overall charge of zero to be achieved, and for the transfer of electrons to be balanced. For example, let's consider the formation of the ionic compound magnesium chloride, which is illustrated in Figure 3.13. We see above that magnesium can donate two electrons to form a doubly charged cation, Mg^{2+}, whereas chlorine can accept one electron to form a singly charged anion, Cl^-. For the overall charge on magnesium chloride to be zero, a single magnesium atom must react with two atoms of chlorine. Figure 3.13 shows how each chlorine atom accepts one electron from the magnesium atom, to give an overall transfer of two electrons. As a result, the two positive charges on the Mg^{2+} ion are

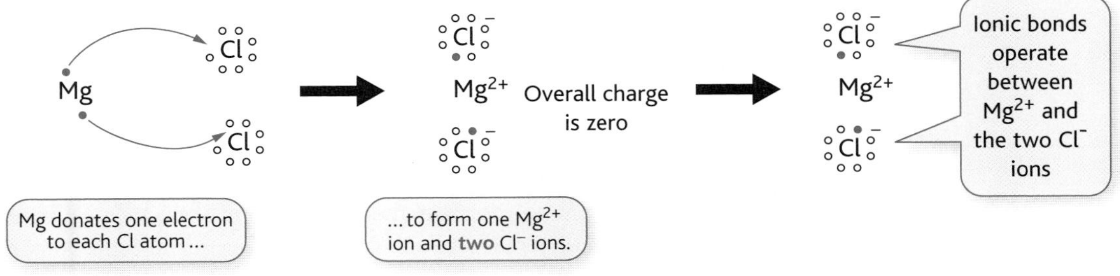

Figure 3.13 The formation of magnesium chloride. Two ionic bonds form between one atom of magnesium and two atoms of chlorine when magnesium donates one electron to each of the chlorine atoms.

balanced by the total of two negative charges carried by the two Cl^- ions to give an overall charge of zero, as required.

Self-check 3.6 The ionic compound sodium oxide is formed from the reaction of sodium and oxygen atoms. What charges do the sodium and oxide ions carry, and how many sodium atoms must react with each oxygen atom to give a neutral compound?

3.5 The chemical formula

An important part of science is being able to share ideas and communicate complicated information in a transparent, reliable way. We see in Section 2.1 how chemical symbols are used to communicate the identity of the chemical elements. A similar type of shorthand is used to communicate the components of compounds. This is called the **chemical formula**.

To describe an ionic compound accurately we need to identify two things:

1. The elements from which the compound is formed
2. The relative number of ions of each element

The elements from which a compound is formed are identified using the same chemical symbols introduced in Chapter 2 (Na, Cl, Mg, etc.).

The *relative number* of ions is indicated by writing a subscript number to the right of the chemical symbol of the respective element. Therefore, the chemical formula for magnesium chloride (which we have seen to comprise two Cl^- ions for every Mg^{2+} ion) is $MgCl_2$.

If there is just a single ion, then *no* subscript number is necessary. Therefore, though we *could* represent sodium chloride (which contains one Na^+ ion for every Cl^- ion) as Na_1Cl_1, there is no need to include the subscript '1': sodium chloride is represented as NaCl. Note that the chemical formula of an ionic compound does not indicate the charges on the component cation and anion. We write NaCl, rather than Na^+Cl^-.

We learn more about sodium chloride and its role in biological systems in Box 3.3.

Self-check 3.7 What is the chemical formula for calcium bromide? Hint: work out the ions formed by atoms of calcium and bromine. How many bromide ions are needed to balance the charge on the calcium ion?

3.6 The covalent bond: sharing electrons

We see how electrons are redistributed to allow covalent bonding to occur in Section 3.3.

Both ionic and covalent bonding arise from the redistribution of electrons. Whereas electrons are completely transferred from one atom to another during ionic bonding, in covalent bonding electrons are **shared** between atoms.

Atoms which are linked by covalent bonds form discrete units called **molecules**. Molecules may comprise atoms of *different* elements (compounds), or atoms of the *same* element.

BOX 3.3 Sodium chloride in everyday life

Sodium chloride, NaCl, is better known as 'table salt' in everyday life, and is a vital part of our diet. Our nervous system in particular requires a sufficiently high concentration of sodium in the body to enable it to transmit nerve impulses effectively. It is the movement of sodium ions into and out of nerve cells, across the cell membrane, which forms the basis of the movement of electrical charge along nerve cells–the nerve impulse.

As with many foods, however, salt can be harmful if consumed in large quantities, particularly to individuals who are at risk of heart or kidney disease. Such conditions leave the body less able to remove excess salt from the body such that its concentration in the body increases. The body retains more water in response to these elevated salt concentrations. And, as a knock-on effect of this higher water retention, the arteries are put under extra pressure, leading to high blood pressure.

In fact, the foodstuffs in a balanced, healthy diet naturally contain enough salt to satisfy our body's demands. Adding extra salt may enhance your enjoyment of a meal, but you're likely to be better off leaving it to Nature to provide all the salt you need.

For example, the most important part of the air we breathe is oxygen gas[3]. When we breathe in oxygen as part of the air that surrounds us (as we are all doing continually) it is not single atoms of oxygen that we're inhaling, but molecules of oxygen–two atoms of oxygen joined by a covalent bond to form O_2. Under ordinary conditions this is the form in which oxygen exists[4].

By contrast, a molecule of the compound glucose comprises *three* elements: carbon, hydrogen, and oxygen. If we took a single molecule of glucose and broke it down into these separate elements, they would no longer behave like glucose. It is the precise combination of six atoms of carbon, twelve atoms of hydrogen, and six atoms of oxygen that make up a molecule of glucose, which confers upon glucose its specific chemical properties.

- ***A molecule is the smallest part of a single element (for example O_2) or a compound (a composite of different elements, such as glucose, $C_6H_{12}O_6$) which can exist alone under ordinary conditions.***

Ionic compounds do not form discrete molecules in the way that covalent compounds do. We see in Section 3.4 how the cations and anions which comprise ionic compounds are arranged into extensive organized lattices. We can represent an ionic compound by its chemical formula (e.g. NaCl). However, while the chemical formula tells us the relative number of cations to anions present in the ionic compound (in the case of NaCl, this is one Na^+ for every Cl^- ion), it does not represent a discrete 'molecule' of NaCl. The extensive lattices of ionic compounds cannot be broken down into discrete molecules.

In contrast, a covalent compound comprises many millions of identical molecules, as depicted in Figure 3.14.

3 Arguably, oxygen is the most important component of air for humans, despite not being the most prevalent component: only 21% of air is oxygen, with the majority (78%) being nitrogen.

4 Unlike humans, plants must absorb carbon dioxide, CO_2, from the air around them to generate energy through the process of photosynthesis. Plants require oxygen for other vital metabolic processes, however, and so depend on oxygen for survival just as much as humans do.

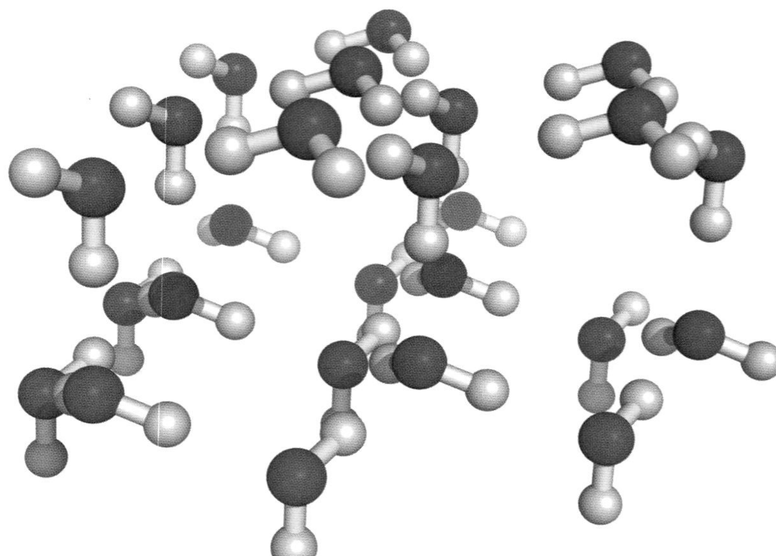

Figure 3.14 Water: an example of a covalent compound. Water comprises many millions of identical molecules. When covalent compounds are in the solid phase (see Section 4.5) the individual molecules are arranged in a regular, repeating way. This figure depicts the arrangement of water molecules when water is in the solid phase–a substance better known to us as ice.

Covalent compounds and electrical charge

Both covalent and ionic compounds are neutral. In the case of ionic compounds, the ions which comprise the compound are charged, but the positively charged cations exactly balance the negatively charged anions so that the compound as a whole carries an electrical charge of zero.

Atoms which comprise covalent compounds do not form ions during the process of bond formation: valence electrons are not fully transferred from one atom to another. As a result, every atom retains its neutrality, and the compound as a whole is neutral.

The molecular formula: identifying the components of a covalent compound

The components of a covalent compound can be represented in the same manner as for ionic compounds. The **molecular formula** tells us the composition of one molecule of a covalent compound:

- It identifies the different elements present in the compound.
- It tells us the number of atoms of each element in one molecule of the compound.

A molecule of glucose comprises six atoms of carbon, twelve atoms of hydrogen, and six atoms of oxygen. Its molecular formula is $C_6H_{12}O_6$. When writing a molecular formula, we list the composite elements in alphabetical order, rather than according to the number of atoms of each element that are present. (So, for example, we write $C_6H_{12}O_6$ rather than $C_6O_6H_{12}$.)

- *The molecular formula tells us the composition of one molecule of a covalent compound.*

Self-check 3.8 (a) A molecule of propane 1-thiol, the compound that gives onions their eye-watering odour, has the molecular formula C_3H_8S. What is the composition of this compound? (b) A molecule of menthol, the mint-scented compound used widely in chest and nasal decongestants, and other medical products, comprises twenty atoms of hydrogen, ten atoms of carbon, and an atom of oxygen. What is its molecular formula?

Covalent bonding and the distribution of electrons

Atoms form bonds with other atoms to achieve full valence shells, with the effect that both atoms achieve a greater degree of stability.

The smallest molecule to exist is hydrogen, H_2, which comprises two hydrogen atoms. A hydrogen atom has a single electron in its 1s orbital, so one more electron must occupy this orbital for a full shell to be attained. Figure 3.15 shows how a full shell is achieved by the two atoms **sharing** their single valence electrons: each atom contributes one electron to a central 'pool'. This pool of two electrons (an electron pair) is shared by both atoms so that they both effectively have a full valence shell.

The sharing of electrons acts as the atomic 'glue' that holds the two hydrogen atoms together to form the hydrogen molecule. When two atoms share a pair of electrons a **covalent bond** forms between them.

Look at Box 3.4 to learn how we can use Lewis structures to represent covalent molecules.

Self-check 3.9 Use the rules for representing covalent bonds and non-bonding pairs in Box 3.4 to draw the Lewis structure of ammonia, NH_3.

Molecular orbitals

In order for two atoms to share a pair of valence electrons, and form a covalent bond, the atomic orbitals containing the valence electrons to be shared must *overlap*. The overlap of atomic orbitals allows the electrons within the atomic orbitals to *mix* with each other. Picture two atoms of hydrogen coming together to form a molecule, as illustrated in Figure 3.17(a). As the atoms move closer together, the two 1s orbitals from the two hydrogen atoms (which each contain one electron) overlap. When overlap occurs, the two electrons become shared between the two atoms.

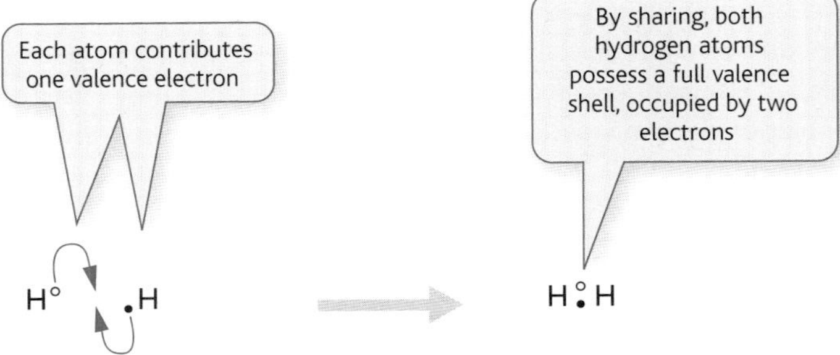
Each atom contributes one valence electron
By sharing, both hydrogen atoms possess a full valence shell, occupied by two electrons
H° •H
H°•H

Figure 3.15 The two atoms that form molecular hydrogen, H_2, share two valence electrons–one from each atom–to form a covalent bond. Both atoms achieve a full valence shell by sharing electrons in this way.

BOX 3.4 Lewis structures

We see in Section 3.2 how we can represent the valence shell of an atom by a Lewis dot symbol. We can combine Lewis dot symbols to generate a Lewis structure, which depicts the molecule formed when two or more atoms are joined by covalent bonds.

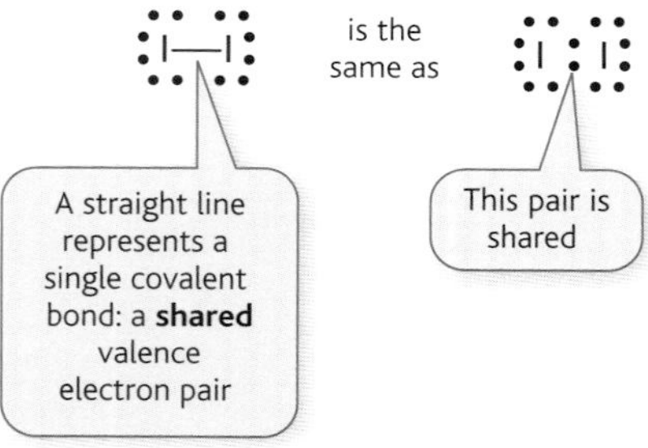
is the same as
A straight line represents a single covalent bond: a **shared** valence electron pair
This pair is shared

Figure 3.16 A molecule of iodine, I_2, represented as a Lewis structure. Notice how a single covalent bond can be represented by a pair of electrons, shared between two atoms.

Figure 3.16 uses the example of a molecule of iodine, I_2, to show the main features of a Lewis structure. Notice how:

- a covalent bond (a pair of electrons that is being shared between two atoms) is depicted by a straight line between the two atoms
- a non-bonding pair of electrons is depicted by a pair of dots.

Similarly, a water molecule (H_2O) can be depicted by the following Lewis structure:

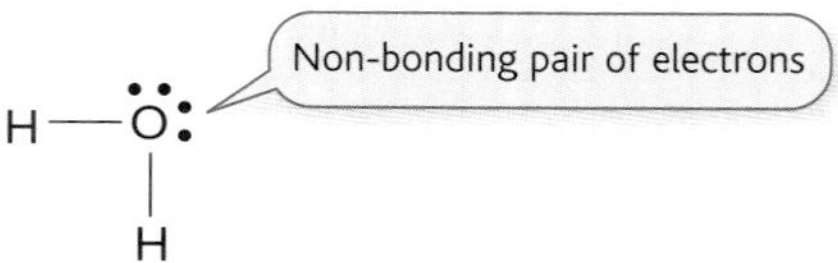
Non-bonding pair of electrons
H
O
H

We can also use Lewis structures to help represent the *shape* of a molecule, as we see in Chapter 8.

New **molecular orbitals** are formed when atomic orbitals overlap. Molecular orbitals are shared by atoms that have joined together to form a molecule, and are occupied by those valence electrons that are shared by the atoms to generate covalent bonds.

Orbital theory is an area of chemistry that is often considered impossibly difficult to understand. This is principally because the properties of orbitals (their energies and shapes) are represented by some really quite challenging maths. However, for our purposes we really don't care about the theoretical chemistry that is used to describe orbitals (and so you won't see any equations here). That said, orbitals are central to determining why molecules behave the way they do. (We see in Chapter 8, for example, how the shapes of molecular orbitals explain quite elegantly the shapes of molecules that we observe.) So it is worth us being aware of some general concepts, which we will explore very briefly here.

Figure 3.17(b) shows how the overlap of two atomic orbitals generates two distinct molecular orbitals: one **bonding** orbital, and one **anti-bonding** orbital[5].

Molecular orbitals–whether bonding or anti-bonding–share one important characteristic with atomic orbitals: they too can hold a maximum of just two electrons. Bonding and anti-bonding orbitals have opposing effects on the formation of covalent bonds. A pair of electrons occupying a **bonding orbital** facilitates covalent bonding, whereas a pair of electrons occupying an **anti-bonding orbital** inhibits covalent bond formation.

5 The rules that govern molecule formation (and which are explored in the field of quantum mechanics) dictate that *all* molecules must have both bonding *and* anti-bonding orbitals. We do not find molecules with just one type or the other.

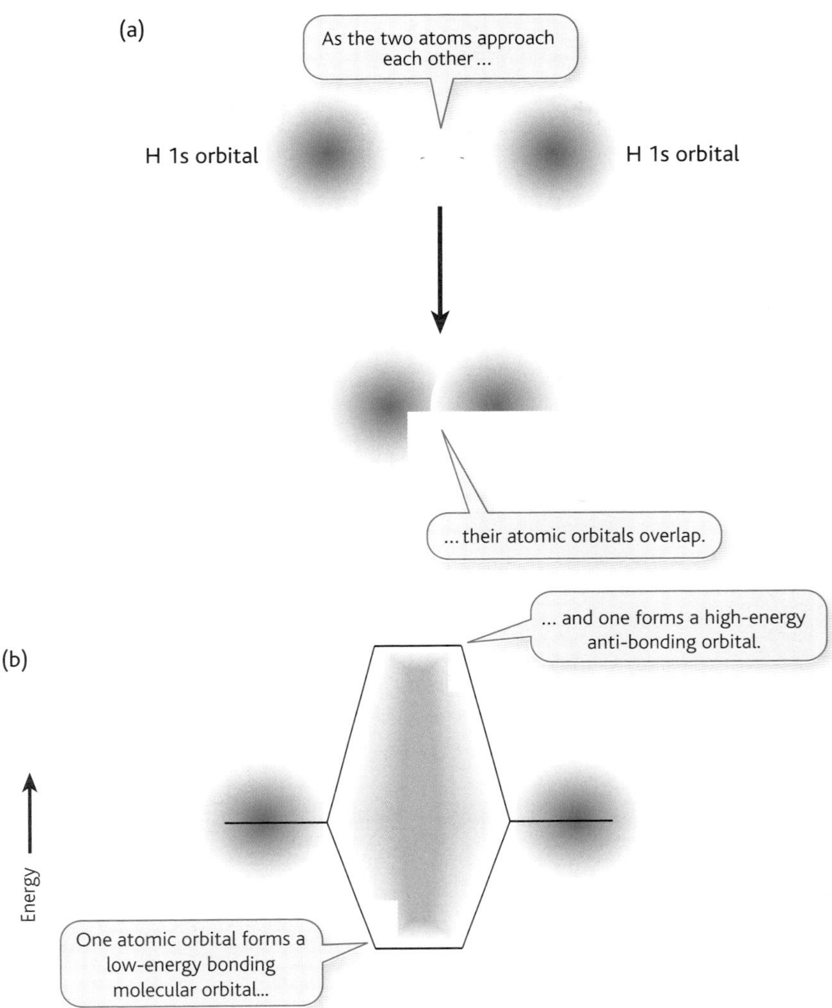

Figure 3.17 The joining of two hydrogen atoms to form a molecule of hydrogen, H_2. (a) When the two atoms join, their atomic orbitals overlap. (b) When the two atomic orbitals combine, they generate two molecular orbitals: one bonding orbital, and one anti-bonding orbital.

There must be more electrons located in a molecule's bonding orbitals than in its anti-bonding orbitals for covalent bonding to occur: electrons in molecular bonding orbitals are those that actively contribute to chemical bonding.

Bonding orbitals possess lower energy than anti-bonding orbitals. It is this difference in energy that accounts for the opposing effect that the two types of orbital have on covalent bond formation. Remember that lower energy correlates with greater stability: a molecule with electrons in a bonding orbital is more stable than a molecule with electrons in an anti-bonding orbital.

- ***Two atomic orbitals combine to form two molecular orbitals.***
- ***There must be more electrons located in a molecule's bonding orbitals than in its anti-bonding orbitals for covalent bonding to occur.***

The distribution of valence electrons between an atom's bonding and anti-bonding orbitals explains why some atoms don't form molecules. For example, helium

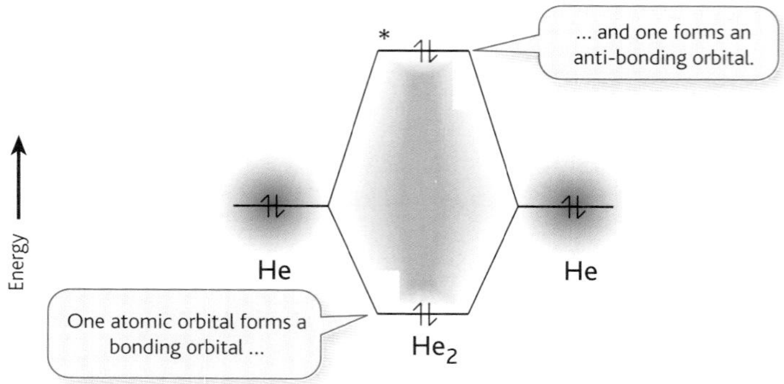

Figure 3.18 When two helium atoms interact to form a filled bonding molecular orbital and a filled anti-bonding molecular orbital, the filled anti-bonding orbital cancels out the effect of the filled bonding orbital, so no bonding occurs.

has the electronic configuration $1s^2$. Look at Figure 3.18, which illustrates what happens when two helium atoms interact in an attempt to form a molecule of helium, He_2. Notice how both the bonding and anti-bonding orbitals, which form when the two helium atoms interact, contain two electrons. The condition that the number of electrons in the bonding orbital must exceed the number of electrons in the anti-bonding orbital is not satisfied, so a covalent bond is not formed. As a result, He_2 does not exist.

Anti-bonding orbitals become important when we study the composition of compounds using a technique called UV–visible spectrophotometry (see p. 365). It is the movement of electrons into anti-bonding orbitals that is detected by this technique.

A third type of molecular orbital, which does not directly influence the formation of covalent bonds, is the **non-bonding** orbital. We explore non-bonding orbitals in more detail in Box 3.5.

During the rest of this book, we're only really interested in the electrons that occupy *bonding* orbitals, so we'll simply use the term 'orbitals' to denote bonding orbitals.

3.7 The formation of multiple bonds

In the above examples, we see how two atoms can share a pair of electrons to form a covalent bond. In some instances, however, two atoms share more than two electrons such that multiple covalent bonds form between them. In fact, up to *three* covalent bonds can form between two atoms. When atoms share increasing numbers of electrons they do so through the filling of a progressively greater number of molecular bonding orbitals–a combination of σ and π orbitals.

Sigma and pi orbitals

There are two types of molecular orbital, sigma (σ) and pi (π). We see in Section 2.3 that there are different types of **atomic** orbital (s, p, d, and f), which each have a distinctive shape. (Take a look at Section 2.3 to refresh your memory.) Similarly, σ and π **molecular** orbitals have contrasting shapes, as shown in Figure 3.20. Indeed, it is just the difference in their shape that distinguishes a σ orbital from a π orbital.

The covalent bond formed from the sharing of a pair of electrons in a sigma orbital is called a sigma bond (or σ bond). The covalent bond formed from the sharing

BOX 3.5 Non-bonding orbitals

When an atom that possesses a non-bonding pair of electrons undergoes covalent bonding with another atom, the non-bonding pair does not participate directly in bonding to form a molecule (Section 3.2). Instead, the non-bonding pair occupies a non-bonding molecular orbital.

Look at Figure 3.19, which shows how nitrogen bonds with three atoms of hydrogen to form ammonia, NH_3. Notice how the non-bonding pair of electrons occupying the 2s orbital in a nitrogen atom occupies a non-bonding molecular orbital in ammonia. By contrast, notice how the unpaired electrons, which occupy nitrogen's three 3p orbitals prior to bonding, move to bonding molecular orbitals when the nitrogen atom forms covalent bonds with the three hydrogen atoms.

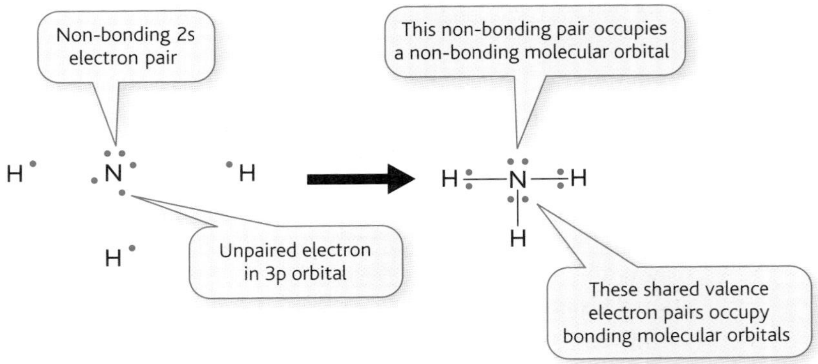

Figure 3.19 The formation of ammonia, NH_3, from one nitrogen atom and three hydrogen atoms. The non-bonding pair from the nitrogen atom occupies a non-bonding molecular orbital when a molecule of ammonia is formed.

of a pair of electrons in a pi orbital is called a pi bond (or π bond). It is important to remember that σ and π bonds are both covalent bonds that are formed from the sharing of one pair of electrons between two atoms. The only key difference is that the orbitals occupied to form the two types of bond have different shapes.

The σ and π symbols are used to denote different molecular orbitals in a manner analogous to the use of s, p, d, and f to denote different atomic orbitals.

- ***A molecular bonding orbital can be one of two types: a sigma orbital or a pi orbital.***

We explore different molecules that possess either σ or π bonds in Section 3.7.

Now that we have been introduced to sigma and pi bonds, we are ready to learn more about the formation of multiple bonds. First, we must ask the question: just

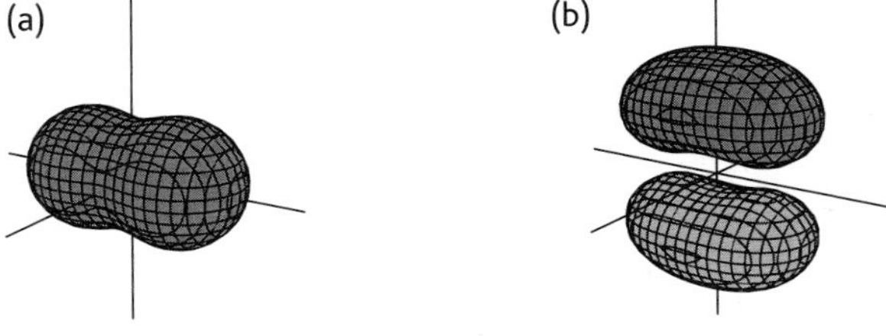

Figure 3.20 The shape of (a) a typical sigma (σ) molecular orbital, and (b) a typical pi (π) molecular orbital. Even though the pi orbital comprises two distinct lobes, it is still just a single orbital.

how many bonds can form between atoms of a given element? To answer this question, we need to consider an atom's valency.

Valency: how many bonds can an atom form?

An atom's **valency** denotes the number of pairs of electrons that it must share with one or more other atoms in order to attain a full valence shell. An atom's valency varies from element to element, reflecting the way that atoms of different elements possess different numbers of valence electrons, and so must share different numbers of pairs of electrons to attain a full valence shell.

As the sharing of one pair of electrons equates to the formation of one covalent bond, an atom's valency also indicates the number of covalent bonds that it can form.

For example, oxygen has a valency of two. Oxygen must share two pairs of electrons to achieve a full valence shell, and so can form a maximum of two covalent bonds. Recall that oxygen shares a pair of electrons with each of two hydrogen atoms to form a molecule of water, H_2O. In so doing, the oxygen atom participates in two covalent bonds, and its valency is satisfied.

In contrast, hydrogen has a valency of one. Hydrogen needs to share just one pair of electrons to fill its valence shell, and can form just one covalent bond with another atom. This situation is again illustrated by the water molecule, in which each hydrogen atom forms a *single* covalent bond with the oxygen atom.

The valencies of several different elements are shown in Figure 3.21.

Self-check 3.10 Ammonia has the molecular formula NH_3. What is the valency of nitrogen?

Let's look at a series of examples to see what happens when atoms share one or more pairs of electrons in order to achieve full valence shells.

Sharing one pair of electrons: the single bond

A single bond is formed when there is one pair of electrons shared between two atoms. This pair of electrons occupies a σ orbital to form a σ bond.

For example, when a H_2 molecule is formed, as illustrated in Figure 3.22(a), the two shared electrons occupy one σ orbital to generate a σ bond. As only one covalent bond has formed, we say that the atoms are joined by a single bond.

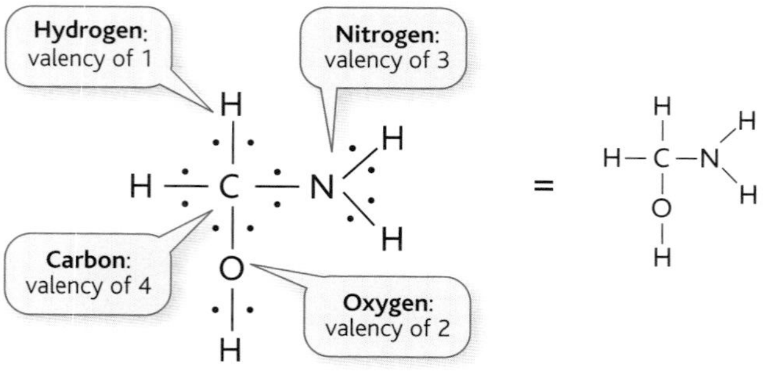

Figure 3.21 The valencies of several different elements. Notice how the different elements form different numbers of covalent bonds in order to achieve full valence shells.

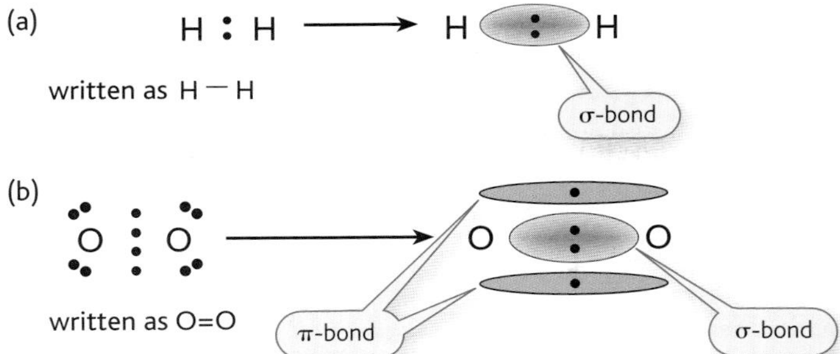

Figure 3.22 (a) The formation of a molecule of hydrogen, H_2. The two hydrogen atoms share one valence electron pair, which occupies a sigma orbital to form a sigma bond. (b) The formation of a molecule of oxygen, O_2. The two oxygen atoms share two valence electron pairs. One pair occupies a sigma orbital to form a sigma bond. The other pair occupies a pi orbital to form a pi bond.

When drawing the structure of a compound, we represent a single bond by a single line joining the two atoms:

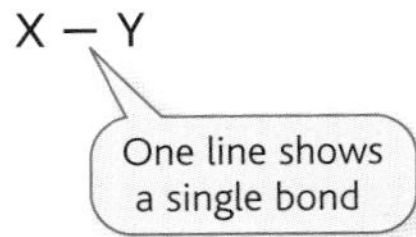

Sharing two pairs of electrons: the double bond

A **double** bond forms when two pairs of electrons are shared between two atoms. One pair of electrons occupy a σ orbital to form a σ bond, while the second pair of electrons occupy a π orbital to form a π bond.

For example, when an O_2 molecule is formed, as depicted in Figure 3.22(b), the four shared electrons occupy two orbitals: one pair occupies a σ orbital, and one pair occupies a π orbital. There are two covalent bonds overall (one σ and one π), so we say that the two oxygen atoms are joined by a double bond.

Double bonds are formed between atoms of elements with a valency ≥ 2, which include the biologically important elements C, N, O, and S. Double bonds are a vital part of many biological molecules, including DNA and amino acids.

We learn more about the structure of DNA and the amino acids in Chapter 7.

When drawing the structure of a compound, we represent a double bond by two parallel lines joining the two atoms:

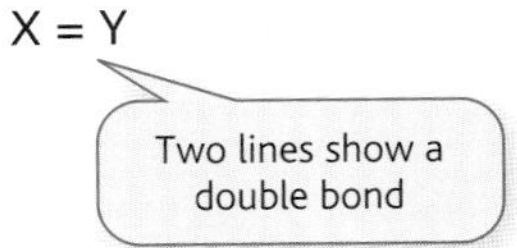

Self-check 3.11 Double bonds can form between which of the following pairs of atoms?

(a) C, H

(b) C, O

(c) O, H

(d) C, C

We explore the chemistry of carbon-containing compounds, which often feature double covalent bonds, in Chapter 5.

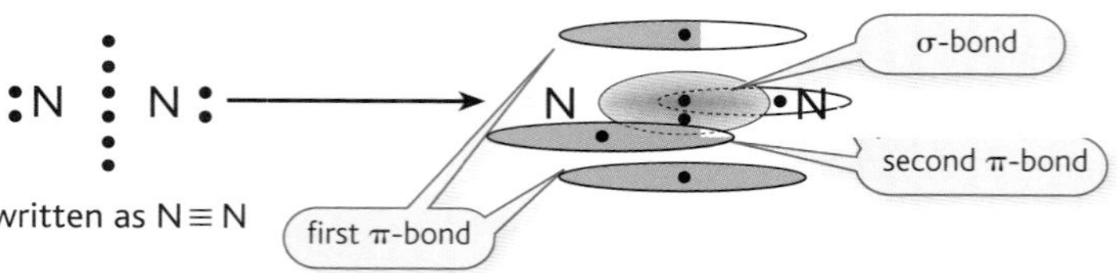

Figure 3.23 A molecule of nitrogen, N_2. The two nitrogen atoms share three valence electron pairs. One pair occupies a sigma orbital to form a sigma bond. The other two pairs occupy two pi orbitals to form two pi bonds.

Sharing three pairs of electrons: the triple bond

A **triple** bond forms when three pairs of electrons are shared between two atoms, and comprises one σ and two π bonds. One pair of electrons occupies a σ orbital to form a σ bond, while two pairs of electrons occupy two π orbitals to form two π bonds.

Figure 3.23 depicts a N_2 molecule. When a N_2 molecule is formed, the three pairs of valence electrons occupy three molecular orbitals: one sigma orbital and two pi orbitals, to form one σ bond and two π bonds. There are three covalent bonds overall, so we can think of the atoms as being joined by a triple bond.

Triple bonds are far rarer than single or double bonds, but occur between atoms of elements that possess a valency ≥ 3. (Both elements must be able to participate in three covalent bonds (one σ and two π) for a triple bond to form between them, and so both must have a valency ≥ 3.)

When drawing the structure of a compound, we represent a triple bond by three parallel lines joining the two atoms:

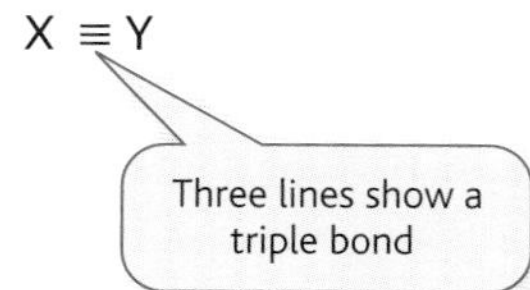

Self-check 3.12 A triple bond can form between which of the following pairs of atoms?

(a) C, N

(b) C, Cl

(c) C, H

(d) C, C

Satisfying valency with multiple bonds

An atom can participate in any combination of single, double, or triple bonds in order to satisfy its valency. For example, carbon, with its valency of four, can form four covalent bonds. These might be four single bonds, two double bonds, a triple bond and a single bond, or a number of other combinations, as illustrated in Figure 3.24. Notice how the valency of carbon is satisfied by a different combination of sigma and pi bonds in each instance. It is this variety that enables carbon, and other

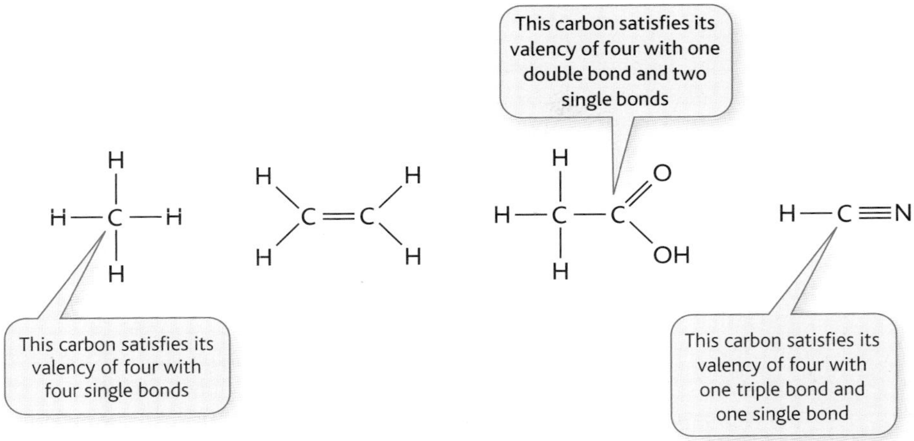

This carbon satisfies its valency of four with one double bond and two single bonds

This carbon satisfies its valency of four with four single bonds

This carbon satisfies its valency of four with one triple bond and one single bond

Figure 3.24 Carbon can satisfy its valency of four by participating in different combinations of single, double, and triple bonds.

high-valency elements, to form a part of such a wide array of compounds. We learn more about carbon, and the range of compounds it can form, in Chapters 5, 6, and 7.

However, an atom's valency also restricts the type of bond it can participate in. For example, hydrogen, with its valency of 1, cannot participate in a double or a triple bond: to participate in a double bond an atom must have a valency of at least 2, and to participate in a triple bond an atom must have a valency of at least 3.

- ***An atom can participate in any combination of single, double, or triple bonds in order to satisfy its valency—but its valency cannot be exceeded.***

Self-check 3.13 Which of the following compounds are drawn to show the correct valency for each atom involved? What are the errors?

(a)

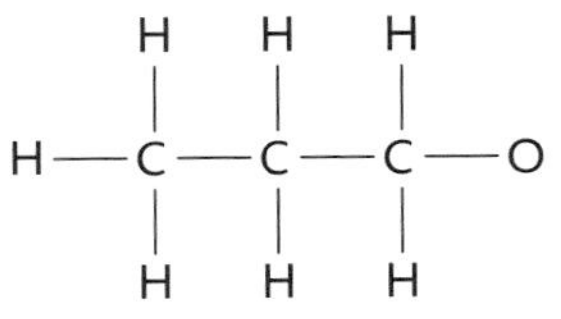

(b)

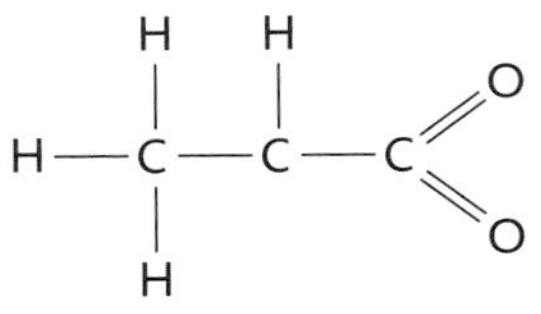

(c)

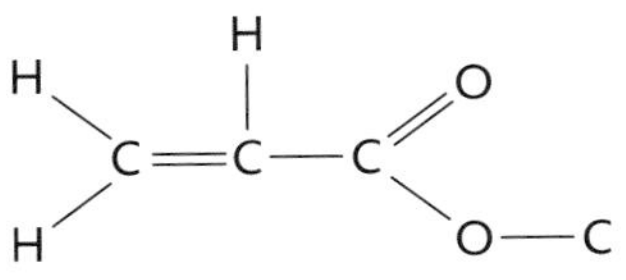

(d)

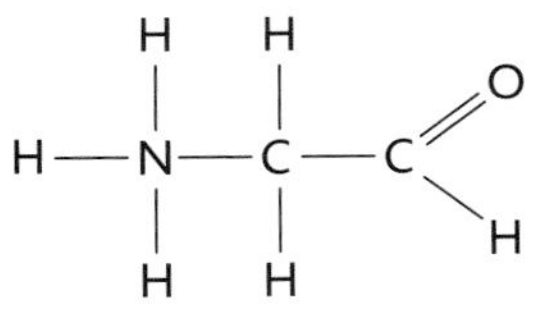

Hypervalency: going beyond the octet rule

We see in Section 3.2 how, as a general rule, we consider an atom's valence shell to be full once it contains eight electrons, as dictated by the octet rule. This eight-electron limit therefore determines an element's valency: it can only share electrons in covalent bonds until its valence shell is full.

However, as we note earlier, there are some elements for which the octet rule does not apply. These elements can possess an expanded valence shell, which may house *more* than eight valence electrons. We say that elements exhibiting such expanded valence shells have undergone **octet expansion**, and species that feature an atom with more than eight electrons are said to be **hypervalent**.

So what determines whether an element sticks to the octet rule, or whether it undergoes octet expansion? The answer may lie in the presence (or absence) of d orbitals in the valence shell of the element being considered. Typically, the elements of period 2, which features the biologically important carbon, nitrogen, and oxygen, obey the octet rule very well. For these elements, the valence shell is shell 2, which has the electronic configuration $2s^22p^6$ when full. (Notice how this electronic configuration represents a total of eight electrons, in keeping with the octet rule.) Beyond period 2, however, we start to see deviations.

For example, let's consider phosphorus, P, which is the central atom of the biologically important phosphate group, PO_4^{3-}. The phosphate group forms the 'backbone' of DNA, and is a vital component of molecules such as ATP that mediate the transfer of energy within the cell.

Look at the structure of the phosphate group when part of the DNA backbone, as depicted in Figure 3.25. Notice how the central phosphorus atom participates in five covalent bonds (one double covalent bond, and three single covalent bonds) such that its valence shell is occupied by a total of *ten* electrons, thereby deviating from the octet rule. Consequently, we see in this instance how phosphorus has a valency of five.

So how does phosphorus accommodate ten electrons in its valence shell? The electronic configuration of its valence shell is $3s^23p^3$. According to the octet rule, P should need just three more electrons to fill its valence shell (such that its electronic

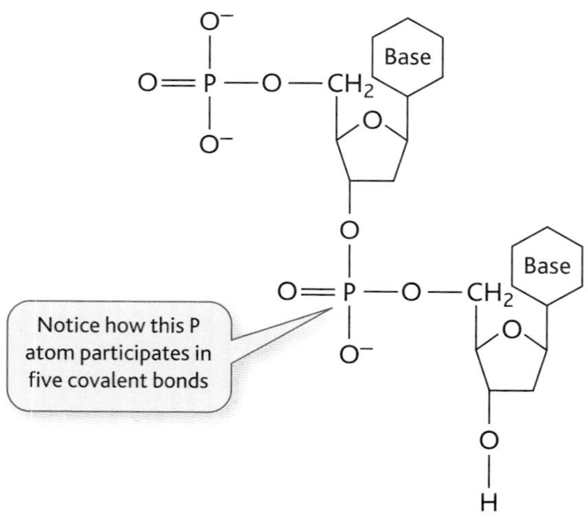

Figure 3.25 The phosphorus atom at the centre of the phosphate group that joins neighbouring nucleotides in a DNA strand participates in five covalent bonds (three single bonds, and one double bond), and so its valence shell is occupied by ten valence electrons.

configuration effectively becomes $3s^2 3p^6$), and could achieve this by participating in just three covalent bonds.

The most straightforward explanation for the observation that phosphorus can participate in five covalent bonds is that P uses its vacant d orbitals to hold some of its valence electrons–that is, it fills its s and p orbitals with eight electrons, and holds the remaining two electrons in the d subshell. (This explanation fits with the observation that elements in period 2 don't typically exhibit hypervalency: the valence shell of these elements is shell 2, which lacks the d subshell. Therefore, the elements of period 2 don't have vacant d orbitals to expand into.) However, it is also possible that the hypervalency exhibited by some elements is achieved by certain electron pairs participating in more than one covalent bond, such that the use of d orbitals as described here isn't so important after all.

Whatever the mechanism, however, the important thing to note is that some elements aren't limited to having just eight electrons in their valence shell, and can consequently form more than four covalent bonds with atoms of other elements.

Taking things one step further, some elements don't just exhibit one particular valency. Phosphorus is an excellent example of an element that can exhibit different valencies while still forming stable compounds. For example, look at Figure 3.26, which shows the structure of phosphorus trichloride, PCl_3. Notice how, in this instance, phosphorus has a valency of just three. In this instance, phosphorus does not need to undergo octet expansion to form a stable compound.

- ***A hypervalent compound is one that features an atom with more than eight valence electrons.***
- ***Some elements can exhibit different valencies, and therefore contribute to different numbers of covalent bonds.***

Sulfur is another important biological element, which exhibits a variable valency. In biological systems, sulfur typically exhibits a valency of 2, as we witness in the amino acid cysteine, (**1**) . However, sulfur can also exhibit a valency of six, as seen in the compound sulfur hexachloride, SF_6 (**2**).

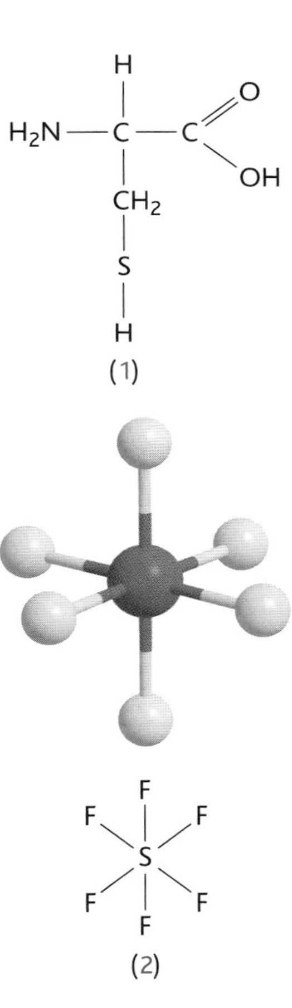

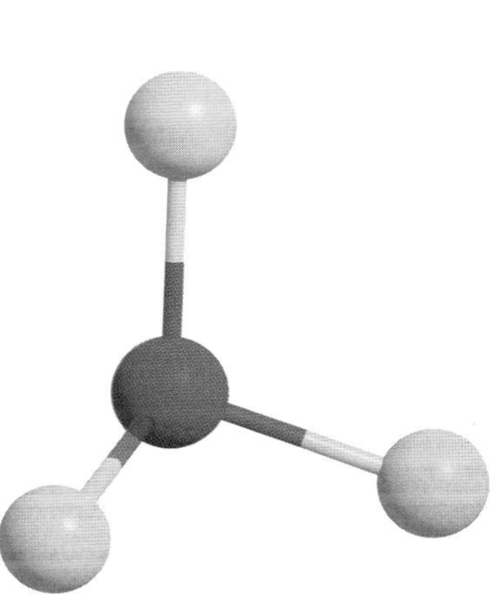

Figure 3.26 The structure of phosphorus trichloride, PCl_3.

Self-check 3.14 Look at structures (1) and (2), while bearing in mind sulfur's electronic configuration of $3s^2 3p^4$. Which of the following statements relates correctly to these two figures?

(a) Sulfur has an expanded octet of electrons in both structures 1 and 2.

(b) Sulfur has an expanded octet of electrons in structure 1, but not structure 2.

(c) Sulfur has an expanded octet of electrons in structure 2, but not structure 1.

(d) Sulfur doesn't have an expanded octet of electrons in either structure.

3.8 Dative bonding: covalent bonding with a twist

In a vast majority of cases, when two atoms share a pair of electrons to form a covalent bond, each of the two atoms contributes one electron to the bond. This isn't an absolute rule, however. In some instances, the pair of electrons that is being shared by two atoms to form a covalent bond comes from just *one* of the two atoms. In these instances, the bond formed is termed a **dative covalent** (or **coordinate**) bond.

It is important to recognize that electrons are still being shared between atoms, and not transferred from one atom to another during dative bond formation. So, it is still a covalent bond (rather than an ionic bond) that is formed, despite both electrons originating from a single atom.

Remember that covalent bonding arises from the overlap of two atomic orbitals, *one* from each of the participating atoms. So, when dative bonding occurs, an orbital containing *two* electrons from one atom must be overlapping with an *empty* orbital from the second atom. The bonding molecular orbital that is formed contains the two electrons donated by the first atom.

Figure 3.27 depicts the Lewis structure for the ammonium ion, NH_4^+. Look at this figure, and notice how this ion comprises three covalent and one dative covalent bond. Three of the covalent bonds arise from the sharing of a pair of electrons between the nitrogen and each of the three hydrogen atoms, with one electron in each pair originating from the N atom, and one originating from the H atom. The pair of electrons which forms the fourth bond, however, originates solely from the

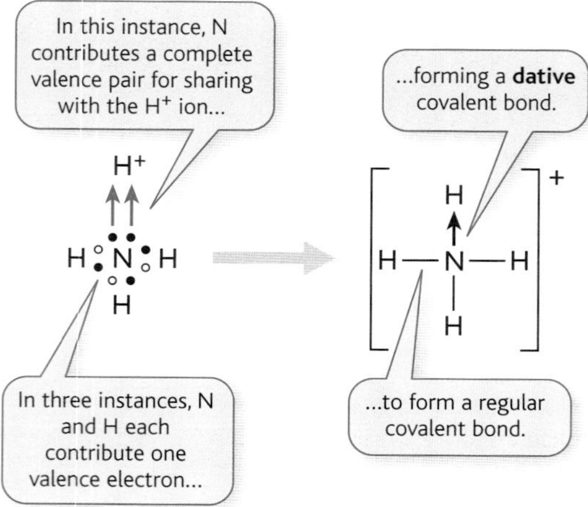

Figure 3.27 An ammonium ion represented as a Lewis structure. Notice how one of the covalent bonds is formed from a pair of electrons, both of which have been donated by the nitrogen atom. We call this a dative covalent bond.

N atom. This bond is a dative covalent bond, and is represented by the addition of an arrowhead to the end of the bond, pointing away from the atom that has donated the pair of electrons.

Polyatomic ions, such as NH_4^+, are explored in more detail in Section 3.10.

A dative covalent bond is chemically identical to any other covalent bond, and arises from the sharing of a pair of electrons between two atoms.

Atoms with non-bonding pairs of electrons are ideally placed to undergo dative covalent bonding. Each non-bonding pair has the potential to be shared with another atom to form a dative covalent bond. We see in Section 3.2 how atoms of the biologically important elements O and N possess non-bonding pairs of electrons. So atoms of both of these elements can form dative bonds with atoms of other elements. These bonds are a vital part of a number of biologically important molecules.

- *A dative covalent bond is formed by the donation of a pair of electrons by a single atom.*

Dative bonds in biological systems

Dative bonds are common in the bonding of some metals to atoms of other elements. Indeed, there are a number of biological molecules whose function is critically dependent upon the association of a metal with the molecule. In many instances, atoms in the molecule that are in close proximity to the metal donate pairs of electrons to form dative covalent bonds with the metal. The site in the molecule at which the metal is found is often called the **reaction centre**, which emphasizes the importance of the metal to the molecule's function.

Haemoglobin is the protein found in red blood cells which is responsible for carrying oxygen from our lungs to all living tissue within the body.

We learn about the impact of carbon monoxide on haemoglobin's oxygen-carrying ability in Box 3.6.

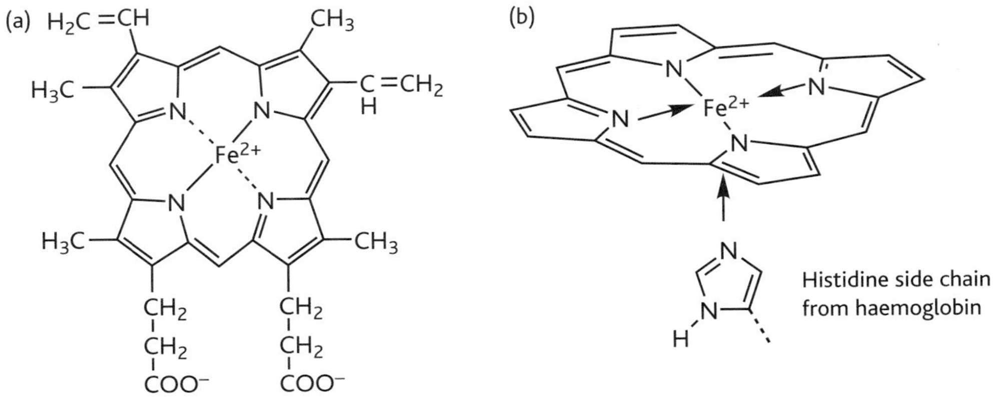

Haemoglobin contains an Fe^{2+} (iron) ion at the site in the protein where oxygen binding occurs. (Without the Fe^{2+} ion oxygen is unable to attach to haemoglobin.)

The Fe^{2+} ion forms part of a **haem group**, a ring-shaped structure depicted in Figure 3.28(a), which gives haemoglobin its oxygen-binding properties.

Figure 3.28 (a) The structure of the haem group, with an Fe^{2+} ion at its centre. (b) Fe^{2+} is held in place at the centre of the haem group by a total of five covalent bonds, three of which are dative covalent bonds.

BOX 3.6 Carbon monoxide: an invisible poison

We all know that we must breathe in oxygen to stay alive. All living tissue, such as our brain, muscle, and liver, requires a constant supply of oxygen to continue to function.

However, it is not just molecules of oxygen that can attach themselves to haemoglobin: carbon monoxide (CO) molecules can too. CO exists as a gas under normal conditions, and is most widely produced from the incomplete combustion of fuels. (Usually, CO_2 gas is produced if a fuel burns completely. If there is insufficient oxygen available in the surroundings, however, or if the equipment which is burning the fuel is poorly maintained, then unwanted CO is also produced.)

Importantly, unlike oxygen molecules, CO molecules bind to haemoglobin irreversibly–once a CO molecule has bound to haemoglobin, the haemoglobin is prevented from binding further molecules of oxygen. If we breathe in excessive amounts of carbon monoxide, all available oxygen binding sites within our red blood cells become blocked by CO. Eventually, our body is no longer able to obtain the levels of oxygen that it needs to survive, with fatal consequences.

Poorly maintained gas fires and gas-fired boilers that burn fuel inefficiently are among the most notorious sources of carbon monoxide. Proper maintenance of such appliances is vital. Failure to do so could, quite literally, cost you your life.

Now look at Figure 3.28(b), which shows how the Fe^{2+} ion is held in place at the centre of the haem group by a total of five bonds, three of which are dative bonds. Notice how the three dative bonds are formed from the donation of electron pairs by three nitrogen atoms, two of which are located in the haem group, and one which is found in an adjacent amino acid, histidine.

3.9 Aromatic compounds and conjugated bonds

Many compounds contain not just single bonds or just double bonds, but a combination of both. In certain circumstances, a combination of both single and double bonds gives rise to a special type of bond whose properties fall part way between those of a single bond and those of a double bond.

Look at the structure of the carbon-based compound benzene:

H C C H C H C H C H C H

Notice how:

- the carbon atoms are joined to form a **ring** structure
- the carbon atoms are joined by an alternating sequence of single and double covalent bonds.

What we see in reality, however, is that the carbon atoms are not joined by alternating single and double bonds, but by a network of bonds that fall halfway between being single and double. So what is happening here? How can there be this 'special' bond network?

Look at Figure 3.29, which illustrates the covalent bonding that occurs in benzene. Figure 3.29(a) shows how each carbon atom shares one valence electron with a hydrogen atom, and one valence electron with each of its neighbouring carbon atoms in the ring–three valence electrons are shared in total to form three σ bonds. Carbon has a valency of four, however, so each carbon atom in benzene's ring has one valence electron that is not directly involved in chemical bonding to form the benzene ring. Look at Figure 3.29(b) and notice how this valence electron occupies a p orbital.

The unique nature of the covalent bonding in benzene stems from the way that the p orbitals (containing the single valence electron) on each of the six carbon atoms in the ring overlap to form a *single* uninterrupted π orbital. This π orbital takes the form of two doughnut-shaped rings sitting above and below the plane of the six-carbon ring, as illustrated in Figure 3.30.

The valence electrons occupying the π orbital are not restricted to being associated with one or two specific carbon atoms upon the formation of this unique π orbital,

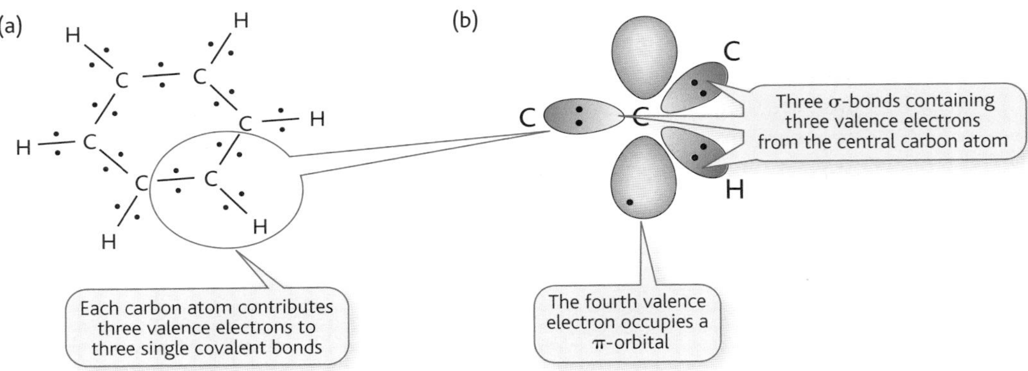

Figure 3.29 (a) Covalent bonding in benzene. Each carbon atom participates in three covalent sigma bonds. (b) Each carbon atom in the benzene ring has one unshared valence electron, which occupies a p orbital.

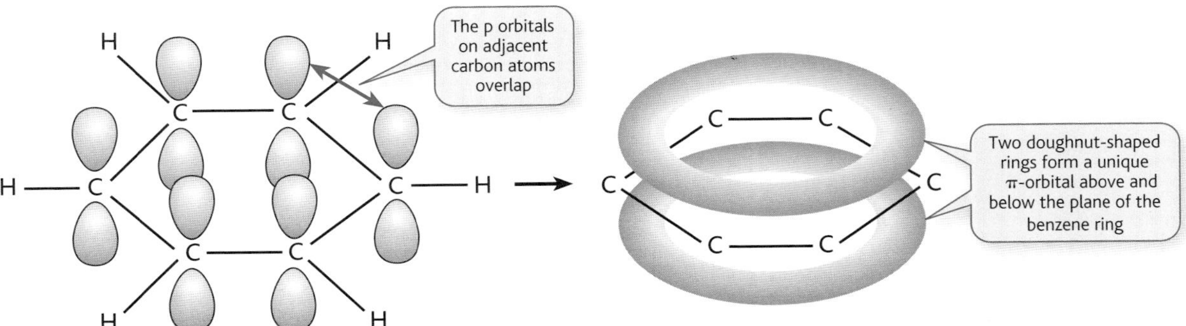

Figure 3.30 The p orbitals of the six carbon atoms that form a benzene ring overlap to form an unusual pi orbital, which takes the form of two doughnut-shaped rings. This pi orbital is occupied by the six unpaired valence electrons which occupied the six p orbitals prior to overlap.

but are free to move *anywhere* within the two doughnut-shaped lobes that form the π orbital–they are said to be **delocalized**[6].

This delocalization means that the bond between carbon atoms in the benzene ring falls halfway between being a single and a double bond. When we measure the distance between carbon atoms in benzene we find it to be 139.5 pm. By contrast, an average single C–C bond measures 154.1 pm, and an average double C=C bond measures 133.7 pm.

We often represent the benzene ring like this:

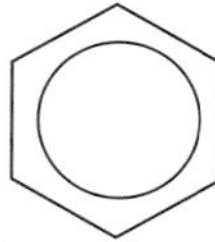

with the circle at the centre of the ring denoting the delocalized system of π electrons.

We call any molecule that includes the ring of delocalized electrons seen in benzene an **aromatic** compound. (By contrast, non-aromatic compounds are what we call **aliphatic**.) Aromatic compounds are widespread in biological systems, and include phenylalanine (an amino acid) and oestradiol (an important female sex hormone). Aromatic hydrocarbons also form the templates for many painkillers, including paracetamol and ibuprofen, depicted in Figure 3.31.

The unique π orbital in benzene is an example of what we call a **conjugated** system, in which the electrons are delocalized. The conjugation of bonds to form a conjugated system–a network of delocalized electrons–can occur wherever two double bonds are separated by *one* single bond, and not just in a ring-shaped molecule like benzene. For example, β-carotene, the pigment that gives carrots their orange colour, contains an extensive network of conjugated bonds:

All eleven double bonds in β-carotene form a conjugated system

Even in linear conjugated systems like this, the system itself arises in the same way as for benzene: from the overlap of adjacent p orbitals. We also find a conjugated system of bonds in the molecule 11-*cis*-retinal, a molecule central to visual perception in humans, as explained in Box 10.3.

HO H N O Paracetamol

CO_2H Ibuprofen

Figure 3.31 The structures of paracetamol and ibuprofen.

6 We call the real structure of benzene a **resonance hybrid**, with six delocalized p electrons shared by all carbon atoms.

BOX 3.7 **Delocalization in non-conjugated systems**

We also see delocalization of electrons in compounds that do not possess a conjugated system of bonds. Electron delocalization occurs in compounds in which two or more different arrangements of bonds can occur.

For example, the three oxygen atoms that comprise ozone, O_3, could join together in two different ways: O–O=O and O=O–O.

In practice, the compound is what we call a resonance hybrid–a hybrid of these two possible structures. The valence electrons occupying the π (double) bond become delocalized between all three atoms. Instead of forming a permanent double bond between a particular pair of oxygen atoms, the valence electrons spread between all three atoms, as seen in the diagram below.

As a result, the two bonds joining the three oxygen atoms don't behave as standard single bonds or double bonds, but have characters that fall part way between both types of bond. For example, the distance between adjacent oxygen atoms in ozone is shorter than a standard O–O bond, but longer than a standard O=O bond.

A particularly important resonance hybrid in biological systems is the peptide bond, the bond that joins together amino acids to form polypeptides. We learn more about the peptide bond–and the effect of it being a resonance hybrid on the structure of polypeptides–on p. 244.

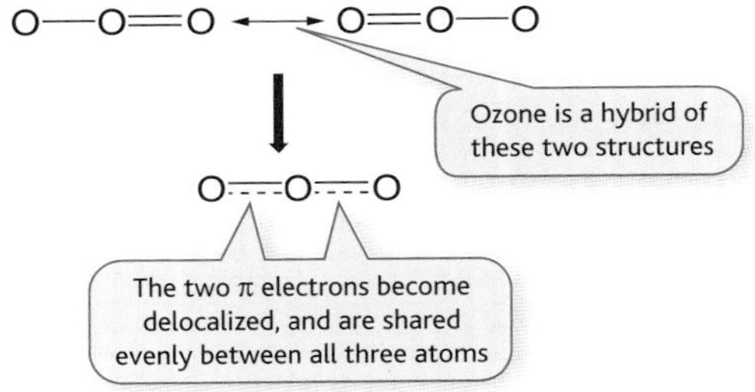
O—O=O ↔ O=O—O
Ozone is a hybrid of these two structures
O=O=O
The two π electrons become delocalized, and are shared evenly between all three atoms

Conjugated compounds are not the only compounds to exhibit delocalization, however, as explained in Box 3.7.

- ***A delocalized system is one in which electrons are shared between more than two atoms.***
- ***A conjugated system arises from the overlap of adjacent p orbitals in any molecule featuring a sequence of alternating single and double bonds.***
- ***Electrons in a conjugated system become delocalized–but delocalization can occur in non-conjugated systems too.***

3.10 Polyatomic compounds

We see in Sections 3.4 and 3.6 how compounds are formed as a result of ionic or covalent bonding between atoms. Compounds are not restricted to containing just one type of bond or the other, however. Some ionic compounds feature both ionic and covalent bonding. These compounds are called **polyatomic ionic compounds**.

In Section 3.4 we see how ionic compounds comprise an anion and a cation, which are held together by electrostatic attraction between the oppositely charged ions. In some cases, the ion may not be a single species (e.g. Mg^{2+} or Cl^-) but may be **polyatomic**, comprising a number of covalently bound species.

For example, calcium carbonate (more commonly known as chalk) has the chemical formula $CaCO_3$. The cation in this ionic compound is the Ca^{2+} ion, while the anion is the CO_3^{2-} (carbonate) ion. The two ions are held together by ionic bonding,

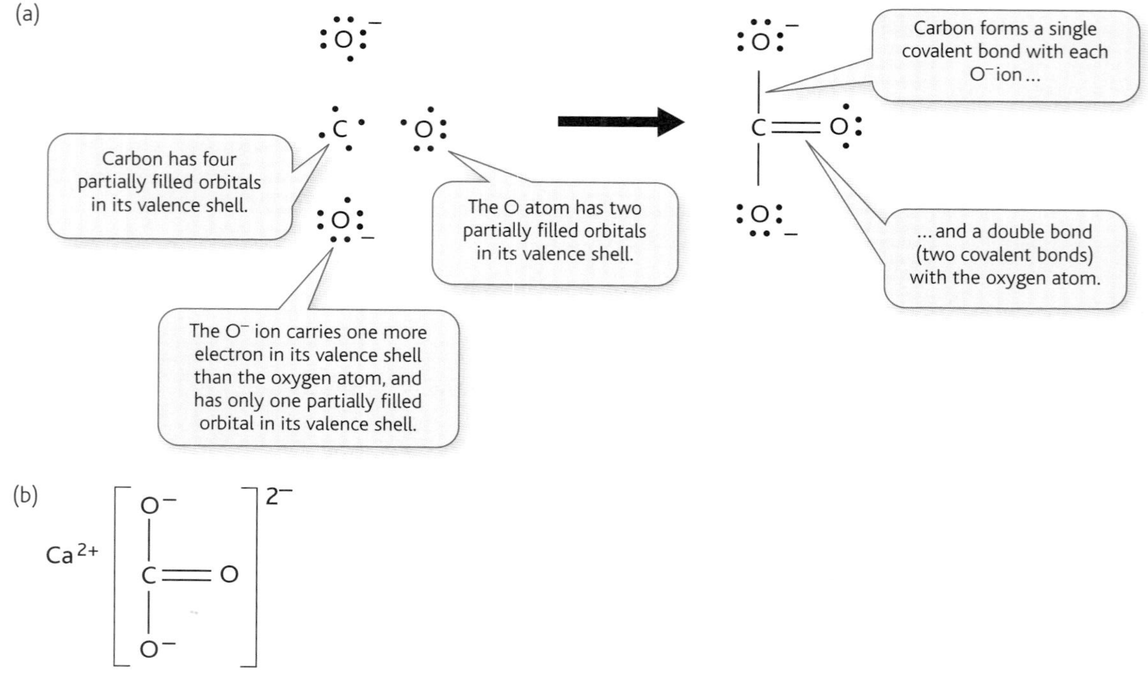

(b) Ca^{2+} $[O^{-}\text{–}C(=O)\text{–}O^{-}]^{2-}$

Figure 3.32 A polyatomic ionic compound comprises two ions, one of which is composed of multiple atoms joined by covalent bonds. For example, the carbonate ion, CO_3^{2-}, comprises four species joined by covalent bonds (a), and can form the polyatomic ionic compound calcium carbonate, $CaCO_3$, (b), in which electrostatic interactions operate between the two oppositely charged ions. Notice how the charges on the two ions, Ca^{2+} and CO_3^{2-}, cancel each other out.

but Figure 3.32 shows how the anion itself comprises two different elements (carbon and oxygen) joined by a total of four *covalent* bonds. Notice how the carbonate ion contains two O^- ions, giving rise to the two negative charges on the CO_3^{2-} ion overall.

Self-check 3.15 Look at Figure 3.32 once again. Based on what you can see in this figure, what is the valency of carbon?

Some other common polyatomic ions are shown in Table 3.2. Look at this table, and notice how virtually all polyatomic ions that we encounter are anions. The only prevalent polyatomic cation is the ammonium ion, NH_4^+. We encounter the ammonium ion in Section 3.8 when we see how one of its covalent bonds is a dative bond.

- ***A polyatomic ionic compound is one in which one of the ions in the compound features atoms held together by covalent bonds.***

The relative number of anions and cations in a polyatomic ionic compound must be balanced in exactly the same way as for any other ionic compound such that the overall charge on the compound is zero.

Ion name	Chemical formula	Charge
Hydroxide	OH^-	−1
Sulfate	SO_4^{2-}	−2
Carbonate	CO_3^{2-}	−2
Phosphate	PO_4^{3-}	−3
Ammonium	NH_4^+	+1
Nitrate	NO_3^-	−1

Table 3.2 The chemical formulae and electrical charges for some common polyatomic ions.

For example, sodium sulfate, Na_2SO_4, comprises the Na^+ cation (which carries one positive charge) and the SO_4^{2-} anion (which carries two negative charges). There must be two Na^+ ions for every SO_4^{2-} ion, as indicated by the chemical formula Na_2SO_4, for the overall charge on the compound to be zero.

Self-check 3.16 What is the chemical formula for lithium carbonate, which comprises the Li^+ and CO_3^{2-} ions?

The chemical formula for a polyatomic ionic compound uses exactly the same notation as any other ionic compound. If there is more than one polyatomic anion per cation, however, the anion is written in parentheses, and the number of anions present per cation denoted by a subscript number outside the parentheses.

For example, magnesium nitrate comprises two NO_3^- ions for every Mg^{2+} ion. The chemical formula for magnesium nitrate is therefore written as $Mg(NO_3)_2$. The polyatomic NO_3^- anion is contained within the parentheses, with the subscript 2 outside the parentheses indicating the presence of two anions for every Mg cation.

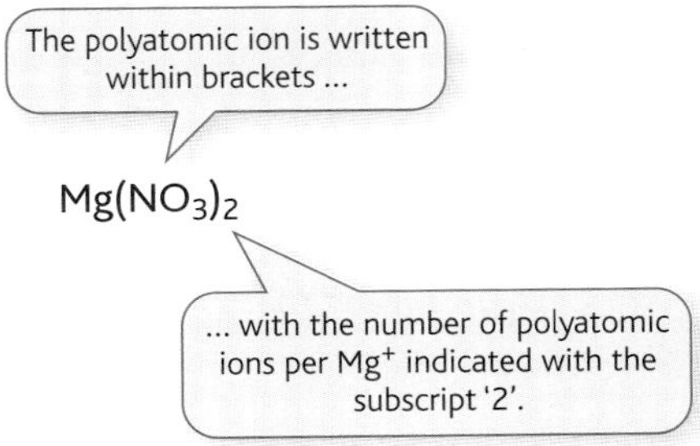
The polyatomic ion is written within brackets ...
Mg(NO3)2
... with the number of polyatomic ions per Mg+ indicated with the subscript '2'.

Self-check 3.17 Calcium hydroxide comprises the Ca^{2+} cation and the OH^- anion. What is the relative number of each ion in the compound and, hence, what is the chemical formula for calcium hydroxide? (Remember: there must be an equal number of positive to negative charges in the compound for it to be neutral overall.)

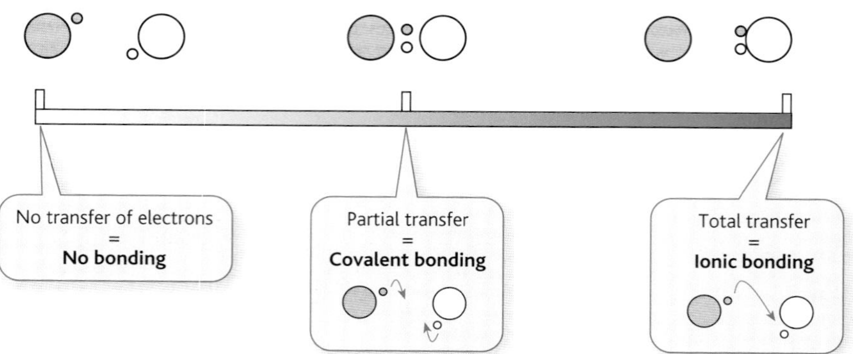

Figure 3.33 The spectrum of chemical bonding. No chemical bonding results if no transfer of electrons occurs. Ionic bonding results if one or more electrons are *totally* transferred. Somewhere in the middle, covalent bonding results if one or more electrons are partially transferred.

3.11 Ionic versus covalent bonding

What is it that determines the type of bonding (either ionic or covalent) that is exhibited by a particular pair of atoms? The answer lies in how readily a particular atom gains or loses electrons.

Look at Figure 3.33. This illustrates that two atoms that have not bonded at all, and two atoms that have undergone ionic bonding are at opposite ends of the same spectrum. At one end of the spectrum, denoting no bonding, no electrons have been transferred between the two atoms. At the other end, denoting ionic bonding, one or more electrons have been *completely* transferred between the two atoms. Somewhere in the middle lies covalent bonding, where electrons could be thought of as having been partially transferred–they are shared between two atoms, spending some of the time associated with one atom, and some of the time associated with the other one.

The type of bonding that exists between atoms of two particular elements depends on how readily electrons are transferred from one atom to another. If transfer occurs very readily then the two atoms are likely to undergo ionic bonding, while those atoms for which transfer occurs less readily undergo covalent bonding. If the electrons of a particular atom are very tightly associated with the atom and cannot interact with neighbouring atoms, then that atom is unable to form *any* bonds–it is chemically **inert**.

Electronegativity: how easily can electrons be transferred?

The readiness with which the transfer of an electron between atoms of two different elements occurs can be assessed by considering the **electronegativity** of the elements in question.

An element's **electronegativity value** (represented by the symbol chi, χ) indicates how strongly an atom of that element can attract an electron. A high value indicates that the atom strongly attracts an electron; a low value indicates that an electron is only weakly attracted. For example, chlorine has an electronegativity value of 3.0, while sodium has an electronegativity of 0.9. Therefore a chlorine atom shows a stronger attraction for an electron than sodium.

Figure 3.34 shows a version of the periodic table which gives the electronegativity values for a range of elements. What do you notice about electronegativity values as you move across a period? Figure 3.35 shows how electronegativity exhibits

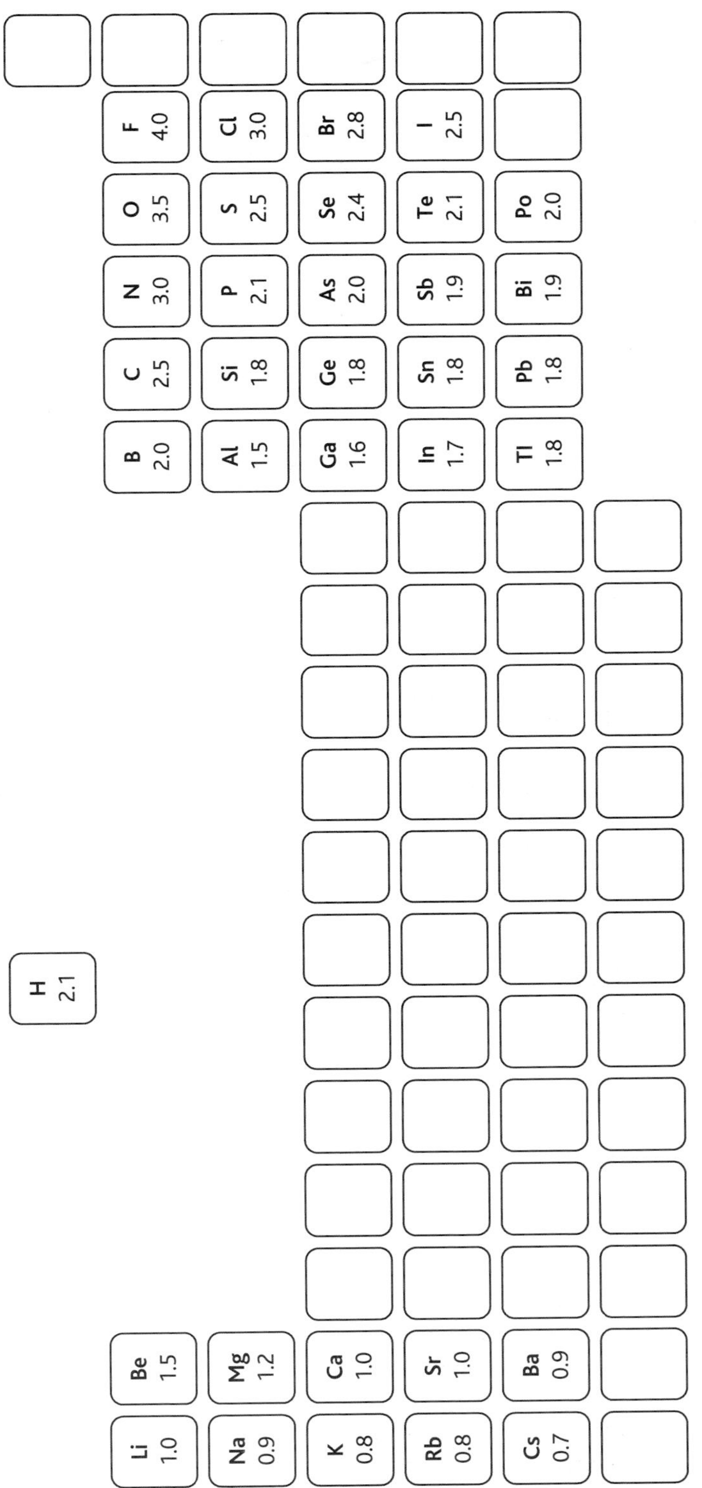

Figure 3.34 The electronegativity values of a range of elements arranged as per the periodic table.

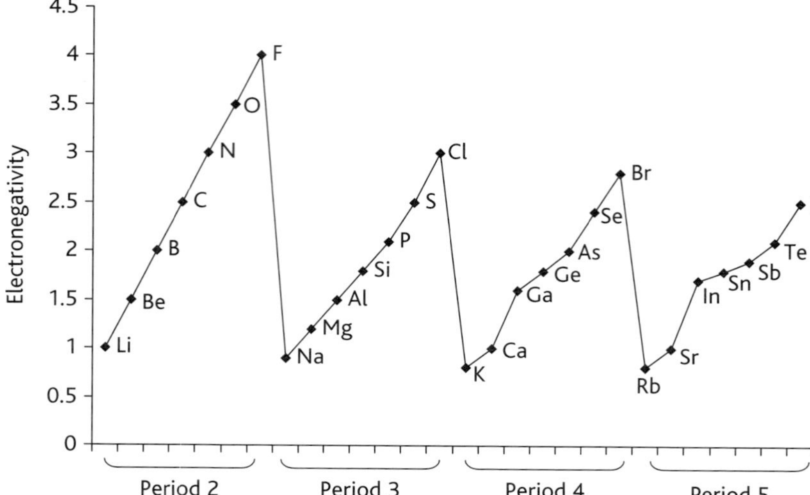

Figure 3.35 A graph showing how electronegativity values vary as we move across the periodic table, from period to period. Notice how electronegativity *increases* as we move across a period.

We learn more about periodicity in Section 2.1

periodicity–increasing as we move across a period, from one side of the periodic table to the other.

How does electronegativity help us to predict whether two atoms undergo ionic or covalent bonding?

If there is a large difference in electronegativity values between two atoms (indicating that one atom has a much stronger attraction for an electron than the other) it is likely that complete transfer of the electron from one atom to the other can occur, and the atoms undergo ionic bonding. Conversely, if electronegativity values are similar, such that the two atoms exert a similar 'pull' on the electron, complete transfer from one atom to another is unlikely, and covalent bonding results. Look at Figure 3.36, which illustrates this concept: where one atom pulls harder than the other (exhibiting greater electronegativity) the electron is pulled strongly towards that atom and becomes fully transferred, resulting in ionic bonding between the two atoms. By contrast, if both atoms pull with an equal force (their electronegativity values are very similar), there is no overall movement: the electron remains shared between both atoms and covalent bonding results.

- ***If the electronegativity values of two atoms are similar they are most likely to undergo covalent bonding.***
- ***If there is a large difference in the electronegativity values of two atoms they are most likely to undergo ionic bonding.***

Generally, if the difference in electronegativity value of two atoms is greater than or equal to 1.7, ionic bonding occurs between the two atoms. If the difference is less than 1.7, covalent bonding is most likely. For example, in the case of NaCl, the difference in electronegativity values is 2.1 (Na = 0.9, Cl = 3.0; difference = 2.1). This difference leads us to predict that sodium and chlorine undergo ionic bonding, which we have already seen is the case.

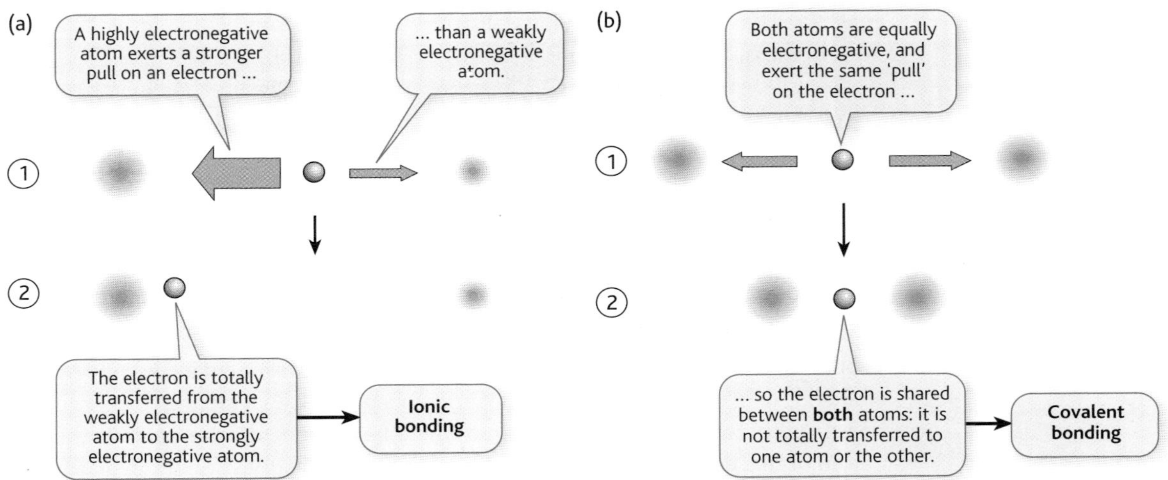

Figure 3.36 The influence of electronegativity on the nature of chemical bonding. (a) If there is a large difference in electronegativity between the two atoms, the strong 'pull' exerted by one atom results in an electron being transferred fully from one atom to another. Ionic bonding results. (b) If there is a small difference in electronegativity between the two atoms, the 'pull' exerted by both atoms is similar, so an electron fails to be transferred fully from one atom to another. Covalent bonding results.

Self-check 3.18 On the basis of electronegativity values, what type of bond would you predict to form between: (a) C and O; (b) Ca and Cl?

The bonding between atoms of a single element is always covalent, as the difference in electronegativity between the atoms is 0.

Ionic and covalent bonding in nature: which is most prevalent?

As we see in later chapters, many of the most biologically important molecules and compounds contain the four elements carbon, nitrogen, hydrogen, and oxygen. Look again at the periodic table showing electronegativities in Figure 3.34. What are the relative electronegativity values for C, N, H, and O, and what are the implications for the type of bonds which form between atoms of these elements?

As we see in the preceding section, ionic bonding is only predicted to occur between those elements that exhibit a large difference in electronegativity. In contrast, carbon, nitrogen, hydrogen, and oxygen all have broadly similar electronegativity values. Atoms of C, N, H, and O readily form covalent bonds with one another. Given the importance of C, N, H, and O in biological molecules, it is apparent that covalent bonding plays a vital part in quite literally holding us all together.

3.12 Blurring the boundaries: polarized bonds

We see in Section 3.6 how a covalent bond involves the equal sharing of electrons between two atoms. In reality, however, the electrons that make up the covalent bond are rarely shared equally between the joined atoms. Instead, the sharing of electrons is more uneven, with the distribution of electrons being skewed towards one of the two atoms that are joined by the covalent bond.

We call a covalent bond in which the electrons are not evenly shared between the two joined atoms a **polar** bond, and we say that the bond is **polarized**[7].

- ***A polar bond is a covalent bond in which the electrons are not evenly shared between the two joined atoms.***

The distribution of electrons in a polar bond is governed by the **electronegativity** of the respective atoms. We see above how we can use an element's electronegativity value, χ, to estimate how strongly an atom of a particular element attracts a shared electron to itself within a bond. If an element has a high electronegativity, then an atom of that element attracts a shared electron to itself very strongly. If an element has a lower electronegativity, then an atom of that element attracts a shared electron to itself only weakly.

The distribution of electrons in a polar bond is skewed towards the atom with the highest electronegativity value (the atom that exerts the strongest attractive force on the shared electrons). The atom with the highest electronegativity value carries a slightly more negative electrical charge than the atom to which it is joined as a consequence of the attraction of electrons to itself.

Look at Figure 3.37, which shows the relative distribution of electrons between an atom of hydrogen and an atom of chlorine when joined by a H–Cl bond in the molecule HCl. Notice how the distribution of electrons is skewed towards the most electronegative atom, chlorine. While HCl carries no electrical charge overall, the unequal distribution of electrons results in the chlorine atom carrying a slight negative charge and the hydrogen atom carrying a slight positive charge, relative to one another. We call these 'slight charges' **partial negative** and **partial positive** charges. We call the difference between these partial positive and negative charges a **dipole moment**.

Partial charges are denoted by the Greek character delta (δ): a partial positive charge is denoted $\delta+$, while a partial negative charge is denoted $\delta-$.

- ***The distribution of the electrons that are shared in a polar bond is skewed towards the most electronegative of the two atoms.***

We need to remember that partial negative and partial positive charges are not equivalent to the *full* negative and positive charges carried by ions. Full positive and negative charges arise only when an electron is *transferred fully* from one atom to another. Figure 3.38 shows how an atom of Na and an atom of Cl undergo ionic bonding to yield a Na^+ cation, carrying a single positive charge, and a Cl^- anion, carrying a single negative charge. By contrast, an atom of H and an atom of Cl undergo covalent bonding to yield a covalently bound molecule in which there is no full charge on either atom.

---

7 We can use the term *polar* to describe any chemical entity in which electrons are distributed unequally.

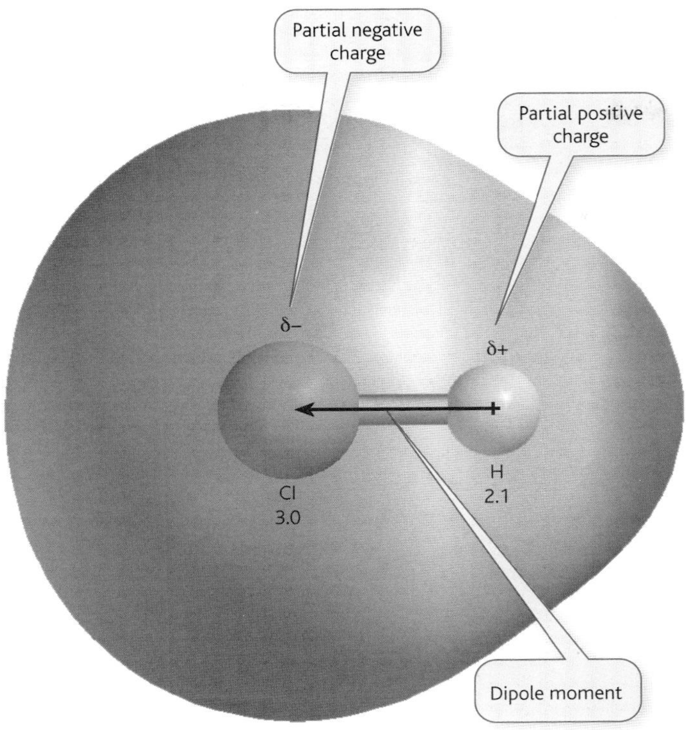

Figure 3.37 The relative distribution of electrons in a molecule of hydrogen chloride, HCl. The distribution of electrons is skewed towards the highly electronegative chlorine atom. The numbers shown are the electronegativity values for each element.

Some of the most widespread polar bonds in biology include the C=O, O–H, and N–H bonds. Notice how these three bonds include atoms of the two highly electronegative elements, oxygen and nitrogen, bonded to less electronegative atoms.

How strongly is a bond polarized?

We can predict how *strongly* polarized a particular bond is (that is, how unequally the electrons are shared between the two joined atoms). We do this by considering the *difference* in the electronegativity of the two atoms.

If there is a large difference in electronegativity, the bond will be highly polarized, with the electrons shared unequally. If there is a small difference in electronegativity, then the bond will be weakly polarized, with the electrons shared more evenly.

Self-check 3.19 Look back at Figure 3.34, which gives the electronegativity values for a range of elements. Based on this information, put the following pairs of atoms in order of bond polarization, starting with the most weakly polarized bond and ending with the most strongly polarized bond:

(a) C–H; (b) C–O; (c) H–Cl; (d) O–H

The comparison of electronegativity values to determine the extent of polarization of a bond is similar to the way we compare electronegativity values to predict whether a bond will be ionic or covalent, as described on p. 72. However, remember that the difference in electronegativity values associated with ionic bonding is large (a difference of at least 1.7). When we compare electronegativity values to predict

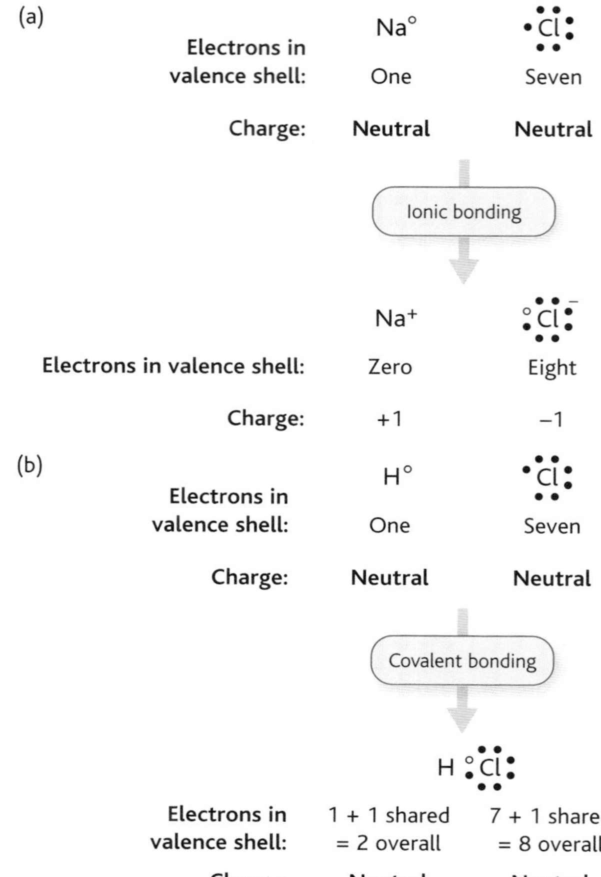

Figure 3.38 (a) An atom of Na and an atom of Cl undergo ionic bonding to yield a Na^+ cation, carrying a single positive charge, and a Cl^- anion, carrying a single negative charge. (b) An atom of H and an atom of Cl undergo covalent bonding to yield a covalently bound molecule in which there is no full charge on either atom.

the polarization of a covalent bond, the differences are much smaller. For example, the difference in electronegativity value between H and Cl is just 0.9.

Some of the most polarized bonds in biological molecules occur between hydrogen and atoms of the highly electronegative elements fluorine, oxygen, and nitrogen. These polar bonds play a critical part in forming one type of non-covalent force that has huge importance in many biological systems: the **hydrogen bond**. We explore the hydrogen bond in more detail in Section 4.4.

Non-polar covalent bonds

If the electrons in a covalent bond *are* genuinely shared equally between the two atoms joined by the bond then we say that the bond is **non-polar**. The equal sharing of electrons occurs when two atoms of the *same* element are joined by a covalent bond, as illustrated in Figure 3.39.

In this instance, both atoms have the same electronegativity and attract the shared electrons with the same strength, such that the electrons remain evenly shared between the two atoms, rather than being drawn towards one atom or the other[8].

8 We call bonds between atoms of the same element a homonuclear bond. A bond between atoms of different elements is a heteronuclear bond. (Homo = same; hetero = different.)

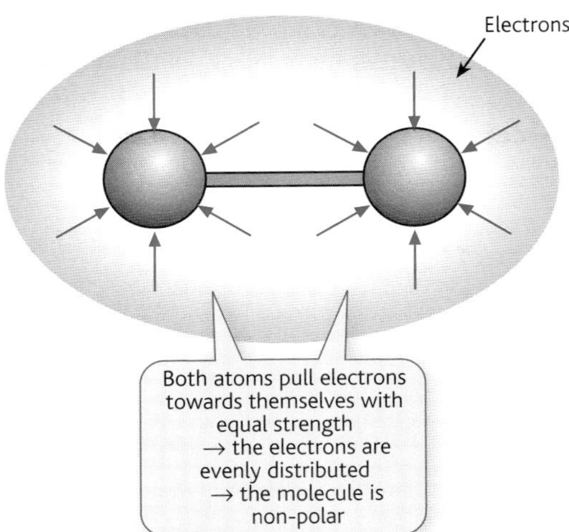
Electrons
Both atoms pull electrons towards themselves with equal strength
→ the electrons are evenly distributed
→ the molecule is non-polar

Figure 3.39 When two atoms of the same element are joined by a covalent bond, electrons are shared equally between the two atoms. The resulting molecule is non-polar.

Because the electrons are equally shared, such that their negative charges are evenly distributed, atoms that are joined by a non-polar covalent bond remain electrically neutral—that is, they do not carry either a partial positive or a partial negative charge. As a consequence, non-polar covalent bonds have a dipole moment of zero.

Check your understanding

To check that you've mastered the key concepts presented in this chapter, review the checklist of key concepts below, and attempt the multiple-choice questions available in the book's Online Resource Centre at **http://www.oxfordtextbooks.co.uk/orc/crowe2e/**.

Checklist of key concepts

The formation of compounds

- A compound is a substance which comprises more than one element

Bond formation

- Chemical bonds hold the components of a compound together
- Chemical bonds are formed by the redistribution of valence electrons between atoms
- According to the octet rule, valence electrons are redistributed so that atoms achieve full valence shells which contain eight electrons
- The valence shells of some elements can contain more than eight electrons
- The valence shell of an atom can be represented by a Lewis dot symbol
- A lone pair of electrons is a pair of valence electrons which fully occupies an atomic orbital
- A lone pair is more usefully called a 'non-bonding pair' when we consider valence electrons in molecules
- The noble gases (Group 18) do not undergo chemical bonding because they possess full valence shells naturally
- Elements which are unable to undergo chemical bonding are chemically inert
- There are two types of chemical bond: ionic and covalent

- An **ionic bond** forms when one or more electrons are totally transferred from one atom to another to generate an ionic compound
- A **covalent bond** forms when one or more pairs of electrons are shared equally between atoms to generate a covalent compound
- A **non-metal** will react with a metal to form an ionic compound, held together by ionic bonds
- Two non-metals will react to form a covalent compound, held together by covalent bonds
- More electrons must occupy a molecule's bonding orbitals than its anti-bonding orbitals for covalent bonding to occur
- A **non-bonding** orbital is occupied by a non-bonding pair of electrons and does not contribute to covalent bonding
- A **dative covalent bond** forms when the pair of electrons in a molecular bonding orbital originates from just one of the two atoms that are bonded together
- Dative bonds are common in the bonding of some metals to atoms of other elements.

The ionic bond

- When atoms undergo ionic bonding they become **ions**
- The attraction between oppositely charged ions holds ionic compounds together
- This attraction is called an **electrostatic interaction**
- Ionic compounds exist as lattice structures in which the anions and cations are arranged in regular repeating patterns
- An ionic compound has **zero charge** overall
- The number of negative charges in an ionic compound must equal the number of positive charges
- The **chemical formula** identifies the elements from which an ionic compound is formed, and the relative number of ions of each element present in the compound

The covalent bond

- Atoms linked by covalent bonds form units called **molecules**
- Molecules may comprise atoms of different elements, or atoms of the same element
- A covalent compound comprises many millions of identical molecules
- A covalent compound carries **zero charge**
- The **molecular formula** tells us the composition of one molecule of a covalent compound
- The composition of a molecule can be represented by a **Lewis structure**
- A covalent bond forms when atomic orbitals on separate atoms overlap and generate **molecular orbitals**
- The overlap of two atomic orbitals generates one **bonding orbital** and one **anti-bonding orbital**
- Electrons in bonding orbitals are those that actively contribute to covalent bonding
- Electrons in anti-bonding orbitals inhibit covalent bond formation

Multiple covalent bonds

- There are two types of molecular orbital, **sigma** (σ) and **pi** (π), which have contrasting shapes.
- A covalent bond can arise from either a pair of electrons occupying a **σ bonding orbital** (to generate a σ bond) or a **π bonding orbital** (to generate a π bond). σ and π orbitals have distinct shapes, but are chemically equivalent
- An element's **valency** tells us how many pairs of electrons an atom of that element must share with other atoms to attain a full valence shell and, therefore, how many covalent bonds it can form
- Atoms of certain elements can form multiple covalent bonds with other atoms
- A **single bond** arises from the sharing of one pair of electrons between two atoms to form one covalent bond: a σ bond
- A **double bond** arises from the sharing of two pairs of electrons between two atoms to form two covalent bonds: one σ bond and one π bond
- A **triple bond** arises from the sharing of three pairs of electrons between two atoms to form three covalent bonds: one σ bond and two π bonds
- A species that features an atom with more than eight valence electrons (and which can participate in more than four covalent bonds) is called **hypervalent**.

Aromatic compounds

- **Delocalized electrons** are valence electrons that are shared between more than one pair of atoms
- Two double bonds separated by a single bond can form a **conjugated** system, in which the π bonding electrons are delocalized throughout the conjugated system
- Bonds in a conjugated system behave in a way that falls part way between being a single bond and a double bond
- An **aromatic compound** is one that includes a ring of delocalized electrons

- If a molecule can be represented by two or more Lewis structures, its actual structure is a hybrid of the possible structures; we call it a resonance hybrid

Polyatomic compounds

- In a polyatomic ionic compound, the anion or cation comprises multiple atoms that are covalently bonded
- A polyatomic ion still carries an overall positive or negative charge

Electronegativity

- An element's electronegativity value, χ, indicates how strongly an atom of that element attracts an electron
- A high value for χ indicates that an electron is strongly attracted; a low value for χ indicates that an electron is weakly attracted
- Whether two atoms are likely to undergo ionic or covalent bonding can be predicted by comparing their electronegativity values
- Two atoms whose electronegativity values differ by 1.7 or more are most likely to undergo ionic bonding
- Two atoms whose electronegativity values differ by less than 1.7 are most likely to undergo covalent bonding
- The principal biological elements have similar electronegativity values, and undergo covalent bonding with one another

Polarized bonds

- A polarized bond is one in which the electrons are shared unequally between the atoms that are joined by the bond
- The distribution of electrons in a covalent bond is dictated by the electronegativity of the atoms joined by the bond: the distribution is skewed towards the most electronegative of the two atoms
- If the difference in electronegativity of two atoms is large, the bond joining them will be highly polarized
- If the difference is small, the bond joining them will be only slightly polarized
- If the electrons in a covalent bond are shared equally, the bond is non-polar

4 Molecular interactions: holding it all together

INTRODUCTION

In Chapter 2, when we look at the formation of bonds, we see how the arrangement of electrons between atoms can result in either ionic or covalent chemical bonds. The resultant distribution of electrons allows individual atoms to associate with one another to form compounds.

If electrons are shared, the bond is covalent. If, however, one or more electrons transfer from one atom to another, two ions form, which are held together by an ionic bond. An ionic bond is defined as a strong electrostatic force that exists between the two oppositely charged species. While ionic bonds hold ions together in large three-dimensional lattices, covalent bonds hold atoms together in discrete units called molecules.

However, the covalent bonds that hold atoms together in molecules are not the only important forces that exist in biological systems. In this chapter we explore non-covalent interactions: the range of other forces that occur within and between molecules. We ask how these forces arise, what effect they have on the way in which molecules interact, and why these forces are so important in biological systems.

4.1 Chemical bonding versus non-covalent forces

Biological systems are assembled from a diverse pool of molecules, including proteins, nucleic acids, sugars, and fats, all of which comprise atoms that are held together in a stable way by networks of covalent bonds.

We explore the chemical composition of some of the major types of biological molecules in Chapter 7.

It is not just the covalent bonds that exist between atoms in a molecule that are vital for the correct operation of a biological system, however. Just as important are the various **non-covalent** interactions that govern how different parts of a molecule, or entirely separate molecules, interact with one another. These molecular interactions lie at the heart of all biological systems, from the assembly of lipid molecules while forming cell membranes, to the hugely condensed packing of our genome into the cell nucleus.

By governing how molecules interact with one another, non-covalent molecular interactions underpin a number of physical properties of molecular substances, including melting points and boiling points[1]. We see in Section 4.5 how non-covalent

1 The melting point is the temperature at which a solid transforms to a liquid. The boiling point is the temperature at which a liquid becomes a gas.

Interaction type	Typical energy (kJ mol^{-1})
Covalent bond	150–1000
Ionic bond	250
Dispersion force	2
Dipole–dipole interaction	2
Hydrogen bond	20

Table 4.1 A comparison of the typical energies of a range of physical interactions.

interactions can help us to understand and explain the physical properties of different molecular substances.

The covalent bonds that exist between atoms in a molecule are strong, and are generally disrupted only when a molecule undergoes a chemical reaction. By contrast, the non-covalent interactions that exist *between* molecules are generally weaker than covalent bonds (typically ten- to a hundred-fold weaker), and can be disrupted more readily.

There are six key types of non-covalent interaction other than formal ionic bonds (the ionic bonds that we encountered in Chapter 3); these six interactions have a vital role in influencing the physical properties of molecules. These are:

1. Dispersion forces
2. Permanent dipolar interactions
3. Steric repulsion
4. Hydrogen bonds
5. Ionic interactions
6. Hydrophobic forces

We explore each of these interactions in more detail in Sections 4.3 and 4.4.

Table 4.1 shows the typical energies of a number of non-covalent interactions that we will encounter in this chapter. For comparative purposes the table includes the *range* of energies typically exhibited by a covalent bond. Notice how covalent and ionic bonds have broadly similar energies, while the other forces are of considerably smaller energy.

We see from Table 4.1 how dispersion forces and permanent dipolar interactions when considered individually are much weaker than either formal ionic or covalent bonds. Indeed, a single non-covalent interaction is so weak as to have a negligible effect on both the ability of two molecules to associate with one another and the resultant stability of these molecules once associated.

The key to the biological significance of non-covalent interactions, however, is that they do not operate as single events (with just one non-covalent interaction occurring between two molecules, for example). Instead, many non-covalent interactions may operate between two molecules, such that the net (overall) effect is large enough for it to be both noticeable and significant, in terms of promoting association between molecules and conferring greater stability.

We learn more about what is happening at the molecular level during melting and boiling in Section 4.5.

Intramolecular versus intermolecular forces

Molecular interactions operate at two levels:

1. **Intramolecular** interactions operate between separate parts of the *same* molecule (intra = within, so 'intramolecular' = within molecules).
2. **Intermolecular** interactions operate between *different* molecules (inter = between, so 'intermolecular' = between molecules).

Both types of interaction are absolutely vital in biological systems, particularly in helping biological molecules to maintain their correct shapes. For example, a protein may comprise several discrete polypeptide molecules, which associate with one another to generate the complete, fully functional protein[2]. At the level of the individual polypeptide molecules, intramolecular interactions help each polypeptide chain to adopt the correct three-dimensional structure, so that the separate molecules can slot together correctly just like the pieces of a jigsaw. Once associated, intermolecular interactions help the separate polypeptide molecules to remain associated in a stable way, and hence ensure that the protein can function as it should.

Look at Box 4.1 to see an example of a biological molecule, which relies on both intramolecular and intermolecular interactions to adopt the correct three-dimensional structure.

We explore the biological significance of intramolecular and intermolecular interactions later in this chapter, while we discover more about the importance of shape in biology (why we talk about a molecule possessing the 'correct' structure, for example) in Chapters 8 and 9.

- ***Intramolecular interactions exist between separate parts of the same molecule.***
- ***Intermolecular interactions exist between different molecules.***

The significance of molecular interactions

Why should we be interested in molecular interactions, particularly if they are relatively weak? The answer is that they confer *stability* upon molecules, most often by facilitating close association between separate molecules, or parts of the same molecule.

Consider a typical camping holiday and its central feature: the tent. What is the most important outcome when pitching a tent? It is to make the tent as stable as possible, so that it does not simply collapse with the lightest gust of wind, or become separated from the ground beneath it. We give the tent stability by using tent pegs: the tent pegs strengthen the association between the tent and the ground and between the separate parts of the same structure, thereby making the overall assembly a stable one.

Molecular interactions are the molecular equivalent of using tent pegs: they strengthen the association between molecules, and confer stability–they hold the molecules in place, instead of the molecules subsequently drifting apart. Indeed, by holding two molecules in place, molecular interactions play an important role in

---

2 Often biologists may refer to such a complex of molecules, which forms the fully functional protein, as a 'molecule'. However, from a chemical point of view, molecules are discrete chemical entities, such as individual polypeptide chains.

BOX 4.1 **How haemoglobin pulls itself together**

The protein haemoglobin comprises four polypeptide subunits, and depends upon both intramolecular and intermolecular interactions to ensure that its structure remains correct.

Figure 4.1 shows the three-dimensional structure of haemoglobin. Look carefully at the four areas with different shading to identify the four discrete molecules that associate to form the complete protein. Intramolecular forces enable the individual polypeptides to adopt the correct structural arrangements, while intermolecular forces hold the four units together as a stable complex.

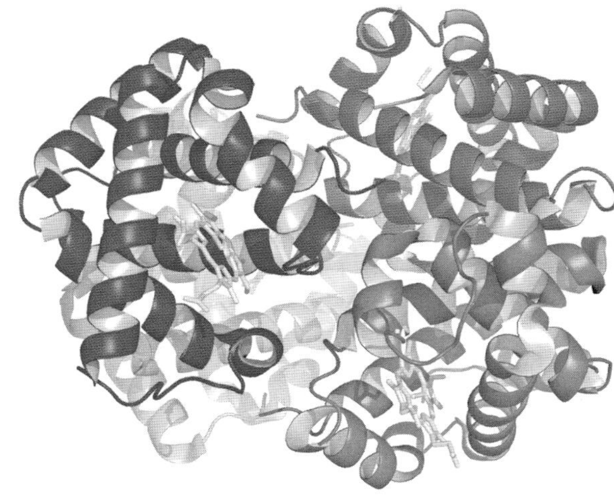

Figure 4.1 The three-dimensional structure of haemoglobin. Haemoglobin comprises four discrete molecules, which are held together by intermolecular forces.

helping biological systems exhibit **specificity**. Many biological systems require the stable association of two or more molecules. For example, our endocrine system (the system of hormones that controls many aspects of our behaviour and physiology) is based upon signals being transmitted through the binding of hormones to their complementary receptors. For a hormone to pass on the right message, it must bind in a stable way solely to its complementary receptor, and not to other receptors. However, these interactions can only occur to a level that makes the association stable if the two interacting molecules have complementary *shapes*.

Figure 4.2 shows how many interactions can occur between two molecules if they have complementary shapes (such that their association is stabilized) whereas few interactions can occur if their shapes are not complementary, so their association is *not* stabilized. This contrast in stability enables a biological system to exhibit specificity, such that only associations between molecules with complementary shapes are stable enough to occur.

We explore the critical link between shape and specificity in more detail in Chapter 9.

4.2 Electrostatic forces: the foundations of molecular interactions

Let's now go on to explore what it is at the physical level that determines whether or not molecular interactions occur. The non-covalent interactions that operate within

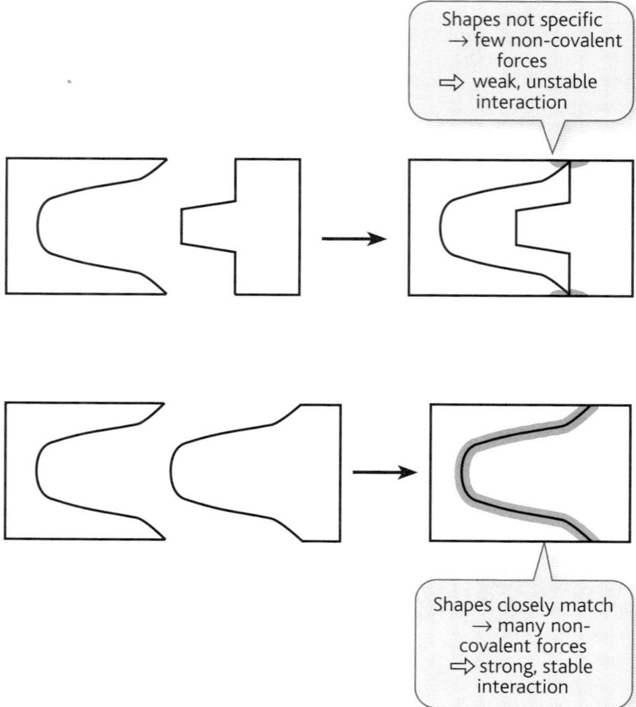
Shapes not specific
→ few non-covalent forces
⇨ weak, unstable interaction
Shapes closely match
→ many non-covalent forces
⇨ strong, stable interaction

Figure 4.2 The association of two molecules with complementary shapes is stabilized by non-covalent interactions. By contrast, two molecules whose shapes aren't complementary do not experience such extensive non-covalent forces; their association is not stabilized.

biological systems have their roots in the complementary principles of attraction and repulsion: opposite charges attract, while like charges repel. A vast majority of molecular interactions are **electrostatic** in nature: they are based on the notion of opposite charges attracting (and, in one case, repelling) one another. So we can see that molecules must possess areas of opposing charge for molecular interactions to arise. But how do these opposite charges occur?

We see in Section 3.12 how the atoms joined by a covalent bond develop partial positive and partial negative charges when the electrons they share within the covalent bond are shared unequally. The bond is polarized, and we call it a **polar bond**. The unequal sharing of electrons doesn't just occur at the level of the bond, however, electrons can also be distributed unequally throughout a molecule as a whole, generating a **polar molecule**[3].

We encounter a simple example of a polar molecule in Section 3.12: a hydrogen chloride molecule features a single, polar H–Cl bond and is therefore polar overall, as depicted in Figure 4.3. Another example of a polar molecule is water, H_2O. Water comprises two hydrogen atoms bonded to a central oxygen atom by two polar O–H bonds, as depicted in Figure 4.4(a). If we look at the overall distribution of electrons within the molecule, as illustrated in Figure 4.4(b) we see how the distribution of electrons across the molecule as a whole is uneven, with most electrons being clustered around the oxygen atom, and fewer electrons extending down towards either hydrogen atom.

Hydrogen chloride and water also both exhibit dipoles. A **dipole** consists of a positive charge and a negative charge which are separated in space within a

3 We can use the term **polar** to describe any chemical entity in which electrons are distributed unequally.

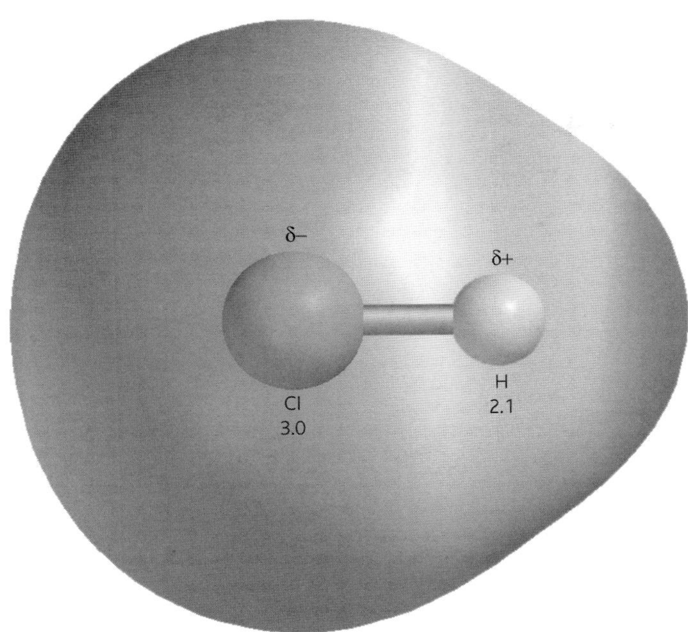

Figure 4.3 The relative distribution of electrons in a molecule of hydrogen chloride, HCl. The distribution of electrons is skewed towards the highly electronegative chlorine atom. The numbers shown are the electronegativity values of hydrogen and chlorine. (Image courtesy Peter Atkins.)

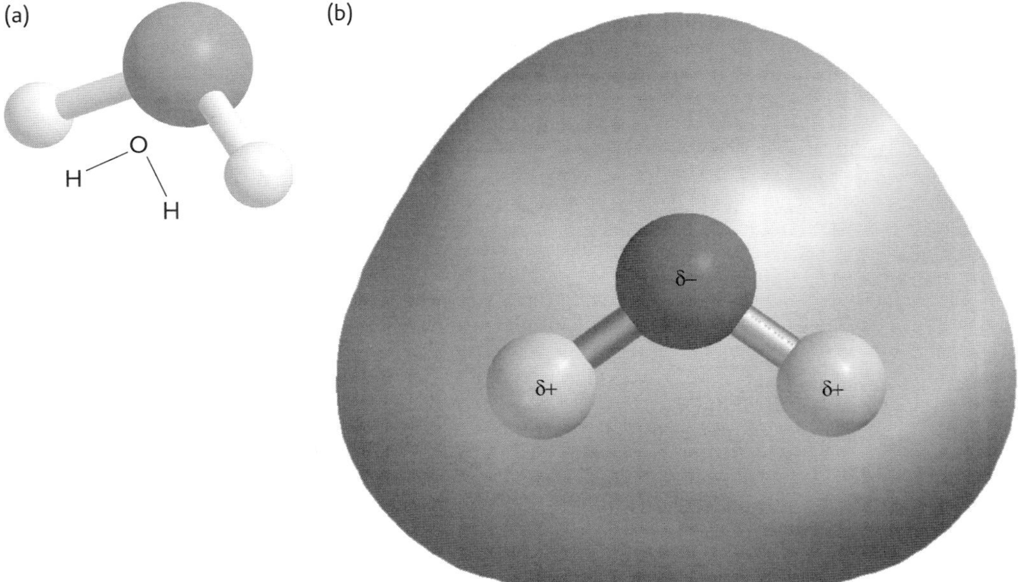

Figure 4.4 The polarity of a water molecule. (a) A water molecule comprises two hydrogen atoms bonded to a central oxygen atom by two polar O–H bonds. (b) Oxygen is more electronegative than hydrogen, so the distribution of electrons in a water molecule is 'skewed' towards the oxygen atom. (Image courtesy Peter Atkins.)

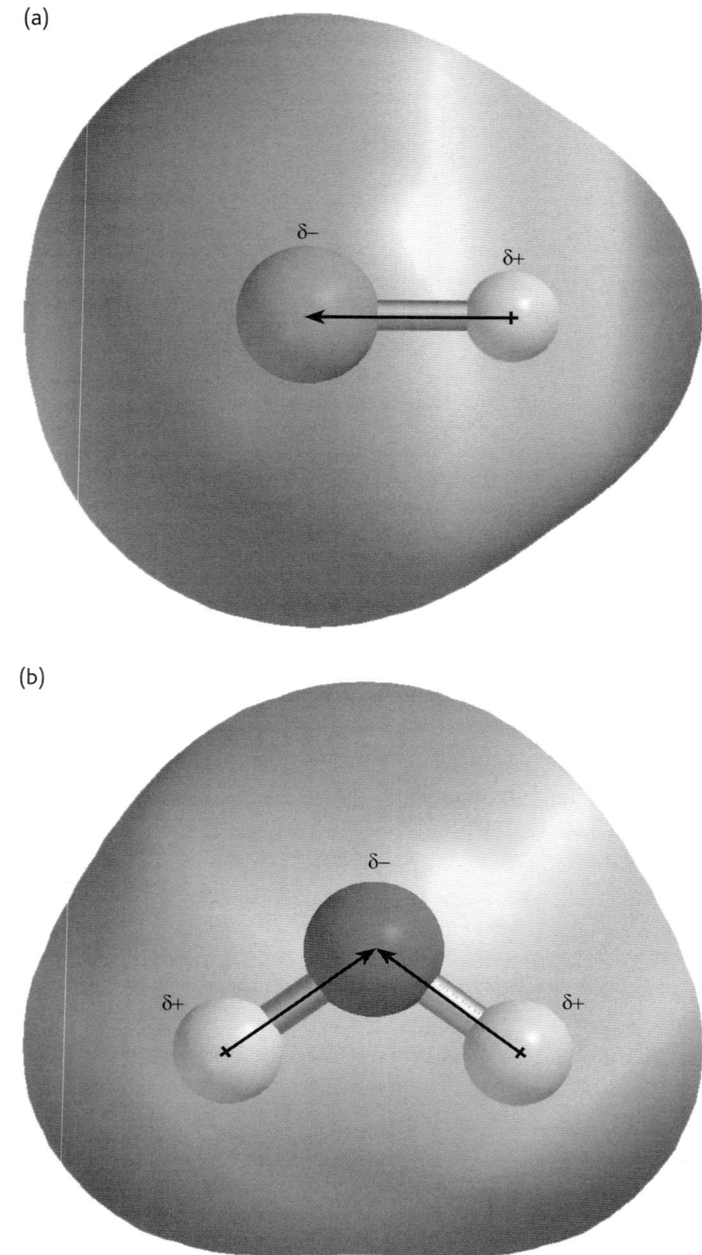

Figure 4.5 The dipole moment in (a) a hydrogen chloride molecule and (b) a water molecule. In each case, the arrowheads point towards the region of highest electron density. In HCl, this is centred upon the electronegative chlorine nucleus. In water, it is centred upon the electronegative oxygen nucleus. (Images courtesy Peter Atkins.)

molecule[4]. This separation of charge between the partial positive and partial negative regions constitutes a dipole. The difference in charge between two ends of a dipole contributes to what is called the **dipole moment**, which is something that

4 The prefix di- = two, so the term 'dipole' literally means 'two poles'. Picture a bar magnet that has two poles, which represent the opposing forces, with a north pole at one end and a south pole at the other. By analogy, a dipole also has opposing forces at its two ends: a partial positive charge at one end and a partial negative charge at the other.

can be measured[5]. The dipole moment can be represented by a particular style of arrow, as illustrated in Figure 4.5; the arrowhead points towards the region of highest electron density.

The uneven distribution of electrons–and the existence of a dipole moment–in a polar molecule (or, indeed, in a polar bond) is permanent, so we refer to a molecule such as hydrogen chloride as a **permanent dipole**. This is in contrast to the temporary dipoles that form the basis of dispersion forces, as we discover in Section 4.3.

- ***A permanent dipole is a molecule that has a permanently partially positively charged region and a permanently partially negatively charged region.***

Polar bonds in non-polar molecules

Hydrogen chloride and water are examples of polar molecules that contain polar bonds. Not all molecules that contain polar bonds are themselves polar, however. A molecule that contains polar bonds will be **non-polar** overall if it comprises identical atoms that are arranged symmetrically relative to one another.

For example, let us consider a molecule of carbon dioxide, CO_2. CO_2 is a linear molecule (that is, its three composite nuclei lie in a straight line), in which each oxygen atom is joined to the central carbon atom by a double bond. Each C=O bond is highly polarized, due to the difference in electronegativity between carbon and oxygen. However, a molecule of carbon dioxide is non-polar overall.

Look at Figure 4.6(a), which depicts a molecule of CO_2: notice how the two oxygen atoms are arranged symmetrically relative to the central carbon atom. While both C=O bonds are polarized, Figure 4.6(b) shows how the highly electronegative oxygen atoms exert their equally strong polarizing effects in exactly opposite directions to one another, and so effectively cancel each other out. As a result CO_2 is non-polar overall. (By analogy, consider a tug-of-war between two teams of exactly the same strength. Both teams exert an identical 'pull' on the rope but in opposite directions. The two 'pulls' cancel each other out, with the result that the middle of the rope remains central.)

By contrast, let's return to water, H_2O. At first glance, CO_2 and H_2O seem very similar, both comprising two identical atoms bonded to a central atom via polar covalent bonds. However, unlike CO_2, water is not linear. Rather, water adopts the V-shaped structure illustrated in Figure 4.4(a). As a result, the two hydrogen atoms are *not* arranged symmetrically relative to the central atom. As a consequence, the dipole moments between the oxygen atom and two hydrogen atoms, which are depicted in Figure 4.5(b), do not cancel each other out, with the result that water remains polar.

Self-check 4.1 Is tetrachloromethane a polar or a non-polar molecule? Why?

5 The formal definition of a dipole moment is (the charges that are separated) × (the distance between the charges).

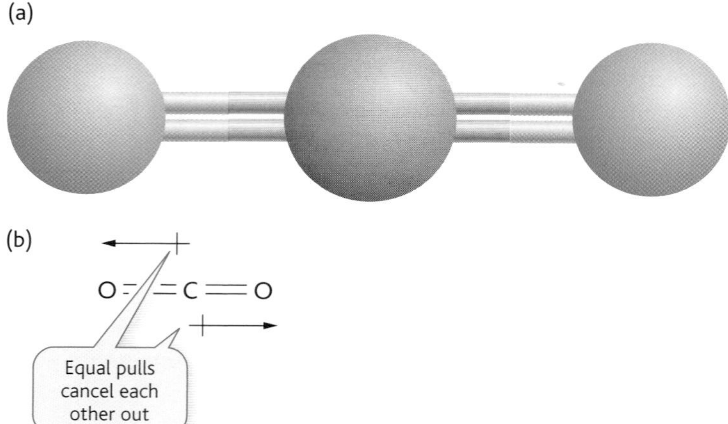

Figure 4.6 The non-polarity of a carbon dioxide molecule. (a) Carbon dioxide features two polar C=O bonds. (b) Despite the individual bonds being polar, carbon dioxide is a non-polar molecule: the two bonds exert equal 'pulls' in opposite directions and cancel each other out.

4.3 The van der Waals interaction

Now that we have seen how charges occur within molecules, it is time for us to consider in more detail what happens when these charges interact. As we mention earlier, these interactions can be either attractive (in the case of opposite charges interacting) or repulsive (if like charges interact). In this section we explore three distinct types of molecular interactions–two attractive and one repulsive–and see how they come together to form an overall interaction called the van der Waals interaction. We start with dispersion forces.

Dispersion forces

Dispersion forces are weak molecular interactions which have the important and unique characteristic of occurring between all molecules, whether or not these molecules carry an electrical charge. We might not expect an electrically neutral molecule (which carries neither full nor partial charges) to experience forces of attraction. As we note above, forces of attraction can only occur between opposite charges. Surely a totally neutral molecule, by definition, lacks regions of electrical charge (either partial or full)?

In fact, an uneven distribution of electrical charge occurs in *all* molecules at an instant in time, and is a consequence of the movement of electrons within them. Electrons in molecules are not static but are in a state of continual motion. While they are constrained to occupy a series of molecular orbitals, there is no restriction placed on where, within an orbital, an electron can reside. (Remember the analogy of a wasp flying round the inside of a jam jar from Chapter 2; the wasp is restricted to flying within the jar, but can fly anywhere within that jar.)

As a result of this movement, the electrons in non-polar molecules are never totally equally distributed. Consider two alternative aerial views of a stretch of free-flowing motorway traffic as depicted in Figure 4.7. We rarely encounter view (a), in which traffic in all three lanes is equally distributed. View (b) is much more typical: some parts of the carriageway contain slightly more traffic than others, due solely to where the traffic happens to be located at a particular moment in time.

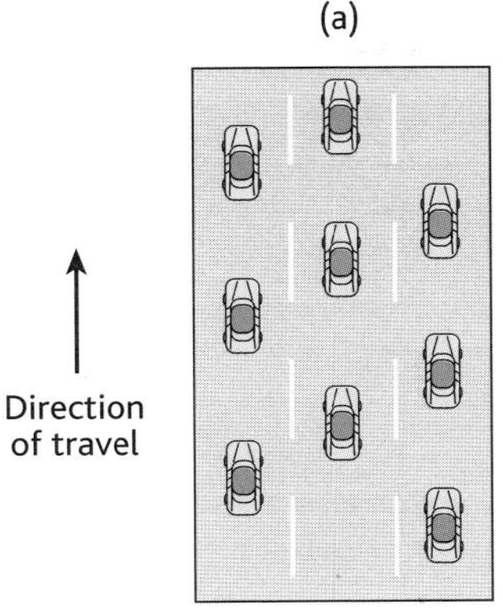

Figure 4.7 Two views of motorway traffic. (a) Traffic in all three lanes is evenly distributed. We rarely see this kind of distribution. (b) Traffic in the three lanes is unevenly distributed. This is a much more typical kind of distribution.

Similarly, if we were to take a 'snapshot' of the location of electrons in a molecule, we would see the same kind of slightly uneven distribution. This uneven distribution gives rise to minor variations in electrical charge throughout a molecule: those parts of a molecule containing slightly more electrons at a particular moment possess a slight negative charge relative to parts of a molecule containing slightly fewer electrons.

The regions of slight negative charge have a knock-on effect in neighbouring areas. The slight negative charge is sufficient to **induce** a slight positive charge in the neighbouring areas, generating what is called an **induced dipole**. Look at Figure 4.8, which shows how a dipole is induced. An area of negative charge repels electrons in its immediate vicinity: a repulsive force exists between the electrons and the region of (like) negative charge. Consequently, the region from which the electrons have been repelled attains a slight positive charge, to form an induced dipole.

A **dispersion force** is the force of attraction that exists between the two areas of opposite charge, which form the induced dipole. Sometimes we see dispersion forces referred to by the alternative name of 'London dispersion forces'[6]. The two terms are interchangeable, but we shall stick with using 'dispersion forces' in this book.

Dispersion forces have three notable characteristics:

1. They are very short-lived. (A dispersion force is so transitory that it exists for just 10^{-16} s.) Areas that are slightly positive one moment may be slightly negative the next as the distribution of electrons shifts slightly from one area to another. Therefore the induced dipoles that give rise to dispersion forces are only temporary.
2. They are very weak. Forces of attraction are proportional to the size of the charges being attracted: full positive and negative charges are attracted to one another more strongly than partial charges, which, in turn, are attracted to one another more strongly than the transient weak charges arising from the movement of electrons in molecules.
3. They operate over very short distances. We see on p. 96 how forces of attraction decrease rapidly as the distance between charges increases. If the force of attraction is weak even at short distances, then it will diminish to practically zero when the charges are moved apart only slightly.

6 The name 'London' here refers to the physicist Fritz London, rather than to the city of that name.

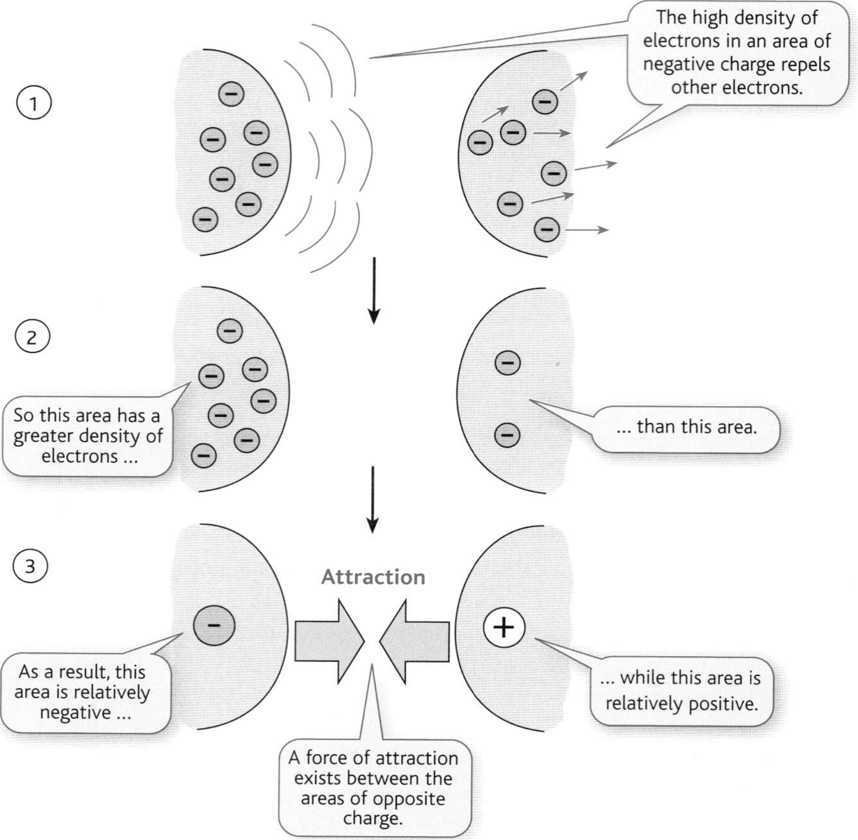

Figure 4.8 The mechanism by which a dipole is induced. Notice how, ultimately, a force of attraction operates between areas of opposite electrical charge. This is a dispersion force.

- ***Dispersion forces are weak, short-lived forces, which are experienced by all molecules.***
- ***The prevalence of dispersion forces between two molecules is influenced by two factors related to the nature of the molecules—their shape and their size.***

Shape

If two molecules can get close together, juxtaposed in three-dimensional space, the dispersion forces existing between them will be stronger. Figure 4.9(a) shows how planar (flat) molecules are able to associate more closely than molecules with a more irregular shape. Because dispersion forces (like other electrostatic forces) are at their strongest when the opposing charges are in close proximity, molecules that are able to associate closely exhibit more extensive dispersion forces than those that associate more weakly.

Size

Large molecules exhibit greater dispersion forces than smaller molecules. The larger a molecule is, the more atoms it comprises, and therefore the more electrons

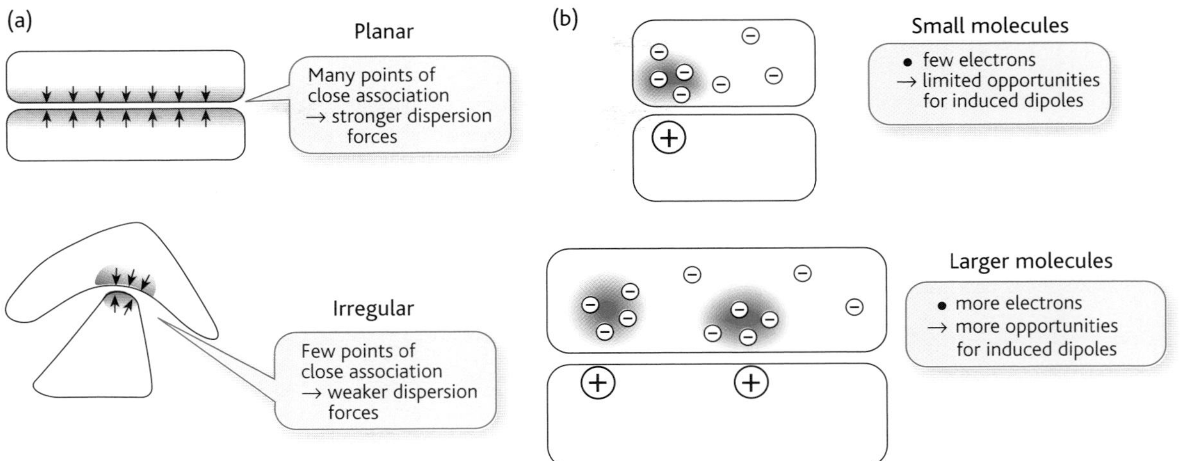

Figure 4.9 (a) Planar molecules are able to associate closely with one another, allowing extensive dispersion forces to occur. By contrast, irregularly shaped molecules cannot associate so closely, so less extensive dispersion forces can occur. (b) Large molecules, with a large number of electrons and more opportunities for induced dipoles to arise, experience greater dispersion forces than smaller molecules, which possess fewer electrons and experience fewer induced dipoles.

it contains. Figure 4.9(b) illustrates how more induced dipoles are possible in a large molecule with more electrons, than in a smaller molecule with fewer electrons. Because more induced dipoles may exist between larger molecules, the dispersion forces (the forces of attraction between the charges forming the dipole) are greater.

Self-check 4.2 Look at the following pair of compounds. Which of the pair experiences the greater dispersion forces, and why?

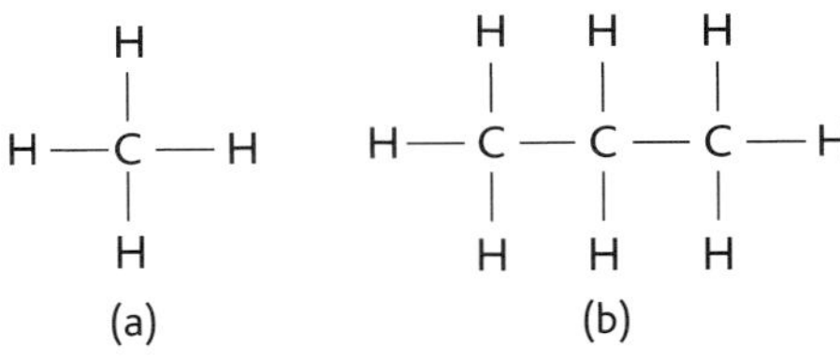

The physical importance of dispersion forces

Look again at Table 4.1 and notice how dispersion forces are among the weakest of the non-covalent forces. Despite being seemingly feeble, however, dispersion forces are a vital component of biological systems. Their significance is due to their ubiquity: dispersion forces operate between *all* molecules. Therefore, while an isolated instance of dispersion forces operating is pretty insignificant, the cumulative effect of many dispersion forces in a biological system makes them very significant indeed. We discuss the cumulative effect of non-covalent forces on the properties of a biological system in Section 4.4.

Indeed, dispersion forces are the only form of non-covalent interaction that can operate between non-polar molecules such as methane (shown in Self-check 4.2), or even neon (as explained in Box 4.2). Molecules of a non-polar liquid, including fuels such as butane or propane, associate solely through the attractive influence of dispersion forces. Given the fact that dispersion forces are weak, however, the individual molecules that comprise non-polar liquids are still held together rather

Fuels such as propane or butane belong to the class of organic compounds called alkanes. We explore the alkanes in more detail in Chapter 5.

BOX 4.2 Faking it: diatomic molecules of the noble elements

We see in Chapter 3 how elements of Group 18–the so-called noble gases–are chemically inert. Each element in the group possesses a full valence shell, and so does not participate in covalent bonds with atoms of other elements, or with other atoms of the same element.

Despite this, diatomic 'molecules' of the Group 18 elements, such as Xe_2, have been detected, occurring very transiently. Their existence is not due to covalent bonding between two atoms, however, but is due solely to dispersion forces–a force of attraction that arises between two neighbouring atoms. As such, entities such as Xe_2 are not molecules in the true sense of the word.

weakly, making these liquids very **volatile**–they readily separate from one another during the transition from liquid to gas, i.e. they have low boiling points.

Permanent dipolar interactions

Permanent dipolar interactions (or **dipole–dipole** interactions) are the attractive forces that exist between opposite partial charges on polar molecules.

We see in Section 3.12 the way in which many covalent bonds are polar: the electrons that participate in the bond are shared unequally between the joined atoms, so a dipole moment (signifying a difference in charge) exists between the two atoms. The most electronegative atom carries a partial negative charge, while the least electronegative atom carries a partial positive charge. Just as forces of attraction exist in an ionic compound (such as NaCl) between full positive and negative charges, so forces also exist between partial positive and partial negative charges. It is this force of attraction that gives rise to dipolar interactions.

Let us consider chloromethane, CH_3Cl, as an example. Chloromethane is a polar molecule, and possesses a permanent dipole moment, as illustrated in Figure 4.10. The partial negative charge attracts the partial positive charges on neighbouring molecules such that a network of dipolar interactions are generated.

Dispersion forces rely on the *temporary* dipoles that arise as the result of electron movement. While such forces are always operating, they only occur transiently in any one region of a molecule at any one point in time. Conversely, dipolar interactions arise between permanent dipoles that result from a *permanently* uneven distribution of electrons between atoms, where the electron distribution in a certain region of the molecule remains essentially the same at all times. Therefore, dipolar interactions are permanent, or last considerably longer than dispersion forces. However, Table 4.1 shows how the typical energy of a permanent dipolar interaction is similar to that of a dispersion force.

- *Dispersion forces are short-lived forces arising from temporary induced dipoles.*
- *Permanent dipolar interactions are long-lived forces arising from permanent dipoles.*

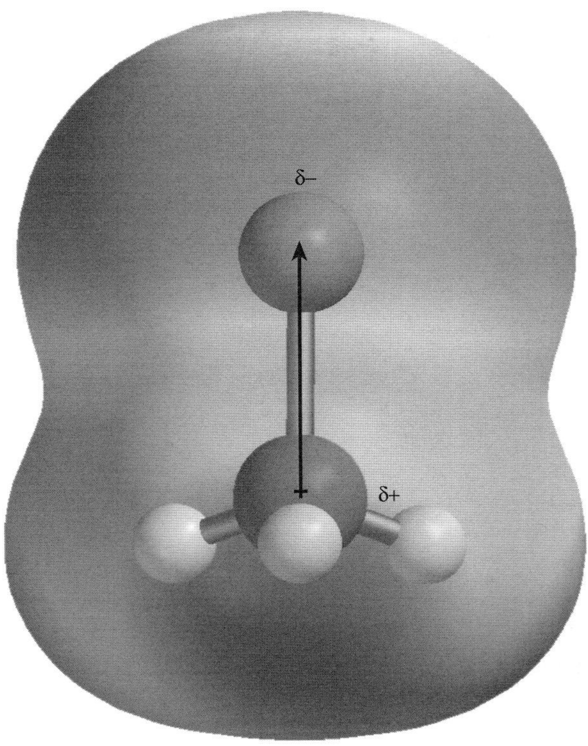

Figure 4.10 Chloromethane possesses a permanent dipole moment, which is generated by the highly electronegative chlorine atom. (Image courtesy Peter Atkins.)

Steric repulsion

We have seen above how dispersion forces and permanent dipolar interactions arise from the forces of attraction that operate between opposite charges, whereby opposite charges attract. However, these attractive forces–which draw two molecules together and stabilize their association–are offset to some extent by the *repulsion* experienced between two areas of like charge, namely the electron clouds that surround every atom.

We see in Chapter 3 how it is that the electron clouds that surround each atom come into contact first as two atoms approach each other. An analogous situation occurs when two molecules interact. A molecule has electrons distributed at its surface, just like an atom. When two molecules approach each other, it is the electrons distributed on their surface that first interact. Because two groups of electrons will both carry the same (negative) charge, the two groups repel each other. We call this **steric repulsion**.

Steric repulsion only operates over a very short distance–that is, two molecules have to be in very close proximity for steric repulsion to come into play. When the molecules are further apart, the attractive dispersion forces and dipolar interactions dominate to bring two molecules together. Once they *are* close together, steric repulsion simply limits just how close to another the two molecules can come.

Balancing attraction and repulsion: the van der Waals interaction

Look at Figure 4.11, which illustrates how the energy between two interacting species varies as the distance between them varies. Notice how, when two molecules

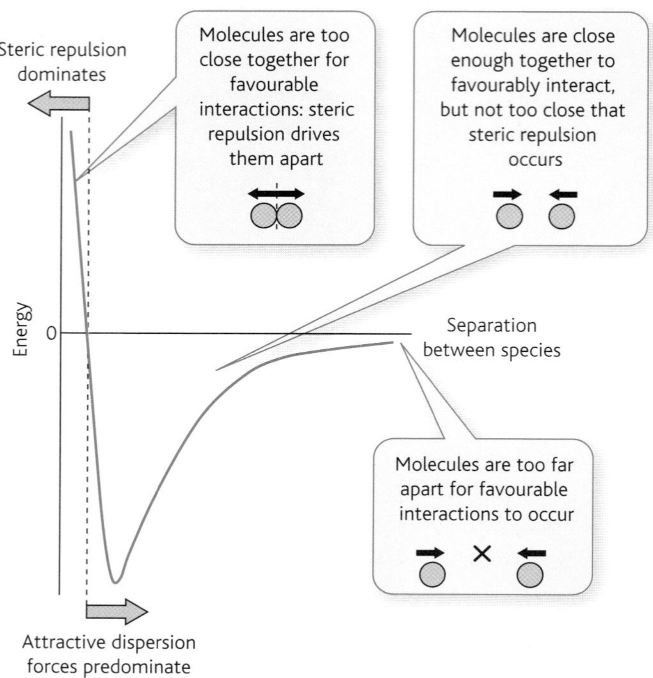
Steric repulsion dominates
Molecules are too close together for favourable interactions: steric repulsion drives them apart
Molecules are close enough together to favourably interact, but not too close that steric repulsion occurs
Energy
0
Separation between species
Molecules are too far apart for favourable interactions to occur
Attractive dispersion forces predominate

Figure 4.11 The variation in the energy of interaction between two species as the distance between them varies. An energy below zero on the graph (y-axis) constitutes a positive (attractive) interaction; an energy above zero constitutes an unfavourable (negative) interaction. When two species are far apart (right side of graph) there is no substantive energy of interaction between them; when two species are very close together, steric repulsion prevents their (favourable) interaction. It is only in the region in the centre portion of the graph that the molecules experience appreciable attractive forces.

are far apart, there is no real energy of interaction; the molecules are not associated. As the molecules move closer together, however, the energy of attraction increases, and the molecules become stably associated. However, as the molecules become very close together, steric repulsion dominates, pushing the two molecules apart to limit just how closely they can interact.

The energy profile depicted in Figure 4.11 sums up the **van der Waals interaction**[7], the term used to describe the overall interaction between two species once both the attractive dispersion forces and dipolar interactions, and the repulsive force of steric repulsion are taken into account. Remember that the van der Waals interaction sums up non-covalent forces, so applies only to two species that aren't covalently bonded to one another.

- ***The van der Waals interaction describes the overall level of interaction between two species once both attractive and repulsive forces have been taken into account.***

The distance between two charges is not the only factor that influences the forces of attraction that arise, however. The medium that the molecules occupy also plays an important role. The strongest forces of attraction occur between two charges in a vacuum. However, molecules in biological systems do not exist in a vacuum. Rather, they occur in aqueous solution–a complicated water-based mix of biochemical compounds. Different substances 'dampen' attractive forces to varying degrees.

7 The van der Waals interaction is more formally termed the van der Waals potential.

For example, the presence of water between two charges decreases the force of attraction between them by about 80-fold, compared to the force that would exist between them in a vacuum.

The different electrostatic interactions that we consider throughout this chapter vary considerably in their relative magnitudes. Nevertheless, all are influenced by how close the two interacting particles are and the medium that they occupy.

4.4 Beyond van der Waals: other biologically essential interactions

We have now explored a number of interactions that exist between all polar molecules (in the case of permanent dipolar interactions) and all molecules, whether polar or non-polar (in the case of dispersion forces and steric repulsion). However, these interactions are not the only ones that are important in biological systems. In this section, we explore a further three interactions that govern how biological molecules interact in a stable way, often fundamentally influencing a molecule's physical properties in the process–hydrogen bonds, ionic forces, and hydrophobic forces.

Hydrogen bonds

The hydrogen bond represents a special class of dipolar interaction, in which electrostatic attractions result in the strong interaction of a hydrogen atom on one molecule to an electronegative atom either on another molecule, or on a spatially distinct part of the same molecule. The electronegative atom must be N, O, or F.

Hydrogen bond formation is critically dependent upon the presence of one of fluorine, oxygen, or nitrogen, in two ways:

1. The hydrogen atom must be bonded to an atom of fluorine, oxygen, or nitrogen.
2. The hydrogen atom can only form a hydrogen bond with a lone pair on an atom of fluorine, oxygen, or nitrogen.

Look at Figure 4.12, which illustrates how a hydrogen bond arises. Notice how the electronegative atom to which the hydrogen atom is bonded causes an uneven distribution of electrons between the two atoms, so the hydrogen atom carries

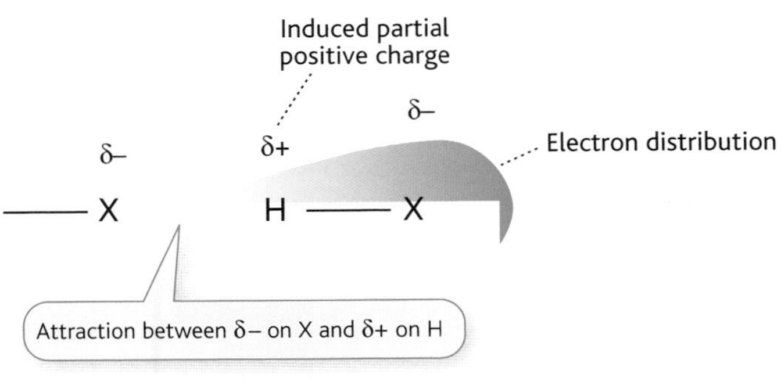

Figure 4.12 The formation of a hydrogen bond. A hydrogen bond forms between an electronegative atom (O, N, or F), and a hydrogen atom which is itself bonded to an atom of O, N, or F. Hydrogen bonds are at their strongest when the three nuclei involved lie in a straight line, as depicted here.

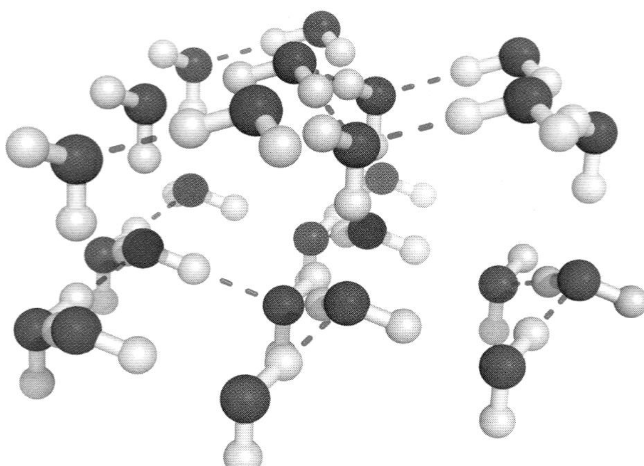

Figure 4.13 Hydrogen bonding in water. Notice how the three nuclei participating in each hydrogen bond (denoted by the blue dashes) lie in a straight line.

a partial positive charge. It is this partial positive charge that facilitates electrostatic attraction between this hydrogen atom and another electronegative atom, either on a different molecule, or on a different part of the same molecule.

Also notice how the nuclei that participate in the hydrogen bond (the H and the two nuclei denoted X in Figure 4.12) lie in a straight line. This geometry is the one in which a hydrogen bond is at its strongest. The arrangement in a straight line of the three nuclei participating in a hydrogen bond can have a big impact on the three-dimensional structure of a compound. For example, if we look back at Figure 3.14 we see how the water molecules in ice are arranged into a regular array. Now look at Figure 4.13, which illustrates the hydrogen bonding in operation between these water molecules. Notice how the three nuclei that are participating in each hydrogen bond are arranged in a straight line.

The strength of a hydrogen bond is also influenced by the distance between the two electronegative atoms (the two atoms denoted X in Figure 4.12). The separation between these two atoms can only lie within certain narrowly defined limits. Again, this has a direct impact upon the arrangement in space of hydrogen-bonded atoms.

- ***A hydrogen bond can only form between a hydrogen atom and an atom of oxygen, fluorine, or nitrogen.***
- ***The hydrogen atom must itself be covalently bonded to an atom of oxygen, fluorine, or nitrogen.***
- ***The three nuclei that participate in a hydrogen bond must lie in a straight line.***

Self-check 4.3 Look again at Figure 4.12. How many chemically distinct hydrogen bonds are possible?

The importance of hydrogen bonds in biology

Hydrogen bonds play a vital role in ensuring that many biological molecules adopt, and subsequently retain, their correct three-dimensional structure. Hydrogen bonds are particularly prevalent in both nucleic acids and proteins.

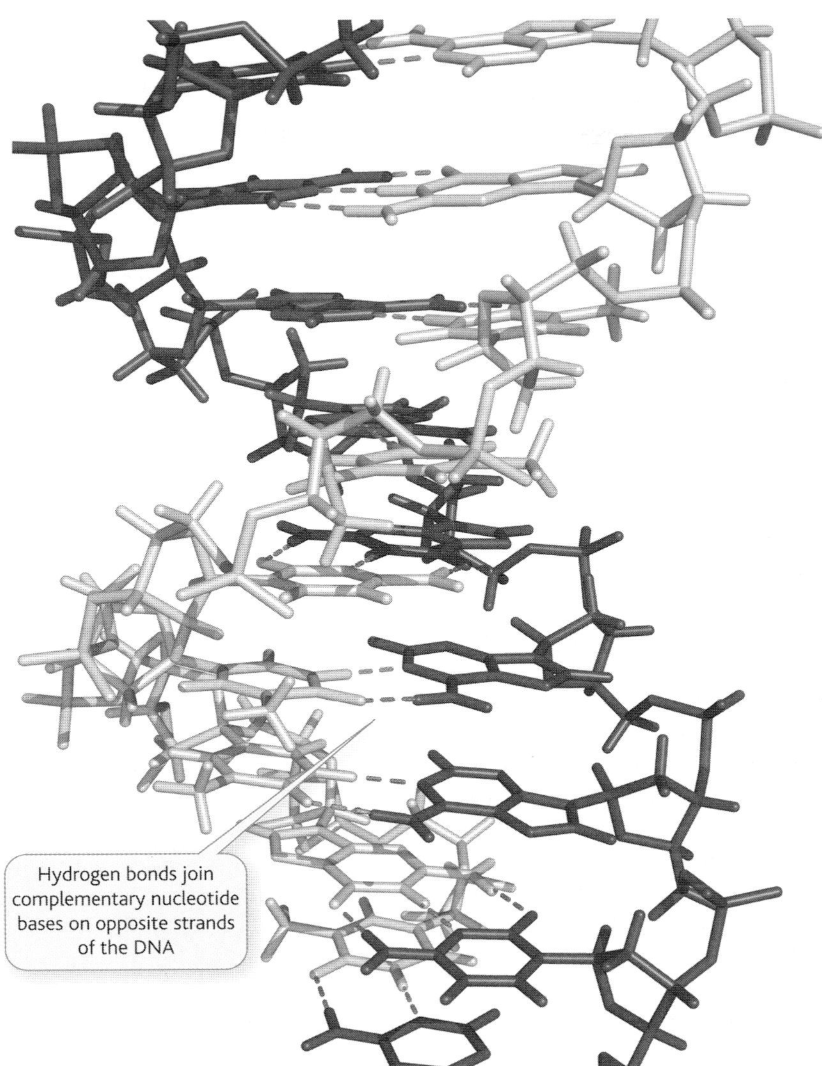
Hydrogen bonds join complementary nucleotide bases on opposite strands of the DNA

Figure 4.14 The two strands of DNA are joined by a network of hydrogen bonds, shown here in blue, which link nucleotide bases on one strand with complementary nucleotide bases on the other strand. The network of hydrogen bonds forms a 'ladder', which extends up the 'inside' of the double-helical structure.

In 1953, James Watson and Francis Crick famously deduced that the structure of DNA was a double helix. Hydrogen bonds play a central role in enabling double-stranded DNA to adopt such a helical structure, and in ensuring its subsequent stability. Figure 4.14 shows how two strands of DNA are joined by a network of hydrogen bonds that span the 'inside' of the helix, linking a nucleotide base on one strand with a complementary nucleotide base on the other to form a series of ladder-like 'rungs'. The base pairs orient themselves, once linked by hydrogen bonds, in such a way that consecutive pairs must twist slightly relative to one another as they stack along the length of the double strand. This twisting gives double-stranded DNA its distinctive double-helical structure.

Indeed, hydrogen bonds mediate the *specific* base pairing which is so fundamental to the conservation of the sequence of DNA during replication. Hydrogen bonding between the double strands of DNA is essential for its structural stability. However, hydrogen bonding can only occur between two specific base pairs: adenine with thymine, and cytosine with guanine. Look at Figure 4.15, which depicts the six

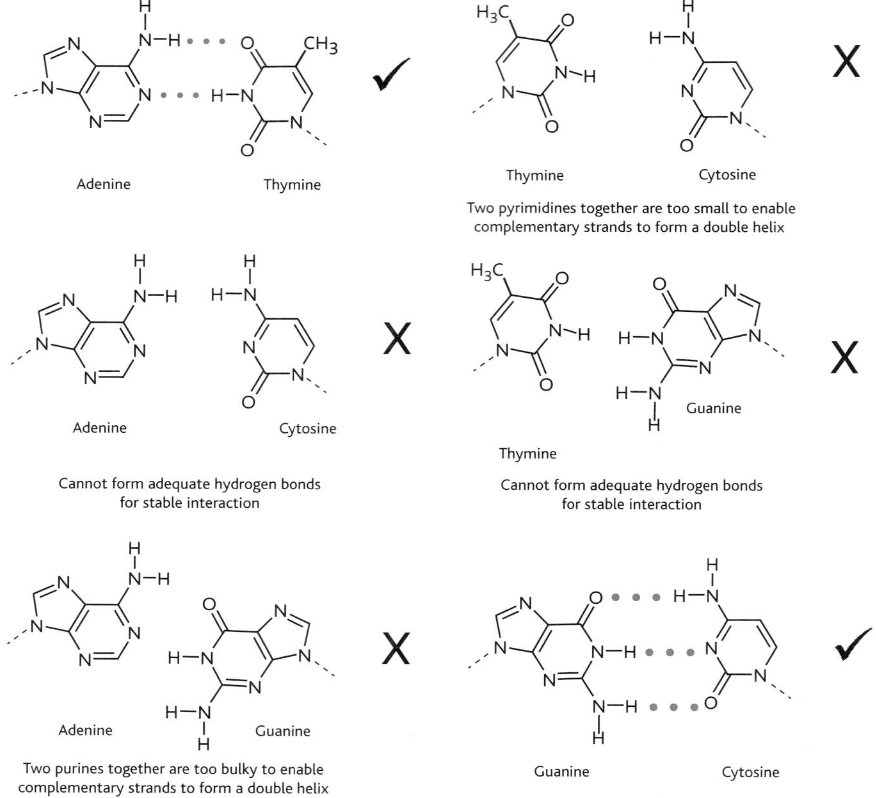

Figure 4.15 Hydrogen bonds only exist between two specific pairs of nucleotide bases: A and T, and C and G. Other base pairings are not possible.

possible ways of combining the four nucleotides that comprise DNA. Notice that hydrogen bonding is only possible for *two* of the possible six combinations. In the other cases, those parts of the nucleotides associated when the two DNA strands are in close proximity do not contain the chemical components required for hydrogen bond formation.

We explore the biological significance of hydrogen bonding again in Chapter 9.

Hydrogen bonding also plays an important part when determining the structure of proteins. Figure 4.16 illustrates how a polypeptide contains the two chemical components required for hydrogen bond formation: a hydrogen atom bonded to a strongly electronegative atom (O, N, or F); and a separate strongly electronegative atom (again, O, N, F). Look at this figure, and notice how the folding of a polypeptide chain brings these components into close enough proximity for non-covalent interactions (in this case, hydrogen bonds) to arise.

Q **Self-check 4.4** Why cannot hydrogen bonds form between parts of a polypeptide chain if the chain is stretched out in a linear fashion?

It is not just the polypeptide backbone that participates in structurally important hydrogen bonding. Instead, the structures of many polypeptides are stabilized further by networks of hydrogen bonds that operate between neighbouring amino acid side chains. A number of polar amino acid side chains have the chemical characteristics that are a prerequisite for hydrogen bonding. Look at Figure 4.17, which shows how a hydrogen bond can form between the side chains of serine and glutamine.

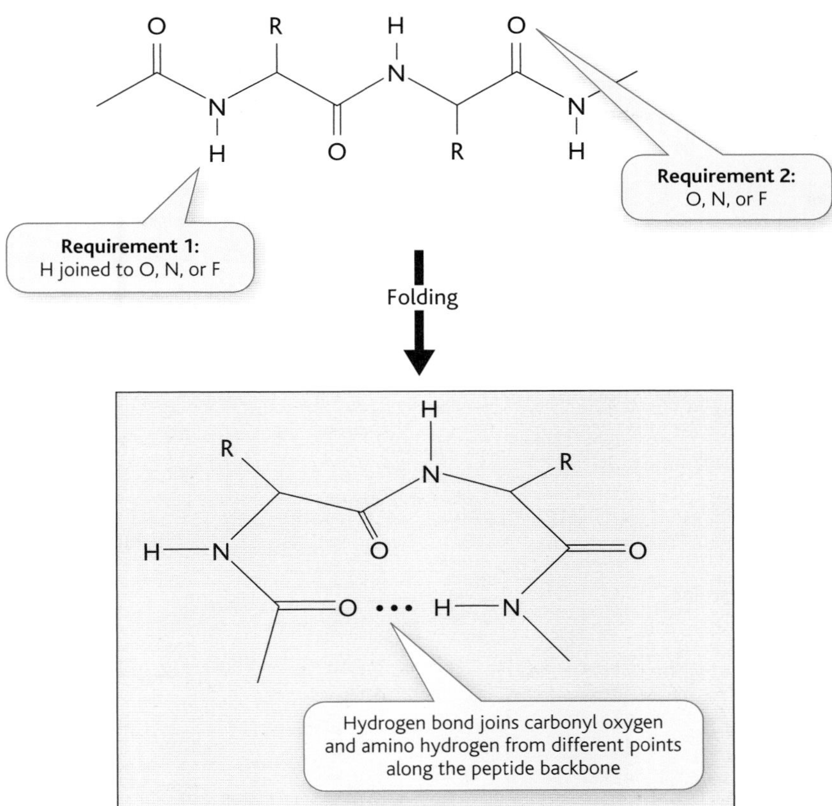

Figure 4.16 A polypeptide contains both the components necessary for hydrogen bond formation. Consequently, hydrogen bonds can form between different regions of a polypeptide chain, or between different polypeptide chains.

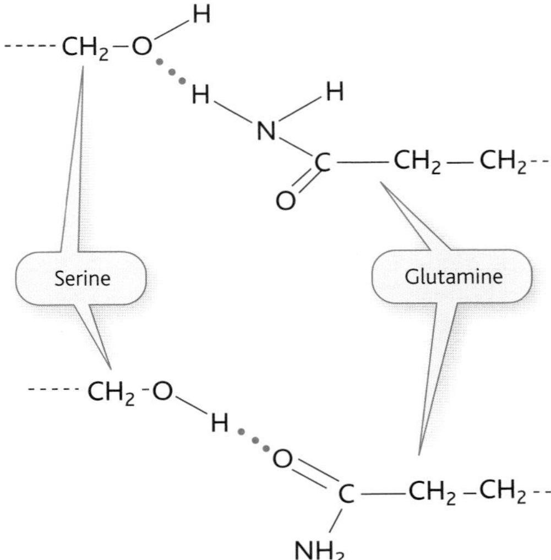

Figure 4.17 Two possible ways in which hydrogen bonds form between the side chains of the amino acids serine and glutamine.

Notice how the same two side chains offer versatility: they can form hydrogen bonds in several different ways.

We see in Chapter 9 how different patterns of hydrogen bond formation give rise to characteristic three-dimensional structures between different regions of a polypeptide chain. These patterns are conserved across many different polypeptides.

- ***Hydrogen bonding stabilizes the structure of several key biological molecules, including DNA and polypeptides.***

Self-check 4.5 Look at the following amino acids. Which have side chains that can participate in hydrogen bonding?

Threonine

```
           O
           ||
H2N—CH-C—OH
       |
      CH-OH
       |
      CH3
```

Asparagine

```
           O
           ||
H2N—CH-C—OH
       |
      CH2
       |
      C=O
       |
      NH2
```

Cysteine

```
           O
           ||
H2N—CH-C—OH
       |
      CH2
       |
      SH
```

Leucine

```
           O
           ||
H2N—CH-C—OH
       |
      CH2
       |
      CH-CH3
       |
      CH3
```

Hydrogen bonds and water solubility

The importance of hydrogen bonds in biological systems extends beyond their structural role described above: they play a vital part in shaping the behaviour of compounds in aqueous systems[8], and particularly whether a compound is **soluble** in a water-based system. Specifically, hydrogen bonds stabilize the association of a water molecule with a water-soluble compound, so that free mixing of the two types of molecule is stable, as illustrated in Figure 4.18(a).

We consider the interaction of molecules with water further on p. 106, when we explore hydrophobic forces.

If a compound *cannot* form hydrogen bonds with water, however, then its molecules cannot interact with water molecules in this stable way. Instead, the most stable arrangement is for the two types of molecule to remain completely separate, and for the water and other compound to form two distinct layers, with no mixing, as depicted in Figure 4.18(b).

8 An aqueous solution is any solution in which water is the solvent. Biological systems are inherently water-based. Our bodies are between 50 and 60% water! (The actual percentage depends on our gender: a greater percentage of an adult male's body is water compared with an adult female.) However, our cells are far from merely being bags of pure water. Rather, they contain water in which is dissolved a complicated mixture of inorganic and organic compounds. This water-based mixture is called the cytoplasm.

The dependence of life upon water is further demonstrated by the way that the presence of water is looked for in space missions as evidence for the possible existence of life.

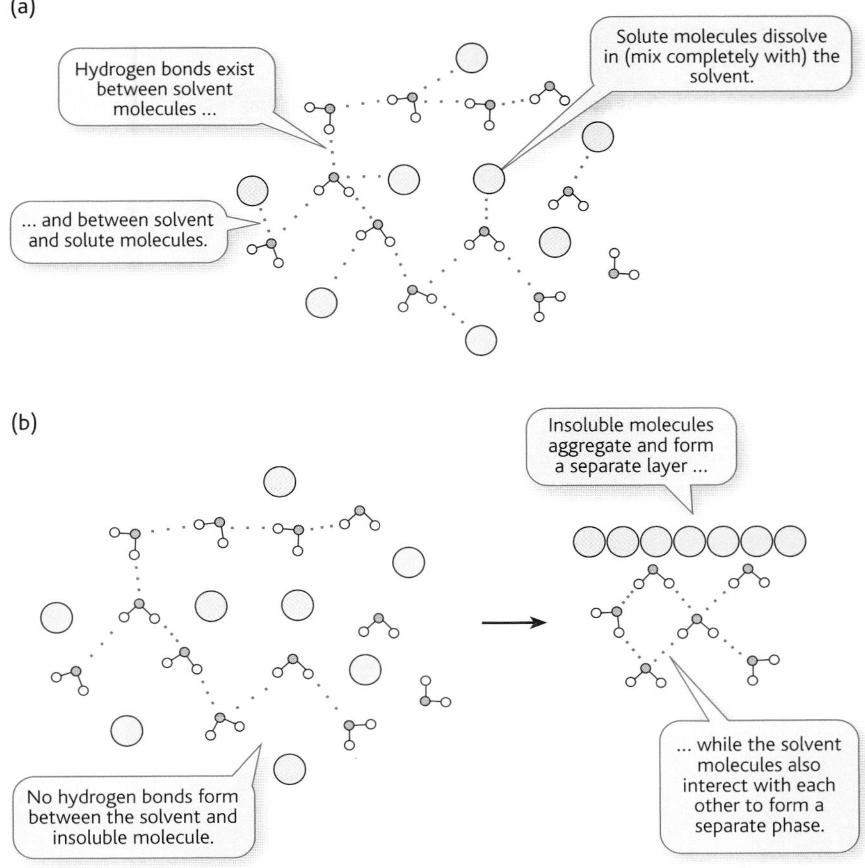
(a)
Hydrogen bonds exist between solvent molecules ...
Solute molecules dissolve in (mix completely with) the solvent.
... and between solvent and solute molecules.
(b)
Insoluble molecules aggregate and form a separate layer ...
... while the solvent molecules also interect with each other to form a separate phase.
No hydrogen bonds form between the solvent and insoluble molecule.

Figure 4.18 (a) Molecules of a water-soluble compound and molecules of water mingle freely with each other: the two types of molecule are able to mix completely. (b) If a compound is insoluble in water, its molecules cannot mix freely with molecules of water. Instead, the two types of molecule remain completely separate. Occasionally, a small amount of the solute dissolves, while the majority floats on top of the solution.

Many small ions are able to mix with water in a stable way. This mixing is not stabilized by hydrogen bonds, however, but by general dipolar interactions between the charged ions and the polar water molecules, as described in Box 4.3.

Ionic forces

Ionic forces are the attractive forces that exist between ionic species–species carrying full positive and negative charges. The ionic bond, which we introduced in Chapter 2, is the prime example of an ionic force. Ionic forces are not confined to ionic compounds, however, but are also an important feature of *covalent* molecules that possess regions carrying a full positive or negative charge, e.g. carboxylate anions or ammonium cations.

Entities possessing a full positive or negative charge are particularly prevalent in proteins, where they are formed by the gain or loss of hydrogen ions by amino acid residues. Figure 4.20 shows how ionic forces may arise between a positively charged amino acid side chain in one part of a polypeptide and a negatively charged amino acid side chain in another part of the same polypeptide. Notice that both side chains carry full charges (they are ionized species), rather than partial charges.

We find out about the movement of hydrogen ions when we explore the properties of acids and bases in Chapter 16.

BOX 4.3 The hydration of ions

Ions mix with water in a process we call **hydration**. Hydration is a specific example of solvation–the attraction and association of molecules of a solvent with molecules or ions of a solute. (In hydration, the solvent molecule is water.) During hydration, the mixing of ions and water molecules is stabilized not by hydrogen bonds, but by general dipolar interactions operating between the charged ions and polar water molecules, as depicted in Figure 4.19. Look at this figure, and notice how an anion is surrounded by a 'shell' of hydrogen atoms, each carrying a partial positive charge; dipolar interactions attract the partial positive charge on the hydrogen atoms to the negative charge on the anion. Conversely, a cation is surrounded by a 'shell' of oxygen atoms; the partial negative charges on the oxygen atoms are attracted to the positive charge on the cation.

It is the electrostatic attraction that occurs between the charge on the ion and the opposite partial charge on the water molecule that stabilizes the association between the ion and the water molecule, and enables small ions and water molecules to mix freely.

The hydration of ions by water molecules is essential to the correct functioning of biological systems. Many aspects of our physiology depend on the presence of ions within and around our cells: our nerve cells conduct nervous impulses by varying the concentrations of Na^+ and K^+ ions across the cell membrane; our kidneys regulate water uptake by our body by varying the concentrations of Na^+, K^+, and Ca^{2+} ions within specific regions of the kidney. If ions could not exist in solution in a stable way, neither of these processes (and many others besides) would be possible.

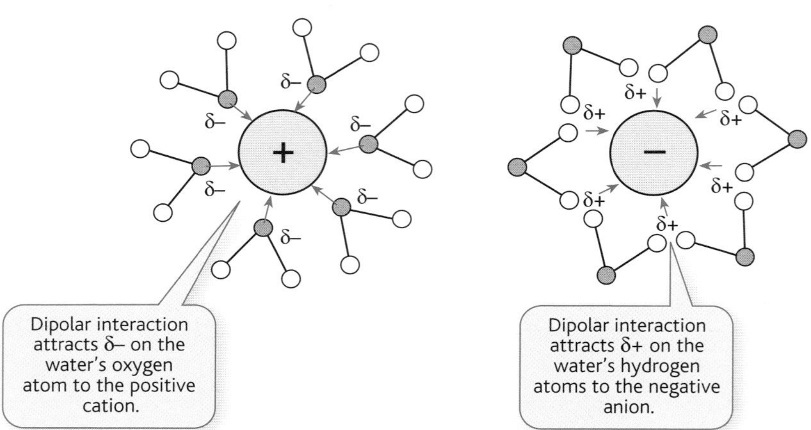

Figure 4.19 The hydration of ions by water molecules. The interaction of ions and water molecules is stabilized by dipolar interactions, which exist between the charge on the ion and the partial charge on the polar water molecule. The partial negative charge on a water molecule's oxygen atom is attracted to a cation's positive charge, while the partial positive charge on a water molecule's hydrogen atom is attracted to an anion's negative charge.

We explore the role played by non-covalent forces in determining the shape of biological molecules in Chapter 9.

The ionic force that operates between two oppositely charged amino acid side chains in a protein has a particular name. We call it a **salt bridge**. Salt bridges are one of a number of vitally important non-covalent forces that help to stabilize the three-dimensional structures of polypeptides and proteins.

Ions form as a result of the *permanent* transfer of electrons from one atom to another. So ionic forces, like dipolar interactions, are consequently permanent rather than transient (in contrast to dispersion forces).

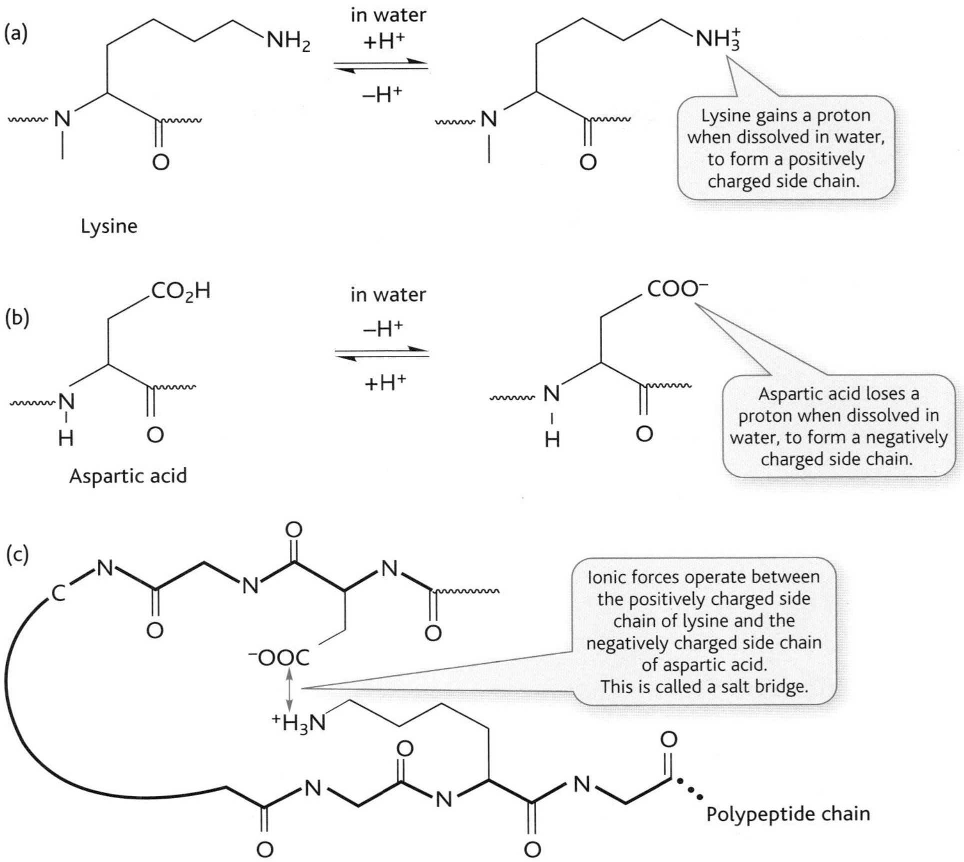

(a) Lysine — in water $+H^+$ / $-H^+$ — NH_2 → NH_3^+

Lysine gains a proton when dissolved in water, to form a positively charged side chain.

(b) Aspartic acid — in water $-H^+$ / $+H^+$ — CO_2H → COO^-

Aspartic acid loses a proton when dissolved in water, to form a negatively charged side chain.

(c) ^-OOC ↔ ^+H_3N

Ionic forces operate between the positively charged side chain of lysine and the negatively charged side chain of aspartic acid. This is called a salt bridge.

Polypeptide chain

Figure 4.20 Ionic forces can operate between a positively charged side chain of one amino acid and a negatively charged side chain of a different amino acid located elsewhere in a polypeptide chain to form what we call a salt bridge.

Self-check 4.6

(a) Look at the three instances of electrostatic interactions operating that are illustrated below. Which is a salt bridge, and why?

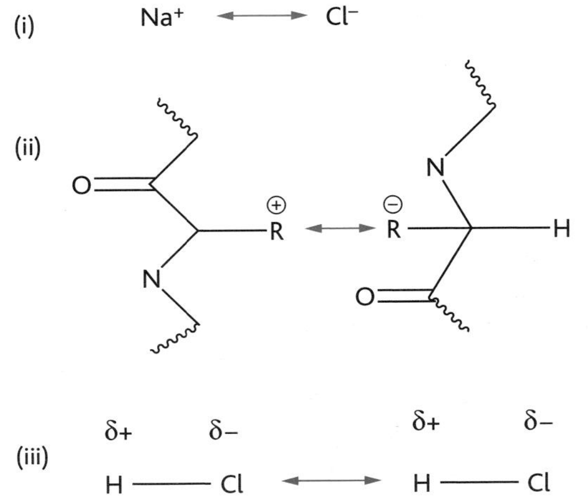

(i) $Na^+ \longleftrightarrow Cl^-$

(ii) $\oplus R \longleftrightarrow \ominus R$

(iii) $\overset{\delta+}{H}—\overset{\delta-}{Cl} \longleftrightarrow \overset{\delta+}{H}—\overset{\delta-}{Cl}$

(b) Consider the pairs of molecules below and state whether the predominant molecular interaction that is likely to occur between each pair is a dispersion force, a permanent dipolar interaction, a hydrogen bond, or an ionic force.

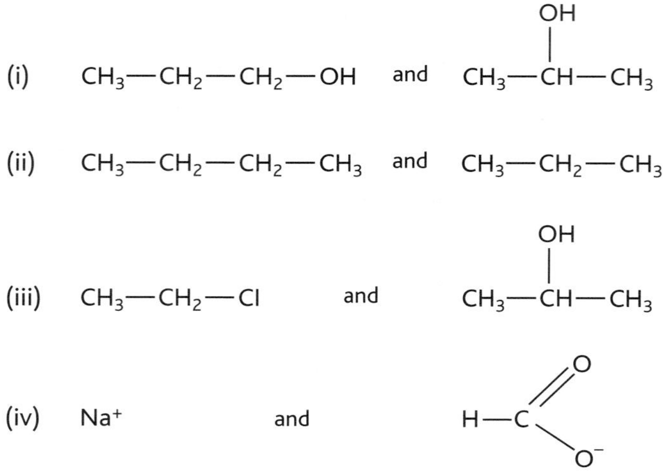

(i) $CH_3-CH_2-CH_2-OH$ and $CH_3-CH(OH)-CH_3$

(ii) $CH_3-CH_2-CH_2-CH_3$ and $CH_3-CH_2-CH_3$

(iii) CH_3-CH_2-Cl and $CH_3-CH(OH)-CH_3$

(iv) Na^+ and $H-C(=O)-O^-$

Hydrophobic forces

We end our exploration of different molecular interactions by considering hydrophobic forces, which are rather different from the other interactions considered so far by not being directly electrostatic in nature. Instead, they are a class of interaction that is driven by a molecule's behaviour when exposed to water. Different chemical entities exhibit differing affinities for water. 'Hydrophilic' ('water-loving') species, which are polar, associate readily with water, and are therefore water-soluble. In contrast, 'hydrophobic' (or 'water-hating') species, which are non-polar, do not interact readily with water, and are therefore water-insoluble[9].

Let us just pause for a moment to consider water solubility a little more. If a compound is soluble in water, then the compound is able to mix completely with water molecules so that the two types of molecule–the water molecule and the compound of interest–mingle freely with one another. For example, picture mixing fruit cordial with water: molecules of both water and the cordial mingle with one another. There is no perceptible difference to us between one type of molecule and the other.

By contrast, if a compound is **insoluble** in water, then the water molecules and molecules of the compound of interest do not mix, but remain separate: the molecules of a water-insoluble compound do not mix with the water molecules, but aggregate to form a separate layer. This describes the behaviour of water and oil: oil is not water-soluble, so its composite molecules aggregate to form a completely separate layer. We observe this as a layer of oil floating on the surface of the water, as depicted in Figure 4.21.

Hydrophobic interactions arise when structural rearrangements occur to keep the hydrophobic portions of a molecule away from their aqueous surroundings. Once clustered together, the hydrophobic portions are stabilized through a network

9 The suffix 'philic' derives from the Greek word 'philos': meaning to love. The suffix 'phobic' derives from the Greek word 'phobos': to fear or hate.

Figure 4.21 Oils are hydrophobic: they do not readily mix with water. Consequently, when we pour oil onto water, the two substances remain separate and form distinct layers.

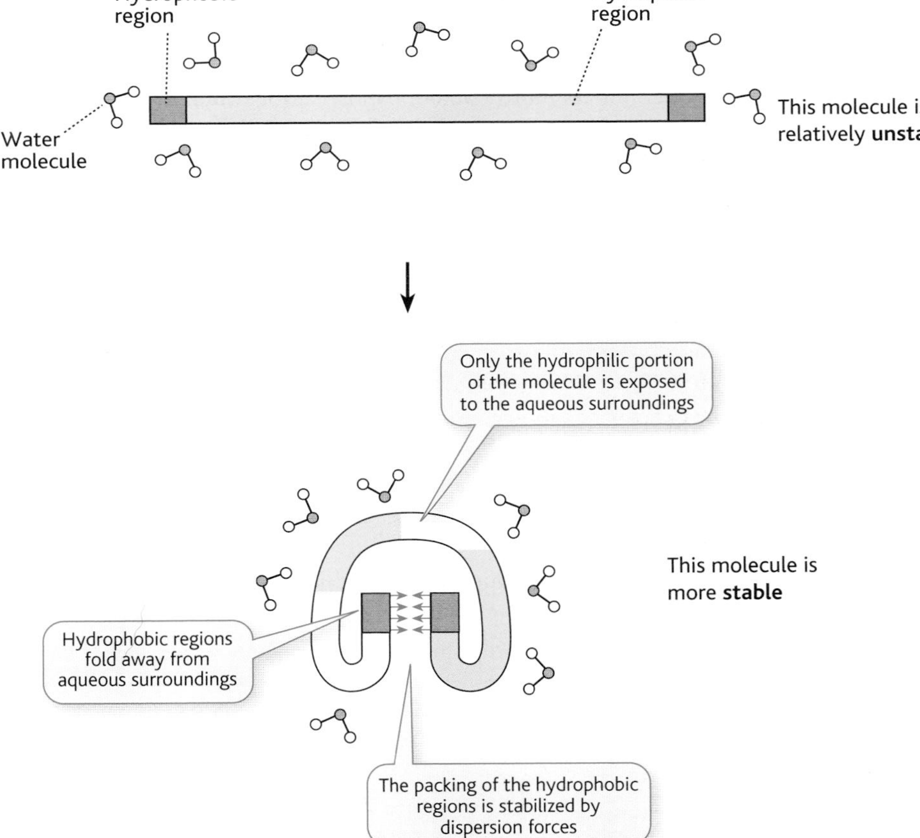

Figure 4.22 The folding of a polypeptide possessing hydrophobic and hydrophilic portions. The darker hydrophobic portions fold away from the aqueous surroundings; this arrangement is stabilized by dispersion forces which operate between the tightly packed hydrophobic portions. By contrast, the hydrophilic portions are exposed to the aqueous surroundings.

of dispersion forces that operate between them. Look at Figure 4.22, which depicts the folding of a polypeptide possessing hydrophobic and hydrophilic portions: the folding occurs in such a way that, after structural rearrangement, the hydrophobic portions are shielded from their aqueous surroundings. Notice how the arrangement of the polypeptide, with the tightly packed hydrophobic portions oriented away from any aqueous surroundings, is stabilized by the dispersion forces that operate between the hydrophobic portions[10].

We learn more about the impact of the hydrophobicity of amino acids on the structure of proteins in Chapter 9.

Hydrophobic forces drive the assembly of various biological structures, including the ubiquitous (and vital) cell membrane, as explained in Box 4.4.

The behaviour of hydrophobic species in the presence of water, i.e. their shielding by structural rearrangement, is driven by a set of **thermodynamic** principles. In essence, thermodynamics is the study of energy, and underpins one of our key themes: that all matter tries to achieve a state of minimum energy and, in consequence, attain its maximum stability. Hydrophobic interactions are one way of increasing stability and, as such, are an example of thermodynamic principles in action.

We explore some of the key thermodynamic principles in more detail in Chapter 13.

Let us briefly consider why hydrophobic interactions confer stability upon hydrophobic molecules. Hydrophobic molecules are more stable in the absence of water: whereas they cannot engage in interactions with water (as we have discovered above), they *can* engage in interactions with each other. Attractive intermolecular forces such as dispersion forces are generally stabilizing; in the case of hydrophobic molecules, such forces help these molecules to remain tightly associated, one with another, and thereby avoid interacting with water.

Water, too, is more stable if exposed to hydrophilic, rather than hydrophobic, molecules. Figure 4.24 shows how hydrophobic molecules disrupt the network of hydrogen bonds that exist in water, and so reduce their stability; this stability is recaptured by the hydrophobic molecules **partitioning**[11] to form a separate layer, which has minimal contact with the water.

It is therefore thermodynamic principles that govern the energy and stability of matter, and that provide the energy and impetus for the folding of molecules to shield their hydrophobic portions from water.

How can we predict whether a molecule is hydrophobic?

The polarity of a molecule gives us a good indication of whether it is hydrophilic or hydrophobic. Broadly speaking:

- polar molecules are hydrophilic
- non-polar molecules are hydrophobic.

Perhaps the most striking example of the hydrophobicity of non-polar molecules is the immiscibility of oil and water[12], as mentioned above. Oils are a class of non-polar organic compound (see Chapter 7); oil and water cannot interact, so when oil is added to water, it separates out to form a layer which floats on top of the water.

---

10 The hydrophobicity (or hydrophilicity) of a particular section of a polypeptide will depend on the amino acids from which it is composed. Some amino acid side chains are hydrophobic, while others are hydrophilic, as we discover in Chapter 7.

11 Partitioning means to divide or separate.

12 Immiscible means unable to mix.

BOX 4.4 The assembly of lipid bilayers

Hydrophobic forces play a vital part in the assembly of each of the many millions of cells from which our bodies form. Every cell of our body is enveloped by a cell membrane, which possesses a structure called a lipid bilayer. As the name suggests, a lipid bilayer comprises two layers of lipid molecules, which lie parallel, one with the other, to form a continuous envelope around the cell's interior. Lipid bilayers also exist *within* cells to form separate compartments called organelles, each of which has its own particular function.

Figure 4.23(a) shows how a lipid molecule has two very distinctive components: a polar head, which is hydrophilic, and a non-polar tail, which is hydrophobic.

The correct assembly of a lipid bilayer, illustrated in Figure 4.23(b), is driven by hydrophobic forces. A lipid bilayer assembles in such a way that the hydrophilic heads of the composite lipid molecules lie on the two outer surfaces of the bilayer. Both the outer and inner layers of the cell membrane are therefore attracted to water and water-based environments.

In contrast, the hydrophobic tails all point inward towards the centre of the bilayer, where they are shielded from the aqueous surroundings. This arrangement is governed by hydrophobic forces that seek to minimize the exposure of hydrophobic entities to water. The lipid bilayer is stabilized by the dispersion forces that operate between the hydrophobic tails of neighbouring lipid molecules, and the hydrogen bonds that operate between the polar heads and surrounding water molecules.

The architecture of the cell membrane, with its hydrophobic core, has important implications for the passage of molecules across it: polar molecules cannot interact in a stable way with a non-polar environment, such as that found in the interior of a lipid membrane. Consequently, special carrier proteins are required to transport polar molecules, such as ions or glucose, across the cell membrane and through the hydrophobic core.

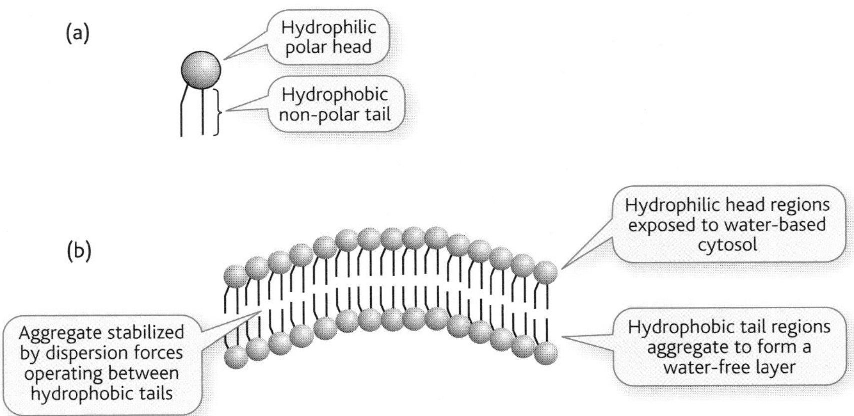

Figure 4.23 (a) The generalized structure of a lipid molecule. A lipid molecule comprises a hydrophilic polar head, and a hydrophobic non-polar tail. (b) Lipid bilayers are formed when the hydrophobic tails of lipid molecules aggregate to generate a water-free environment, leaving the hydrophilic heads exposed to the aqueous surroundings.

But what makes non-polar molecules hydrophobic? For a molecule to be water-soluble it must be able to experience dipolar interactions to undergo hydration, as explained in Box 4.3, or form hydrogen bonds with water. Non-polar molecules can do neither of these things: they are not polar, so they cannot participate in dipolar interactions. They also lack the electronegative atom (e.g. O, N, or F) required in order for hydrogen bonding to occur.

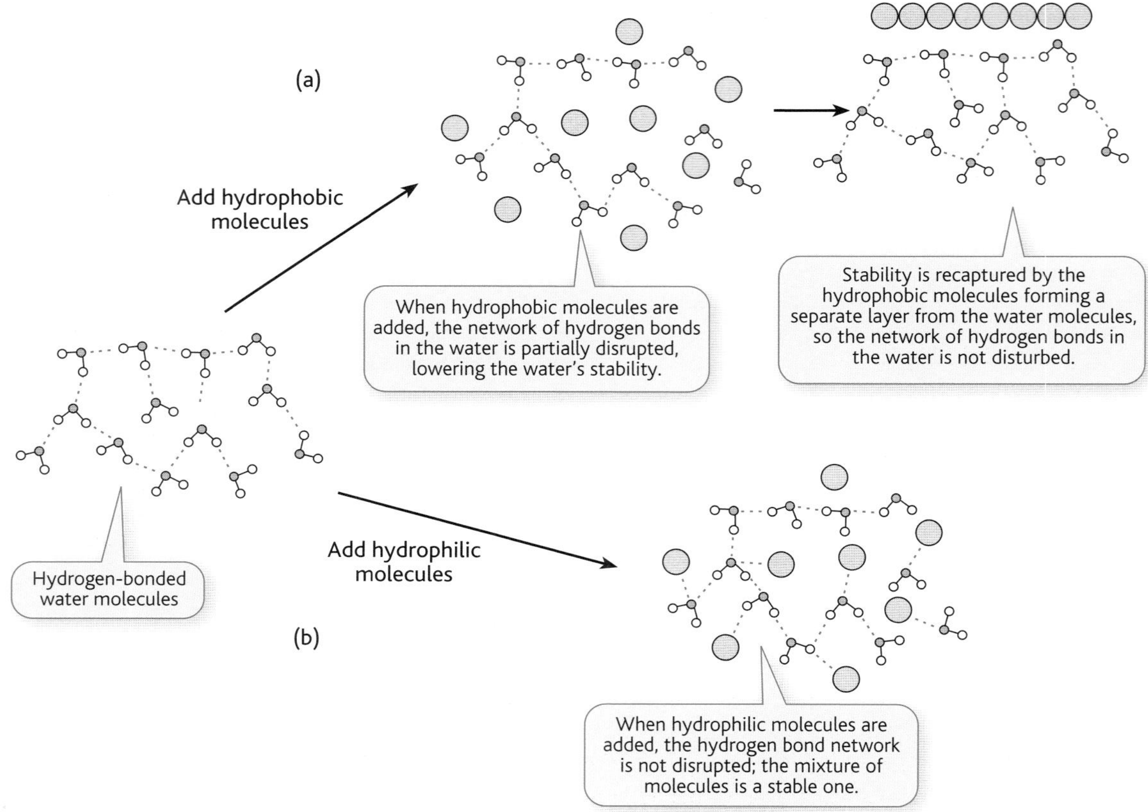

Figure 4.24 (a) Hydrophobic molecules disrupt the network of hydrogen bonds that exist in water. Consequently hydrophobic molecules partition to form a separate layer (just like oil forms a separate layer which floats on water). (b) Hydrophilic molecules can integrate into a network of hydrogen bonds, and so can mix fully with water.

Look at Figure 4.25 which shows how polar molecules can participate in dipolar interactions (and maybe also hydrogen bonding), whereas a non-polar molecule cannot. Notice in particular how both hydrogen bonds *and* dipolar forces can operate between ethanol and water, which explains the relatively high solubility of ethanol in water.

Self-check 4.7 Are the following molecules most likely to be hydrophobic or hydrophilic?

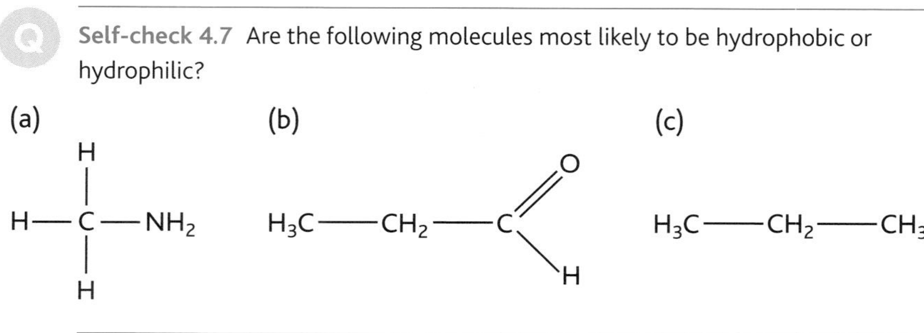

Holding it together: non-covalent interactions in biological molecules

Throughout this chapter we have seen how different non-covalent interactions operate within and between biological molecules. Figure 4.26 summarizes the

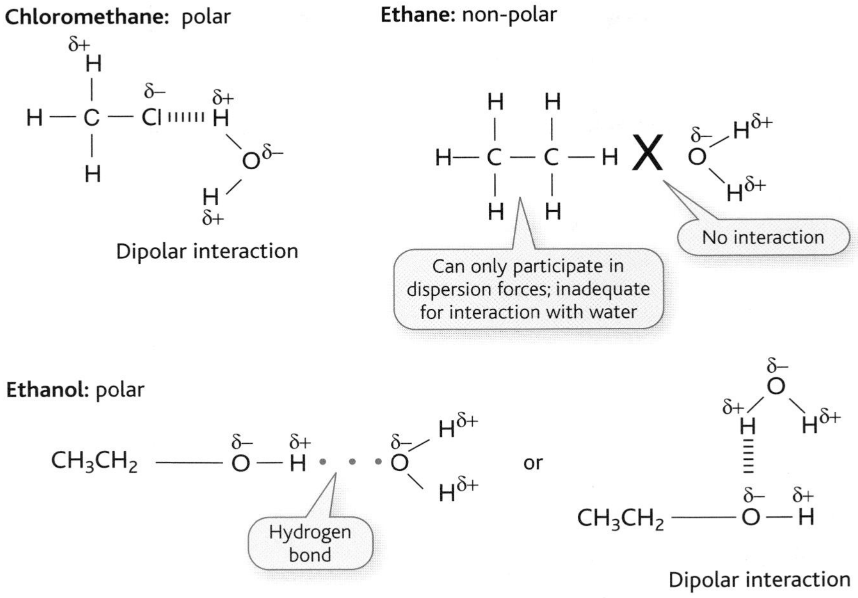

Chloromethane: polar

δ+ H — C — Cl δ− ⋯ H δ+ — O δ− — H δ+

Dipolar interaction

Ethane: non-polar

H — C — C — H X O δ− H δ+ H δ+

Can only participate in dispersion forces; inadequate for interaction with water

No interaction

Ethanol: polar

CH_3CH_2 — O δ− — H δ+ • • • O δ− H δ+ H δ+

Hydrogen bond

or

CH_3CH_2 — O δ− — H δ+ ⋯ H δ+ — O δ− — H δ+

Dipolar interaction

Figure 4.25 For a molecule to be water-soluble it must be able to participate in dipolar interactions or hydrogen bonds with water. Polar molecules can participate in dipolar interactions (and, in some cases, hydrogen bonds) and so are water-soluble; they are hydrophilic. By contrast, non-polar molecules cannot participate in dipolar interactions or hydrogen bonds, and so are not water-soluble; they are hydrophobic.

different types of non-covalent force that we have encountered, and shows how each force may operate within a single biological molecule, in this case a polypeptide. While certain interactions (including salt bridges and hydrogen bonds) occur only between specific portions of the molecule, remember that dispersion forces operate throughout the entire molecule, wherever different parts of the molecule come into close enough proximity for the dispersion forces to operate.

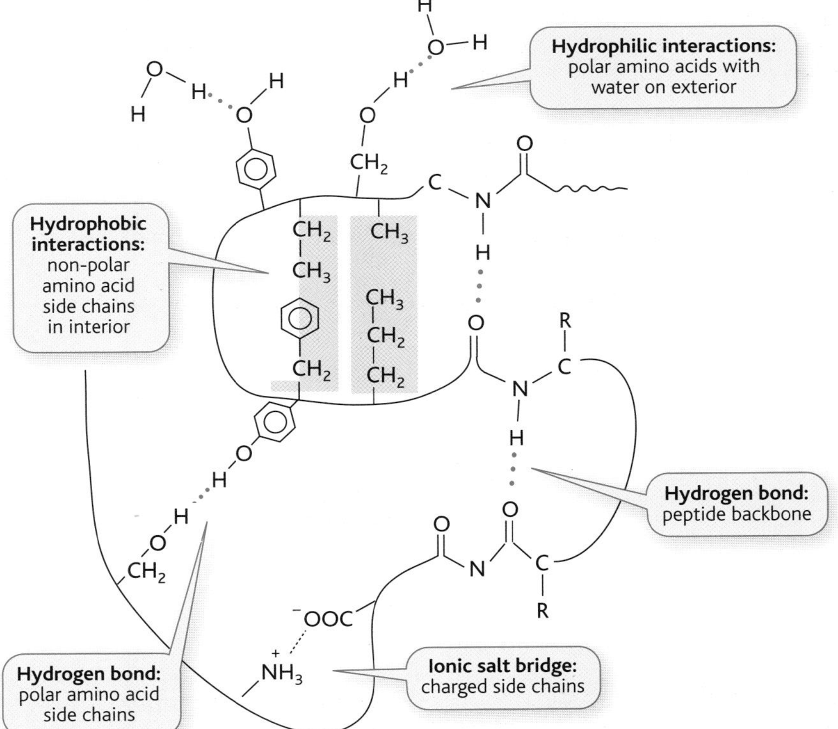

Figure 4.26 The various non-covalent interactions that can operate in a biological molecule, such as a polypeptide.

4.5 Breaking molecular interactions: the three states of matter

In this chapter we consider non-covalent forces and their effect on how molecules interact, and in this section we see how these interactions influence whether a substance is a solid, liquid, or gas at a particular temperature.

If many non-covalent interactions exist between a group of molecules, then the molecules are tightly associated. If fewer non-covalent interactions exist, then association is weaker. It is the *extent* of the non-covalent forces that exist between molecules that dictates a substance's physical state: whether it is a solid, liquid, or gas.

Figure 4.27 shows how molecular solids, liquids, and gases are characterized by the number of non-covalent interactions that exist between their composite molecules.

When molecules form **solids** they experience many non-covalent interactions, and are hence very tightly associated[13]. The shape of a solid is defined: the composite molecules are held together so tightly that they are effectively locked in position and are unable to move relative to one another. A **liquid** possesses fewer non-covalent interactions than a solid; hence the molecules are held together less tightly and are able to move around more freely.

Consider an ice cube and water in a glass: an ice cube (solid water) has a defined shape because the composite water molecules are held together very tightly. If we move the ice cube, it retains its defined shape. In contrast, water (liquid) is fluid–if we tilt the glass, the water moves to adopt a new shape. In this instance, not all the water molecules are held together by non-covalent interactions, so molecules are able to move past one another such that the liquid can adopt new shapes.

In contrast to both a solid and a liquid, the molecules in a **gas** experience no intermolecular interactions (covalent or ionic)–there is essentially nothing to hold the

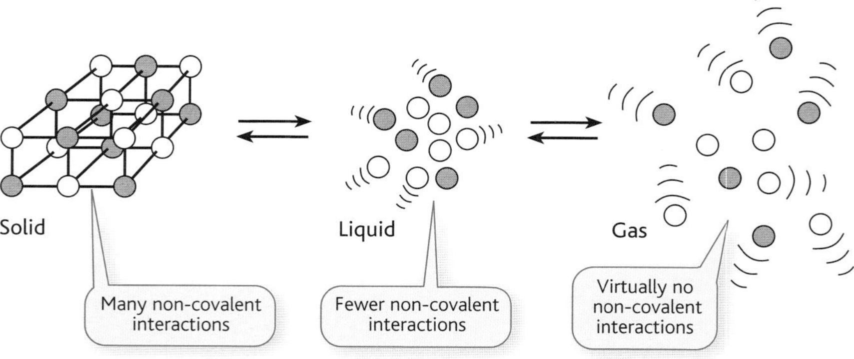

Figure 4.27 Molecular solids, liquids, and gases are characterized by the number of non-covalent interactions that exist between their composite molecules.

13 We need to note that not all solids comprise *molecules*. For example, some solids comprise atoms of a *single* element joined together by a network of solely covalent bonds or comprise ions joined by ionic bonds, as explained in Box 4.5. Some covalently bonded solids also comprise multiple elements, as in the case of silicon dioxide, SiO_2, in which each silicon atom is covalently bonded to four oxygen atoms via a strong Si–O bond.

molecules together, so they can move around freely. (There are, of course, still interactions between the *atoms* in gaseous molecules such as CO_2 or methane.)

The relative characteristics of molecular solids, liquids, and gases are summarized in Table 4.2.

	Number of molecular interactions	Shape of substance	Degree of movement of molecules	Relative energy
Solid	Many	Fixed	Virtually none	Low
Liquid	Few	Variable	Moderate	Medium
Gas	Virtually none	Unrestricted	High	High

Table 4.2 A comparison of the physical properties of typical molecular solids, liquids, and gases.

BOX 4.5 **The world of non-molecular solids**

Not all solids comprise discrete ***molecules*** held together by non-covalent interactions. Some solids comprise a large array of atoms of a single element, which are joined by covalent bonds.

For example, we see in Chapter 10 how naturally occurring carbon exists in several forms, including graphite and diamond. These two solids comprise arrays of carbon atoms joined together by covalent bonds, as illustrated in Figure 4.28(a) and (b). There are no individual molecules making up the solid.

Because covalent bonds are so strong, compared with non-covalent bonds, it takes a huge amount of energy to turn a solely covalently bonded solid into a liquid. This explains why we don't routinely see liquid diamond: diamond has a melting point of 3550 °C.

Figure 4.28(c) shows how solid NaCl comprises a network of alternating sodium and chloride ions, which are held together *wholly* by non-covalent interactions, i.e. ionic bonds. There are *no* covalent bonds in this compound.

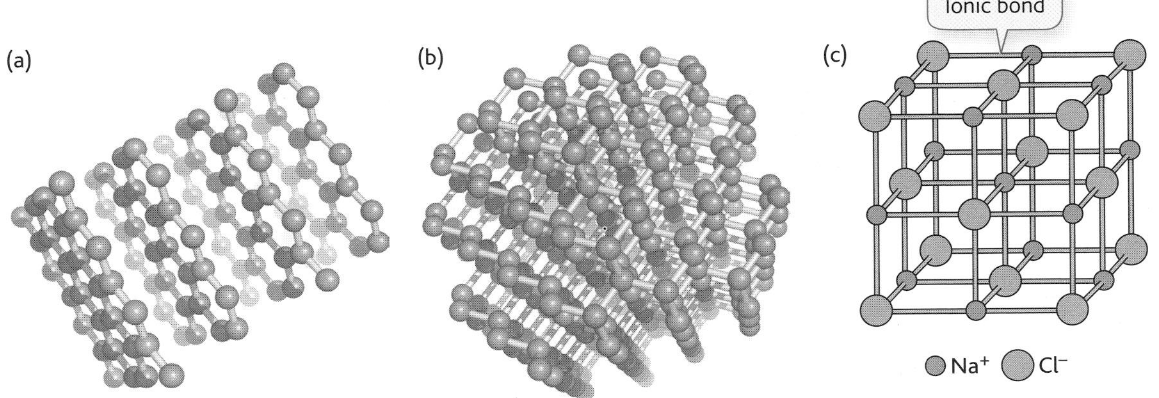

Figure 4.28 Solid carbon can take several forms including: (a) graphite, and (b) diamond. In both instances, neighbouring carbon atoms are joined by covalent bonds to form an extensive 3D network. By contrast, solid sodium chloride (c) comprises Na^+ and Cl^- ions, which are joined by ionic bonds to form an extensive lattice.

Changing states

The physical state of a substance can be changed by altering the number of non-covalent interactions that exist between the substance's composite molecules: if we decrease the number of interactions then the composite molecules become more weakly associated and a substance changes from solid to liquid, or from liquid to gas; if we increase the number of interactions then the composite molecules become more strongly associated and a substance changes from gas to liquid, or from liquid to solid, or direct from gas to solid.

Look again at Figure 4.27, and notice how the number of non-covalent interactions affects the physical state of a substance. How can we alter the number of non-covalent forces that occur?

One of our essential concepts is that change involves **energy**. (For example, if we want a car to move then we have to put in energy by using the engine to burn petrol.) Consistent with this essential concept, changing the physical state of a substance requires us to alter the energy of that substance. We usually do this by changing its temperature: if we increase the temperature (heat up), then we increase the substance's energy; if we decrease the temperature (cool down), then we decrease the substance's energy.

Look again at Table 4.2, and consider the relative energy possessed by solids, liquids, and gases. Notice how solids are low-energy, liquids have intermediate energy, and gases are high-energy. Now consider the extent of movement exhibited by the molecules in a solid, liquid, and gas respectively. Molecules in a solid show little movement; molecules in a liquid show a reasonable degree of movement, while molecules in a gas show a great deal of movement.

Notice how there is a **positive relationship** between the energy of a substance, and the degree of movement exhibited by its composite molecules: as the energy of a substance increases, so the extent of movement of its molecules increases also. Consider the process of boiling water. When the temperature is low, the water remains virtually motionless; as the temperature increases, however, the water moves with ever-increasing vigour until it boils[14].

In contrast, notice how there is a **negative relationship** between the energy of a substance and the number of interactions between molecules of a substance: as the energy of a substance *increases* so the number of interactions *decreases*.

As we increase the energy of a substance by heating it, its composite molecules exhibit a greater degree of movement. If we increase the energy to a high enough level, the movement of the molecules will be enough to overcome the attractive forces that are holding the molecules together. By analogy, consider a Velcro™ fastening: if we pull only gently on the fastening then it holds firm; as we increase the force with which we pull, however, we eventually reach a point where the fastening will rip apart: we have overcome the 'force of attraction' holding the two parts of the fastening together.

14 If we look closely at water in a saucepan as it reaches boiling point, we notice that bubbles originate from the bottom of the saucepan. It is those water molecules nearest to the source of heat–and the bottom of a saucepan–that heat up the quickest and so enter the gas phase most rapidly. We witness this transition from liquid to gas as the bubbling of water as it boils–the bubbles contain gaseous molecules of water.

It is at the point at which this threshold level of overcoming intermolecular forces is reached that we see a change in physical state. These 'threshold levels' are better known to us as a compound's **melting point** and **boiling point**.

Returning to the example of boiling water, as the temperature increases, so the water molecules gain more and more energy, and exhibit increasing vigour, until they are able to break the attractive forces holding them together. At this point, the water molecules enter the gas phase.

The transition between states

Look again at Figure 4.27, and notice how the changes in state are reversible. While an increase in energy results in an increase in movement, and a decrease in non-covalent interactions, a *decrease* in energy results in a decrease in movement, and a concomitant increase in non-covalent interactions. So, if we cool a gas, so that its composite molecules exhibit less movement, the molecules are able to associate for long enough for intermolecular forces to arise. As a result of increased associations, the molecules make the transition from the gas to the liquid phase, a process we call **condensation.**

Each change in physical state associated with an increase or decrease in the number of non-covalent interactions is given a specific name, as depicted in Figure 4.29:

- **Melting**: the transition from solid to liquid, associated with a decrease in intermolecular forces.
- **Vaporization**: the transition from liquid to gas, associated with a further decrease in intermolecular forces.
- **Condensation**: the transition from gas to liquid, associated with an increase in intermolecular forces.
- **Freezing**: the transition from liquid to solid, associated with a further increase in intermolecular forces.

BOX 4.6 Why the misty outlook?

Those of us who have travelled in a car on a cold morning are familiar with the process of condensation. While the windscreen may be clear when we first get in the car, within a few moments the windscreen mists up.

This misting is caused by the process of condensation. As we exhale, we breathe out molecules of water vapour (gaseous water). The moment the water vapour comes into contact with the cold glass of the windscreen it rapidly cools down. As they cool, and their extent of movement is reduced, the water molecules are able to form an increased number of interactions with one another and hence, change from the gas phase to the liquid phase.

As a result of this condensation process, a thin film of liquid water forms on the windscreen. The only way to remove it–and regain visibility through the window–is to 'demist' the window. We reverse the condensation process by heating the water molecules, so they re-enter the gas phase–the process of 'vaporization'. We usually achieve the heating up of water molecules required for vaporization (and demisting) by using hot air from the car engine or, in the case of the rear window, a built-in electrical heating element.

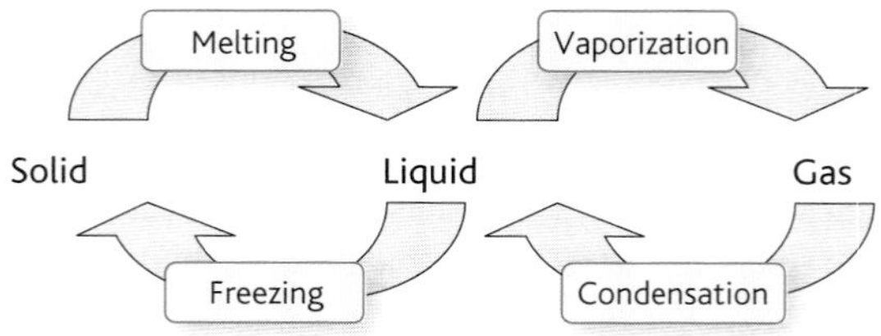

Melting Vaporization

Solid Liquid Gas

Freezing Condensation

Figure 4.29 Each change in physical state associated with an increase or decrease in the number of non-covalent interactions is given a specific name.

The impact of non-covalent interactions on melting and boiling points

Different compounds can participate in different numbers of non-covalent interactions. All molecules (both polar and non-polar) experience dispersion forces, while only polar molecules participate in dipolar interactions, and just a subset of polar molecules (those carrying O, N, or F) can form hydrogen bonds.

We have already seen how an increase in the number of non-covalent interactions in a molecular compound leads to an increase in its melting and boiling points. As a consequence, molecules which experience different numbers of non-covalent interactions have different melting and boiling points:

- Molecules that participate in *few* non-covalent interactions have *low* melting and boiling points, as little energy is needed to separate the molecules.
- Molecules that participate in *many* non-covalent interactions have *high* melting and boiling points, as more energy is needed to separate them.

Look at Table 4.3, which summarizes the types of molecular interactions that different types of molecule can experience. Notice how polar molecules possessing atoms of O, N, or F experience the highest number of non-covalent interactions (dispersion forces, dipolar interactions, and hydrogen bonds), while extremely non-polar molecules experience the fewest (only dispersion forces).

As a consequence, polar molecules generally have higher melting and boiling points than non-polar molecules: polar molecules are held together by a more extensive network of non-covalent interactions, and so more energy is required to separate the molecules, and effect the transition from solid to liquid, or from liquid to gas.

- ***Polar molecules have high melting and boiling points.***
- ***Non-polar molecules have low melting and boiling points.***

Table 4.3 The 'palette' of non-covalent interactions available to non-polar molecules, polar molecules, and polar molecules carrying O, N, or F atoms.

	Dispersion	Dipolar	Hydrogen bond
Non-polar	✓	X	X
Polar (no O, N, F)	✓	✓	X
Polar **with** O, N, or F	✓	✓	✓

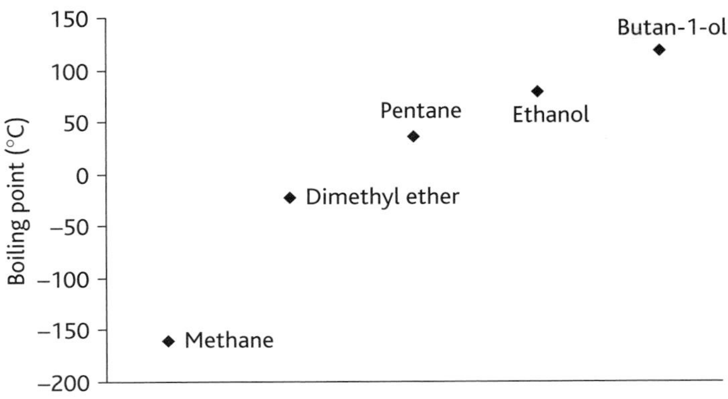

Figure 4.30 A comparison of the boiling points of a range of polar and non-polar compounds.

Look at Figure 4.30, which compares the boiling points of a range of compounds, both polar and non-polar. We notice that a non-polar compound such as methane, CH_4, experiences only dispersion forces, and has a relatively low boiling point.

We can also see the effect of an increase in the range of non-covalent interactions upon the boiling point of a compound by comparing polar molecules that can and cannot participate in hydrogen bonding.

Both ethanol, CH_3CH_2OH, and dimethyl ether, CH_3OCH_3, are polar molecules, and so both experience dipolar interactions. However, only ethanol can form hydrogen bonds with itself. The hydrogen bonding that occurs in ethanol enhances the association between ethanol molecules in a way that is not possible between molecules of dimethyl ether, so more energy is required to separate the ethanol molecules, compared to dimethyl ether molecules. As a consequence, ethanol has a higher boiling point than dimethyl ether, as shown in Figure 4.30.

The different types of non-covalent interaction also possess different energies relative to one another, which accentuate the differences in melting and boiling points exhibited by non-polar and polar compounds. Look again at Table 4.1 for a reminder of the relative energies of the different types of non-covalent interaction. Not only are non-polar compounds held together solely by dispersion forces, but these forces are relatively weak (relative to hydrogen bonds), and so relatively easy to overcome. By contrast, hydrogen bonds are of higher energy, and are more difficult to overcome.

As a final example of how the different types of non-covalent interaction possess different energies, we see in Table 4.1 how the strongest electrostatic interaction by far is the ionic force, which operates within ionic compounds such as sodium chloride. Given the link between the energy of interactions and melting points, we might predict that the melting point of an ionic compound would be much higher than that of a covalent compound (comprising molecules held together by weaker non-covalent forces). This prediction is borne out by reality: ionic sodium chloride has a melting point of 801 °C–much higher than the kinds of covalent compound shown in Figure 4.30. For example, ethanol has a melting point of −114 °C, while methane has a melting point of −182.5 °C.

Self-check 4.8

(a) Look at the following compounds and place them in order of their melting and boiling points, working from lowest to highest. Why have you chosen that order?
(i) $CH_3CH_2CH_3$; (ii) H_2O; (iii) CH_3Br

(b) Which compound has the highest boiling point, ethanol (CH_3CH_2OH) or ethane (CH_3CH_3)?

Check your understanding

To check that you've mastered the key concepts presented in this chapter, review the checklist of key concepts below, and attempt the multiple-choice questions available in the book's Online Resource Centre at **http://www.oxfordtextbooks.co.uk/orc/crowe2e/**.

Checklist of key concepts

Molecular interactions

- Non-covalent interactions operate between parts of one molecule, or between separate molecules
- Intermolecular interactions operate between *different* molecules
- Intramolecular interactions operate between different parts of a *single* molecule
- Non-covalent molecular interactions are primarily electrostatic in nature
- Non-covalent interactions only operate over short distances
- Individual non-covalent interactions are weak, but many interactions may operate between two molecules, with a large overall effect
- Molecular interactions can act to stabilize a molecule, or the association of neighbouring molecules

Polarity and polarization

- Polarization is the process by which electrons are unevenly distributed within a molecule
- The uneven distribution of electrons throughout a molecule gives rise to partial positive and partial negative charges
- The difference between partial positive and partial negative charges contributes to the dipole moment
- A dipole is a molecule, or part of a molecule, that possesses a region of partial negative charge and a region of partial positive charge
- A permanent dipole is one in which the uneven distribution of electrons is permanent
- If electrons in a covalent bond are shared equally, the bond is non-polar
- Molecules containing polar bonds may be non-polar overall

The van der Waals interaction

- The van der Waals interaction is the overall interaction between two species once both attractive and repulsive interactions are taken into account
- The extent of the van der Waals interaction operating between two molecules is determined by the distance between the two molecules, and the medium in which the molecules are found.

Dispersion forces

- Dispersion forces are weak forces of attraction that operate over short distances between all covalent molecules
- Dispersion forces exist because electrons are always slightly unevenly distributed within molecules, generating areas of partial negative and partial positive charge, which form an induced dipole
- Induced dipoles are only temporary

Permanent dipolar interactions

- Permanent dipolar interactions are the forces of attraction that exist between opposite partial charges on polar molecules
- These interactions arise between permanent dipoles, and are less transient than dispersion forces

Steric repulsion

- Two areas of like charge experience repulsion, which acts to offset attractive interactions to some extent
- Steric repulsion operates over only very short distances

The hydrogen bond

- The hydrogen bond is a special type of dipolar interaction in which a hydrogen atom on one molecule interacts strongly with one of three electronegative atoms either on another molecule, or on a spatially distinct part of the same molecule
- Hydrogen bond formation is governed by two rules:
    - The hydrogen atom must itself be bonded to an atom of oxygen, fluorine, or nitrogen
    - The hydrogen atom can only form a hydrogen bond with an atom of oxygen, fluorine, or nitrogen
- The three nuclei participating in a hydrogen bond must lie in a straight line, and the distance between the two electronegative atoms must fall within a narrow range of values
- Hydrogen bonds facilitate the formation of the double-helical structure of DNA, and underpin the specific base pairing upon which the semi-conservative replication of DNA depends
- Hydrogen bonds operate between different parts of the peptide backbone of polypeptides to generate conserved three-dimensional motifs
- Hydrophilic interactions operate between polar molecules and water molecules to make polar molecules soluble in aqueous media; they are mediated by hydrogen bonds between the polar molecule and water

Ionic forces

- Ionic forces operate between ionic species carrying full positive and negative charges
- Ionic forces may operate between parts of a covalent molecule that possess full positive and negative charges
- The ionic force that operates between oppositely charged amino acid side chains in a protein is called a salt bridge

Hydrophobic forces

- Hydrophobic forces arise when hydrophobic entities cluster together to shield themselves from water
- The clustering of hydrophobic entities during hydrophobic interactions is stabilized by dispersion forces
- Non-polar molecules are hydrophobic, and are immiscible with water
- Polar molecules are hydrophilic, and interact readily with water
- Hydrophobic interactions can be explained by thermodynamic principles

Physical states

- The extent of the non-covalent forces that exist between molecules dictates a substance's physical state
- A solid comprises molecules that possess many non-covalent interactions
- A liquid possesses fewer non-covalent interactions than a solid
- A gas comprises molecules that experience virtually no non-covalent interactions
- As the energy of a molecule increases, the number of non-covalent interactions that it experiences decreases
- The melting point is the temperature at which a compound makes the transition from a solid to a liquid
- The boiling point is the temperature at which a compound makes the transition from a liquid to a gas
- Melting is the transition from solid to liquid, associated with a decrease in intermolecular forces
- Vaporization is the transition from liquid to gas, associated with a further decrease in intermolecular forces
- Condensation is the transition from gas to liquid, associated with an increase in intermolecular forces
- Freezing is the transition from liquid to solid, associated with a further increase in intermolecular forces
- Non-polar molecules experience few non-covalent interactions and have low melting and boiling points
- Polar molecules experience more non-covalent interactions than non-polar molecules, and have higher melting and boiling points
- Hydrogen bonds contribute more to the elevation of a compound's melting and boiling point than dispersion forces or dipolar interactions

5 Organic compounds 1: the framework of life

INTRODUCTION

One of the essential concepts we encounter in Chapter 1 is that complexity can have very simple origins. The variety that we see in the world around us is the outcome of simple building blocks coming together in diverse ways. For example, all of life is generated from a pool of fewer than 30 chemical elements, while just four different nucleotide bases generate a huge range of DNA sequences – those genetic templates that make every individual on the planet unique (with the notable exception of identical twins).

Carbon lies at the heart of the building blocks that give life its variety, stability, and energy. It is, in effect, the scaffolding upon which many vitally important biological molecules are built – from the ATP that is broken down in mitochondria to give us energy to the lipids that form our cell membranes.

In this chapter we discover why we could consider carbon the central biological element. We see how just two elements, carbon and hydrogen, combine to form the framework for an array of different compounds, and learn how this framework is built upon to generate different families of organic compound, many of which lie at the heart of life itself.

5.1 Organic chemistry

Organic chemistry is the field of chemistry that describes the chemistry of carbon-containing compounds. Over 20 million **organic compounds** are known; of these, many are naturally occurring **bioorganic** compounds, while others are **synthetic**– they do not occur naturally but are produced in the laboratory.

'**Synthetic**' is a generic term used to describe any compound that is made artificially. Many of our clothes are made from a combination of synthetic and naturally occurring materials: for example, cotton is naturally occurring, while polyester is synthetic.

The two types of compound are not mutually exclusive. Some naturally occurring compounds are also made artificially in the lab. For example, vitamin C is naturally present in many fruits and vegetables. However, the vitamin C supplements which many of us may take (for example, as once-a-day tablets) contain vitamin C that has been mass-produced artificially in the lab, rather than expensively extracted from fresh fruit or vegetables.

Self-check 5.1 What other chemical compounds can you think of in everyday use that occur naturally, but which are also widely available in synthetic form?

Before we explore some of the many different compounds that are encompassed by organic chemistry, let's look more at carbon, the one element that is common to all of these compounds.

Carbon: its defining features

There are only a few aspects of our life in which carbon isn't present in some guise or another, whether it be as a fuel (coal, petrol, oil), as an artist's material (pencil lead, charcoal), or even as a strengthening agent (carbon fibre). Indeed, have you ever used a water filter and noticed tiny specks of black appear in the base of the filter? This is carbon, in the form of 'activated' charcoal, whose affinity for absorbing pollutants accounts for its widespread use in filters.

Pure ('elemental') carbon exists in several forms, including graphite, diamond, and the fullerenes. The three types of compound listed here consist solely of carbon atoms, but vary in the way in which the carbon atoms are joined to one another. (We say that they are **allotropes** of carbon.) We learn more about these three forms of carbon in Box 10.1.

Little of the carbon in the environment is in one of those 'pure' forms described above, however. Carbon exists as carbonate (CO_3^{2-}) in rocks, as hydrogen carbonate (HCO_3^-) in the seas, and as carbon dioxide (CO_2) and methane (CH_4) gases in the atmosphere. Carbon moves between the earth, sea, and air through a series of natural processes that, taken together, form the **carbon cycle**, as depicted in Figure 5.1.

Those of us who drive cars, or have gas-fired central heating, rely upon a particular part of the carbon cycle: the death of living organisms and their subsequent fossilization over millions of years. The decay of organisms (whose cells contain carbon in the form of many important biomolecules, as we see throughout this book)

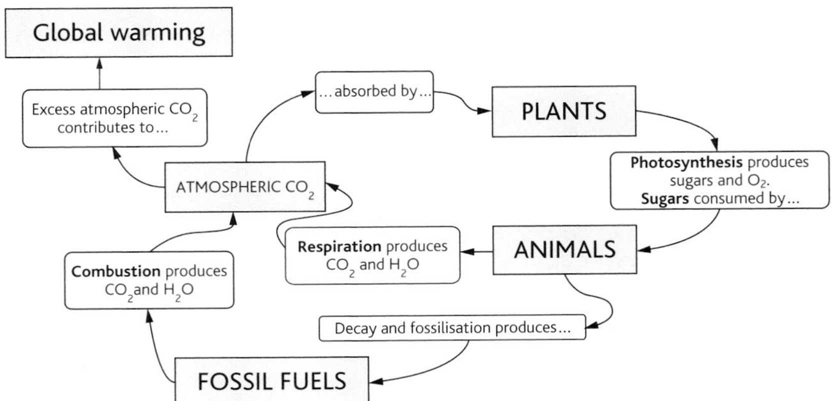

Figure 5.1 A representation of the carbon cycle, showing the natural processes by which carbon cycles through the earth, sea, and air.

returns much carbon to the earth over a relatively short timescale (weeks, months, and years). Composting is one way in which we give the carbon cycle a helping hand, by transferring decaying organic matter to the soil. On a far longer timescale, however, dead organisms become fossilized. Deposits of fossilized marine algae and plants, which were laid down millions of years ago, form the reserves of **fossil fuels** on which we are so dependent today. Natural gas and oil are extracted from deposits of fossilized marine algae, while coal is extracted from deposits of fossilized plants[1].

However, the form in which carbon is most vital to us is as the central component of the array of biological molecules on which we depend for life: the food we eat; the tissues from which our bodies are formed; and the hormones that regulate the intricate network of chemical reactions that keep us alive. These biological molecules are members of the huge group of organic compounds that we explore during the rest of this chapter, and encounter again throughout this book.

Carbon: the central biological element?

We learn more about the valency of different elements in Chapter 3.

Carbon's central role in biological compounds is facilitated by its valency. Carbon has a valency of four, telling us the following:

- An atom of carbon must share four pairs of electrons to attain a full valence shell, and satisfy the 'octet rule'.
- An atom of carbon can therefore form covalent bonds with up to four other atoms.
- An atom of carbon can form multiple covalent bonds: it has a sufficiently high valency to form both double and triple bonds (see section 3.7).

These characteristics confer upon carbon considerable chemical versatility, enabling an atom of carbon to form chemical bonds with many different combinations of atoms of both carbon, and other elements.

Think of carbon as the central piece of a jigsaw on which all other pieces rely. While pieces at the edge of a jigsaw can connect with just two or three other pieces, those at the centre can connect with the most number of other pieces. Similarly, carbon is the central atom in many biological compounds: it acts as the hub to which other atoms anchor themselves.

The key components of organic compounds

We can envisage organic compounds as having two key physical features:

1. A simple framework, which provides structure and stability
2. A series of more diverse and specialized components, which attach to the simple framework and confer upon the compound its functionality.

We could consider an organic compound to be analogous to the human body. The body has a simple framework: the skeleton, on which the rest of the body is built. The skeleton gives the body structural stability and durability.

1 The extraction of natural gas and oil from fossilized marine algae explains the location of oil rigs far out at sea. These oil rigs drill deep into the sea bed to extract natural gas and oil from fossil deposits which were laid down millions of years ago.

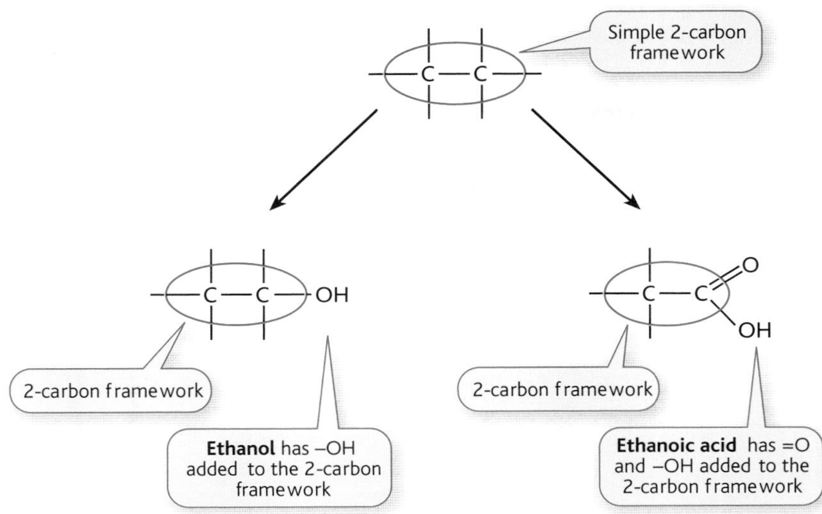

Figure 5.2 Ethanol and ethanoic acid are based on the same two-carbon framework.

The body also has a wide range of specialized components–heart, lungs, skin, muscle–all of which are connected to the skeleton in some way. The body's various components have the vital role of giving the body its functionality. Without muscles, our body could not move; without a brain, we could not instruct our muscle to contract and produce that movement.

The specialized components that attach to the framework of an organic compound are what we call **functional groups**.

Organic compounds are grouped into families according to the functional groups that they possess. Compounds from different families may share the same structural framework, but belong to the different families because they possess different functional groups. For example, both ethanol, CH_3CH_2OH, and ethanoic (acetic) acid, CH_3COOH, are based on the same two-carbon framework, as illustrated in Figure 5.2. Ethanol and ethanoic acid have contrasting functions, however. While ethanol puts the 'alcohol' into alcoholic beverages, ethanoic acid is the substance that gives vinegar its sharp taste. We see these differences in chemical properties because ethanol and ethanoic acid possess different functional groups: ethanol carries a hydroxyl (–OH) group, while ethanoic acid carries a carboxyl (–COOH) group.

All members of a particular family of organic compound share the same functional group. For example, the alcohols are a family of organic compound that possess the hydroxyl (–OH) functional group, whereas the carboxylic acids possess the carboxyl (–CO_2H) functional group. However, members of the same family of organic compound differ according to the nature of the structural framework they possess. Look at Figure 5.3; the compounds we see in this figure are all alcohols, but they are all *different* alcohols. Look at these structures closely, and notice how the framework to which the hydroxyl group is attached differs in each case.

We learn more about the most common functional groups, including the hydroxyl and carboxyl groups, in Chapter 6.

- ***A functional group comprises a specific, characteristic set of atoms, like –OH or –CO_2H.***
- ***All members of a particular family of organic compound share the same functional group(s).***

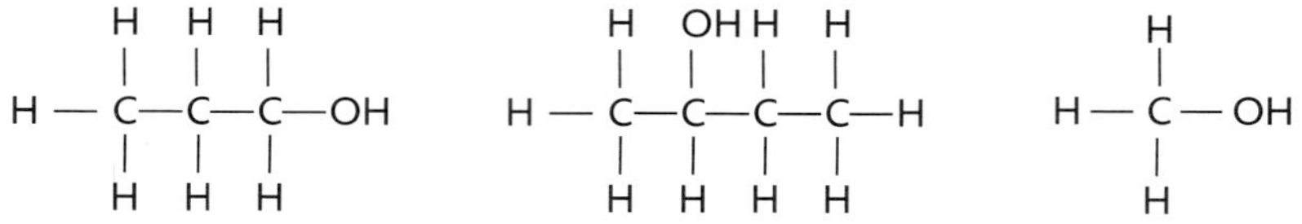

Propan-1-ol Butan-2-ol Methanol

Figure 5.3 These compounds are all alcohols; they all possess the –OH functional group. However, they all have different carbon frameworks.

Carbon has its central role in the chemistry of biological systems as the key component of the **framework** of an organic compound, as we discover in the next section.

5.2 The framework of organic compounds

Look again at the structural 'skeleton' of each of the alcohols shown in Figure 5.3. What do you notice? The framework of an organic compound is composed of atoms of just two elements, **carbon** and **hydrogen**.

Specifically, the framework of an organic compound comprises a carbon 'backbone', a chain of carbon atoms joined to one another, to which hydrogen atoms are attached.

The simplest organic compounds comprise nothing more than a simple carbon + hydrogen skeleton, with no other functional groups attached. These simple organic compounds belong to a family called the **hydrocarbons**. Despite being composed solely of carbon and hydrogen, the hydrocarbons comprise a large number of compounds, ranging in composition from just four atoms, to tens or even hundreds.

The hydrocarbons are split into three groups, which are distinguished by the nature of the covalent bonds between the carbon atoms that form the backbone of the compound:

- An **alkane** contains only **single** carbon–carbon bonds

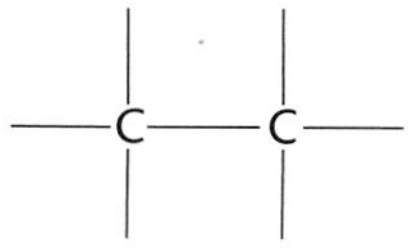

- An **alkene** contains at least one **double** carbon–carbon bond

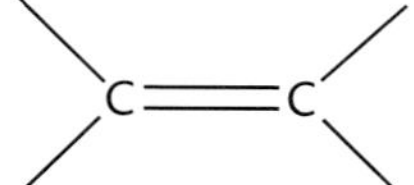

- An **alkyne** contains at least one **triple** carbon–carbon bond

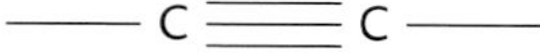

Alkanes, alkenes, and alkynes are members of two different groups of compounds, characterized by the type of chemical bond they contain. **Saturated** compounds contain only single covalent bonds, while **unsaturated** compounds contain double or triple covalent bonds. Alkanes are saturated compounds: they contain only *single* carbon–carbon bonds. By contrast, alkenes and alkynes are **unsaturated** compounds, and contain some *double* or *triple* covalent bonds[2].

We often use the word 'saturated' in the context of the absorption of water. If we are caught in a heavy downpour of rain, then we might complain that our clothes have become 'saturated'–they have absorbed rainwater to the point that they can absorb no more. Similarly, alkanes are saturated with hydrogen atoms–they have 'absorbed' as many hydrogen atoms as they possibly can (given the constraints placed on an atom in terms of the number of covalent bonds it can form).

By contrast, alkenes are not saturated with hydrogen atoms: two further hydrogen atoms could (in theory) be added wherever there is a double carbon–carbon bond. Similarly, four hydrogen atoms could be added wherever there is a triple carbon-carbon bond.

- ***Saturated compounds contain only single covalent bonds.***
- ***Unsaturated compounds contain at least one multiple bond.***

The hydrocarbons have few direct roles in biological systems. However, the alkane-like carbon + hydrogen structure forms the framework for many important biological molecules.

Before we learn more about the alkanes, however, let's consider how we represent organic structures visually.

Representing chemical structures: the structural formula

We see in Chapter 3 how a compound's molecular formula tells us the composition of one molecule of a covalent compound. However, they do not give us enough information to establish the *structure* of a compound, because they fail to tell us the *connectivity* of atoms, that is, how the atoms comprising the molecule are joined to each other. For this, we use **structural formulae**.

A molecule's structural formula tells us how the different atoms in the molecule are connected to one another. There are several ways of representing the structural formula of a molecule; each one reveals different levels of detail about how the atoms are arranged. Look at Figure 5.4, which shows the most commonly used formats for representing the structural formula.

We use structural formulae throughout the rest of this book, to help us explore the way compounds are formed, and how the structure of a compound can often have a major influence on its function.

---

2 We have probably all seen television adverts for butter-substitute spreads, which drive home the point that these spreads are 'low in saturates, and high in unsaturates'. Such adverts describe the chemical nature of the fats that make up the spread, telling us that the spread contains a high proportion of unsaturated molecules (molecules that contain double carbon–carbon bonds) and a low proportion of saturated molecules (molecules that only contain single carbon–carbon bonds).

Pentane:

C_5H_{12}

Molecular formula

$CH_3CH_2CH_2CH_2CH_3$

Structural formulae

Ethanol:

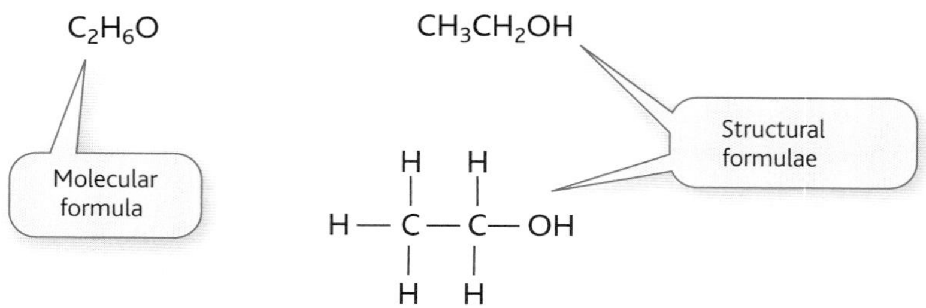

Figure 5.4 Some of the most commonly used formats for representing the structural formula. Compare the information given in a structural formula with that given in a molecular formula. Notice how the structural formula gives us much more detail about the compound–in particular, showing how atoms in the compound are arranged relative to one another.

- ***A molecular formula indicates the relative number of atoms present in one molecule of a given compound.***
- ***A structural formula tells us how the different atoms in the molecule are connected to one another.***

Simplifying the structural formula

The full structural formula can become quite complicated and visually cluttered, particularly when we've considering large molecules. To make it easier for us to visualize the structure of the carbon framework we can represent the framework using a **simplified** structural formula.

The simplified structural formula shows the bonds linking atoms, but does not show the carbon and hydrogen atoms that comprise the framework itself. For example, when using the simplified structural formula we represent propane as

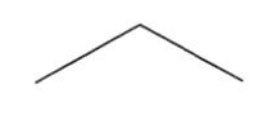

rather than

$$H_3C—CH_2—CH_3$$

Notice how the simplified structural formula communicates to us the *shape* of the compound somewhat more clearly than when the carbon and hydrogen atoms are also shown.

We can use the simplified structural formula to represent compounds with double bonds in the same way. For example, here are two ways of drawing pent-2-ene:

is the same as $H_3C—CH{=}CH—CH_2—CH_3$

Self-check 5.2 Draw the following 'expanded' structural formulae in simplified form.

```
    H  H  H  H  H           H  H  H  H            H  H  H
    |  |  |  |  |           |  |  |  |            |  |  |
H — C— C— C— C— C— H    H — C— C= C— C— H     H — C— C— C— OH
    |  |  |  |  |           |        |            |  |  |
    H  H  H  H  H           H        H            H  H  H
         (a)                    (b)                  (c)
```

Remember: there is a carbon atom at the end of each straight line in a simplified structural formula. Each carbon has a sufficient number of hydrogen atoms attached to satisfy carbon's valency of 4.

Let's now examine the alkanes in more detail.

The alkanes: the backbone of organic chemistry

An alkane is a hydrocarbon that contains carbon atoms which are joined solely by single covalent bonds. The simplest alkane (and indeed the simplest hydrocarbon) is **methane**, CH_4, which comprises four hydrogen atoms covalently bonded to a central carbon atom (1).

We learn more about the biological impact of methane in Box 5.1

```
     H
     |
 H — C — H
     |
     H
    (1)
```

The members of the alkane family are differentiated by the number of carbon atoms they contain. Methane, CH_4, contains one carbon atom. The next largest alkane (adding one carbon to the 'carbon backbone' of methane) is **ethane**, CH_3CH_3 (2), which contains two carbon atoms and six hydrogen atoms:

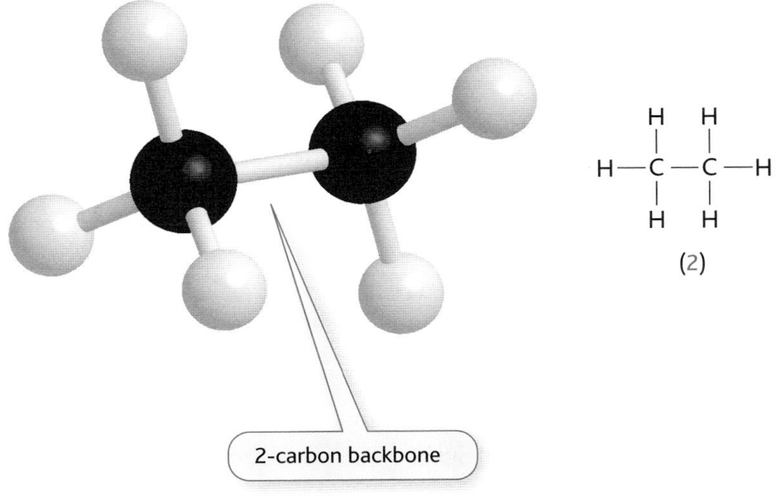


```
    H   H
    |   |
H — C — C — H
    |   |
    H   H
     (2)
```

BOX 5.1 Danger in the air

Believe it or not, cows could be considered to be an environmentalist's greatest enemy. A cow produces on average a staggering 600 litres of methane, CH_4, a day, enough to fill roughly 120 party balloons. Methane is one of the so-called greenhouse gases–those gases that contribute to global warming by forming an insulating layer around the Earth. When solar radiation (light from the sun) reaches the Earth, a proportion is reflected back into the atmosphere. The insulating layer of greenhouse gases prevents this reflected solar radiation from escaping the Earth's atmosphere. Instead it traps this heat within the atmosphere, leading to the warming of our climate which we are now experiencing.

Perhaps surprisingly, the greenhouse gases comprise around just 1% of the gases in our atmosphere. The vast majority comprises nitrogen (about 78%) and oxygen (21%). Of the greenhouse gases themselves, carbon dioxide, CO_2, is probably the most widely recognized. However, like-for-like, methane produces over 20 times more warming than carbon dioxide. So shouldn't we be more worried about it? Perhaps we should, if we consider the natural sources of methane, and how these might impact on atmospheric levels in the future.

Beyond the cows mentioned above (which are an all-too-real source of the gas), the largest natural source of methane is from global wetlands. For example, vast reservoirs of methane are usually trapped under a layer of permafrost in regions like Greenland's tundra. However, as warming takes hold, this permafrost is thawing, allowing these reserves of methane to escape into the atmosphere.

There is a danger that this process of thawing and methane release could become a vicious circle, whereby warming causes thawing, which releases more methane, leading to more warming, more thawing, and yet more methane release–a continual (and potentially destructive) loop.

So what is likely to happen in the future? At this stage, it is difficult to predict. Concentrations of methane in the atmosphere since around 1750 (the time of the Industrial Revolution) have increased by over 100%, significantly more than the rise in carbon dioxide (which has increased by 30%). Despite this increase, methane still only accounts for ~20% of the global warming caused by greenhouse gases, compared to the ~60% contribution made by carbon dioxide (because the concentration of methane in the atmosphere is so much lower than that of carbon dioxide, despite the increased levels of methane we have witnessed). So carbon dioxide would still seem to be the key threat. Will this remain true in the future? Only time will tell.

Figure 5.5 depicts the structures of propane, butane, and pentane, whose carbon backbones comprise three, four, and five carbon atoms respectively.

Naming the alkanes

A hydrocarbon's name has two components. The first part (the root) tells us the number of carbon atoms forming the longest unbroken chain in the molecule. The second part (the suffix) tells us whether the hydrocarbon is saturated or unsaturated, and so identifies the hydrocarbon group to which it belongs.

The first part of the hydrocarbon's name is based on the Greek language (pent- = 5; oct- = 8, etc.). Look at Table 5.1 to find the roots for carbon backbones containing one to ten carbon atoms.

The suffix '-ane' tells us that the compound is an alkane, and comprises just carbon and hydrogen atoms, joined by *single* covalent bonds. It is a **saturated** compound.

By contrast, the suffix -ene indicates that the hydrocarbon is an alkene, and contains at least one *double* carbon–carbon bond, while the suffix -yne indicates that the hydrocarbon is an alkyne, and contains at least one *triple* carbon–carbon bond.

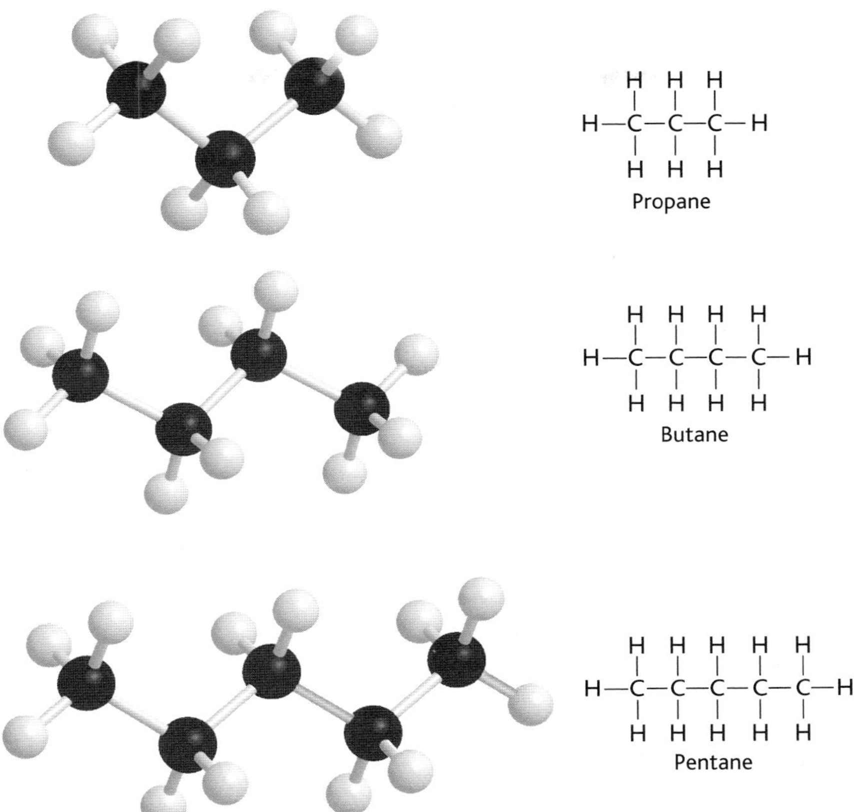

Figure 5.5 The structures of (a) propane, (b) butane, and (c) pentane.

So, this alkane is called pentane:

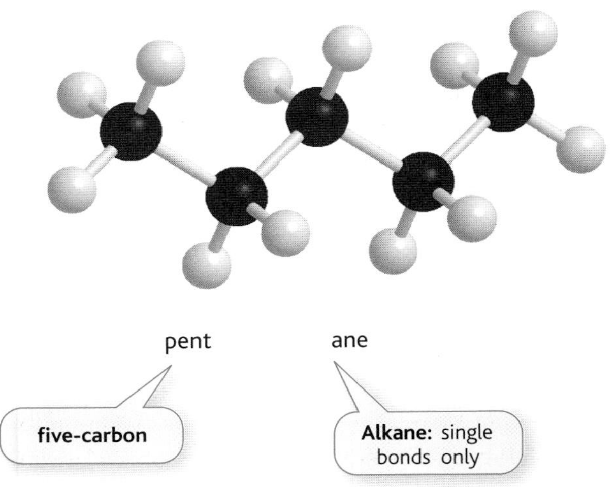

Table 5.1 The roots for carbon backbones containing one to ten carbon atoms.

Stem	Number of carbon atoms
Meth-	1
Eth-	2
Prop-	3
But-	4
Pent-	5
Hex-	6
Hept-	7
Oct-	8
Non-	9
Dec-	10

This alkene (which contains a double bond) is called propene:

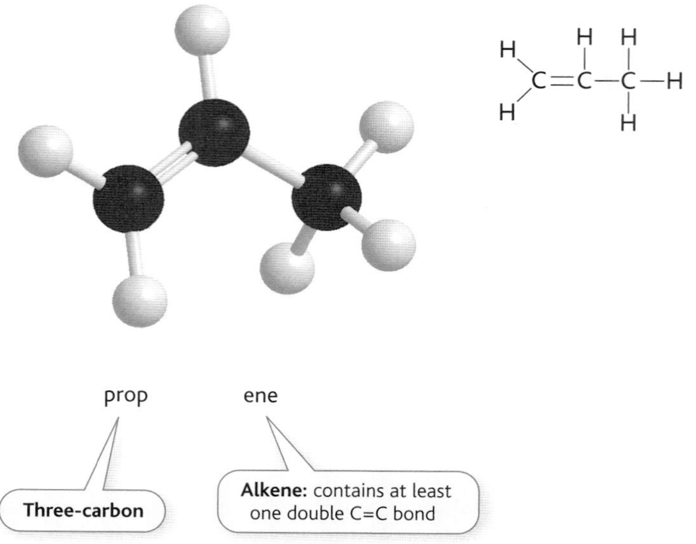

Self-check 5.3 What is the name of the alkane with an eight-carbon framework?

The alkanes mirror one of our essential concepts (Section 1.2), that of complicated structures being generated from the joining of simple building blocks. In the case of the alkanes, the simple building block is the CH_2 group. Look again at Figure 5.5 and notice how butane has an extra CH_2 group compared to propane, and pentane has one more CH_2 group than butane.

The shape of organic compounds

Carbon's valency of four (its ability to form single covalent bonds with four other atoms) enables organic compounds to possess a variety of shapes. The carbon atoms of an alkane may join together in different ways to generate backbones with three distinctive shapes:

- straight chain
- branched chain
- cyclic (ring-shaped).

Figure 5.6 shows how the three types of structure compare.

Straight-chain hydrocarbons comprise carbon atoms which are joined together to form a continuous chain. This is much like a group of actors taking a bow during the curtain call at the end of the show, where the actors arrange themselves in a single line, and join hands to form an unbroken chain.

Branched-chain hydrocarbons comprise a carbon backbone which branches to form two separate chains, a situation analogous to a road splitting into two where a fork in the road occurs. Look at Figure 5.6(b). There are five carbon atoms altogether, just as there are in Figure 5.6(a), but they are not all connected to form a continuous (linear) chain.

(a)

Straight-chain
forms linear chain

(b)

Branched-chain
chain of carbon atoms splits into two at this atom

(c)

Cyclic
ends of chain join to form a sealed ring

Figure 5.6 (a) A straight-chain, (b) a branched-chain, and (c) a cyclic molecule.

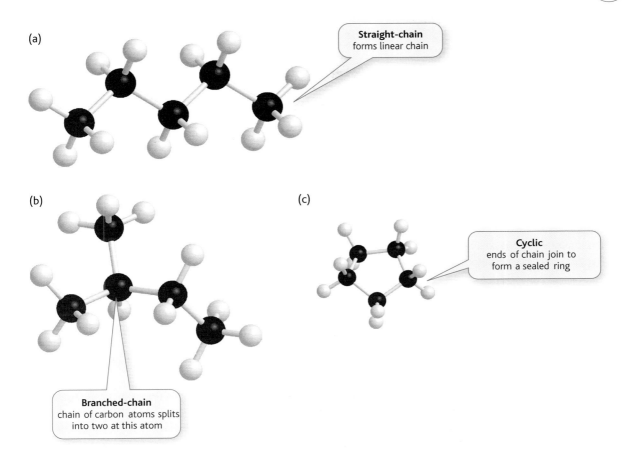

Cyclic hydrocarbon rings arise when the ends of a straight carbon backbone join together to form a cyclic or ring structure, as illustrated in Figure 5.6(c).

Notice in this figure how hydrocarbon rings comprise two fewer hydrogen atoms than the corresponding straight-chain hydrocarbon. When the ends of a straight-chain hydrocarbon join together to form a ring structure, the carbon atoms at the two ends of the chain satisfy their valency of four by bonding to one another, rather than to hydrogen atoms, as depicted in Figure 5.7.

Hydrocarbon rings are named by attaching the prefix 'cyclo–' to the name of the straight-chain hydrocarbon that has a corresponding number of carbon atoms. For example, cyclohexane is the cyclic alkane comprising *six* carbon atoms, and cyclopropene is a cyclic alkene comprising five carbon atoms.

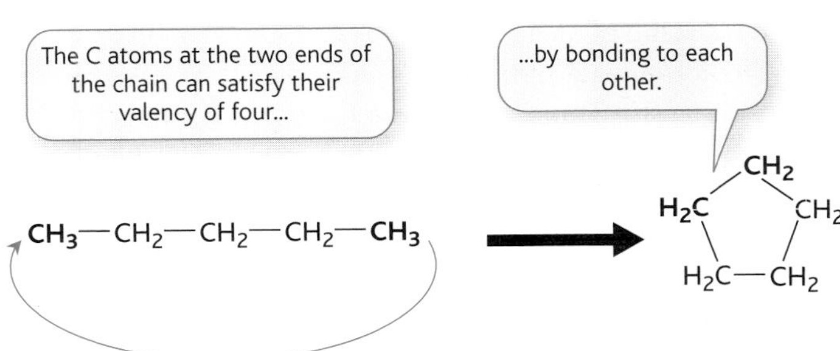

Figure 5.7 The two carbon atoms at the end of a hydrocarbon chain can satisfy their valence of four by bonding to one another, rather than to hydrogen atoms.

BOX 5.2 The multi-ringed steroid template

An unusual cyclic alkane provides the template for the **steroids**, a vitally important class of biological molecule. The steroid template comprises a distinctive four-ringed hydrocarbon whose structure is shown in Figure 5.8, and features three cyclohexane rings and one cyclopentane ring joined together. We say they are 'fused'.

The steroid template is the basis of a number of organic compounds that have important roles in biology; it is featured in the sterol cholesterol, and the steroid sex hormones testosterone and progesterone.

The steroids have earned rather a notorious reputation through their misuse by athletes: anabolic steroids are used to build muscle and give athletes extra power (and an unfair advantage over their competitors!).

We find out more about the steroids in Chapter 7.

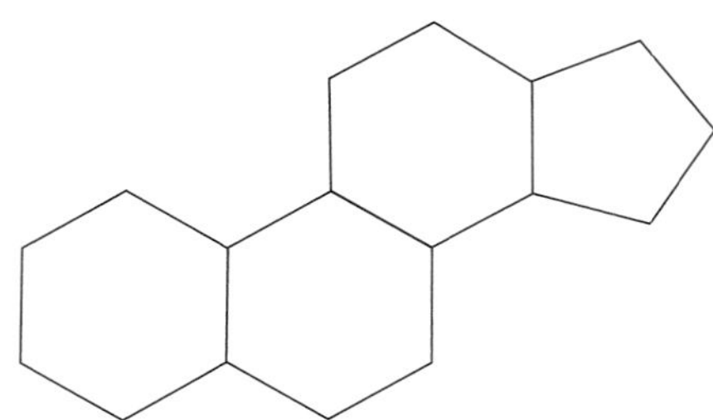

Figure 5.8 The steroid template. This structure features three cyclohexane rings and one cyclopentane ring fused together.

Look at Box 5.2 to find out about the steroid template–a cyclic hydrocarbon template, which comprises *four* ring structures.

Physical properties of the alkanes

The compounds within all three groups that form the hydrocarbon family (the alkanes, alkenes, and alkynes) are **non-polar** molecules (Section 4.2): they comprise carbon and hydrogen atoms held together by covalent bonds, and neither the carbon atoms nor the hydrogen atoms are sufficiently electronegative to cause significant polarity within the molecule.

We see in Chapter 4 how very few non-covalent forces exist between non-polar molecules. The lack of polarity means these molecules cannot participate in strongly dipolar or ionic interactions; they also lack the ability to form hydrogen bonds[3].

Dispersion forces are very weak dipolar forces. We explore dispersion forces on p. 90.

Therefore the hydrocarbons experience only **dispersion forces**, the transitory attraction between areas of opposite charge, which is a consequence of the continual movement of electrons in a molecule.

Consequently, the hydrocarbons have both low melting points and low boiling points: it requires relatively little energy to overcome the weak dispersion forces that attract neighbouring alkane molecules to one another, to effect the transition from solid to liquid, or from liquid to gas. In other words, melting and boiling occur at relatively low temperatures.

As the size of the alkanes increases, so their melting and boiling points increase also. This trend is a result of the increase in the magnitude of dispersion forces that operate between molecules as the molecules increase in size. (We see on p. 93 how the number of electrons in a molecule increases as the molecule gets larger, and that an increase in the number of electrons results in stronger dispersion forces.)

3 Remember: molecules of a particular compound can only form hydrogen bonds with each other if they possess a hydrogen atom attached to one of three electronegative atoms–oxygen, fluorine, or nitrogen. Look at p. 97 to find out more about hydrogen bonding.

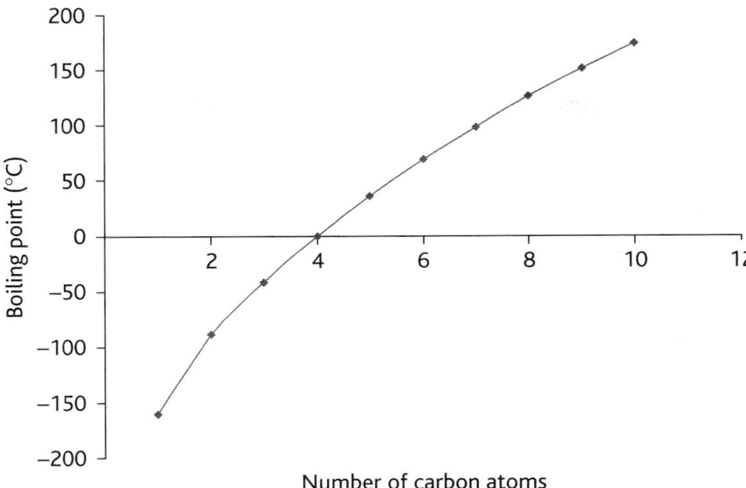

Figure 5.9 A graph depicting the boiling points of alkanes of increasing carbon chain length. Notice how the boiling points of the alkanes increase as their carbon chains increase in length.

Look at Figure 5.9, which shows the boiling points for alkanes as a function of increasing chain length (i.e. with progressively longer carbon backbones). Notice how the boiling points increase as the length of the carbon chain increases. More energy is needed to drive the transition from liquid to gas as the chain length increases in order to overcome the increasing dispersion forces that hold these non-polar molecules together.

Solubility in water

We see in Chapter 4 how a covalent compound is soluble in water only if it can form hydrogen bonds with water molecules. Given their lack of a suitably electronegative atom, the hydrocarbons are unable to form hydrogen bonds with water. Hence they are **insoluble** in water.

This observation explains in part why the alkanes have few direct roles in biological systems: biological systems are aqueous (water-based), yet the alkanes cannot dissolve in water. By contrast, many other biologically important organic compounds *can* dissolve in water. It is the functional groups that are attached to the otherwise insoluble alkane-like skeleton of these organic compounds which facilitate dissolving in water.

The physical properties of the alkanes are summarized in Table 5.2. Notice how certain physical compositions dictate the physical properties that the alkanes exhibit.

Table 5.2 The physical property checklist for the alkanes.

Does the alkane experience...		Consequence
Dispersion forces?	✓	Low melting and boiling points
Dipolar interactions?	×	
Hydrogen bonding between molecules of the same compound?	×	
Hydrogen bonding with water?	×	Insoluble in water

Chemical properties of the alkanes

The carbon and hydrogen atoms from which the alkanes are formed are connected by a series of strong single covalent C–C and C–H bonds. A relatively large amount of energy is required to break these bonds, so they generally stay intact: the alkanes are very stable. However, with stability comes a lack of reactivity. Because the bonds joining the atoms of an alkane are relatively hard to break, alkanes rarely react readily with other compounds. The alkanes do, however, undergo **combustion** readily[4].

We exploit the combustion of alkanes by using them as a component of many fuels, from the propane gas used in camping stoves, to the petrol that powers many of our cars, as explained in Box 5.3.

Once again, we see a stark contrast between the behaviour of the alkanes and the behaviour of many biologically important organic compounds. If all organic compounds were as chemically unreactive as the alkanes then we would not exist! Chemical reactions occur in virtually every cell of our bodies every second of the day: they generate energy from food; they bring our senses to life. The alkane-like skeleton of organic compounds provides robustness and stability, but they need something else to provide the 'spark'–to provide chemical reactivity. During the remainder of this chapter we see how **functional groups** provide the spark to make organic compounds chemically reactive. We learn about the different types of functional group that exist, and how they make many organic compounds very different, in terms of physical and chemical properties, from the rather mundane alkanes.

- ***The alkanes are very stable compounds, and hence are chemically relatively unreactive.***

BOX 5.3 Burn baby burn....

Petrol (or gasoline) is a complicated mixture of hydrocarbons whose carbon backbones comprise between five and twelve carbon atoms. A majority of the hydrocarbons present are alkanes. The performance of petrol in an engine depends on the relative amounts of heptane (containing seven carbon atoms; Figure 5.10(a)) and isooctane (containing eight carbon atoms; Figure 5.10(b)) present.

The petrol in an engine cylinder needs to burn at a constant rate, and completely, for optimum performance. If there is too little isooctane present, however, the petrol ignites too early, leading to 'knocking': the firing of cylinders out of sequence, and a subsequent loss of power. (We also hear an audible 'knock' from the engine.)

Petrol is given a rating in the form of an **octane number**, which indicates how resistant the petrol is to 'knocking', and so also reflects its performance and efficiency. The higher the octane number, the greater the amount of isooctane relative to heptane present in the petrol, and the greater the fuel's performance. (An octane number of 100 is assigned to 100% (pure) isooctane; an octane number of 0 is assigned to 100% heptane.) Most regular unleaded petrol in the UK has an octane number of 95; 'super-unleaded' (high performance) petrol typically has an octane number of 98.

'Isooctane' is a common (or 'trivial') name. The IUPAC accepted name for isooctane is 2,2,4-trimethylpentane:

4 Combustion is the generic name given to any chemical reaction where a compound reacts with oxygen to produce carbon dioxide, water, and energy. When a substance undergoes combustion we often describe it by saying that the substance burns.

BOX 5.3 **Continued**

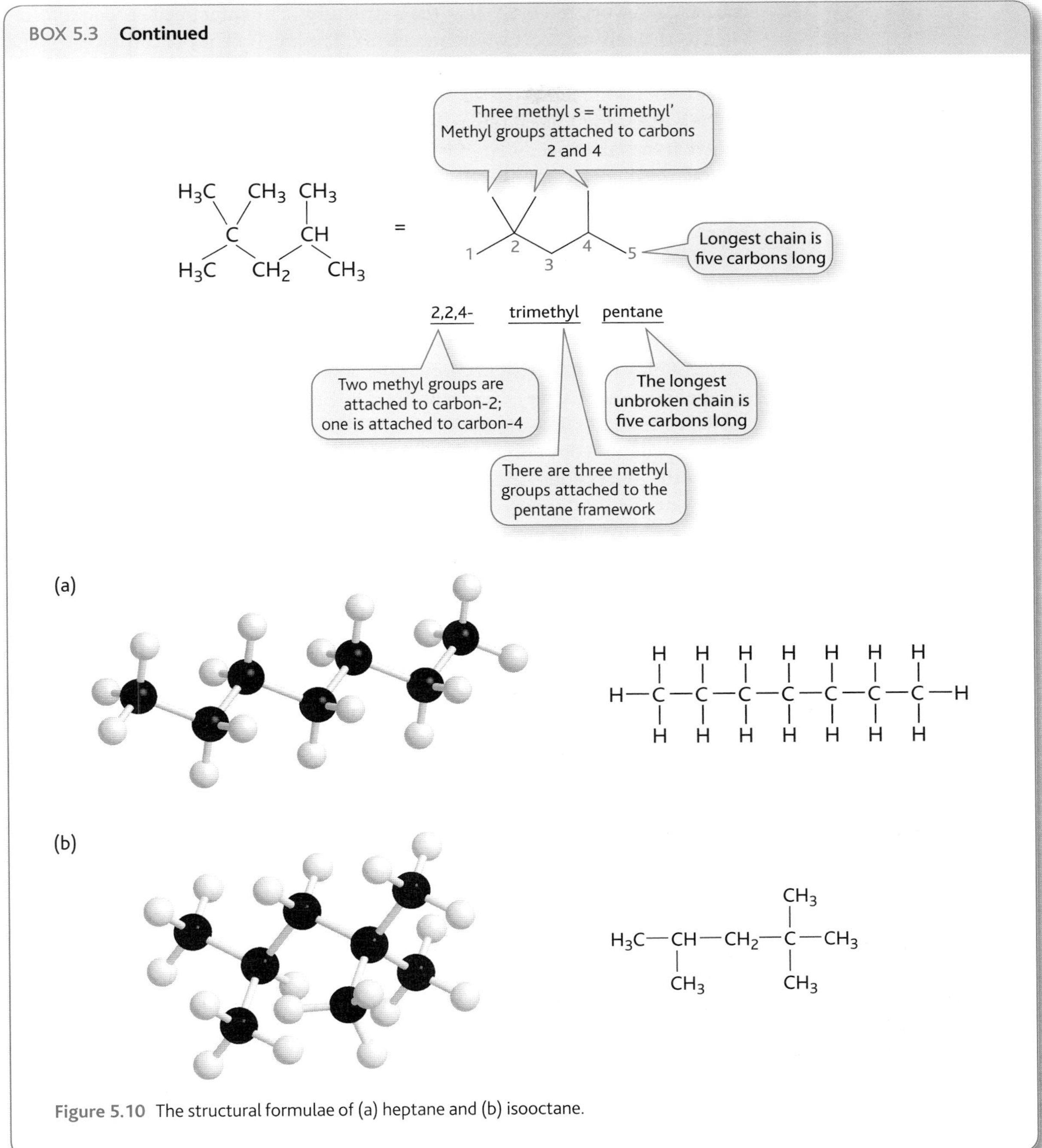

Figure 5.10 The structural formulae of (a) heptane and (b) isooctane.

5.3 Functional groups within the carbon framework

The chemical stability of the alkanes explains why their structure and composition is utilized by the framework of many organic compounds. Our skeleton would be no use to us if it wasn't robust; similarly, organic compounds require a stable, robust framework to ensure that their structures are maintained–a function fulfilled by their alkane-like carbon framework. The high valency of carbon offers the potential for the attachment of a diverse range of atoms (just as the skeleton offers a framework upon which muscle, skin, internal organs, etc., can be attached), while the carbon backbone itself acts as a durable, stable skeleton on which chemical construction can take place.

Functional groups are the chemical components that are added to the simple skeleton of an organic compound to generate chemical diversity and functionality. Functional groups add character to the 'blank canvas' of the alkane-like skeleton of an organic compound just as an artist uses different coloured paints to enliven and add character to a blank canvas. In effect, the different functional groups represent the components of a chemical 'toolbox' (just as the artist's paints form a palette). Different functional groups have different characteristics, and confer upon the recipient molecule different properties. By combining different components from the chemical toolbox, we generate molecules with contrasting properties.

We see on p. 123 how organic compounds that share the same functional group are grouped into common families. For example, those organic compounds possessing the –OH group are all grouped into a single family called the **alcohols**. Because a functional group confers specific physical and chemical properties upon the molecule to which it is attached, those molecules belonging to the same family of organic compounds (and, hence, sharing the same functional group) generally exhibit very similar physical and chemical properties.

- ***Functional groups confer specific physical and chemical properties upon the molecule to which they are attached.***
- ***Molecules possessing the same functional group belong to the same family of organic compound.***
- ***Because they share the same functional group, members of the same family of organic compound exhibit very similar physical and chemical properties.***

Most functional groups comprise a specific grouping of atoms. However, we begin our exploration of functional groups not by looking at molecules formed by adding atoms to the carbon skeleton, but by looking at the double bond, a functional group formed by taking atoms away.

The double bond

A double bond is formed when two pairs of electrons are shared by two atoms. One pair of electrons occupies a σ bonding orbital to form a σ bond; the second pair of electrons occupies a π bonding orbital to form a π bond. In essence, a double bond acts like a reserve of valence electrons. It contains two more electrons than the atoms joined by the bond actually need in order to be held together. A single σ bond is sufficient to hold two atoms together; the additional π bond in a double bond acts as reinforcement, strengthening the linkage between the two atoms. Therefore, the π bond can be dispensed with, freeing up two valence electrons, without disrupting the link between the two atoms.

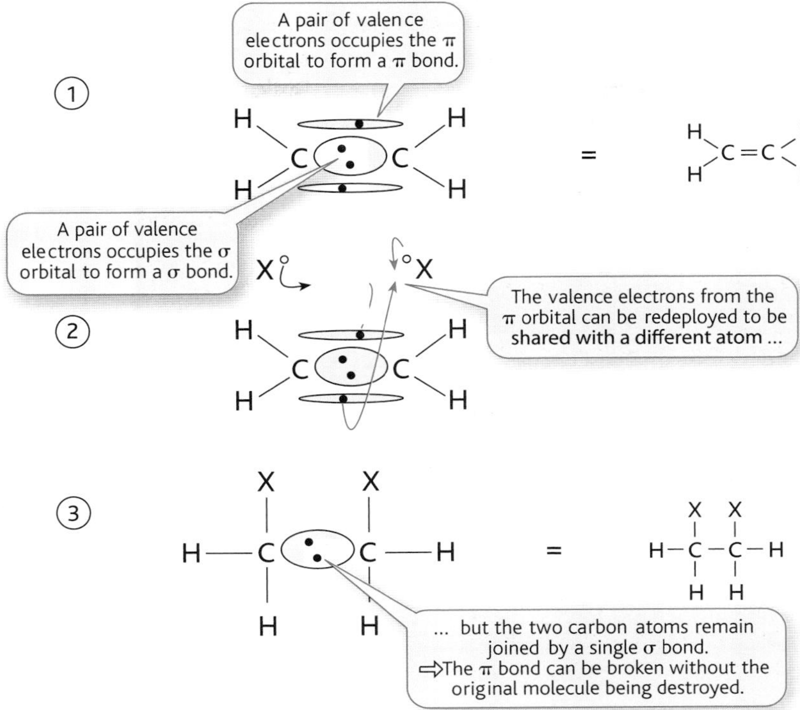

Figure 5.11 A pair of valence electrons occupying a π bond can be redeployed to occupy a different molecular orbital and, hence, form a new σ bond. Here we see a π bond joining two carbon atoms being broken to form two new σ bonds. However, the two carbon atoms remain joined by an existing σ bond.

It is this 'reserve' of valence electrons that provides those organic compounds which possess a double bond with their source of chemical reactivity. Chemical reactions occur merely by the redistribution of valence electrons. Figure 5.11 shows how the pair of valence electrons occupying a π bond can be redeployed to occupy a different molecular orbital and, hence, form a new σ bond. By contrast, the valence electrons occupying the σ bonding orbital remain in the same location, leaving the σ bond intact. As a result, the pair of atoms that was originally joined by the double bond remains connected by the (single) σ bond.

Double bonds occur in a wide array of compounds and can join atoms of any elements that have a valency ≥ 2. However, for now, let us focus on double bonds occurring between two carbon atoms.

We call the reaction depicted in Figure 5.11 an **addition** reaction. We learn more about different types of chemical reactions involving organic compounds in Chapters 17 and 18.

The alkenes: hydrocarbons with a double carbon–carbon bond

We see in Section 5.2 how the alkanes, alkenes, and alkynes are three groups of compound that belong to the hydrocarbon family, a family of organic compounds that consist solely of carbon and hydrogen atoms. The alkenes are differentiated from the alkanes by their possession of at least one double **carbon–carbon** bond.

The simplest alkene is ethene, C_2H_4, which comprises two CH_2 groups joined by a double carbon–carbon bond, (3).

The naming of alkenes uses the same convention as that for the alkanes (p. 128), comprising two parts.

H₂C=CH₂ structure: H, H, C=C, H, H

(3)

- The first part tells us the number of carbon atoms forming the longest unbroken chain in the compound.

- The suffix tells us whether the compound is saturated or unsaturated. The suffix '-ene' tells us that the compound is unsaturated and contains at least one double bond.

The presence of the double bond makes alkenes more chemically reactive than their saturated cousins, the alkanes. We explore some chemical reactions that are characteristic of the alkenes in Chapter 17.

Unlike the alkanes, several alkenes have important biological roles, acting as hormones in plants and animals, as explained in Box 5.4. Also, several plant-produced alkenes have aromas with which we're very familiar: geraniol (4) gives roses their delicate aroma, while limonene (5) gives oranges and lemons their refreshing citrus scent. (These alkenes are now mass produced synthetically at industrial chemical plants. Therefore, while a citrus-scented air freshener may smell to us of fresh lemons, the source of the smell is likely to be a synthetic copy of the naturally occurring compound.)

OH

(4)

(5)

Physical properties of alkenes

We see on p. 132 how all three members of the hydrocarbon family share similar physical properties, because they all comprise atoms of just two elements, carbon and hydrogen. The alkenes, like the alkanes, are non-polar molecules and so experience only dispersion forces. However, Figure 5.12 shows how alkenes have two fewer electrons than those alkanes possessing a corresponding number of carbon atoms. As a result, an alkene experiences weaker dispersion forces than its equivalent alkane. Consequently, the melting and boiling points of the alkenes are a little lower than those of the corresponding alkanes (i.e. those with the same length carbon chain) as illustrated in Table 5.3.

Self-check 5.4 Look at Table 5.3 and find the boiling points of pentane and ethene. Which has the lower boiling point, and why?

BOX 5.4 **Savour the flavour...**

The development of fruit typically follows a two-step process: the growth of the fruit to its full size, and subsequent ripening. Ripening encompasses a range of changes in the fruit, including softening, changes in colour, and changes in sugar content (which leads to the development of flavour). Ripening has two drawbacks, however. While ripening makes fruit edible, it also signals the start of a gradual breakdown of the fruit towards decay and rot, meaning losses for fruit producers. A ripe fruit is also more prone to bruising during transit, whereas fruit producers want their stock to reach the consumer in tip-top condition.

Ripening is naturally regulated by the alkene ethene, C_2H_4 (more commonly known as ethylene). Ethene acts as a plant hormone: it is released by the plant, and triggers the onset of the molecular and physiological processes that characterize ripening. To many fruit producers, however, the notion of *delaying* ripening is an appealing one. Not only does it minimize losses through spoilage, but it also minimizes damage during transit–meaning greater yields, and greater profits. But can you switch off the effect of ethene? The answer seems to be: yes.

Genetic engineering is now being used to delay fruit ripening by preventing the production of ethene by the plant. (Genetic engineering is used to block the action of enzymes which are essential to the production of ethene by the plant.) With ethene's ripening signal turned off, fruit grows to its full size but fails to ripen, leaving it in an ideal condition for transport and extending its shelf life. Once the fruit has reached its destination, ethene can be supplied artificially to trigger the ripening process, leaving the fruit ready for the consumer.

2-carbon alkane

$$\begin{array}{ccccc} & H & & H & \\ & | & & | & \\ H- & C & - & C & -H \\ & | & & | & \\ & H & & H & \end{array}$$

7 single bonds
= 14 valence electrons

2-carbon alkene

$$H_2C{=}CH_2$$

4 single bonds + 1 double bond
= 12 valence electrons

Figure 5.12 Alkenes have two fewer valence electrons than those alkanes possessing a corresponding number of carbon atoms.

Self-check 5.5 Compounds (a), (b), and (c) have the structures shown below. The possible boiling points of these compounds are –42 °C, –6 °C, or –0.5 °C. If we know that structure (c) has a boiling point of –0.5 °C, what are the boiling points of structures (a) and (b)?

(a) $H_3C{-}CH_2{-}CH_3$

(b) $H_3C{-}CH{=}CH{-}CH_3$

(c) $H_3C{-}CH_2{-}CH_2{-}CH_3$

The alkenes also lack an electronegative atom (oxygen, fluorine, or nitrogen) to enable them to hydrogen-bond with water. Therefore, the alkenes share the other hydrocarbons' insolubility in water.

The presence of the double bond in alkenes makes them more chemically reactive than their corresponding alkanes. We note above how the double bond acts as a 'reserve' of valence electrons–electrons that can be used to participate in new covalent bonds with other atoms. We learn much more about the way the valence electrons within double bonds are used during chemical reactions in Chapter 17, when we explore chemical reaction mechanisms.

- ***The presence of a double bond makes an alkene more chemically reactive than the corresponding alkane.***

Double bonds are perhaps the most common functional group in biological molecules. They do not occur solely between pairs of carbon atoms in alkenes, but

Table 5.3 The boiling points of alkanes and alkenes comprising equivalent carbon frameworks.

Number of carbon atoms	Alkane boiling point (°C)	Alkene boiling point (°C)
2	–89	–104
3	–42	–47
4	–0.5	–6
5	36	30

also between carbon and oxygen atoms as a component of other functional groups, as we see in Section 5.4. By contrast, triple bonds are seen only rarely in biological molecules. Despite their limited presence in biological systems let us briefly consider the final hydrocarbon family, the alkynes, which are characterized by their possessing a C≡C bond.

The alkynes: hydrocarbons with a triple carbon–carbon bond

An alkyne is a hydrocarbon in which one or more pairs of carbon atoms are joined by a triple bond. The high-energy triple bond confers upon the alkynes much lower stability than the alkanes, but far greater reactivity, like the double carbon–carbon bond in alkenes.

Alkynes are identified by the suffix -yne, and their names use the same convention as that for the alkanes and alkenes, comprising two parts.

- The first part tells us the number of carbon atoms forming the longest unbroken chain in the compound.
- The suffix tells us whether the compound is saturated or unsaturated. The suffix '-yne' tells us that the compound is unsaturated and contains at least one triple bond.

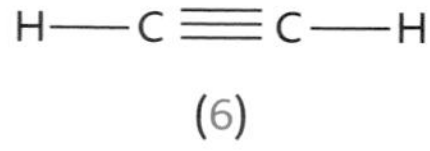

(6)

The simplest alkyne is ethyne, C_2H_2 (6)

The alkynes aren't as widespread in nature as the alkenes. Like both the alkanes and alkenes, however, the alkynes burn readily in air, leading to some useful applications in everyday life. Ethyne is also known by its common name of acetylene. This gives a clue as to one of ethyne's most common uses: in oxyacetylene (welding) torches.

A mixture of ethyne and oxygen gases burns to generate a flame with a temperature of roughly 3300 °C–one of the highest temperatures achieved from a mixture of combustible gases. A flame at this temperature can be used to cut through metals quite literally like a hot knife through butter.

Self-check 5.6 The following hydrocarbons contain how many carbon atoms, and are members of which hydrocarbon group?

(a) hexene (b) ethyne (c) decane

5.4 Adding functional groups to the carbon framework

Let's now find out more about the principal functional groups that occur in biological molecules, and how they can transform the rather unreactive alkane-like hydrocarbon skeleton into much more dynamic structures.

Different organic compounds possess different degrees of chemical sophistication, depending on the functional groups that are added to their basic alkane-like skeletons, just as a computer can be upgraded by installing new software to add extra functionality and sophistication to the basic operating system.

We see in Section 5.2 how the carbon atoms, which are joined by a series of single covalent bonds to form the backbone of an alkane molecule, satisfy their valency of four by sharing additional pairs of valence electrons with hydrogen atoms. Figure 5.13

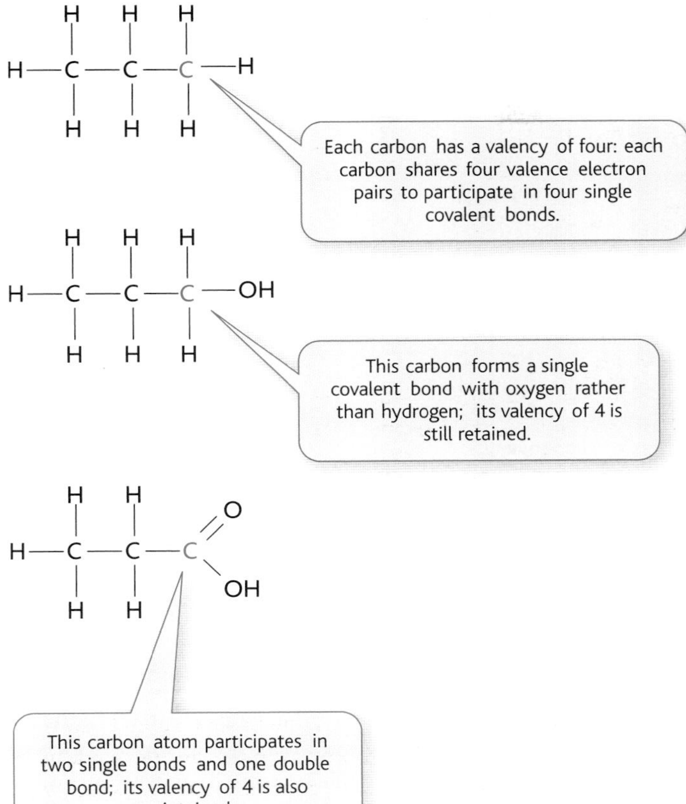

Each carbon has a valency of four: each carbon shares four valence electron pairs to participate in four single covalent bonds.
This carbon forms a single covalent bond with oxygen rather than hydrogen; its valency of 4 is still retained.
This carbon atom participates in two single bonds and one double bond; its valency of 4 is also maintained.

Figure 5.13 Carbon atoms that form the backbone of more complicated organic molecules than the alkanes satisfy their valency of four by sharing their valence electrons with both hydrogen atoms and atoms that form part of functional groups.

shows how the carbon atoms that form the backbone of more complicated organic compounds satisfy their valency of four by sharing their valence electrons with both hydrogen atoms *and* functional groups to create a diverse 'mix-and-match' range of molecules.

The alkane-like skeleton that forms the backbone of many organic compounds, and to which functional groups attach, is what we call an **alkyl group**. Let's now look at alkyl groups in a bit more detail before we consider the functional groups themselves.

Alkyl groups

Alkyl groups are hydrocarbon fragments–they consist solely of carbon and hydrogen atoms. Alkyl groups resemble alkane molecules, but with a hydrogen atom removed from a terminal carbon atom (the carbon atom at the beginning or end of a carbon chain) to enable the molecule to join to something else, as illustrated in Figure 5.14. Look at this figure, and notice how a majority of the alkyl group is identical in structure and composition to the alkane shown next to it; the only difference is that it lacks a hydrogen atom–creating both space on the framework for another atom (or functional group) to attach, and the necessary electron to facilitate the formation of the necessary bond.

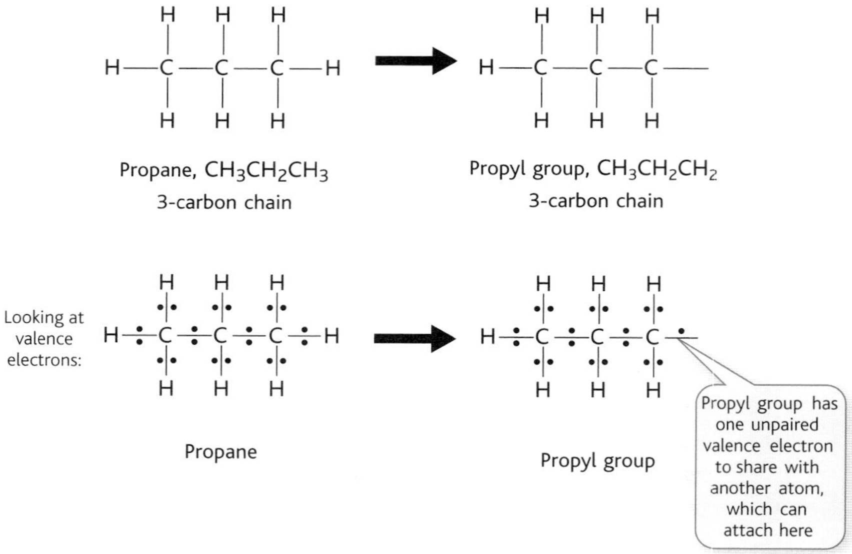

Propane, $CH_3CH_2CH_3$
3-carbon chain

Propyl group, $CH_3CH_2CH_2$
3-carbon chain

Looking at valence electrons:

Propane

Propyl group

Propyl group has one unpaired valence electron to share with another atom, which can attach here

Figure 5.14 Alkyl groups resemble alkane molecules, but with a hydrogen atom removed from a terminal carbon atom. In this figure we see how the propyl group is generated by the removal of a hydrogen atom from propane.

Don't mistake the -yl ending of a terminal alkyl group for the -yne suffix of an alkyne.

The alkyl group formed from the removal of an H atom from methane is called a methyl group ($-CH_3$). Other common alkyl groups are shown in Table 5.4. The alkyl groups are named by removing the '-ane' suffix from the name of the parent alkane, and replacing it with the suffix '-yl'. When we see the suffix '-yl' we know that we are looking at a group of atoms which is able to bond to another atom to form a new, larger molecule.

Self-check 5.7 What is the name and structural formula of the alkyl group formed from the alkane heptane?

The symbol R is often used within structural formulae to represent *any* alkyl group. While each alkyl group has a specific name (methyl, butyl, etc.), R is used within structural formulae to indicate that *any* alkyl group could be present and still form a chemically viable compound. So, for example, ROH could be used to represent

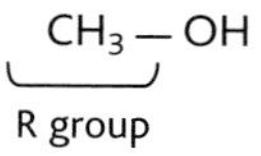

$$\underbrace{CH_3}_{\text{R group}} - OH$$

Table 5.4 Some common alkyl groups.

Alkyl group name	Parent alkane	Structural formula
Methyl	Methane	$-CH_3$
Ethyl	Ethane	$-CH_2CH_3$
Propyl	Propane	$-CH_2CH_2CH_3$
Butyl	Butane	$-CH_2CH_2CH_2CH_3$

but could equally well represent

$$\underbrace{CH_3CH_2CH_2}_{\text{R group}} \text{—— OH}$$

We say that a molecular formula such as ROH, which uses the general symbol R, represents the **generic formula** for a group of compounds. It depicts the atoms that are common to all members of the group (in this instance, the -OH group), and the atoms that vary between members of the group (in this instance, the alkyl group, which could be one of a number of different groups of atoms).

- ***An alkyl group is a fragment of an alkane.***
- ***The symbol R is used to denote a generic alkyl group–that is, we can interpret it as representing any alkyl group of our choosing.***

Alkyl groups often have functional groups added to them to generate molecules belonging to distinct families of organic compound. For example, we see above how an –OH can be added to a methyl group to generate the alcohol methanol, CH_3–OH. However, despite not being chemically reactive themselves, alkyl groups are more than mere anchor sites for more chemically exciting groups of atoms. Indeed, methyl groups, in particular, play a vital biological role in the control of gene expression, as explained in Box 5.5.

BOX 5.5 Methyl groups: laying down a new blueprint for life?

The chemical simplicity of the methyl group–just one carbon atom and three hydrogen atoms–belies the central role that it plays both in the regulation of gene expression, and in the modification of the genetic information that is passed on from generation to generation.

The genetic information stored in an organism's genome directs the way in which that organism develops and behaves–that is, it determines an organism's phenotype. Traditionally, the genetic information was thought to be encoded solely in the sequence of bases from which an organism's DNA is formed (that is, its genome sequence). However, the field of **epigenetics** opens up strong challenges to this traditional view of the DNA sequence being the sole dictator of phenotype.

The ascribing of a single, robust definition to the term epigenetics remains controversial: it is a term that tends to mean different things to different people. However, in broad terms, epigenetics refers to the way that the behaviour of the genome–and the phenotype it determines–can be modified not by changes in DNA sequence, but by chemical modifications either to the biochemical structure of the DNA itself, or to the special group of proteins–called histones–that associate with DNA to form chromatin within the cell nucleus. Further, such epigenetic modifications seem not only to affect the way the genome is expressed in an existing cell (or population of cells), but also in later generations, when the existing cells replicate. That is, epigenetic modifications appear to be heritable–they can be passed on from generation to generation.

So what has epigenetics to do with the lowly methyl group? It is now clear that the addition of methyl groups to both DNA and histone proteins is a vital part of the epigenetic mechanisms that operate in the cell. For example, the methylation (addition of a methyl group) to the cytosine residue of CG or CNG DNA sequences (where N represents any of the four bases) appears to result in the repression of transcription of the stretch of DNA along which the methylation has occurred. In other words, the act of adding a methyl group switches off genes, and prevents them from being transcribed.

The methylation of histone proteins seems to be even more intriguing (and challenging). While it was originally thought that methylated histone proteins resulted in the repression of

BOX 5.5 Continued

gene activity, it now seems to be the case that these modified proteins can both activate *or* repress transcription, depending on the biological context. (For example, the triple methylation of lysine 4 in the histone called histone H3 has been linked to the induction of transcription, whereas the triple methylation of lysine 9 within the *same* histone protein has been linked to regions of DNA that are transcriptionally repressed.)

Methyl groups aren't the sole players in epigenetic modification: acetylation (the addition of an acetyl group) and ubiquitination (the addition of an ubiquitin protein) are among other types of chemical modification that help to mediate epigenetic changes to the genome.

However our understanding of the field of epigenetics evolves in the future it is clear that something as simple as the presence or absence of a methyl group has far-reaching consequences for the way genomes control biological phenotypes, to the extent that it is changing the way we think about DNA as the blueprint for life.

The aryl group: a special hydrocarbon group

An **aryl group** is a functional group derived from an aromatic hydrocarbon by the removal of one or more hydrogen atoms.

For example, benzene comprises an aromatic six-carbon ring in which delocalized electrons occupy a distinctive π orbital that follows the circumference of the ring structure (p. 67). Each carbon atom in benzene has one hydrogen atom attached, to give the structural formula shown in Figure 5.15.

We explore the concept of delocalization on p. 68.

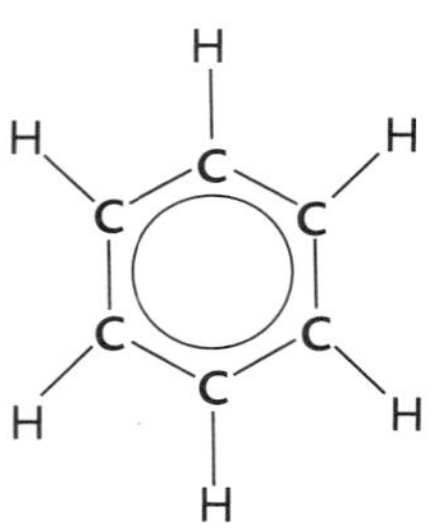

Figure 5.15 The structural formula of benzene.

Benzene forms an aryl group called a **phenyl group** when a hydrogen atom is removed from the benzene ring. The aryl group can then join to another atom or group of atoms. For example, the alcohol phenol comprises the phenyl radical attached to an -OH group, as depicted in Figure 5.16.

The group of atoms to which an aryl group attaches may be a small functional group, or may be part of a larger, more complicated molecule. Aryl groups occur in many biological molecules, including the amino acid phenylalanine, and the hormone adrenaline. Look at the structural formulae of these two molecules in Figure 5.17 and locate the aryl group in each molecule.

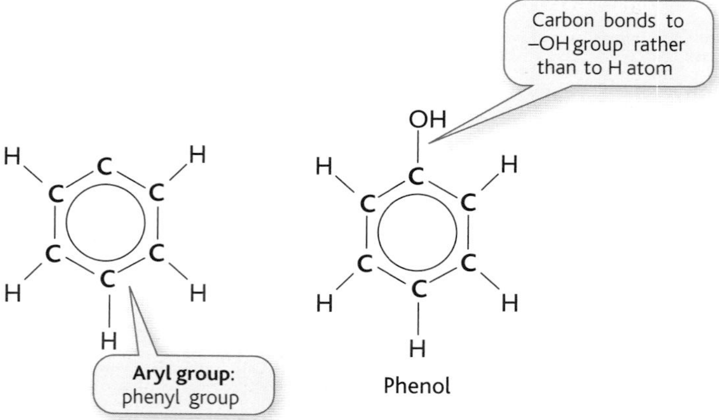

Figure 5.16 The alcohol phenol comprises the aryl group attached to a hydroxyl group.

Phenylalanine

Adrenaline

Figure 5.17 The amino acid phenylalanine and the hormone adrenaline both possess aryl groups.

The generic symbol R can also be taken to represent an aryl group as well as the alkyl groups that we've learned about above. Therefore, the generic formula ROH could equally well represent structure (7) as it could CH_3CH_2OH.

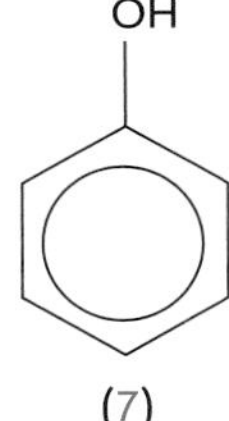

- *An aryl group is an aromatic alkyl group.*

Functional groups and the properties of organic compounds

Table 5.5 shows those functional groups that occur most often in organic compounds, and which we'll explore in more detail during Chapter 6. Look at the elements that form the functional groups shown in Table 5.5. What do you notice?

Hydrogen, carbon, oxygen, and nitrogen make up nearly 97% of the body mass as a whole (see Section 2.1). Therefore it's little surprise that these four elements are the key components of a majority of those functional groups that add chemical spice to the somewhat bland alkane-like hydrocarbon skeleton of many organic compounds[5].

We note in Section 5.3 that functional groups confer upon molecules specific physical and chemical properties. The alkane-like framework of many organic compounds is chemically unreactive and non-polar; functional groups modify these characteristics. Oxygen and nitrogen are highly electronegative atoms; they can induce polarity in parts of a molecule surrounding them (or in adjacent molecules), and both can participate in hydrogen bonds.

The two characteristics conferred by oxygen and nitrogen–the polarization of a molecule, and its ability to participate in hydrogen bonds–have two key effects on the physical properties of that molecule.

Effect 1: We see an increase in melting and boiling points

We see in Chapter 4 how a compound's melting and boiling points are influenced by the magnitude of non-covalent forces that exist between molecules of that

5 Functional groups are another example of our key theme of complexity arising from simplicity: a majority of biologically important functional groups arise from a pool of just four elements: C, H, O, and N.

Table 5.5 The key functional groups that occur in organic compounds.

Functional group	Structural formula	Class of organic compound
Double bond	$>C=C<$	Various unsaturated
Hydroxyl	—OH	Alcohols
Alkoxy (ether)	—O—R	Ethers
Carbonyl	—C(=O)H	Aldehydes
"	—C(=O)R	Ketones
Carboxyl	—C(=O)OH	Carboxylic acids
"	—C(=O)O—R	Esters
Amino	$—NH_2$	Amines
Amide	$—C(=O)NH_2$	Amides
Halo	—X where X = F, Br, Cl, or I	Halogeno-hydrocarbons
Thiol	—SH	Thiols

compound. If extensive non-covalent forces exist, more energy is required to separate the molecules, meaning the compound has a high melting and boiling point. If few non-covalent forces exist, little energy is required to separate the molecules, and the compound has a low melting and boiling point. The dipolar interactions experienced by a polarized molecule, and the hydrogen bonding that may be possible, both contribute to greater non-covalent interactions in a molecule possessing an electronegative atom, compared to a molecule without an electronegative atom. Greater non-covalent interactions equate to elevated melting and boiling points.

Effect 2: We see an increased solubility in polar solvents, like water

Water is a polar solvent. We see in Chapter 4 how the solubility of a covalent compound in water depends on whether the compound can form hydrogen bonds with water. If it *can* form hydrogen bonds with water, then its solubility in water is

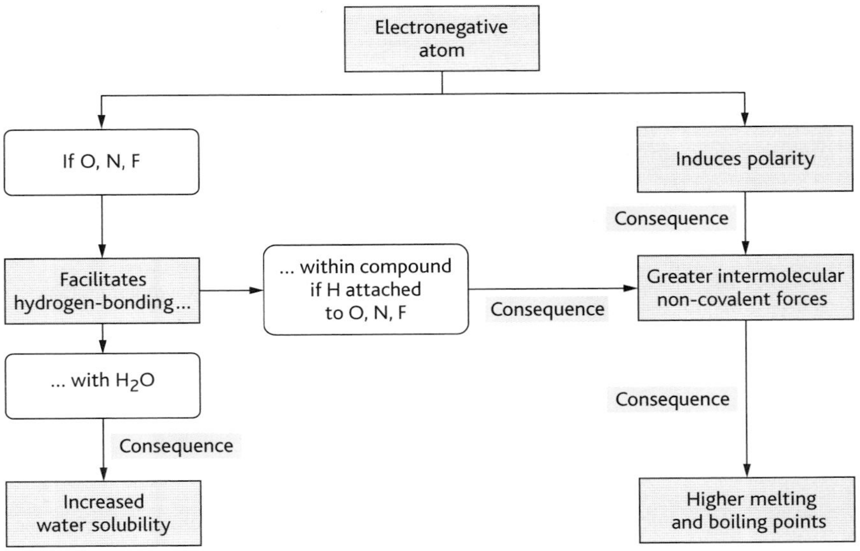

Figure 5.18 A summary of how an electronegative atom can influence the physical properties of a molecule of which it is part.

enhanced; if it *can't* form hydrogen bonds, it is insoluble in water. The presence of an oxygen or nitrogen atom in a covalent compound increases the solubility of that compound in water by promoting hydrogen bond formation between molecules of the compound, and molecules of water.

Look at Figure 5.18, which summarizes how an electronegative atom can influence the physical properties of a molecule; notice how it can have more than one effect depending on its chemical identity, and the atoms to which it is attached.

Look out for oxygen and nitrogen as we explore the different functional groups in more detail in Chapter 6, and particularly note how the electronegativity of these two elements influences the physical properties of the organic compounds that we consider.

- ***Functional groups often increase melting and boiling points, and increase solubility, relative to a corresponding hydrocarbon with no functional groups.***

Functional groups versus the carbon framework: a balancing act

The ability of a functional group to modify the inherently non-polar character of the carbon framework of organic compounds (i.e. its insolubility in water, and weak non-covalent interactions leading to low melting and boiling points) depends on the *size* of the carbon framework. The non-polar nature of a small carbon framework (typically one to four carbon atoms in size) can be offset by the addition of an electronegative functional group, such that solubility is increased and melting and boiling points are elevated. As the carbon framework increases in size, however (typically beyond a four-carbon chain), its stubbornly non-polar nature begins to outweigh the effect of the electronegative functional group, such that solubility in water in particular is far less pronounced.

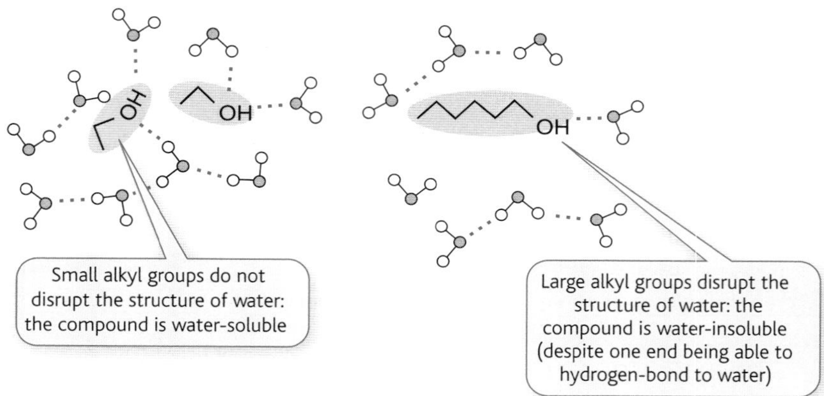
OH
OH
OH
Small alkyl groups do not disrupt the structure of water: the compound is water-soluble
Large alkyl groups disrupt the structure of water: the compound is water-insoluble (despite one end being able to hydrogen-bond to water)

Figure 5.19 There comes a chain length above which the hydrophobic nature of an alkyl group outweighs the hydrophilic nature of a hydroxyl group because it disrupts the hydrogen bonding in water too greatly. Alcohols whose alkyl groups are above a certain size are therefore insoluble in water.

For example, while those alcohols (Section 6.1) possessing a carbon framework of one to four carbon atoms are very soluble in water, alcohols whose carbon frameworks comprise more than four carbon atoms are much less water-soluble. Figure 5.19 shows how there comes a size beyond which the inability of the alkyl group (the carbon framework) to hydrogen-bond with water disrupts the structure of water so much that the alcohol molecule and water can no longer mix; it is at this point that the alcohol becomes partially or totally water-insoluble.

Therefore we face a balancing act: when the carbon framework of an organic compound is small, its physical properties will be most strongly influenced by its functional group; when the carbon framework is larger, its physical properties will be most strongly influenced by the carbon framework itself, e.g. the alkyl or aryl group.

The insolubility of some organic compounds is central to their biological function. For example, many biologically important fats and lipids depend on their insolubility in water to function correctly. It is the insolubility of lipids that enables them to aggregate into layers (or **membranes**), which divide cells up into different compartments or organelles (a feature which is crucial to the cells' existence).

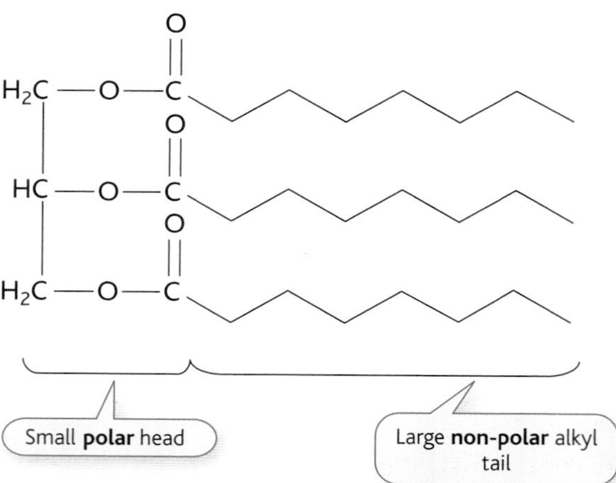
H_2C—O—C
O
HC—O—C
O
H_2C—O—C
O
Small **polar** head
Large **non-polar** alkyl tail

Figure 5.20 A generalized structure of a fat molecule, showing its polar head region, and large non-polar alkyl tail. The large, non-polar tail outweighs the hydrophilic tendencies of the polar head region, making a fat molecule insoluble in water.

If lipids were soluble in water they would be unable to aggregate in this manner, and our cells would have no way of being compartmentalized.

Many fats and lipids have long alkyl group tails attached to small polar heads, as illustrated in Figure 5.20. The large non-polar tails greatly outweigh the solubilizing influence of the polar heads to render such molecules insoluble in water, and ensure that the fats and lipids can function as intended.

In the next chapter we look at the key families of organic compound, and the functional groups that they possess. To make the comparison of different organic compounds easier, when we learn about the solubility of different organic compounds in the next chapter, we consider the behaviour of compounds with small alkyl groups (carbon frameworks), in which the functional group holds more sway over the physical properties of the compound than does the alkyl group.

Check your understanding

To check that you've mastered the key concepts presented in this chapter, review the checklist of key concepts below, and attempt the multiple-choice questions available in the book's Online Resource Centre at **http://www.oxfordtextbooks.co.uk/orc/crowe2e/**.

Checklist of key concepts

Organic chemistry

- Organic chemistry is the chemistry of carbon and carbon-containing compounds
- Carbon has an important structural role in organic compounds because of its high valency
- Organic compounds have two key physical features: a simple framework, to which are attached more specialized groups of atoms
- These specialized groups of atoms are called functional groups
- Organic compounds are grouped into families according to the functional group that they possess
- Members of a family possess the same functional group, but different carbon frameworks

The framework of organic compounds

- The framework of an organic compound comprises just carbon and hydrogen
- Organic compounds that contain just carbon and hydrogen are what we call the hydrocarbons
- There are three different types of hydrocarbon
    - An alkane contains only single carbon–carbon bonds
    - An alkene contains at least one double carbon–carbon bond
    - An alkyne contains at least one triple carbon–carbon bond
- Saturated compounds contain only single covalent bonds
- Unsaturated compounds contain multiple covalent bonds
- A molecule's structural formula tells us how the different atoms in the molecule are connected to one another
- The simplified structural formula shows the bonds linking atoms, but does not show the carbon and hydrogen atoms that comprise the framework itself
- The framework of organic compounds can adopt one of three different shapes: straight-chain (linear), branched-chain, or cyclic
- The hydrocarbons are non-polar and are insoluble in water
- They experience few intermolecular forces and have low melting and boiling points
- The alkanes are chemically unreactive

Adding functional groups to the carbon framework

- Functional groups enhance the physical and chemical properties of organic compounds
- Members of the same family of organic compound exhibit similar physical and chemical properties
- The double bond acts as a reserve of valence electrons, and increases the chemical reactivity of the molecule that possesses it
- An alkyl group is identical to an alkane from which a hydrogen atom has been removed
- We can represent an alkyl group with the generic symbol R
- An aryl group is derived from an aromatic hydrocarbon by the removal of one or more hydrogen atoms
- Most biological functional groups contain either oxygen or nitrogen
- Functional groups may increase the solubility and melting and boiling points of the molecules to which they are attached
- Alkyl groups above a certain size override the effect of an attached functional group, causing a relative decrease in solubility, and melting and boiling points.

Organic compounds 2: adding function to the framework of life

INTRODUCTION

What do computers and organic compounds have in common? The answer: they both comprise a simple, stable 'skeleton', with additional features that add variety and functionality. Every computer has the same simple skeleton: a hard drive, a keyboard, a processing unit, and an operating system. Beyond this, however, computers are far from identical: they feature different pieces of software and hardware that enable us to use them to fulfil many different functions–to key in and format reports, essays, and publications, to edit and print photographs, to build websites, and even to broadcast images and sound via the Internet.

Similarly, organic compounds possess a chemically simple framework that comprises only atoms of carbon and hydrogen. Different organic compounds are given their contrasting properties, however, by the range of functional groups that are attached to the simple framework. These functional groups mean that different organic compounds fulfil many different functions, just as we use computers with different hardware and software to fulfil different functions in everyday life.

In this chapter we explore the different functional groups that give organic compounds their chemical variety. We discover the components of some of the most biologically important functional groups, and ask how functional groups influence the physical and chemical properties of organic compounds.

6.1 Organic compounds with oxygen-based functional groups

In this section we focus on the four key functional groups that contain oxygen, and the six families of organic compound that feature these functional groups.

The alcohols: the hydroxyl group

Generic formula: ROH

The simplest oxygen-containing functional group is the **hydroxyl** group: –O–H

The presence of the hydroxyl group is the defining characteristic of a particular class of organic compound with which we're all already familiar: the **alcohols**.

The alcohols comprise a hydrocarbon skeleton to which one or more hydroxyl groups are attached. The hydroxyl group attaches to the hydrocarbon skeleton via a single covalent bond between the hydroxyl group's oxygen atom and a carbon atom:

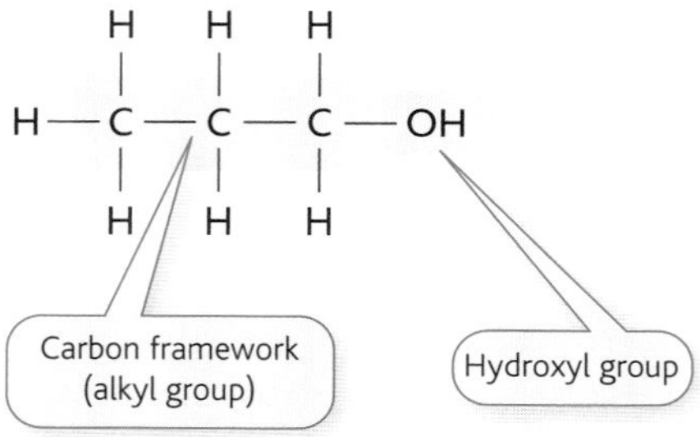

```
     H   H   H
     |   |   |
  H—C—C—C—OH
     |   |   |
     H   H   H
```

Carbon framework (alkyl group)

Hydroxyl group

The simplest alcohol is **methanol**, CH_3OH. When we use the word 'alcohol' in everyday use (usually in the context of alcoholic drinks), we are actually referring to **ethanol**, CH_3CH_2OH. Ethanol can have a powerful effect on our body, not least to lift inhibitions (which can be the cause of a spectrum of behaviours, from amorous advances, to unprovoked violence and depression). Look at Box 6.1, p. 155 to find out more about ethanol and its interaction with the human body.

- ***Alcohols possess the –OH functional group.***

Naming the alcohols

The alcohols are named by removing the '-e' from the parent alkane name, and adding the suffix '-ol'.

CH_3CH_2OH has a two-carbon framework, and is based on the parent alkane ethane. To name this alcohol:

1. remove -e from the parent alkane name:

   ethane → ethan

2. add the suffix '-ol':

   ethan → ethan**ol**.

Self-check 6.1 What is the name of the alcohol with the following structural formula?

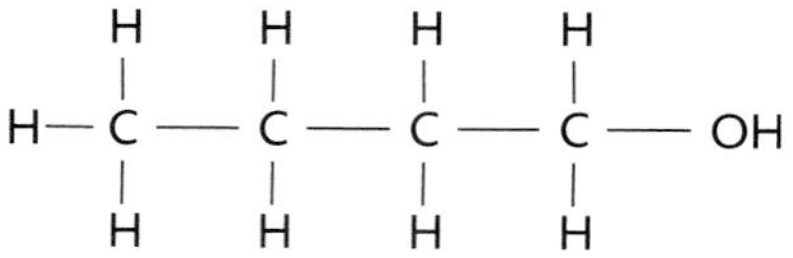

```
     H   H   H   H
     |   |   |   |
  H—C—C—C—C—OH
     |   |   |   |
     H   H   H   H
```

A vitally important alcohol in biological systems is glycerol, a three-carbon alcohol which possesses three hydroxyl groups, as depicted in Figure 6.1. Just as

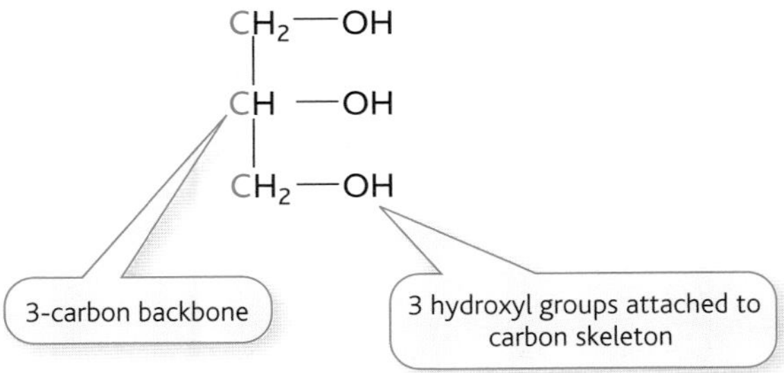

Figure 6.1 The structure of glycerol. Glycerol comprises three hydroxyl groups attached to a three-carbon backbone.

an alkyl skeleton forms the framework for a diverse range of biological molecules, so glycerol forms the framework for a particular class of fats and oils known as **triacylglycerols**. We find out more about the triacylglycerols, and other fats and oils, in Section 7.3.

Read Box 6.1 to find out how our liver protects our bodies from the potentially harmful effects of alcohol consumption.

Physical properties of alcohols

Table 6.1 shows the physical property checklist for the alcohols. The presence of a highly electronegative oxygen atom in the hydroxyl functional group means that the alcohols are **polar** molecules, as illustrated in Figure 6.2, and experience dipolar interactions, as well as dispersion forces.

Table 6.1 Physical property checklist for the alcohols.

Do alcohols experience...		Consequence
Dispersion forces?	✓	High melting and boiling points
Dipolar interactions?	✓	
Hydrogen bonding between molecules of same compound?	✓	
Hydrogen bonding with water?	✓	Soluble in water, if small

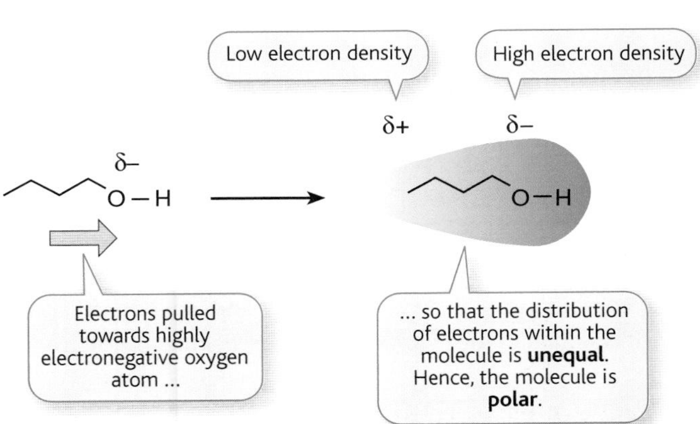

Figure 6.2 The alcohols are polar molecules. Electrons are withdrawn towards the highly electronegative oxygen atom, distorting the distribution of electrons throughout the molecule as a whole. Consequently, the molecule is polar.

Table 6.2 The boiling points of alkanes and alcohols of varying carbon chain length.

Number of C atoms in carbon backbone	Boiling point (°C)	
	Alkane	Alcohol
1	−164	65
2	−89	78
3	−42	82
4	−0.5	118

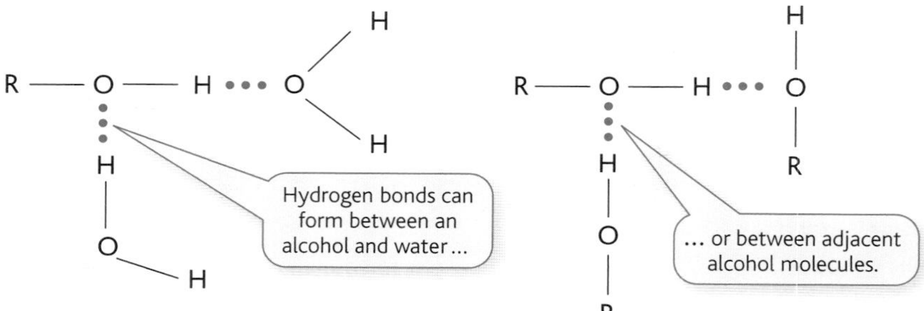

Figure 6.3 Molecules of an alcohol can form hydrogen bonds both with water, and with each other.

The combination of a hydrogen atom joined to a highly electronegative oxygen atom also enables the hydroxyl group to participate in hydrogen bonds. Look at Figure 6.3 to see how molecules of an alcohol can form hydrogen bonds both with water, and with each other.

As a result of their polarity *and* their ability to form hydrogen bonds with one another, the alcohols experience very strong non-covalent forces and, consequently, exhibit higher melting and boiling points than alkanes of corresponding carbon chain length. Look at Table 6.2, which shows the boiling points of alkanes and alcohols possessing carbon skeletons of varying length. Notice how the boiling points of the alcohols are consistently higher than those of the corresponding alkanes, as a result of the more extensive non-covalent forces that operate between alcohol molecules.

Their ability to form hydrogen bonds with water molecules also makes short-chain alcohols soluble in water[1]. Contrast this with the non-polar alkanes, which, as we see on p. 133, are insoluble in water.

The ethers: the alkoxy group

Generic formula: ROR′

The **alkoxy group** is a combination of an alkyl group and an oxygen atom (**alk**yl + **oxy**gen = alkoxy). It is similar to the hydroxyl group, except that the hydrogen

1 A word of caution here: we see on p. 147 how an *increase* in the size of a non-polar alkyl group causes a *decrease* in the solubility of the compound of which it is part. Only **small** alcohols are soluble in water.

BOX 6.1 **The liver: protection from the 'demon drink'**

The liver is responsible for metabolizing alcohol (or, more specifically, ethanol–the alcohol found in alcoholic drinks), breaking it down into substances that can be more easily removed from the body.

The breakdown of ethanol is a three-step process, as illustrated in Figure 6.4. Firstly ethanol is converted into ethanal (an aldehyde–see p. 157), which itself is converted into ethanoic acid (a carboxylic acid–see p. 163). Ethanoic acid can then be broken down into carbon dioxide and water.

Some people experience a lack of mental and physical coordination after consuming alcohol: elated mood twinned with dizziness, and a remarkable inability to walk in a straight line, despite their best efforts. It is not ethanol that causes these characteristics, however, but the ethanal by-product generated as the alcohol is metabolized. (This may explain the delay we experience between consuming a drink and feeling its effects.) Ethanol itself affects electrical activity in the brain, acting as both a stimulant and a depressant.

Both the ethanal and ethanoic acid intermediates formed during ethanol metabolism in the liver are damaging to it. This explains why prolonged heavy drinking can lead to cirrhosis of the liver, a disease in which the damage caused during alcohol metabolism results in the severe impairment of liver function.

It is critically important that the liver can get alcohol out of our system quickly despite the damaging side effects that the liver faces as a result of alcohol metabolism. If levels of ethanol in the bloodstream reach more than 500 mg per 100 ml we risk slipping into a coma, and possible death. (To put this amount in context, the legal blood alcohol limit in the UK, above which driving is illegal, is 80 mg per 100 ml.)

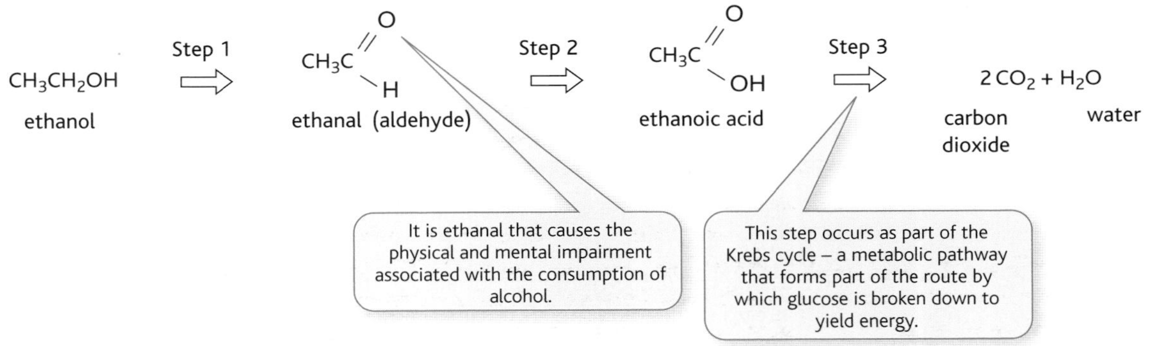

Figure 6.4 The breakdown of ethanol by the body is a three-step process.

atom is replaced with an alkyl group (such as the methyl or ethyl group that we encounter on p. 141):

Hydroxyl group	Alkoxy group
—O–H	—O–R

The alkoxy group attaches to the hydrocarbon skeleton via a single covalent bond linking the oxygen atom of the ether group with a carbon atom from the hydrocarbon skeleton, a situation analogous to the attachment of the hydroxyl group.

The family of organic compounds that is characterized by their possession of an alkoxy group is the **ethers**. In effect, ethers comprise two alkyl groups linked by a bridging oxygen atom, as illustrated in Figure 6.5.

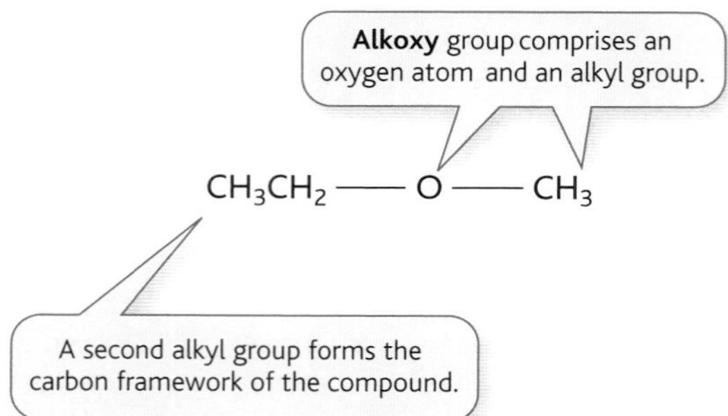

Figure 6.5 Ethers comprise two alkyl groups linked by a bridging oxygen atom.

Multiple alkyl groups can be represented in a general way by modifying the R symbol with the ′ ('prime') character. For example, if a compound contains two different alkyl groups, they can be denoted by the symbols R and R′, while three alkyl groups on a single molecule are denoted R, R′, and R″.

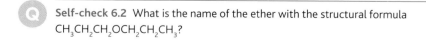

- *Ethers possess the –OR functional group.*

Naming the ethers

The ethers are named by listing the two alkyl groups in order of length, shortest first, with the following modifications:

- Removing the –yl suffix from the shortest alkyl group and replacing it with –oxy
- Using the name of the parent alkane that corresponds to the longest alkyl group, rather than the 'alkyl' version
- $CH_3OCH_2CH_3$ features the methyl ($-CH_3$) and ethyl ($-CH_2CH_3$) groups. To name this ether:

1. Identify the shortest alkyl group. This is the methyl group, $-CH_3$
2. Remove the -yl suffix from the alkyl group name:

   methyl → meth
3. Add the suffix '-oxy':

   meth → meth**oxy**
4. The longest alkyl group is the ethyl group. The corresponding parent alkane is ethane. So we add this name to 'methoxy' to get the full ether name: methoxyethane.

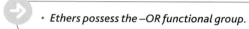

Self-check 6.2 What is the name of the ether with the structural formula $CH_3CH_2CH_2OCH_2CH_2CH_3$?

The same rules apply even if an ether is composed of two identical alkyl groups. We just don't have to think about which group is shortest or longest! Therefore CH_3OCH_3, which features two methyl groups, is called methoxymethane.

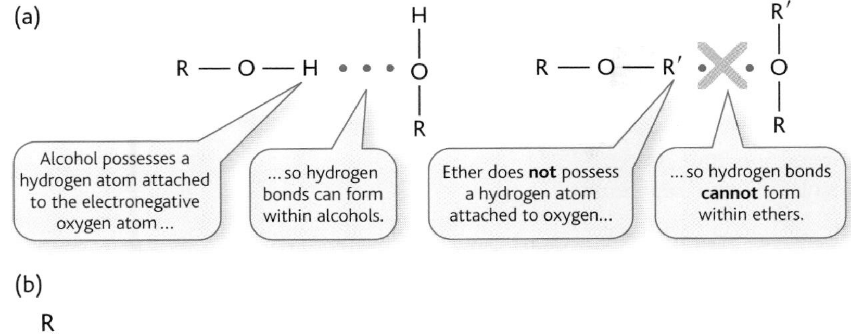
(a)
R — O — H • • • O
H
R
R — O — R′ • • O
R′
R
Alcohol possesses a hydrogen atom attached to the electronegative oxygen atom...
... so hydrogen bonds can form within alcohols.
Ether does **not** possess a hydrogen atom attached to oxygen...
... so hydrogen bonds **cannot** form within ethers.

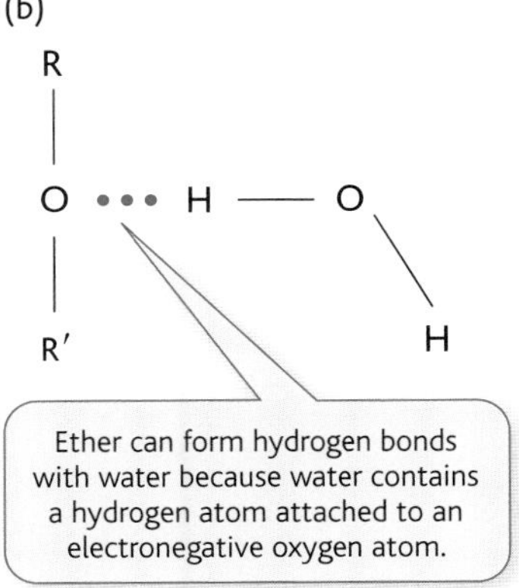
(b)
R
O • • • H — O
R′
H
Ether can form hydrogen bonds with water because water contains a hydrogen atom attached to an electronegative oxygen atom.

Figure 6.6 Hydrogen bonding within alcohols and ethers
(a) Alcohols feature the molecular requirements for hydrogen bonds to be able to form within them. By contrast, ethers lack the required molecular features, so hydrogen bonds cannot form within them.
(b) Ethers can, however, form hydrogen bonds with water molecules.

Physical properties of ethers

The ethers are relatively polar molecules because of the presence of the electronegative oxygen atom in the alkoxy group. However, ethers are unable to form hydrogen bonds to the same extent as alcohols. Remember that a hydrogen bond forms between an electronegative atom (oxygen, nitrogen, or fluorine) and a hydrogen atom, which is itself joined to oxygen, nitrogen, or fluorine. Look at Figure 6.6(a), and notice how an alcohol can form hydrogen bonds with itself because it contains a hydrogen atom joined to the electronegative oxygen atom. By contrast, the ether cannot form hydrogen bonds with itself: while it possesses an electronegative oxygen atom, it lacks a hydrogen atom bonded directly to this oxygen.

Ethers can, however, form hydrogen bonds with water, despite being unable to form hydrogen bonds with themselves, as depicted in Figure 6.6(b). Ethers are therefore more soluble in water than their corresponding alkanes (but are still only slightly soluble).

Look at the physical property checklist for the ethers shown in Table 6.3. If we compare this checklist to Table 6.1 we notice how ethers experience fewer non-covalent forces than alcohols (they experience dispersion and dipolar forces, but not hydrogen bonding), and so exhibit melting and boiling points which are lower than for alcohols of corresponding carbon skeleton length.

The aldehydes and ketones: the carbonyl group

Generic formula:

Aldehydes	Ketones
–RCHO	–RCOR′

Table 6.3 Physical property checklist for the ethers.

Do ethers experience...		Consequence
Dispersion forces?	✓	Intermediate melting and boiling points
Dipolar interactions?	✓	
Hydrogen bonding between molecules of same compound?	×	
Hydrogen bonding with water?	✓	Slightly soluble in water

Carbonyl groups possess an oxygen atom that is joined to a carbon atom[2] by a *double* covalent bond (in contrast to the hydroxyl and ether groups where the oxygen is joined by a single covalent bond):

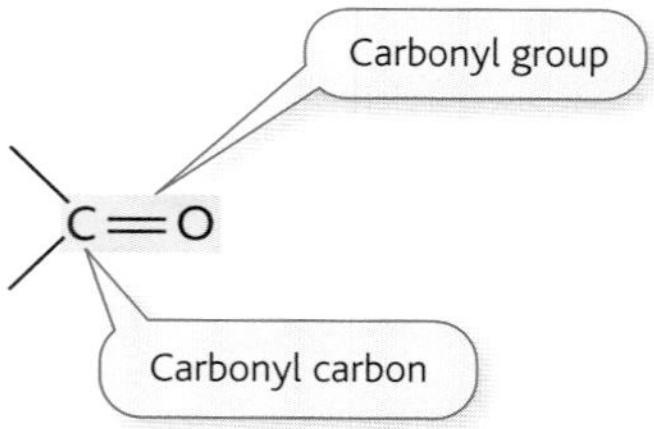

The carbonyl group is a particularly important functional group. Notice how it contains both an electronegative oxygen atom and a double bond. The electronegative atom leads to enhanced physical properties–it induces polarization, and confers the ability to participate in hydrogen bond formation–while the double bond serves as a reserve of electrons, and leads to enhanced chemical reactivity[3].

The two simplest classes of organic compound containing the carbonyl group are the **aldehydes** and the **ketones**. We distinguish aldehydes from ketones by considering the number of carbon atoms to which the carbonyl group is attached:

- In aldehydes, the carbonyl group is attached to *one* carbon atom (and a hydrogen atom):

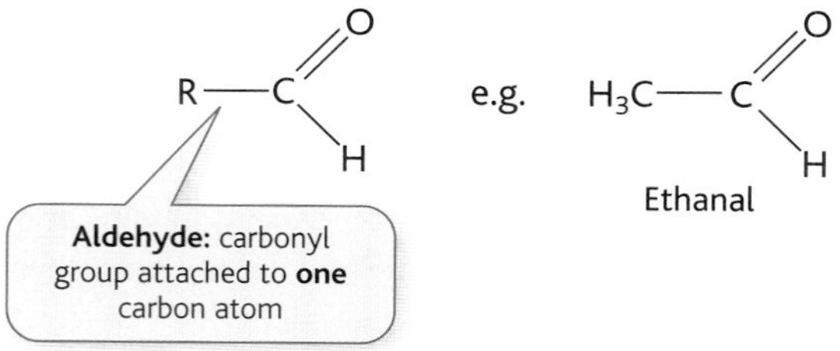

2 We call the carbon atom of a carbonyl group to which the oxygen atom is attached the **carbonyl carbon**.

3 In fact, we see in Chapter 17 how an electronegative atom contributes as greatly to the chemical reactivity of compounds as does the presence of a double bond.

- In ketones, the carbonyl group is attached to *two* carbon atoms (often two alkyl groups):

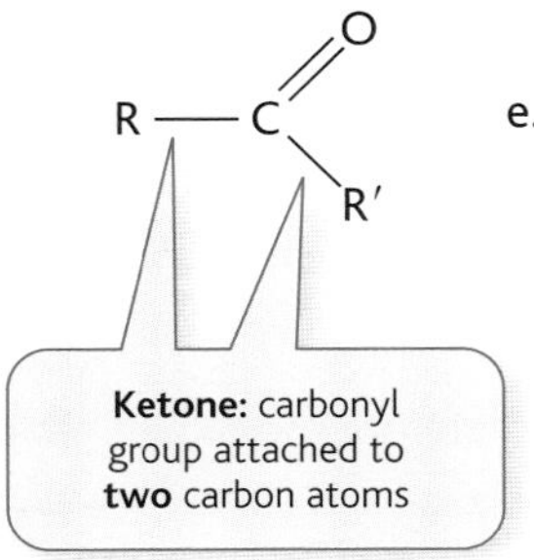

Remember: the alkyl group is represented in general terms by the letter R; a second alkyl group is represented by R′.

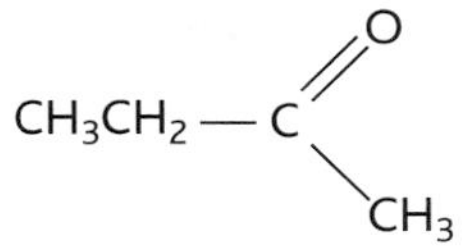

R—C(=O)—R′ e.g. CH_3CH_2—C(=O)—CH_3

Ketone: carbonyl group attached to **two** carbon atoms

Butanone

The aldehydes and ketones are used widely as solvents, and generate the potent, penetrating smell that we associate with substances such as paint stripper or nail polish remover. The simplest aldehyde is methanal, CH_2O, based on the parent alkane methane:

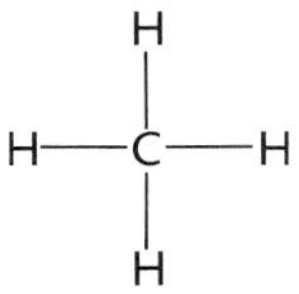

1-carbon alkane
methane

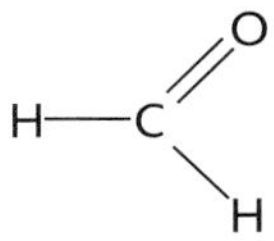

1-carbon aldehyde
methanal

Methanal is often referred to by its **common name**[4], formaldehyde. The simplest ketone is propanone, C_3H_6O, though we are probably more familiar with its common name, acetone. (Acetone is the key ingredient of the nail polish removers referred to above.)

- Aldehydes and ketones possess the carbonyl group, C = O.
- In aldehydes, the carbonyl group is attached to one carbon atom and one hydrogen atom.
- In ketones, the carbonyl group is attached to two carbon atoms.

Naming aldehydes and ketones

Aldehydes and ketones are named by taking the parent alkane name, removing the -e, and adding the suffix -al to denote an aldehyde, or -one to denote a ketone.

---

4 Every compound has a formal chemical name, which is determined by IUPAC rules on nomenclature (p. 7), and reflects the chemical structure and composition of the compound. However, some compounds also have a **common** (or 'trivial') **name**–a name that is used on a day-to-day basis in place of the proper chemical name. The major disadvantage of a common name is that, unlike the IUPAC name, it does not tell us anything about the compound's chemical structure and composition.

For example, consider the aldehyde with the following structural formula:

```
     H   H   H    O
     |   |   |   //
  H—C———C———C———C
     |   |   |    \
     H   H   H     H
```

4-carbon backbone

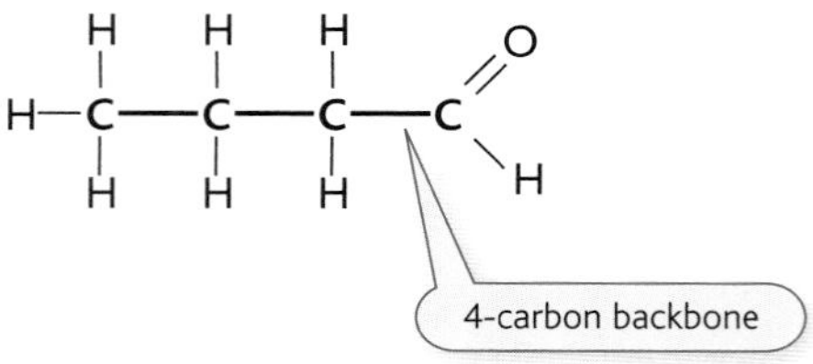

This has a four-carbon backbone, and therefore the parent alkane is butane. To name this aldehyde:

1. remove the -e from the parent alkane name:

   butane → butan
2. add the suffix -al:

   butan → butan**al**

   Similarly, the ketone with the following structural formula:

```
      H   O   H
      |   ‖   |
  H — C — C — C — H
      |       |
      H       H
```

3-carbon framework

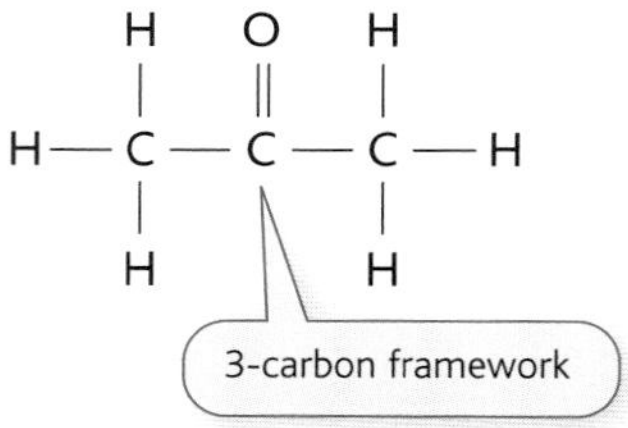

is named propanone, because it's based on the 3-carbon parent alkane, propane:

1. propane → propan
2. propan → propan**one**

Self-check 6.3 Identify whether the following compounds are aldehydes or ketones. What is the name of each?

```
        H   H   H   H   O   H
        |   |   |   |   ‖   |
(a) H — C — C — C — C — C — C — H
        |   |   |   |       |
        H   H   H   H       H

        H   H   H   H   O
        |   |   |   |   ‖
(b) H — C — C — C — C — C — H
        |   |   |   |
        H   H   H   H

        H   H   O   H
        |   |   ‖   |
(c) H — C — C — C — C — H
        |   |       |
        H   H       H
```

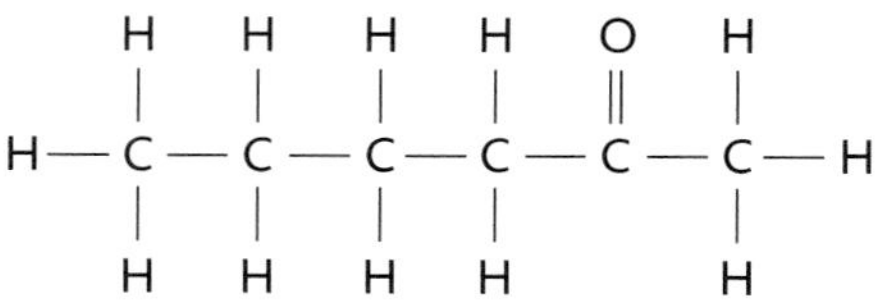

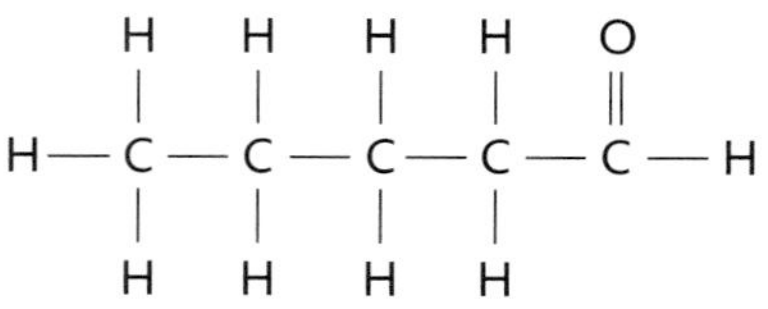

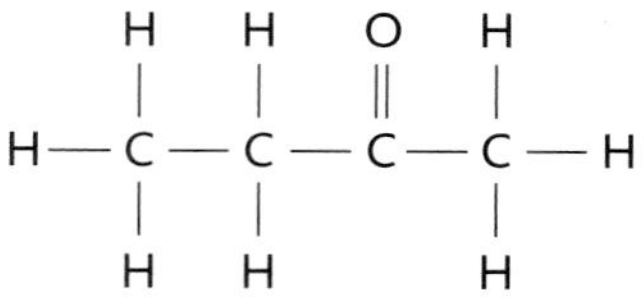

Aldehydes and ketones feature most prominently in biological systems as **intermediates** within metabolic pathways[5].

We see an example of a short metabolic pathway in Box 6.1, where we find out how ethanol is broken down by the liver in a series of steps to effect its removal from the body. This metabolic pathway includes the aldehyde ethanal as an intermediate species.

Aldehydes also play a vital function in enabling us to study the structure of biological tissues. Look at Box 6.2 to learn how aldehydes are used as fixatives to preserve biological specimens prior to microscopic examination.

Self-check 6.4 Why is the simplest ketone not ethanone?

Physical properties of aldehydes and ketones

Figure 6.8 shows how the carbonyl group is a highly electronegative group, which makes both aldehydes and ketones polar in nature. Look at this figure and notice how the distribution of electrons is skewed towards the electronegative carbonyl group.

Now look at Figure 6.9 which shows how the presence in the carbonyl group of the electronegative oxygen atom also means that both aldehydes and ketones can form hydrogen bonds with water molecules. Notice, however, that neither

BOX 6.2 Getting in a fix

Aldehydes are widely used to preserve biological specimens in a process we call fixation. Fixation is an important part of the process of preparing biological samples for microscopic examination; it helps to stabilize the structure of cells to prevent damage, decay, or distortion, so that what we see through the microscope is, in fact, an accurate representation of how the tissue appears in the living organism.

A well-known proponent of the preservative properties of aldehydes is the British 'conceptual' artist Damien Hirst. Hirst is renowned for his artistic installations, some of which comprise aldehyde-filled glass tanks in which preserved biological specimens are displayed; one of Hirst's more controversial installations displays a cow that has been cut in half, giving the viewer a rare insight into the cow's internal anatomy.

One of the most common fixatives (the name given to compounds that are used for fixation) is the aldehyde methanal (1), known more commonly as formaldehyde. But how does such a fixative work? Formaldehyde is firstly dissolved in water to form an aqueous solution before being applied to the biological sample. (Formaldehyde in water is sometimes called formalin.) Once exposed to the biological sample, formaldehyde reacts with specific parts of any proteins that are present in the sample. Figure 6.7 shows how formaldehyde forms a 'bridge' between the side chain of a lysine residue and the nitrogen atom of a nearby peptide linkage. We call this structure a methylene bridge.

The methylene bridge acts as a brace, holding the protein in a rigid–or fixed–structure. Formaldehyde can effectively stabilize a whole tissue sample by forming many methylene bridges throughout the sample.

H—C(=O)—H

(1)

5 An **intermediate** represents the product of one step in a multi-step chain of reactions through which a starting substance is gradually converted into an end product. We learn more about intermediates and reaction pathways in Chapter 13.

BOX 6.2 Continued

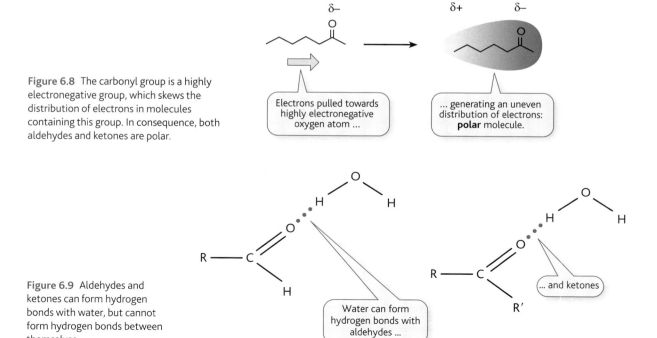

Figure 6.7 Formaldehyde (methanal) acts as a fixative by forming a 'bridge' between the side chain of a lysine residue and the nitrogen atom of a nearby peptide linkage.

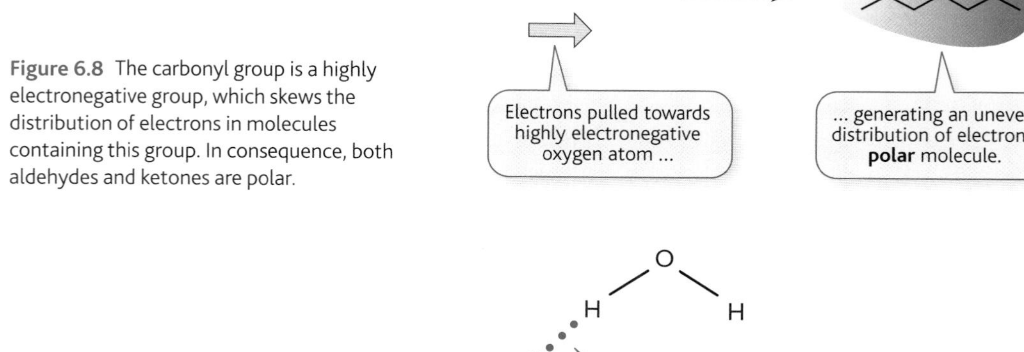

Figure 6.8 The carbonyl group is a highly electronegative group, which skews the distribution of electrons in molecules containing this group. In consequence, both aldehydes and ketones are polar.

Figure 6.9 Aldehydes and ketones can form hydrogen bonds with water, but cannot form hydrogen bonds between themselves.

Do aldehydes and ketones experience...		Consequence
Dispersion forces?	✓	Intermediate melting and boiling points
Dipolar interactions?	✓	
Hydrogen bonding between molecules of same compound?	×	
Hydrogen bonding with water?	✓	Soluble in water (if small)

Table 6.4 Physical property checklist for the aldehydes and ketones.

compound has a hydrogen atom directly connected to the oxygen atom; consequently, neither compound can form hydrogen bonds between its own molecules.

Table 6.4 shows the physical property checklist for aldehydes and ketones. Their melting and boiling points are lower than for the corresponding alcohols (hydrogen bonds do occur in alcohols), but higher than the corresponding alkanes (which experience only dispersion forces). The similar general compositions of aldehydes and ketones are reflected in the way they can exist as structural isomers, as explained in Chapter 10.

The carboxylic acids: combining the hydroxyl and carbonyl groups

Generic formula: RCOOH

We now look at the **carboxyl** group, a functional group that incorporates two functional groups that we encounter above, and which is the defining feature of the **carboxylic acids**. The carboxyl group has two components, as illustrated in Figure 6.10.

Carboxylic acids play a vital part in our metabolism, and include several intermediate compounds that are formed during the chain of biochemical reactions through which our body generates energy from the sugars in our diet. For example, during **glycolysis** (glyco = sugar; lysis = splitting) a molecule of glucose is split into two molecules of the carboxylic acid pyruvic acid:

H O
C
H—C—OH
HO—C—H
H—C—OH
H—C—OH
CH_2OH
Glucose

As part of the process by which our body generates energy from food, glucose is converted into two molecules of the carboxylic acid, pyruvic acid, in a ten-step process.

O OH
C
C=O
CH_3
O OH + $4H^+$
C
C=O
CH_3
2 pyruvic acid

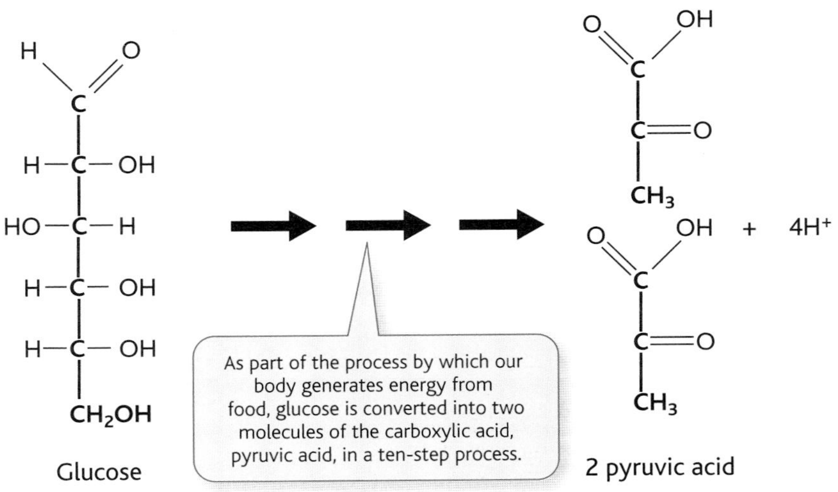

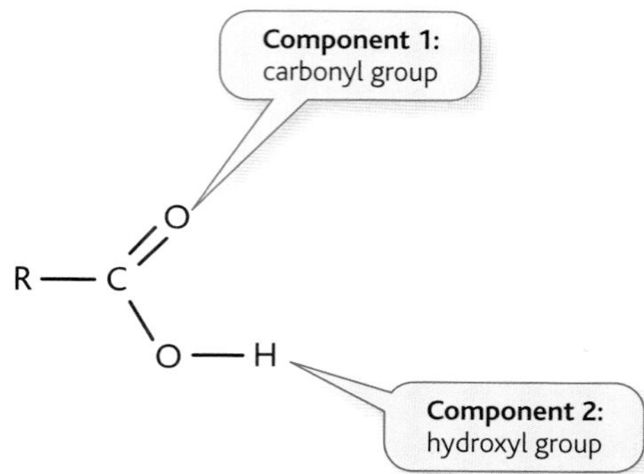

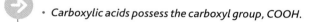

Figure 6.10 The carboxyl group has two components: a carbonyl group and a hydroxyl group.

$$^{-}OOC{-}C(=O){-}CH_3$$

(2)

In fact, to be totally accurate, the product of glycolysis is the **pyruvate** ion (2).

Pyruvate is formed when pyruvic acid becomes ionized (forms an ion) by giving up a proton in the form of a hydrogen ion.

- *Carboxylic acids possess the carboxyl group, COOH.*

We encounter the citric acid cycle in more detail in Chapter 18, when we explore some biological reaction mechanisms.

Carboxylic acids also feature often in the **citric acid** (or **Krebs**) cycle, which takes place later in the process of generating energy from sugars.

BOX 6.3 No pain, no gain...?

Many of us may experience a burning sensation in our muscles during exercise. This pain is caused by the build up of lactic acid, a carboxylic acid:

$$CH_3{-}CH(OH){-}C(=O){-}OH$$

During normal levels of activity our body gets its energy by breaking down glucose (sugar), a process which involves the oxygen that we inhale when we breathe. This process is known as **aerobic respiration** (aerobic = oxygen present).

During periods of strenuous exercise, however, the body needs energy more quickly than the rate at which we are able to get oxygen into the cells. (This mirrors the situation faced by firemen on board steam trains in the early twentieth century, who fervently stoked the train's fire with shovel after shovel of coal; the coal would burn almost the instant it entered the fire.) As a result the body has to break down sugar without oxygen–this is known as **anaerobic** respiration. In the absence of oxygen, glucose gets broken down into lactic acid, which builds up in those cells that are burning energy most rapidly–the muscle cells–causing the burning sensation with which we're familiar.

Anaerobic respiration is not unique to humans. Bacteria also break down sugar to lactic acid under anaerobic conditions. The outcome isn't always pleasant: when bacteria break down the sugars that are naturally present in milk, lactic acid is formed–and the lactic acid proceeds to turn the milk sour, giving it the stomach-churning smell we associate with milk that has gone off.

Those carboxylic acids that possess long alkyl chains form a family of compounds called the **fatty acids**. The fatty acids typically have between 10 and 20 carbon atoms in their carbon backbone, which may be saturated (comprising only single carbon–carbon bonds) or unsaturated (including at least one double or triple carbon–carbon bond). We explore the fatty acids in more detail in Section 7.3. Read Box 6.3 to see how one carboxylic acid, lactic acid, is the cause of the burning we sometimes feel in our muscles when we exercise.

Naming the carboxylic acids

Carboxylic acids are named by removing the -e from the parent alkane name, and adding the suffix '-oic acid'. For example, butanoic acid (3),

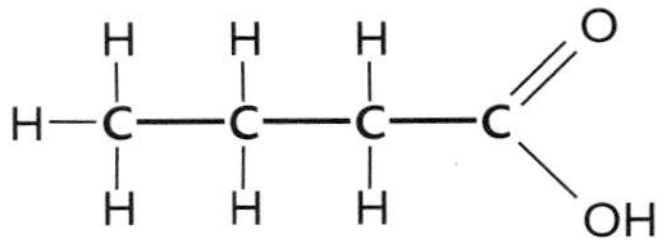

```
     H   H   H      O
     |   |   |    //
  H—C———C———C———C
     |   |   |    \
     H   H   H     OH
```

(3)

which smells of goats, contains a four-carbon backbone, and is named from the corresponding four-carbon alkane, butane:

1. remove -e from parent alkane name:

   butane → butan

2. add the suffix '-oic acid':

   butan → butan**oic acid**.

Self-check 6.5 What is the name of the carboxylic acid represented by the following structural formula?

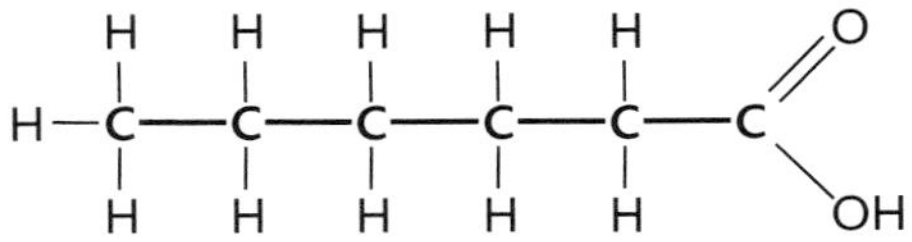

```
     H   H   H   H   H      O
     |   |   |   |   |    //
  H—C———C———C———C———C———C
     |   |   |   |   |    \
     H   H   H   H   H     OH
```

Physical properties of carboxylic acids

Look at Table 6.5, which shows the physical property checklist for the carboxylic acids. Notice that there are two ticks in the 'Experiences hydrogen bonding between

Do carboxylic acids experience...		Consequence
Dispersion forces?	✓	Very high melting and boiling points
Dipolar interactions?	✓	
Hydrogen bonding between molecules of same compound?	✓✓	
Hydrogen bonding with water?	✓	Soluble in water

Table 6.5 Physical property checklist for the carboxylic acids.

molecules of the same compound?' column, and that the carboxylic acids are noted as exhibiting very high melting and boiling points. Why is this?

The carboxyl group possesses two electronegative oxygen atoms, one of which has a hydrogen atom attached. *Both* entities are able to participate in hydrogen bonds. As a consequence, a relatively extensive network of hydrogen bonds exists between carboxylic acid molecules, so a great deal of energy is required to pull neighbouring molecules apart. Therefore, the carboxylic acids have *unusually high* melting and boiling points. Look at Table 6.6, and notice how high the boiling points of the carboxylic acids are compared to the alcohols, which can also form hydrogen bonds with themselves, but to a lesser extent than the carboxylic acids.

The ability of a carboxylic acid molecule to participate in two hydrogen bonds also means that gaseous carboxylic acids can join together in pairs to form a structure called a dimer, as illustrated in Figure 6.11. Notice how two hydrogen bonds form: one between the carbonyl group of one carboxylic acid molecule and the hydroxyl group of another carboxylic acid; and the second between the hydroxyl group of the first carboxylic acid and the carbonyl group of the other. This dimerization results in a stable association between neighbouring carboxylic acid molecules. However, dimerization only occurs when a carboxylic acid is in the gas phase.

Self-check 6.6 Would you predict ethanoic acid to be more or less soluble in water than ethane? Why?

The esters: a modified carboxyl group

Generic formula: RCOOR′

The **esters** are a family of organic compounds which are defined by the presence of a *modified* form of the carboxyl group, the functional group that is characteristic of the carboxylic acids. Figure 6.12 shows how in esters the hydroxyl group is modified to an **alkoxy** (–OR) group, whereas carboxylic acids possess a regular carboxyl group as we see above.

Esters are formed from the joining through a chemical reaction of an alcohol and a carboxylic acid. The carbonyl carbon of the carboxylic acid joins with the oxygen

Table 6.6 Boiling points for carboxylic acids and alcohols with the same-sized carbon framework.

Carboxylic acid	Boiling point (°C)	Alcohol	Boiling point (°C)
Methanoic acid, HCOOH	101	Methanol	65
Ethanoic acid, CH_3COOH	118	Ethanol	78
Propanoic acid, CH_3CH_2COOH	141	Propan-1-ol	82
Butanoic acid, $CH_3CH_2CH_2COOH$	164	Butan-1-ol	118

Figure 6.11 In the gas phase, carboxylic acids are able to form **dimers**.

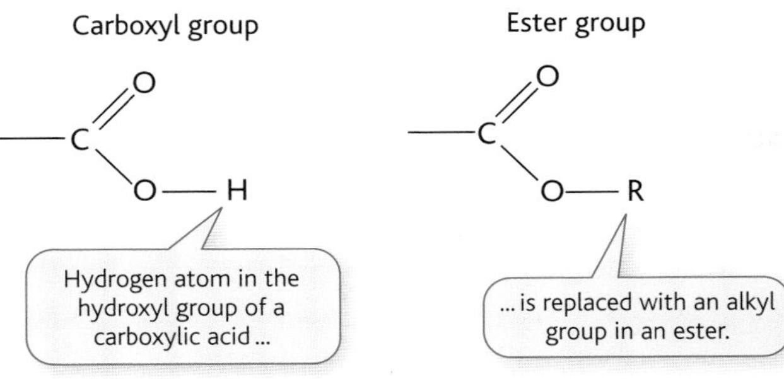

Figure 6.12 Esters possess a modified carboxyl group, in which the hydroxyl group is modified to an alkoxy group.

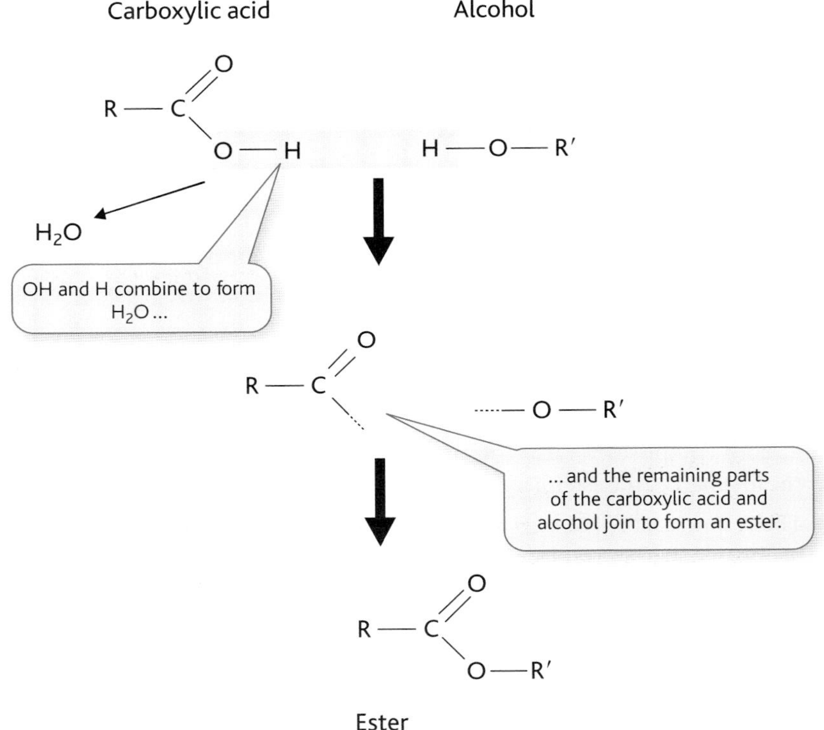

Figure 6.13 Carboxylic acids and alcohols can react to form esters, with the release of water.

atom of the alcohol's hydroxyl group to form the ester, as illustrated in Figure 6.13. The process by which an ester is formed from a carboxylic acid and an alcohol is called **esterification**, and is an example of a **condensation reaction**.

We explore the different types of chemical reaction (such as condensation) that biological molecules undergo in more detail in Chapter 17.

- *The esters possess a modified carboxyl group, COOR.*

Naming the esters

The name of an ester consists of two parts:

1. The first reflects the name of the alcohol.
2. The second reflects the name of the carboxylic acid.

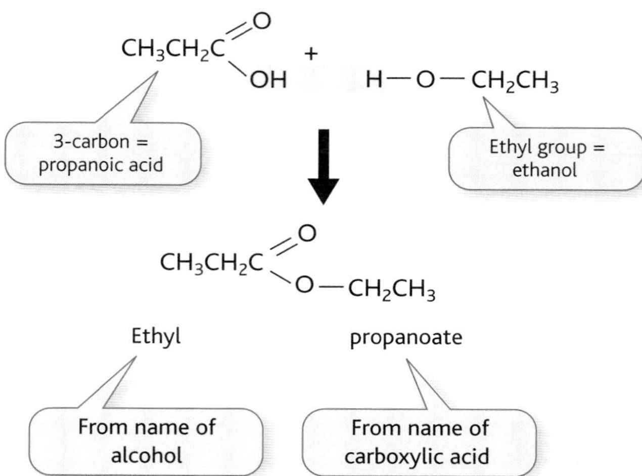

Figure 6.14 The formation of the ester ethyl propanoate from ethanol and propanoic acid.

Specifically:

1. The first part takes the name of the **alkyl** group on which the alcohol is based.
2. The second part takes the carboxylic acid name, and substitutes -ic acid with the suffix **-ate**.

As an example, let's consider the formation of an ester from ethanol and propanoic acid, as illustrated in Figure 6.14:

- Ethanol comprises the **ethyl** group with a hydroxyl group attached. Therefore, the first part of the ester name is **ethyl**.
- The second part of the ester name is generated by removing the '-ic acid' from propanoic acid, and replacing it with the suffix '-ate'. Therefore the second part of the ester name is **propanoate**.

Taken together, the ester formed from the joining of ethanol and propanoic acid is named **ethyl propanoate**.

Self-check 6.7 What is the name of the ester formed from the joining of propanol and butanoic acid? What is its structural formula?

Some of the most biologically important esters are the **triacylglycerols**, a group of fats and oils based on the alcohol glycerol, which possesses three ester groups, as depicted in Figure 6.15. We find out more about the triacylglycerols in Section 7.3. Some esters also have medicinal value. Look at Box 6.4 to find out about a ubiquitous ester-based medicine, aspirin.

Physical properties of esters

Unlike carboxylic acids, esters lack the ability to form hydrogen bonds with themselves. (Look again at the structure of the carboxyl group of an ester, compared to that of a carboxylic acid (Figure 6.12). Notice that an ester does *not* have a hydrogen atom connected to an electronegative oxygen atom, and so cannot form hydrogen bonds with itself.) Consequently, esters have noticeably lower melting and boiling

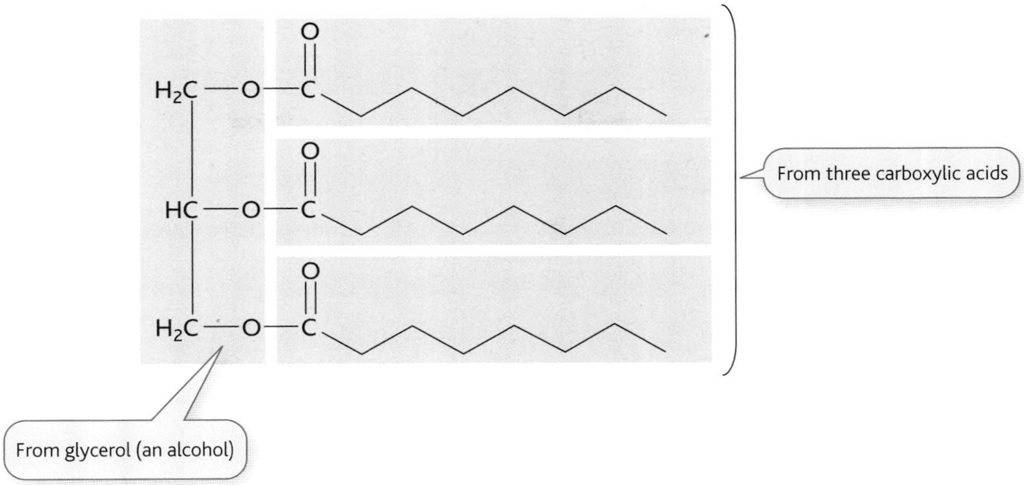

From three carboxylic acids

From glycerol (an alcohol)

Figure 6.15 Triacylglycerols are based on the alcohol glycerol, and possess three ester groups.

BOX 6.4 An ester a day keeps the headache away

The ester with which we are probably all most familiar is acetylsalicylic acid, though we are most likely to recognize its common name: aspirin.

At first sight, aspirin seems to contradict what we have just seen about the formation of esters from a carboxylic acid and an alcohol: aspirin is formed from two carboxylic acids. However, this isn't really breaking the rules: one of the carboxylic acids contains a separate hydroxyl group (the characteristic component of an alcohol) and it is this part of the carboxylic acid that is involved in the formation of aspirin, as shown in the reaction below.

Salicylic acid occurs naturally in willow bark, whose pain-relieving properties have been known for centuries. However, salicylic acid causes irritation of the stomach lining, a symptom that was found to be eased by joining salicylic acid with acetic acid to form aspirin.

Aside from its use in pain relief, aspirin is also recommended to those with heart conditions because it helps to reduce the risk of blood clots, and the heart attacks that they can trigger. Blood clots are a normal and important part of our body's repair strategy, and are formed by the packing together of platelets, a particular type of blood cell. However, the formation of blood clots in those blood vessels supplying the heart with blood, the risk of which is elevated in individuals with a narrowing of the blood vessels, can lead to a heart attack. Aspirin prevents platelets from binding to one another and, hence, inhibits clot formation, simultaneously reducing the risk of a heart attack in those susceptible.

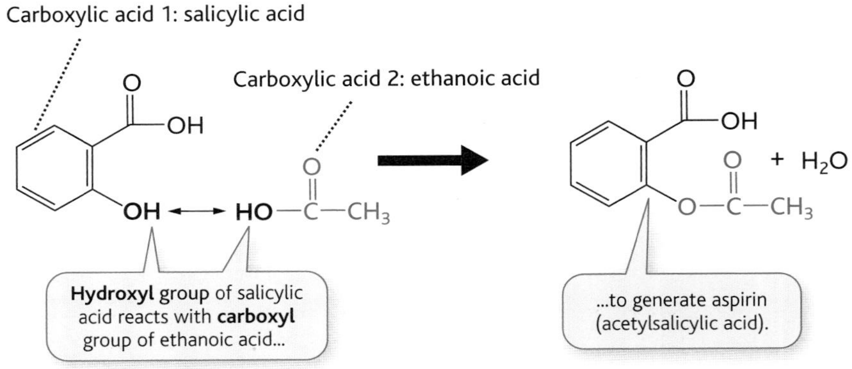

Carboxylic acid 1: salicylic acid

Carboxylic acid 2: ethanoic acid

Hydroxyl group of salicylic acid reacts with **carboxyl** group of ethanoic acid...

...to generate aspirin (acetylsalicylic acid).

Table 6.7 Physical property checklist for the esters.

Do esters experience...		Consequence
Dispersion forces?	✓	Intermediate melting and boiling points
Dipolar interactions?	✓	
Hydrogen bonding between molecules of same compound?	×	
Hydrogen bonding with water?	✓	Soluble in water (if small)

points than equivalent carboxylic acids. For example, the ester methyl methanoate (methyl formate), (4) has a boiling point of 32 °C, while the equivalent carboxylic acid, ethanoic acid, (5) has a boiling point of 118 °C.

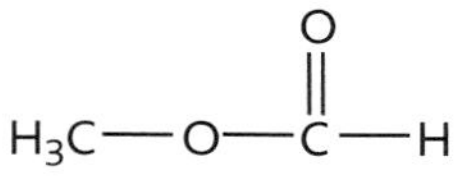

$H_3C—O—C(=O)—H$

(4)

$H_3C—C(=O)—OH$

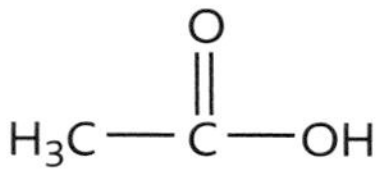

(5)

The physical property checklist for esters is shown in Table 6.7. Notice that, in common with the other organic compounds featured in this section, the esters can form hydrogen bonds with water; small esters are therefore water-soluble.

6.2 Organic compounds with nitrogen-based functional groups

Let's now turn our attention to nitrogen, the second electronegative element that is prevalent in biological systems. Nitrogen is a component of two main functional groups: the **amino** and **amide** groups.

The amines: the amino group

Generic formula: RNH_2

The amines are a family of nitrogen-containing organic compounds that are characterized by the attachment to the carbon framework of an **amino group**. The amines are based on the compound ammonia, (6), but in which one or more hydrogen atoms are replaced with an R group.

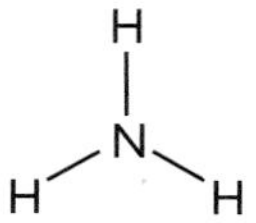

NH_3 (H—N(—H)—H)

(6)

Consequently, the amino group–with a nitrogen atom as its central 'hub'–can take one of three different forms:

- The nitrogen atom can bond with one R group and two hydrogen atoms:

We call an amine in which one hydrogen atom of ammonia is replaced by *one* R group a **primary amine**.

- The nitrogen atom can bond with one hydrogen atom and two R groups:

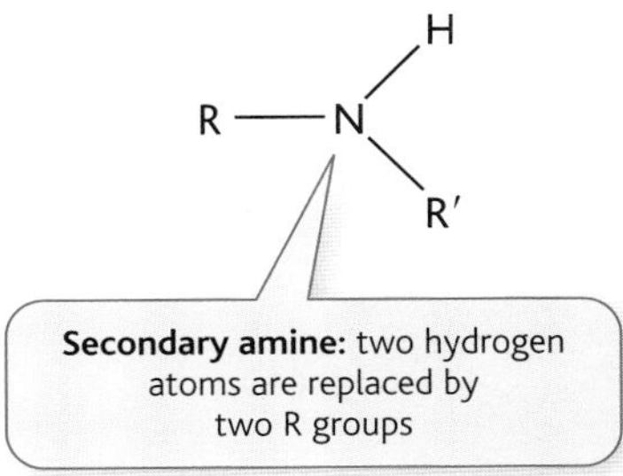
H
R
N
R′
Secondary amine: two hydrogen atoms are replaced by two R groups

We call an amine in which the two hydrogen atoms of ammonia are replaced by *two* R groups a **secondary amine**.

- The nitrogen atom can bond with three R groups:

Remember: the symbol 'R' indicates either an alkyl *or* an aryl (aromatic) group.

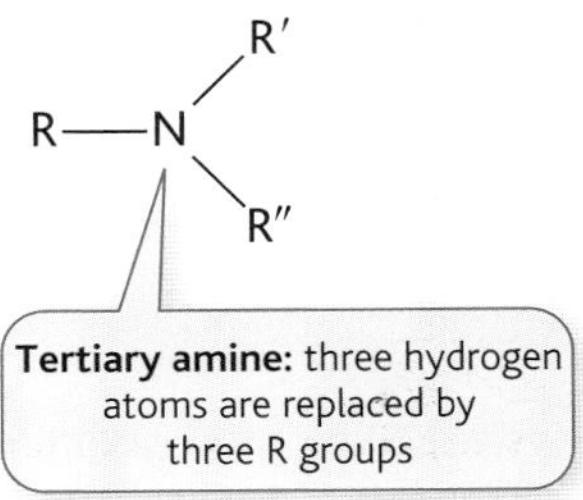
R′
R
N
R″
Tertiary amine: three hydrogen atoms are replaced by three R groups

We call an amine in which the three hydrogen atoms of ammonia are replaced with *three* R groups a **tertiary amine**.

Notice how nitrogen satisfies its valency of three in each type of amino group.

- ***A primary amine comprises a molecule of ammonia in which one hydrogen atom is replaced with one R group.***
- ***A secondary amine comprises a molecule of ammonia in which two hydrogen atoms are replaced with two R groups.***
- ***A tertiary amine comprises a molecule of ammonia in which all three hydrogen atoms are replaced by R groups.***

Naming amines

We name amines by:

1. taking the name of the alkane that has an equivalent number of carbon atoms (Section 5.2)
2. removing the -e from the suffix and replacing it with the suffix **-amine**.

For example, the amine $CH_3CH_2CH_2NH_2$ has a three-carbon framework, and so is named **propanamine**, after the three-carbon alkane **propane**. However, amines are also commonly named by removing the -ane from the parent alkane name and replacing it with -ylamine.

For secondary and tertiary amines, we find the longest carbon chain, and use this to identify the corresponding alkane and, hence, to generate the amine name:

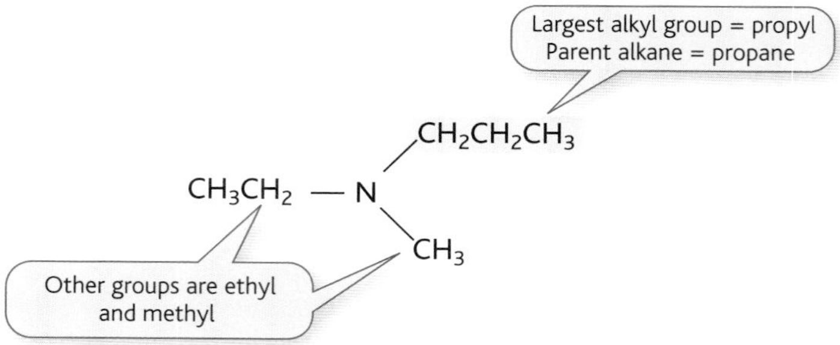

Largest alkyl group = propyl
Parent alkane = propane

$CH_2CH_2CH_3$

CH_3CH_2 — N

CH_3

Other groups are ethyl and methyl

We then prefix this name with the names of the (smaller) alkyl groups attached to the nitrogen atom, listed in alphabetical order, and each prefixed with N-:

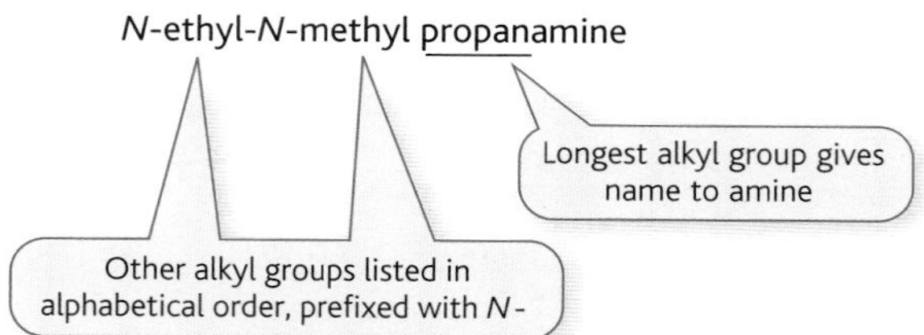
N-ethyl-*N*-methyl propanamine
Longest alkyl group gives name to amine
Other alkyl groups listed in alphabetical order, prefixed with *N*-

The 'N-' prefix simply tells us that the smaller substituent groups are attached to the nitrogen of the amino group.

Q

Self-check 6.8 What are the names of these amines?

(a) $CH_3CH_2CH_2CH_2NH_2$

(b) CH_3CH_2N (with two CH_3 groups on N)

(c) $CH_3CH_2CH_2N$ (with H and CH_2CH_3 on N)

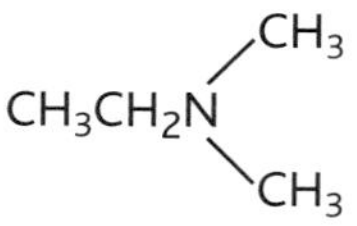

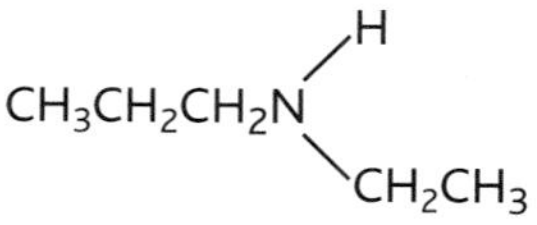

Our nervous system depends on a range of amines to function properly. Biologically important amines include several hormones (including adrenaline and noradrenaline) that help our bodies respond to stress. For example, a surge in adrenaline (7) triggers the 'fight or flight' response whereby the body is put on a state of alert. Stored sugars are broken down, and the heart rate quickens to transport oxygenated blood to tissues such as muscle. This enables the muscle to rapidly convert the released sugars into energy–and prime our body for sudden energetic movement.

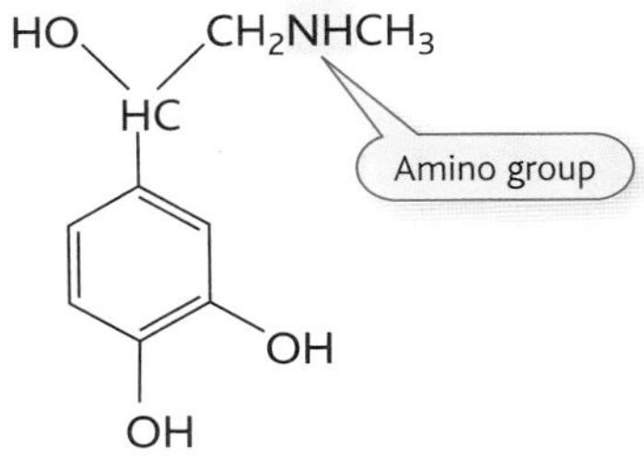

HO, CH_2NHCH_3, HC, Amino group, OH, OH

(7)

Amines are also vital for our nervous system to exert proper control over our movements. Dopamine (8) is a neurotransmitter with an important role in the coordination and control of movement.

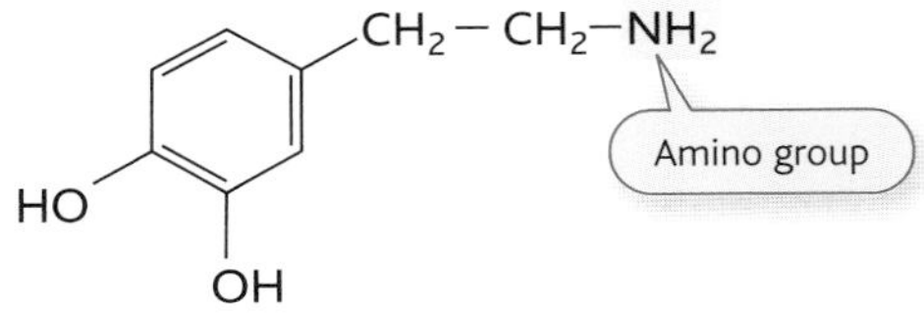

$CH_2-CH_2-NH_2$, Amino group, HO, OH

(8)

A deficiency in dopamine contributes to the neurological disorder Parkinson's disease, in which the control of movement is impaired, leading to involuntary, and often incapacitating, movement of the limbs.

Not all amines have such a beneficial effect on our health and well-being, however. We see in Box 6.5 how a family of naturally occurring amines are something of a mixed blessing.

Physical properties of amines

Look at Table 6.8, which shows the physical property checklist for primary amines. Notice how primary amines can form hydrogen bonds with themselves, and with water. Hence, they have relatively high boiling points, and are soluble in water.

Why do we specify that we are considering primary amines here? Primary, secondary, and tertiary amines have different physical properties because they can form hydrogen bonds to different extents, as illustrated in Figure 6.16.

Look at Table 6.9, which shows the boiling points of a primary, secondary, and tertiary amine with the same-sized carbon framework.

Notice how there is a correlation between the number of hydrogen bonds that can form in the amine, and its boiling point. Two hydrogen bonds can form in primary amines; they have a relatively high boiling point. One hydrogen bond can form in secondary amines; they have an intermediate boiling point. No hydrogen bonds can form in tertiary amines; they have a low boiling point.

BOX 6.5 Alkaloids: a natural wake-up call . . .?

The alkaloids are a group of amines that occur naturally in plants. Alkaloids are something of a paradox: they include compounds that are amongst the most beneficial to our lives, but also those that have the potential to be the most harmful.

The alkaloids with which we are probably most familiar are those that act as stimulants. These include caffeine (**9**) the alkaloid that gives coffee its 'wake up' properties, and nicotine, (**10**).

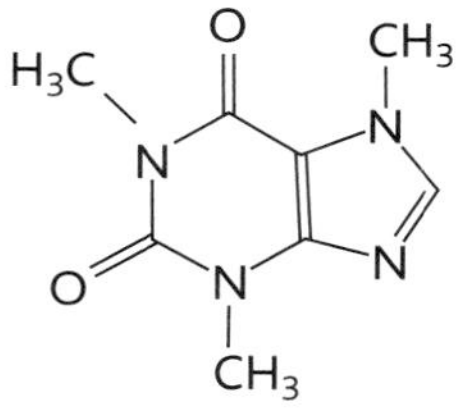

(9)

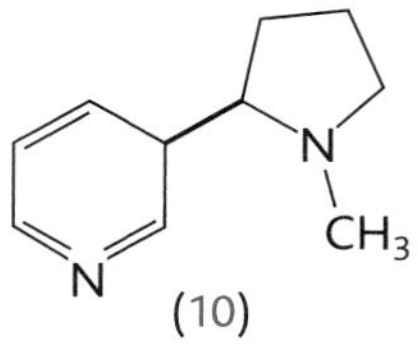

(10)

Nicotine is a potent stimulant: it causes the body to be on a high state of alert by stimulating an increase in the levels of adrenaline in the blood. Adrenaline primes the body for sudden bursts of activity by increasing the heart rate and blood pressure. It is the adrenaline boost triggered by the intake of nicotine that gives smokers the 'kick' that they crave and which, for many, makes smoking so enticing as a quick pick-me-up. There is a problem, however: our body isn't designed to be regularly primed for action. The high blood pressure caused by the elevated levels of adrenaline stimulated by nicotine can cause heart and lung damage.

The alkaloids also include a number of painkilling drugs that belong to a sub-class of alkaloids called **opiates** (which are produced naturally by poppies). These drugs include codeine (**11**) and morphine (**12**)

The most negative feature of the alkaloids is that they are habit-forming. It is the nicotine in tobacco that makes cigarettes addictive; caffeine can make even coffee addictive if consumed in very large quantities. Also, long-term users of morphine find themselves subject to debilitating withdrawal symptoms if administration of the drug is stopped; we see in Chapter 10 (p. 315) how methadone is used as a substitute to help rehabilitate those addicted to both morphine and its chemically similar cousin heroin.

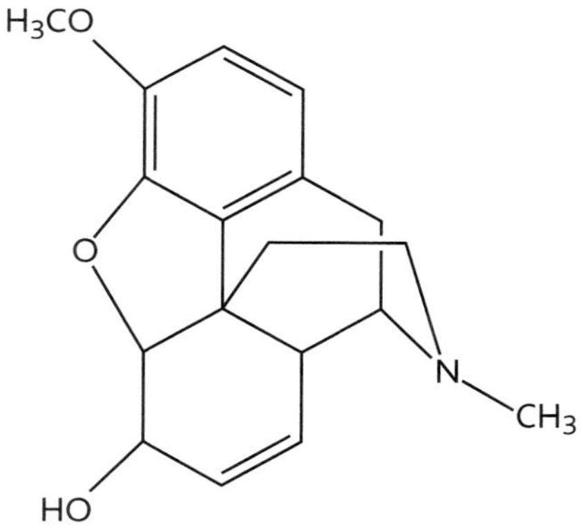

(11)

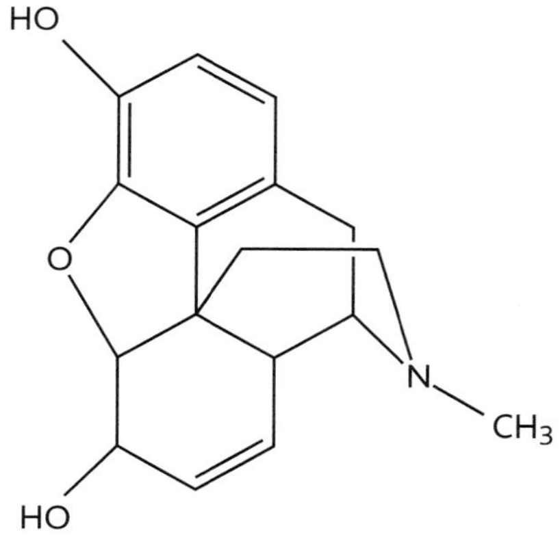

(12)

Table 6.8 Physical property checklist for the primary amines.

Do the primary amines experience...		Consequence
Dispersion forces?	✓	High melting and boiling points
Dipolar interactions?	✓	
Hydrogen bonding between molecules of same compound?	✓	
Hydrogen bonding with water?	✓	Soluble in water

Structural formula	Boiling point of amine (°C)
Primary amine: $CH_3—CH_2—CH_2—NH_2$	48
Secondary amine: $CH_3—CH_2—NH—CH_3$	36
Tertiary amine: $CH_3—N(CH_3)—CH_3$	3

Table 6.9 The boiling points of primary, secondary, and tertiary amines.

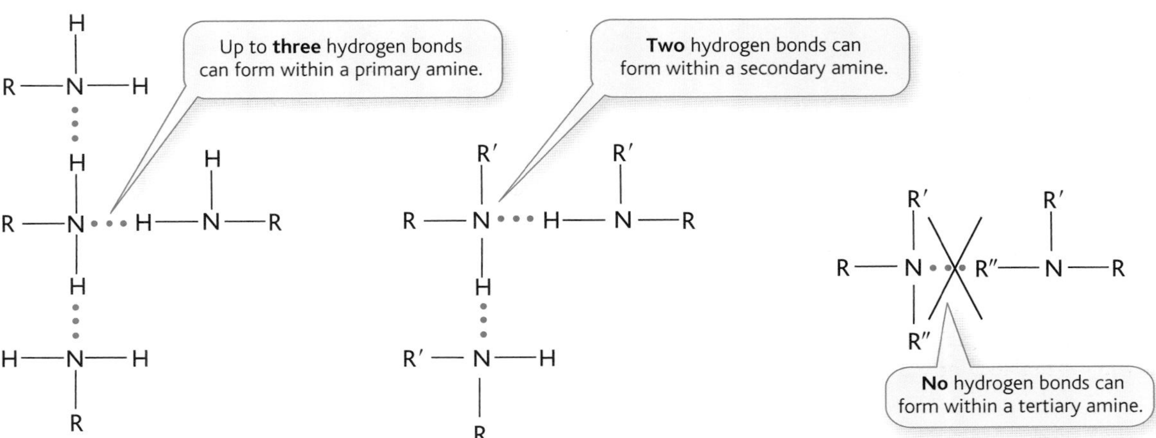

Figure 6.16 Primary, secondary, and tertiary amines can form hydrogen bonds to themselves to different extents. Up to three hydrogen bonds can form within primary amines; no hydrogen bonds can form within tertiary amines.

Primary Secondary Tertiary

Figure 6.17 Primary, secondary, and tertiary amines can all form hydrogen bonds with water.

However, Figure 6.17 shows how all types of amine–whether primary, secondary, or tertiary–can form hydrogen bonds with water, so most amines are water-soluble.

Now look at Table 6.10, which compares the boiling points of several alcohols and primary amines of corresponding carbon framework size. What do you notice?

Amines have *lower* melting and boiling points than the corresponding alcohols because nitrogen has only one non-bonding pair of electrons, and so can form one hydrogen bond, while oxygen has *two* non-bonding pairs of electrons, and so can form *two* hydrogen bonds, as depicted in Figure 6.18.

Table 6.10 The boiling points of corresponding alcohols and amines.

Number of carbon atoms	Boiling point of alcohol, ROH (°C)	Boiling point of amine, RNH_2 (°C)
1	65	−7
2	79	17
3	97	48

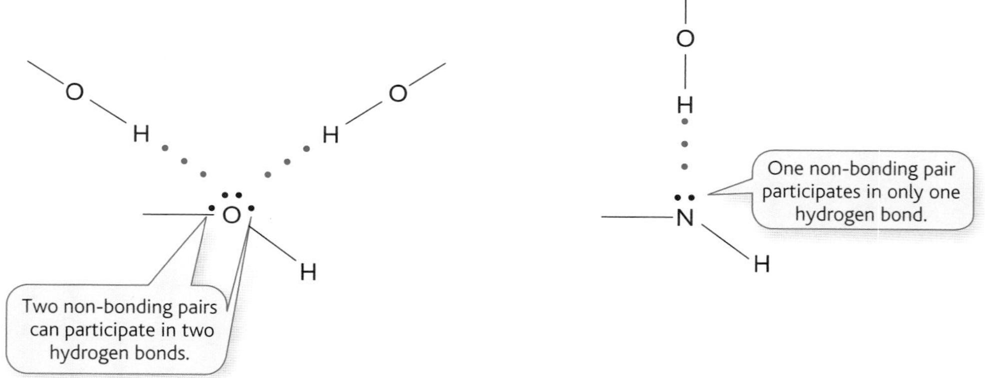
O
H
H
O
O
H
Two non-bonding pairs can participate in two hydrogen bonds.
O
H
N
H
One non-bonding pair participates in only one hydrogen bond.

Figure 6.18 Amines form fewer hydrogen bonds than the corresponding alcohols.

In addition, oxygen is *more* electronegative than nitrogen. Oxygen causes a greater degree of polarization than does nitrogen as a result of this greater electronegativity. Oxygen also forms stronger hydrogen bonds for the same reason.

Consequently, the non-covalent forces that are induced by oxygen are greater than those induced by nitrogen; we observe these greater non-covalent forces as higher melting and boiling points in alcohols compared to amines.

Self-check 6.9 Arrange these compounds in order of their boiling points, starting with the compound with the lowest boiling point.

(a) $CH_3CH_2OCH_3$

(b) $CH_3CH_2CH_3$

(c) $CH_3CH_2NH_2$

(d) CH_3CH_2OH

The amides: the amide group

Generic formula: $RCONH_2$

The **amides** are a family of organic compounds that contain the amide group (13)

(13)

Notice how the amide group closely resembles the carboxyl group (14) of carboxylic acids (Section 6.1).

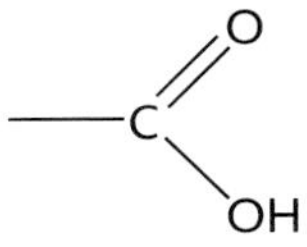

(14)

Both the carboxyl and amide group contain the carbonyl (C=O) group. However, in an amide, the hydroxyl portion of the carboxyl group is replaced with an **amino group**, $-NH_2$.

We see above how an amine may be classed as primary, secondary, or tertiary, depending on the number of R groups that are attached to the nitrogen atom. Amides are also classed as primary, secondary, and tertiary by applying exactly the same criteria to the amine portion of their amide group:

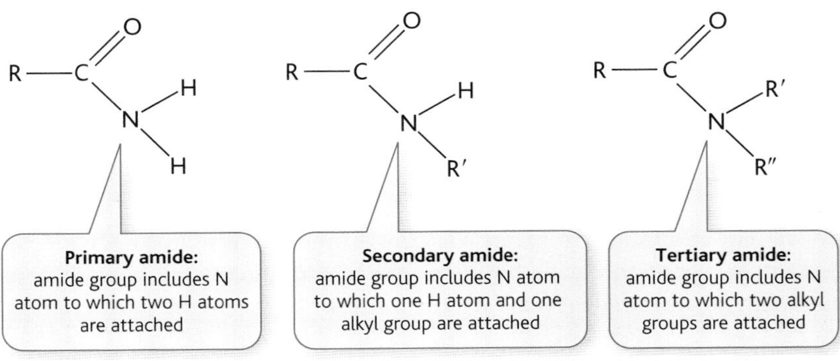
Primary amide:
amide group includes N atom to which two H atoms are attached
Secondary amide:
amide group includes N atom to which one H atom and one alkyl group are attached
Tertiary amide:
amide group includes N atom to which two alkyl groups are attached

- ***Amides possess the amide functional group comprising both a carbonyl group and an amino ($-NH_2$, $-NHR$, $-NR_2$) group.***

Naming amides

We name an amide by taking the equivalent carboxylic acid name, removing the suffix -oic acid, and replacing it with the suffix -amide. (The 'equivalent' carboxylic acid is the one that has exactly the same structure and composition as the amide but with a hydroxyl group attached to the carbonyl carbon, not an amino group.)

For example, the carboxylic acid that is equivalent to the amide represented by structure (**15**)

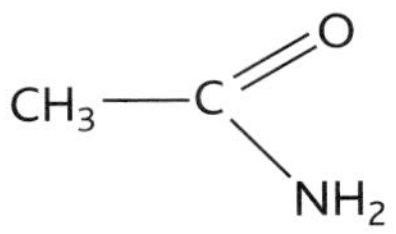

(15)

is ethanoic acid (**16**)

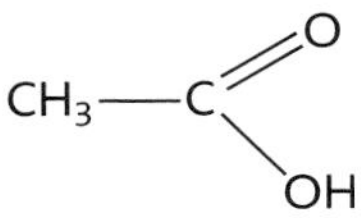

(16)

We therefore name the amide by:

1. removing the suffix **-oic acid** from the carboxylic acid name:
   ethanoic acid → ethan
2. replacing it with the suffix –amide:
   ethan → ethan**amide**.

Self-check 6.10 What is the name of this amide?

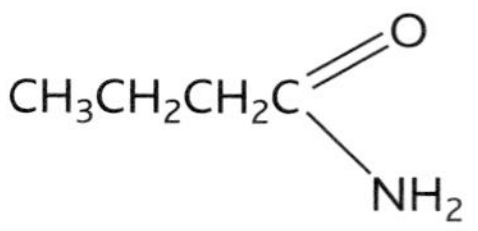

$CH_3CH_2CH_2C(=O)NH_2$

Physical properties of amides

Look at Table 6.11, which shows the physical property checklist for **primary** amides.

We see the same decrease in melting and boiling points as we move from primary amide to secondary amide to tertiary amide as we do for amines (p. 173). The reason is the same: Figure 6.19 shows how more hydrogen bonds can form in primary amides than in secondary amides, while no hydrogen bonds at all can form in tertiary amides.

Both amine and amide groups are a feature of biologically important molecules. For example, the amino acids histidine, lysine, and arginine all contain amino groups. Amino and amide groups are also a feature of one of the most important

Table 6.11 Physical property checklist for the primary amides.

Do the primary amides experience...		Consequence
Dispersion forces?	✓	High melting and boiling points
Dipolar interactions?	✓	
Hydrogen bonding between molecules of same compound?	✓	
Hydrogen bonding with water?	✓	Soluble in water

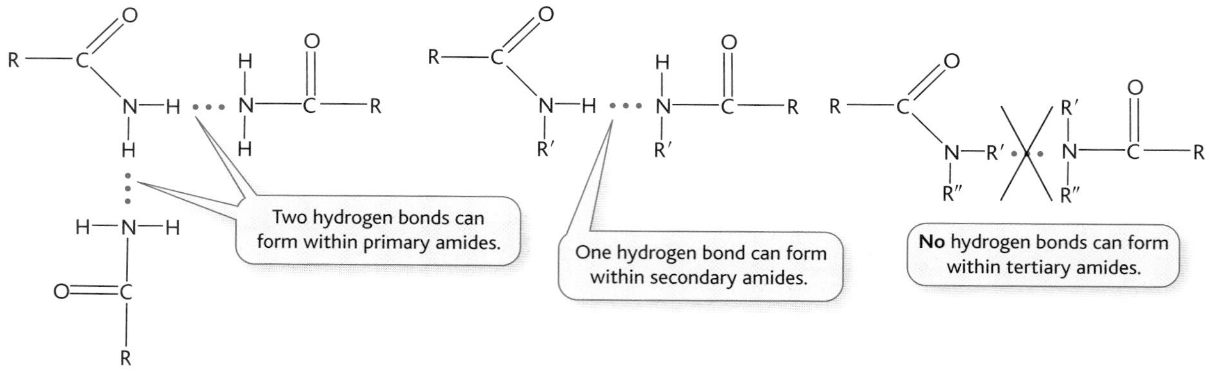

Two hydrogen bonds can form within primary amides.
One hydrogen bond can form within secondary amides.
No hydrogen bonds can form within tertiary amides.

Figure 6.19 More hydrogen bonds can form in primary amides than in secondary amides, while no hydrogen bonds at all can form in tertiary amides.

groups of biological molecule, the nucleic acids. We learn more about the amino acids and nucleic acids in Chapter 7.

6.3 Other functional groups

We conclude this chapter by considering two types of functional group that have as their central 'hub' atoms other than oxygen or nitrogen, the thiol group and the halo group.

The thiols and the sulfur-based functional group

The key sulfur-based functional group is the **thiol** group. The thiol group is structurally similar to the hydroxyl group (Section 6.1), but with a sulfur atom substituted for the oxygen atom:

Hydroxyl	Thiol
—O–H	—S–H

The thiol group does not impart the same functionality upon a molecule as those functional groups containing oxygen or nitrogen: sulfur is only as electronegative as carbon, so it does not induce additional polarity; neither is it able to facilitate hydrogen bond formation.

However, the thiol group plays a very important structural role in biological macromolecules, particularly proteins. Neighbouring thiol groups are able to join together as illustrated in Figure 6.20(a) to form a strong **disulfide bridge**. Disulfide bridges have a vital role in stabilizing the structure of proteins by firmly linking together amino acids from different parts of the protein to help 'lock' the protein in place. For example, Figure 6.20(b) shows how disulfide bridges hold together the two polypeptide chains of insulin into a single, stable complex.

We explore the role of disulfide bridges further in Chapter 9, where we find out more about the importance of the shape of biological molecules, and how shape can be stabilized and maintained.

The **thiols**–the class of organic compounds that contain the thiol group–have one very distinctive characteristic: their smell. Thiols are responsible for the pungent aromas of various common foods, including cheese and garlic. For example, 1-propanethiol (17) is the compound that gives onions their potent smell.

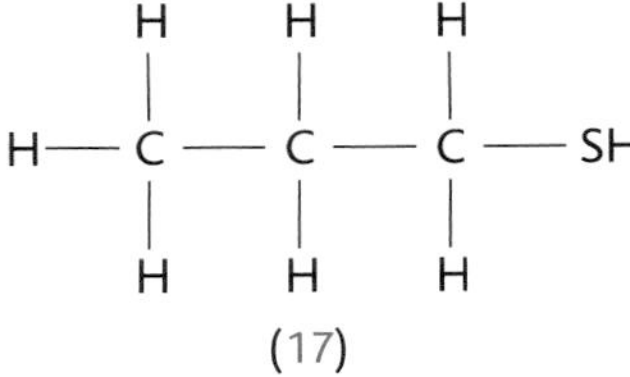

(17)

We may curse the thiols as we stand weeping over our chopping boards, trying to wield a razor-sharp kitchen knife through a blur of tears, but their penetrating smells can be exploited to our advantage. When we smell natural gas from the gas main, it's not actually the gas itself that we can smell, but small amounts of the thiol ethanethiol (shown in Figure 6.21) which are added to the gas supply. Natural gas (methane) is odourless; without the addition of some kind of distinctive smell to trigger our senses–and grab our attention–many gas leaks could go unnoticed for long periods of time, with potentially catastrophic consequences.

The haloalkanes and the halogen-based functional group

A final group that may augment the alkane-like framework of an organic compound is the halogen-based **halo**- group. The halogens, group 17 in the periodic table, are

(a)

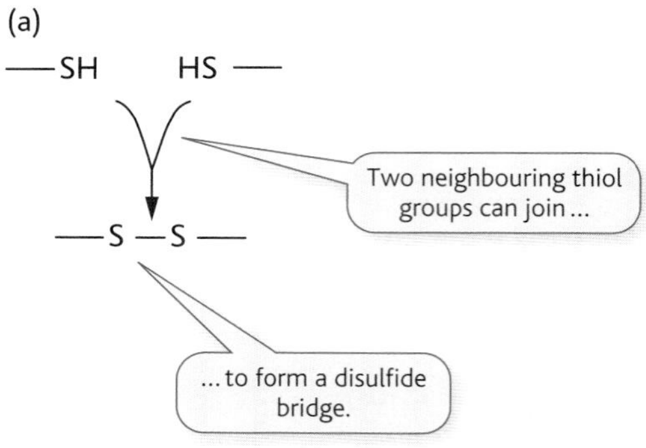
—SH HS—
—S—S—
Two neighbouring thiol groups can join ...
... to form a disulfide bridge.

(b)

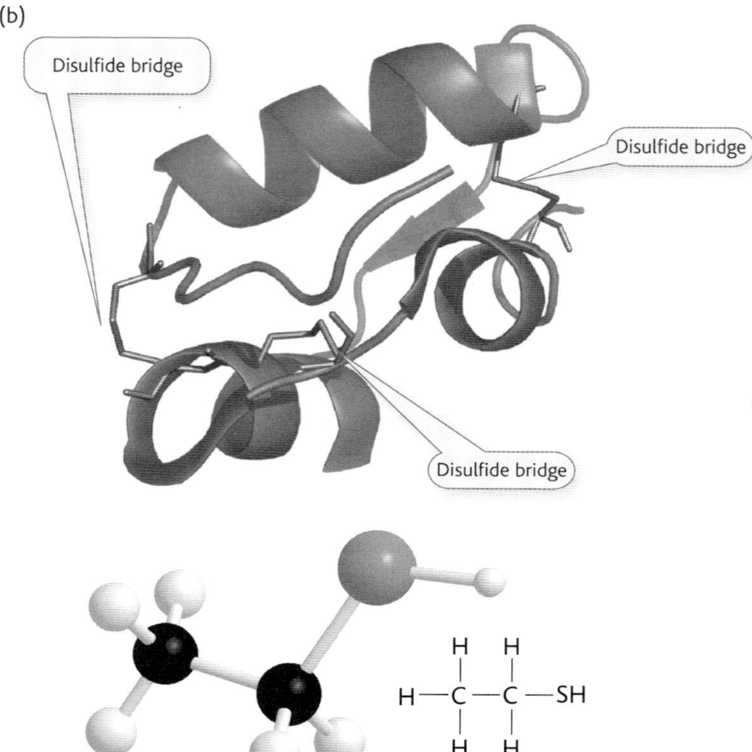
Disulfide bridge
Disulfide bridge
Disulfide bridge

Figure 6.20 (a) The structure of a disulfide bridge. A disulfide bridge is formed when neighbouring thiol groups join together as depicted here. (b) Disulfide bridges are important structural components of proteins. For example, disulfide bridges covalently join the two polypeptide chains that comprise the hormone insulin, as shown here.

Figure 6.21 The structure of ethanethiol.

highly electronegative, and readily form ionic compounds by strongly attracting an electron from a neighbouring atom to the extent the electron is fully transferred to the halogen atom, ensuring that the valence shell of the halogen becomes fully occupied (Section 3.1).

However, a halogen joins to the carbon backbone of an alkane by forming a single *covalent* bond with a carbon atom. The attachment of a halogen atom to the alkane-like framework of an organic compound generates a **haloalkane**. The most common haloalkanes contain chlorine, bromine, or iodine.

Naming the haloalkanes

The haloalkanes are named by prefixing the name of the parent alkane with a shortened version of the halogen name: **fluoro** for fluorine, **chloro** for chlorine, and **bromo** for bromine.

For example, the following haloalkane, in which a hydrogen in methane is replaced with chlorine is chloromethane:

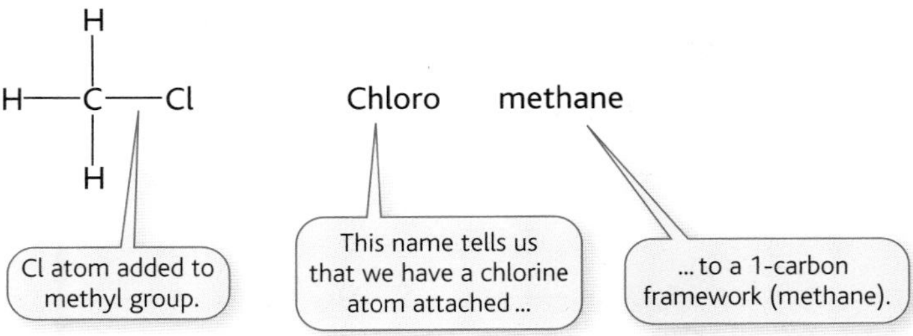

Similarly, the addition of bromine to ethane generates bromoethane:

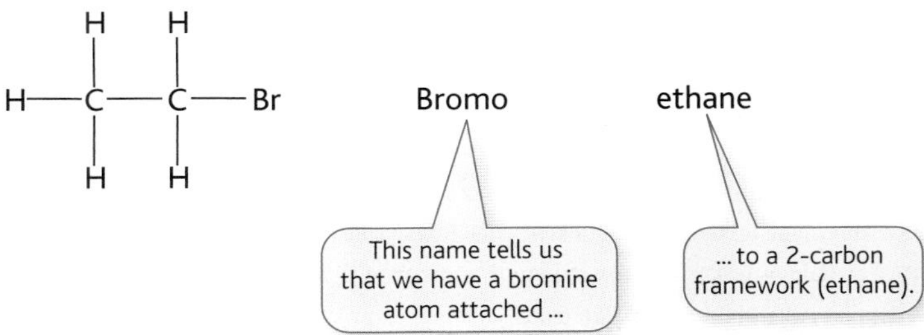

A number is inserted at the start of the name of the haloalkane to identify the point of attachment of a halogen atom part way along the carbon chain of a haloalkane. For example, structure (18) is named 2-bromobutane.

```
     H   Br  H   H
     |   |   |   |
 H — C — C — C — C — H
     |   |   |   |
     H   H   H   H
```

(18)

Q **Self-check 6.11** What are the names of the following haloalkanes?

(a)
```
     H
     |
 H — C — Br
     |
     H
```

(b)
```
     H   F   H   H   H
     |   |   |   |   |
 H — C — C — C — C — C — H
     |   |   |   |   |
     H   H   H   H   H
```

(c)
```
      H   H   H   H
      |   |   |   |
 Cl — C — C — C — C — H
      |   |   |   |
      H   H   H   H
```

The biological impact of haloalkanes

A haloalkane may comprise more than one halogen atom, which may attach to a single atom, or several different carbon atoms along the carbon backbone. For example, halothane (19) the general anaesthetic in most widespread use today, comprises a two-carbon backbone to which an atom of bromine, an atom of chlorine, and three atoms of fluorine are attached.

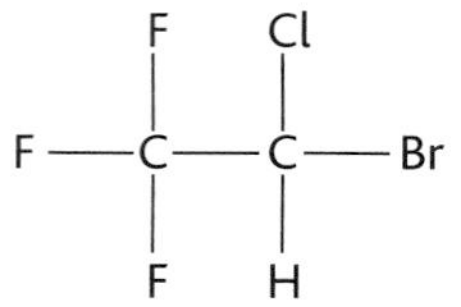

F Cl
F—C—C—Br
F H

(19)

Indeed, the anaesthetic properties of the haloalkanes are their most significant biological characteristic. Chloroform, $CHCl_3$, was once widely used as a general anaesthetic. However, it was found to be toxic, and hence withdrawn from use. Halothane is far less harmful to the body, and is now most widely used.

Perhaps the most infamous haloalkanes, however, are the chlorofluorocarbons (CFCs)–small molecules with central carbon atoms bonded to atoms of chlorine and fluorine only, as depicted in Figure 6.22.

CFCs were widely used as refrigerants until a link was discovered between the presence of these compounds in the atmosphere and the depletion of the ozone layer. The ozone layer is the protective shell of ozone gas (O_3) that shields us from the Sun's harmful ultraviolet rays, and helps to maintain the stability of the Earth's climate. It lies in the stratosphere, a volume of space that forms a band around the Earth between 15 and 50 km above its surface. The ozone layer works by absorbing ultraviolet radiation from the Sun and preventing it from reaching the surface of the Earth. It is thought that the depletion of the ozone layer is a contributory factor to the global warming which many environmental scientists believe we are now experiencing: a thinner ozone layer traps less ultraviolet radiation, allowing more to reach the Earth's surface, where the radiation causes warming.

CFCs do not destroy the ozone layer directly; rather, it is the action of chemical species, formed as CFCs decay in the stratosphere, that are thought to be to blame. Figure 6.23 shows how these chemical species–principally atomic chlorine (Cl) and chlorine monoxide (ClO)–cause the breakdown of ozone into molecular oxygen (O_2), which lacks the ability of ozone to absorb UV radiation.

The production of CFCs is now limited under the terms of the Protocol on Substances that Deplete the Ozone Layer (known more widely as the Montreal Protocol), which came into force on 1 January 1989. Proof that concerted global action can

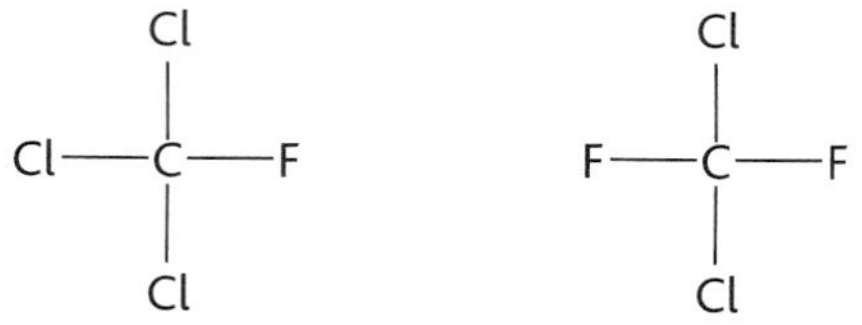

Cl
Cl—C—F
Cl

Trichlorofluoromethane
(Freon 11)

Cl
F—C—F
Cl

Dichlorodifluoromethane
(Freon 12)

Figure 6.22 The structure of two chlorofluorocarbons, trichlorofluoromethane and dichlorodifluoromethane.

Atomic chlorine · Ozone · Molecular oxygen

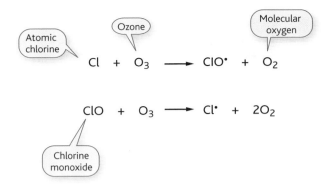

$$Cl + O_3 \longrightarrow ClO^{\bullet} + O_2$$

$$ClO + O_3 \longrightarrow Cl^{\bullet} + 2O_2$$

Chlorine monoxide

Figure 6.23 Atomic chlorine and chlorine monoxide can react with ozone to generate molecular oxygen, O_2. O_2 lacks the ability of ozone to absorb UV radiation, such that more of this radiation enters Earth's atmosphere, where it causes warming.

make a difference, the Protocol has reduced CFC production by over 95%. Indeed, if the Protocol (which has undergone seven revisions since its introduction in 1989) is adhered to, it is believed that the ozone layer could recover by 2050.

However, it is not all good news. In many cases, CFCs have been replaced by hydrofluorocarbons (HFCs), another family of haloalkanes which contain only fluorine (and not both chlorine and fluorine, as is the case with CFCs). Examples include trifluoromethane (or fluoroform) (**20**)

$$CHF_3$$

(20)

and fluoromethane (**21**)

$$CH_3F$$

(21)

While HFCs do not damage the ozone layer in the way that CFCs were found to do, HFCs are far from environmentally friendly: they are potent greenhouse gases, whose warming effect is thought to be thousands of times greater than that of carbon dioxide. Their warming effect has not gone unnoticed, however: a commitment to reducing levels of atmospheric HFCs forms part of the Kyoto Protocol, which came into force in 2005.

Check your understanding

To check that you've mastered the key concepts presented in this chapter, review the checklist of key concepts below, and attempt the multiple-choice questions available in the book's Online Resource Centre at **http://www.oxfordtextbooks.co.uk/orc/crowe2e/**.

Checklist of key concepts

Oxygen-based functional groups

- The alcohols carry the hydroxyl functional group, –OH, and have the generic formula ROH
- Small alcohols are readily soluble in water
- Alcohols experience large intermolecular non-covalent forces, and so have high melting and boiling points
- Ethers have the generic formula ROR′
- Small ethers are soluble in water, but have lower melting and boiling points than the corresponding alcohols
- Aldehydes and ketones possess the carbonyl group, C=O
- Aldehydes have the generic formula RCHO
- Ketones have the generic formula RCOR′–they contain two alkyl groups whereas aldehydes contain only one
- Small aldehydes and ketones are soluble in water, but have lower melting and boiling points than the corresponding alcohols
- Carboxylic acids carry the carboxyl functional group, COOH, and have the generic formula RCOOH
- Small carboxylic acids are very soluble in water because they can form multiple hydrogen bonds
- The esters possess a modified carboxyl group in which the hydroxyl group (–OH) is replaced with an alkoxy group (–OR′)
- Small esters are soluble in water

Nitrogen-based functional groups

- The amines are based on ammonia, NH_3, in which one or more hydrogen atoms are replaced with R groups
- A primary amine has one hydrogen atom replaced by an R group
- A secondary amine has two hydrogen atoms replaced by R groups
- A tertiary amine has three hydrogen atoms replaced by R groups
- Primary, secondary, and tertiary amines have different physical properties because they can form different numbers of hydrogen bonds
- Amines have lower melting and boiling points than the corresponding alcohols because nitrogen is less electronegative than oxygen, and so generates weaker intermolecular non-covalent forces
- All three types of amine are water-soluble
- The amides possess the amide functional group, comprising both a carbonyl group and an amino ($-NH_2$, –NHR, $-NR_2$) group
- Amides may also be primary, secondary, or tertiary

Other functional groups

- The thiols contain the sulfur-based thiol group, –SH
- The thiol group is particularly important in the structural biology of proteins because two thiol groups on the side chains of adjacent amino acid residues can join together to form a strong disulfide bridge
- The haloalkanes possess the halo group (a halogen atom)
- Haloalkanes are formed from the joining of a halogen atom to an alkyl group

7

Biological macromolecules: providing life's infrastructure

 INTRODUCTION

Organic compounds are amongst the key materials from which all living organisms are made. The cells from which our bodies are constructed comprise lipids, proteins, and sugars, just as a builder uses bricks, mortar, and plaster to construct the walls of a house. Organic compounds also play an essential role in regulating life: they act as messengers to pass signals from cell to cell; they store and transmit genetic information to ensure organisms develop as they should, from generation to generation.

In Chapter 6 we see how functional groups add chemical 'character' to the otherwise simple carbon skeleton of an organic compound. In this chapter we see how some of the most important organic compounds in nature comprise more than one functional group, and how these compounds are the building blocks for the range of biological macromolecules on which we depend for survival. We also see how organic compounds do not have a monopoly on the chemistry of life, but that metals also play a vital role.

7.1 Amino acids and proteins

Amino acids are the building blocks of proteins, the macromolecules that have diverse and vital functions in biological systems. Although many different amino acids exist, only twenty occur in those proteins found in humans, and we focus on these here.

The composition of amino acids

The 'hub' of an amino acid is a central carbon atom, to which four different components are attached:

1. An amino group ($-NH_2$)
2. A carboxyl group ($-COOH$)

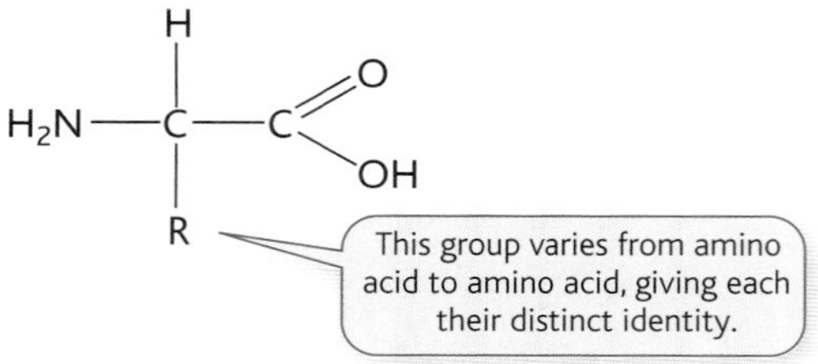

Figure 7.1 The general structure of an amino acid.

3. A hydrogen atom
4. A variable side chain (denoted 'R').

Taken together, these components generate the general structure of an amino acid depicted in Figure 7.1.

Notice how an amino acid has *two* functional groups: an amino group and a carboxyl group.

The twenty amino acids each have contrasting chemical characteristics that are determined by the composition of the variable side chain, R. Look at Figure 7.2, which shows the structural formulae of the side chains of the twenty amino acids. Notice how the side chain may comprise an alkyl or aromatic group, or may comprise an additional functional group. For example:

- The side chains of serine, threonine, and tyrosine each contain a **hydroxyl** group.
- The side chain of cysteine contains a **thiol** group.
- The side chains of asparagine and glutamine contain an **amide** group.
- The side chains of aspartic acid and glutamic acid contain an additional **carboxyl** group.

Notice too the unusual cyclic structure of proline, in which the side chain is 'fused' to the amino nitrogen.

The amino acids in Figure 7.2 are divided into two groups: those that are hydrophobic, and those that are hydrophilic. Why is this grouping important? The hydrophobicity (or hydrophilicity) of amino acids in a given polypeptide has a big impact on how that polypeptide folds–that is, the three-dimensional shape that the polypeptide adopts. Hydrophobic amino acids will preferentially cluster on the inside of a folded polypeptide (where they will be shielded from the aqueous surroundings); by contrast, hydrophilic amino acids will preferentially lie on the surface of a folded polypeptide, fully exposed to the aqueous surroundings.

We learn more about the impact of amino acids on the three-dimensional structure of proteins in Chapter 9.

If we look again at Figure 7.2 we also notice how several of the amino acids exist in ionized form. For example, the –OH groups of the carboxyl side chains of aspartic and glutamic acid are ionized to give the charged $–O^-$ species. Ionized amino acid side chains can make an important contribution to the biological activity of a protein: such groups can often participate directly in biochemical reactions occurring at the active site of an enzyme, for example.

We learn more about the behaviour of acids and bases in Chapter 16.

Self-check 7.1 Disulfide bridges are important structural components of proteins, and are formed from the joining of two adjacent thiol groups on different amino acids. Which amino acid has an important role in disulfide bridge formation?

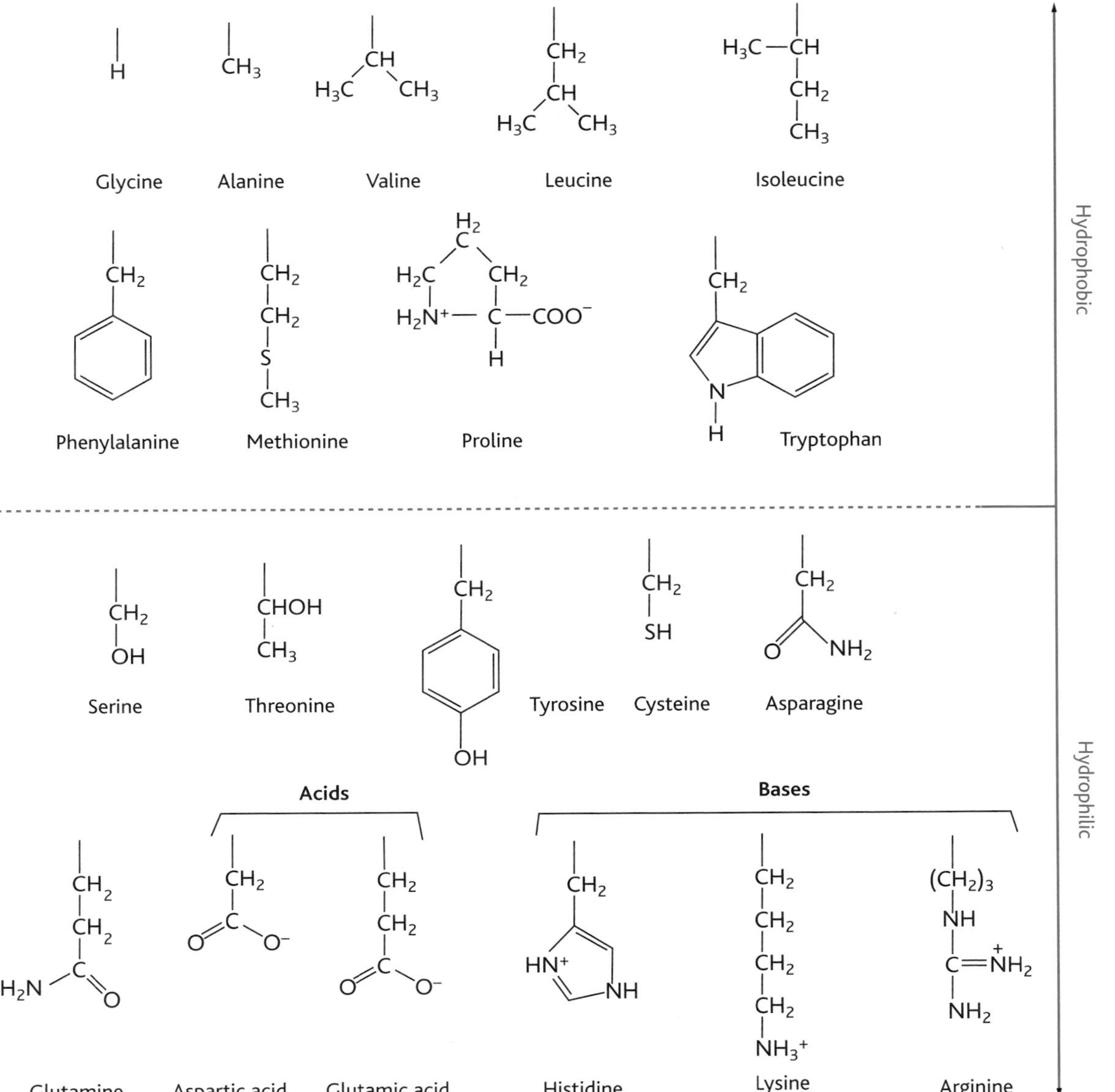

Figure 7.2 The structural formulae of the side chains of the twenty naturally occurring amino acids. Notice how these amino acids are split into two groups: those that are hydrophobic, and those that are hydrophilic.

The simplest amino acid is glycine; look at Figure 7.2 and notice how the R group in glycine is a hydrogen atom.

Formation of polypeptides

Proteins are **polymers** of amino acids[1]: many amino acids are joined together to form a long chain. These long chains are properly termed **polypeptides**; a protein

1 A polymer comprises many basic building blocks joined together (poly = many). The basic building blocks of a polymer are termed monomers (mono = one).

BOX 7.1 Tasting the difference: the foodie's fifth Beatle

We are all familiar with four of the basic tastes–bitter, sweet, sour, and salty. But what about umami? Umami is the food equivalent of the Fifth Beatle–the one-time member of that well-known band that most of us have never heard of (a label that, in the context of the Beatles themselves, could be attached to either Stuart Sutcliffe or Pete Best, the original bassist and drummer respectively).

Umami is described, quite generically, as a savoury taste; it is characteristic of protein-heavy foods, including meats, cheese, and stocks. Its association with protein-heavy foods gives a clue to its origin: the umami taste comes from the presence in a food of the amino acid glutamate.

Umami is a Japanese word meaning 'savoury', but the oriental link goes further than this: glutamate–as the root of this taste–is widespread in oriental foods, including soy sauce and fish sauce.

The savoury taste of foods can often be enhanced by adding a source of glutamate to them, to enrich the umami experience. This is often achieved by the use of the somewhat controversial monosodium glutamate (MSG), whose widespread use in Chinese cooking gives us yet another oriental strand to the umami story.

MSG's controversial status stems from the way that some individuals have been found to exhibit short-term symptoms (together often called 'Chinese Restaurant Syndrome') after consuming MSG. These symptoms include numbness, burning sensation, tingling, facial pressure or tightness, chest pain, headache, nausea, rapid heartbeat, drowsiness, and weakness. Those most at risk of developing these symptoms include individuals consuming large quantities of MSG on an empty stomach and those with severe asthma.

However, a number of research groups–among them the Federation of American Societies for Experimental Biology (FASEB)–have concluded that MSG is perfectly safe for consumption by the general population if consumed at usual levels (less than 0.5 mg per meal). So, as is often the case, moderation is the key.

may comprise a single polypeptide chain, or may comprise several polypeptides that associate closely to form an overall functional unit. For example, the hormone insulin comprises two polypeptide chains, while haemoglobin (which we encounter in Section 4.1) comprises *four* subunits, each subunit being a single polypeptide chain.

Insulin is synthesized in the body as an inactive 84 amino acid polypeptide chain, **proinsulin**. The active form of insulin is generated by the cleavage (removal) of 33 of the amino acid residues that lie in the central part of the inactive polypeptide chain, leaving two separate polypeptide segments that are held together by disulfide bridges, as shown in Figure 7.3.

We explore in Chapter 17 the different chemical reactions that underpin the joining of molecules, and learn about the three-dimensional structure of proteins in Chapter 9.

A polypeptide chain forms when the carboxyl group of one amino acid reacts with the amino group of another amino acid to form an amide group, as illustrated in Figure 7.4. The amide link between adjacent amino acids in a polypeptide chain is called the **peptide bond**. It is the presence of the carboxyl and amino groups, and the chemical properties that they possess, which confer upon amino acids the ability to form peptide bonds and, hence, join together to form long chains.

Once amino acids are joined together via a peptide bond we refer to them as amino acid **residues**. A peptide comprises only a few amino acid residues, whereas a polypeptide comprises many amino acid residues.

- *Polypeptides are polymers of amino acids.*
- *Neighbouring amino acid residues in a polypeptide chain are joined by a peptide bond.*

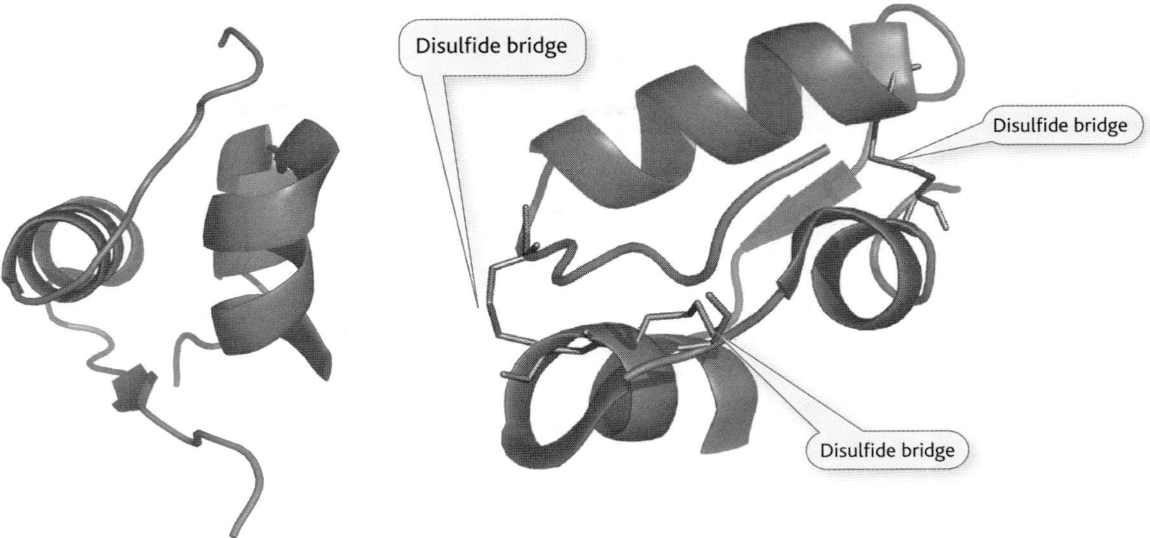

Figure 7.3 The three-dimensional structure of insulin. (a) This side-on view shows how insulin comprises two separate polypeptide chains. (b) The two polypeptide chains are held together by a series of disulfide bridges, which form between the thiol groups of cysteine residues that lie close together in space.

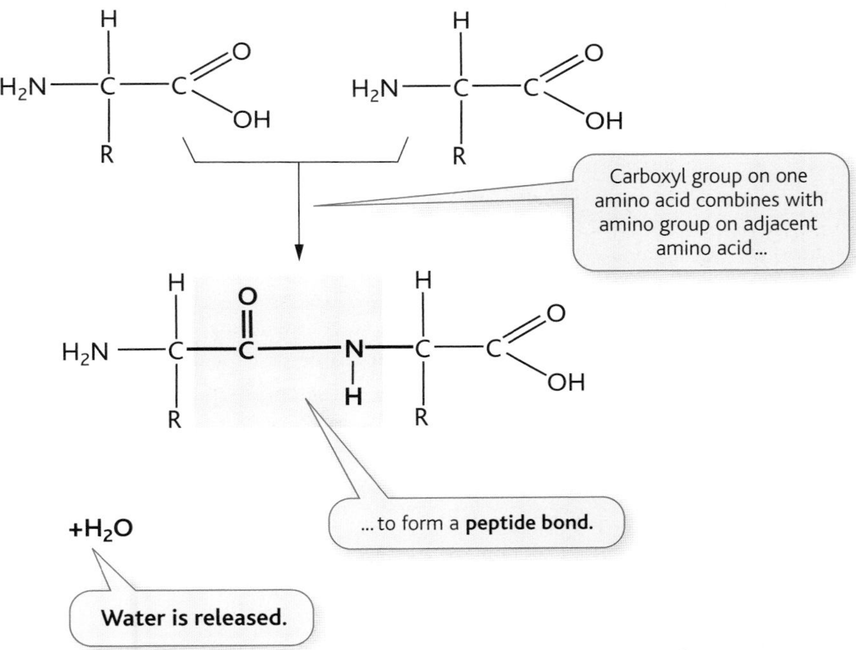

Figure 7.4 A polypeptide chain forms when the carboxyl group of one amino acid reacts with an amino group of another amino acid to form a peptide bond.

BOX 7.2 The bitter taste of sweeteners

Have you ever noticed food packaging that bears the label 'contains a source of phenylalanine', and wondered why this warning needs to be given? Amino acids may be the vital building blocks of proteins, which, in turn, are vital to our well-being, but in certain circumstances some amino acids can do more harm than good.

Phenylketonuria (PKU) is a rare genetic disease in which the body is unable to metabolize the amino acid phenylalanine. Normally, the enzyme phenylalanine hydroxylase (PAH) is responsible for converting phenylalanine into tyrosine, another amino acid:

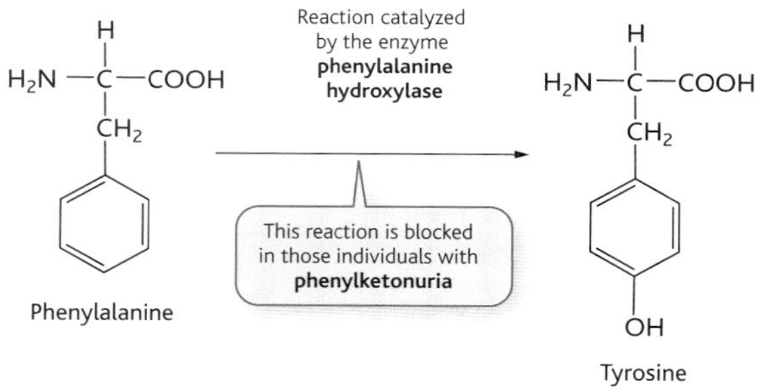

Reaction catalyzed by the enzyme **phenylalanine hydroxylase**

This reaction is blocked in those individuals with **phenylketonuria**

Phenylalanine

Tyrosine

Individuals suffering from phenylketonuria, however, carry a mutation in the gene encoding PAH, such that the enzyme is less efficient, or even completely inactive. As a result, phenylalanine fails to be converted to tyrosine and accumulates in the body instead.

This accumulation of phenylalanine causes serious health problems, including mental retardation and organ damage. The prognosis need not be bleak though: babies born with PKU can be diagnosed at birth using an appropriate blood test, and the onset of health problems associated with the accumulation of phenylalanine can be avoided by the careful control of the amount of phenylalanine in the diet. For example, high-protein foods contain phenylalanine by their very nature, and must be avoided.

The majority of foodstuffs carrying the warning 'contains a source of phenylalanine' include those containing the artificial sweetener aspartame. These foodstuffs typically include carbonated drinks, 'health' (low-sugar) drinks, and artificial sweetener tablets.

Figure 7.5 shows the structural formula of aspartame. Look at Figure 7.5 and notice how the central portion of the molecule is derived from phenylalanine. When aspartame is metabolized in the body, phenylalanine is produced. Aspartame is the 'source of phenylalanine' that such labels warn against, and must not be consumed by individuals with phenylketonuria. The phenylalanine produced from the breakdown of aspartame will simply accumulate in the body.

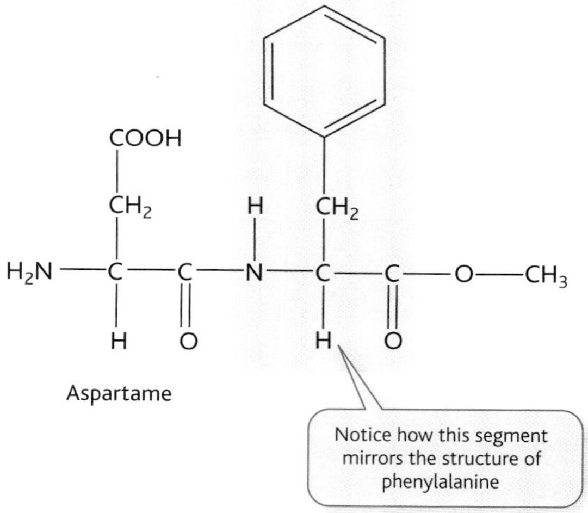

Aspartame

Notice how this segment mirrors the structure of phenylalanine

Figure 7.5 The structural formula of aspartame.

7.2 Carbohydrates

We all know how a sugary snack can help to give us an energy boost: glucose is a fuel for the chemical reactions which power the human body just as petrol is a fuel for the chemical reactions which power the car engine.

The carbohydrates are known more simply as 'sugars', and provide us with a vital source of energy. A marathon runner often consumes a 'high carb' diet in the weeks before a race–a diet heavy in carbohydrate-rich foods, such as pasta and potatoes. The plentiful supply of carbohydrates enables the body to build up reserves of energy-rich glycogen in the body tissue, which is then broken down during the race to help power the marathon runner and help them complete the full gruelling distance.

We rely on carbohydrates for our survival. During the process of **photosynthesis**, plants use carbon dioxide, water, and energy (from sunlight) to form glucose (a carbohydrate) and oxygen. The opposite process occurs in animals: glucose and oxygen are consumed to form carbon dioxide, water, and energy: a process known as **respiration**. Look at Figure 7.6 to see the interrelationship between these two processes.

There are three key classes of sugar: **monosaccharides, disaccharides**, and **polysaccharides**. These forms of sugar are characterized by the number of sugar (saccharide) units from which they are formed. The monosaccharides are single sugar units–the simple building blocks from which disaccharides and polysaccharides are formed. Disaccharides comprise two monosaccharide units joined together, while polysaccharides have multiple (many) monosaccharides joined in a long chain.

Notice how the simple building block of sugars (the monosaccharide) polymerizes to form long-chain polysaccharides in an analogous fashion to the simple building block of proteins (the amino acid) polymerizing to form long-chain polypeptides. This is another example of one of our essential concepts: simple building blocks can join together to form larger, more complicated structures.

- *Mono = one; Monosaccharide = one sugar unit*
- *Di = two; Disaccharide = two sugar units joined together*
- *Poly = many; Polysaccharide = many sugar units joined together*

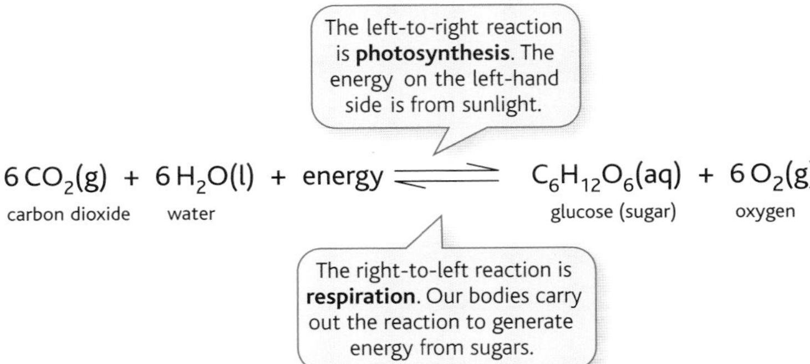

The left-to-right reaction is **photosynthesis**. The energy on the left-hand side is from sunlight.
$6\,CO_2(g) + 6\,H_2O(l) + \text{energy} \rightleftharpoons C_6H_{12}O_6(aq) + 6\,O_2(g)$
carbon dioxide
water
glucose (sugar)
oxygen
The right-to-left reaction is **respiration**. Our bodies carry out the reaction to generate energy from sugars.

Figure 7.6 The interrelationship between photosynthesis and respiration. Note that very different molecular components carry out the two reactions even though, at the level of chemistry, they are two sides of the same coin.

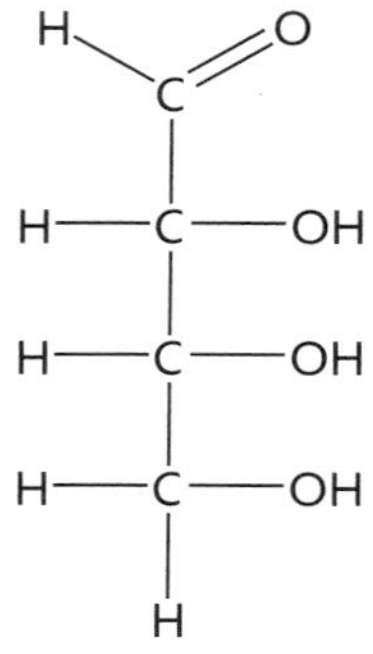

H–C=O
H–C–OH
H–C–OH
H–C–OH
H

Figure 7.7 The structure of a typical monosaccharide, erythrose.

The composition of monosaccharides

The single sugar unit represented by a monosaccharide possesses two functional groups: a carbonyl group and a hydroxyl group. Figure 7.7 shows how a monosaccharide has a single carbonyl group, but possesses multiple hydroxyl groups.

The carbon framework on which a monosaccharide is constructed comprises between three and eight carbon atoms. The most important monosaccharides are glucose, galactose, and fructose, which possess six-carbon backbones. Look at Figure 7.8 and notice how all three monosaccharides possess six carbon atoms, five hydroxyl groups, and one carbonyl group, but differ very slightly in the way in which these functional groups are attached to the carbon backbone.

The monosaccharides are able to adopt two distinct structures: a **straight-chain** (or **open-chain**) structure, and a **cyclic** (ring) structure. The transition to a cyclic structure occurs when the hydroxyl group on the fifth carbon atom in the carbon backbone reacts with the carbonyl group (either the first or second carbon atom in the backbone). Figure 7.9 shows how the reaction between these two functional groups generates a cyclic structure in which carbon 1 and carbon 5 become linked by a bridging oxygen atom.

We learn more about the conformation of six-carbon sugars on p. 244.

In nature, most of the glucose molecules exist in the cyclic form.

Self-check 7.2 Look at Figure 7.9, and use this to help when drawing the cyclic structure of galactose.

When adopting a cyclic structure, many sugars exhibit a distinct functional group called the **hemiacetal** group. A hemiacetal group is characterized by a carbon atom to which is attached both a hydroxyl (–OH) group, and an OR group (that is, an alkoxy group), as illustrated in Figure 7.10(a). Look at the cyclic structure of glucose in Figure 7.10(b), and notice how the blue carbon is the central atom of a hemiacetal group. Also notice how the –OR and R′ groups in this instance are joined to one another as part of the overall cyclic structure.

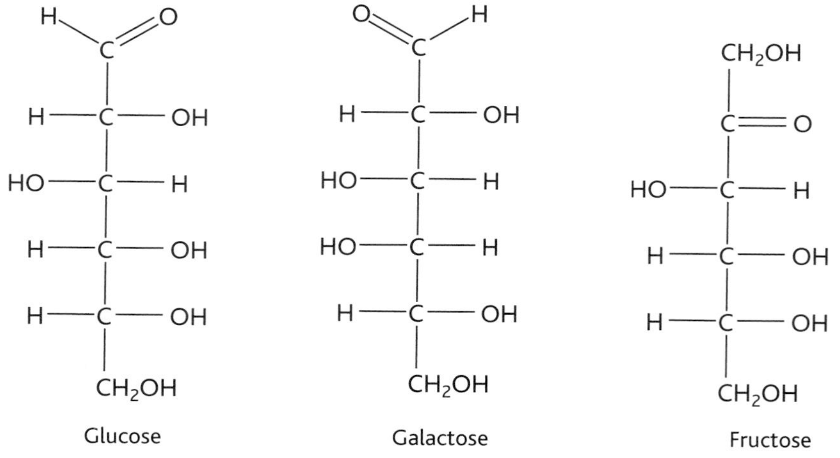

Glucose	Galactose	Fructose
H–C=O	H–C=O	CH_2OH
H–C–OH	H–C–OH	C=O
HO–C–H	HO–C–H	HO–C–H
H–C–OH	HO–C–H	H–C–OH
H–C–OH	H–C–OH	H–C–OH
CH_2OH	CH_2OH	CH_2OH

Figure 7.8 The structures of the monosaccharides glucose, galactose, and fructose.

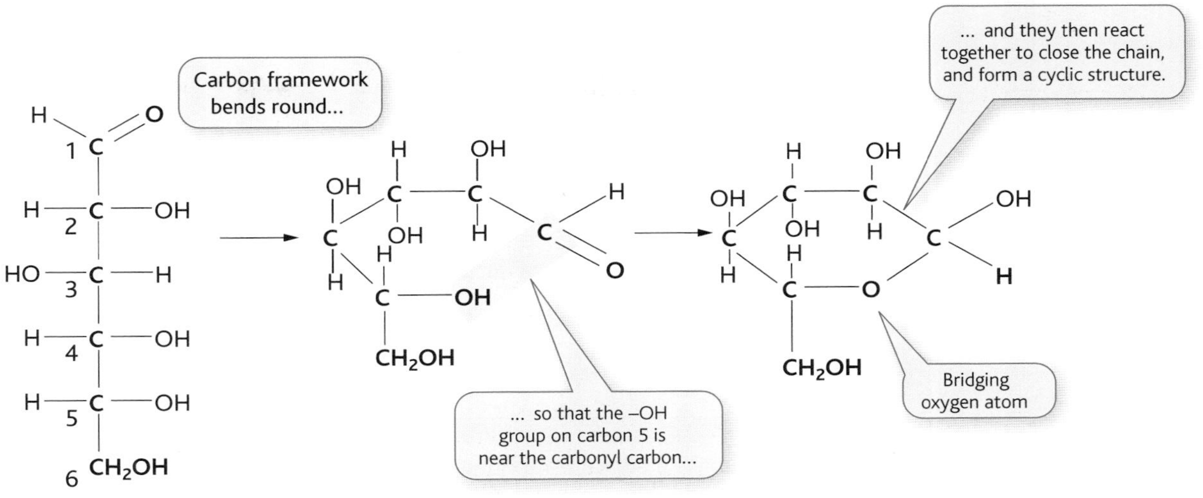

Figure 7.9 The cyclic form of monosaccharides occurs when the two ends of the straight chain form join together to form an unbroken ring.

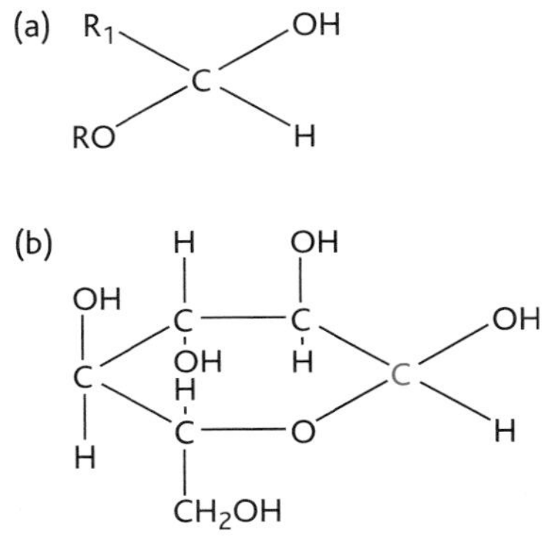

Figure 7.10 (a) A hemiacetal group is characterized by a central carbon atom to which both an alkoxy group (–OR) and a hydroxyl group (–OH) are attached. (b) In this cyclic form of glucose, the carbon shown in blue is the central atom of a hemiacetal group.

If we now compare Figure 7.10(b) to the straight-chain version of glucose in Figure 7.8 we notice how none of the carbon atoms in the straight-chain glucose molecule is part of a hemiacetal group. (Check this for yourself: none of the C atoms have –OH, –OR, H, and R′ groups all attached.) So we see how the hemiacetal group is only present when glucose adopts a cyclic structure.

Monosaccharides can be divided into two groups according to the carbon atom to which the carbonyl group is attached (when the molecule adopts a straight-chain conformation). If the carbonyl group is attached to the terminal (end) carbon atom, the monosaccharide is termed an **aldose**. If the carbonyl atom is attached to the second carbon atom, the monosaccharide is termed a **ketose**. For example, the carbonyl group in glucose is attached to the terminal carbon, so glucose is an aldose. In contrast, fructose possesses a carbonyl group attached to the second carbon atom; fructose is a ketose.

Another biologically vital monosaccharide is the five-carbon sugar ribose, a central component of those nucleic acids that store and transmit genetic information in all living organisms.

Ribose

We explore the composition of nucleic acids in more detail in Section 7.4.

Self-check 7.3 Look at the structure of galactose in Figure 7.8. Is galactose an aldose or a ketose?

The link between adjacent monosaccharide units is called a **glycosidic bond**. A glycosidic bond forms when a hydroxyl group on one monosaccharide unit joins with a second hydroxyl group on another monosaccharide unit.

Figure 7.11 shows how the hydroxyl group attached to carbon 1 of galactose joins with the hydroxyl group attached to carbon 4 on glucose to form the disaccharide lactose. Look at this figure, and notice how the glycosidic bond takes the form of a bridging oxygen atom that links the two monosaccharide rings.

Self-check 7.4 Which monosaccharides form the disaccharide sucrose?

Sucrose

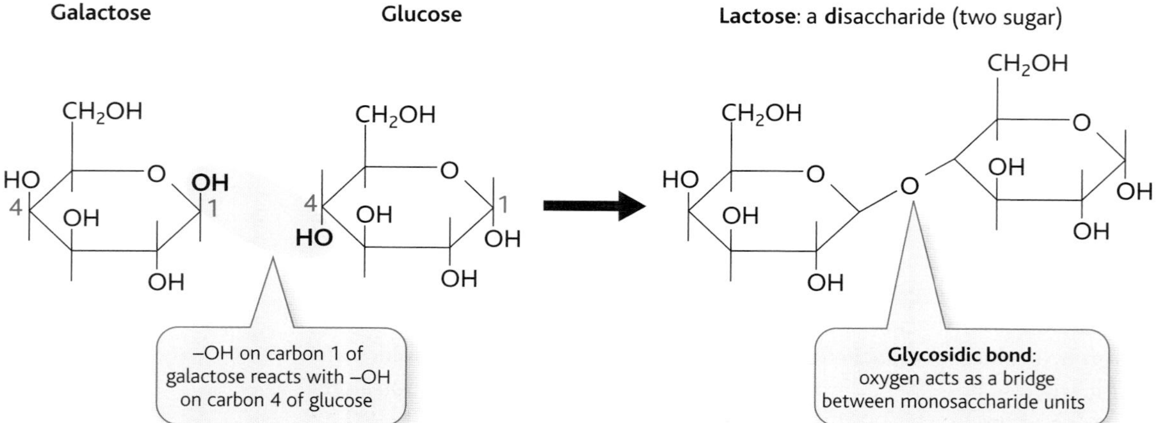

Figure 7.11 The structure of the disaccharide lactose. Lactose is formed when the monosaccharides galactose and glucose are joined by a glycosidic bond.

Once again, we see how it is the presence of specific functional groups (in this instance the carbonyl and hydroxyl groups) that confer on sugars the ability to join to one another to form larger, more complicated molecules.

7.3 Lipids

Lipids have many vital biological functions and mediate biological processes upon which we all depend for survival. The three most important types of lipid in biological systems are the **steroids**, the **triacylglycerols**, and the **glycerophospholipids**. Let's look at each of these in turn.

Steroids

We see in Chapter 5 how the basic skeleton of a steroid is a series of fused alkyl rings (1).

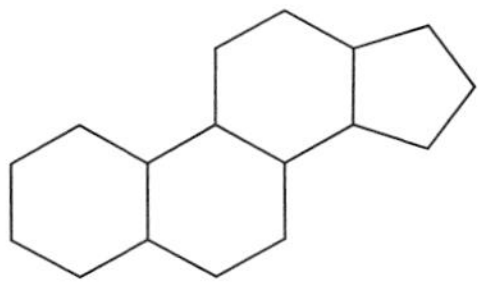

(1)

An array of functional groups can be added to the basic steroid skeleton to generate compounds with a range of functions. Compounds built upon the steroid template include the sex hormones testosterone (2), oestradiol (3) and cholesterol.

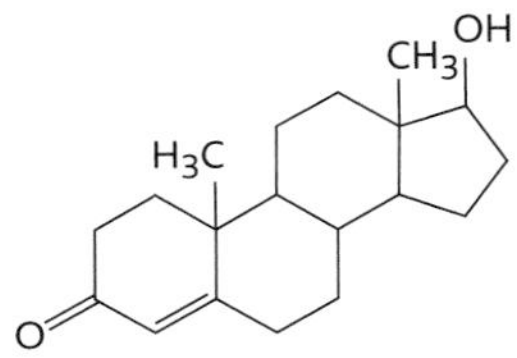

H_3C CH_3 OH O

(2)

CH_3 OH HO

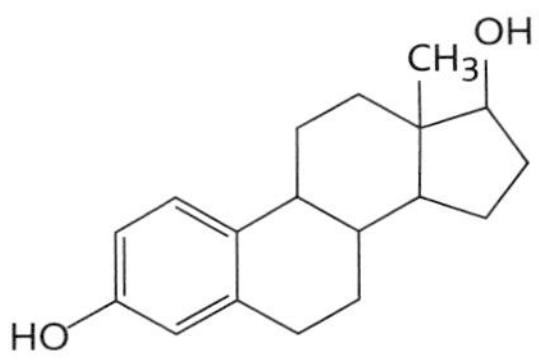

(3)

Our proper development and physiological well-being are being threatened by synthetic chemicals in the environment that are disrupting the effect of important sex hormones in the body, as explained in more detail in Box 7.3.

BOX 7.3 The chemical gender-benders

There are some obvious physical characteristics that help differentiate the male of our species from the female: facial hair, deeper voice, lack of breasts. These sexual characteristics are regulated by the sex hormones, the male-determining androgens (including testosterone), and a relative lack of the female-determining oestrogens (including oestradiol).

It is not the presence or absence of these hormones in the body that make us either typically male or female (the bodies of males and females contain both androgens and oestrogens, and the presence of both types of hormone is essential for proper development). Rather, it is the *relative* amounts of these hormones–the balance that exists between them–that are important.

The sex hormones form part of our **endocrine system**, a carefully controlled and balanced network of hormones that regulates the operation of our body. The relative amounts of a hormone dictate whether we feel awake or sleepy, they control the growth and renewal of damaged and worn-out tissues, and even impart sexual urges and tell us to feel hungry or thirsty.

It is now thought, however, that the balance of the endocrine system is being disrupted by chemicals known as **endocrine disrupters**, which are present in substances as diverse as pesticides, plastics, dyes, and deodorants. The endocrine disrupters work in two ways. Firstly, they may possess structures that mirror those of hormones so closely that they are able to mimic the hormones' activity–in other words, they are able to have the same effect on the human body as the hormone itself. Alternatively, they may block the activity of hormones, stopping them from communicating their intended messages within the body.

Of particular concern are those endocrine disrupters that upset the balance of the sex hormones in the body. The influence of endocrine disrupters has already been detected in marine organisms: the chemical tributyltin (TBT) (**4**) has been linked to the development of male sexual organs in a female marine snail; and synthetic oestrogen-mimicking chemicals in aquatic systems have been linked to the production of female-specific proteins by male fish.

We have yet to establish whether endocrine disrupting chemicals are having a definite effect on humans, but there is evidence of falling sperm density in the western world, which some are attributing to the effect of endocrine disrupters on the balance of androgens and oestrogens in males. It is also now believed that hormones from the female contraceptive pill are excreted in a woman's urine, and are ultimately transferred into drinking water. There is now much evidence that the presence of these female hormones in drinking water affects men.

The debate over the true effect of endocrine disrupters is set to continue, but it certainly looks set to add something of a sinister twist to the classical nature/nurture debate.

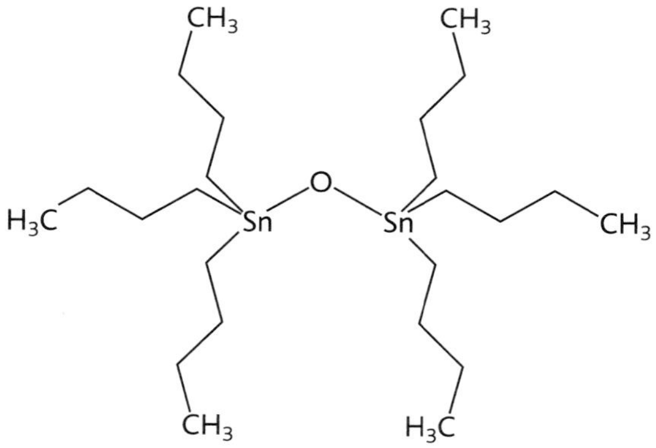

CH_3 CH_3

H_3C Sn O Sn CH_3

CH_3 H_3C

(4)

Cholesterol

Cholesterol (5) is a non-polar **sterol** (based on the steroid template), which is insoluble in water, and other water-based substances such as blood. Cholesterol is synthesized in the liver, and is needed in tissues throughout the body. The main network for transporting materials round the body is the blood supply. So how does cholesterol get where it's needed if it is insoluble in the blood? We discover the answer in Box 7.4.

H_3C CH_3 CH_3 CH_3 CH_3 HO

(5)

Triacylglycerols

The triacylglycerols are a family of organic compounds that are better known to us as **fats** and **oils**. The triacylglycerols are derived from two compounds:

1. The three-carbon alcohol, **glycerol** (6)
2. **Fatty acids**, those carboxylic acids possessing long carbon backbones (p. 165)

OH HO HO

(6)

BOX 7.4 The transport of cholesterol in the blood

Cholesterol is transported in the bloodstream when associated with a soluble protein, named **lipoprotein**. Different lipoproteins act as 'tags' that direct the transport of cholesterol to different locations in the body (just as we write an address on a letter so it gets delivered to the right place). There are two key members of the lipoprotein family: low-density lipoprotein (LDL) and high-density lipoprotein (HDL), which form two distinct complexes with cholesterol, **LDL-cholesterol** and **HDL-cholesterol**.

HDL directs the transport of excess cholesterol to the liver, where it can be broken down safely (metabolized). Therefore HDL-cholesterol has come to be known as 'good cholesterol'. By contrast, LDL directs the transport of cholesterol to the arteries, where it is deposited in the artery walls. The depositing of cholesterol may contribute to the condition **atherosclerosis**, that is, the narrowing and hardening of the arteries. Specifically, cholesterol accumulates as fatty 'hot spots' known as plaques. Not only do these plaques disrupt the blood flow (leading to chest pain), but they can also rupture, releasing debris into the artery, which may cause a heart attack or blockage in the brain – a condition leading to a stroke.

Blood clots may also form at the site of a plaque, further heightening the risk of a heart attack or stroke. If a blood clot becomes dislodged, it may travel through the arterial system to one of the major organs, where it can cause a blockage, restricting the blood supply to that organ. The restriction of blood to the brain can trigger a stroke; lack of blood to the heart can cause a heart attack; a blood clot in the lung is termed a pulmonary embolism, and is often fatal.

LDL-cholesterol has earned the name 'bad cholesterol' as it promotes plaque formation. Accordingly, the amount of LDL-cholesterol in the bloodstream should be kept to a minimum.

Ironically, cholesterol itself is not harmful. Indeed, it has been found to have antioxidant properties (antioxidants inhibit a type of cellular damage that may lead to the onset of cancer and ageing). Instead it is the association with lipoprotein that leads cholesterol astray–with the potentially fatal consequences outlined above.

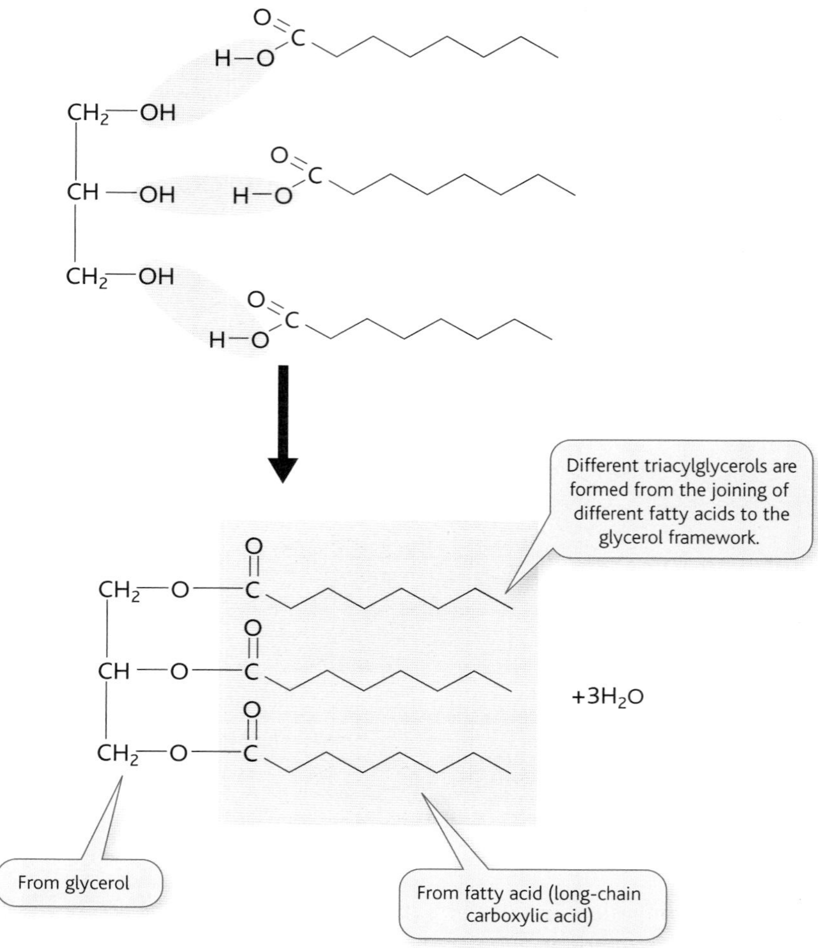

Figure 7.12 The formation of triacylglycerols. Triacylglycerols occur when the carboxyl groups of three fatty acids react with the three hydroxyl groups on glycerol to generate three ester groups.

Triacylglycerols occur when the carboxyl groups of three fatty acids react with the three hydroxyl groups on glycerol to generate three ester groups, as illustrated in Figure 7.12. This feature explains the first part of their name, 'tri-acyl'.

Fatty acids fall into one of three groups:

- **Saturated**–the alkyl chain contains only single carbon–carbon bonds.
- **Monounsaturated**–the alkyl chain contains one carbon–carbon double bond.
- **Polyunsaturated**–the alkyl chain contains more than one carbon–carbon double bond.

Table 7.1 shows the structural formulae of some common saturated, monounsaturated, and polyunsaturated fatty acids. Notice how the melting point of each type of fatty acid increases as the number of carbon atoms increases.

Self-check 7.5 (a) Look at Table 7.1 and find the melting points for lauric acid, palmitic acid, and arachidic acid. Why do these fatty acids have progressively higher melting points? (b) Identify the six fatty acids shown in Table 7.1 as saturated, monounsaturated, or polyunsaturated.

Table 7.1 The structural formulae and melting points of some common saturated, monounsaturated, and polyunsaturated fatty acids.

Name	Structural formula	Melting point (°C)
Arachidonic acid	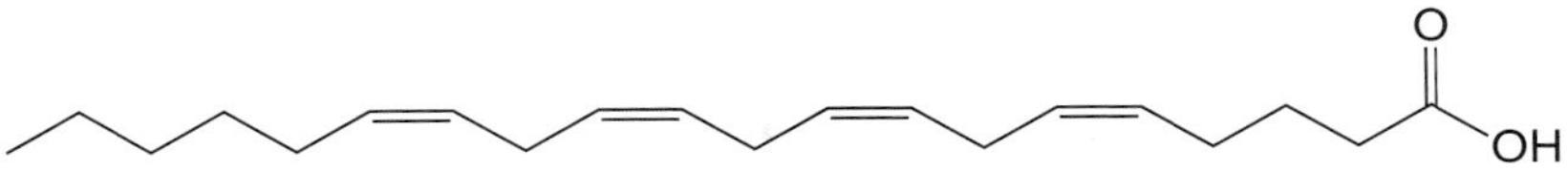	−50
Linoleic acid	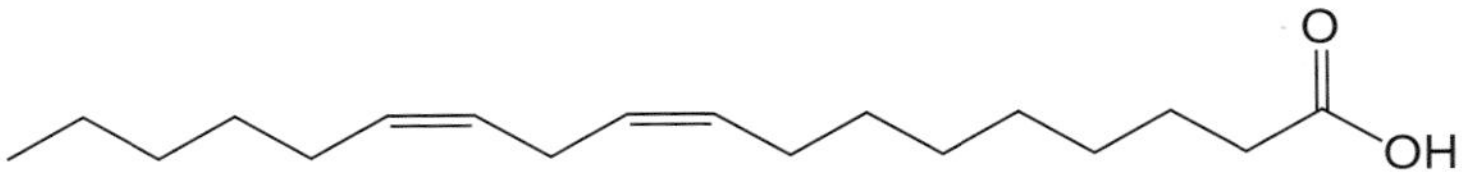	−9
Oleic acid	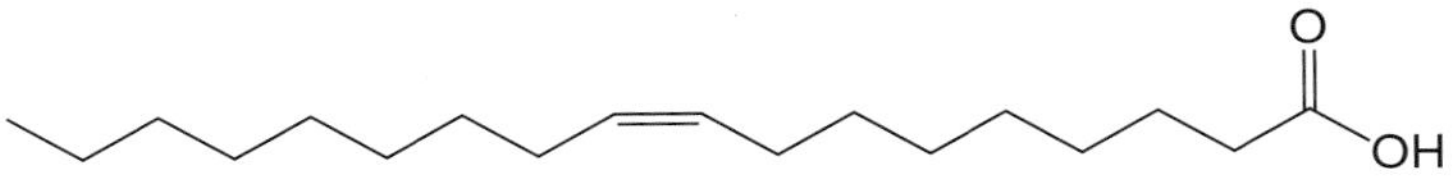	13
Lauric acid	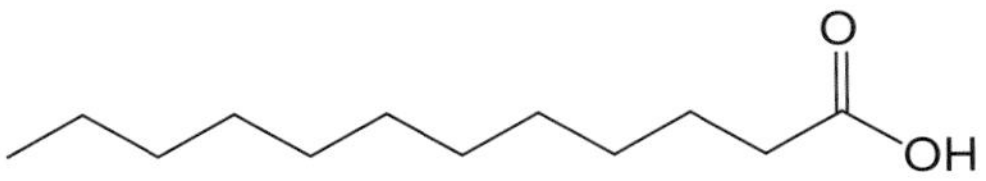	43
Palmitic acid	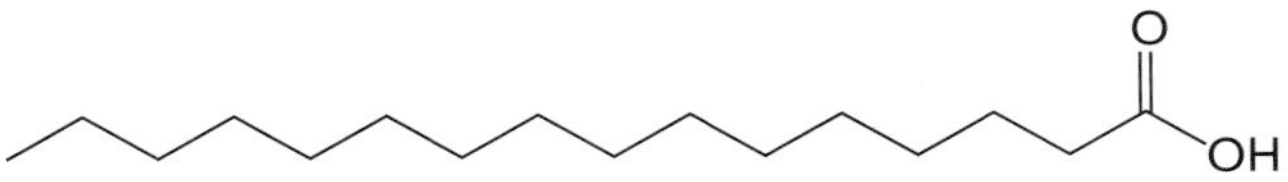	62
Arachidic acid	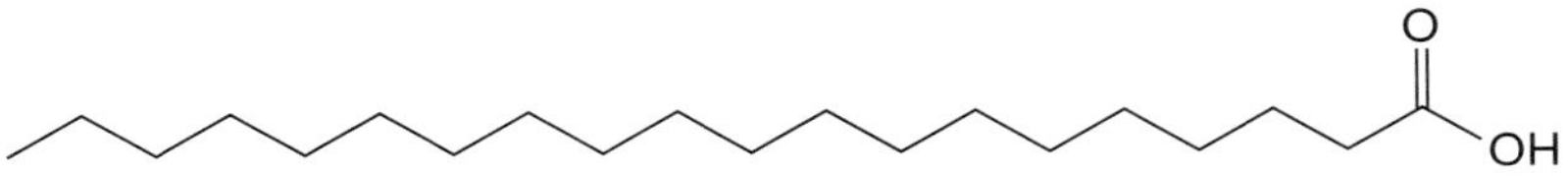	76

A saturated compound has a higher melting point and a higher boiling point than an equivalent unsaturated compound. Saturated compounds are able to pack together more closely than unsaturated compounds; intermolecular forces are consequently higher, and melting and boiling points are elevated relative to unsaturated compounds.

It is the **fatty acid** component of triacylglycerols that give different fats and oils their characteristic properties.

Triacylglycerols with a high saturated fatty acid content are classed as fats[2]. They have high melting and boiling points, and exist as **solids** at room temperature. For example, butter has a melting point of 32 °C and is a solid at room temperature.

In contrast, triacylglycerols with high unsaturated fatty acid content are classed as oils[3]. They have low melting and boiling points, and exist as **liquids** at room temperature. For example, both olive oil and sunflower oil are liquids at room temperature, with melting points of −6 °C and −18 °C respectively.

Taken at face value, it seems our health can be directly influenced by something as simple as the presence or absence of one carbon–carbon double bond. For some time now, experts on nutrition have advised us to maintain diets that contain a low

2 Triacylglycerols with a high saturated fatty acid content are generally termed **saturated fats**.

3 Despite being classed as oils, triacylglycerols with high unsaturated fatty acid content are generally termed **unsaturated fats**.

BOX 7.5 **Good fat? Bad fat?**

The common concept of unsaturated fats being 'good fats' might not be as universally applicable as we once thought. One class of unsaturated fat–the so-called *trans* fats–are thought to be even more dangerous to our health than saturated fats.

We see in Chapter 10 how double bonds exist in either *cis* or *trans* configurations. The designation '*cis*' or '*trans*' depends on the conformation of the groups around the double bond. A vast majority of naturally occurring unsaturated fats (including the mono- and polyunsaturated fats mentioned above) possess the *cis* conformation, and pose no threat to health.

Trans fats, however, have been found to increase 'bad' LDL-cholesterol levels while lowering 'good' HDL-cholesterol levels. As a result, less cholesterol is taken to the liver for safe removal, while more cholesterol is targeted for depositing in the arteries–the worst-case scenario as far as our body is concerned.

Only very limited amounts of *trans* fats occur in nature. They are generated artificially, however, through the process of **hydrogenation**: an industrial process in which unsaturated fats are solidified by decreasing the number of double bonds they contain. (Solidified fats have found wider uses in the food industry than their unsaturated cousins, and also possess longer shelf lives, resulting in less food spoilage, and therefore more profits for the food manufacturer.)

Critically, hydrogenation can also convert double bonds from a *cis* to a *trans* conformation in addition to decreasing the extent of unsaturation. The *trans* double bond conformation confers upon *trans* fats a shape that closely mimics that of saturated fats. One of our essential concepts is that physical form (e.g. shape) dictates function. So it should come as little surprise that, given their similar shapes, *trans* fats and saturated fats both have similar negative effects on the respective levels of LDL- and HDL-cholesterol.

The discussion above helps to show the importance of understanding chemical structure: one double bond marks the difference between saturated and monounsaturated fats, yet the two types of fat have opposite effects on LDL-cholesterol. Equally, a switch in conformation around a double bond, from *cis* to *trans*, can have equally opposing effects on amounts of cholesterol and, ultimately, our health. Proof, if needed, that chemistry really is central to life.

proportion of saturated fats (containing no double bonds), and a higher proportion of unsaturated fats (containing at least one double bond).

Research has found that high levels of saturated fats in the diet are linked with elevated levels of the cholesterol–protein complex LDL-cholesterol, which has been found to greatly increase the risk of atherosclerosis and heart disease (see Box 7.4).

Polyunsaturated fats (containing more than one double bond) have been found to lower both LDL- and HDL-cholesterol such that, while less cholesterol is deposited in the arteries (by LDL), less cholesterol is also directed to the liver (by HDL) for safe disposal.

Monounsaturated fats (containing just one double bond) represent the ideal situation: they have been found to lower LDL-cholesterol but preserve levels of the 'good' HDL-cholesterol. As a result, while the amount of cholesterol deposited in the arteries is lowered (reducing the risk of heart disease), more of the excess cholesterol is targeted to the liver for safe removal.

It is now becoming clear, however, that the argument against saturated fats in favour of unsaturated fats is far less clear-cut, as explained in Box 7.5.

Glycerophospholipids

Glycerophospholipids are the central component of **cell membranes**, the structures that envelop the contents of cells and define their exterior boundaries (just as the walls of a house enclose the contents of the house and delineate one house from its neighbour).

Triacylglycerols

Glycerophospholipids

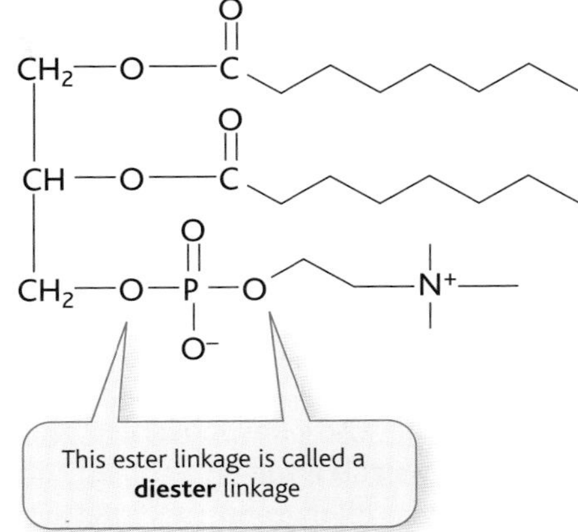

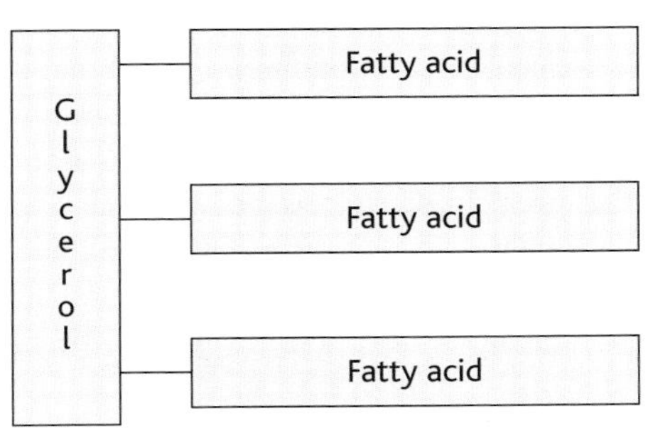

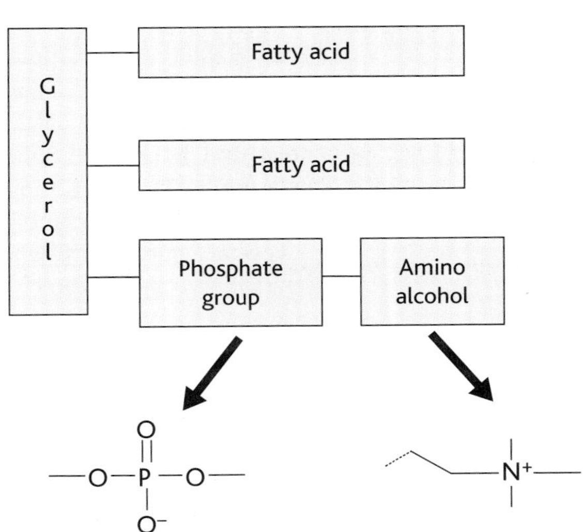

Figure 7.13 A comparison of the general structures of triacylglycerols and glycerophospholipids.

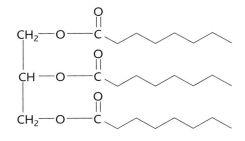

The basic framework of the glycerophospholipids comprises a glycerol backbone and, as such, mirrors that of the triacylglycerols. One of the hydroxyl groups in a glycerophospholipid is attached to a **phosphodiester group**, however (in contrast to the triacylglycerols, in which each hydroxyl group on the glycerol backbone has reacted with a fatty acid to form three ester groups). We compare the general structures of glycerophospholipids and triacylglycerols in Figure 7.13.

The phosphodiester group has two components: a **phosphate** group and an **amino alcohol**.

The **phosphate group** acts as the bridge between the amino alcohol and the glycerol backbone. The phosphodiester group contains one of three amino alcohols: serine, choline, or ethanolamine. Look at Figure 7.14 and notice how they each contain a hydroxyl functional group and an amino group. Notice too how choline has a *tertiary* amino group, while both serine and ethanolamine have *primary* amino groups.

Self-check 7.5 Only one of the amino alcohols occurring in glycerophospholipids is also found in proteins. Which one?

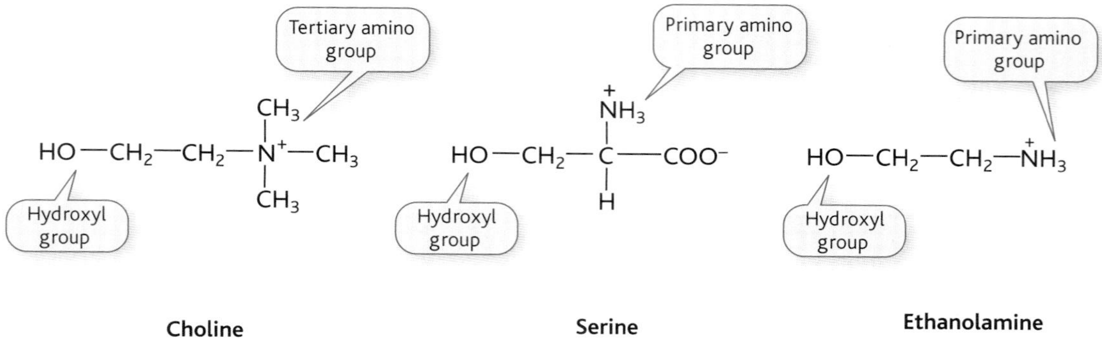

Tertiary amino group

$HO{-}CH_2{-}CH_2{-}N^{+}(CH_3)_3$

Hydroxyl group

Choline

Primary amino group

$HO{-}CH_2{-}CH(\overset{+}{N}H_3){-}COO^{-}$

Hydroxyl group

Serine

Primary amino group

$HO{-}CH_2{-}CH_2{-}\overset{+}{N}H_3$

Hydroxyl group

Ethanolamine

Figure 7.14 The structures of choline, serine, and ethanolamine.

The phosphodiester group carries the 'diester' tag because both the amino alcohol and glycerol/fatty acid portions of the glycerophospholipid are recognized as ester groups, as noted in Figure 7.13.

Glycerophospholipids and the formation of the lipid bilayer

The suitability of glycerophospholipids as the central component of the cell membrane is derived from its possessing both **polar** and **non-polar** components, as illustrated in Figure 7.15(a). The long alkyl tails of the fatty acid groups are non-polar and hence hydrophobic, while the phosphodiester group is polar and hydrophilic. These opposing characteristics hold the key to a glycerophospholipid's ability to form a cell membrane.

Look at Figure 7.15(b), which shows how the glycerophospholipids arrange themselves in two sheets, each with a hydrophilic and a hydrophobic face. Notice how

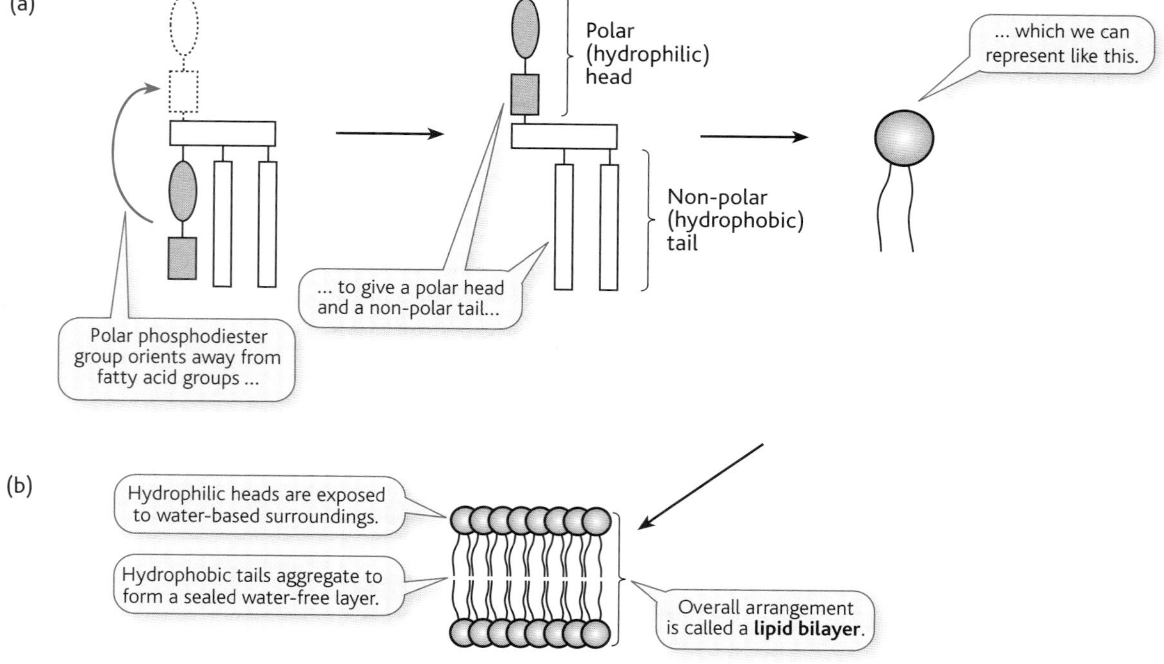
(a)
Polar (hydrophilic) head
... which we can represent like this.
Non-polar (hydrophobic) tail
... to give a polar head and a non-polar tail...
Polar phosphodiester group orients away from fatty acid groups ...
(b)
Hydrophilic heads are exposed to water-based surroundings.
Hydrophobic tails aggregate to form a sealed water-free layer.
Overall arrangement is called a lipid bilayer.

Figure 7.15 (a) A glycerophospholipid has a hydrophilic head group (a phosphodiester group), and a hydrophobic tail, formed from fatty acid groups. (b) Glycerophospholipids arrange themselves in two sheets, each with a hydrophilic and a hydrophobic face. The hydrophobic faces of the two sheets lie against one another, forming a sealed, water-free layer.

the hydrophobic faces of the two sheets (comprising the fatty acid tails) lie against one another, forming a sealed, water-free layer that is shielded from the aqueous environment in and around the cell. In contrast, the hydrophilic faces of the two sheets comprising the phosphodiester groups (which interact readily with water) lie exposed to the aqueous surroundings. The overall two-sheet assembly is termed a **lipid bilayer.**

The lipid bilayer is arguably one of the most important structures found in living organisms. The lipid bilayer divides an organism into individual cells, and the interior of cells into distinct organelles, each with contrasting properties and functions, just as the walls of a house delineate rooms with different functions. The lipid bilayer helps impart a physical shape to all living things. Without the lipid bilayer we would have no physical shape. In fact, we would be little more than a puddle of chemicals on the floor.

However, the lipid bilayer doesn't solely comprise glycerophospholipids. The lipid bilayer of cell membranes also features a range of proteins with various roles, including transmembrane (membrane-spanning) proteins. Some form pores to allow the transport of molecules into and out of the cell or organelle. Others act as receptors for signalling molecules, mediating the passage of messages from the cell exterior to the interior (and vice versa), and hence allowing vital communication between neighbouring cells (and, indeed, disparate parts of the body).

Another essential component of the lipid bilayer is cholesterol. In recent years, cholesterol has gained a notorious reputation due to its association with heart disease (Box 7.4). However, the presence of cholesterol in our body is essential for good health. Figure 7.16 shows how cholesterol molecules are intercalated (inserted) at intervals between glycerophospholipid molecules on the outer face of a lipid bilayer (the outermost sheet of glycerophospholipids), and are thought to prevent the glycerophospholipid molecules packing too tightly together, i.e. helping to maintain the fluidity of the cell membrane. Beyond its role in cell membranes, cholesterol is also a precursor (ingredient) in the synthesis of a number of steroid hormones and vitamins.

7.4 Nucleic acids

The nucleic acids are perhaps the most fundamental class of biological molecule. They enable genetic information to be used to direct the assembly and function of biological systems. They also enable the same information to be stored, and transmitted from one generation to the next.

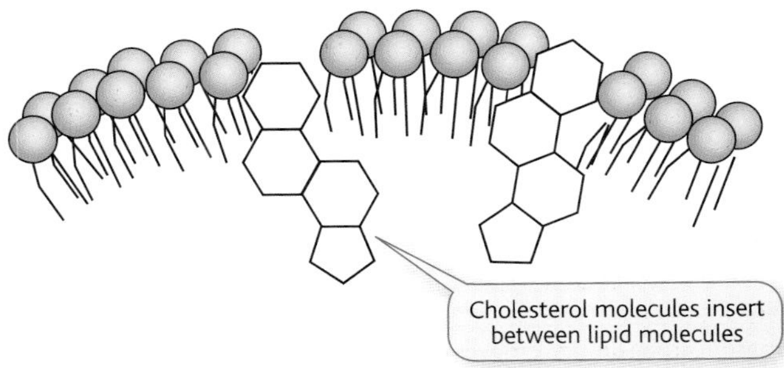
Cholesterol molecules insert between lipid molecules

Figure 7.16 Cholesterol molecules are inserted at intervals between glycerophospholipid molecules on the outer face of a lipid bilayer, helping to maintain the fluidity of the cell membrane.

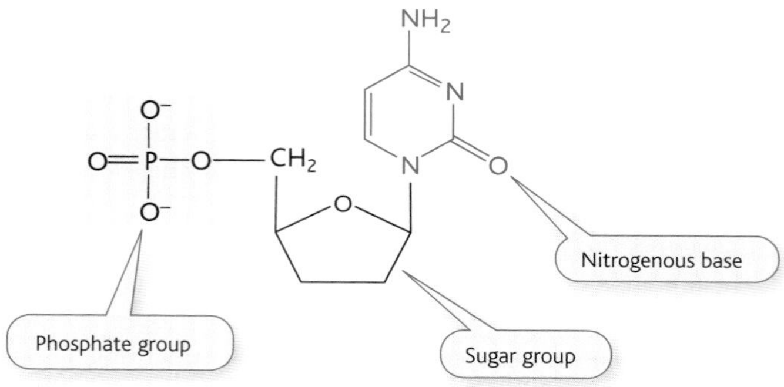

Figure 7.17 The general structure of a nucleotide.

Nucleotides and their composition

Two nucleic acids occur in biological systems, **deoxyribonucleic acid** (DNA) and **ribonucleic acid** (RNA). Both DNA and RNA are polymers of a family of compounds called **nucleotides**. Nucleotides comprise three components:

1. A nitrogen-containing (nitrogenous) base
2. A sugar
3. A phosphate group.

The three components join together to give the nucleotides the general structure depicted in Figure 7.17.

Five different nitrogenous bases appear most widely in DNA and RNA; three of these (**adenine**, **cytosine**, and **guanine**) appear in both DNA and RNA, while **thymine** appears only in DNA, and **uracil** appears only in RNA. Isomers (structural variants) of these five molecules occur in small amounts in both nucleic acids.

The nitrogenous bases each have as their 'template' one of two types of ring structure, a **pyrimidine** ring or a **purine** ring. Figure 7.18 shows how cytosine, thymine, and uracil are based on the pyrimidine ring, while adenine and guanine are based on the purine ring. In each case the pyrimidine or purine rings are modified to give each base a distinct chemical identity. Both the amino and amide groups help create these distinct identities. Look at Figure 7.18 and notice how cytosine, adenine, and guanine have an amino group attached to the pyrimidine ring, while cytosine, guanine, thymine, and uracil all possess amide groups.

Self-check 7.7 Look again at Figure 7.18. What chemical difference distinguishes uracil from thymine?

The amino groups, in which hydrogen atoms are bonded to an electronegative nitrogen atom, also play a vital part in the hydrogen bonding through which the complementary base pairing of the two DNA strands in the DNA double helix is facilitated, as we see in Section 9.2.

RNA is distinguished from DNA by a slight difference in the sugar component of their nucleotides. RNA comprises nucleotides that contain the sugar **ribose**, a five-carbon sugar (Section 7.2). By contrast, the nucleotides from which DNA is formed contain the sugar **deoxyribose**. The prefix 'deoxy' means 'without oxygen'[4].

4 It is the lack of an oxygen atom on carbon 2 of deoxyribose that gives DNA its name, deoxyribonucleic acid: that is, ribonucleic acid without oxygen.

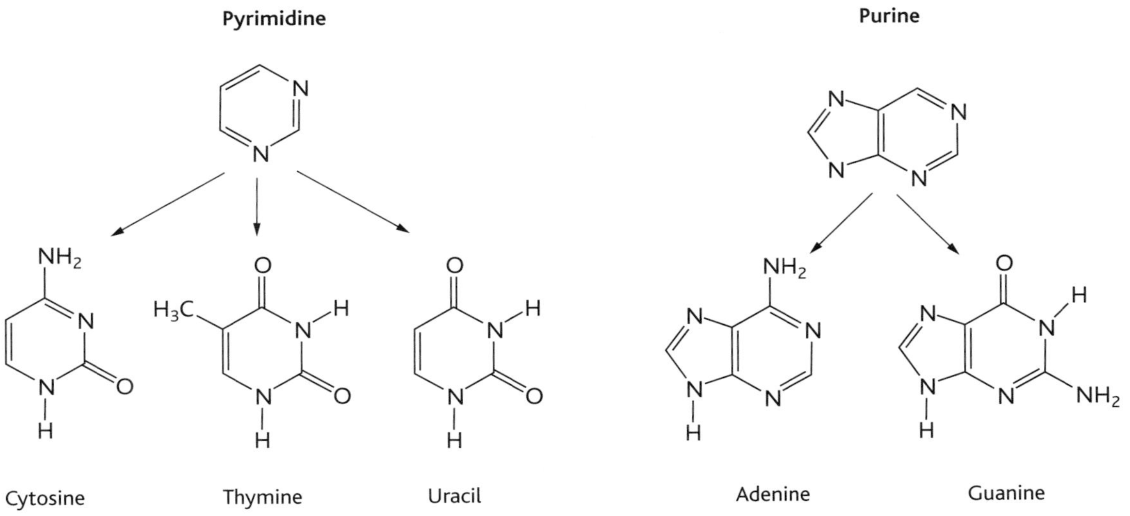

Pyrimidine

Purine

Cytosine Thymine Uracil Adenine Guanine

Figure 7.18 Cytosine, thymine, and uracil are based on the pyrimidine ring, while adenine and guanine are based on the purine ring.

While deoxyribose is also a five-carbon sugar, it has a hydrogen atom attached to carbon 2, rather than a hydroxyl group–a difference of one oxygen atom. Look at Figure 7.19 and compare the structures of ribose and deoxyribose. Notice how the chemical compositions of the two sugars are identical, except for the lack of one oxygen atom attached to carbon 2 of deoxyribose.

The nucleotides of both DNA and RNA contain *identical* phosphate groups, which join to the ribose or deoxyribose sugar via a **phosphate ester** bond.

The structural formulae of the nucleotides from which DNA is formed are illustrated in Figure 7.20. Remember: the nucleotides that comprise RNA are identical to those found in DNA (with one exception–see Self-check 7.8), except the sugar group has an –OH group and not an H atom attached to carbon 2.

- *Nucleotides have three key components:*
  1. *a nitrogenous base*
  2. *a sugar group*
  3. *one or more phosphate groups.*

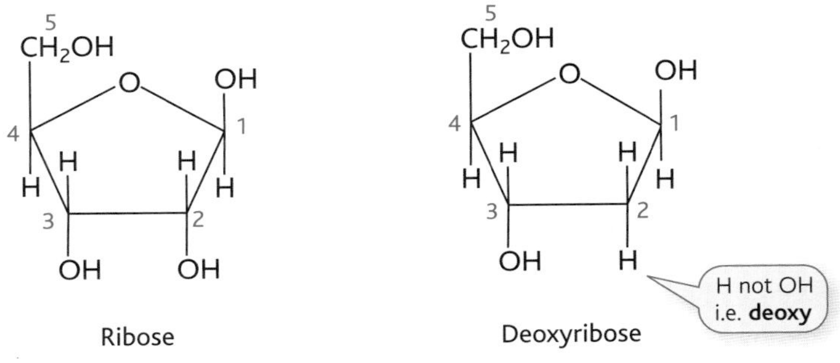

Ribose Deoxyribose

H not OH i.e. **deoxy**

Figure 7.19 The structures of ribose and deoxyribose.

Deoxyadenosine 5' -monophosphate

Deoxyguanosine 5' -monophosphate

Deoxycytidine 5' -monophosphate

Deoxythymidine 5' -monophosphate

Figure 7.20 The structural formulae of the nucleotides from which DNA is formed.

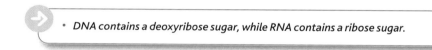

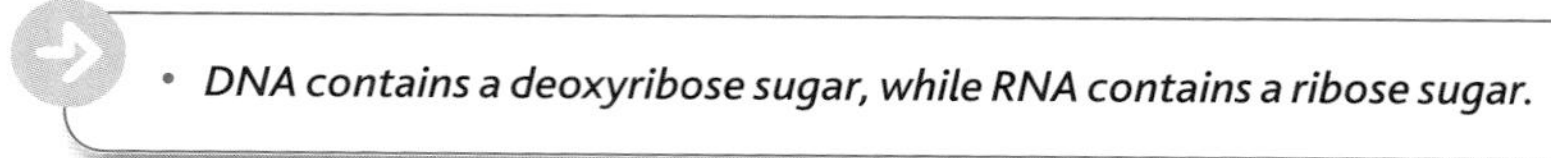

• *DNA contains a deoxyribose sugar, while RNA contains a ribose sugar.*

Self-check 7.8 Draw the structure of the nucleotide uridine 5′-monophosphate, which is found only in RNA. The nitrogenous base found in uridine 5′-monophosphate is uracil, whose structure is shown in Figure 7.18.

Formation of nucleic acids

We learn about the *mechanism* of this reaction on p. 616.

DNA and RNA are formed from the joining together of nucleotides in a long chain. A human chromosome typically contains 130 million nucleotides, while an RNA strand is much shorter, comprising from 75 to several thousand nucleotides. Look at Figure 7.21, which shows how nucleotides form a nucleic acid chain by the joining of the phosphate group of one nucleic acid with a hydroxyl group of an adjacent nucleotide.

When the phosphate groups join together, a **phosphodiester bond** forms between them. It is the joining of the different nucleotides in a different order that gives rise to different DNA and RNA **sequences**. It is the variation in DNA sequence between individuals that makes each of us unique. Paradoxically, however, much attention has been paid in the past few years to determining the typical ('average')

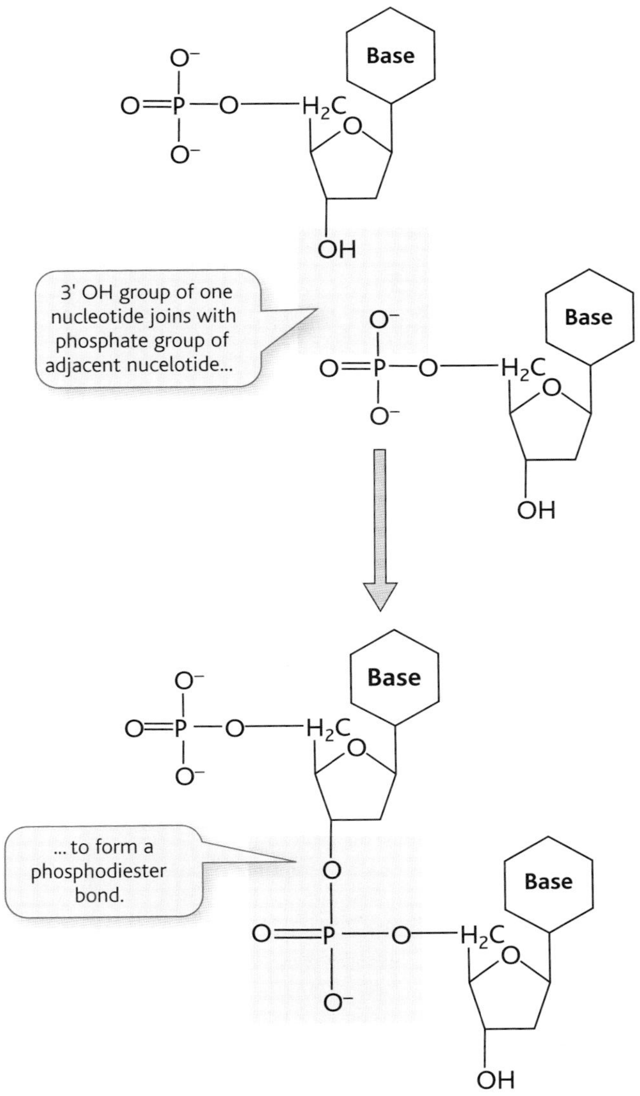

Figure 7.21 Nucleotides form a nucleic acid chain by the joining of the phosphate group of one nucleotide with a hydroxyl group of an adjacent nucleotide.

DNA sequence of a human. This was the aim of the so-called **Human Genome Project**, as explained in Box 7.6.

The shape of nucleic acids

Chromosomal DNA occurs as a two-stranded double helical structure, as depicted in Figure 7.23. The physical properties of the sugar, phosphate, and nitrogenous base components of the nucleotides are central in determining the shape of the DNA helix. Both the sugar and phosphate groups of a nucleotide are relatively hydrophilic, while the nitrogenous base component is relatively hydrophobic. As a consequence, the long chain of nucleotides that comprise DNA is arranged in such a way that the hydrophilic sugar and phosphate groups are exposed to the aqueous environment of the cell structure, while the hydrophobic nitrogenous bases are shielded from it.

We learn about hydrophobic and hydrophilic entities in Chapter 4.

We find out more about how the chemical composition of the nucleotides influences the shape of DNA in Chapter 9.

Look at Figure 7.23 and notice how the sugar and phosphate groups lie on the *outside* of the helical structure (exposed to the aqueous surroundings of the cell) while the nitrogenous bases lie *inside* the helix, where they are shielded from the cell's aqueous environment.

The sugar and phosphate groups form a 'framework' (or 'backbone') for the nucleic acids in a fashion analogous to carbon forming the framework of many other organic compounds.

Nucleotides: nature's energy stores

The contribution of nucleotides to biological systems is not limited to the formation of the nucleic acids. Nucleotides also play a vital role in the storage and release of chemical energy in cells. The key players in the storage and release of energy are adenosine monophosphate (AMP), one of the nucleotides found in RNA, and its sister compounds adenosine diphosphate (ADP) and adenosine triphosphate (ATP). As their names imply, these three compounds differ according to the number of phosphate groups they possess. Look at Figure 7.24, which shows the structures of AMP, ADP, and ATP, and notice how AMP has one phosphate group (**mono**phosphate), ADP has two phosphate groups (**di**phosphate), and ATP has three phosphate groups (**tri**phosphate).

BOX 7.6 The DNA sequence of Mr Average: the Human Genome Project

In June 2000, the Prime Minister of the UK, Tony Blair, and the President of the USA, Bill Clinton, announced the completion of the draft of the human genome–the sequence of deoxyribonucleic acid comprising the 24 chromosomes present in humans. (Females have only 23 different chromosomes–they carry two identical 'X' chromosomes. By contrast, males carry 24 different chromosomes – they carry both an 'X' and a 'Y' chromosome.)

The purpose of the Human Genome Project was to establish the sequence of nucleotides present in human DNA. The only chemical difference between consecutive nucleotides in a DNA strand is the nitrogenous base that each nucleotide possesses: adenine, guanine, cytosine, or thymine. The identity of the nitrogenous base within each of the nucleotides in a strand of DNA therefore defines the DNA's sequence.

Sequencing determines the *order* in which the nucleotides appear in the DNA strand. In practice, we do this by identifying the base each nucleotide contains. But how do we tell the different bases apart? Each base is labelled with a different fluorescent dye, so that each base gives a different 'signal' as we proceed along the DNA strand.

An example of a DNA sequence is shown in Figure 7.22. Look at this figure, and notice how peaks with different line weights on the profile correspond to different bases (the darkest line shows G, the palest line shows A).

The Human Genome Project involved the exact sequencing of a staggering 3 billion nucleotides which, together, encode the roughly 35 000 genes that define the human being. The Human Genome Project used DNA samples from many individuals throughout the world, in order to generate an 'average' sequence, representing an 'average' human.[5]

Of course, the genomes of any two individuals will not match exactly with the published human genome sequence. Our genomes all vary slightly from the 'average' sequence. The similarity means that we are all human; the difference means that we each have a different genome.

We find out more about DNA profiling in Box 11.3.

5 In fact, two sequencing projects ran in parallel: the privately funded Celera Genomics project, and the publicly funded International Human Genome Sequencing Consortium project. Both projects published draft sequences simultaneously, in February 2001.

BOX 7.6 **Continued**

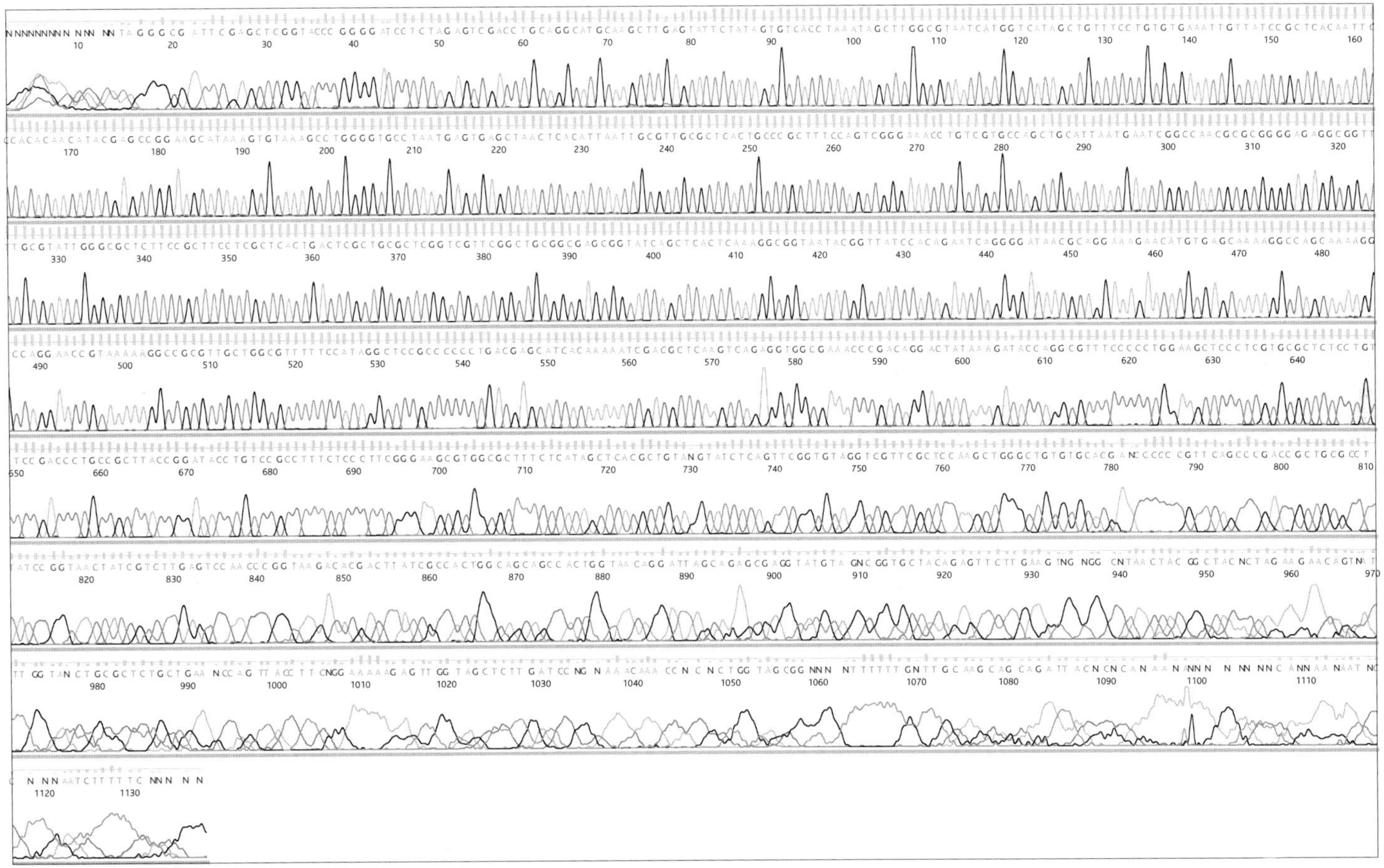

Figure 7.22 An example of a DNA sequence. (Courtesy: A. M. Lesk.)

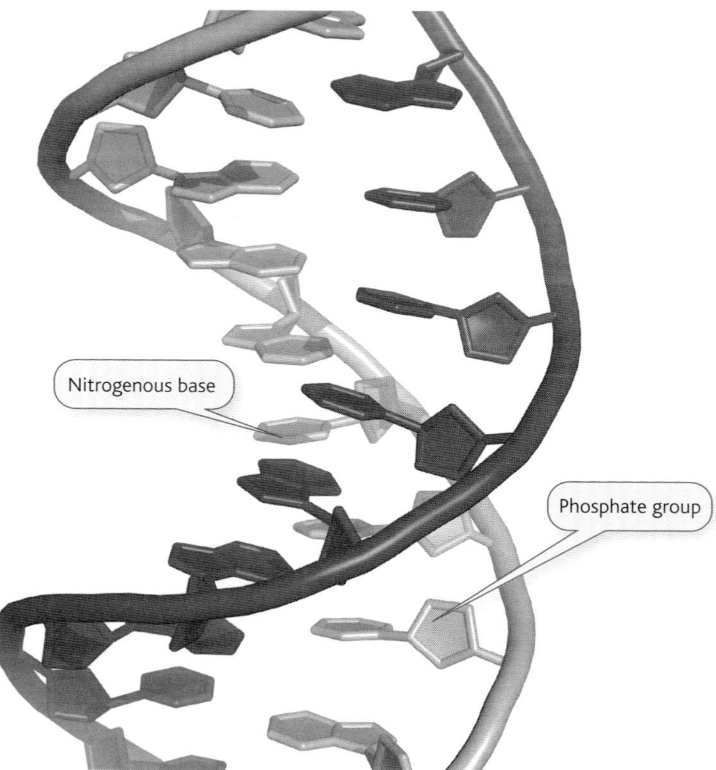

Figure 7.23 The two-stranded double helical structure of DNA.

We find out what it is about these three compounds that makes them so well suited to the job of storing and releasing energy in Chapter 13, when we explore the link between chemical bonds and changes in energy.

7.5 Metals in biology

Throughout the last three chapters we have focused heavily on organic compounds, and the roles they play in biological systems. However, organic compounds are far from having exclusivity when it comes to the chemistry of life. Instead, compounds that have metal components play vital roles, as we discover in the final section of this chapter.

Many biological compounds, most notably proteins, feature metal ions bound within them. For example, we see in Chapter 3 (p. 65) how each of the polypeptide chains making up the oxygen-carrying protein haemoglobin possesses an Fe^{2+} ion to which an oxygen molecule can bind. The Fe^{2+} ion is held in place by five covalent bonds formed with surrounding nitrogen atoms, including three dative covalent bonds.

We learn about dative covalent bonds in Chapter 3.

The presence of metal ions within biological compounds can contribute to both the structure and the function of these compounds. Though we won't explore metal-containing compounds in detail here, let's consider a couple of specific examples that illustrate how metal ions contribute to both structure and function.

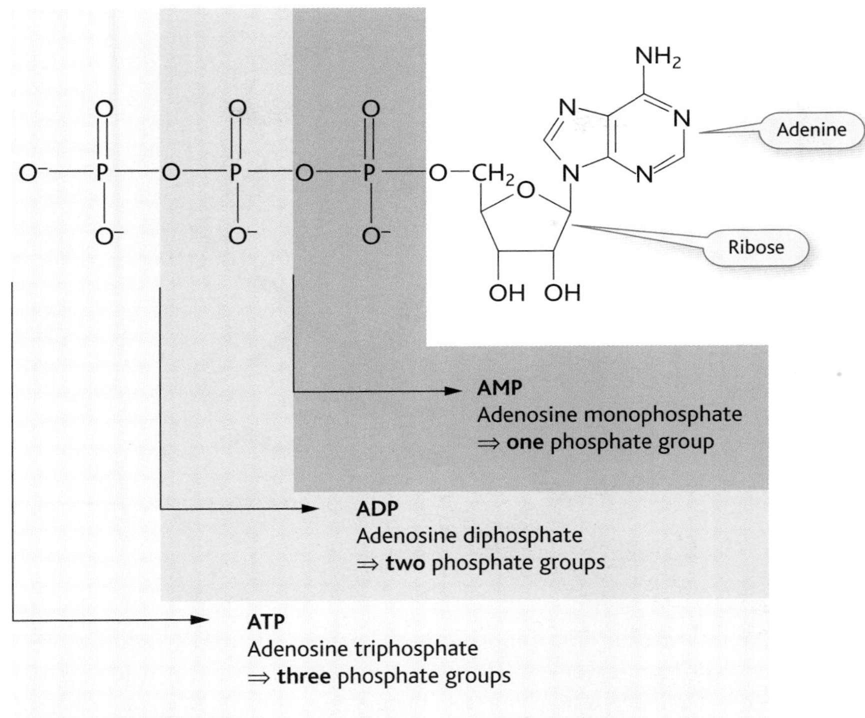

Figure 7.24 The structures of AMP, ADP, and ATP.

Metals and molecular structure: the zinc finger motif

The zinc finger motif is an amino acid sequence found in a wide range of DNA-binding proteins. The motif has the characteristic sequence shown in Figure 7.25. Look at this figure, and notice how the zinc finger motif has a number of conserved residues; four of these–two cysteine and two histidine–are of particular importance. It is these four residues that are vital to the formation of the zinc finger motif: they form four bonds with a central zinc atom, as shown in Figure 7.26.

The coordination of the zinc atom by the four conserved amino acid residues means that the polypeptide chain, of which the conserved residues are part, is held in a characteristic three-dimensional structure, illustrated in Figure 7.26. Essentially, the amino acid sequence containing the four conserved residues forms a finger-like protrusion around the zinc atom, which sits within the 'finger'.

—(Tyr, Phe)—X—Cys—$X_{2\text{-}4}$—Cys—X_3—Phe—X_5—Leu—X_2—His—$X_{3\text{-}5}$—His—

Figure 7.25 The characteristic amino acid sequence for the zinc finger motif. 'X' denotes a variable amino acid. The four residues in bold are the conserved cysteine and histidine residues that coordinate the zinc atom, which lies within the finger-like protrusion of this motif.

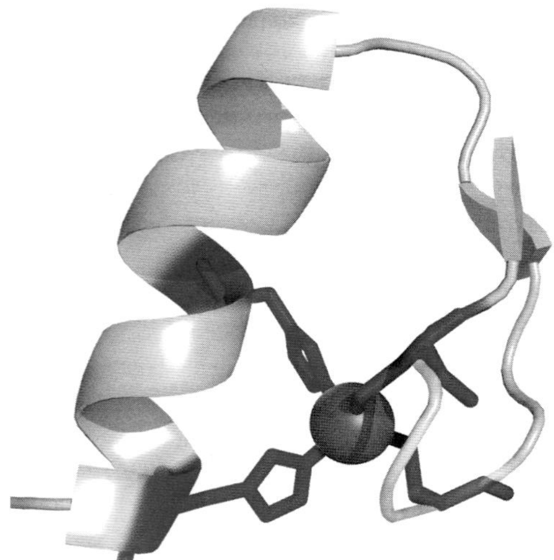

Figure 7.26 The three-dimensional structure of the zinc finger motif, showing the position of the zinc atom (dark grey sphere) within the fold of the finger. The zinc atom is held in place with dative covalent bonds from the four conserved amino acid residues depicted in Figure 7.25, shown here in blue.

We note above that the zinc finger motif is characteristic of DNA-binding proteins. Why is this? The answer lies in the way that the finger-like protrusion fits into the major groove of a DNA molecule, as illustrated in Figure 7.27, giving proteins bearing the zinc finger motif their DNA-binding ability.

Zinc finger motifs are particularly characteristic of gene regulatory proteins, which have their effect by binding to DNA, affecting the expression of genes encoded by the DNA as a result.

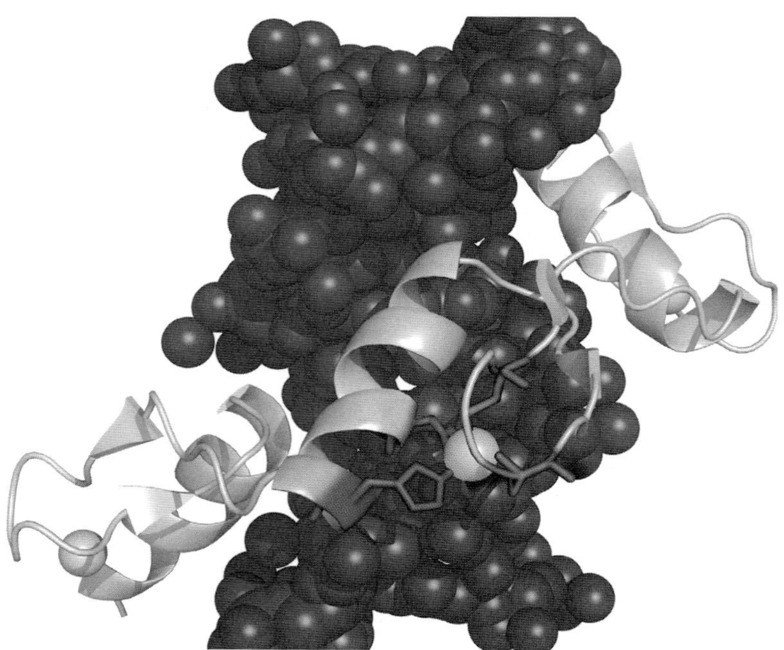

Figure 7.27 A protein containing a zinc finger domain binding to the major groove of a double-stranded DNA molecule. The zinc finger domain is the same as that illustrated in Figure 7.26, but now depicted as part of a larger protein molecule. Notice how it sits snugly in the major groove.

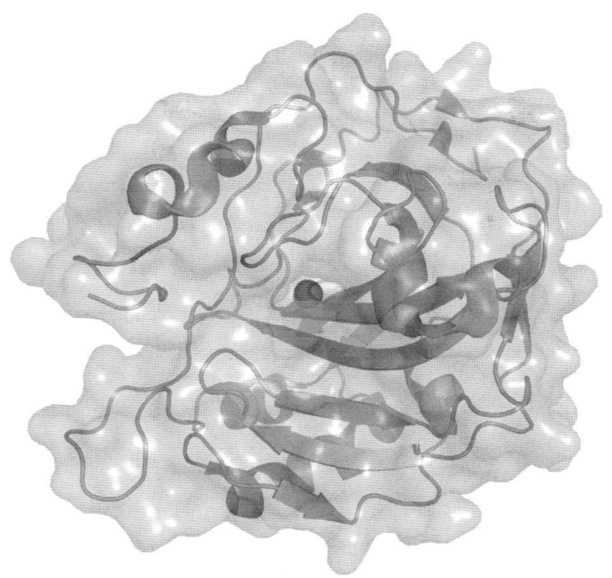

Figure 7.28 The three-dimensional structure of the enzyme carbonic anhydrase, showing the location of zinc (blue sphere) at the centre of the molecule. This image shows the surface of the protein to emphasize the way that the zinc sits in a hollow pocket in the protein structure.

Metals and biological function: metalloproteins

The zinc finger motif gives a nice example of how metals are central to some biological molecules adopting their correct three-dimensional shape. However, metals are also often central to proteins being able to *function* as they should. (Indeed, we see in Chapter 9 how having the correct shape is intrinsically linked to being able to perform the correct function.)

Metalloproteins are proteins that contain one or more metal ions. Many metalloproteins are enzymes, whereby the metal ion contributes directly to the enzyme's catalytic function (for example, participating in oxidation and reduction reactions). A range of metal ions act as components of metalloproteins; these include zinc (Zn), iron (Fe), magnesium (Mg), manganese (Mn), copper (Cu), and nickel (Ni).

Zinc is often found in enzymes that catalyze acid–base reactions. For example, Figure 7.28 shows the location of zinc in the active site of the enzyme carbonic anhydrase, which catalyzes the conversion of carbon dioxide to the bicarbonate ion[6]:

$$CO_2 + H_2O \rightleftharpoons HCO_3^- + H^+$$

In this enzyme, the zinc ion is held in place by dative covalent bonds with three nitrogen atoms on surrounding histidine residues. Notice a commonality with the zinc finger motif here: again, we see histidine playing an important part in metal ion binding. Each histidine has a nitrogen atom with a non-bonding pair of electrons, which can participate in a dative covalent bond with the metal ion.

6 This reaction is a highly important one in biological systems, as it provides a means of regulating the pH of the blood. We learn more about the regulation of blood pH in Chapter 16.

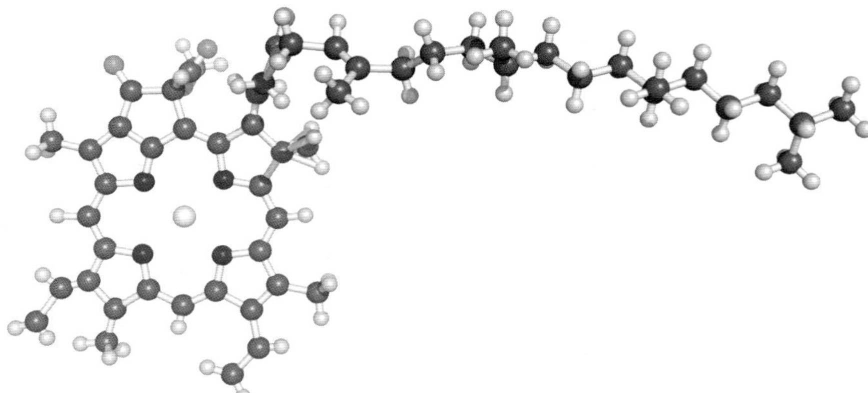

Figure 7.29 The three-dimensional structure of chlorophyll, with its prominent porphyrin ring. Notice how the porphyrin ring has a magnesium atom at its centre.

Metals are also components of other biological compounds, which mediate processes that are essential to life. For example, photosynthesis in plants occurs in specialized organelles called chloroplasts. The first stage of photosynthesis is the 'harvesting' of light, which is ultimately used to power the synthesis of sugars. The light itself is absorbed by the pigment chlorophyll, which is present in the chloroplast membrane as part of giant light-harvesting antennae. Each chlorophyll molecule has a magnesium atom at its centre, as illustrated in Figure 7.29. Look at this figure, and notice how the central ring of the chlorophyll is a porphyrin ring, which is very similar to the haem ring found in haemoglobin (which is illustrated in Figure 3.28). The key difference is that the haem ring has an iron ion at its centre, whereas the porphyrin ring of chlorophyll has a magnesium atom.

Needless to say, we could go on to describe many other metalloproteins that play important and diverse roles in biological systems. The examples here are just intended to give a quick taster–and to show how cells and organisms depend on more than just purely organic compounds to function as they should.

During this section–and, indeed, throughout this chapter as a whole–you could be forgiven for thinking that only large, complicated compounds have important functions in biological systems. Just to put things in a little more perspective, look at Box 7.7, which explains how some biologically important molecules are really quite small indeed.

Check your understanding

To check that you've mastered the key concepts presented in this chapter, review the checklist of key concepts below, and attempt the multiple-choice questions available in the book's Online Resource Centre at **http://www.oxfordtextbooks.co.uk/orc/crowe2e/**.

BOX 7.7 **Life is a gas: the 'gasotransmitters'**

For many of us, hydrogen sulfide gas (H_2S) is associated solely with the smell of rotting eggs – and certainly not a compound that we'd think to be of value to us. However, it is now thought to belong to a group of so-called 'gasotransmitters', small molecule gases, which have important signalling functions within the body.

The most well-known gaseous signalling molecule is nitric oxide, NO. Originally thought to be little more than a pollutant – nitric oxide in car exhausts reacts with oxygen to produce smog – it is now known to regulate brain activity, and the activity of the lungs, liver, kidneys, stomach, and other organs; it also plays an important role in controlling blood circulation. Indeed, it was the relaxation-inducing effect that nitric oxide has on the muscle of blood vessels that first brought it to the attention of scientists, before the realization that it acts as a key player in the physiology of most organs and tissues. The relaxation of blood vessels can help to ease hypertension, and can even help to overcome impotence. (The erectile dysfunction drug Viagra has its effect by enhancing the release of nitric oxide in the body.)

Now, scientists are finding that hydrogen sulfide has a similarly beneficial impact on blood vessels: mice carrying the mutant form of an enzyme known to be responsible for hydrogen sulfide production in bacteria were shown to produce lower levels of the gas, and were found to suffer from hypertension as a result.

How do such small molecules have such a potent physiological effect? Nitric oxide often has its effect by covalently modifying amino acid residues in proteins, inhibiting their activity as a result. The process, known as *S*-nitrosylation, results in the addition of a nitric oxide group to the sulfur atom of a cysteine side chain:

It is currently thought that hydrogen sulfide acts in a similar way – though, instead of a nitric oxide group being added to the S atom of a cysteine residue, the –SH group becomes a sulfur–sulfur–hydrogen group, as shown in the reaction schemes below.

It is still early days in terms of characterizing the physiological role of hydrogen sulfide, though some are already saying that it could have even more impact than nitric oxide. Whatever the case, the gasotransmitters are proof that, when it comes to the chemistry of life, small really can be beautiful.

$$\underset{\text{Cysteine}}{\text{-----}\overset{\displaystyle H}{\underset{\displaystyle CH_2-SH}{C}}\text{-----}} + ONOO^- \rightleftharpoons \text{-----}\overset{\displaystyle H}{\underset{\displaystyle CH_2-S-NO_2}{C}}\text{-----} + OH^-$$

$$\underset{\text{Cysteine}}{\text{-----}\overset{\displaystyle H}{\underset{\displaystyle CH_2-SH}{C}}\text{-----}} + H_2S \rightleftharpoons \text{-----}\overset{\displaystyle H}{\underset{\displaystyle CH_2-S-S-H}{C}}\text{-----} + 2H^+$$

Checklist of key concepts

Amino acids and proteins

- An amino acid has two functional groups: an amino group, and a carboxyl group
- An amino acid also possesses a variable side chain, the composition of which determines the identity of the amino acid
- The variable side chain may also possess an additional functional group
- Amino acids can be classified as either hydrophobic or hydrophilic
- Some amino acid side chains can act as acids or bases
- Amino acids join together to form polymers named polypeptides
- The link between amino acids in a polypeptide chain is called a peptide bond

Carbohydrates

- The carbohydrates are also known as sugars
- There are three main classes of sugar
  - Monosaccharides comprise a single sugar unit
  - Disaccharides comprise two sugar units
  - Polysaccharides comprise many sugar units
- Sugars possess two types of functional group: a carbonyl group, and multiple hydroxyl groups
- The monosaccharides may adopt an open or cyclic structure
- If the carbonyl group is attached to the terminal carbon of an open-chain sugar, the monosaccharide is called an aldose
- If the carbonyl atom is attached to the second carbon atom, the monosaccharide is called a ketose
- The link between adjacent monosaccharide units is called the glycosidic bond

Lipids

- The basic skeleton of a steroid is a series of four fused alkyl rings
- Triacylglycerols are better known as fats and oils
- Triacylglycerols comprise two components: glycerol and three fatty acids
- A fatty acid is a carboxylic acid with a long alkyl chain
- Triacylglycerols are classed as saturated, monounsaturated, or polyunsaturated, depending on the chemical composition of their fatty acid groups
- A saturated fat comprises fatty acids containing only single carbon–carbon bonds
- A monounsaturated fat comprises fatty acids containing a single double bond
- A polyunsaturated fat comprises fatty acids containing more than one double bond
- Triacylglycerols with high saturated fatty acid content are classed as fats, and exist as solids at room temperature
- Triacylglycerols with high unsaturated fatty acid content are classed as oils, and are liquids at room temperature
- Glycerophospholipids comprise two hydroxyl groups and one phosphodiester group, joined to a glycerol skeleton
- A phosphodiester group comprises a phosphate group and an amino alcohol group
- Glycerophospholipids contain both polar and non-polar regions, which enable them to form a lipid bilayer

Nucleic acids

- The two key natural nucleic acids are deoxyribonucleic acid (DNA) and ribonucleic acid (RNA)
- Nucleic acids are polymers of nucleotides
- A nucleotide comprises three components: a nitrogenous base, a sugar, and a phosphate group
- In DNA, the sugar is deoxyribose
- In RNA, the sugar is ribose
- In DNA, the nitrogenous bases are adenine, guanine, cytosine, and thymine
- In RNA, the nitrogenous bases are adenine, guanine, cytosine, and uracil
- Cytosine, thymine, and uracil are pyrimidines, while adenine and guanine are purines
- Nucleotides join to form nucleic acids via phosphodiester bonds
- Phosphodiester bonds link the phosphate groups of neighbouring nucleotides
- The sugar and phosphate groups are hydrophilic, and lie on the outside of the DNA helix

- The nitrogenous base is hydrophobic, and lies on the inside of the DNA helix
- The ordering of nucleotides in a nucleic acid determines its sequence
- Nucleotides also play an important biological role in energy storage and release

Metals in biology

- Metals contribute to the correct structure and function of biological molecules
- The zinc finger domain is a characteristic amino acid sequence, which confers upon a protein the ability to bind to the major groove of DNA
- Metalloproteins are proteins that contain one or more metal atoms
- Many enzymes are metalloproteins, in which the metal ion participates directly in the catalytic activity of the protein

Molecular shape and structure 1: from atoms to small molecules

INTRODUCTION

Everything around us has some kind of shape and structure, from the uniform hexagonal array of a honeycomb, to the more irregular, yet still striking, design of a spider's web. The shape of an object rarely arises solely by chance. Instead, shape and structure are often carefully designed with a specific function in mind. The chair that you are sitting in while reading this book will have been designed to have a shape that puts the seat at the right height for you to sit comfortably; the hands with which you are holding this book have evolved with the protrusions we know as fingers, which are shaped and structured to grasp objects, and hold them firm.

Throughout chemistry, too, shape and structure are not a matter of chance, but are driven by a number of important factors–how atoms join to one another; how small molecules connect to form larger macromolecules; how molecules then fold up to arrange their component atoms in characteristic three-dimensional ways.

In this chapter we explore the factors that influence the shape and structure of chemical substances at the scale of atoms and small molecules. We ask what it is that gives different chemical substances their different shapes and structures, how different shapes and structures are generated, and why a substance's structure is linked so intimately to its function.

8.1 The link between structure and function

The relationship between **structure** and **function** pervades biology: consider, for example, the teeth of either a carnivore (a meat eater), such as a lion, or a herbivore (a plant eater), such as a cow. The primary function of teeth is to break down food to make it more readily digestible, so the shape of the teeth in each animal reflects its diet. While the cow has large, flat molars, which are perfect for grinding plant-based material such as grass, the lion possesses pronounced and sharp canine teeth, which they use to tear up their prey's flesh.

The intrinsic link between structure and function can have negative implications, as well as positive ones: if a substance has the wrong shape or structure, then it cannot perform its function correctly. A striking example of this is the disease sickle-cell anaemia. Individuals suffering from the disease carry red blood cells which have an unusual crescent (or 'sickle') shape, as opposed to the characteristic doughnut shape of a healthy red blood cell.

The sickle shape is caused by abnormal sickle-cell haemoglobin molecules carried by the red blood cells. Unlike normal haemoglobin molecules, sickle-cell haemoglobin molecules stick to one another to form long, rod-like structures inside the red blood cells. It is these rod-like structures that distort the shape of the red blood cell, forcing it into a rigid, sickle-shaped conformation.

While sickle-cell haemoglobin can still bind oxygen, sickle-shaped red blood cells cannot move through blood vessels as easily as normal shaped red blood cells. Instead they are prone to clumping together and causing blockages (particularly in small blood vessels), preventing the oxygen carried by these red blood cells from reaching its target tissue. People with sickle-cell anaemia experience severe pain as a result of the oxygen starvation of tissue and, in time, experience damage to organs, including the liver, kidneys, lung, and heart[1].

8.2 The shape of small molecules

In this chapter and the chapter that follows, we explore the shape and structure of molecules at two distinct scales–at the level of small molecules, and at the level of large macromolecules. These two scales are closely related, as large macromolecules (particularly biological macromolecules) are often generated from the joining of many small molecules. Shape and structure at these two different scales are influenced by rather different factors; it is the factors influencing the shape and structure of small molecules that we explore first.

One of our essential concepts is that many complicated molecules are generated from the joining of simple, small subunits or building blocks. (We see in Chapter 7, for example, how polypeptides are generated from the joining of multiple small subunits, amino acids.) But what factors influence the shape and structure of small molecules, such as amino acids?

Three key factors determine the shape and structure exhibited by a molecule, as illustrated in Figure 8.1. These are:

- bond lengths
- bond angles
- bond rotation.

We consider each of these factors in turn, in the following three sections.

Bond lengths

The **bond length** is the distance between the nuclei of two covalently bonded atoms. Atoms in molecules are not all positioned at equal distances apart. Rather, bond lengths vary, and are dictated by two things:

1 We may think that the abnormal haemoglobin molecules that give rise to sickle-cell anaemia give us no benefit, and should have been selected against as humans evolved. Paradoxically, however, sickle-cell anaemia confers upon the sufferer immunity to malaria. To some extent, therefore, the gene encoding the abnormal haemoglobin responsible for sickle-cell anaemia offers a benefit to our health, hence the fact that it has remained part of the human genome as we have evolved.

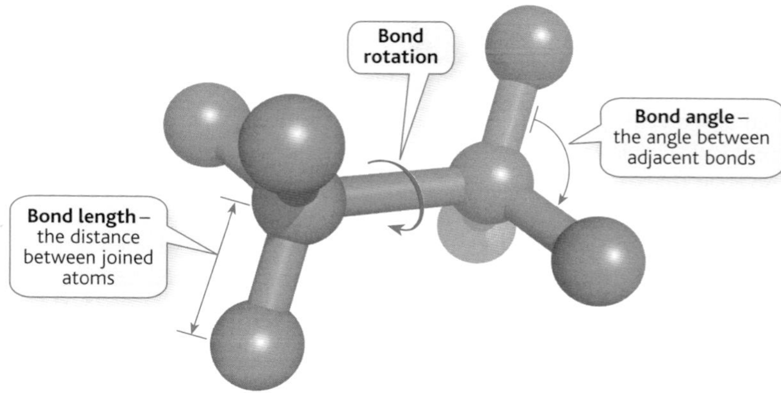

Figure 8.1 There are three key factors that determine the shape and structure of a molecule: bond lengths, bond angles, and bond rotation.

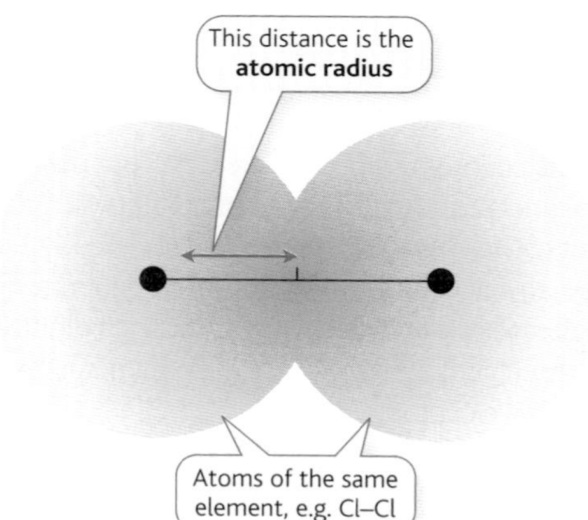

Figure 8.2 The atomic radius is the distance from the nucleus to the outer boundary of the atom, which is itself delineated by the volume of space occupied by the valence electrons.

1. The atomic radii of the atoms that are joined by the bond (and, hence, the chemical identity of the atoms)
2. The type of covalent bond (whether it is a single, double, or triple bond).

Atomic radii

The length of a single covalent bond is equivalent to the sum of the **atomic radii** of the two bonded atoms.

The **atomic radius** is usually defined as half the distance between two covalently bonded atoms *when the two atoms are identical*–that is, when the covalently bonded atoms are of the same element–as depicted in Figure 8.2.

Look at Figure 8.3 which depicts the atomic radii of a number of biologically important elements. If we look carefully at this figure we notice some general trends:

- As the valence shell number increases, so the atomic radius increases. (Compare the atomic radii of H, C, and P, whose valence shells are 1, 2, and 3 respectively. Notice how there is an increase in atomic radius as we move from H to C, and from C to P.) Why is this? As the valence shell number increases, valence electrons are occupying orbitals at increasing distances from the nucleus, so the atomic radius increases accordingly.

Element	H	C	N	O
Atomic radius (picometres, 10^{-12} m)	37	77	74	73
Electronic configuration of outer shell	$1s^1$	$2s^22p^2$	$2s^22p^3$	$2s^22p^4$

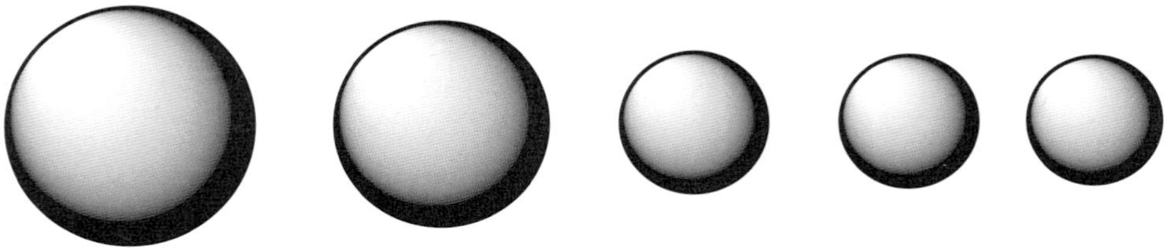

Element	Na	Mg	P	S	Cl
Atomic radius (picometres, 10^{-12} m)	186	160	110	103	100
Electronic configuration of outer shell	$3s^1$	$3s^2$	$3s^23p^3$	$3s^23p^4$	$3s^23p^5$

Figure 8.3 The atomic radii of a number of biologically important elements.

- As an increasing number of valence electrons enter the *same* valence shell, the atomic radius decreases slightly. (Compare the atomic radii of C, N, and O, which possess four, five, and six valence electrons respectively. Notice how the atomic radius *decreases* slightly as we move from C to N, and from N to O.) This slight decrease is caused by an increase in attraction between the positively charged nucleus, and the negatively charged valence electrons as a valence shell becomes filled. The increase in attraction draws the valence electrons closer to the nucleus, so that the distance from nucleus to the valence shell (the outer boundary of the atom) decreases.

We explore electronic configurations and valence shells in more detail in Chapter 2.

Bearing in mind that the length of a single covalent bond is equivalent to the sum of the atomic radii of the two bonded atoms, the length of a C–H bond is the sum of the atomic radii of atoms of C and H:

$$\begin{aligned}\text{C–H bond length} &= (\text{atomic radius of C}) + (\text{atomic radius of H})\\ &= 77\text{ pm} + 37\text{ pm}\\ &= 110\text{ pm}\end{aligned}$$

Remember: 1 pm = 10^{-12} m.

Atoms with large atomic radii exhibit longer bond lengths than those with smaller atomic radii. For example, if we consider the C–Cl bond:

$$\begin{aligned}\text{C–Cl bond length} &= (\text{atomic radius of C}) + (\text{atomic radius of Cl})\\ &= 77\text{ pm} + 100\text{ pm}\\ &= 177\text{ pm}\end{aligned}$$

Cl has a larger atomic radius than H so consequently the C–Cl bond length is greater than the C–H bond length.

- *The bond length is the sum of the atomic radii of two covalently-bonded atoms.*
- *The atomic radius is half the distance between the nuclei of two identical atoms, when joined by a single covalent bond.*

Self-check 8.1 What are the bond lengths of the following single covalent bonds?
(a) C–S; (b) O–H; (c) C–N

Bond type

As the number of electrons shared by two atoms increases–to form a double or triple bond–the bond length decreases.

We see in Chapter 3 how double and triple bonds arise between two atoms from the sharing of two and three pairs of valence electrons respectively. This increase in the number of shared electrons represents an increase in electron density: there are more electrons within the volume of space between the two atoms. Figure 8.4 shows how the increased electron density exerts a greater attractive pull on the positive nuclei of both atoms, drawing the atoms closer together and, hence, reducing the bond length.

Look at Figure 8.5, which depicts the lengths of the C–C, C=C, and C≡C covalent bonds. Notice how a double bond between two atoms is shorter than a single bond between the same two atoms, while a triple bond is shorter than the equivalent double bond.

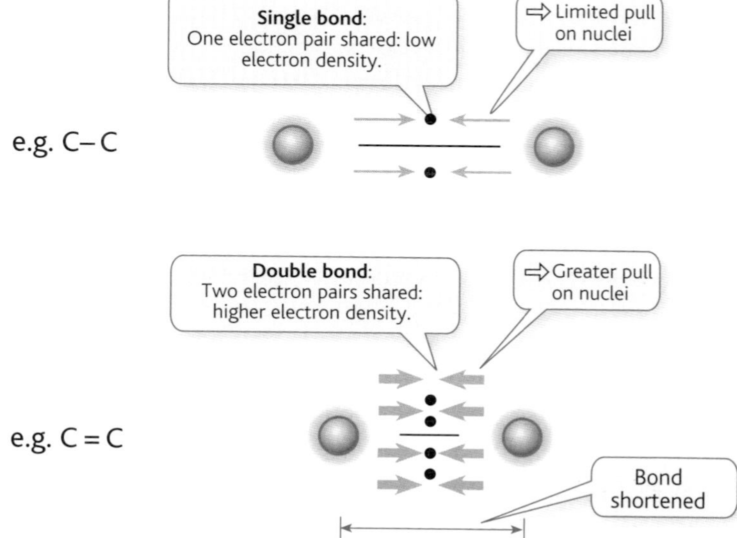

Figure 8.4 Double and triple bonds exhibit shorter lengths than single bonds joining atoms of the same elements. In this figure we see how double bonds have a greater electron density than single bonds, and so exert a greater 'pull' on the atoms' positive nuclei, shortening the distance between the atoms.

Self-check 8.2

(a) Identify the shortest bond in each of these pairs:

(i) H–H; C–H

(ii) C–O; C=O

(iii) N=N; N≡N

(b) The C=O bond has a length of 122 pm. Use the information in Figure 8.3 to calculate the length of the C–O bond, and hence verify your answer to (a) (ii) above.

C—C	154 pm
C=C	134 pm
C≡C	121 pm

Figure 8.5 The relative lengths of single, double, and triple bonds.

8.3 Bond angles

While bond lengths describe the distance between two bonded atoms, bond angles describe the three-dimensional arrangement of three covalently bonded atoms.

A bond angle is the angle between two bonds that are joined by a common central atom:

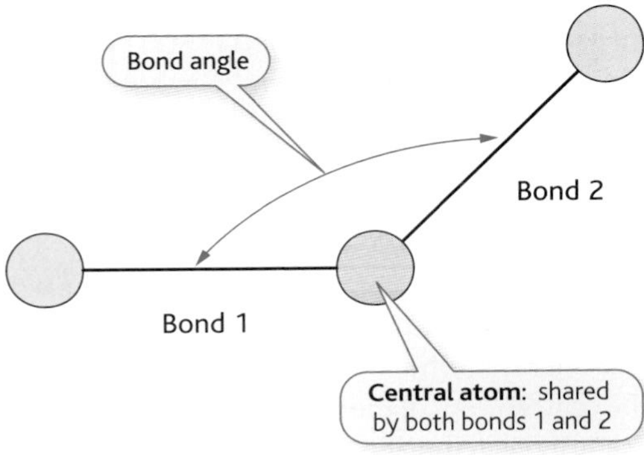

A bond angle may vary from 60° to 180°. A bond angle of 180° indicates that the atoms are arranged in a linear fashion:

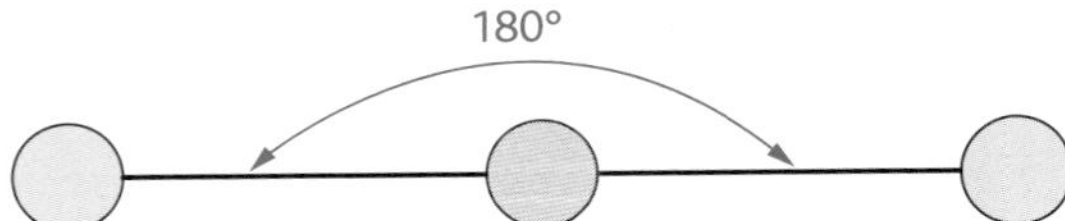

The bond angle arising between a particular group of three atoms is not random, however, but is influenced heavily by the chemical identity of the atoms that are joined together. Different groups of atoms arrange themselves in particular orientations, which can be described in two ways:

1. By the valence shell electron pair repulsion (VSEPR) theory
2. By the hybridization of atomic orbitals.

Let us now consider each of these in turn.

Valence Shell Electron Pair Repulsion (VSEPR)

Every covalently bonded atom possesses at least one valence electron pair. These valence electrons may be bonding pairs (shared between two atoms as a covalent bond), or non-bonding pairs (not shared between two atoms). For example, when their valence shells have been filled by covalent bonding with other atoms, hydrogen possesses one valence pair (a bonding pair); carbon has four valence pairs (four bonding pairs); oxygen also has four valence pairs (two bonding and two non-bonding pairs).

The valence pairs surrounding an atom arrange themselves in such a way that they minimize the repulsive forces that operate between them. This behaviour is described by the VSEPR theory. The VSEPR theory allows us to predict the arrangement of valence electrons around a single chosen atom (the **central** atom)[2]. By applying VSEPR theory sequentially to a number of atoms, we can then build up a bigger picture of valence electron arrangements within larger molecules.

Let's begin by considering a single atom, and seeing how VSEPR theory can help us to predict the arrangement of valence electrons around that atom.

In order to minimize repulsive forces, valence electron pairs on an atom move as far apart as is physically possible. Look at Figure 8.6, which illustrates how valence electron pairs surrounding an atom repel one another and, in so doing, influence how the atoms which are sharing the valence electrons orient themselves relative to one another. Notice that a non-bonding pair behaves in a similar way to a bonding pair: it repels valence pairs that are close to it, and so still influences the orientation of nearby atoms, despite not being directly involved in bonding.

There are a limited number of ways in which valence electrons can arrange themselves in order to minimize interactions between them. The most common arrangements are illustrated in Figure 8.7. Notice how each arrangement adopts a specific, defined geometry (tetrahedral, trigonal planar, etc.)

So the precise arrangement of a set of bonded atoms depends on the total number of valence electron pairs that surround the central atom. We can apply VSEPR theory to predict molecular shape by counting the total number of valence electron pairs surrounding the central atom of the molecule and selecting the corresponding geometric arrangement indicated in Table 8.1. It is through this arrangement that

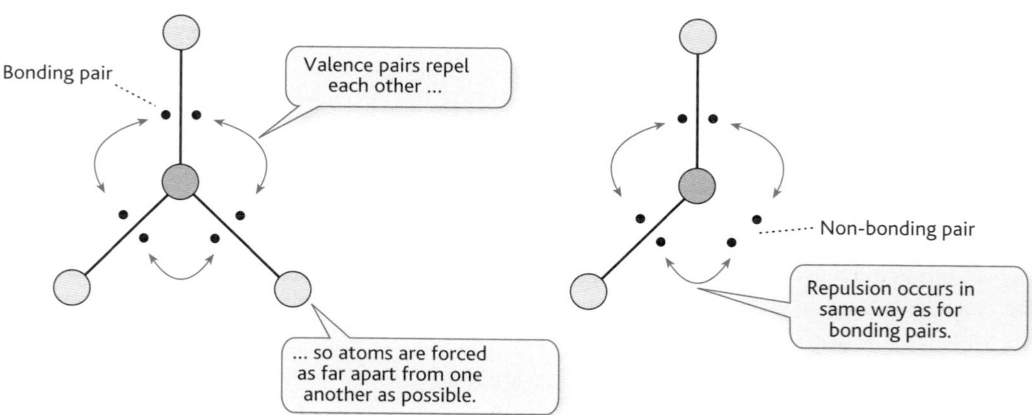

Figure 8.6 Valence electron pairs surrounding an atom repel one another and, in so doing, influence how the atoms which are sharing the valence electrons orient themselves relative to one another. Notice how both bonding and non-bonding pairs are involved in repulsion.

2 We use VSEPR theory to predict the arrangement of valence electrons around one atom at a time. We call the atom under consideration the central atom.

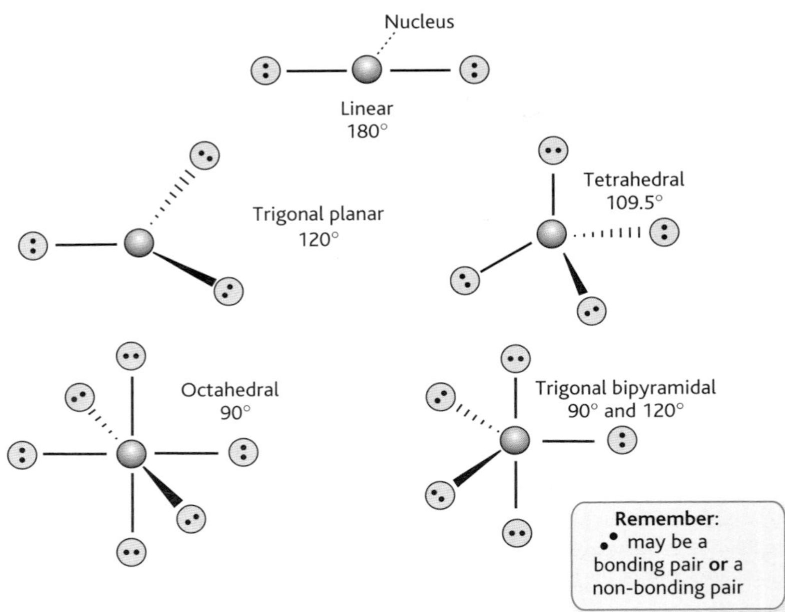

Figure 8.7 The most common geometries adopted by valence electrons. The geometry adopted depends on the number of valence electron pairs that surround the nucleus.

Table 8.1 The geometries and associated bond angles exhibited by different numbers of valence electron pairs around a central atom

Number of valence electron pairs	Geometry of molecule	Associated bond angle
2	Linear	180°
3	Trigonal planar	120°
4	Tetrahedral	109.5°
5	Trigonal bipyramidal	90° and 120°
6	Octahedral	90°

the number of valence pairs in question can orient themselves as far apart from one another as possible.

For example, we could predict that a central atom surrounded by three valence pairs would adopt a trigonal planar geometry, whereas a central atom surrounded by five valence pairs would adopt a trigonal bipyramidal arrangement.

Look at Figure 8.8, which shows how VSEPR theory can be used to predict the shapes of a range of small molecules, based on the number of valence electron pairs surrounding the central atom of each molecule. We need to remember that, when applying VSEPR theory, a valence pair can be either a bonding pair or a non-bonding pair. So, we see in Figure 8.8 how the non-bonding pairs on oxygen and nitrogen influence the geometry of water and ammonia respectively, even though these non-bonding pairs are not actively involved in formal covalent bonding.

When applying the VSEPR theory, we are considering the geometric arrangement of all valence pairs around the central atom, including the non-bonding pairs that aren't actually involved in a covalent bond with another atom. That is, the number of valence pairs we count may be more than the number of atoms present. For example, water has three atoms, yet we count four valence pairs for the purposes of VSEPR

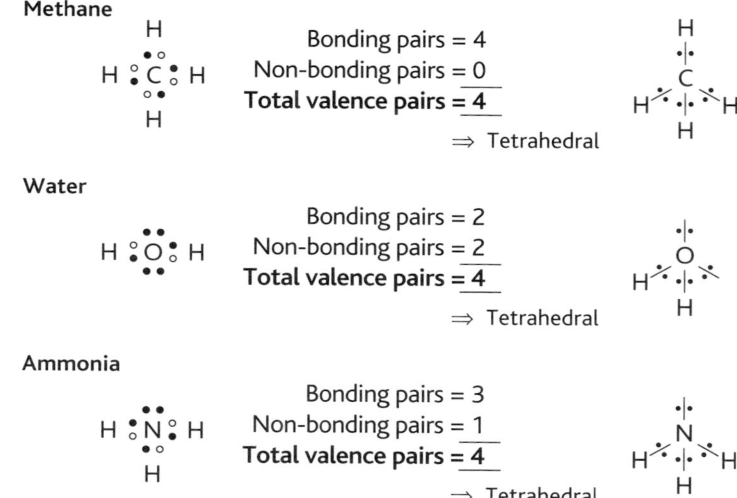

Figure 8.8 The use of VSEPR theory to predict the shapes of a range of small molecules. Lewis structures are shown in the left column, while actual geometries are illustrated on the right.

theory. Consequently, we describe a water molecule as 'V-shaped' (see Figure 4.7) because we're only 'seeing' the O and two H atoms and not the non-bonding pairs, but we say that water has a tetrahedral geometry.

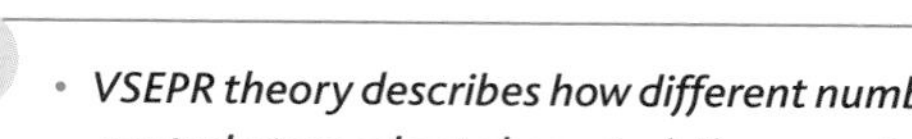

- *VSEPR theory describes how different numbers of valence electrons around a central atom adopt characteristic geometries.*
- *To use VSEPR theory, we count the number of pairs of bonding electrons* plus *the number of non-bonding pairs around the central atom.*

It is important to take account of both bonding pairs *and* non-bonding pairs when predicting the structure of molecules. If, for example, we consider only the bonding pairs in the case of the shape of a water molecule we might expect the molecule to adopt a linear structure:

H : O : H

Bonding pairs	= 2
Non-bonding pairs	= 0
Total valence pairs	**= 2**

⇒ Linear

Wrongly ignoring two non-bonding pairs on oxygen

Similarly, if we discount the lone pair on nitrogen when predicting the structure of ammonia, we might expect it to adopt a trigonal planar structure:

Bonding pairs	= 3
Non-bonding pairs	= 0
Total valence pairs	**= 3**

⇒ Trigonal planar

Wrongly ignoring one non-bonding pair on nitrogen

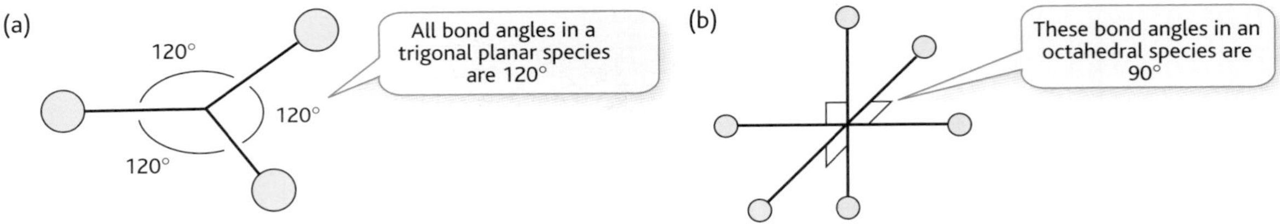

Figure 8.9 Different geometric arrangements have associated with them defined bond angles. For example, a trigonal planar arrangement is associated with a bond angle of 120° (a), whereas an octahedral arrangement is associated with a bond angle of 90° (b).

Both of these predictions are incorrect; only by taking into account both non-bonding pairs and bonding pairs does VSEPR predict molecular shape in a reliable way.

Each of the defined geometric arrangements has associated with it a defined bond angle. Look at Figure 8.9 and notice how each of the bond angles in a trigonal planar arrangement is identical at 120°. Similarly, each bond angle in an octahedral arrangement is 90°.

Non-bonding pairs: distorting the expected geometry

The force with which repulsion occurs depends on the nature of the valence electrons: non-bonding pairs repel more strongly than do bonding pairs. (Non-bonding pairs occupy more space than bonding pairs, and so surrounding electron pairs have to move slightly further away in order to evade the repulsive force from the non-bonding pair.) The relatively larger 'push' exerted by non-bonding pairs results in a slight distortion of the bond angles relative to those that we see in Figure 8.8.

For example, for water, the H–O–H bond angle is not 109.5° as predicted by VSEPR theory on the basis of a genuine tetrahedral arrangement. Look at Figure 8.10 (a). Notice how the two non-bonding pairs on the oxygen atom exert a greater push on the hydrogen atoms than would two bonding pairs, squeezing the H–O–H bond angle to a tighter 104.5°. We see a similar squeezing of bond angle for ammonia (Figure 8.10(b)), which is again predicted to give an H–N–H bond angle of 109.5° according to VSEPR theory. In reality, the additional repulsion from the non-bonding pair of electrons on the nitrogen squeezes the H–N–H bond angle to a tighter 106.6°.

- ***Non-bonding pairs repel neighbouring valence electrons to a greater extent than bonding pairs, making bond angles smaller than we would otherwise expect for a given geometry.***

VSEPR theory and the shape of molecules with multiple bonds

We see in Chapter 3 how molecules may possess single, double, or triple bonds, comprising one sigma (σ) bond, one σ bond and one pi (π) bond, and one σ bond and two π bonds respectively.

We can also use VSEPR theory to help us predict the shape of molecules containing both double and triple covalent bonds. A double bond comprises two valence

Remember:
- a single bond = 1 σ bond
- a double bond = 1 σ bond + 1π bond
- a triple bond = 1 σ bond + 2π bonds

For more details see p. 56, Chapter 3.

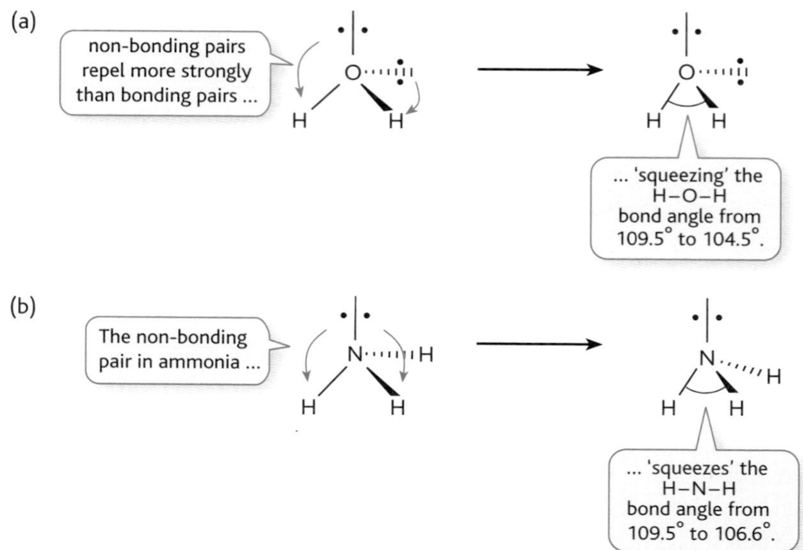

Figure 8.10 Non-bonding pairs exert a relatively large 'push' on surrounding atoms, leading to a distortion of bond angles relative to those that we' predict from the standard geometric arrangements. For example, the two non-bonding pairs on the oxygen atom in a water molecule squeeze the H–O–H bond angle to a tighter 104.5° (a), whereas the non-bonding pair of electrons on the nitrogen in ammonia squeezes the H–N–H bond angle to a tighter 106.6° (b).

electron pairs (one pair forms the sigma bond; one pair forms the pi bond), while a triple bond comprises three valence electron pairs (one pair forms the sigma bond; two pairs form the two pi bonds). For the purposes of using VSEPR theory to predict molecular shape, however, we only need to consider the *number of valence electron pairs occupying sigma bonds*, and can ignore the valence electron pairs occupying pi bonds.

- ***When using VSEPR theory to predict molecular shape, we only need to consider the number of valence electrons occupying sigma bonds.***

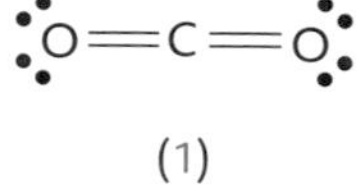

(1)

As an example, let's consider the structure of carbon dioxide, CO_2 (1).

To predict the structure of CO_2 using the VSEPR theory, we count the number of valence electron pairs around the central carbon atom. Look at Figure 8.11 and notice how the carbon atom possesses four bonding pairs of valence electrons and no non-bonding pairs. However, we only need to count those pairs of bonding electrons that form sigma bonds. The valence pairs on carbon participate in two sigma bonds, and two pi bonds; we ignore the two pairs of valence electrons participating in the two pi bonds, and so have a total of just *two* valence pairs.

If we look at Table 8.1, we see that a central atom possessing two valence pairs is predicted to adopt a **linear** structure. Therefore, we can predict that CO_2 is linear. And it is.

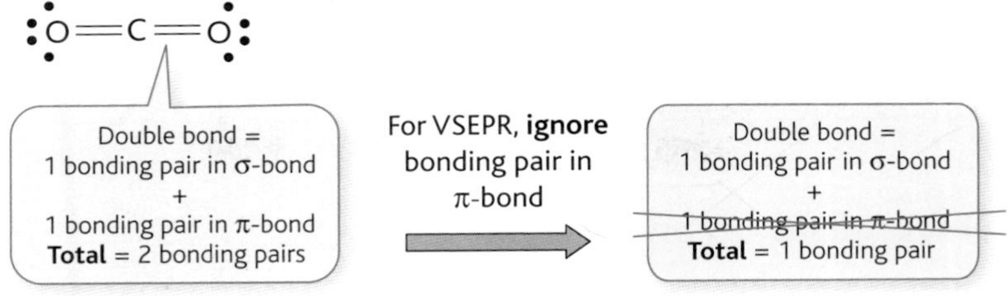

Figure 8.11 When we use VSEPR theory to predict the structure of a molecule with multiple bonds, we only need to consider those valence electron pairs that form sigma bonds. For example, a molecule of CO_2 has four valence electron pairs, two in each double bond. However, we only need to consider the two pairs which occupy the two sigma bonds.

Let's take another example, ethene (CH_2CH_2) (2).

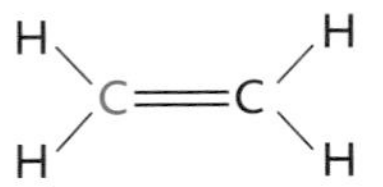

(2)

EXAMPLE

Problem: Predict the arrangement of valence electrons around the central C atom of ethene, indicated in blue in the structure shown here.

Strategy: We count the number of valence electron pairs (bonding pairs + non-bonding pairs) around the central C atom, but exclude those valence pairs which participate in pi bonds.

Solution:

- The central carbon atom has four valence electron pairs: four bonding pairs and no non-bonding pairs. The bonding pairs participate in two single covalent bonds and one double covalent bond:
  - 2 single bonds → 2 × sigma bonds (2 pairs)
  - 1 double bond → 1 × sigma bond (1 pair) and 1 × pi bond (1 pair).
- We ignore the valence pair which participates in the pi bond.
- We are left with a total of *three* valence electron pairs.
- Table 8.1 shows that a central atom with three valence electron pairs adopt a trigonal planar structure.

Answer: We predict that the three valence electron pairs around the central C atom in ethene adopt a trigonal planar structure.

Look again at the Lewis structure for ethene. Notice how both carbon atoms are chemically equivalent–they both participate in two single bonds with two H atoms, and one double bond with each other. Therefore, *both* C atoms in ethene adopt a trigonal planar structure.

The trigonal planar arrangement of valence pairs around the central carbon atom in ethene is the reason why we draw the structural formula as:

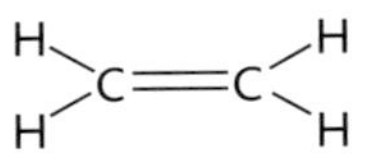

rather than

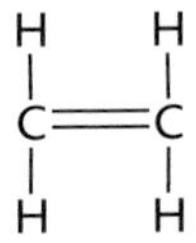

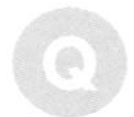

Self-check 8.3 What geometry would you predict the atoms in a molecule of hydrogen cyanide, H—C≡N, to adopt?

8.4 Hybridization and shape

We see above how interaction between neighbouring valence electron pairs governs the configuration of atoms that are bonded together, and that the configuration which is adopted around a given atom can be predicted by VSEPR theory. The arrangement of valence electrons predicted by VSEPR theory occurs as a result of the **hybridization** of atomic orbitals: a change in the shape of the orbitals, which allows the electrons they contain to occupy new locations.

Consider what happens when two human hands 'overlap' during the process of shaking hands. Before the hands actually meet, both participants must move their arms so that their hands come into close proximity. By analogy, we can consider hybridization to be the repositioning of atomic orbitals so that valence electrons are placed in the optimal locations for sharing between atoms during covalent bond formation (the equivalent of atoms 'joining hands').

It is worth pausing for a moment to recall that all electrons are restricted to occupying spatially defined orbitals. If valence electrons adopt a tetrahedral arrangement, then they can only have done so if the molecular orbitals in which they are located have also adopted a tetrahedral arrangement.

Methane has a tetrahedral structure, as predicted by VSEPR theory: the valence electron pairs that surround the central carbon atom in methane adopt a tetrahedral arrangement such that the hydrogen atoms to which the carbon atom is covalently bonded are forced into positions which correspond to the corners of a tetrahedron.

Specifically, carbon forms single covalent bonds with four hydrogen atoms; each single bond comprises one bonding pair of valence electrons. Each bonding pair occupies a molecular bonding orbital, which, according to VSEPR theory, must adopt a tetrahedral arrangement around carbon's nucleus.

Now look at Figure 8.12, which depicts the four atomic orbitals that comprise carbon's valence shell prior to covalent bond formation (shell 2: one s orbital and three p orbitals). Notice that these atomic orbitals do *not* adopt a tetrahedral arrangement. How do the orbitals that comprise carbon's valence shell make the transition from the arrangement shown in Figure 8.12 to the tetrahedral arrangement of methane?

When carbon bonds with four atoms of hydrogen, the four atomic orbitals, which comprise carbon's valence shell (one 2s and three 2p orbitals), merge together to form four new **hybrid** orbitals, which exhibit shapes that are different to the atomic orbitals from which they are formed. Because the hybrid orbitals arise from the hybridization of **one s** and **three p** orbitals, we call the hybrid orbitals **sp^3 orbitals**, and the overall process **sp^3 hybridization**.

- *Hybridization is the name we give to the process of atomic orbitals merging together to form new hybrid orbitals.*

We see in Chapter 2 how each atomic orbital has a characteristic shape; the hybrid orbitals exhibit shapes which fall part way between the shapes of the atomic orbitals

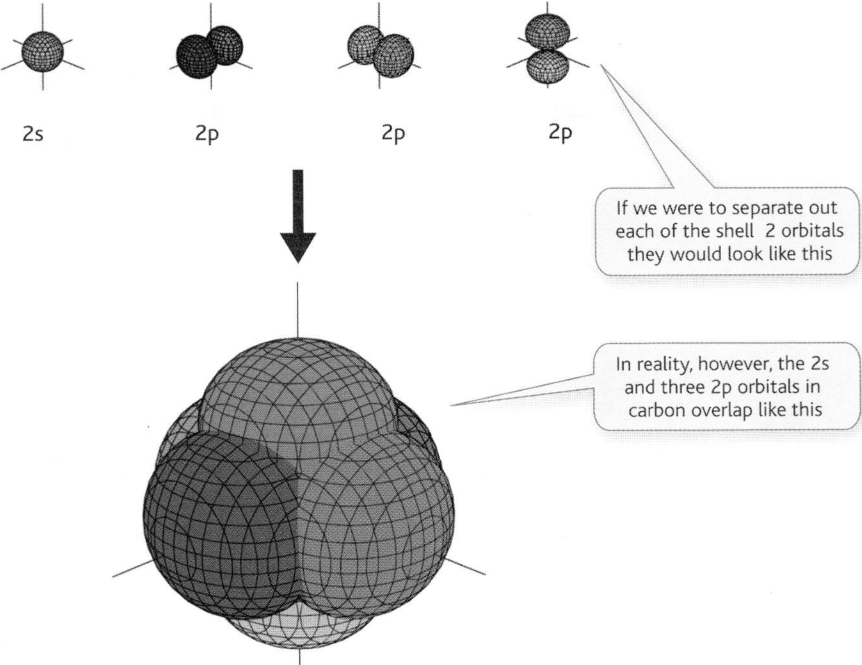
2s
2p
2p
2p
If we were to separate out each of the shell 2 orbitals they would look like this
In reality, however, the 2s and three 2p orbitals in carbon overlap like this

Figure 8.12 Shell 2 of carbon comprises one 2s orbital and three 2p orbitals. In reality, these orbitals overlap as shown here. These orbitals do not adopt a tetrahedral arrangement.

that have hybridized to form them. Look at Figure 8.13 and compare the shapes of the orbitals in carbon (a) pre-hybridization and (b) post-hybridization. Notice how the four sp^3 orbitals retain the double-lobed structure of a p orbital, but that one lobe is now much smaller than it was before hybridization, reflecting the influence of the s orbital during hybridization.

Figure 8.14 shows how the shape of methane reflects a tetrahedral geometry (with angles of 109.5° between each of the sp^3 orbitals) through the overlap of four 1s orbitals on the hydrogen atoms with the four sp^3 hybrid orbitals on the carbon atom, resulting in the formation of four sigma bonds.

The number of hybrid orbitals formed is always equal to the number of atomic orbitals that undergo hybridization: in the case of methane, notice how *four* orbitals (one s and three p) hybridize to form *four* sp^3 hybrid orbitals.

- ***The number of hybrid orbitals formed after hybridization is equal to the number of atomic orbitals undergoing hybridization.***

Hybridizing different numbers of orbitals

Hybridization can also explain the shape of molecules in which the central atom participates in fewer than four sigma bonds.

The number of atomic orbitals that hybridize is generally equal to the number of sigma bonds that occur between the central atom and other atoms in the molecule

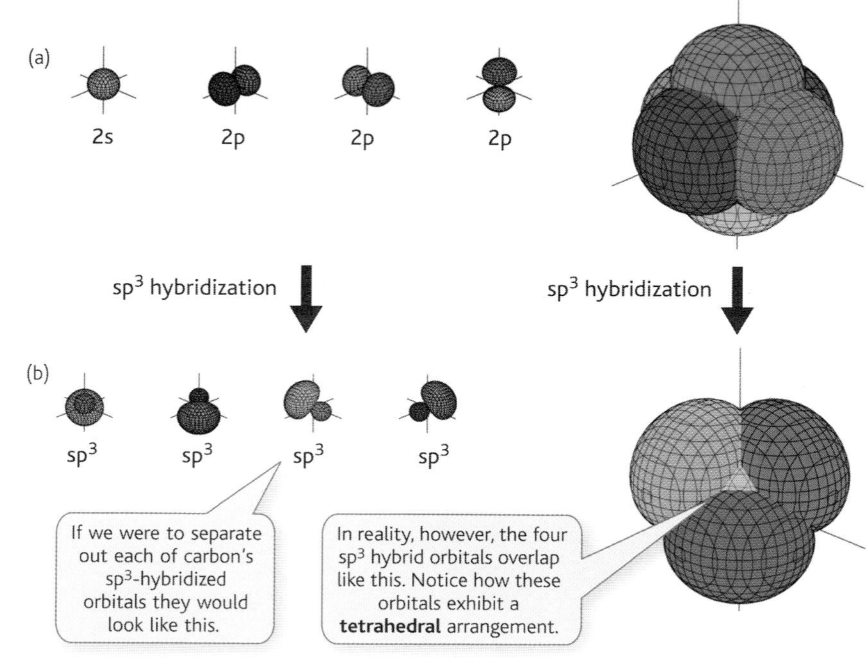

Figure 8.13 sp^3 hybridization. One s orbital and three p orbitals hybridize to form four sp^3 orbitals. Notice how the sp^3 hybrid orbitals exhibit a tetrahedral geometry whereas the unhybridized orbitals do not.

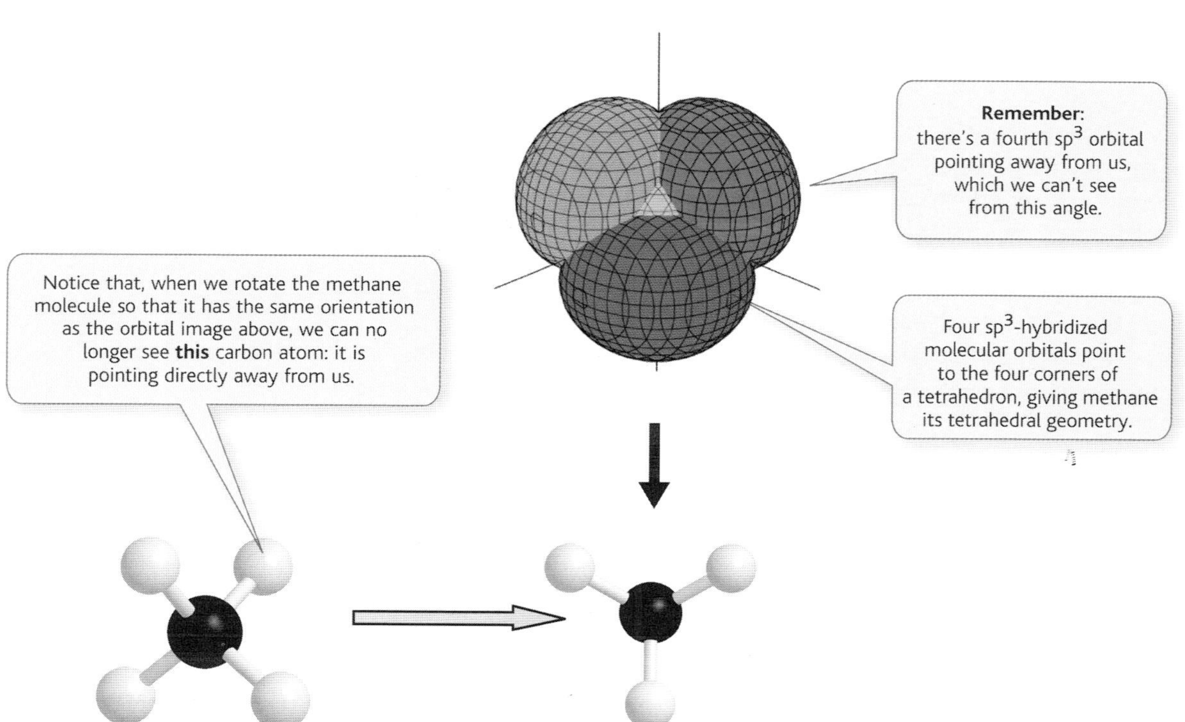

Remember: there's a fourth sp^3 orbital pointing away from us, which we can't see from this angle.
Notice that, when we rotate the methane molecule so that it has the same orientation as the orbital image above, we can no longer see **this** carbon atom: it is pointing directly away from us.
Four sp^3-hybridized molecular orbitals point to the four corners of a tetrahedron, giving methane its tetrahedral geometry.

Figure 8.14 The shape of methane reflects a tetrahedral geometry (with angles of 109.5° between each of the sp^3 orbitals) through the overlap of four 1s orbitals on the hydrogen atoms with the four sp^3 hybrid orbitals on the carbon atom, resulting in the formation of four sigma bonds.

to which it is contributing. Therefore, if an atom in a molecule participates in three sigma bonds with neighbouring atoms, then three of its atomic orbitals must have become hybridized prior to covalent bond formation. Similarly, if an atom participates in just two sigma bonds, then two of its atomic orbitals become hybridized.

Let's return to the example of ethene. We see above that each carbon atom in ethene contributes to three sigma bonds and one pi bond. The number of atomic orbitals that hybridize must be equal to the number of sigma bonds that occur once covalent bonding has occurred. Therefore, three atomic orbitals must hybridize for three sigma bonds to exist between each carbon atom and its neighbouring atoms in ethene.

In this instance, **one s** orbital and only **two p** orbitals on carbon hybridize to form **three sp^2** hybrid orbitals, as illustrated in Figure 8.15. Notice, once again, that the number of hybrid orbitals formed is equal to the number of atomic orbitals that become hybridized.

Look at Figure 8.15, and notice how, consistent with the orientation of valence pairs according to VSEPR theory, the three sp^2 hybrid orbitals adopt a **trigonal planar** geometry, with an angle of 120° between each sp^2 orbital.

Why is it that one 2s and two 2p orbitals, rather than three 2p orbitals, from carbon's valence shell become hybridized to form three sp^2 orbitals? The answer lies

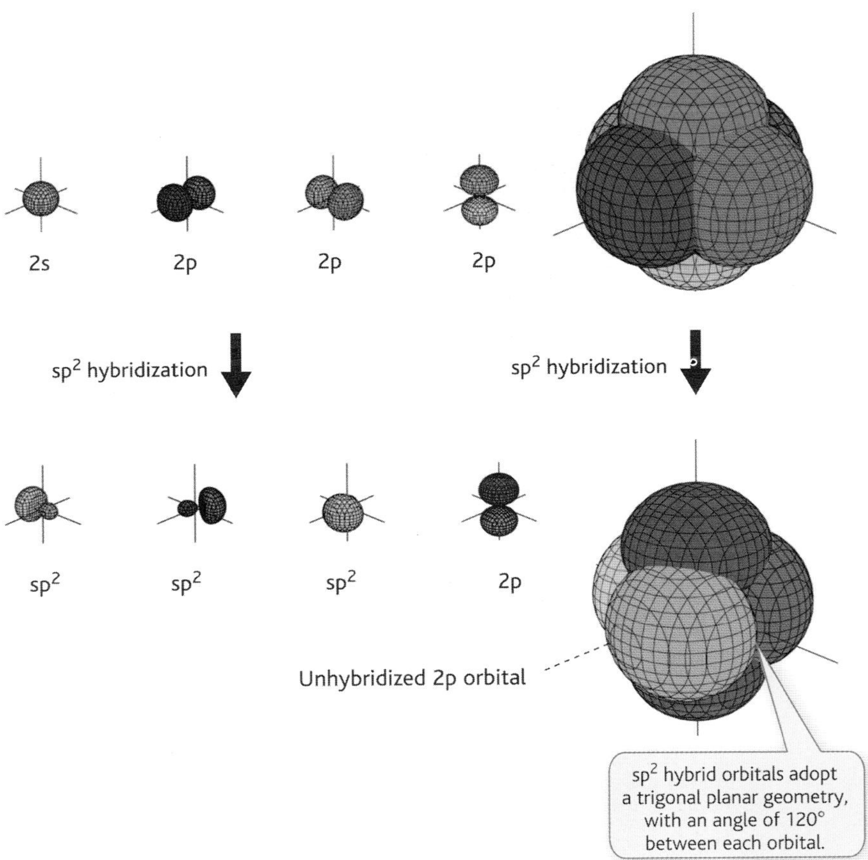

Figure 8.15 sp^2 hybridization. One s orbital and only two p orbitals on carbon hybridize to form three sp^2 hybrid orbitals. One 2p orbital remains unhybridized. Notice how the three sp^2 hybrid orbitals adopt a trigonal planar geometry.

in one of our essential concepts: the tendency to adopt the most stable (that is, the lowest-energy) option. We see in Chapter 2 how an s orbital has lower energy than a p orbital. The 2s orbital in carbon's valence shell becomes hybridized in preference to the third 2p orbital, as it represents the lowest energy (and therefore most stable) option.

This process is analogous to the order of filling of atomic orbitals with electrons, in which we see s orbitals filled before p orbitals (see p. 25).

Now look at Figure 8.16(a), which shows how two sp^2-hybridized carbon atoms and four hydrogen atoms come together to form ethene: two sp^2 orbitals on each carbon atom overlap with the 1s orbitals of two hydrogen atoms to form two sigma

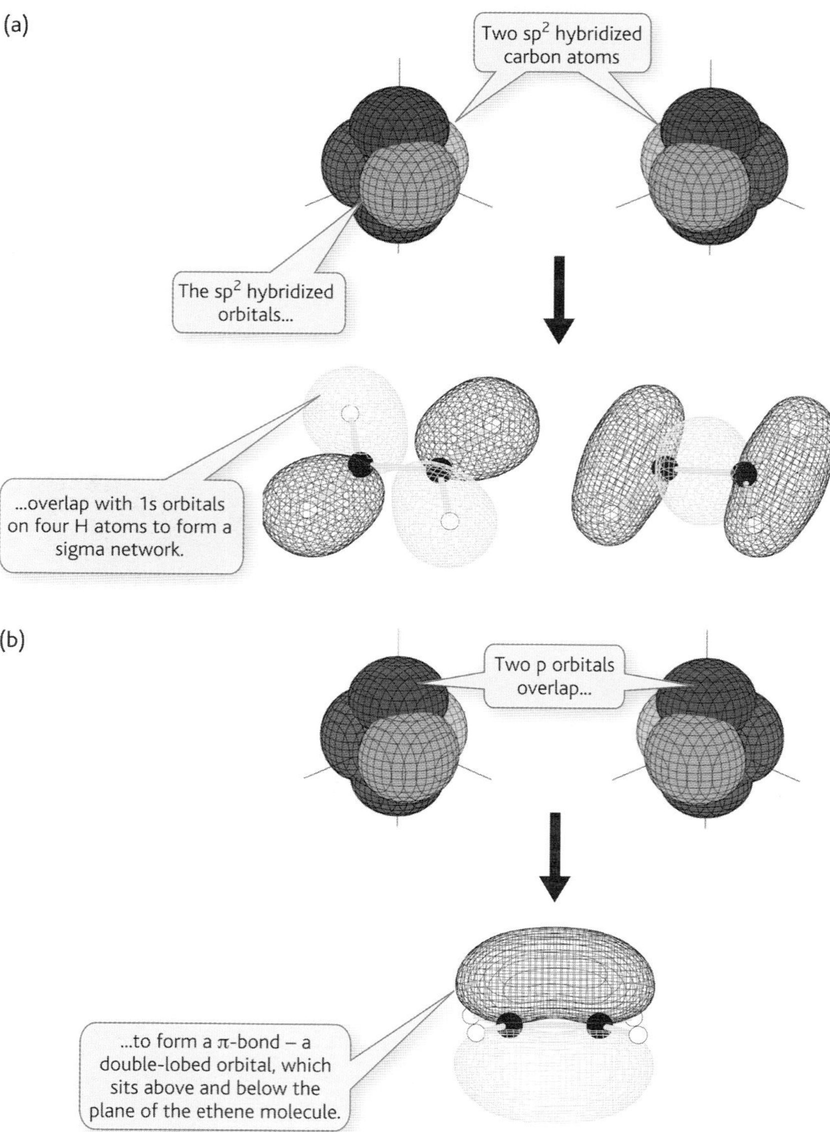

Figure 8.16 In ethene, two sp^2-hybridized carbon atoms and four hydrogen atoms come together. (a) The sp^2 hybrid orbitals overlap with 1s orbitals on four hydrogen atoms to form a sigma framework (four C–H bonds, and one C–C bond). (b) The two unhybridized p orbitals on the two carbon atoms overlap to form the pi bond.

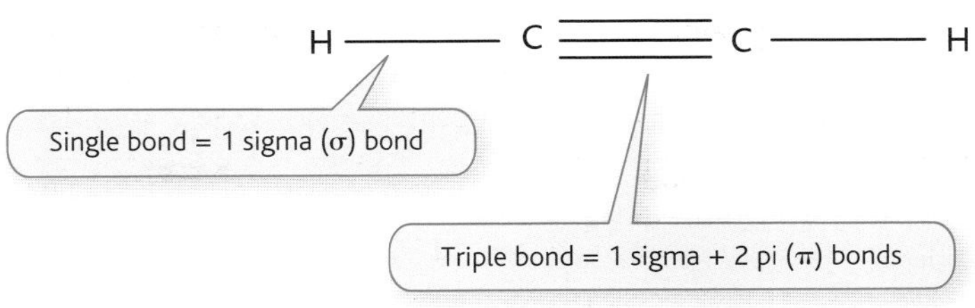

Figure 8.17 Each carbon in ethyne contributes to two sigma bonds and two pi bonds.

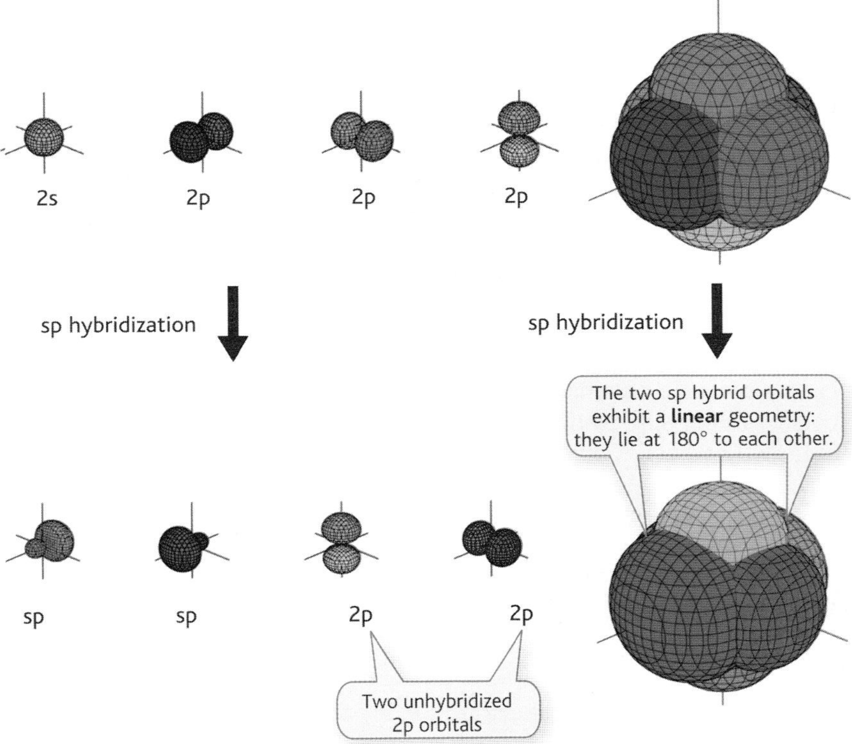

Figure 8.18 sp hybridization. One s orbital and one p orbital come together to form two sp hybrid orbitals. *Two* 2p orbitals remain unhybridized. Notice how the two sp hybrid orbitals adopt a linear geometry.

bonds, while the third sp^2 orbital on each carbon atom overlaps with the other to form a sigma bond between the two carbon atoms. Therefore a total of three sigma bonds are formed.

Note that one orbital in the valence shell of carbon–a 2p orbital–remains unhybridized. It is this orbital that overlaps with the equivalent orbital on the other carbon atom to form the pi bond between the two carbon atoms, as shown in Figure 8.16(b).

Finally, let us consider an example where a central carbon atom participates in just two sigma bonds: ethyne. Figure 8.17 shows how each carbon in ethyne contributes to two sigma bonds and two pi bonds. Two atomic orbitals on carbon must become hybridized for two sigma bonds to exist between carbon and its neighbouring atoms in ethyne.

This time, one s orbital and one p orbital in carbon's valence shell hybridize to form two sp hybrid orbitals[3], as illustrated in Figure 8.18. These two orbitals adopt a linear geometry, as predicted by VSEPR theory. The two sp orbitals lie at 180° to one another.

In this instance, two orbitals in carbon's valence shell–two 2p orbitals–remain unhybridized. The two 2p orbitals on both carbon atoms overlap with each other to form the two pi bonds that, together with the sigma bond formed by overlap of two sp hybrid orbitals, generates a triple covalent bond. This situation is called **sp hybridization.**

In summary:

- One s and three p orbitals undergo sp^3 hybridization to form **four sp^3** hybrid orbitals with tetrahedral geometry.
- One s and two p orbitals undergo sp^2 hybridization to form **three sp^2** hybrid orbitals with trigonal planar geometry.
- One s and one p orbital undergo sp hybridization to form **two sp** hybrid orbitals with linear geometry.

Non-bonding pairs and orbital hybridization

Often, atomic orbitals occupied by non-bonding pairs of valence electrons undergo hybridization in the same way as those that become occupied by bonding pairs of electrons after covalent bond formation.

Let's consider the carbonyl group, (3).

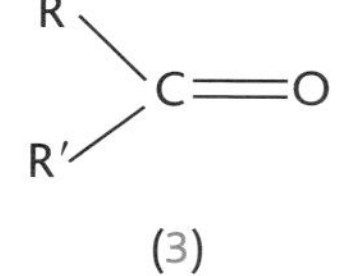
R
C
O
R′
(3)

Look at Figure 8.19. Notice how the central carbon atom in a carbonyl group behaves in the same way as the carbon atom in ethene: it participates in three sigma bonds, and so undergoes sp^2 hybridization to form three sp^2 orbitals, which adopt a trigonal planar geometry.

By contrast, the oxygen atom contributes to just one sigma bond (which, together with a pi bond, forms a double bond between the oxygen atom and the carbon atom). However, unlike the carbon atom, the oxygen atom carries two **non-bonding pairs** of electrons.

Earlier in this section, we see how we apply VSEPR theory to the arrangement of valence electron pairs around a central atom by adding the number of those bonding pairs which form sigma bonds to the number of non-bonding pairs of electrons. During the hybridization of atomic orbitals we can often treat non-bonding pairs of

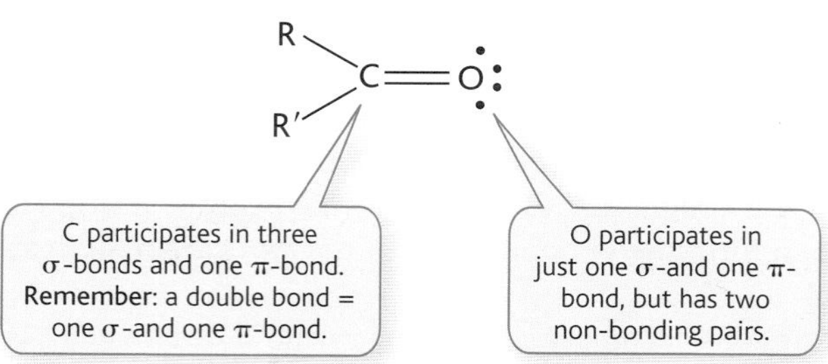
R
C
O
R′
C participates in three σ-bonds and one π-bond. Remember: a double bond = one σ-and one π-bond.
O participates in just one σ-and one π-bond, but has two non-bonding pairs.

Figure 8.19 The central carbon atom in a carbonyl group behaves in the same way as the carbon atom in ethene: it participates in three sigma bonds, and so undergoes sp^2 hybridization to form three sp^2 orbitals, which adopt a trigonal planar geometry.

3 Once again, we see one s and one p orbital becoming hybridized, rather than two p orbitals, because the hybridization of an s, rather than a p, orbital represents the most stable (again, the lowest-energy) option.

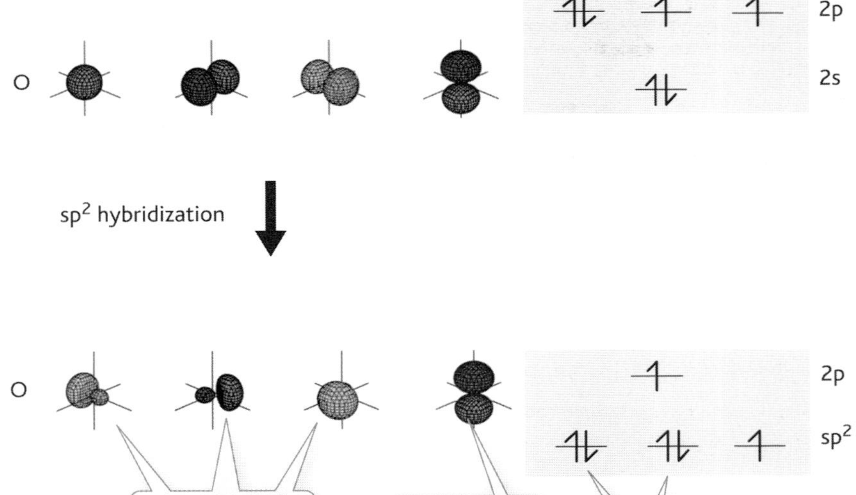

Figure 8.20 The oxygen atom of a carbonyl group is sp^2-hybridized. One 2s orbital and two 2p orbitals in oxygen's valence shell (shell 2) become hybridized to form three sp^2 hybrid orbitals, which adopt a trigonal planar geometry.

electrons in an analogous way: as being equivalent to a sigma bond (i.e. a bonding pair of electrons). (There are some exceptions to this rule, but treating non-bonding pairs and sigma bonds as equivalent entities works well enough for our exploration of hybridization.[4])

The oxygen atom of a carbonyl group contributes to one sigma bond and has two non-bonding pairs of electrons: this is equivalent to the oxygen atom contributing to *three* sigma bonds.

Therefore, like the carbon atom, three of its atomic orbitals must hybridize to form three hybrid orbitals. Figure 8.20 shows how one 2s orbital and two 2p orbitals in oxygen's valence shell (shell 2) become hybridized to form three sp^2 hybrid orbitals, which adopt a trigonal planar geometry. Notice how the non-bonding pairs, which occupy the 2s and one 2p orbital prior to hybridization, occupy two sp^2 hybrid orbitals after hybridization. Notice too that one of the three 2p orbitals does *not* undergo hybridization.

Remember: sp^3 orbitals adopt a tetrahedral geometry; sp^2 orbitals adopt a trigonal planar geometry; and sp orbitals adopt a linear geometry.

Now look at Figure 8.21 to see how orbitals on the sp^2-hybridized oxygen atom overlap with orbitals on the sp^2-hybridized carbon atom to form the carbonyl group: one sp^2 hybrid orbital on the oxygen atom overlaps with an sp^2 orbital on the carbon atom to form a sigma bond. The remaining two sp^2 orbitals on the oxygen atom are occupied by the two non-bonding pairs of valence electrons. The remaining unhybridized 2p orbital on the oxygen atom overlaps with the unhybridized 2p orbital on the carbon atom to form a pi bond. This pi bond, together with the sigma bond formed from sp^2 overlap, generates the double bond that joins the oxygen and carbon atoms.

4 Take care here. We are *not* saying that a non-bonding pair is the chemical equivalent of a sigma bond (or that a non-bonding pair participates in a sigma bond). We are merely saying that, *for the purposes of understanding how atomic orbitals hybridize*, non-bonding pairs of valence electrons and bonding pairs of valence electrons (sigma bonds) can be treated as being equivalent.

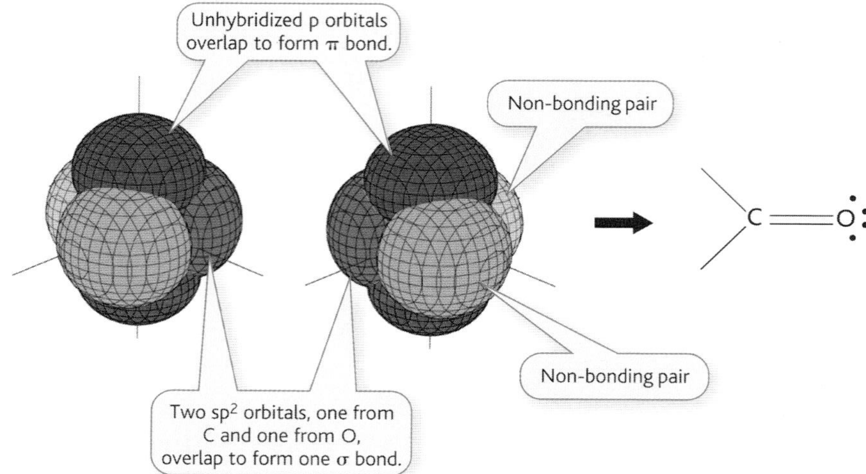

Figure 8.21 The overlap of an sp^2-hybridized oxygen atom with orbitals on the sp^2-hybridized carbon atom to form a carbonyl group.

- ***For the purposes of understanding how atomic orbitals hybridize, non-bonding pairs of valence electrons and bonding pairs of valence electrons (sigma bonds) can be treated as being equivalent.***

Self-check 8.4 Consider a molecule of water, H_2O.
How many non-bonding pairs does oxygen have in this molecule?
What kind of hybridization occurs when an oxygen atom bonds with two hydrogen atoms to form water, H_2O?

8.5 Bond rotation and conformation

We now know how single atoms arrange themselves spatially. Their 3D structure is principally influenced by the repulsion of valence electron pairs, such that these valence electron pairs–and the atoms that share them–adopt specific geometries, i.e. shapes. Underpinning these geometries is the mixing of atomic orbitals to generate new shapes through the process of hybridization, to allow valence electrons to occupy new regions of space.

While bonded atoms are locked into precise configurations which are governed by the repulsion of valence electron pairs, **groups** of bonded atoms may adopt different **conformations** relative to each other. A molecule's conformation describes the way in which connected atoms are arranged relative to one another at a particular moment in time. The arrangement of joined atoms is variable; this variability is a consequence of **bond rotation**.

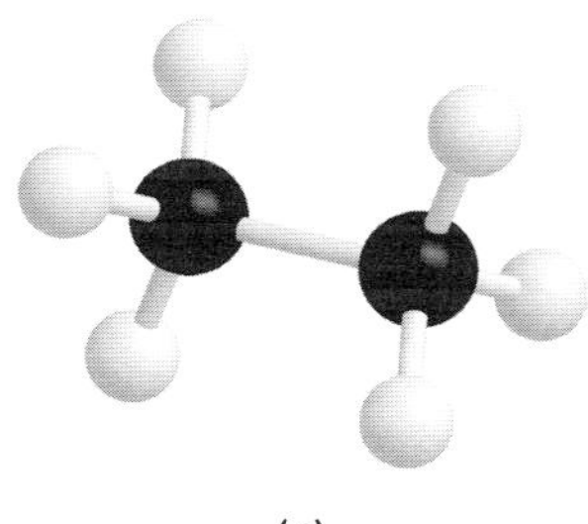

(4)

Consider a molecule of ethane, C_2H_6, (4). The four atoms positioned around each carbon atom (three H and one C in each case) adopt a tetrahedral geometry. This shape is preferred as a means of minimizing charge repulsion between the valence electrons that are shared to form the covalent C–H and C–C bonds. The four pairs of

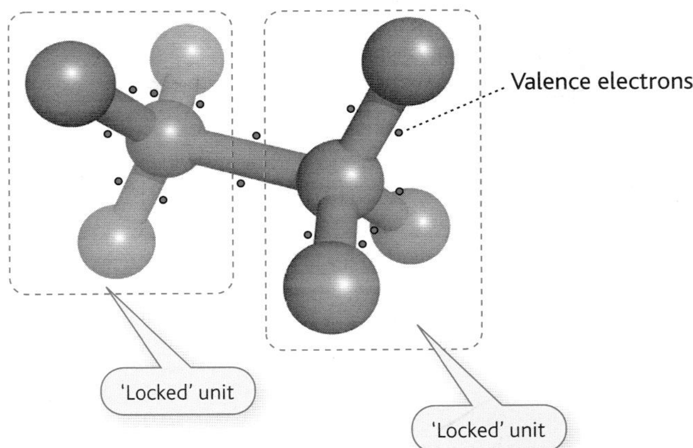
Valence electrons
'Locked' unit
'Locked' unit

Figure 8.22 The four pairs of valence electrons surrounding each carbon atom in ethane must remain as far apart from each other as possible, forming a 'locked' tetrahedral unit.

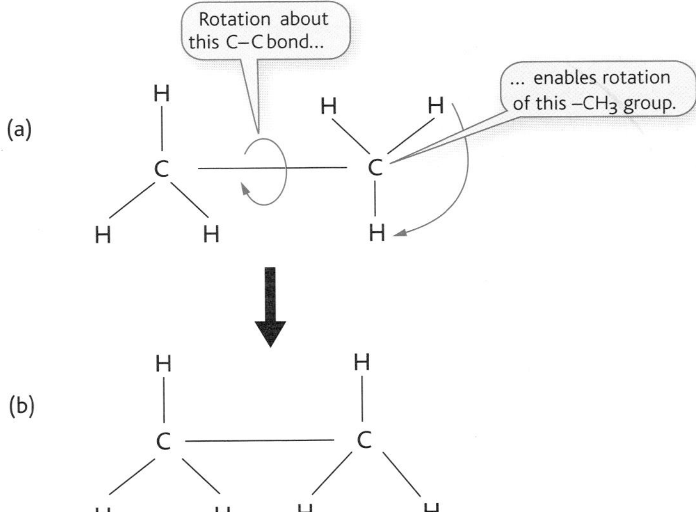
Rotation about this C–C bond...
... enables rotation of this $-CH_3$ group.
(a)
(b)

Figure 8.23 While the configuration of atoms around both carbon atoms in ethane may be 'locked', the two methyl ($-CH_3$) groups are able to rotate relative to each other by rotation about the central C–C bond.

valence electrons surrounding each carbon atom must remain as far apart from each other as possible, forming a 'locked' tetrahedral unit, as depicted in Figure 8.22.

However, while the configuration of atoms around both carbon atoms may be 'locked', the two methyl ($-CH_3$) groups are able to rotate relative to each other by rotation about the central C–C bond. Look at Figure 8.23; notice that the right-hand $-CH_3$ group is able to rotate such that the orientation of the $-CH_3$ group in space mirrors that of the $-CH_3$ group at the other end of the C–C bond. Now look again at the position of the valence electrons when the ethane molecule adopts conformations (a) and (b). Notice how the valence electrons are at the same fixed distances apart (they retain the same configuration with the 109.5° bond angle characteristic of a tetrahedral geometry) whether the molecule adopts conformation (a) or (b).

But what is the difference between 'configuration' and 'conformation'? Let's pause for a moment to find out.

Self-check 8.5 The C–C sigma bond in ethane is the result of the overlap of two hybrid orbitals–one from each of the two C atoms. Each carbon atom in ethane contributes to four sigma bonds. What type of hybridization occurs between the valence shell orbitals of each C atom? Hence, what types of hybrid orbital overlap to form the C–C bond?

Conformation versus configuration

A molecule's **configuration** describes how its composite atoms are joined together in three-dimensional space. A molecule's configuration is **fixed**, and can only be changed by breaking bonds.

In the case of ethane, both $-CH_3$ groups are locked in a specific tetrahedral configuration. Repulsion between valence electron pairs dictates that the atoms comprising the $-CH_3$ groups must remain fixed in this tetrahedral arrangement to minimize the repulsive forces.

A molecule's **conformation** describes its overall three-dimensional structure at a particular moment in time. A molecule's conformation is **variable**.

Consider a wooden-framed futon, as depicted in Figure 8.24. A futon can typically exhibit two conformations: a folded conformation (in which the futon may be used as a sofa), and an extended conformation (in which the futon may be used as a bed). Both conformations utilize the same structural components, which are attached to each other in exactly the same way. The overall shapes are very different however (as are the functions that the futon may have when arranged in either conformation).

In the case of ethane, rotation about the central C–C bond means that it can adopt different conformations, in which its composite atoms arrange themselves differently relative to one another without changing their connectivity (that is, how they are joined to one another). Look at Box 8.1 to find out *why* bond rotation about the C–C bond is possible.

- ***A molecule's configuration describes how its composite atoms are joined together in three-dimensional space, and is fixed.***
- ***A molecule's conformation describes how its composite atoms are arranged at a particular moment in time, and is variable.***

We see in Chapter 5 that the alkanes are a family of hydrocarbons, the members of which comprise atoms of carbon and hydrogen that are joined to form chains of

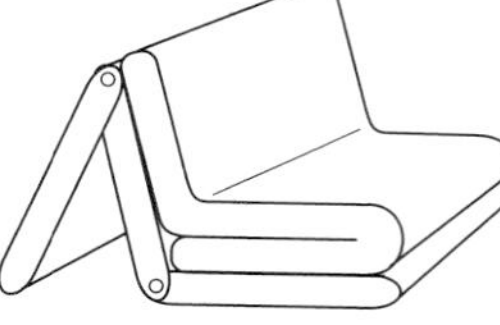

'Sofa' conformation

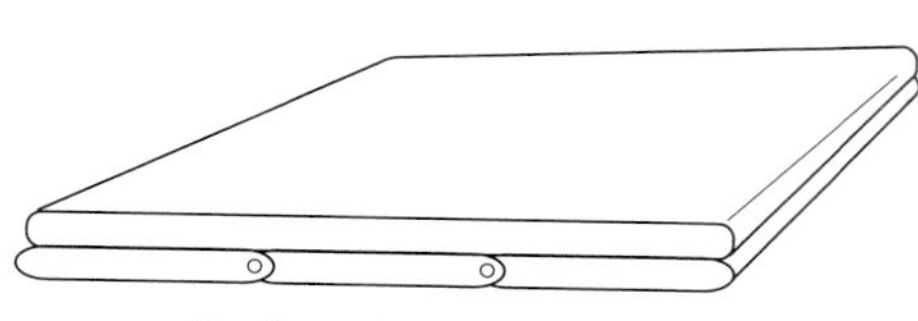

'Bed' conformation

Figure 8.24 A futon can typically exhibit two conformations: a folded conformation (in which the futon may be used as a sofa), and an extended conformation (in which the futon may be used as a bed).

BOX 8.1 Why is rotation about the C–C bond possible?

Rotation can occur about a single covalent bond such as the central C–C bond in ethane because it does not disrupt the overlap between the two bonding orbitals that comprise the single bond.

Single bonds such as the C–C bond occur as a result of the 'head-to-head' overlap of two bonding orbitals. In the case of ethane, the C–C bond forms from the overlap of two sp^3 hybrid orbitals, as depicted in Figure 8.25(a). The two orbitals are symmetrical about the C–C bond axis. That is, if we imagine looking down the length of the C–C bond while rotating the two orbitals, the shape we see remains the same, as depicted in Figure 8.25(b). Therefore, we can say that the bond is symmetrical about its axis (its length). It doesn't matter how the orbitals rotate relative to one another, their end-on shapes (the parts that are in contact) remain the same. Therefore, the overlap of the two orbitals is unaffected by the rotation, and the sigma bond remains unaltered.

Rotation is only possible if the two overlapping orbitals meet head-to-head, i.e. they lie along a straight line, and are symmetrical about this line.

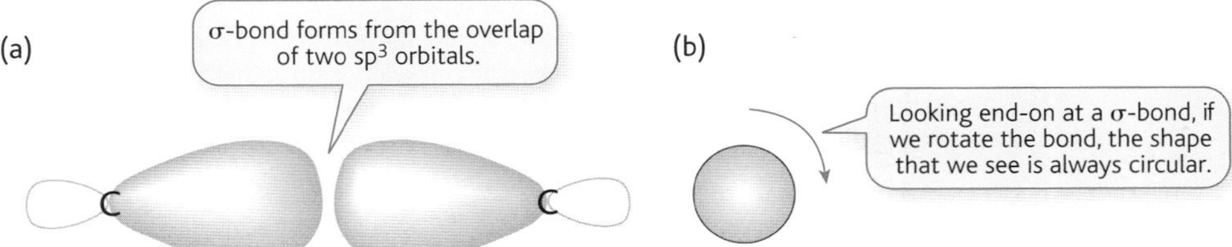
(a)
σ-bond forms from the overlap of two sp³ orbitals.
C
C
(b)
Looking end-on at a σ-bond, if we rotate the bond, the shape that we see is always circular.

Figure 8.25 (a) The C–C bond in ethane forms from the overlap of two sp^3 hybrid orbitals. (b) If we imagine looking down the length of the C–C bond while rotating the two orbitals, the shape we see remains the same: the two orbitals are symmetrical about the C–C bond.

varying length. For example, butane has a four-carbon backbone (Figure 8.26(a)) whereas octane has an eight-carbon backbone:

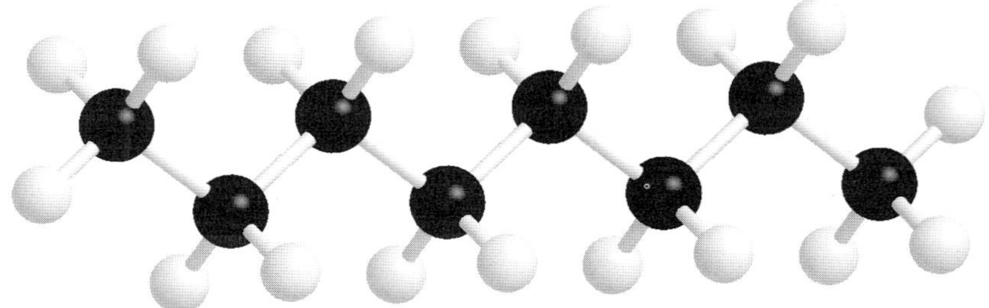

Bond rotation is possible about each C–C bond in the carbon backbone, such that numerous conformations of a molecule such as butane or octane are possible.

Look at Figure 8.26, which shows two of the possible conformations that butane can adopt due to rotation about the middle C–C bond. In one conformation, the four carbon atoms lie in a zigzag type arrangement; in the other, the atoms form a 'boat-shaped' structure. However, butane can also adopt millions of shapes between these two extremes.

While both conformations are possible, only one conformation–the zigzag structure–is favoured in reality. Look at the two methyl groups, which lie at opposite ends of the carbon chain: the zigzag structure is favoured because it places them as far apart as possible. We see in Section 8.3 how VSEPR theory predicts the arrangement

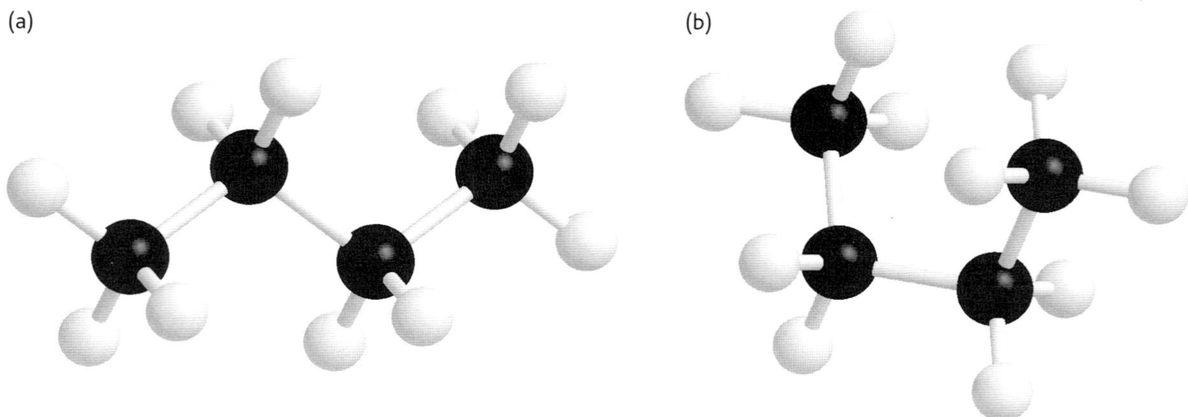

Figure 8.26 Two of the conformations that butane can adopt due to rotation about the middle C–C bond. In one conformation, the four carbon atoms lie in a zigzag type arrangement (a); in the other, the atoms form a 'boat-shaped' structure (b). Notice that the configurations within butane–how the atoms are connected around each carbon atom–are identical in each of these conformations.

of valence electron pairs around a central atom, also placing them as far apart from one another as possible. (This physical separation minimizes the repulsive forces that operate between the valence pairs and hence maximizes their stability.) The same principle operates here, making the zigzag conformation the m ost stable.

Indeed, a zigzag conformation is favoured for all straight-chain alkanes, because it represents the most stable way of arranging the groups of atoms that form these chains. (Look again at the structure of octane above; notice how this has the same zigzag structure as exhibited by butane in Figure 8.26(a).)

Limitations on bond rotation

Rotation is only possible about *single* covalent bonds. Both double and triple bonds have **fixed** conformations. This inability to rotate follows from the orbital overlaps required for pi bond formation: overlap is broken if the two atoms joined by the pi bond rotate relative to one another.

We see above how two atomic orbitals, which are overlapping to form a sigma bond, may rotate about the bond axis without orbital overlap being disrupted. However, pi bond formation can only occur if the overlapping orbitals are held in fixed positions relative to one another. In this context, look at Figure 8.27, which shows how a pi bond forms from the overlap of two p orbitals aligned parallel to one another. If one atom rotates, such that the two p orbitals lie perpendicular (at right angles) to one another, orbital overlap is broken, and the pi bond cannot form.

While rotation about a double or triple bond is never possible without breaking the bond, there are circumstances under which rotation about a single bond is not possible either. Rotation about a single bond may often be blocked simply because different groups of atoms physically get in the way of one another. We call this phenomenon **steric hindrance**. The blocking of rotation about a single bond is particularly important in influencing the conformation of polypeptide chains, as we discover in Box 8.2.

(a) **No bond rotation:**

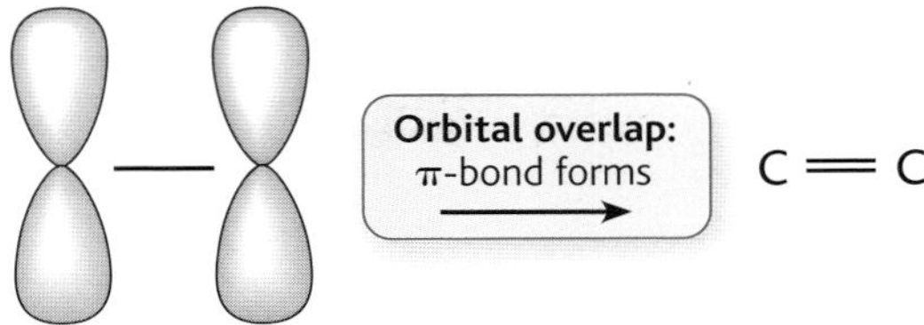
Orbital overlap:
π-bond forms
C ═ C

(b) **Rotation about double bond:**

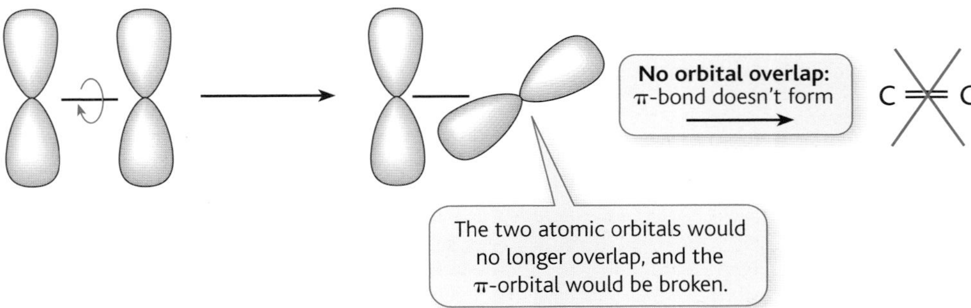
No orbital overlap:
π-bond doesn't form
C ═ C
The two atomic orbitals would no longer overlap, and the π-orbital would be broken.

Figure 8.27 The effect of bond rotation on the formation of a pi bond. (a) A pi bond forms from the overlap of two p orbitals aligned parallel to one another. (b) If one atom rotates, such that the two p orbitals lie perpendicular (at right angles) to one another, orbital overlap is broken, and the pi bond cannot form.

Rotation about a single bond is also not possible if the bond is part of a **delocalized** system of electrons. We see in Chapter 3 how delocalized electrons are π-bonding electrons that are shared between more than two atoms. When delocalization occurs, the bonds joining the atoms between which the π-bonding electrons are shared behave in a way that falls part way between being a single bond and double bond. As a consequence, single bonds acquire the double bond characteristic of being unable to rotate.

We learn more about the delocalization of electrons on p. 68.

One pair of bonds in which delocalization occurs is the C=O and C–N bonds that comprise a peptide bond (the bond that joins adjacent amino acid residues in polypeptides), as depicted in Figure 8.28. Consequently, the C–N bond does not behave like a 'standard' single bond, and is unable to rotate. This lack of rotation has a significant impact on the structure of polypeptides, as we see in Box 8.2.

(a) Peptide bond

(b) Delocalization occurs within these two bonds.

Figure 8.28 Delocalization in the peptide bond. (a) The structure of the peptide bond. (b) Delocalization of electrons occurs within the C=O and C–N bonds, as denoted by the dotted line here.

BOX 8.2 Peptide bonds and the three-dimensional shape of polypeptides

Polypeptides are formed from the sequential joining of amino acids to form long chains, often tens or even hundreds of amino acid residues long. We see in Chapter 7 how the 'backbone' of the polypeptide chain forms from the joining together of the carbonyl group of one amino acid with an amine group of an adjacent amino acid to form a peptide bond. Figure 8.29 shows how the polypeptide backbone comprises a chain of alternating peptide, $C–C^{\alpha}$, and $C^{\alpha}–N$ single bonds. While rotation is not possible about the peptide bond (as explained above), rotation is possible about the two other single bonds that form the repetitive three-bond sequence. We call the angles through which these two bonds are able to rotate **dihedral angles**. It is the particular dihedral angles exhibited by an amino acid residue which dictate the conformation of the peptide chain at that location in the chain.

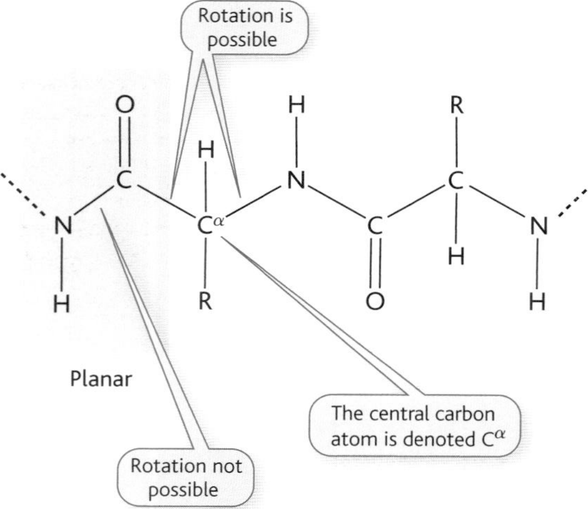

Figure 8.29 The polypeptide backbone comprises a chain of alternating peptide, $C–C^{\alpha}$ and $C^{\alpha}–N$ single bonds. Bond rotation is not possible about the peptide bond, so it forms a planar unit.

However, even rotation about these two bonds is restricted to some extent. Each amino acid is given its particular identity by the chemical composition of the side chain which is bonded to the central carbon atom. Glycine, for example, has the simplest side chain: –H, whereas leucine has the side chain:

$$—CH_2—CH(CH_3)_2$$

Such a side chain is often rather bulky, and so gives rise to steric hindrance: it blocks the rotation of the C–C and C–N bonds by physically restricting movement of different parts of the peptide chain relative to one another.

As a result of the steric hindrance, the $C–C^{\alpha}$ and $C^{\alpha}–N$ bonds are only able to rotate within certain narrow ranges, i.e. the peptide chain adopts conformations which possess only a small number of different dihedral angles. Consequently, peptide chains exhibit just a small range of different conformations; these different conformations confer upon the peptide chain characteristic shapes.

We find out more about the characteristic shapes adopted by peptide chains in Section 9.2.

- *Rotation is not possible about double or triple bonds, or where delocalization occurs.*
- *The restriction of movement due to a lack of space between neighbouring atoms is called* **steric hindrance.**

Restricted bond rotation: cyclic structures

We see in Chapter 5 how the carbon chains of hydrocarbon molecules can form not only linear structures, but that the end of the chain can join to generate cyclic (ring) structures, as illustrated in Figure 8.30(a).

While bond rotation is possible about a single bond in a linear molecule, bond rotation is *not* possible about a single bond in a cyclic structure. Figure 8.30(b) shows how the ends of a linear molecule are untethered, giving the molecule freedom to twist along its length without constraint. In contrast, when the ends of a linear structure become tethered to one another to form a cyclic structure, rotation becomes constrained.

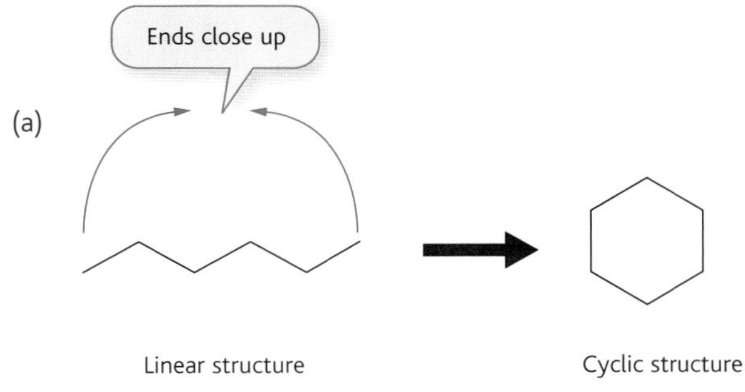
(a)
Ends close up
Linear structure
Cyclic structure

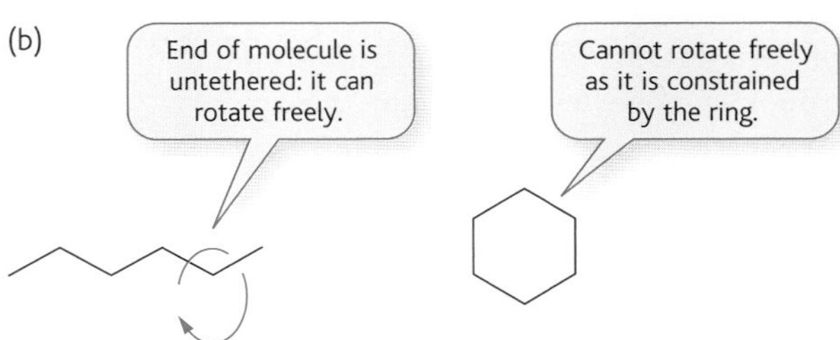
(b)
End of molecule is untethered: it can rotate freely.
Cannot rotate freely as it is constrained by the ring.

Figure 8.30 (a) Linear structures can become cyclic if the two ends of the chain become joined. (b) The ends of a linear molecule are untethered, giving the molecule freedom to twist along its length without constraint. In contrast, when the ends of a linear structure become tethered to one another to form a cyclic structure, rotation becomes constrained.

The movement of atoms in a cyclic structure is not altogether impossible, however. While unable to rotate totally freely, a ring structure does possess enough flexibility to allow it to move into, and adopt, different conformations. The most characteristic conformations are those adopted by cyclic molecules comprising a six-carbon framework, which includes the sugar glucose.

Such molecules may adopt one of two characteristic conformations–either the 'boat' conformation or the 'chair' conformation, as illustrated in Figure 8.31. Look at this figure, and notice how flexibility in the cyclic structure allows the connected atoms to rotate into distinct conformations.

Self-check 8.6 Bond rotation is possible about which of the following highlighted bonds?

(a) H–C(H)(H)–C(OH)(H)–C(H)(H)–H

(b) O, H_3C

(c) H_3C—CH=CH—CH=CH—CH_3

Having explored the different factors that affect the shape of molecules at the scale of the joined atoms, we are now ready to consider what happens when many atoms join together to form large, complicated molecules. What determines the shape of

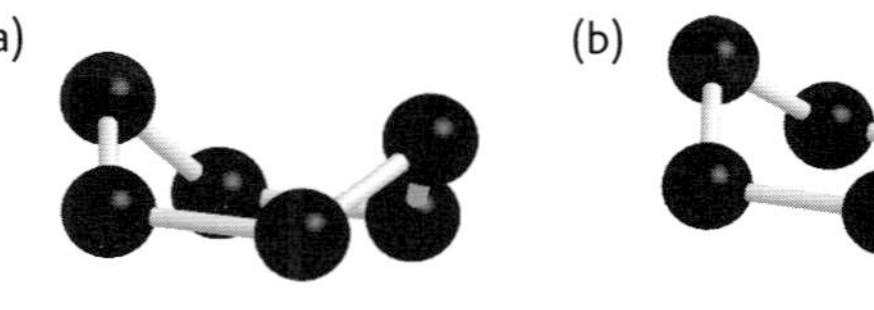

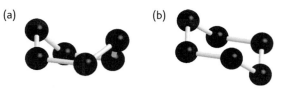

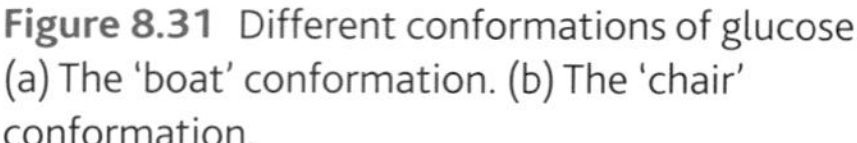
Figure 8.31 Different conformations of glucose. (a) The 'boat' conformation. (b) The 'chair' conformation.

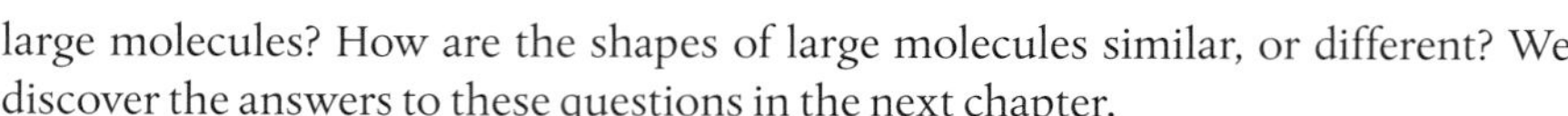
large molecules? How are the shapes of large molecules similar, or different? We discover the answers to these questions in the next chapter.

Check your understanding

To check that you've mastered the key concepts presented in this chapter, review the checklist of key concepts below, and attempt the multiple-choice questions available in the book's Online Resource Centre at **http://www.oxfordtextbooks.co.uk/orc/crowe2e/**.

Checklist of key concepts

The shape of small molecules

- The bond length is the distance between two covalently bonded atoms
- The atomic radius is half the distance between two covalently bonded atoms of the *same* element
- As the number of electrons shared by two atoms increases, the bond length decreases
- The bond angle is the angle between two bonds that are joined by a central atom
- Valence Shell Electron Pair Repulsion (VSEPR) theory allows us to predict the arrangement of valence electron pairs around a central atom

Geometric arrangements

- Four valence electron pairs adopt a tetrahedral geometry, with a bond angle of 109.5°
- Three valence electron pairs adopt a trigonal planar geometry, with a bond angle of 120°
- Two valence electron pairs adopt a linear geometry, with a bond angle of 180°
- Both bonding pairs and non-bonding pairs contribute to the structure of a molecule
- We consider only bonding pairs comprising sigma bonds when applying VSEPR theory

Hybridization

- Hybridization is the process of atomic orbitals merging together to form new hybrid orbitals
- One s and three p orbitals undergo sp^3 hybridization to form four sp^3 hybrid orbitals
- One s and two p orbitals undergo sp^2 hybridization to form three sp^2 hybrid orbitals
- One s and one p orbital undergo sp hybridization to form two sp orbitals
- sp^3 orbitals adopt a tetrahedral geometry
- sp^2 orbitals adopt a trigonal planar geometry
- sp orbitals adopt a linear geometry

Bond rotation

- Bond rotation is possible about a single bond only
- A molecule's configuration is a description of how its composite atoms are joined to one another in three-dimensional space
- A molecule's conformation describes the way in which connected atoms arrange themselves relative to each other at a particular moment in time
- A molecule's configuration is fixed; a molecule's conformation is variable
- Steric hindrance is the restriction of movement due to a lack of space between neighbouring atoms

9

Molecular shape and structure 2: the shape of large molecules

INTRODUCTION

We are surrounded every day by large, complicated structures, which are formed from small subunits. The homes we live in are often constructed from hundreds of bricks; the clothes we wear may be made from polymers comprising repetitive subunits. Yet, when we consider the bigger picture, there are ways in which even large structures are similar. Houses may have different layouts; they may be semi-detached or terraced. However, the materials from which they are constructed are actually arranged in similar ways to yield characteristic structures: walls, ceilings, and roofs.

In this chapter we explore the factors that affect the shape and structure of large molecules to give different molecules different shapes. We also consider what it is about large molecules that make them similar: what are the recurring motifs that are mirrored by different compounds, even if their structure at the atomic scale differs greatly. We begin by considering how large molecules are constructed from smaller subunits, and the factors that affect shape and structure as we move from small to big.

9.1 Constructing larger molecules

One of our essential concepts from Chapter 1 is that complexity can arise from the joining of many small, simple units. Many important–and complicated–biological molecules, including proteins and sugars, are generated from the joining of smaller subunits. In general terms, we call these small subunits **monomers**, and the larger substances which the small subunits form are called **polymers**.

- *A polymer is formed from the joining together of many small repeating units called monomers.*

Three key factors dictate the shape and structure of a polymer:

1. The geometry of joined atoms
2. The sequence in which monomers join to one another
3. The physical nature of the bonds that join neighbouring monomers.

We explore each of these factors in turn below.

The geometry of joined atoms

VSEPR theory focuses on the arrangement of valence electrons around a single central atom. We can predict the conformation of a large molecule by considering the arrangement of valence electrons (and hence connected atoms) around each individual atom in turn.

We see in Section 8.3 how VSEPR theory dictates that valence electron pairs must arrange themselves around a central atom in a way that minimizes interactions between the valence pairs, so that they are positioned as far away from one another as possible. It is the positioning of valence electrons that dictates the position of the atoms that are covalently bonded to the central atom–the connected atoms must occupy the same geometry as the valence electrons that they share with the central atom, because they are attached via the orbitals which the valence electrons occupy.

Look at Figure 9.1, which depicts the structure of butane, C_4H_{10}. As we see in Section 8.3, VSEPR theory leads us to predict that the valence electrons around a central carbon atom adopt a tetrahedral geometry, and therefore those atoms that are covalently bonded to a carbon atom by single (sigma) bonds adopt a tetrahedral geometry also. We can apply VSEPR theory not to just one carbon atom in butane, but to each of the carbon atoms. Look at Figure 9.1 and notice how each of the four carbon atoms displays a tetrahedral geometry with respect to the four atoms joined to it.

The sequence of monomers within a polymer

The three-dimensional structure of a large polymer is influenced by the identity, and sequence of joining, of its composite monomeric units.

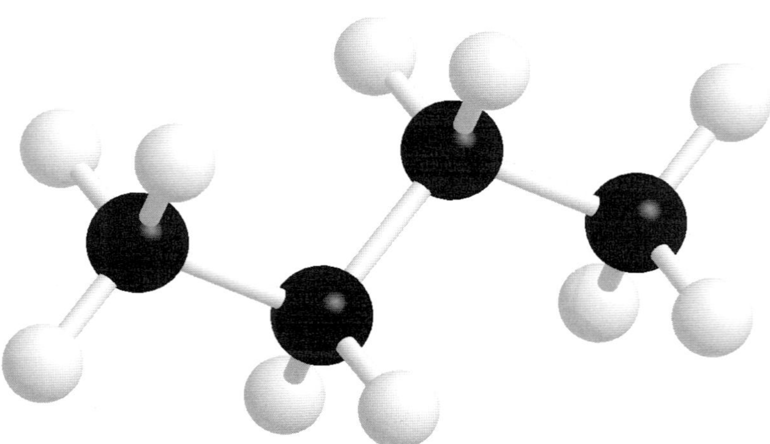

Figure 9.1 The structure of butane, C_4H_{10}. Notice how the atoms around each carbon atom adopt a tetrahedral geometry.

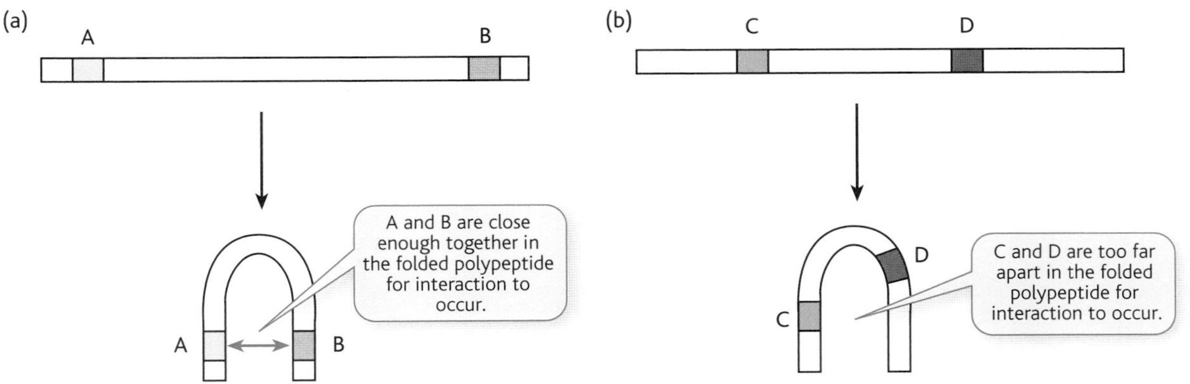

Figure 9.2 Side chains that are in close proximity when a polypeptide chain adopts its three-dimensional, folded structure may be some distance apart in the polypeptide chain itself. Look at this polypeptide chain, and notice how amino acids A and B are some distance apart in the *sequence* of the polypeptide chain, but actually come into close proximity when the chain is folded.

The link between the identity of monomeric units, and the structure of the polymer as a whole, are particularly strong in polypeptides. We see in Chapter 4 how several types of non-covalent interaction can occur between side chains of amino acids: ionized side chains may interact to form salt bridges; polar side chains possessing atoms of O or N may associate closely through the formation of hydrogen bonds; non-polar side chains may cluster together as a consequence of their hydrophobic behaviour.

The ability of side chains to interact in this way, however, depends on the correct amino acid side chains occurring in close proximity in the folded polypeptide chain. Side chains that are in close proximity when the polypeptide chain adopts its three-dimensional, folded structure may be some distance apart in the polypeptide chain itself. They are only brought into close proximity when folding occurs.

Look at Figure 9.2, which illustrates this point. Notice how amino acids A and B are some distance apart in the *sequence* of the polypeptide chain, but actually come into close proximity when the chain is folded. In the same way, the two handles of a skipping rope can actually touch despite the length of rope between them. By contrast, amino acids C and D are closer to one another in the unfolded polypeptide chain but do not come into close enough proximity for interaction to occur when the chain is folded.

Having the right amino acid residues at the correct locations in the polypeptide chain is vital if a polypeptide is to adopt its correct structure[1]. Let us imagine that amino acids A and B in Figure 9.2 are serine and glutamine. The side chains of serine and glutamine are able to participate in a hydrogen bond, which would serve to stabilize the polypeptide chain in the folded conformation illustrated in Figure 9.2(a). If, however, amino acid B was glycine rather than glutamine, no hydrogen bonds could form between the two side chains, so the folded polypeptide chain would be less stable.

1 An amino acid **residue** is an amino acid that has joined with at least one other amino acid via a peptide bond to form a peptide (that is, only a few amino acid residues long) or polypeptide (that is, many amino acid residues long).

We discover how non-covalent interactions, especially hydrogen bonds, influence the adoption by biological molecules (such as polypeptides) of characteristic structures and shapes in Section 9.2 below.

Bonding between monomers

In some biological molecules, monomers are able to join together through the formation of different covalent bonds, which generate polymers of contrasting structure and shape.

Let's consider the sugar glucose, $C_6H_{12}O_6$. Glucose is able to form four distinct polymers[2]: amylose (an unbranched starch), glycogen and amylopectin (branched starches), and cellulose. All four polymers comprise many molecules of glucose joined together by what we call **glycosidic bonds**. However, it is the *way* the neighbouring glucose units are joined that makes each polymer distinct.

- *A glycosidic bond is the name given to the C–O–C bond that joins adjacent glucose units in starch, cellulose, and glycogen.*

Look at Figure 9.3, which illustrates the differences in structure between amylose, glycogen, and cellulose. Notice how amylose and glycogen are structurally very similar–they both comprise glucose residues which are linked by the same class of glycosidic bond (an α-1, 4-glycosidic bond)[3]. The conformation of the α-1, 4-glycosidic bond is such that the polysaccharide chain of amylose can adopt helical (spiral-like) arrangements. However, the two polysaccharides are distinguished by the existence of α-1, 6-glycosidic bonds at intervals along the polysaccharide chain in glycogen, which generates its branched structure.

By contrast, the glucose units in cellulose are joined by a different class of glycosidic bond to that found in either amylose or glycogen: the β-1, 4-glycosidic bond. The conformation of the β-1, 4-glycosidic bond is such that the unbranched chains of glucose molecules in cellulose form long parallel rows, rather than helical coils. The parallel rows are held in tight association with one another through a network of hydrogen bonds to form the rigid structure characteristic of cellulose.

Amylopectin and glycogen share the same branched structure, generated by α-1, 6-glycosidic bonds. However, branching occurs more frequently along the polysaccharide chain of glycogen (about once every 10–15 glucose units) than in amylopectin (about once every 25 glucose units).

While all four polysaccharides of glucose are composed of the same basic unit, the difference in bonding between glucose units exhibited by the four polysaccharides is sufficient to give each of them distinct physical properties, and very different biological functions.

Amylose and **amylopectin** make up **plant starch**, the form in which glucose is stored by plants. Plant starch (a mixture of 20% amylose and 80% amylopectin) is found principally in potatoes, rice, wheat, and cereals. The starch in plants is a vital source of energy for us: our saliva contains the enzyme amylase which

---

2 We call a polymer of a sugar monomer unit a **polysaccharide**.

3 Glycosidic bonds are often named using two components that describe the specific nature of the bond. The prefixes α and β indicate the conformation of the bond–that is, how it is oriented in space relative to the monomers it is joining together. The numbers 1 and 4 or 6 are then used to denote the position of the carbon atoms within the sugar rings that are joined by the bond, as explained in Figure 9.3.

Amylose

α-1, 4-glycosidic bond joins 1-carbon on one glucose unit with the 4-carbon on an adjacent glucose unit.

Glycogen

α-1, 6-glycosidic bond joins 1-carbon on one glucose unit with the 6-carbon on an adjacent glucose unit.

Cellulose

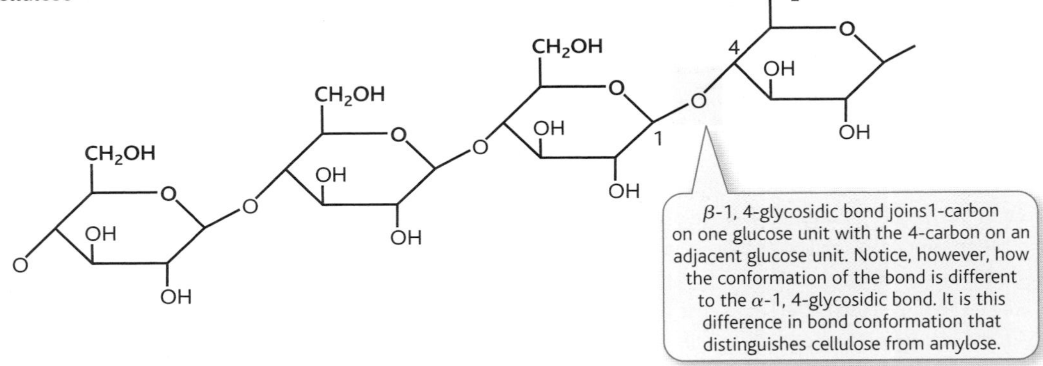

Figure 9.3 The structures of amylose, glycogen, and cellulose. Notice how the glucose subunits that form each compound are joined together in different ways, making the three compounds distinct.

breaks down starch into its composite glucose units, which our body then uses to generate energy.

Cellulose is another plant-based polymer of glucose. Cellulose is the major building material of plants, and is used to give rigidity to plant cell walls. Unlike starch, however, cellulose is not a good source of energy for us. Our bodies cannot digest cellulose–in other words, we cannot break it down to release its constituent glucose molecules. (While we produce an enzyme to break down starch, we do not produce an enzyme capable of breaking down cellulose.)

In contrast, ruminants such as cows *can* digest cellulose, and can live on a diet of plant-based matter such as grass[4].

Despite its lack of nutritional value to humans, however, cellulose is an important component of our diet. We are often advised to eat plenty of fibre-rich foods–foods rich in cellulose, such as vegetables and cereal. If our diet contains too little fibre we run the risk of having a sluggish digestive system, with side-effects such as constipation and bad breath.

We can help to avoid these symptoms by exploiting the fact that our bodies cannot digest cellulose. Because it cannot be digested, cellulose passes straight through our digestive system untouched. While this might seem like a pointless exercise, it actually helps to ensure that our digestive system functions smoothly, and that a regular passage of material through the digestive system is maintained (hence avoiding the discomfort of constipation).

Glycogen is the form in which glucose is stored in our body. We need a continual supply of energy–and hence a continual supply of glucose–in order to survive. Rather than having to eat non-stop, however, reserves of glycogen are built up in our liver and muscle through the polymerization of glucose units. (The glucose may be from sugary foods, or from starch-based plant foodstuffs, the starch having first been broken down into glucose monomers by our digestive system.)

When our body detects that glucose levels are lower than the energy requirements of the body, hormones trigger the breakdown of glycogen in our liver and muscle, converting it back into glucose monomers, which can be metabolized to generate the required energy.

The levels of glycogen and glucose in the body are regulated by two hormones, glucagon and insulin. Glucagon stimulates the breakdown of glycogen to glucose when the body detects that blood glucose levels are too low, thereby increasing glucose levels; insulin stimulates the transport of glucose out of the blood stream when the body detects that blood glucose levels are too high. But what happens if this control mechanism breaks down? We discover the answer in Box 9.1.

BOX 9.1 Life's not so sweet: blood sugar levels in diabetes

Diabetes is a medical condition caused by there being too much glucose in the blood. In 2008, an average of 3.86% of the UK's population–around 2.5 million people–were known to be diagnosed with the condition.

The elevated levels of glucose in the blood arise because of a deficiency in the activity of the hormone insulin. In an individual without diabetes, insulin stimulates the transport of glucose out of the blood, and its storage as glycogen in the liver and muscle. In an individual with diabetes, however, insulin activity is either depleted or absent altogether.

There are two types of diabetes: in those with **type 1** diabetes, the body fails to produce any insulin. In **type 2** diabetes, insulin activity is too low for correct control of blood sugar to occur, either because too little insulin is produced, or because the body does not respond to the insulin in the way that it should. Of the two types, type 1 diabetes is much rarer: people with type 1 diabetes make up only 5–15% of all people with diabetes.

An elevation in blood sugar levels, as seen in those with untreated diabetes, is called **hyperglycaemia**. Symptoms include thirst, frequent urination, tiredness, and increased susceptibility to infections, such as thrush. Prolonged hyperglycaemia can lead to weight loss (as the body tries to remove excess glucose through your urine) and blurred vision (as the high blood sugar changes the shape of the lenses in your eyes). Longer-term, heart disease, kidney damage, circulation problems in the legs, and nerve damage to the feet can occur.

4 In fact, like us, cows do not produce cellulose-digesting enzymes. Rather, their stomachs contain bacteria which secrete the necessary enzymes.

BOX 9.1 **Continued**

While diabetes cannot be cured, it can be treated: blood sugar levels can be actively managed to keep them to within safe limits. In the case of type 1 diabetes, in which insulin fails to be produced by the body, insulin preparations are injected, giving the body the supply of insulin that it needs (but cannot supply itself) to regulate blood glucose levels.

In the case of type 2 diabetes, in which insulin is produced by the body, it is possible to control symptoms through eating a healthy diet. However, insulin treatment–such as that needed in type 1 diabetes–can sometimes be required.

What puts individuals at risk of developing diabetes? In most cases, type 1 diabetes is thought to be the result of an auto-immune disease, whereby the body's immune system inappropriately attacks its own cells. In the case of type 1 diabetes, it is thought that the immune system damages the cells of the pancreas–the organ that produces insulin–preventing insulin from being produced. We don't know for sure what causes this inappropriate immune response, but infection with a particular virus may be one cause.

In the case of type 2 diabetes, the most common cause is thought to be obesity: type 2 diabetes is linked to having excess body fat, and over 80% of those with type 2 diabetes are overweight, and tend not to get much exercise. However, ethnic origin can also be a strong factor: you are at least five times more likely to develop type 2 diabetes if you are African-Caribbean or of south Asian origin and living in the UK, compared with someone who is white.

So, as with so many medical conditions, there are many contributing factors that can increase your risk of developing diabetes, some of which–particularly for type 1 diabetes–may be outside of your control. Risk factors are most likely to have both genetic and environmental origins–though the greater prevalence of type 2 diabetes, and the major part that obesity seems to play in the onset of this particular condition, suggests that something as simple as a healthy diet could prevent a large proportion of the population developing diabetes later in life.

9.2 The shape of larger molecules

It is not just small molecules that have characteristic, predictable structures. Large molecules, too, have characteristic structural features, which may be shared by families of molecules, and arise through the arrangement of joined atoms in regular and predictable ways.

The complicated, three-dimensional shapes of many biological macromolecules are governed, to a large extent, by the stabilizing effect of non-covalent interactions that operate between different parts of biological polymers. In this section we see how non-covalent interactions govern the formation of characteristic structural motifs in many biological polymers.

The shapes of many biological molecules are built up over four levels of complexity: the **primary structure**, **secondary structure**, **tertiary structure**, and **quaternary structure**[5]. As we see in this section, the nature of one level of structure influences the nature of the next level of structure in the hierarchy. So primary structure influences secondary structure, secondary structure influences tertiary structure, and so on.

- ***The structure of biological molecules is built up on four levels: primary structure, secondary structure, tertiary structure, and quaternary structure.***

5 We use the terms primary, secondary, tertiary, and quaternary structure most often with reference to protein structure. However, the concept of structure being built up in a hierarchical way applies to several different types of biological molecule, not just proteins.

Primary structure

The primary structure of a biological polymer describes the sequence in which monomers have joined together to form the polymer.

The primary structure of a polypeptide tells us the identity of the amino acids that have joined to form the polypeptide, and the *order* in which they have joined. For example, a very small polypeptide may have the primary structure:

Gly-Ala-Gly-Pro-Asp-Leu-Val-Leu-Cys-Lys-Arg-Pro

or, using the single-letter codes for each amino acid:

GAGPNLVLCKRP

This primary structure tells us not only the identity of the twelve amino acids that have joined together to form the polypeptide, but the order in which they have joined.

Similarly, the primary structure of a nucleic acid tells us the sequence of nucleotide bases that have joined together to form the nucleic acid. For example, a short RNA strand may have the primary structure:

AUUCGCCGAUUA

We usually refer to the primary structure of a nucleic acid as its **sequence**.

The Human Genome Project mentioned on p. 208 had as its aim the identification of the primary structure of the entire human genome (the sequence of nucleotide bases of all of the DNA strands that comprise the 24 different chromosomes of the human genome).

Unlike other levels in the structural hierarchy, primary structure is mediated solely by covalent bonds, which join together adjacent monomers to yield the primary structure.

Secondary structure

A biological polymer's secondary structure describes the way a polymer's primary structure folds to yield characteristic three-dimensional conformations.

There are two predominant secondary structures exhibited by biological molecules: both proteins and nucleic acids possess regions of primary structure that form **helices**, while only regions of proteins form **sheets**. Unlike the primary structure, both types of secondary structure are mediated by **non-covalent interactions**.

The secondary structure of proteins

We see in Chapter 4 how hydrogen bonds occur between different parts of the peptide chain backbone of polypeptides, joining the C=O group on one part of the peptide chain with the N–H group on another part of the chain. The uniform, repetitive structure of the peptide chain backbone (the N–H group, central C atom, and carbonyl group depicted in Figure 7.1) facilitates its folding into two characteristic structures, **α-helices** and **β-sheets**. Both of these structures are stabilized by regular, repeating networks of hydrogen bonds between different parts of the peptide backbone.

α-helices arise when a hydrogen bond forms between the C=O group of one amino acid and the N–H group of another amino acid four residues along the polypeptide chain, as illustrated in Figure 9.4. (Notice how several hydrogen bonds stabilize each turn of the helix.)

In contrast, β-sheets form from the interaction of two adjacent polypeptide strands, which are held in close association by hydrogen bonds between the C=O

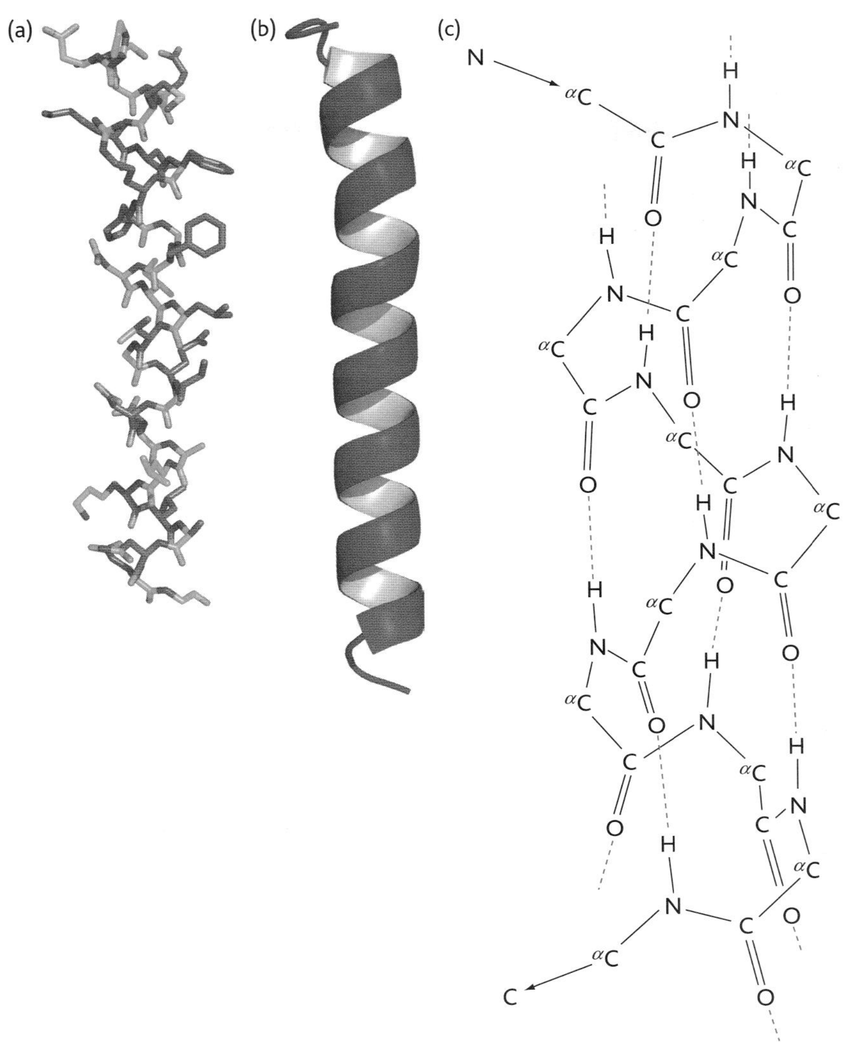

Figure 9.4 The α-helix. (a) The three-dimensional structure of an α-helix, showing how the polypeptide chain folds into a distinctive three-dimensional shape. (b) An α-helix is often denoted in this simpler cartoon form. (c) α-helices arise when a hydrogen bond forms between the C=O group of one amino acid and the N–H group of another amino acid four residues along the polypeptide chain, as shown in blue here (reproduced from Price and Nairn, 2009).

and N–H groups on the two strands, as depicted in Figure 9.5[6]. β-sheets aren't as harmless as they may first seem. Look at Box 9.2 to find out how β-sheets are thought to play an important part in the onset of several neurodegenerative diseases.

- ***The two characteristic secondary structures exhibited by proteins are the α-helix and the β-sheet.***

The simplest secondary structure in polypeptides, however, is the β-turn (or reverse turn), which causes the chain to double back on itself, as depicted in Figure 9.6. The β-turn comprises just three or four amino acid residues, and is stabilized by hydrogen bonds occurring between the carbonyl oxygen of one amino acid residue, and the amide N–H of the residue three places further along the polypeptide chain. Notice how, having doubled back on itself, the polypeptide chain now comprises

6 To be precise, this figure depicts an antiparallel β-sheet arrangement. The arrows in Figure 9.5(b) point in an N-terminal to C-terminal direction, such that strands lying adjacent to each other run in opposite directions. In a parallel β-sheet all the arrows would be lined up to point in the *same* direction.

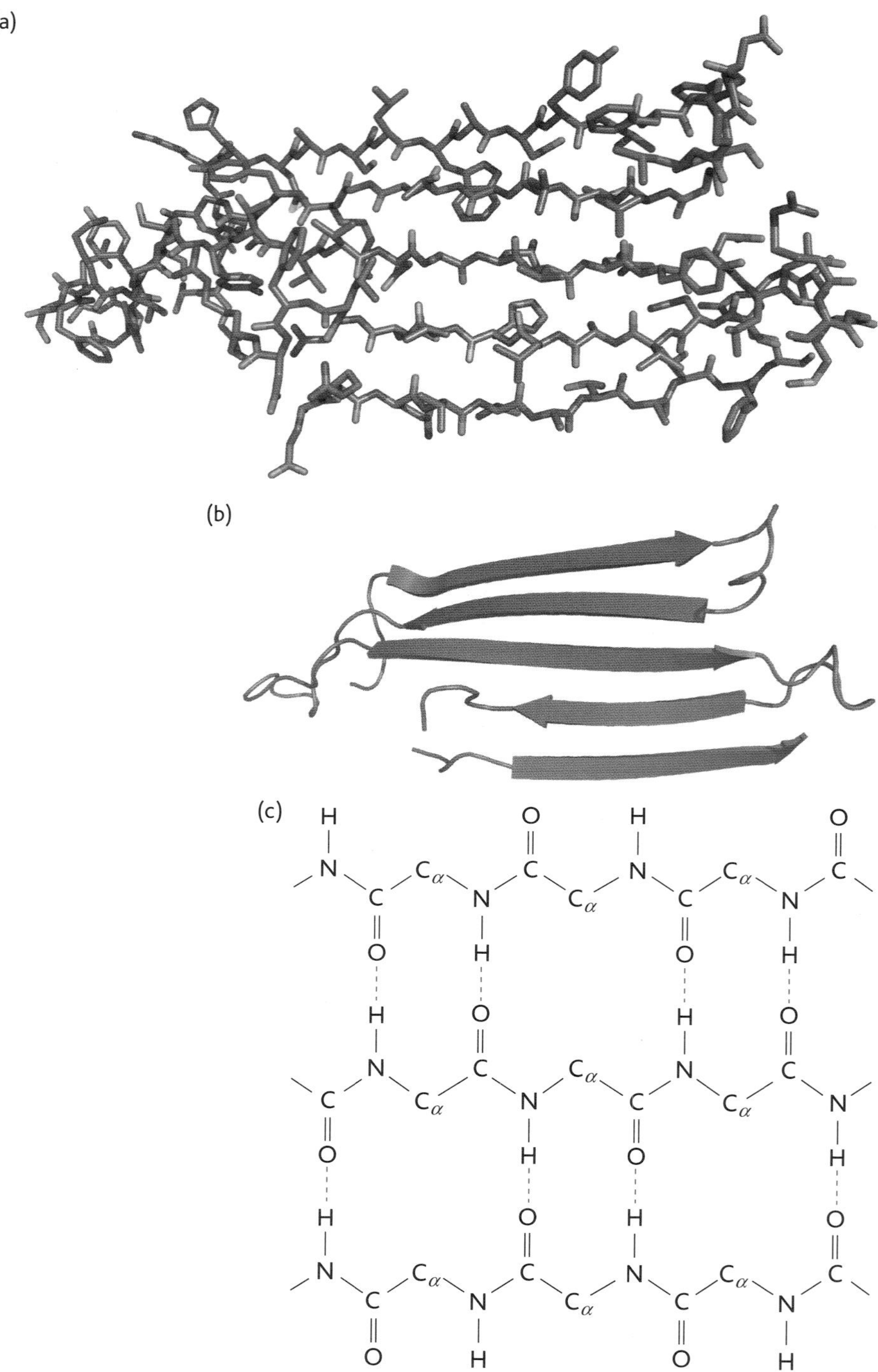

Figure 9.5 The beta sheet. (a) The three-dimensional structure of a β-sheet. (b) β-sheets may be denoted in this simpler cartoon form. (c) β-sheets form from the interaction of two adjacent polypeptide strands, which are held in close association by hydrogen bonds between the C=O and N−H groups on the two strands, as shown in blue here (reproduced from Price and Nairn, 2009).

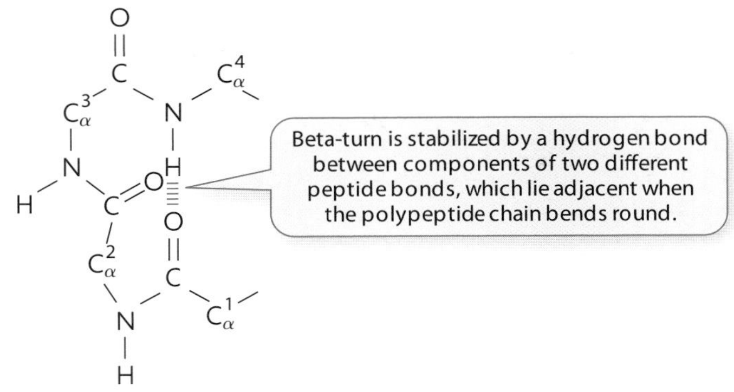

Beta-turn is stabilized by a hydrogen bond between components of two different peptide bonds, which lie adjacent when the polypeptide chain bends round.

Figure 9.6 The structure of a β-turn. When a β-turn occurs in a polypeptide chain, it causes the chain to double back on itself. (Remember: the atoms denoted C_α are the central carbon atoms to which the amino acid residue side chains are attached.)

two adjacent strands. These strands can be stabilized by hydrogen bonds to form the β-sheet structure described above.

While the secondary structures described above are primarily stabilized by hydrogen bonds occurring between different parts of the peptide backbone–the nature of which is invariant, regardless of the specific amino acid residues forming the polypeptide chain–the tendency of a particular region of polypeptide to adopt a particular secondary structure is influenced by the amino acid residues present in that region of the chain. That is, the tendency of a polypeptide to adopt a particular secondary structure is influenced by its primary structure.

For example, leucine, methionine, glutamine, and glumatic acid are found most often in α-helices: regions of polypeptides whose primary structure contains these amino acid residues are particularly likely to adopt an α-helical structure. By contrast, valine, isoleucine, and phenylalanine are most often found in β-sheets. Further, glycine and proline are commonly found in β-turns. Glycine is particularly well suited to being located in a β-turn: its lack of a bulky side chain means that glycine has the flexibility to adopt a wider range of conformations, and can fit in the relatively compact space required for a β-turn to successfully form.

By contrast, both glycine and proline are rarely found in α-helices. Indeed, these residues are often called **helix disrupters**, because their presence does not favour α-helix formation. Instead, when proline is present in an α-helix, it produces a kink

BOX 9.2 The darker side of β-sheets

A range of neurodegenerative diseases, including Bovine Spongiform Encephalopathy (BSE) in cows (known more familiarly as 'mad cow disease'), and the related human form Creutzfeldt–Jakob disease (CJD), are thought to be caused by the presence a particular (and unusual) type of protein called a prion.

Prions are believed to have different structural conformations. The normal form of the prion (denoted PrPC) is present in healthy cells, and has no disease-causing effects. However, research has found that PrPC can be transformed into a disease-causing form, denoted PrP^{Sc}. The transformation from normal PrPC to pathogenic PrP^{Sc} appears to be the result of a change in the structural conformation of the prion molecule.

Structural studies have shown that a far greater proportion of the PrP^{Sc} molecule adopts a β-sheet conformation, compared to the PrPC molecule: PrPC has a 42% α-helix content, but just a 3% β-sheet content. In contrast, PrP^{Sc} has a lower α-helix content (21%) but a much higher β-sheet content (43%). This difference in structural conformation–from α-helix to β-sheet–seems to be sufficient to convert the prion protein from its harmless form to a form capable of contributing to the onset of CJD.

Indeed, studies in 2005 demonstrated that a particular β-sheet region is essential for the infectivity of one particular prion, HET-s (the prion protein of the yeast, *Podospora anserine*): when a characteristic β-sheet region comprising four β-strands was disrupted through mutation, the infectivity of HET-s was strongly affected. These studies seem to confirm the central role that the β-sheet plays in conditions like CJD, a particularly debilitating, distressing–and fatal–human disease.

Table 9.1 The preferences of the twenty amino acids for different secondary structural elements. A value of 1.0 means that the amino acid appears in a given secondary structural element at random. A value greater than 1.0 means that an amino acid shows a preference for that secondary structure; a value less than 1.0 means that an amino acid does not favour that secondary structure. Reproduced from Petsko and Ringe, *Protein Structure and Function* (Oxford University Press, 2004), with data from Williams, R. W. *et al.*, *Biochim Biophys Acta* 1987, **916**:200–204.

Amino acid	Secondary structure		
	α-helix	β-sheet	β-turn
Glutamine	**1.59**	0.52	1.01
Alanine	**1.41**	0.72	0.82
Leucine	**1.34**	1.22	0.57
Methionine	**1.30**	1.14	0.52
Glutamate	**1.27**	0.98	0.84
Lysine	**1.23**	0.69	1.07
Arginine	**1.21**	0.84	0.90
Histidine	**1.05**	0.80	0.81
Valine	0.90	**1.87**	0.41
Isoleucine	1.09	**1.67**	0.47
Tyrosine	0.74	**1.45**	0.76
Cysteine	0.66	**1.40**	0.54
Tryptophan	1.02	**1.35**	0.65
Phenylalanine	1.16	**1.33**	0.59
Threonine	0.76	**1.17**	0.90
Glycine	0.43	0.58	**1.77**
Asparagine	0.76	0.48	**1.34**
Proline	0.34	0.31	**1.32**
Serine	0.57	0.96	**1.22**
Aspartate	0.99	0.39	**1.24**

in the helix: its side chain cannot participate in the normal pattern of hydrogen bonding found in α-helix formation and instead disrupts it, causing the kink.

Look at Table 9.1, which illustrates the secondary structures that the twenty amino acids preferentially participate in. Notice how some amino acids show a strong, clear preference for a particular secondary structure (indicated by values in this table markedly greater than 1).

Self-check 9.1 A polypeptide chain is particular rich in valine, tyrosine, and threonine. What secondary structure is this polypeptide chain most likely to adopt?

- ***Different amino acid residues favour the formation of different secondary structures.***

The secondary structure of nucleic acids

We see on p. 99 how hydrogen bonds are vital in giving DNA its ability to maintain a characteristic **double-helical** structure. In the case of DNA, the nucleotide bases that covalently bind to one another to generate the DNA sequence (the primary structure of DNA) twist relative to one another to adopt a helical conformation in the presence of a complementary strand. DNA is a double-stranded molecule in which *both* strands adopt a helical conformation. These two helices interweave along a common axis to form the double helix itself.

- ***The characteristic secondary structure exhibited by DNA is the double helix.***

Remember: the α-helical structure of polypeptides is structurally different from the double-helical structure of DNA. Both secondary structures are stabilized by hydrogen bonds, but in different ways.

While hydrogen bonding is important in both the α-helical structure of polypeptides, and the double-helical structure of DNA, the *nature* of the hydrogen bonding in both instances is very different. Figure 9.7 shows how the hydrogen bonding, which maintains the α-helical structure of polypeptides, is **intramolecular**, while the hydrogen bonding which maintains the double-helical structure of DNA is **intermolecular.**

DNA is not the only nucleic acid capable of adopting a helical secondary structure, however. RNA molecules also exhibit regions of helical secondary structure. For example, transfer RNA (or tRNA) is a form of nucleic acid which helps ribosomes translate mRNA sequences into the primary sequences of polypeptide chains by carrying amino acids to the ribosome for attachment to the growing polypeptide chain. A tRNA molecule has a characteristic L-shaped structure, as depicted in

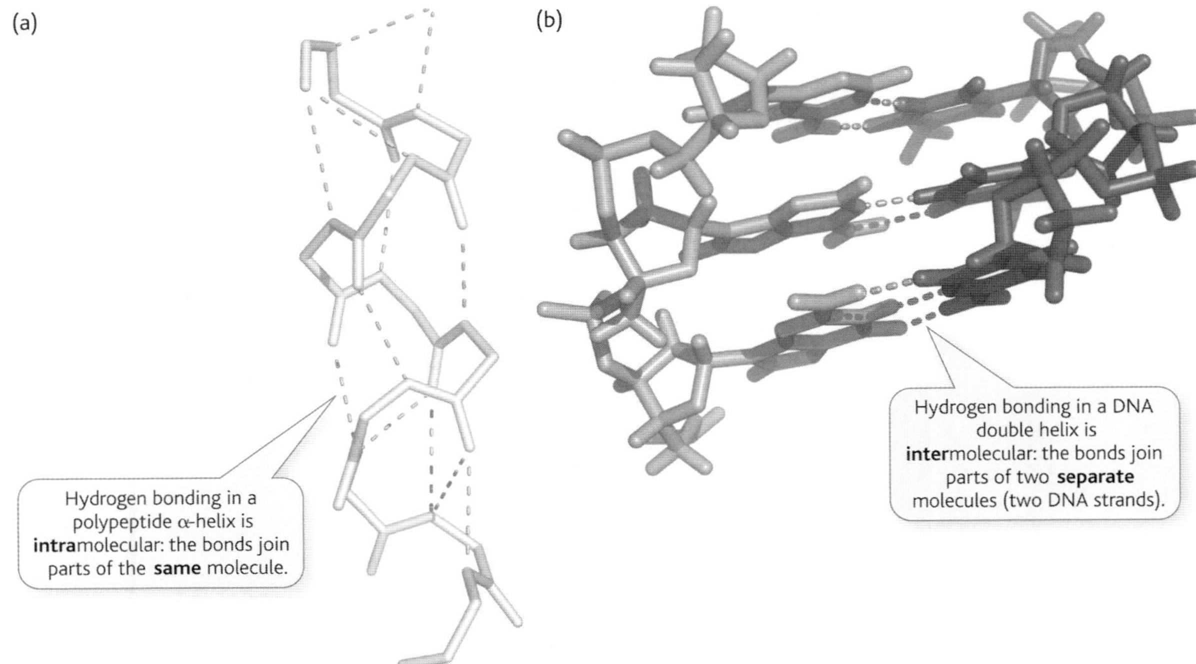
(a)
(b)
Hydrogen bonding in a polypeptide α-helix is **intra**molecular: the bonds join parts of the **same** molecule.
Hydrogen bonding in a DNA double helix is **inter**molecular: the bonds join parts of two **separate** molecules (two DNA strands).

Figure 9.7 The α-helices of polypeptides (a) and the double helices of DNA (b) vary in the nature of the hydrogen bonding they exhibit. Polypeptide α-helices are stabilized by *intramolecular* hydrogen bonds, while DNA double helices are stabilized by *intermolecular* hydrogen bonds.

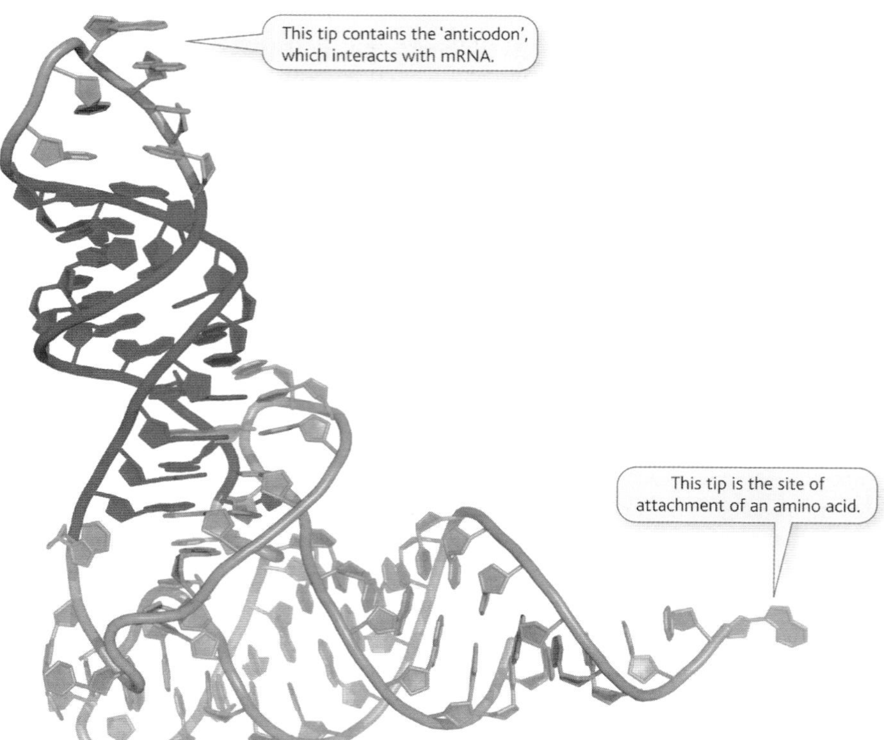

Figure 9.8 The characteristic L-shaped structure of a tRNA molecule. Notice how the region shown in blue here adopts a characteristic double-helical structure.

Figure 9.8. The amino acid attaches to one tip of the molecule, while the other tip interacts with the mRNA that is being 'decoded' by the ribosome. Each 'arm' of the L-shaped molecule comprises two RNA strands which undergo base pairing in an analogous fashion to DNA to form a short double-helical structure (shown in blue in Figure 9.8). The double helices of each arm are joined by a so-called 'hinge' region.

Base pairing between RNA strands is base-specific, just like base pairing between DNA strands. Figure 9.9 shows how a cytosine base on one strand forms a hydrogen bond to a guanine base on the adjacent strand, while adenine on one strand forms a hydrogen bond to uracil on the other strand. No other base pairing is possible.

Self-check 9.2 Look again at Figure 9.9. Notice how three hydrogen bonds form between guanine and cytosine, but only two hydrogen bonds form between adenine and uracil. Why is this?

Guanine Cytosine Adenine Uracil

Figure 9.9 Base pairing between RNA strands is base-specific. Cytosine forms hydrogen bonds with guanine, while adenine forms hydrogen bonds with uracil.

The secondary structure of nucleic acids is also dependent upon them possessing appropriate primary structures. For example, the double-helical structure of DNA can only be achieved if the two nucleic acid strands possess *complementary* primary structures. We see in Chapter 4 how complementary DNA strands undergo specific base pairing, mediated by the hydrogen bonding of nitrogenous bases on adjacent DNA strands. Only two base pairs are possible: A with T, and C with G. Other base pair combinations (for example, A with C, and G with T) are not favoured due to space restrictions: the bases do not come into close enough proximity for a full complement of hydrogen bonds to exist, and steric hindrance inhibits the packing of the bases into the conformation required for helix formation. These alternative base pairings may occur (albeit infrequently), but often lead to mutation in the genome. It is the A-T and C-G base pairings that are the most stable, and are favoured.

Tertiary structure

A biological polymer's tertiary structure describes the way in which the secondary structure is packed to form regions of defined three-dimensional shape. In proteins, the tertiary structure may comprise solely α-helices, solely β-strands, or a combination of both. Look at Figure 9.10, which illustrates the tertiary structures of four different polypeptides. Notice how the tertiary structure of some of the polypeptides comprises just one type of secondary structure, whereas others comprise both α-helices and β-sheets.

Look again at Figure 9.10(b). This illustrates the tertiary structure of the membrane protein FhuA. Notice how the β-sheets that form the tertiary structure of this protein have curled round to form a barrel. We call this characteristic tertiary structure a **β-barrel**; it is stabilized by hydrogen bonds, which operate between the first and last β-strands to maintain the barrel-like shape.

Hydrogen bonds aren't the only non-covalent interaction to play an important part in the stabilization of tertiary structures. Instead, hydrophobic interactions also play a vital role. We see in Chapter 7 how polypeptides comprise both hydrophobic, non-polar amino acid side chains, and hydrophilic polar side chains. This combination of both hydrophobic and hydrophilic portions within a polypeptide chain is central to the correct three-dimensional folding of proteins to give a particular tertiary structure. While non-polar hydrophobic amino acid side chains are buried at the centre of a folded polypeptide, to minimize their exposure to water, hydrophilic side chains often occur on the outer surfaces of a folded polypeptide, where they can form hydrogen bonds with water molecules in the surrounding aqueous medium. These hydrophilic interactions help to stabilize the polypeptide in its correct three-dimensional shape.

We learn about hydrophobic interactions in Chapter 4.

- ***A molecule's tertiary structure describes the way its secondary structural elements fold into distinct three-dimensional arrangements.***

Let's consider the protein α-keratin, which illustrates how hydrophobic interactions play a part in tertiary structure formation. α-keratin is a structural protein, which is ubiquitous in many body tissues, including hair and nails. α-keratin has a primary structure that favours α-helix formation. However, its primary structure has one other important feature: non-polar (hydrophobic) amino acid residues

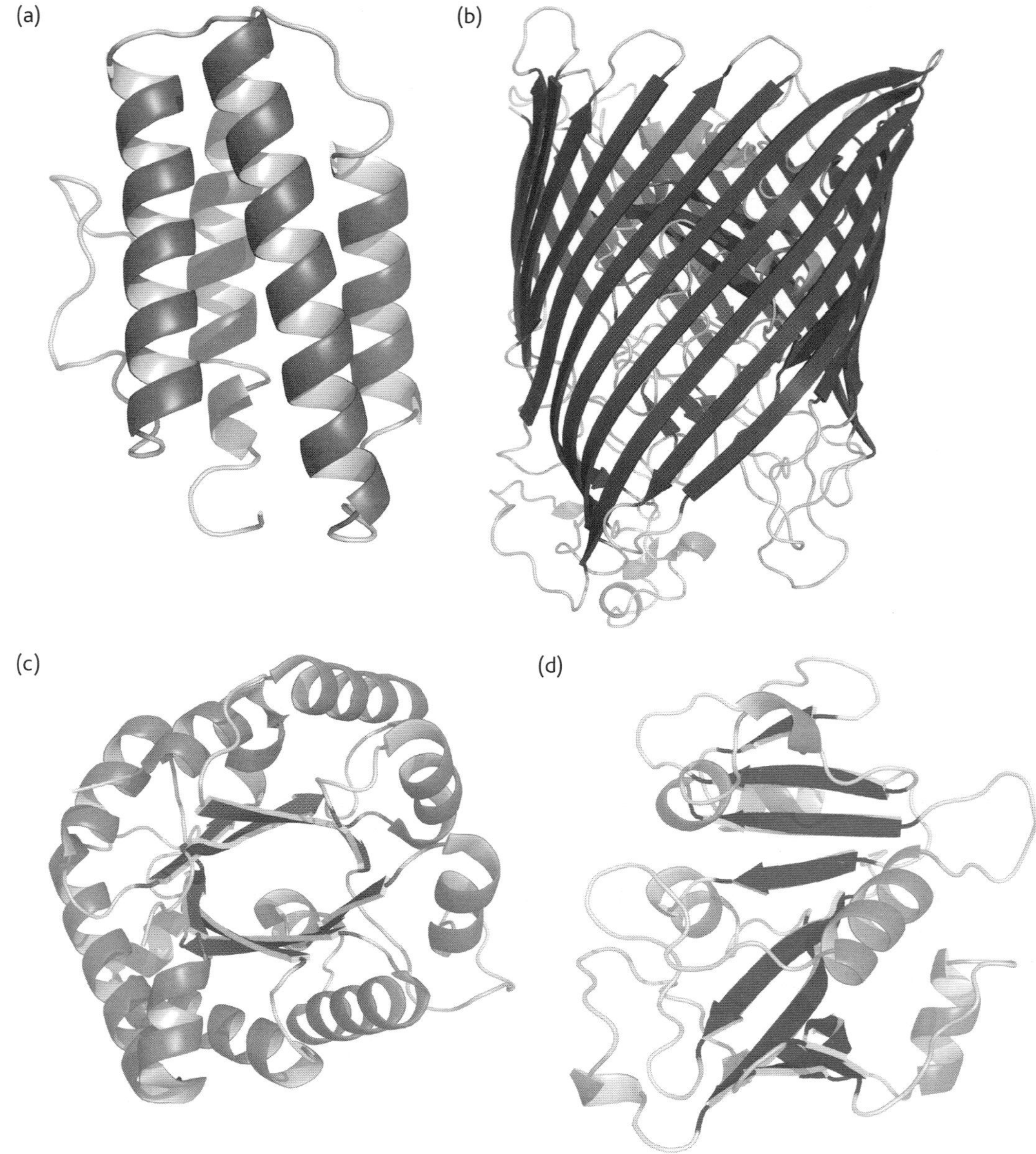

Figure 9.10 The tertiary structure of proteins. (a) Myohemerythrin, composed solely of α-helices. (b) The membrane protein FhuA. Notice how the beta strands form a barrel, which spans the membrane as a pore structure. (c) The enzyme triosphosphate isomerase (TIM), with a β-barrel at the centre of the molecule and α-helices around it. (d) The enzyme dihydrofolate reductase (DHFR), with a β-sheet running down its interior.

fall at regular intervals within those regions of the primary structure that fold to form α-helices. Figure 9.11 shows how hydrophobic amino acid residues are located at regular intervals within the primary structure of α-keratin. Consequently, these hydrophobic amino acid residues are positioned in a repetitive, uniform way within the α-helix that is formed when the polypeptide chain of α-keratin folds.

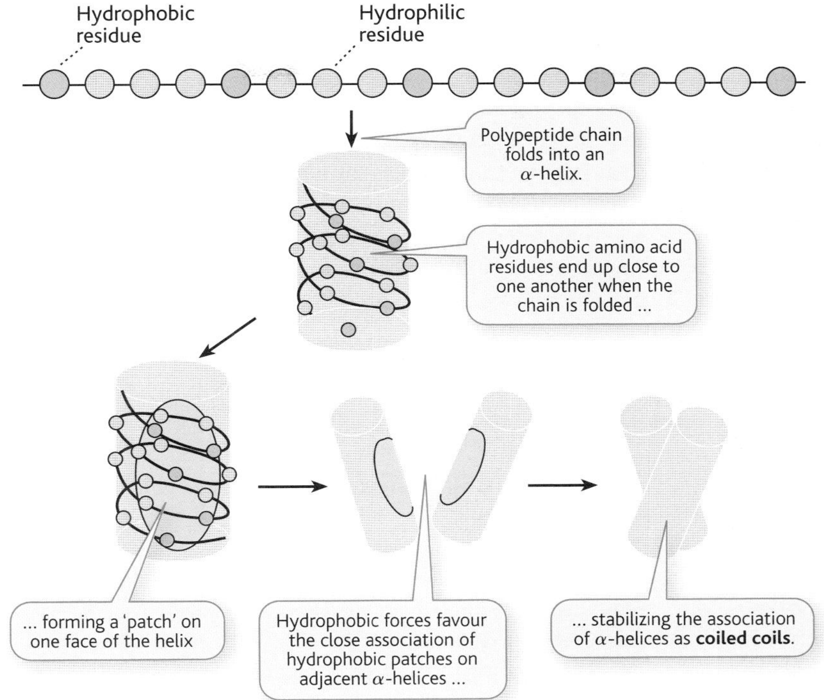

Figure 9.11 The regular positioning of hydrophobic amino acid residues in the primary sequence of α-keratin generates a hydrophobic 'patch' around one face of the α-helix when the polypeptide adopts its folded, three-dimensional structure (adapted from C. Dobson, J. Gerrard, and A. Pratt (2001)).

Look at Figure 9.11, and notice how the hydrophobic residues form a hydrophobic 'patch' around one face of the α-helix.

An α-keratin molecule possesses numerous α-helices, which lie adjacent to one another when the molecule packs into its native three-dimensional conformation: its tertiary structure. When packing occurs, hydrophobic forces drive the tight association of the patches of hydrophobic residues on neighbouring α-helices. The neighbouring helices intertwine as a result of their close interaction to form regions of characteristic tertiary structure which we call **coiled coils**.

It is the intertwining of the adjacent α-helices to form a 'coiled coil' tertiary structure that makes α-keratin such an excellent structural protein. Just as rope is given its strength through the intertwining of fibres, so α-keratin achieves its structural strength in an analogous fashion: the intertwining α-helices, stabilized by hydrophobic interactions, constitute robust structures which are the perfect ingredients for body tissues which require physical strength, such as hair and nails.

Coiled coils are characteristic of a range of proteins, including tropomyosin, a protein found in muscle (another body tissue that must exhibit structural robustness).

Not all interactions that stabilize tertiary structures are non-covalent, however. We see in Box 9.3 how one particular covalent interaction, the **disulfide bond**, plays an important role in conferring structural stability upon proteins.

Quaternary structure

The final level in the structural hierarchy of biological polymers is quaternary structure. The function of many biological molecules derives from their position within large **complexes**. Each molecule represents one **subunit** of the overall complex, but the complex as a whole requires all subunits to be present, and to associate with one another in the correct way, for the complex to function as it should.

By analogy, consider a car engine. A car engine has many component parts, each of which can be isolated: the carburettor, radiator, gearbox, clutch, etc. However, the overall complex–the car engine–can only function properly if all components–or subunits–are present, and connected to one another in the correct way.

The quaternary structure describes the way in which the component subunits of a large biological macromolecule assemble to form the fully functioning complex. We see in Section 4.1 how haemoglobin comprises four separate subunits;

BOX 9.3 Covalent contributions to tertiary structure: the disulfide bond

A disulfide bond is generated when a covalent bond forms between the thiol groups of two adjacent cysteine residues, as depicted in Figure 9.12. The disulfide bond acts as a cross link between the two amino acid residues. As such, the disulfide bond has a vital role in ensuring the structural stability of proteins–it may link together adjacent parts of a single polypeptide strand, or tether together two separate polypeptides, like rungs on a rope ladder.

For example, disulfide bonds hold together the two polypeptide strands that form the hormone insulin (see Figure 6.20b); they also exist within and between the heavy and light chains that form immunoglobulin molecules, as illustrated in Figure 9.13.

Disulfide bonds are particularly essential to the hairdresser, as they are crucial to the process of hair **perming**. Hair is rich in the protein **keratin**, which contains numerous cysteine residues. The cysteine residues are cross-linked via disulfide bonds to hold adjacent keratin molecules in fixed positions relative to one another.

Perming is a two-stage process. Firstly, an alkaline perming lotion is applied to the hair. This lotion penetrates the hair and breaks some of the pre-existing disulfide bonds that link neighbouring keratin strands. Disulfide bond breakage allows adjacent keratin strands to slide past one another; we experience this as a physical softening of the hair.

In the second stage in the process, the softened hair is manipulated into the desired style and a neutralizing lotion applied. This neutralizing lotion reverses the effect of the perming lotion, encouraging the formation of new disulfide bonds between cysteine residues. The new disulfide bonds tether adjacent keratin molecules to one another, locking them into new conformations, which gives the hair its new style and shape.

The name 'perm' reflects the characteristic conformation of permed hair: a **perm**anent wave.

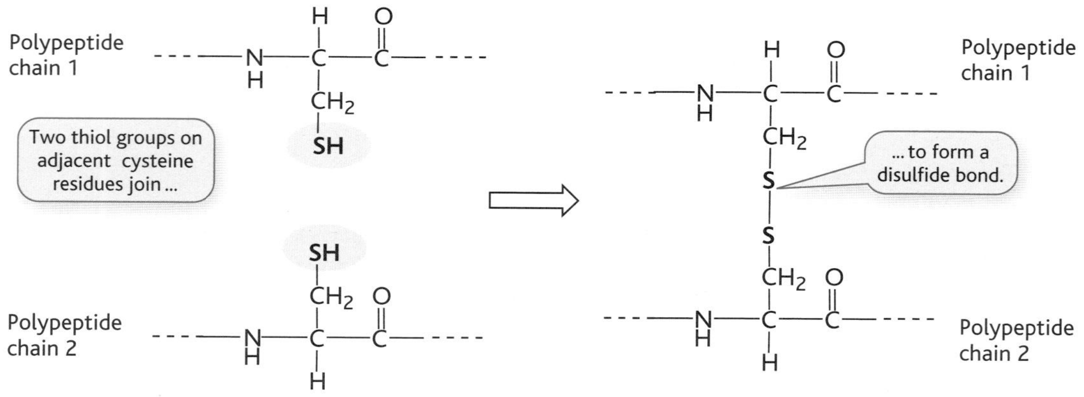

Figure 9.12 The structure of a disulfide bond. A disulfide bond is generated when a covalent bond forms between the thiol groups of two adjacent cysteine residues.

BOX 9.3 **Continued**

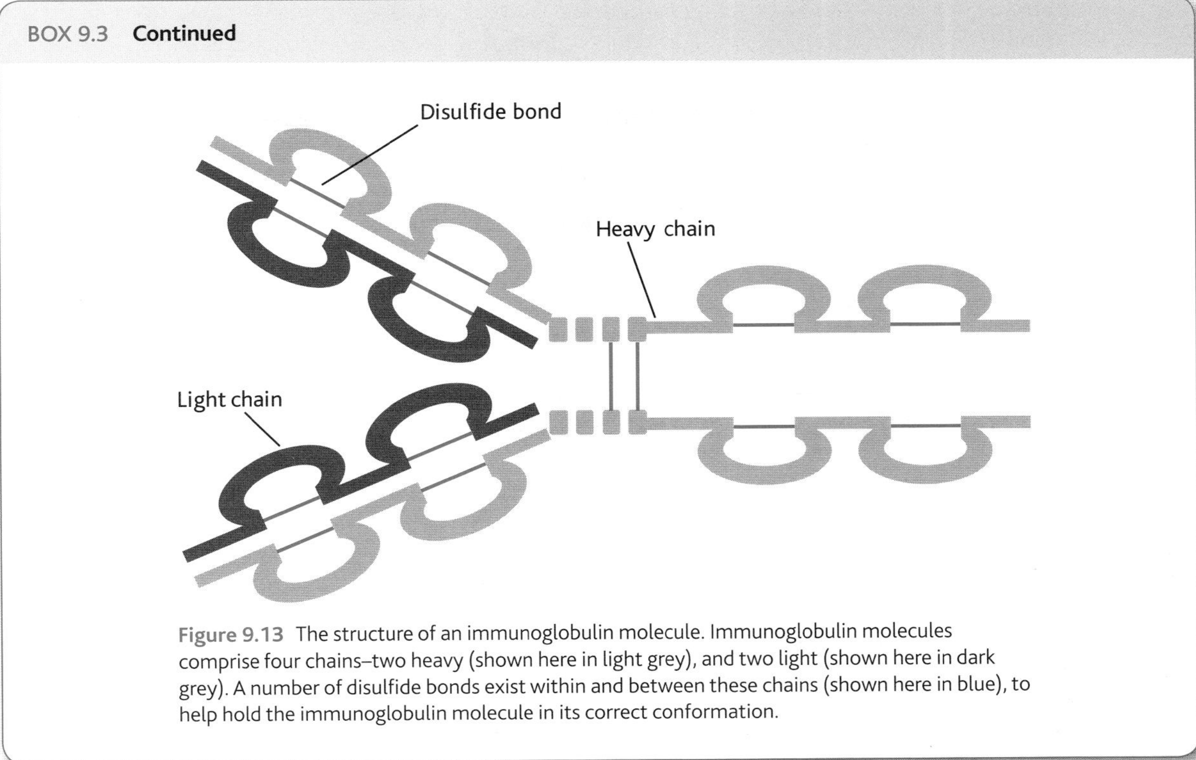

Figure 9.13 The structure of an immunoglobulin molecule. Immunoglobulin molecules comprise four chains–two heavy (shown here in light grey), and two light (shown here in dark grey). A number of disulfide bonds exist within and between these chains (shown here in blue), to help hold the immunoglobulin molecule in its correct conformation.

the quaternary structure of haemoglobin describes how the four subunits associate with one another to form the complete protein.

Another more complicated biological assembly is the enzyme ATP synthase, which synthesizes ATP from ADP. In mitochondria, the F_0F_1 ATP synthase has two portions: the F_0 portion is embedded in the mitochondrial inner membrane, while the F_1 portion lies above the membrane. Each portion comprises multiple subunits, which come together to form an overall molecular assembly. In bacteria, the F_1 portion of the equivalent enzyme comprises five different subunits–three alpha, three beta, one gamma, one delta, and one epsilon–to give nine subunits overall. By contrast, the F_0 portion comprises three different subunits–one a, one b, and around 10–15 c. The mitochondrial F_0F_1 ATP synthase is more complicated, however, and contains at least 17 different subunits.

Look at Figure 9.14, which depicts the quaternary structure of the bovine heart F_1 ATP synthase. Notice how this complex comprises a series of distinct molecules, which have come together in a stable assembly to form the overall structure.

The interaction of subunits to form complete, complicated biological macromolecules is stabilized by networks of non-covalent interactions, just like the interaction of neighbouring monomers to form secondary structures, and the interaction of adjacent secondary structures to form tertiary structures. The association of all neighbouring molecules is stabilized by dispersion forces, while hydrogen bonds and hydrophilic and hydrophobic interactions may exist between molecules possessing appropriate chemical characteristics.

We explore the chemical characteristics that must be exhibited by a molecule for it to participate in hydrogen bonding, or hydrophobic or hydrophilic interactions, in Chapter 4.

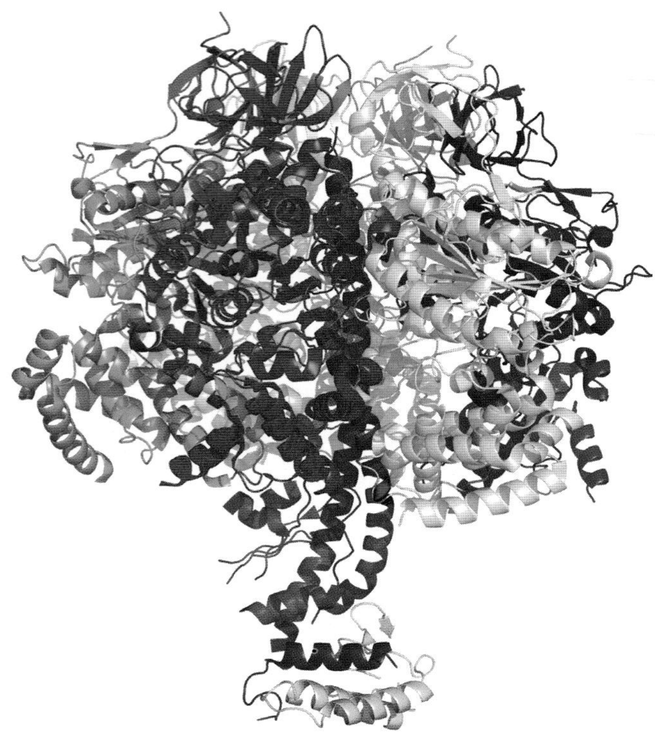

Figure 9.14 The bovine heart F1 ATP synthase. The F1 subunit forms part of the overall ATP synthase complex. Each individual molecule (polypeptide chain) in this assembly is shown in a different colour.

- ***The quaternary structure describes the way multiple individual molecules associate to form a larger complex.***

Self-check 9.3 Is it possible for a coiled coil to be an example of both a tertiary and a quaternary structure? Why or why not?

The hierarchy of biological structure: a summary

We see throughout this section how the complexity of biological shape and structure is built up on four different levels, and how each level impacts upon the level of complexity above it in the hierarchy. We can view biological shape and structure in two ways: we can 'peel off' the different levels just like peeling off the layers of an onion, starting with the highest level of complexity (quaternary structure) and ending with the lowest (primary structure); or we can *build up* complexity, starting with the simple primary structure and building up layer after layer of complexity until we reach quaternary structure.

Look at Figure 9.15(a), which summarizes the layering of structural complexity. Notice, once again, the interplay that occurs between the different levels of complexity: only by understanding the chemical factors that underpin each layer can we then understand how something as seemingly simple as the sequence of amino acids in the primary sequence of a polypeptide chain influences the overall structure of a biological macromolecule. Now look at Figure 9.15(b), which depicts how

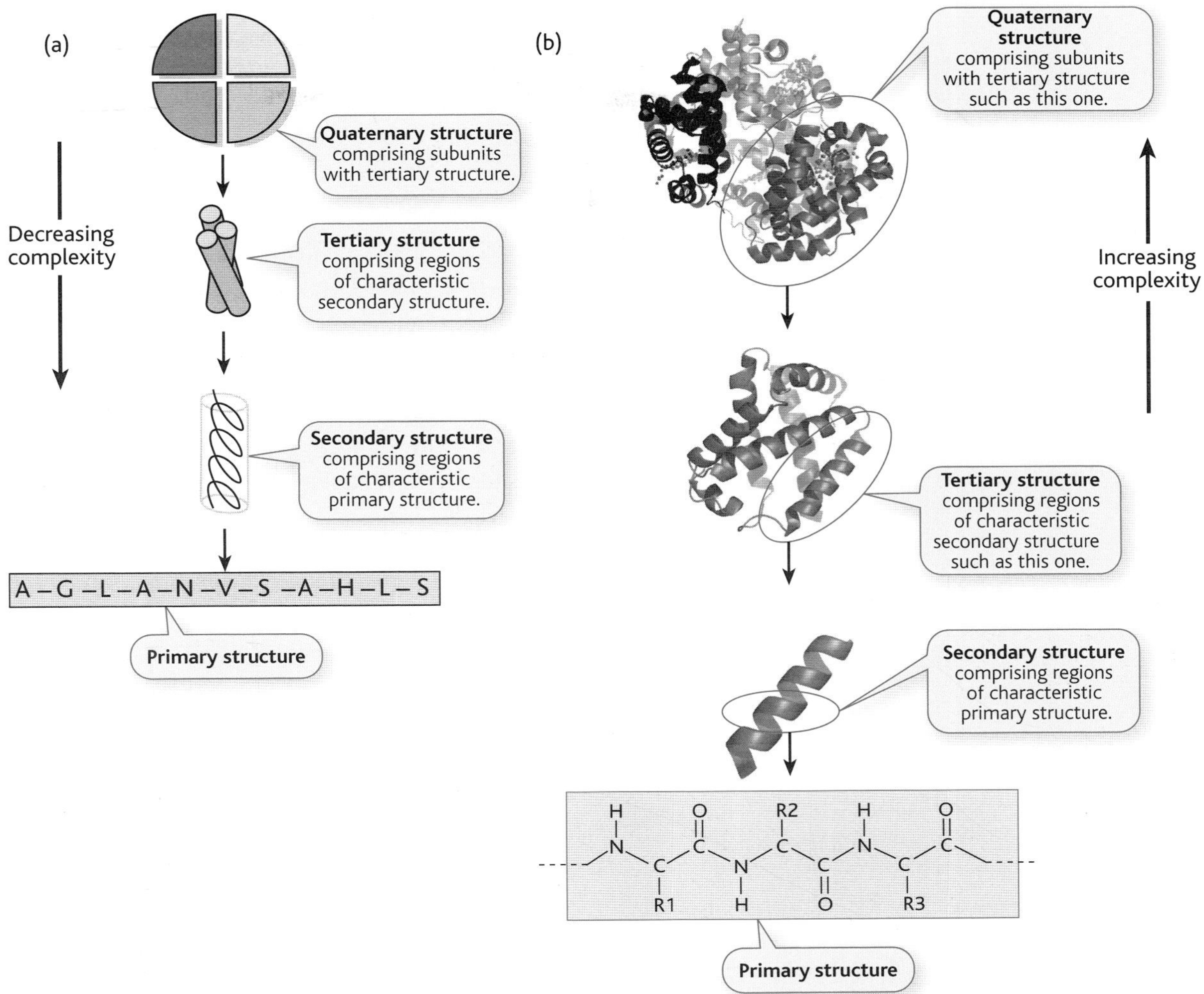

Figure 9.15 (a) An illustration of the structural complexity of biological molecules. (b) The layering of structural complexity in the protein haemoglobin. Notice how this specific example mirrors the general scheme shown in (a).

the same structural complexity arises in a real biological molecule, the protein haemoglobin.

Perhaps the most stunning example of structural complexity is the packing of DNA into chromosomes. We explore this amazing biological feat in Box 9.4.

9.3 Maintaining shape, and allowing flexibility

One of our key themes is that physical structure underpins biological function. A molecule must often fulfil two criteria for it to function properly:

1. It must be held in the correct shape.
2. It must exhibit structural flexibility.

At first glance, these criteria seem to be antagonistic: a molecule must be held in the correct shape, but it must also possess flexibility. Can a molecule be held in a particular shape and still possess flexibility? The answer is: yes. In this section we see how it is the underlying chemistry of a molecule that allows a molecule to satisfy both these criteria at once.

Consider the human body. We require structural rigidity to maintain our overall shape, but we also require flexibility–we must maintain an overall shape that is recognizably human, but still be able to bend, to stretch, to walk, to kneel, to sit. Our skeleton provides rigidity–it is a network of bones that form a robust framework on which the rest of our body is assembled. Our skin and other biological materials operate throughout different locations around the body to hold organs in place, to connect different tissues together–in other words, to complete the structural composition that is characteristically human.

However, while our skeleton fixes our overall configuration–it fixes the length and position of our limbs, torso, and head–it also allows flexibility. A network of joints, muscles, tendons, and cartilage enables us to bend our arms and legs, to rotate our head, to grasp with our hands.

A molecule possesses characteristics that reflect the human body almost exactly. A molecule has a skeleton: a network of covalent bonds that defines the molecule's configuration. The bonds specify the relative location of atoms, and how far apart in the molecule the atoms are positioned. Different parts of large molecules are also held in close associations through networks of non-covalent interactions.

BOX 9.4 Structural hierarchy and the packing of DNA

Each cell of the human body contains roughly 2–3 m of DNA. This DNA is packaged into the cell nucleus, an organelle with a diameter of just 10–20 μm. This packaging is possible only because of the highly condensed way that DNA is folded. Figure 9.16 illustrates how DNA undergoes various levels of packing, from the coiling of DNA into the characteristic double-helical structure, through to the tightly packed chromatin of chromosomes in the nucleus.

The way in which DNA is packaged has a huge impact on its biological function. Tightly packaged DNA–called heterochromatin–cannot be accessed by the regulatory molecules required to activate a gene, or the enzymes required to transcribe it. Consequently, the genes within heterochromatin cannot be expressed: we say that the DNA is transcriptionally silent. By contrast, less tightly packed DNA is called euchromatin. Such DNA *can* be accessed by the molecules required for the gene to be actively expressed.

In fact, the nature of packaging of DNA–either as heterochromatin or euchromatin–is used by the cell as a way of actively regulating gene expression. But what determines whether a stretch of DNA exists as tightly packed (and functionally silent) heterochromatin, or more active euchromatin? The answer appears to lie–at least in part–with the histone proteins that form the nucleosome–the protein-DNA complex formed when the DNA strand wraps itself around the histone proteins as part of the packaging process. Specifically, the addition of chemical groups or small proteins to the side chains of different histone proteins appears to influence the structure of chromatin.

For example, highly acetylated histone proteins (proteins that have had an acetyl group, $-COCH_3$, added) are associated with euchromatin, while heterochromatin tends to exhibit very low levels of histone acetylation. Other modifications to histone proteins, including methylation (the addition of a methyl, $-CH_3$, group), phosphorylation (addition of a phosphate group), and ubiquitination (addition of the ubiquitin protein) also modify the behaviour of chromatin according to modification patterns that we are only just beginning to understand. What is clear, however, is that gene regulation goes beyond the activity of regulatory proteins and transcription enzymes, but is influenced by the very nature of DNA itself.

BOX 9.4 **Continued**

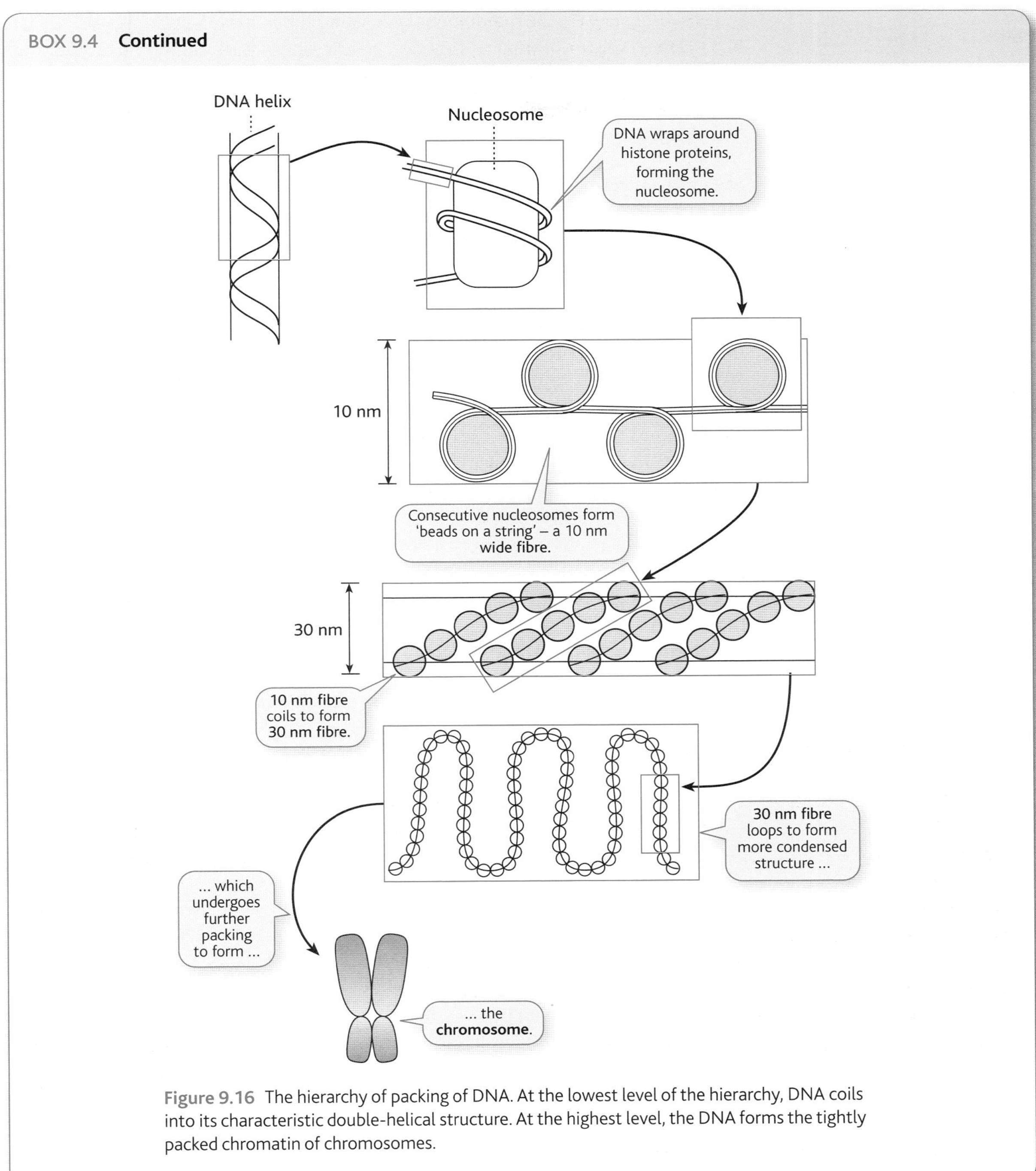

Figure 9.16 The hierarchy of packing of DNA. At the lowest level of the hierarchy, DNA coils into its characteristic double-helical structure. At the highest level, the DNA forms the tightly packed chromatin of chromosomes.

However, molecules are flexible. Rotation is possible about single bonds, allowing atoms to twist relative to one another (just as joints allow arms and legs to twist and rotate). Just as skin helps to maintain overall shape, but does not constrict movement (it does not 'lock' the body into a rigid position), so non-covalent forces allow molecules to adjust conformation, while stabilizing the molecule in its overall shape.

Let's discover why the maintenance of overall shape, and structural flexibility, are both important for biological function by considering a couple of examples: the contraction of muscle, and the activity of enzymes.

The importance of structural flexibility: muscle contraction

Skeletal (or striated) muscle is the type of muscle responsible for powering the movement of our limbs. 'Striated' muscle is so called due to its characteristic striated (or stripy) appearance when viewed under a microscope. Striated muscle principally comprises two types of protein: actin and myosin. Both these fibres exist as thin and thick filaments respectively. Figure 9.17(a) shows how the thin actin filaments and thick myosin filaments are arranged in striated muscle. Look at this figure, and notice how it is the particular arrangement of thin and thick filaments that gives striated muscle its stripy appearance.

When a muscle contracts, it shortens in length. This shortening occurs when the alternate layers of actin and myosin filaments slide over one another, as illustrated by the change between Figures 9.17(a) and (b). But what powers this shortening? Research has shown that the shortening of muscle fibre during muscle contraction occurs as a result of the repeated attachment and release of the thick myosin filament from the neighbouring thin actin filament. Figure 9.17(c) shows how, in effect, the myosin filament 'claws' its way along the length of the actin filament (much like a mountaineer uses ice picks to claw their way up the face of a snow-covered mountain).

The myosin filament joins to the actin filament through a region that we call the myosin 'head'. It is the movement of this head region that enables the myosin filament to claw its way along the actin filament. This movement can only occur because the myosin molecule is **flexible**. It does not possess a rigid, unchanging structure, but is able to change shape in order to carry out a particular function (in this case, effecting the shortening of muscle fibre).

Structural research has allowed us to visualize the movement in the myosin head that is associated with muscle contraction. Look at Figure 9.18, which shows the structure of the myosin head before and after muscle contraction. Notice how the protruding helical strand undergoes a marked change in conformation (relative to the rest of the myosin head) during contraction: it is this change in conformation that is believed to drive the movement of the myosin filament along the actin filament and, hence, drive the shortening of muscle required for muscle contraction.

So, while the overall shape of a myosin molecule must be correct (it must possess the characteristic head region required for binding actin, and must form the thick filaments seen in striated muscle fibre), it must also exhibit flexibility if myosin is to perform its appropriate biological function.

The importance of structural flexibility: enzymes

The enzymes are the class of protein to which both the maintenance of correct overall shape and a degree of structural flexibility are of perhaps the most importance.

In general, an enzyme interacts with a particular starting material (the **substrate**) and mediates its conversion into a new end **product**, as depicted in Figure 9.19.

Enzymes must be of the correct shape in order to exhibit **specificity**: they must be able to recognize and bind the correct substrate. Specificity is mediated by the enzyme having a complementary shape to its substrate. Recognition and binding

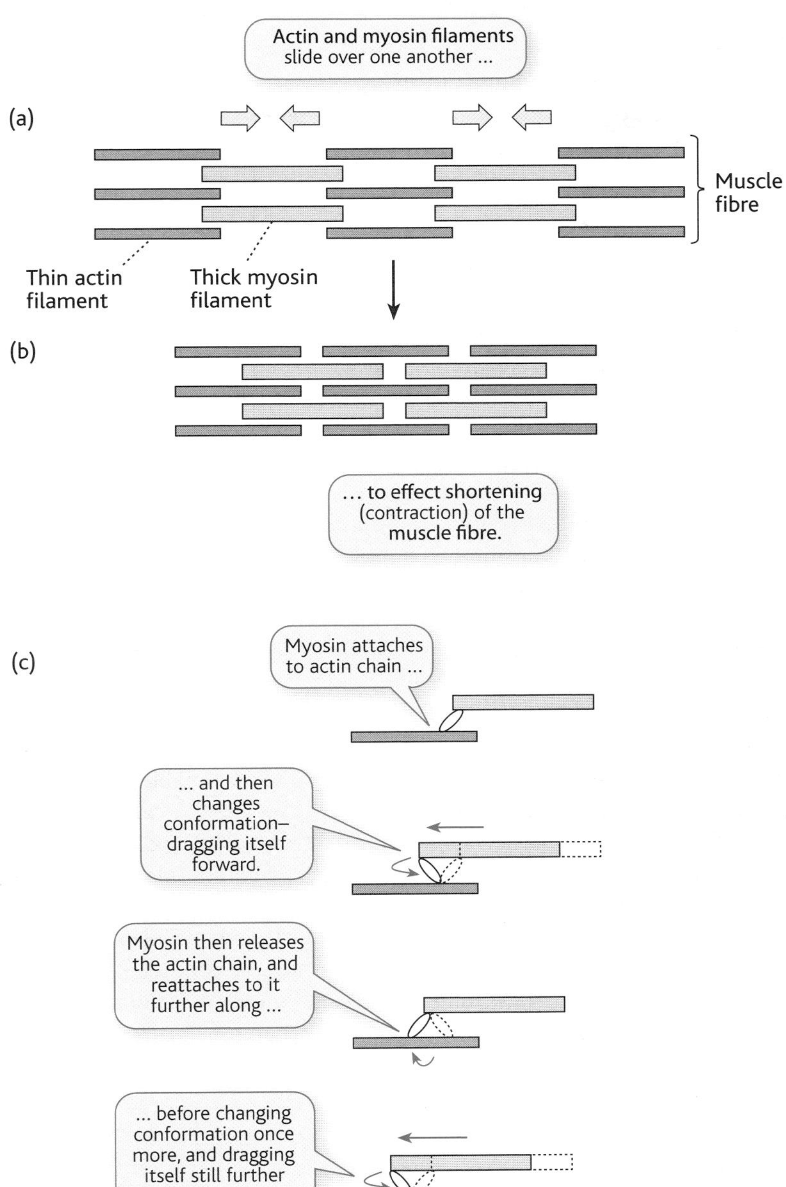

Figure 9.17 A schematic illustration of the arrangement of actin and myosin filaments in striated muscle. (a) Actin and myosin filaments slide over one another; (b) this movement effects shortening of the muscle fibre. (c) The shortening is driven by the 'clawing' of myosin along the length of the actin filament.

can only occur if the shapes of the enzyme and its substrate are complementary, as depicted in Figure 9.20[7]. Because they have complementary shapes, the enzyme and substrate can interact closely enough for extensive non-covalent interactions to

7 The binding of substrate and enzyme follows the same principle of complementarity of shape as that behind a lock and key: a key will only open a lock if its shape is complementary. The similarity has given its name to a model that describes the way in which an enzyme binds its substrate: the **lock and key** model.

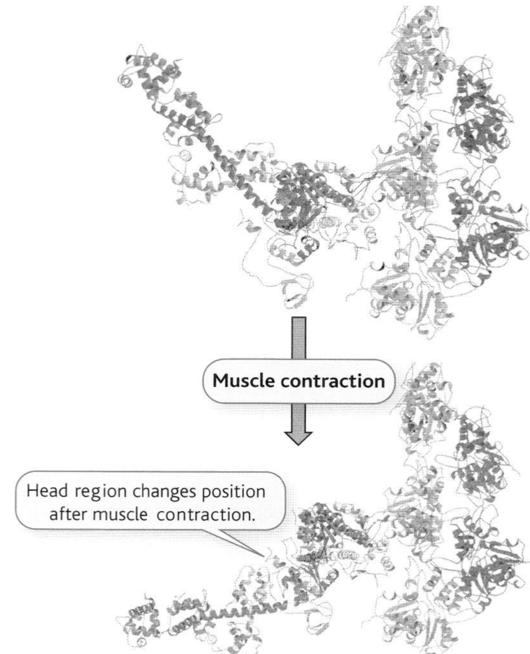
Muscle contraction
Head region changes position after muscle contraction.

Figure 9.18 The structure of the myosin head region before and after muscle contraction. Notice how the head region has changed position relative to the rest of the polypeptide structure. This structural flexibility is central to myosin's correct function. (Reprinted, with permission of Prof. M. Geeves and Prof K. Holmes, from the *Annual Review of Biochemistry*, Volume 68 ©1999 by Annual Reviews, **http://www.annualreviews.org/**.)

operate between the two molecules, stabilizing their association long enough for the ensuing catalysis to occur.

However, an enzyme's structure must also be flexible. Usually, the shape of the end product of the reaction catalyzed by the enzyme is different from that of the substrate, even if only subtly. Indeed, as it makes the transition from substrate to product, a substrate molecule passes through a state called a transition state, which we could consider a halfway house between the substrate and product, and which has a structure that is distinct from either the substrate or product. What impact does this have on the enzyme?

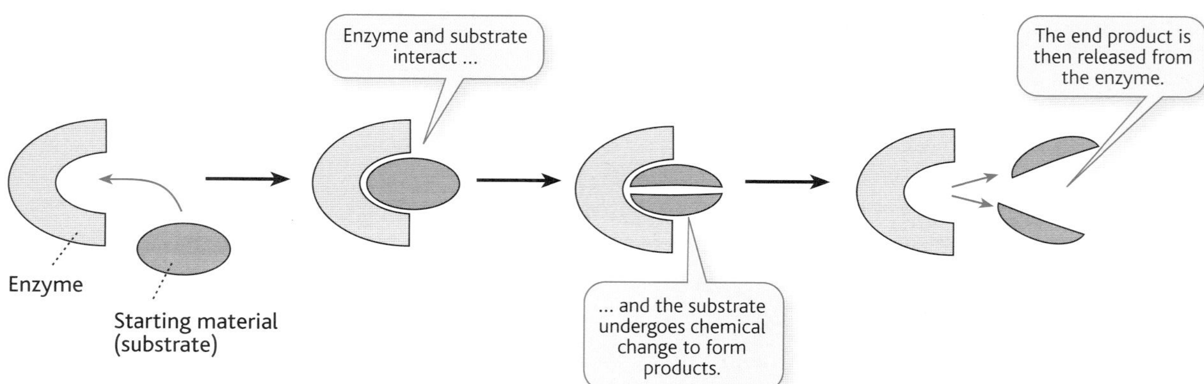
Enzyme and substrate interact ...
The end product is then released from the enzyme.
Enzyme
Starting material (substrate)
... and the substrate undergoes chemical change to form products.

Figure 9.19 The general action of an enzyme. An enzyme interacts with a particular starting material (the substrate) and mediates its conversion into a new **end product**.

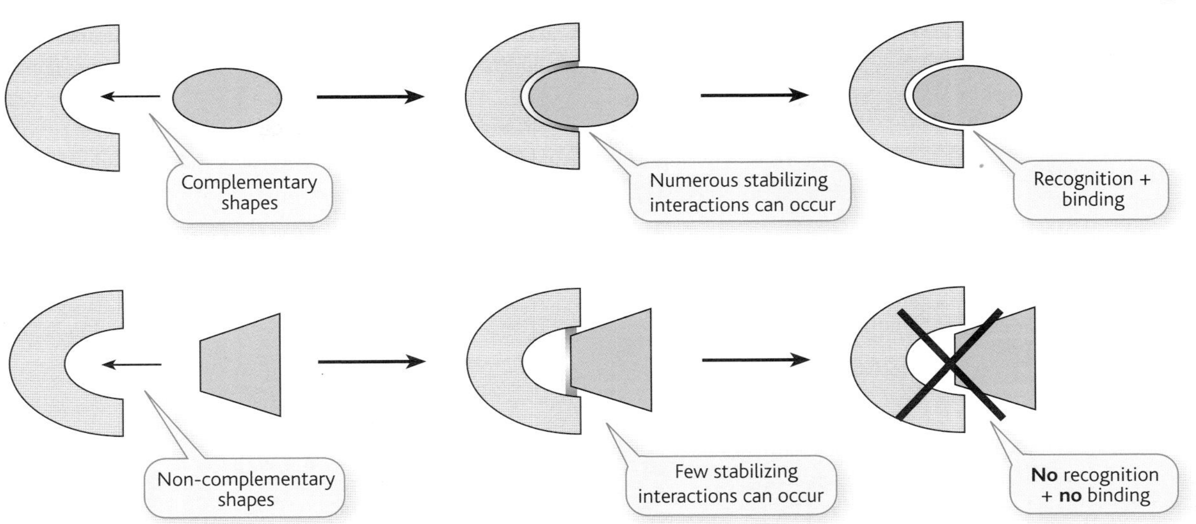

Figure 9.20 Recognition and binding can only occur if the shapes of the enzyme and its substrate are complementary. Complementary shapes allow more extensive interactions, which stabilize the binding of enzyme and substrate.

The answer is that the enzyme must be flexible enough to 'mould' to the structure of the transition state. Indeed, it is the change in structure of the enzyme that *induces* a change in the substrate so that it enters the transition state. (We call this the **induced fit model** of enzyme catalysis: the enzyme facilitates the reaction by inducing the substrate to become the transition state by changing its own shape). Further, the enzyme must be flexible enough that it doesn't bind so tightly to the product that the product can't diffuse away once the reaction is complete.

Figure 9.21 shows how an enzyme's structure must be flexible if the catalysis of a reaction from substrate to product is to proceed. Look at Figure 9.21 and notice how, if the enzyme is totally rigid, it cannot induce a change in the shape of the

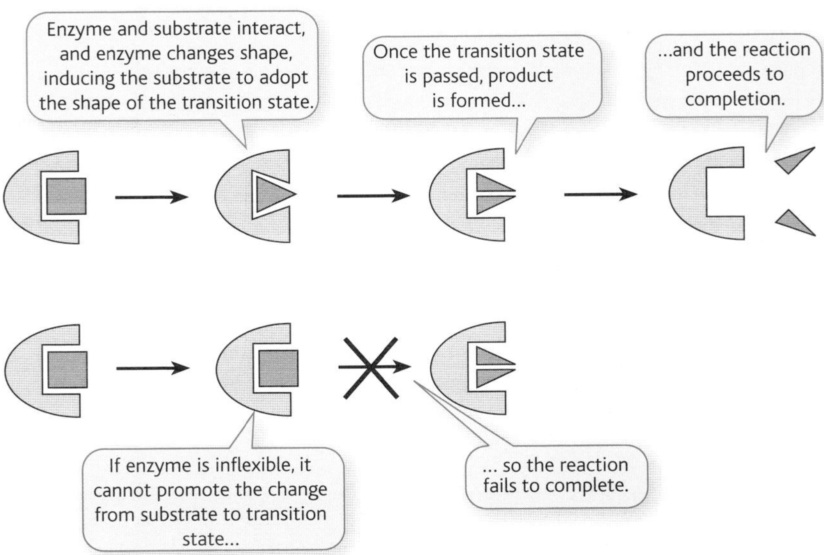

Figure 9.21 An enzyme's structure must be flexible if it is to induce a change in the shape of the substrate, and promote progress to the transition state.

We explore the biological significance of enzymes in more detail in Chapter 14 when we consider chemical kinetics–the rate at which chemical reactions occur.

substrate, and the reaction fails to occur. By contrast, if the enzyme exhibits slight flexibility, the substrate is induced to adopt the transition state, and the reaction proceeds to completion.

But what of proteins that must recognize and bind to many different target molecules–that is, where they don't show such tight specificity as most enzymes? For example, ubiquitin, (1), is a regulatory protein that recognizes and attaches to many different proteins, tagging them for destruction. Ubiquitin appears to operate by shuffling rapidly between many different conformations; these different conformations enable it to associate (and tag) a range of different binding partners.

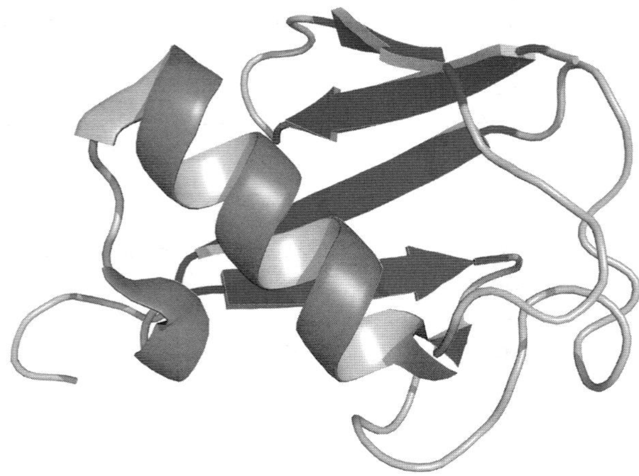

A degree of flexibility can actually enhance the performance of a protein, as explained in Box 9.5, which explores the concept of **allostery**.

BOX 9.5 It's good to talk: communicating protein-style

Haemoglobin exhibits a remarkable property: the binding of oxygen to the haem group of one of its four subunits makes it more efficient at binding further oxygen molecules through its other three subunits. In effect, when oxygen binds to a particular subunit, the subunit 'communicates' to its neighbouring subunits that binding has occurred, making them more receptive to the binding of oxygen. But how can subunits communicate with one another? The answer lies in tiny–yet significant–changes in the structure of the subunits of haemoglobin when oxygen binding occurs.

We see in Section 3.8 (p. 65) how oxygen binds to the Fe^{2+} ion, which is held at the centre of a haem group. When oxygen is not attached, the Fe^{2+} ion sits just below the plane of the porphyrin ring (the ring-shaped structure that constitutes the majority of the haem group). However, Figure 9.22 shows how, when oxygen binds to the Fe^{2+} ion, the Fe^{2+} ion shifts position very slightly, so that it lies in the *plane* of the porphyrin ring. This movement pulls the histidine residue, to which the Fe^{2+} ion is coordinate-bonded, towards the haem group, a movement which, in itself, triggers a cascade of conformational changes throughout the polypeptide subunit.

The conformational changes triggered by the binding of oxygen are transmitted to the neighbouring polypeptide subunits of haemoglobin, such that their haem groups adopt conformations which can bind oxygen more readily.

BOX 9.5 **Continued**

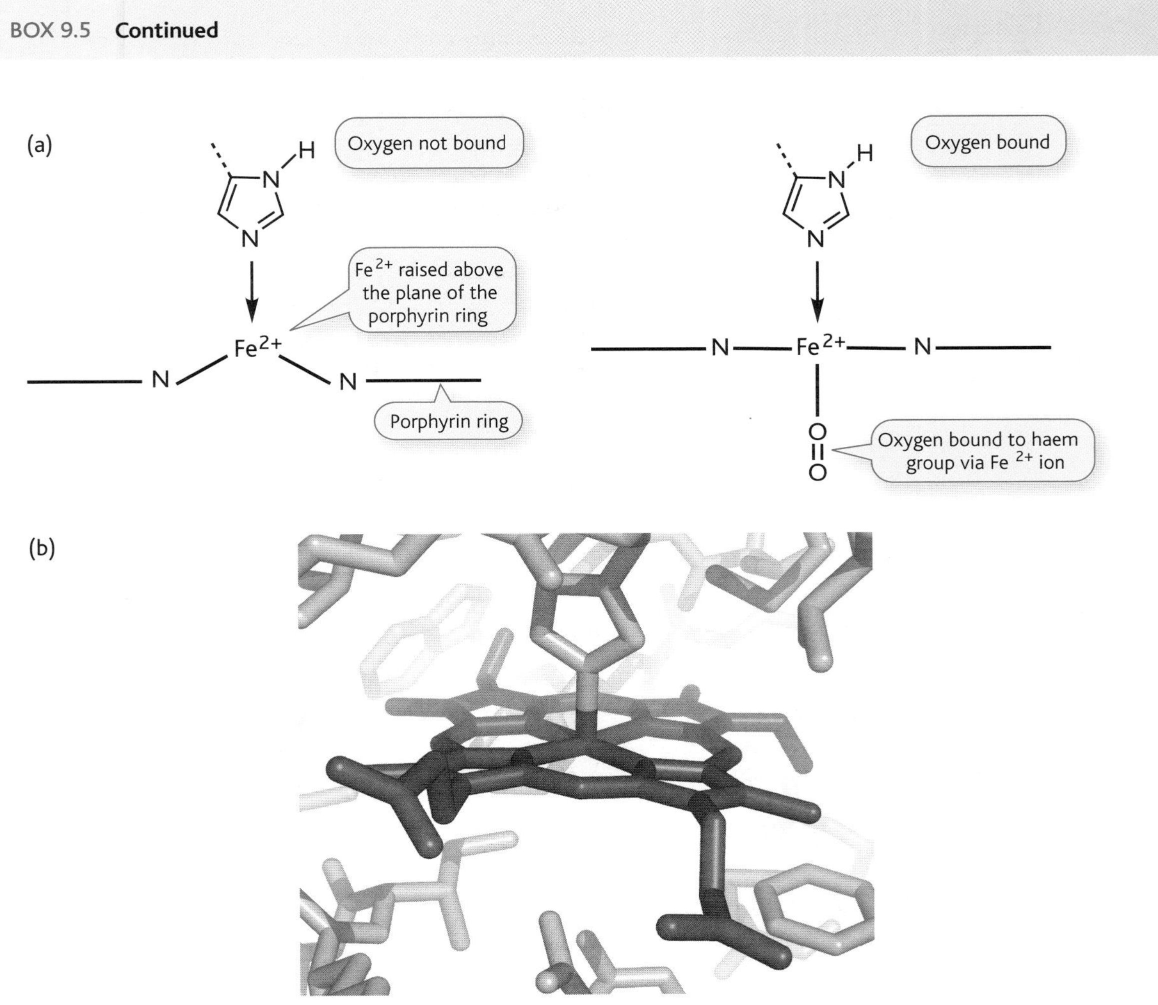

Figure 9.22 (a) When oxygen binds to the Fe^{2+} ion at the centre of one of the haem groups found in haemoglobin, the Fe^{2+} ion shifts position very slightly, so that it lies in the plane of the porphyrin ring. This movement triggers a cascade of conformational changes throughout the polypeptide subunit. (b) The porphyrin ring joined, via a histidine residue, to the rest of the haem group.

We give the name allostery to the phenomenon of the modification in the behaviour of a protein through conformational change triggered by the binding of a molecule to the protein. Allostery lies at the heart of many biological processes, including cell signalling (communication within and between cells). For example, allostery allows messages to be passed across the often impenetrable cell membrane. Figure 9.23 shows how allostery facilitates cell-to-cell signalling: a change in structure of a membrane-bound receptor molecule upon the binding of a 'signal' molecule can trigger a signalling pathway on the inside of the cell, ensuring that the message isn't blocked by the cell membrane, and can reach its target.

BOX 9.5 **Continued**

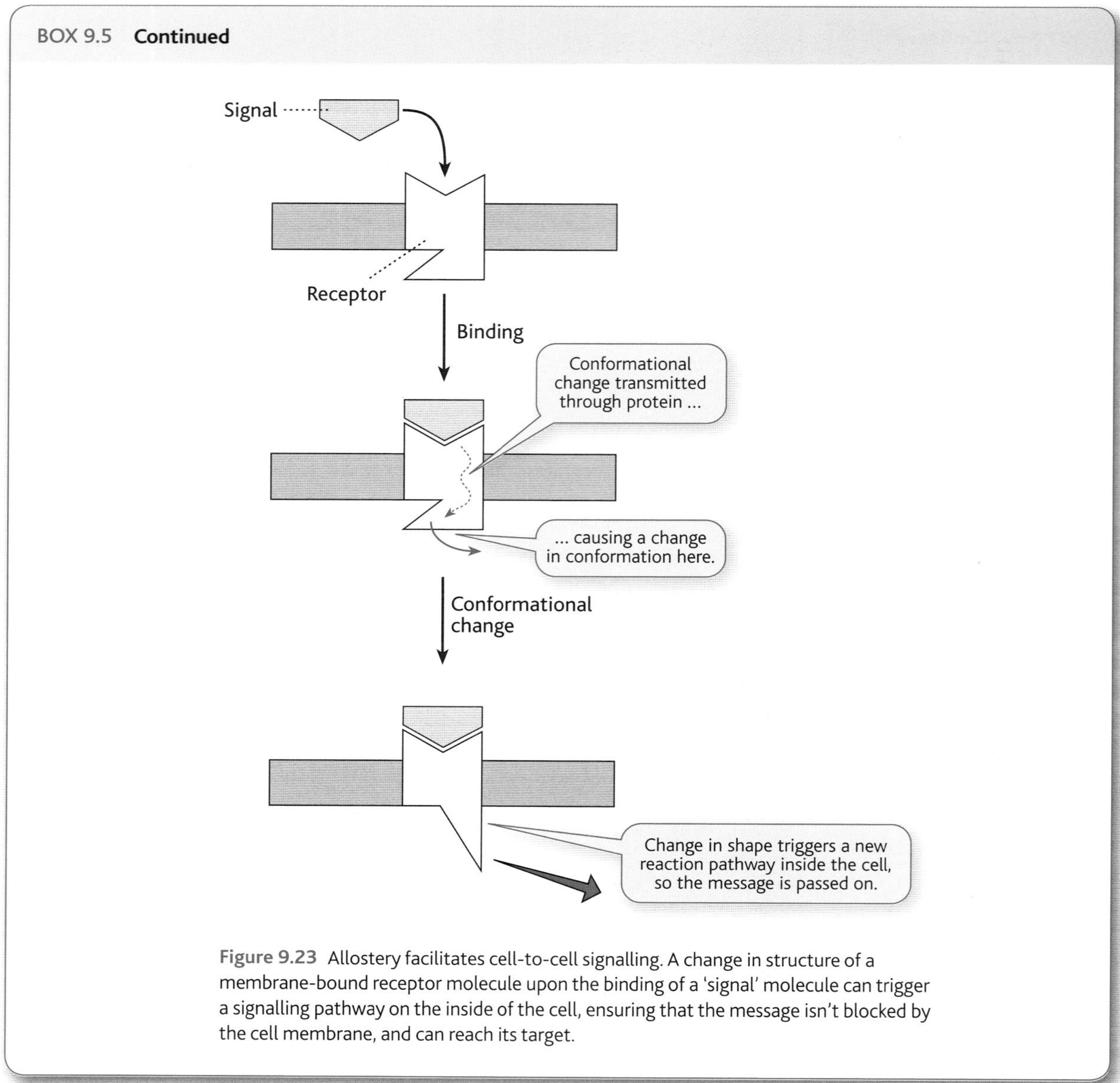

Figure 9.23 Allostery facilitates cell-to-cell signalling. A change in structure of a membrane-bound receptor molecule upon the binding of a 'signal' molecule can trigger a signalling pathway on the inside of the cell, ensuring that the message isn't blocked by the cell membrane, and can reach its target.

We have now seen how molecular shape and structure is critically important to the correct function of biological molecules. In the next chapter, we go on to discover how even the most subtle of changes in three-dimensional shape can affect a molecule's biological role, as we explore the concept of isomerism.

Check your understanding

To check that you've mastered the key concepts presented in this chapter, review the checklist of key concepts below, and attempt the multiple-choice questions available in the book's Online Resource Centre at **http://www.oxfordtextbooks.co.uk/orc/crowe2e/**.

Checklist of key concepts

The shape of larger molecules

- Polymers are formed from the joining together of many small repeating monomers
- The structure of a polymer is influenced by the geometry of the joined atoms; the sequence in which monomers join to one another to form the polymer; and the physical nature of the bonds that join neighbouring monomers
- Sugar units are joined by glycosidic bonds to give polysaccharides
- Starch, cellulose, and glycogen are all polymers of glucose

Structural hierarchies

- The structure of a biological molecule builds up sequentially over a number of levels of complexity
- Primary structure describes the sequence of joining of monomers to form a biological polymer
- Secondary structure describes the folding of lengths of primary structure into characteristic three-dimensional conformations
- Secondary structure is stabilized by hydrogen bonds
- α-helices are regions of coiled polypeptide
- β-sheets are stacked 'sheets' of polypeptide
- β-turns link the stacked sheets of a β-sheet
- Secondary structures can only form if the polymer has an appropriate primary structure: different amino acids favour the formation of different secondary structures
- The characteristic secondary structure of DNA is the double helix
- Tertiary structure describes the packing of regions of secondary structure into a defined three-dimensional shape
- A β-barrel is formed from the curling round of β-sheets
- A coiled coil forms from the packing of neighbouring α-helices, and is stabilized by hydrophobic interactions between the helices
- Disulfide bonds form between the thiol groups of adjacent cysteine residues
- Disulfide bonds play an important role in stabilizing the structure of proteins
- Quaternary structure describes the way in which the component subunits of a large biological macromolecule assemble to form the complete, functional entity

Shape and flexibility

- A molecule must possess the correct shape to exhibit its desired biological function
- Many biological molecules must exhibit structural flexibility in order to carry out their correct function
- Enzymes may act to catalyse a reaction by changing shape to induce the substrate to adopt the shape of the transition state. This is called the induced fit model of enzyme catalysis
- Allostery is the modification in the biological activity of a protein due to conformational change driven by the binding of a molecule to that protein

10 Isomerism: generating chemical variety

INTRODUCTION

Many children's toys are based on the premise of taking a small group of objects and combining them in various ways to create objects of different shapes or structures. The most basic building blocks are designed to be stacked in different ways, creating arrangements with differing shapes. At a more sophisticated level, construction kits such as Lego™ are still based on the same premise: assembling the same 'pool' of pieces in different ways creates an array of different structures.

Chemistry, too, involves creating diversity from a finite range of components: the chemical elements. Sometimes this diversity arises from simply combining a single selection of atoms in different ways, just as a child may take a Lego™ set and build a model house one day, a model car the next, and so on.

In this chapter we explore the concept of isomerism–the joining together of identical groups of atoms to form different molecules. We discover the different types of isomer that can arise, asking how isomers differ from one another, and find out why isomerism is so fundamental to understanding how biological systems operate.

10.1 Isomers

The chemical elements come together in a huge number of different combinations to generate compounds that exhibit contrasting physical and chemical properties, and structures. However, two molecules do not have to comprise different atoms to exhibit contrasting physical and chemical properties, or structures.

Isomers are groups of compounds that comprise exactly the same atoms (that is, they have the same chemical composition), but which have different structures, and often different physical and chemical properties too[1].

Isomerism generates chemical variety from a single group of atoms–it could be considered the chemical equivalent of origami. Origami can transform a single sheet of paper into many different intricate shapes, just through folding the paper in different ways. Similarly, isomerism takes a specific pool of atoms–the chemical

1 The word 'isomer' means 'same units'. ('Iso' = same', and 'mer' = 'unit'.) Compare this to two other words that we find in this book: 'polymer' = 'many units', and 'monomer' = 'one unit'.

equivalent of the sheet of paper–and arranges them in different ways to produce distinct structures.

There are two important types of isomer: **structural isomers** and **stereoisomers**. We examine both types of isomer in turn in the following sections.

10.2 Structural isomers

Structural isomers are molecules that have the same chemical composition (they are made from the same chemical 'ingredients') but the atoms are linked together in different ways[2]. We say the atoms exhibit different **connectivities**. This is analogous to a child using a single box of Lego™ to build different models: the component parts are the same, but the end results are different. Look at Box 10.1 to see how different connectivities between atoms of carbon result in pure carbon existing in a number of distinct forms.

We see in Chapter 3 how a molecular formula tells us the chemical composition of a molecule. It identifies the different elements from which the molecule is composed, and how many atoms of each element are present in one molecule. Structural isomers share the *same* molecular formula: they have exactly the same chemical composition.

Structural isomerism is particularly prevalent among organic compounds, e.g. the hydrocarbons. For example, look at Figure 10.1, which depicts three hydrocarbons that share the molecular formula C_6H_{14}. These molecules are all structural isomers of the alkane hexane–they all contain six carbon atoms, and 14 hydrogen atoms. Notice, however, that the atoms are arranged in different ways in each case; while structure (a) is a straight chain, both structures (b) and (c) are branched chains.

We learn more about straight- and branched-chain alkanes in Section 5.2 (p. 130).

- ***Structural isomers comprise the same atoms, joined together in different ways.***
- ***Structural isomers share the same molecular formula.***

(a) $CH_3—CH_2—CH_2—CH_2—CH_2—CH_3$

hexane

(b) $CH_3—CH(CH_3)—CH_2—CH_2—CH_3$

2-methylpentane

(c) $CH_3—CH_2—CH(CH_3)—CH_2—CH_3$

3-methylpentane

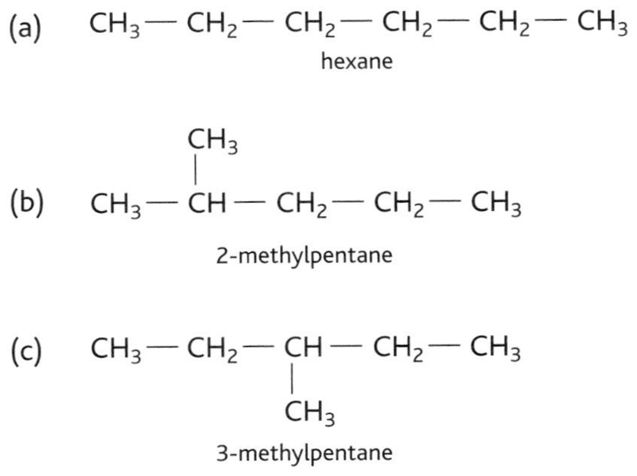

Figure 10.1 These three hydrocarbons all share the molecular formula C_6H_{14}. They are all structural isomers of the alkane hexane.

2 You may also see structural isomers referred to as **constitutional isomers**. The two terms are interchangeable.

Distinguishing structural isomers

Remember: a molecular formula indicates the relative number of carbon and hydrogen atoms present in one molecule of a given hydrocarbon. A structural formula tells us how the different atoms in the molecule are connected to one another.

We see in Chapter 5 how molecular formulae do not give us enough information to differentiate between different structural isomers because they fail to tell us the connectivity of atoms. Instead, we use **structural formulae** to tell us about this connectivity.

Structural formulae tell us how atoms are connected, and so allow us to distinguish between different structural isomers.

Let's compare the molecular formula and structural formula of ethanol. The molecular formula, C_2H_6O, tells us only that one molecule of ethanol contains two carbon atoms, six hydrogen atoms, and one oxygen atom. We cannot deduce, for example, whether the atoms are joined together like this:

```
     H    H
     |    |
H—  C  — C — OH
     |    |
     H    H
```

BOX 10.1 Varying connectivities: the allotropes of carbon

Elemental ('pure') carbon occurs in a surprising array of forms, of which perhaps the best-known are diamond, graphite, and the fullerenes (hollow, ball-like structures such as buckminster fullerene). In common with all the different forms of carbon, these three differ in the way that the carbon atoms are linked to one another–their connectivities–as illustrated in Figure 10.2.

Figure 10.2(a) shows how every carbon atom in diamond is joined to four other atoms to create a giant 3D lattice. This structure, which is very difficult to disrupt, confers upon diamond rigidity and hardness, and explains why diamond is used so widely in cutting. Diamond-tipped cutting tools and drill bits are very hard-wearing, and can be used on substances which can't be cut or drilled by tools made of softer, less durable, materials.

Each carbon atom in graphite bonds to three others to form giant sheet-like structures, which then stack up on top of one another, as illustrated in Figure 10.2(b). Graphite is a much weaker structure than diamond: the sheets of carbon atoms in graphite simply slip over one another when pressure is applied. This slippage is useful, and is exploited in the use of graphite as a lubricant.

The fullerenes exhibit similar bonding to graphite: each atom is covalently bonded to three other carbon atoms. Rather than forming a flat sheet, however, the bonded atoms curl up to form a hollow ball or tube, as illustrated in Figure 10.2(c). The fullerenes are formed from defined numbers of atoms, unlike diamond and graphite, which comprise vast arrays of many millions of carbon atoms. The best-known member of this class is buckminster fullerene, which comprises 60 carbon atoms. More than 1000 other fullerenes are known. (Some have been predicted, but not yet synthesized.)

The fullerenes aren't the only family of compounds to show great variety: there are at least 10 different sorts of graphite-like carbon too. These graphite-like compounds all share a common building block: graphene. Graphene comprises a sheet of carbon atoms just one atom thick, in which the atoms are joined in hexagonal arrays. Individual sheets of graphene were first isolated in 2004. Since then there has been huge interest in its application within the field of ultrafast (and ultra-small) electronics. Graphene conducts electricity in an unusually organized way. Typically, electrons move erratically when they pass through a material that conducts electricity–but not so with graphene. Beyond this, it is also thought that graphene has the potential to conduct electricity at very high speeds.

Graphene brings with it two further advantages: it is mechanically strong, and can be flexed into different shapes. (Indeed, carbon nanotubes are effectively rolled-up sheets of graphene.) In addition, its one-atom thickness means that it is transparent. All of this has led the BBC News website to describe the research into graphene to be 'the most promising path to bendy, invisible electronics'–a path that, if followed, could have some truly fascinating applications.

Compounds comprising atoms of the *same element* that are connected to each other in different ways are what we call allotropes. Oxygen is another element that exists as different allotropes: molecular oxygen, O_2, and ozone, O_3.

BOX 10.1 **Continued**

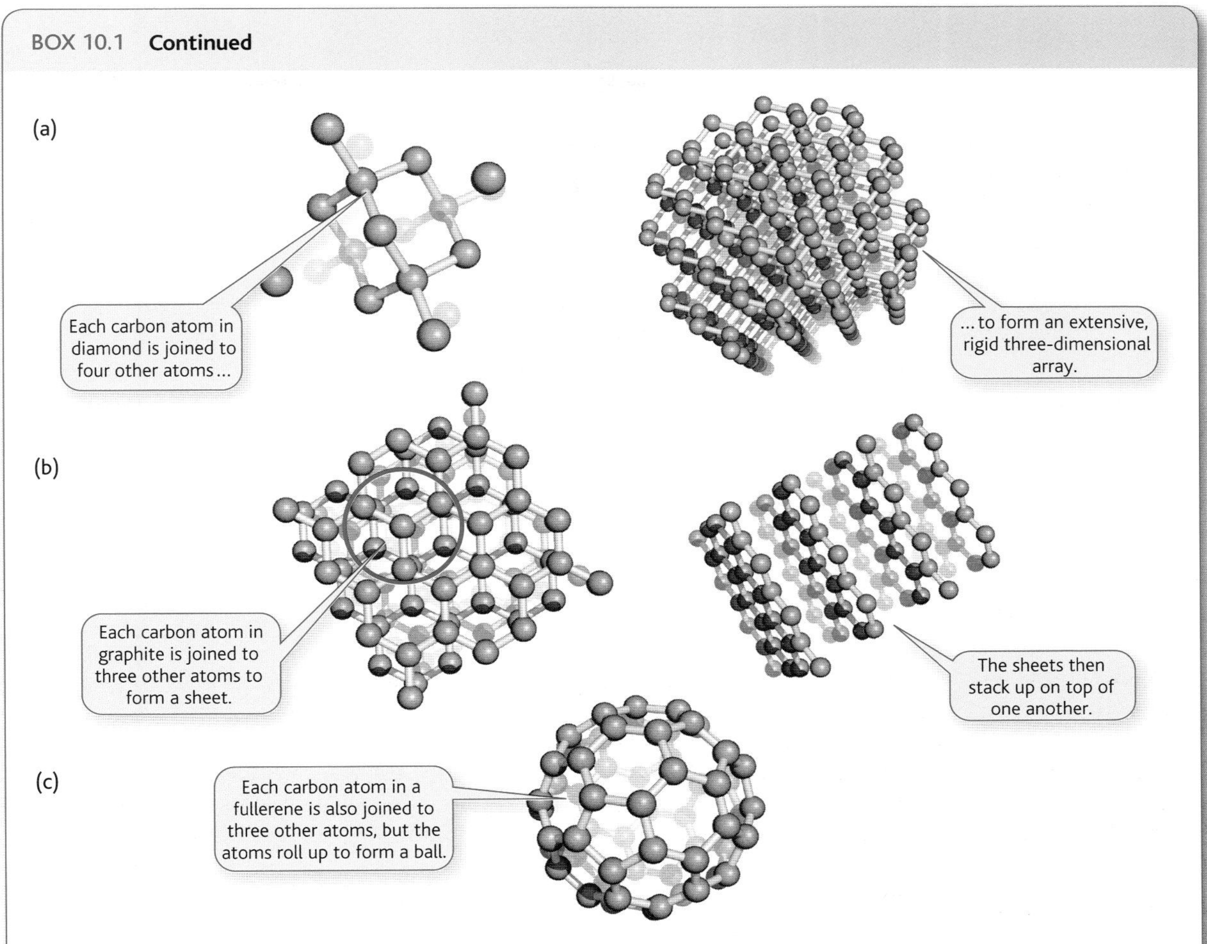

Figure 10.2 Graphite, carbon, and the fullerenes all comprise only carbon atoms. However, they vary in how the carbon atoms are joined to one another–their connectivities. (a) Every carbon atom in diamond is joined to four other carbon atoms. (b) Each carbon atom in graphite bonds to three others to form giant sheet-like structures, which then stack up on top of one another. (c) Each carbon atom in a fullerene is covalently bonded to three other carbon atoms. Rather than forming a flat sheet, however, the bonded atoms curl up to form a hollow ball or tube.

which does indeed give ethanol, or like this:

```
     H       H
     |       |
  H—C—O—C—H
     |       |
     H       H
```

to give a completely different compound, methoxymethane ('dimethyl ether').

By contrast, the structural formula tells us not only the same information as the molecular formula, but depicts how the atoms are connected to one another in the molecule, removing the kind of uncertainty about the molecule's structure that is associated with looking at the molecular formula alone.

Self-check 10.1 Look at structure (1).

$$
\begin{array}{ccccccc}
 & & CH_3 & & CH_3 & & \\
 & & | & & | & & \\
H & — & C & — & C & — & H \\
 & & | & & | & & \\
 & & CH_3 & & CH_3 & &
\end{array}
$$

(1)

This structure is not a structural isomer of structures (a)–(c) in Figure 10.1. True or false? Explain the reason for your choice.

Structural isomerism and the shape of the carbon framework

The number of different structural isomers that are possible increases dramatically as we consider alkanes with hydrocarbon chains of increasing length. While there is only one configuration possible for methane (CH_4), ethane (C_2H_6), and propane (C_3H_8), there are two structural isomers of butane (C_4H_{10}):

$$CH_3 — CH_2 — CH_2 — CH_3$$

$$
\begin{array}{ccc}
 & CH_3 & \\
 & | & \\
 & CH & \\
H_3C & & CH_3
\end{array}
$$

three structural isomers of pentane (C_5H_{12}):

$$CH_3 — CH_2 — CH_2 — CH_2 — CH_3$$

$$
\begin{array}{ccccccc}
 & & CH_3 & & & & \\
 & & | & & & & \\
H_3C & — & C & — & CH_2 & — & CH_3 \\
 & & H & & & &
\end{array}
$$

$$
\begin{array}{ccccc}
 & & CH_3 & & \\
 & & | & & \\
H_3C & — & C & — & CH_3 \\
 & & | & & \\
 & & CH_3 & &
\end{array}
$$

and five structural isomers of hexane (C_6H_{14}):

$$CH_3—CH_2—CH_2—CH_2—CH_2—CH_3$$

$$
\begin{array}{ccccccccc}
 & & CH_3 & & & & & & \\
 & & | & & & & & & \\
CH_3 & — & C & — & CH_2 & — & CH_2 & — & CH_3 \\
 & & H & & & & & &
\end{array}
$$

$$
\begin{array}{ccccccccc}
CH_3 & — & CH_2 & — & CH & — & CH_2 & — & CH_3 \\
 & & & & | & & & & \\
 & & & & CH_3 & & & &
\end{array}
$$

$$
\begin{array}{ccccccc}
 & & CH_3 & & & & \\
 & & | & & & & \\
CH_3 & — & CH & — & CH & — & CH_3 \\
 & & & & | & & \\
 & & & & CH_3 & &
\end{array}
$$

$$
\begin{array}{ccccccc}
 & & CH_3 & & & & \\
 & & | & & & & \\
CH_3 & — & C & — & CH_2 & — & CH_3 \\
 & & | & & & & \\
 & & CH_3 & & & &
\end{array}
$$

Notice how some of the structural isomers possess a straight-chain carbon framework, while others possess a branched-chain framework.

We learn more about the shape of the carbon framework of organic compounds on p. 130.

Table 10.1 shows the number of structural isomers that are possible for alkanes possessing an increasing number of carbon atoms. Look at this table, and notice how the number of structural isomers does not increase by a regular amount as the number of carbon atoms increases, but increases in larger and larger 'jumps'.

Self-check 10.2 Write down the structural formulae of methane, ethane, and propane to confirm to yourself that only one structure is possible for each of these molecules.

Table 10.1 The number of structural isomers that are possible for alkanes possessing an increasing number of carbon atoms.

Number of possible atoms in alkane	Number of carbon structural isomers
1–3	1
4	2
5	3
6	5
7	9
8	18
9	35
10	75
15	4 347
20	366 319

Structural isomerism and the positioning of functional groups

We see in Chapters 5 and 6 how the variety of organic compounds is made possible through the addition of functional groups to the simple hydrocarbon framework upon which all organic compounds are built. Many structural isomers arise from the positioning of functional groups at different positions on the hydrocarbon framework so that, while the molecules are composed of the same atoms, the atoms are connected in different ways.

Let's consider the alkene pentene, C_5H_{10}. We see in Chapter 5 how alkenes are unsaturated hydrocarbons: they possess at least one carbon–carbon double bond.

Straight-chain pentene possesses one double bond. However, there are two different locations in the carbon framework of pentene in which the double bond could appear, as illustrated in Figure 10.3(a). Notice how both molecules have the molecular formula C_5H_{10}: both molecules are structural isomers of pentene. We call these two molecules pent-1-ene and pent-2-ene. The numbers '-1-' and '-2-' tell us where, along the five-carbon backbone of pentene, the double bond is located, as explained on p. 285.

We might think that there are *four* structural isomers of straight-chain pentene, as depicted in Figure 10.3(b). However, when we stop for a moment and consider these structures, we see how structures (1) and (4) are the same molecule, and structures (2) and (3) are again the same molecule, the only difference being that molecule (1) has been 'flipped' through 180° in the plane of the page to give (4), and likewise for (2) to give (3). To verify this similarity, imagine looking at structures (3) and (4) from the other side of this page. What would they look like? They would be identical to structures (2) and (1) respectively.

Let us consider another family of organic compounds, the alcohols. The alcohols are characterized by the presence of the **hydroxyl** functional group, –OH. The alcohol propanol has the molecular formula C_3H_8O, and can exist as the two structural

(a) $H_2C{=}CH{-}CH_2{-}CH_2{-}CH_3$ $H_3C{-}CH{=}CH{-}CH_2{-}CH_3$

Pent-1-ene   Pent-2-ene

(b) $H_2C{=}CH{-}CH_2{-}CH_2{-}CH_3$ (1) $H_3C{-}CH{=}CH{-}CH_2{-}CH_3$ (2)

$H_3C{-}CH_2{-}CH{=}CH{-}CH_3$ (3) $H_3C{-}CH_2{-}CH_2{-}CH{=}CH_2$ (4)

Figure 10.3 Structural isomers of pentene. (a) There are two locations in the carbon framework of pentene in which the double bond could appear. (b) It may appear that there are four structural isomers of pentene. However, structures (1) and (4), and (2) and (3) form identical pairs.

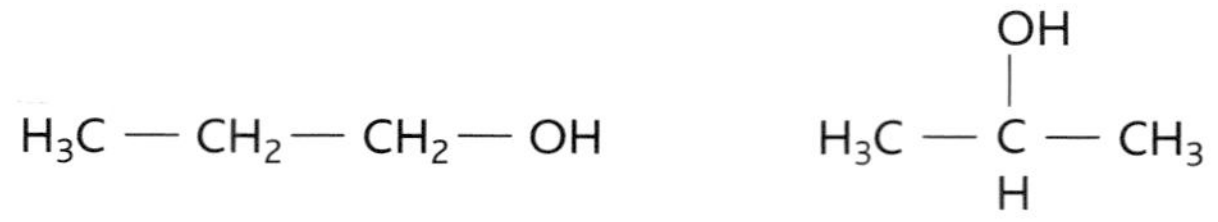

Figure 10.4 The two structural isomers of propanol.

$H_3C — CH_2 — CH_2 — OH$

$H_3C — CH(OH) — CH_3$ (OH above and H below the central C)

isomers shown in Figure 10.4. Look at these structures and notice how they vary solely in the positioning of the hydroxyl functional group.

Self-check 10.3 Draw structural formulae for the structural isomers of the alcohol butanol, $C_4H_{10}O$. How many structural isomers are there?

The above examples show how writing the names of compounds in a simple way (such as 'propanol' or 'pentene') does not tell us all we need to know in order to accurately predict the structures of these compounds. Structural isomerism means that such simple names are ambiguous: does 'propanol' have the hydroxyl group located at the end of the hydrocarbon chain or in the middle? Does 'pentene' have the C=C bond between the first and second carbons, or the second and third carbons in the hydrocarbon chain? Is it a straight chain at all? We can distinguish between structural isomers by using the correct rules for naming a compound–the correct **nomenclature**.

- *There are two possible sources of structural isomerism: the position of functional groups, and the structure of the carbon framework.*

Removing ambiguity: using nomenclature to distinguish between structural isomers

We are all familiar with the system of using street numbers to identify the relative locations of individual houses in a street. House number 1, in a street of 50 houses is likely to be found at one end of the street, while number 50 is likely to be found at the opposite end of the street. Number 25 is likely to be located roughly halfway along the street.

We use a system of numbering analogous to that of numbering the houses in a street to help us to identify where, along the carbon 'backbone' of an organic compound, functional groups are attached. We number sequentially the carbon atoms that form the carbon backbone:

C—C—C—C—C---
1 2 3 4 5

These numbers give us a point of reference to help identify the location of functional groups along the carbon backbone.

Look at Figure 10.5, which shows the structural isomers of the alcohol pentanol. Notice how there are several structural isomers of pentanol, which differ according to where, along the carbon backbone, the hydroxyl group is attached.

Remember: the prefix 'pent-' tells us that the carbon backbone of pentanol is *five* carbon atoms long.

We distinguish between the structural isomers by inserting a number between 'pentan-' and '-ol', to indicate the carbon atom to which the hydroxyl group is attached.

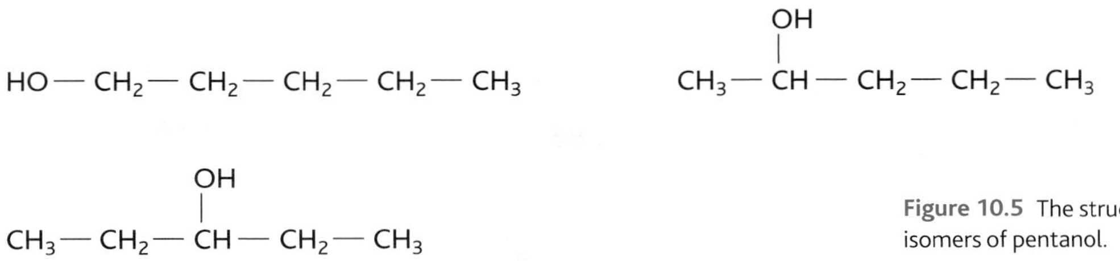

$$HO—CH_2—CH_2—CH_2—CH_2—CH_3$$

$$CH_3—CH(OH)—CH_2—CH_2—CH_3$$

$$CH_3—CH_2—CH(OH)—CH_2—CH_3$$

Figure 10.5 The structural isomers of pentanol.

Pentan-2-ol has the hydroxyl group attached to the *second* carbon atom along the five-carbon backbone of pentanol:

$$C—C(OH)—C—C—C$$
1 2 3 4 5

Pentan-2-ol
–OH is attached to the second carbon in the carbon chain.

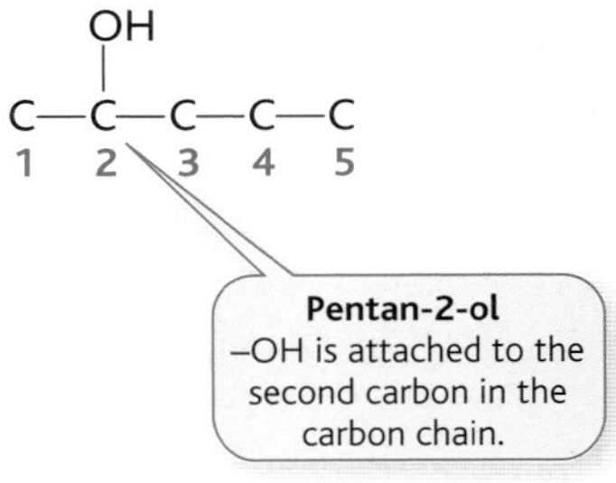

while pentan-3-ol has the hydroxyl group attached to the third carbon atom:

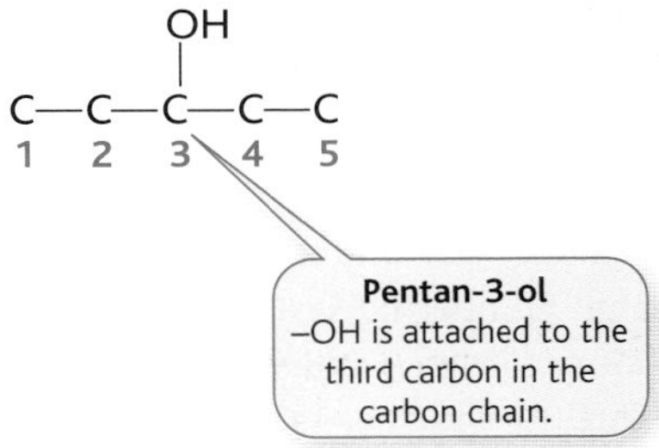

$$C—C—C(OH)—C—C$$
1 2 3 4 5

Pentan-3-ol
–OH is attached to the third carbon in the carbon chain.

We could have numbered the carbon backbone of pentanol from right to left, rather than left to right, giving us the name pentan-4-ol rather than pentan-2-ol:

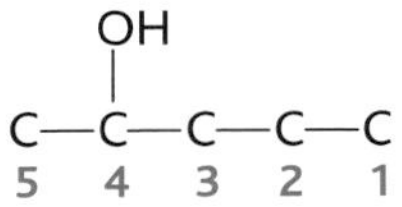

$$C—C(OH)—C—C—C$$
5 4 3 2 1

However, we always number the carbon chain so that the carbon atom to which the functional group is attached is assigned the lowest possible number. Therefore, structure (2)

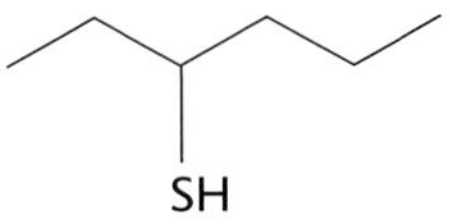

SH

(2)

is called hexane-3-thiol and not hexane-4-thiol.

We also use chemical nomenclature to help resolve any ambiguity regarding the positioning of double and triple bonds within a molecule. We see in Chapter 5 how the double and triple bonds of alkenes and alkynes link two carbon atoms. We identify the location of such bonds by numbering the carbon backbone and identifying the **lowest number** carbon atom that is joined by the double bond.

(3)

For example, the double bond in the isomer of pentene shown in the margin (**3**) joins the second and third carbons in the carbon backbone. We identify the location of this double bond by choosing the lowest number carbon atom. Hence we name this isomer of pentene as pent-2-ene, rather than pent-3-ene.

- ***We indicate the position of a functional group along the carbon backbone by assigning the lowest possible number to the carbon atom to which the functional group is attached.***

Self-check 10.4 What are the names of these three compounds?

(a) (b) (c)

Denoting multiple functional groups

What happens if there is more than one functional group attached to a carbon backbone? We use the same numbering system to specify the location along the backbone of *each* carbon atom to which a functional group is attached. Look at the alcohol depicted by structure (**4**), which has a four-carbon backbone. This molecule has *two* hydroxyl functional groups attached. We specify the locations of both of these functional groups along the four-carbon chain:

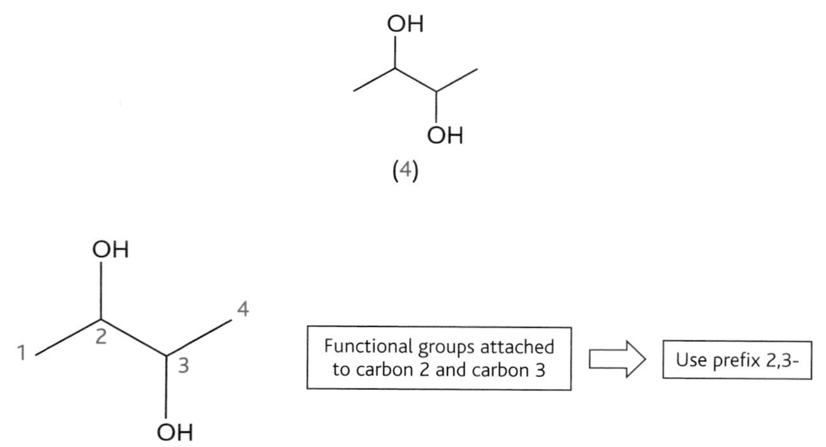

We also modify the name of the alcohol to indicate that there are *two* hydroxyl groups attached:

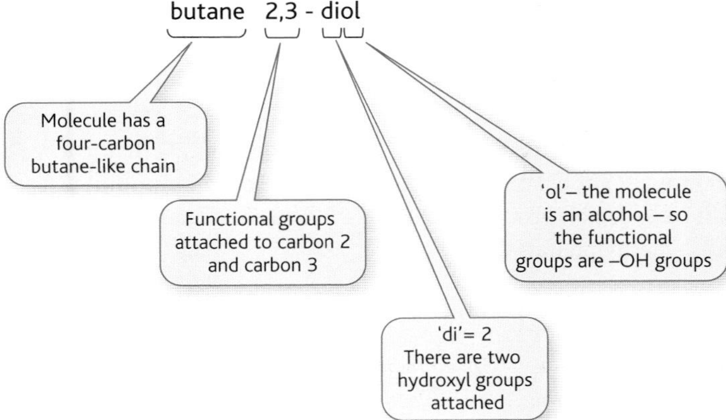

Let's consider one more example. This six-carbon alkene also has two functional groups attached, so we name the molecule as follows:

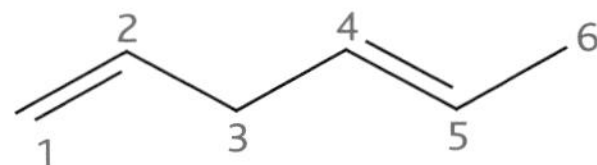

Two double bonds start at carbon 1 and carbon 4

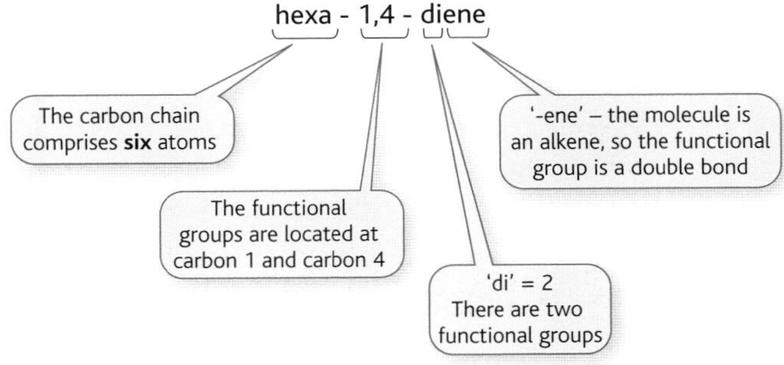

Notice how this molecule is named hexa-1,4-diene, and not hexa-2,5-diene, following the rule that the numbering must assign the lowest possible numbers to the carbon atoms to which the functional groups are attached.

Self-check 10.5 What are the names of these two compounds?

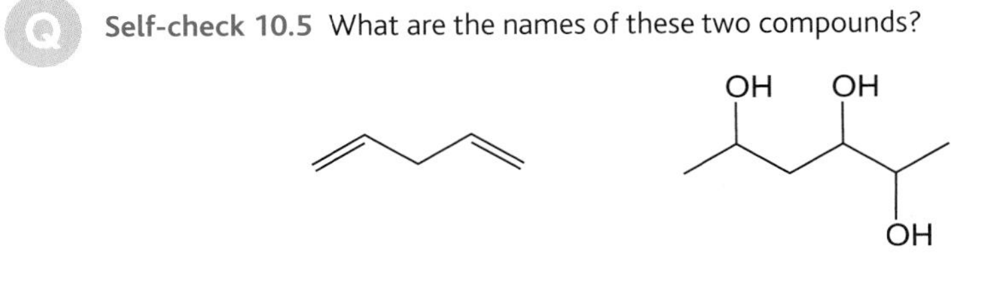

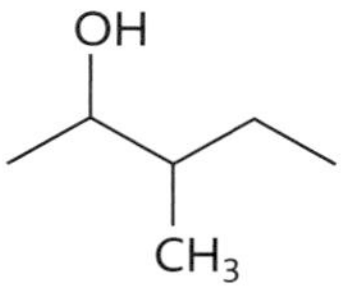

OH

CH_3

(5)

We use exactly the same rules when naming compounds that feature different functional groups. For example, structure (5) denotes a compound featuring both a methyl and a hydroxyl group attached to a five-carbon framework. We name this compound 3-methylpentan-2-ol:

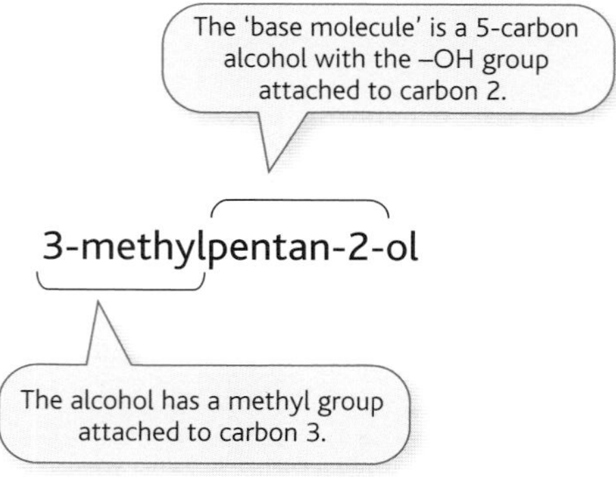
The 'base molecule' is a 5-carbon alcohol with the –OH group attached to carbon 2.
3-methylpentan-2-ol
The alcohol has a methyl group attached to carbon 3.

Self-check 10.6 What are the names of these two compounds?

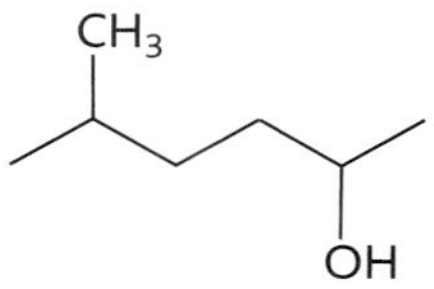

CH_3 OH

(a)

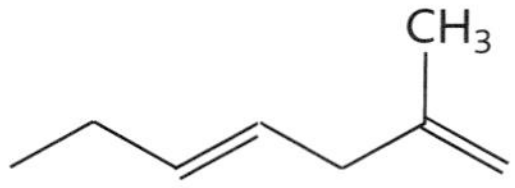

CH_3

(b)

Structural isomerism: unifying chemical families

In the sections above we have seen examples of structural isomers that belong to the same family of compounds (for example, structural isomers that are all alkanes, or all alcohols). Structural isomers do not always belong to the same family of compounds, however, but may span different chemical families.

We see in Chapter 6 how several families of organic compound possess very similar functional groups. For example, both alcohols and ethers are characterized by the presence of an oxygen atom. In an alcohol the oxygen atom forms part of a hydroxyl group:

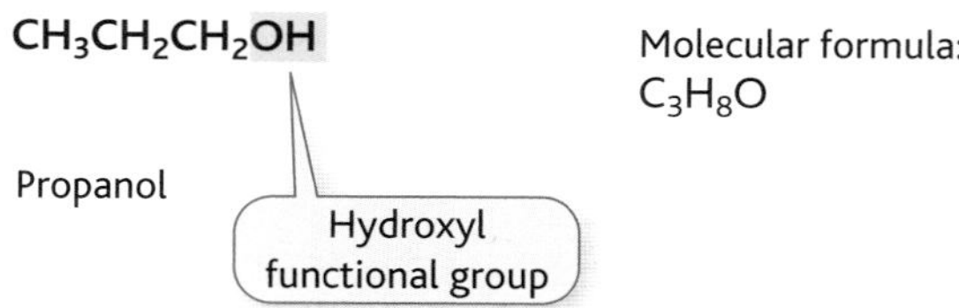
$CH_3CH_2CH_2OH$
Propanol
Hydroxyl functional group
Molecular formula: C_3H_8O

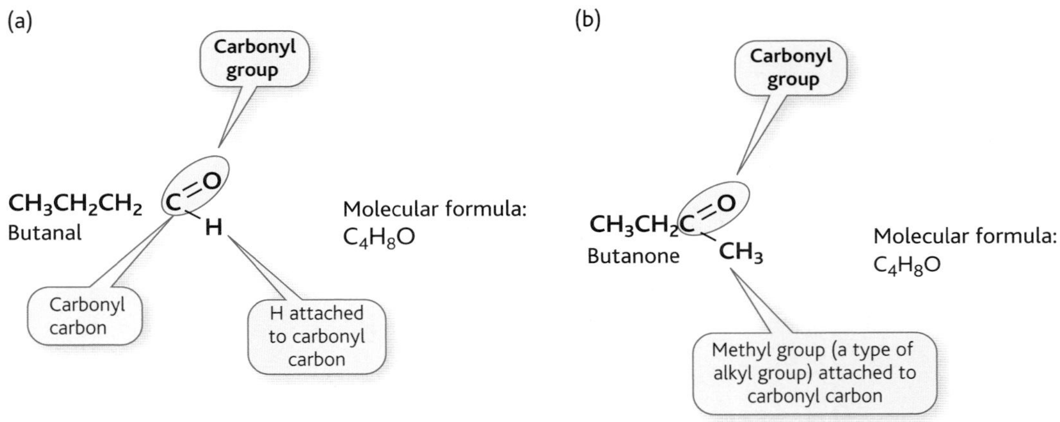

Figure 10.6 Aldehydes and ketones can exist as structural isomers. Both butanal (a) and butanone (b) have the same molecular formula. They are therefore structural isomers.

In an ether, the oxygen atom acts as a linkage between two alkyl groups:

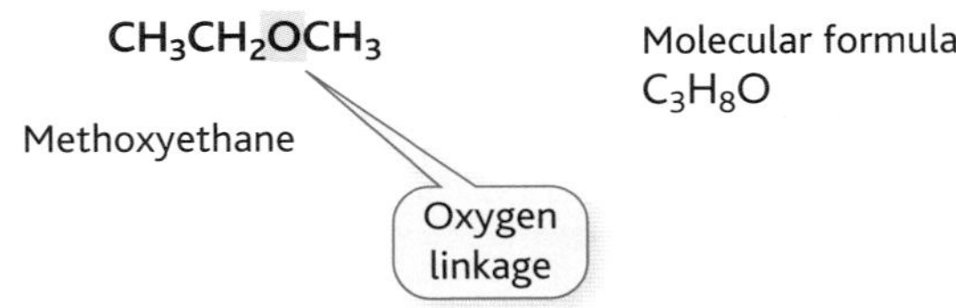

Remember: two molecules are structural isomers if they share the same molecular formula. The two molecules need *not* belong to the same family of compounds.

Now look at the molecular formulae of propanol and methoxyethane. Although propanol is an alcohol and methoxyethane is an ether, both share the *same* molecular formula, C_3H_8O. They are therefore structural isomers.

Similarly, aldehydes and ketones possess similar functional groups. Both types of compound possess a carbonyl group. However, whereas an aldehyde also has a hydrogen atom attached to the carbonyl carbon a ketone has an alkyl group attached, as depicted in Figure 10.6.

We discuss aldehydes and ketones in more detail in Chapter 6.

Look at the molecular formulae of butanal and butanone. Notice that, once again, the molecular formulae are the same. Therefore butanal and butanone are structural isomers.

- ***Structural isomers may belong to different families of organic compound.***

Self-check 10.7 (a) The following four compounds comprise two pairs of structural isomer. Identify the two pairs.

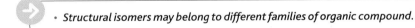

O OH O

O

(a) (b) (c) (d)

(b) Draw the structural formula for the aldehyde of which pentan-3-one is a structural isomer:

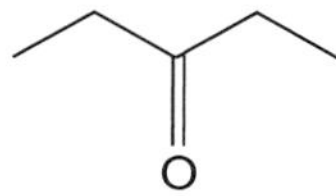

Having now explored structural isomers, and learned how to use chemical nomenclature to distinguish between them, let us now consider the second class of isomer, the stereoisomers.

10.3 Stereoisomers

Stereoisomers possess the same atoms, which exhibit the same connectivities (they are joined to one another in exactly the same way), but the composite atoms are oriented differently in space—we say they have different **configurations**.

- ***Stereoisomers possess the same atoms, which are connected to one another in the same way. It is the configurations of the molecules (how their atoms are arranged in three-dimensional space) that differ.***

There are two classes of stereoisomer, which we explore during the rest of the chapter:

- Geometric (*cis-trans*) isomers
- Enantiomers

Geometric isomers

Geometric isomers are molecules that share the same atoms, which are joined to one another in the same way. However, geometric isomers have different **configurations**.

Look at the two molecules of but-2-ene shown in Figure 10.7. Notice how both molecules possess the same atoms, connected to one another in exactly the same way.

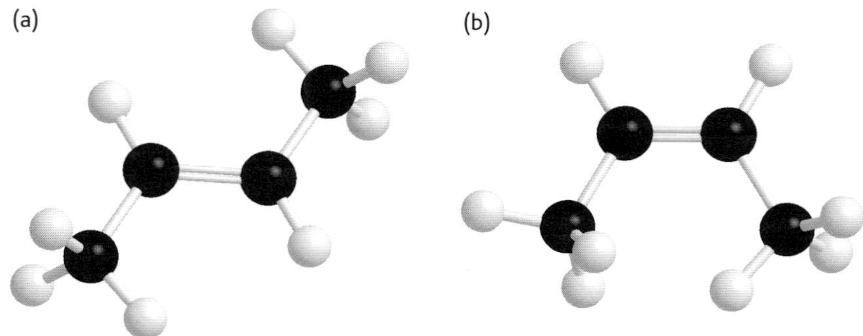

Figure 10.7 Two geometric isomers of but-2-ene. Notice how, in (a), one methyl group points 'up' and the other points 'down', while, in (b), both methyl groups point 'down'.

However, the two molecules have different overall configurations–the atoms occupy different locations in space. Notice how, in (a), one methyl group points 'up' and the other points 'down', while, in (b), both methyl groups point 'down'. Neither molecule is able to switch to a different conformation: each is locked into the configuration we see here.

The locking of atoms into fixed configurations is a feature exhibited by molecules that have one of two characteristics:

1. They possess at least one double bond.
2. They are cyclic: they possess a ring-shaped carbon framework.

Look at Section 8.5, p. 240 for further details about the difference between **configuration** and **conformation**.

In both instances, the 'locking' of groups of atoms so that different conformations cannot be adopted is a consequence of the way that no bond rotation is possible about double bonds, or about bonds that comprise a ring-shaped structure.

Let's first consider geometric isomers that possess a double bond.

The double bond and cis/trans isomerism

Look at Figure 10.8, which demonstrates the way bond rotation about a single bond allows the conformation of atoms joined by the bond to change (so that atoms may occupy different positions relative to one another). By contrast, no bond rotation is possible about a double bond. Therefore, the atoms joined by the double bond cannot adopt different conformations. (Compare a futon to a regular bed. While the different components of a futon are able to move relative to one another, so that the futon can adopt different conformations, a regular bed frame has a fixed structure: it is not possible to move different parts of the bed frame into new positions (conformations).)

The most important class of compound containing the double bond are the alkenes, the C=C-containing hydrocarbons (see Section 5.3).

Remember: bond rotation is not possible about a double *or* a triple bond. We explore bond rotation in more detail in Section 8.5 (p. 238).

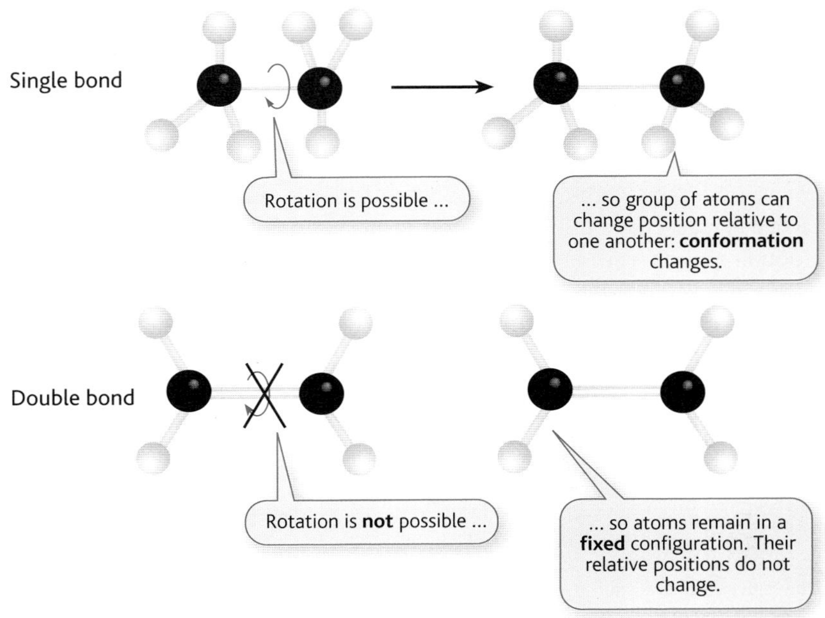
Single bond
Rotation is possible ...
... so group of atoms can change position relative to one another: **conformation** changes.
Double bond
Rotation is **not** possible ...
... so atoms remain in a **fixed** configuration. Their relative positions do not change.

Figure 10.8 Bond rotation about a single bond allows the conformation of atoms joined by the bond to change (so that atoms may occupy different positions relative to one another). Bond rotation is *not* possible about a double bond, however.

As we see above, the alkene but-2-ene can exist as two geometric isomers[3].

The methyl groups and hydrogen atoms can be attached to the carbon–carbon double bond in but-2-ene in two distinct ways:

1. In a *cis* conformation, with the methyl groups on the *same* side of the C=C bond:

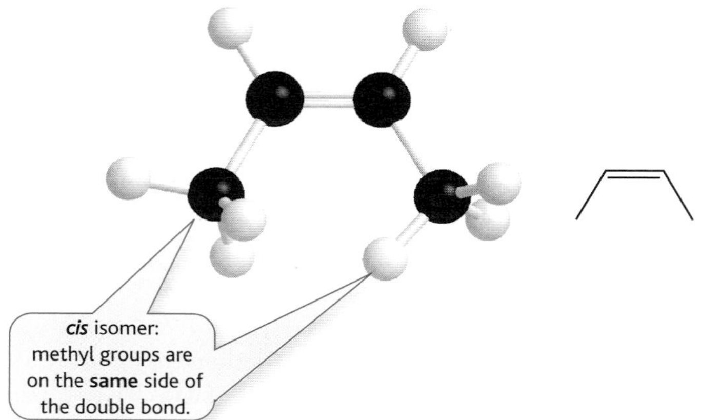
cis isomer:
methyl groups are
on the **same** side of
the double bond.

2. In a *trans* conformation, with the methyl groups on *opposite* sides of the C=C bond:

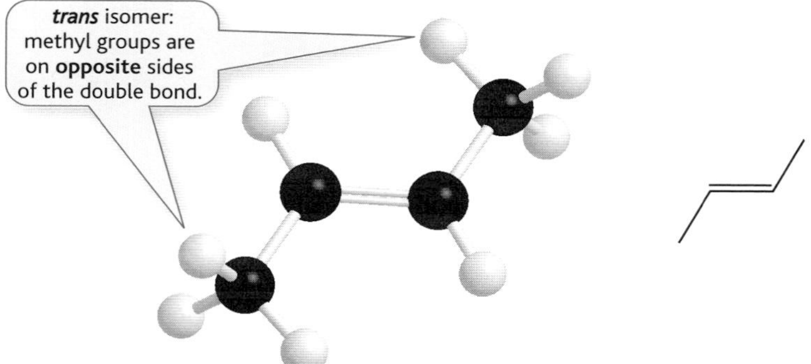
trans isomer:
methyl groups are
on **opposite** sides
of the double bond.

Both molecules contain exactly the same atoms joined together in exactly the same way (a carbon–carbon double bond, with a hydrogen atom and a methyl group attached at each end), but with the atoms *positioned differently* in space. (Remember: the double carbon–carbon bond does not allow rotation in the same way that a single carbon–carbon bond does. A specific molecule of but-2-ene cannot switch the position of its methyl groups: they are locked by the C=C bond into one of the two configurations shown here.)

So, in the case of geometric isomers with a double bond, the labels *cis* and *trans* describe the conformation of atoms connected to the double bond. In a ***cis*** isomer the groups lie on the *same* side of the double bond. In a ***trans*** isomer the groups lie on *opposite* sides of the double bond.

3 The '2' between 'but-' and '-ene' in but-2-ene indicates the carbon–carbon double bond is located between the second and third carbon atoms in the carbon backbone.

- *In a* cis *isomer the groups lie on the* same *side of the double bond.*
- *In a* trans *isomer the groups lie on* opposite *sides of the double bond.*

Self-check 10.8 How many pairs of geometric isomers are there within this group of compounds? Identify the compounds forming these pairs.

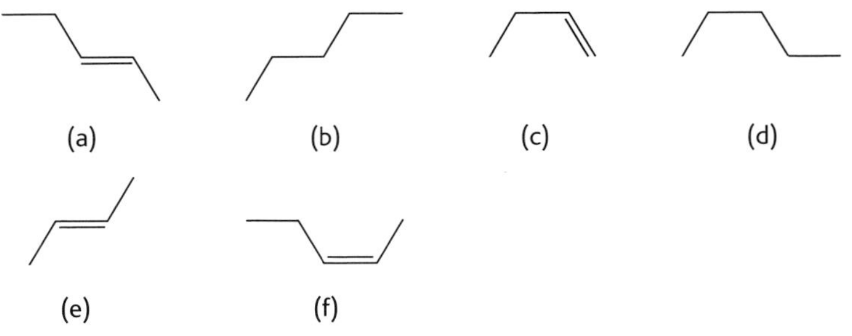

Self-check 10.9 Here are two geometric isomers of hex-3-ene:

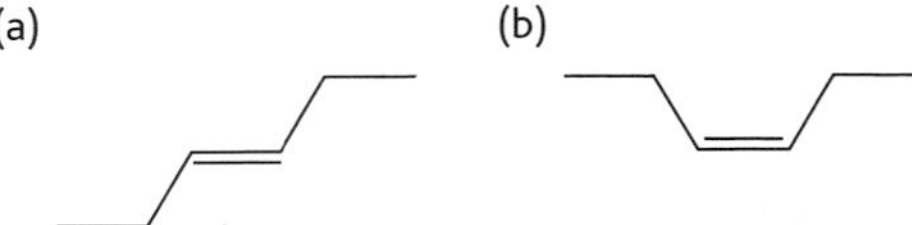

Which is the *cis* isomer, and which is the *trans* isomer?

The labels *cis* and *trans* are useful until we have to distinguish between isomers in which four different groups are attached across a carbon–carbon double bond. In such cases, we use the E/Z nomenclature, as explained in Box 10.2.

BOX 10.2 Zusammen oder entegen? The *E*/*Z* nomenclature

Look at the two compounds in Figure 10.9. Notice how these compounds are geometric isomers, but with four different groups attached across the carbon–carbon double bond. In such instances, we cannot use the labels *cis* and *trans* to distinguish between the two isomers. Instead, we use the labels *E* and *Z*.

The *E*/*Z* nomenclature relies on giving the different groups that are attached to the double bond a priority ranking, accord-

(a) HO(Cl)C=C(CH$_2$CH$_3$)(CH$_3$)

(b) HO(Cl)C=C(CH$_3$)(CH$_2$CH$_3$)

Figure 10.9 Two geometric isomers with four different groups joined across the double bond. We cannot use the labels *cis* and *trans* to distinguish between such isomers.

BOX 10.2 Continued

ing to their atomic number. This ranking follows what is called the **Cahn–Ingold–Prelog convention.** So how do we perform this ranking? First, let us label the structure in Figure 10.9(a) to give us some points of reference:

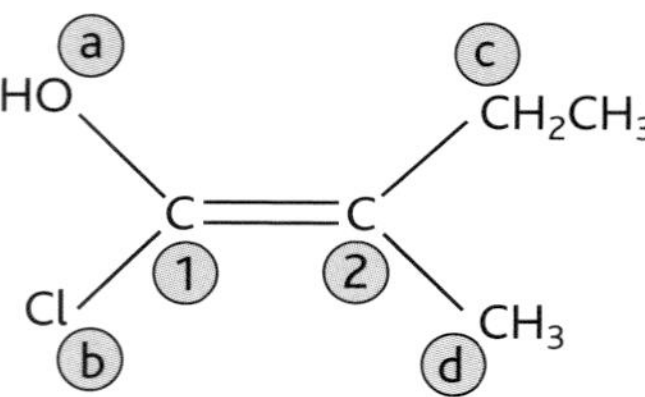

Let's start with carbon (1). This has two atoms directly attached: O (as part of group a) and Cl (group b), with atomic numbers 8 and 17 respectively. Of these, Cl has the highest atomic number, and so group (b) is given the highest priority out of the two attached groups.

Now let's move to carbon (2). The two atoms directly attached to carbon (2) are both C, and so have the same atomic number, and so equal priority. Consequently, we must then consider the atoms to which these two carbons are joined, and identify which of these atoms has the highest atomic number.

The first carbon in group (c) is attached to a C and two Hs; the atom with highest atomic number here is the C, with a value of 12. By contrast, the carbon in (d) is attached to three Hs, all of whom have an atomic number of 1. Therefore, by considering the atoms two along from carbon (2) we find that group (c) has a higher priority than group (d): the highest atomic number in the second position for group (c) is 12, versus an atomic number of just 1 for group (d).

So, we have determined that the two highest-priority groups are groups (b) and (c). If we look again at the whole molecule, we see that the two highest-priority groups lie on *opposite* sides of the double bond. We therefore call this the *E* isomer (E comes from the German *entgegen*, meaning opposite).

The other isomer, with the two highest-priority groups on the *same* side of the double bond (Figure 10.9(b)) is denoted the *Z* isomer (from the German *zusammen*, meaning together).

So, to apply the *E*/*Z* nomenclature we:

1. Prioritize the groups attached at each end of the double bond according to atomic number using the Cahn-Ingold-Prelog convention
   a. If the atoms attached directly to the double bond carbons are the same, we consider the atoms attached to *these* atoms
2. If the highest-priority groups at each end of the double bond lie on the *same* side of the bond, it is the *Z*-isomer
3. If the highest-priority groups at each end of the double bond lie on opposite sides of the bond, it is the *E*-isomer.

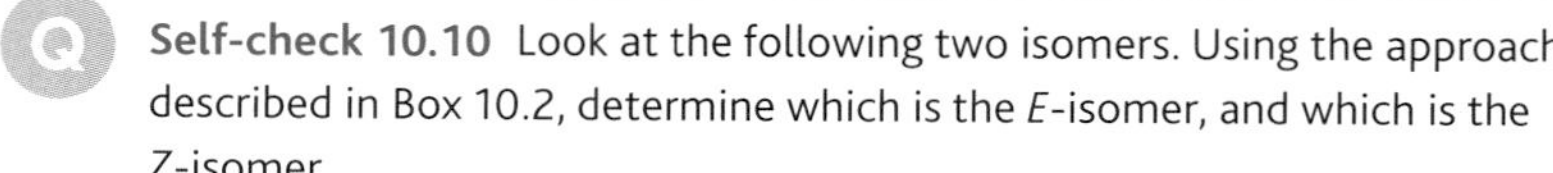

Self-check 10.10 Look at the following two isomers. Using the approach described in Box 10.2, determine which is the *E*-isomer, and which is the *Z*-isomer.

(a) HO, H_3CO — C=C — H, CH_3

(b) HO, H_3CO — C=C — CH_3, H

The example of but-2-ene is one in which there are identical atoms joined by the C=C (both carbon atoms, attached by the double bond, have a hydrogen atom and a methyl group attached). However, geometric isomers do not always possess the same groups at *both* ends of the carbon–carbon bond.

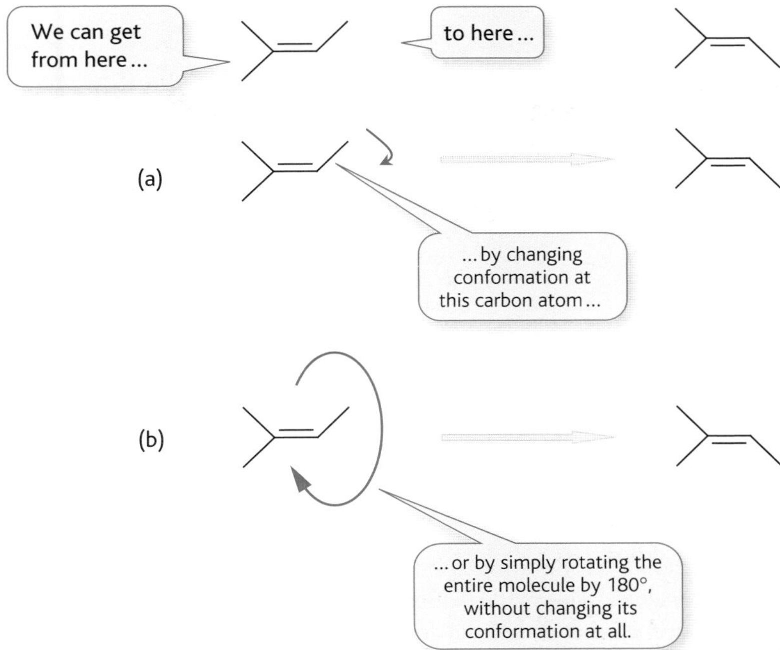

Figure 10.10 2-methylbut-2-ene does *not* exist as a pair of geometric isomers.

For example, pent-2-ene also exists as two geometric isomers:

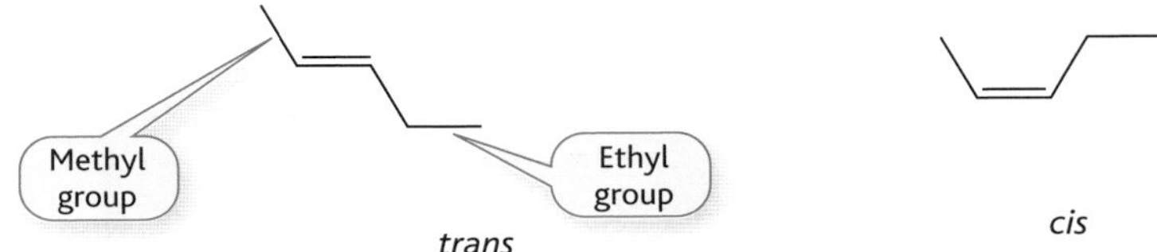

Notice how there is a methyl group attached to one end of the double bond, and an ethyl group attached to the other. However, pent-2-ene still exists in both *cis* and *trans* forms, with the methyl and ethyl groups on the same side and on opposite sides of the C=C bond respectively[4].

Not all alkenes exist as geometric isomers, however. Geometric isomerism occurs only when the carbon atom at each end of the double bond has *different* groups attached to it. Look at the structure of 2-methylbut-2-ene (6).

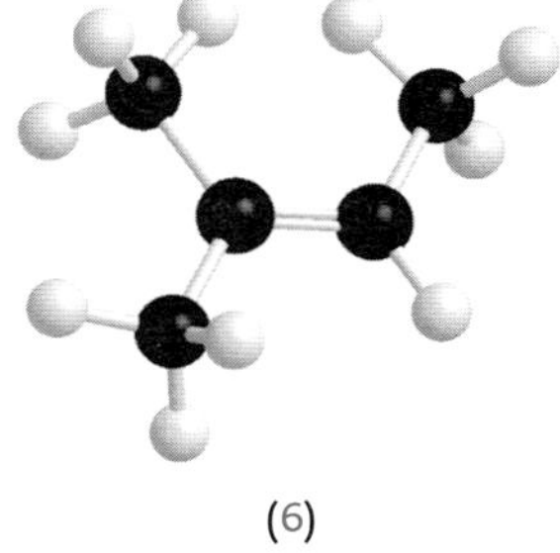

(6)

4 It is only when we have three or four different groups attached to the carbon–carbon double bond that the use of *cis* and *trans* becomes difficult, and we use the *E*/*Z* notation instead.

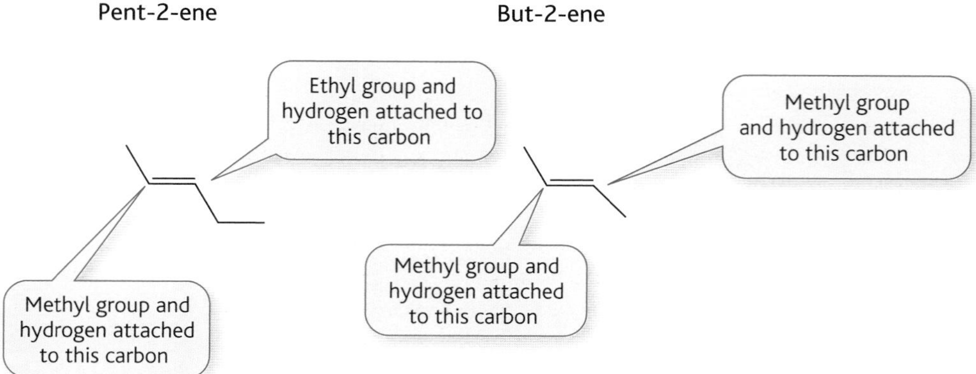

Figure 10.11 Both but-2-ene and pent-2-ene can exist as pairs of geometric isomers because they have different groups attached at each end of the double bond.

Notice how a carbon atom at one end of the double bond has two identical methyl groups attached. Figure 10.10 shows how it doesn't matter if the methyl group attached at the other end of the double bond changes conformation, the resulting molecule remains structurally unchanged overall. Therefore 2-methylbut-2-ene does *not* exist as a pair of geometric isomers.

By contrast, Figure 10.11 shows how the carbon atom at each end of the double bond in both but-2-ene and pent-2-ene has different groups attached to it. And, as we see above, both but-2-ene and pent-2-ene can exist as pairs of geometric isomers.

Self-check 10.11 Which of the following compounds can exist as a pair of geometric isomers? It might help you to draw out the full structural formulae first.

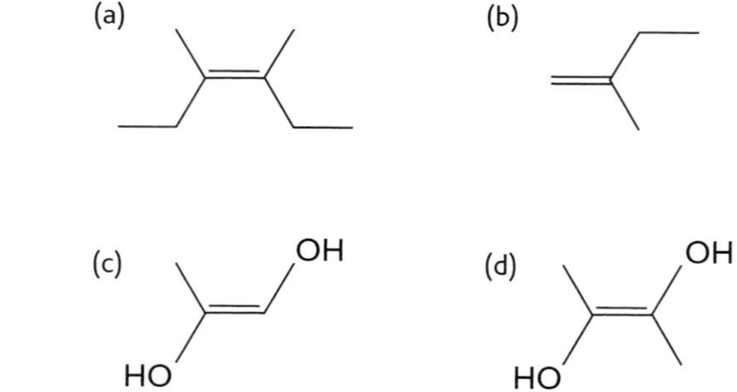

Read Box 10.3 to see how geometric isomerism lies at the heart of visual perception: that is, our ability to see.

Cyclic structures and geometric isomerism

Cis-trans isomerism is not solely a feature of molecules possessing double bonds. Bonds comprising the ring-shaped framework of cyclic structures are also unable to rotate freely, leading to the existence of further geometric isomers.

BOX 10.3 Seeing clearly: the geometric isomers of rhodopsin

We see in Box 2.4 how visual perception (the detection of light by our eyes, and perception of this light by our brain) is mediated by the pigment rhodopsin. Rhodopsin comprises the protein opsin, to which is attached the molecule 11-*cis*-retinal.

When rhodopsin absorbs light, the energy absorbed triggers a change in the 11-*cis*-retinal molecule: its structure changes to that of its geometric isomer, 11-*trans*-retinal, in a process known as **photoisomerism**. This process is illustrated in Figure 10.12.

('Photo' = light; photoisomerism is light-driven isomerism–the change from one isomer to another.)

This geometric isomerism is driven by a change in configuration at a single double bond–from the *cis* configuration to the *trans* configuration, as indicated in Figure 10.12.

It is this change in configuration that triggers dissociation (release) of the opsin protein from the rhodopsin complex. The free opsin protein then triggers a cascade of biochemical and electrochemical reactions, which result in the brain 'perceiving' the light.

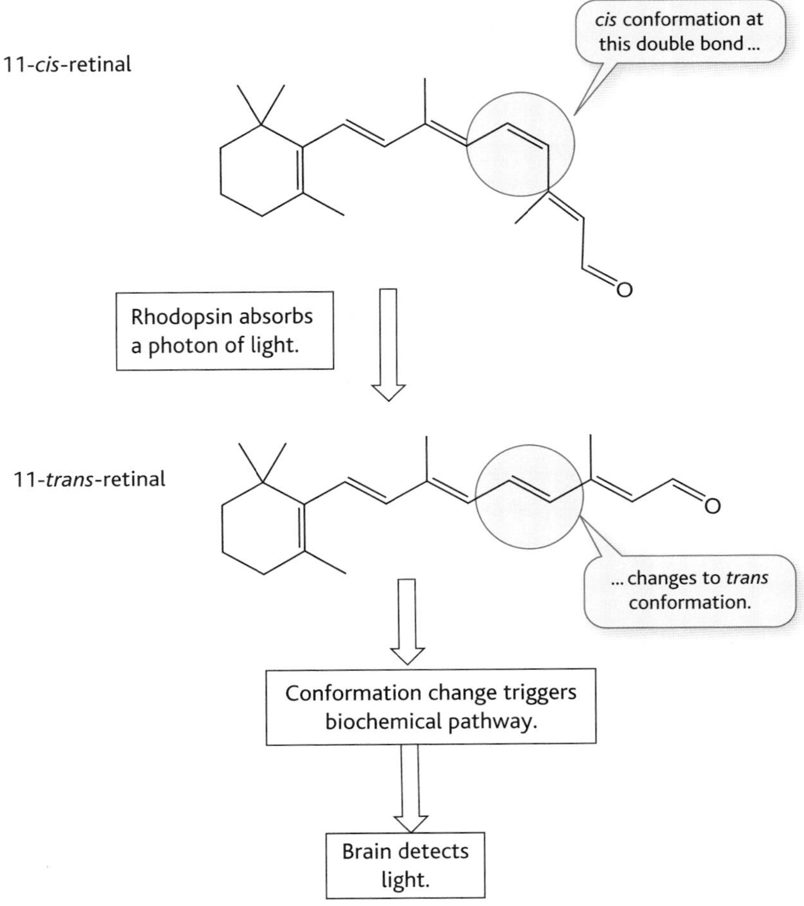

Figure 10.12 When the pigment rhodopsin absorbs light, the structure of 11-*cis*-retinal changes to that of its geometric isomer, 11-*trans*-retinal. Notice how this change brings about a marked difference in the overall shape of the molecule.

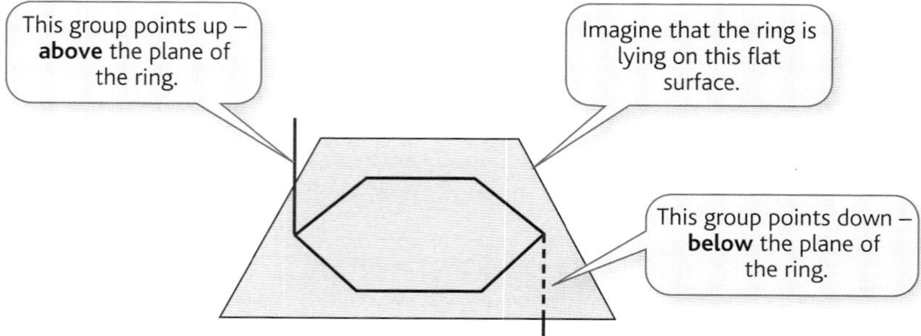

Figure 10.13 Groups attached to a cyclic molecule may lie either *above* or *below* the plane of the ring.

Figure 10.13 shows how groups attached to a cyclic molecule may lie either above or below the plane of the ring. (Imagine that the ring is a dinner plate lying on a flat surface: the group attached above the plane of the ring would be positioned above the flat surface; the group attached below the plane of the ring would be positioned below the flat surface.)

When we consider *cis-trans* isomerism across a double bond, the *cis* isomer possesses groups that are attached to the same side of the double bond while the *trans* isomer possesses groups that are attached to opposite sides of the double bond. We distinguish between cyclic *cis-trans* isomers in a similar way.

Look at Figure 10.14, which shows how the *cis* isomer possesses groups that are oriented in the *same way* in relation to the plane of the ring, whereas the *trans* isomer has groups that are oriented in *opposite* ways in relation to the plane of the ring.

For example, the cyclic alcohol cyclohexane 1,4-diol exists as the following *cis-trans* isomers[5]:

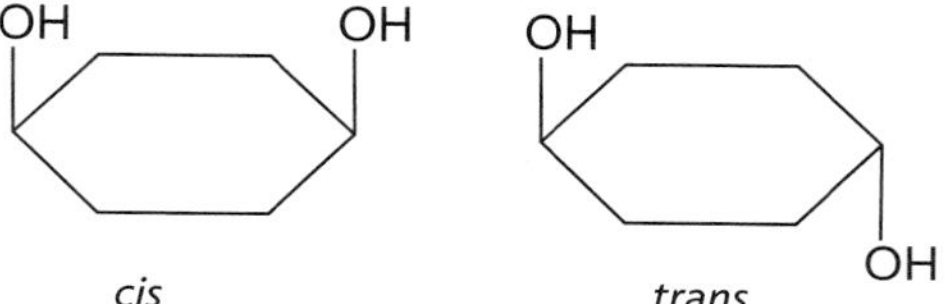

Enantiomers

Enantiomers are pairs of molecules that also fit the definition of a stereoisomer: molecules whose atoms are the same, are joined together in exactly the same way, but which occupy different positions in space.

For example, Figure 10.15 shows how the carboxylic acid 2-methylbutanoic acid exists as a pair of enantiomers[6]. Look carefully at these two molecules. How do they differ?

5 Notice the nomenclature used here: there are two hydroxyl groups attached to carbons 1 and 4 in the six-member ring, hence '1,4-diol'.

6 Notice how this figure makes use of wedged and dashed bonds to help represent the three-dimensional conformation of a molecule. A wedged bond is used to denote a bond that is pointing forward from the plane of the page; a dashed bond is used to denote a bond that is pointing backwards, behind the plane of the page. Regular bonds (that is, straight lines) are used to denote bonds that are lying flat in the plane of the page.

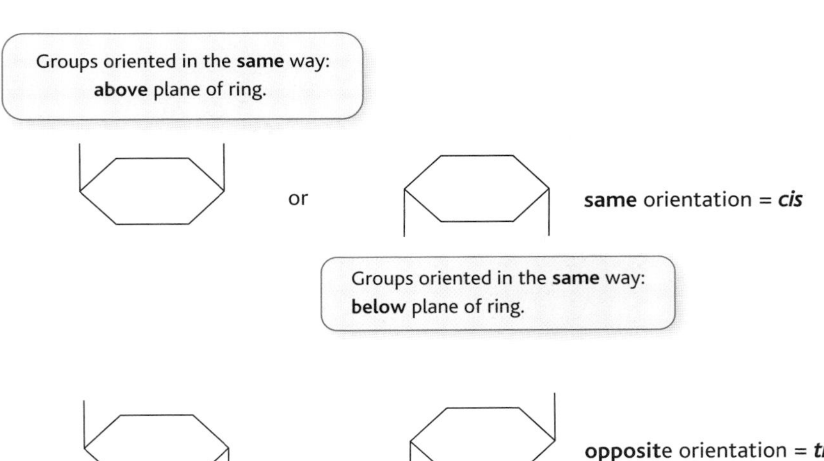
Groups oriented in the **same** way: **above** plane of ring.
or
same orientation = ***cis***
Groups oriented in the **same** way: **below** plane of ring.
opposite orientation = ***trans***
Groups oriented in **opposite** ways: one group above the plane of the ring, one group below.

Figure 10.14 When we consider isomers of cyclic structures, the *cis* isomer possesses groups that are oriented in the *same way* in relation to the plane of the ring, whereas the *trans* isomer has groups that are oriented in *opposite* ways in relation to the plane of the ring.

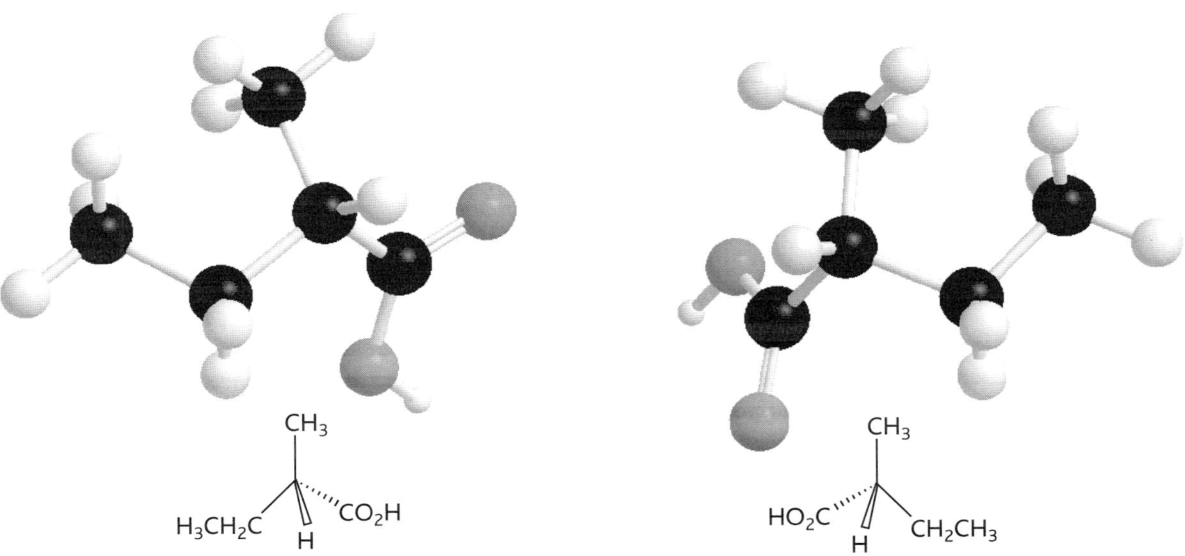
CH_3
H_3CH_2C
H
CO_2H
CH_3
HO_2C
H
CH_2CH_3

Figure 10.15 The two enantiomers of 2-methylbutanoic acid.

While geometric isomers differ because they feature groups of atoms that are locked into contrasting configurations, everything about enantiomers is the same, other than their atoms being arranged in such a way as to make them **non-superimposable mirror images** of one another.

- ***Enantiomers are pairs of molecules that are non-superimposable mirror images of one another.***

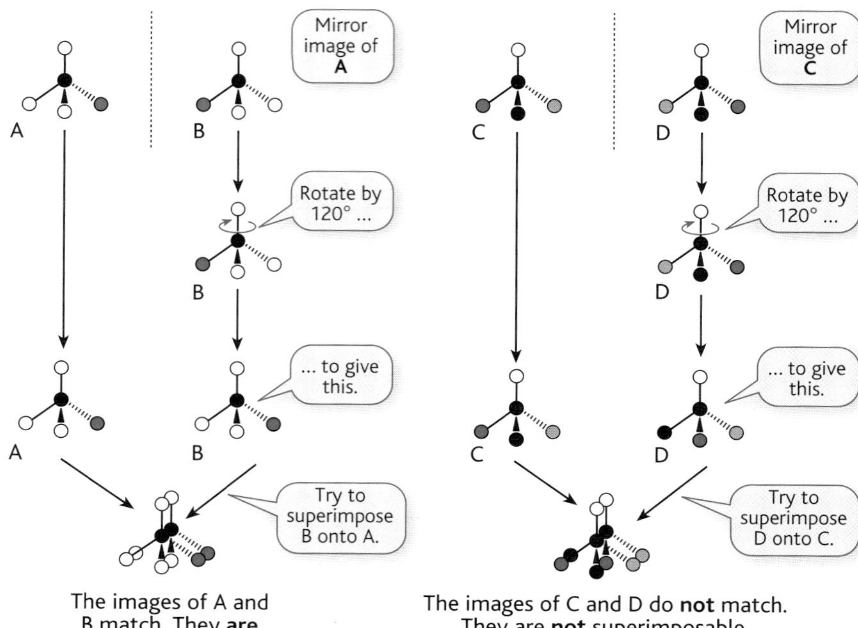

Figure 10.16 Two molecules and their mirror images. Molecule A and its mirror image (image B) are superimposable. However, molecule C and its mirror image (D) are *not* superimposable. Molecules C and D are therefore a pair of enantiomers.

What do we mean by a 'non-superimposable' mirror image? We mean that, if we take an object and its mirror image, and try to overlay one image on top of the other (in other words, to **superimpose** the two images), the two images do not match with one another exactly.

We are surrounded every day by pairs of enantiomers. Our hands and feet are both pairs of enantiomers. Put your hands out in front of you, palms facing downwards. Notice how they are mirror images of one another. (To convince yourself of this, simply place your left hand against a mirror; the reflection looks like your right hand.) Now try superimposing one hand directly on top of the other. Notice how they don't match up–they are not superimposable. (Try doing the same with your feet; you'll notice that they are non-superimposable too.)

Now look at Figure 10.16, which depicts two molecules and their mirror images (A and B, and C and D). Notice how molecule A and its mirror image (image B) are superimposable but molecule C and its mirror image (D) are *not* superimposable. Molecules C and D are therefore a pair of enantiomers; molecules A and B are not.

Structures that cannot be superimposed upon their mirror image are what we call **chiral**, and enantiomers are examples of **chiral** molecules. Chirality is an important feature of biological systems, so we explore chirality further in the next section.

10.4 Chirality

Our lives are filled with chiral objects. Suppose we put on a pair of shoes and get into our car. The shoes and the car are all chiral objects: they cannot be superimposed on their mirror image.

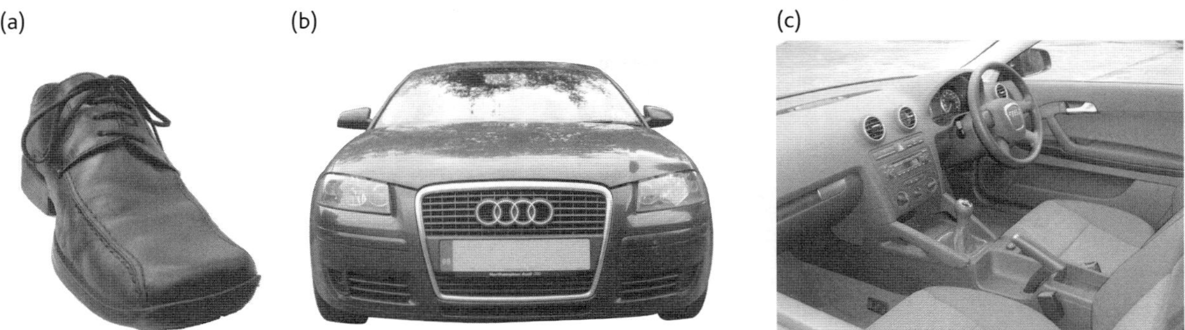

Figure 10.17 Both a shoe and a car are asymmetric objects. A car may seem to be symmetrical from the outside, but if we look inside we notice that the interior is not arranged symmetrically: the car is therefore asymmetric.

A chiral object has a particular feature: it lacks a plane of symmetry (it is **asymmetric**[7]).

Think about both a shoe and a car: Figure 10.17 shows how both are asymmetric.

A **chiral molecule** is one that has four *different* atoms or groups attached to a central atom. Only molecules that have four different groups attached to a central atom lack a plane of symmetry. We call a central atom that has four different atoms or groups attached a **chiral centre**.

Self-check 10.12 Identify the chiral centre in each of the following enantiomers:

O (a) | NH_2 (b) | OH, OH (c) | OH (d)

Don't forget that carbon has a valency of 4, so some atoms are also attached to a hydrogen.

Let's look again at the structural formula of 2-methylbutanoic acid (7).

CH_3 | HO_2C–C–CH_2CH_3 | H

(7)

Notice how the carbon atom at the centre of this molecule has four *different* groups attached to it. Therefore, 2-methylbutanoic acid is a chiral molecule. Notice also that

7 If an object is **asymmetric** it means that it has **no symmetry**.

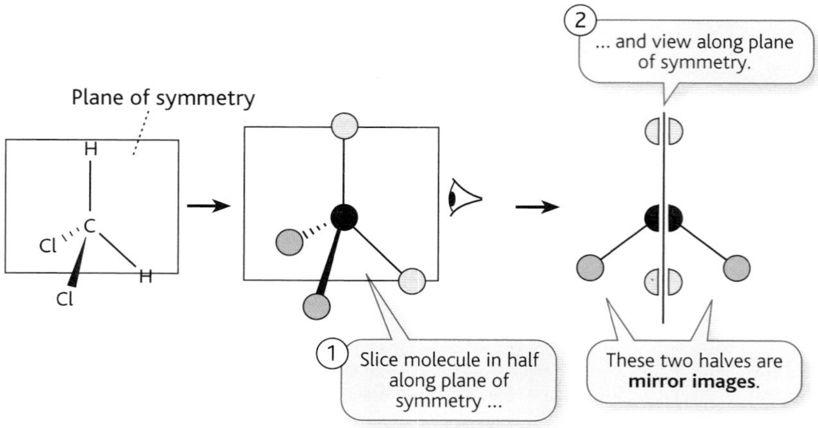

Figure 10.18 If we split dichloromethane along one of its planes of symmetry, the two resulting half-molecules are mirror images. Consequently, dichloromethane is *not* chiral, and does *not* exist as a pair of enantiomers.

2-methylbutanoic acid is **asymmetric** (it has no symmetry: there is no way of splitting the molecule in half such that the two halves are identical).

By contrast, dichloromethane (8) has only two different groups attached to the central carbon atom.

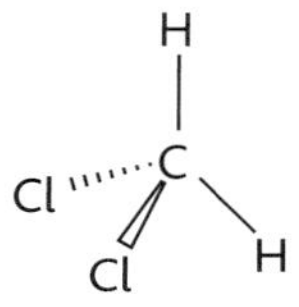

(8)

It also has several planes of symmetry. Look at Figure 10.18. Notice how, if we split the molecule along one of its planes of symmetry, the two resulting half-molecules are mirror images. Consequently, dichloromethane is *not* chiral, and does *not* exist as a pair of enantiomers.

Let's return to the molecules from Figure 10.16 to show how chiral molecules have no plane of symmetry. Look at Figure 10.19, which compares molecule A (not chiral; not an enantiomer) and molecule C (chiral enantiomer). Notice that, while molecule A is symmetrical (it can be divided in half along the plane of the page to generate two half molecules, which are identical), the enantiomer, molecule C, is asymmetric. (It cannot be divided in half along the plane of the page, or along any other plane, to generate two identical half molecules.)

Let's summarize the important features of enantiomers:

- Enantiomers exist as pairs of molecules that are non-superimposable mirror images.
- Non-superimposable mirror images are chiral structures.
- A chiral structure lacks a plane of symmetry.
- For a molecule to lack a plane of symmetry it must have four different groups attached to a central atom.

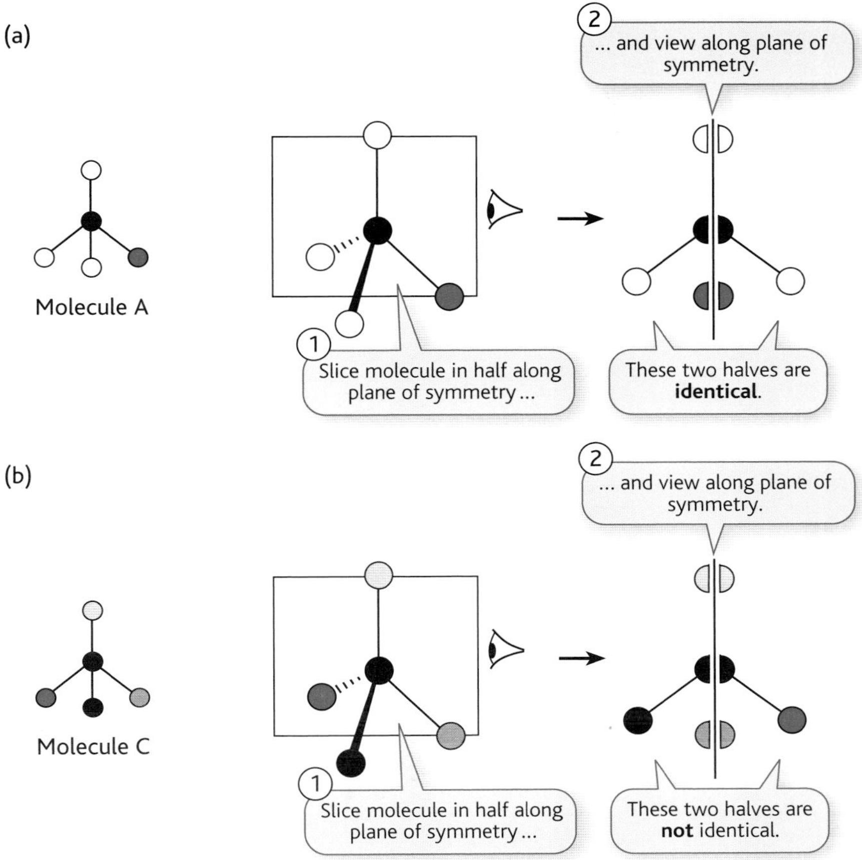
(a)
Molecule A
1
Slice molecule in half along plane of symmetry ...
2
... and view along plane of symmetry.
These two halves are **identical**.
(b)
Molecule C
1
Slice molecule in half along plane of symmetry ...
2
... and view along plane of symmetry.
These two halves are **not** identical.

Figure 10.19 (a) Molecule A is symmetrical: it can be divided in half along the plane of the page to generate two half molecules, which are identical. (b) However, the enantiomer, molecule C, is asymmetric. It cannot be divided in half along the plane of the page, or along any other plane to generate two identical half molecules.

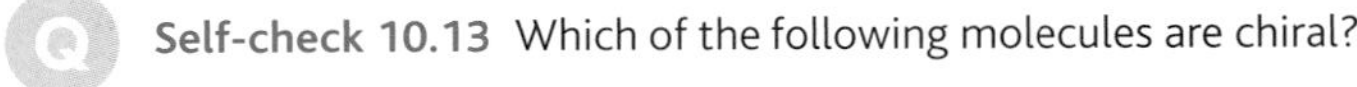

Self-check 10.13 Which of the following molecules are chiral?

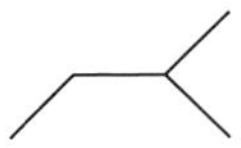

HO OH

H_2N

(a) (b) (c)

OH

HO_2C NH_2

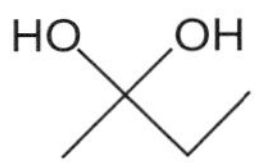

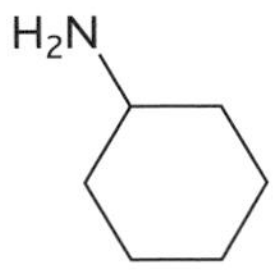

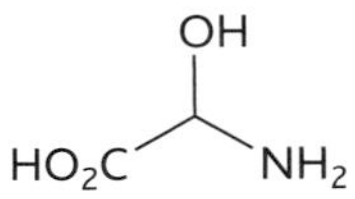

(d) (e)

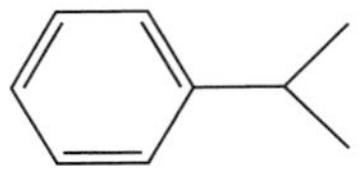

BOX 10.4 Enantiomers and chiral centres

Many biological molecules possess more than one chiral centre. Consider the structural formula of glucose:

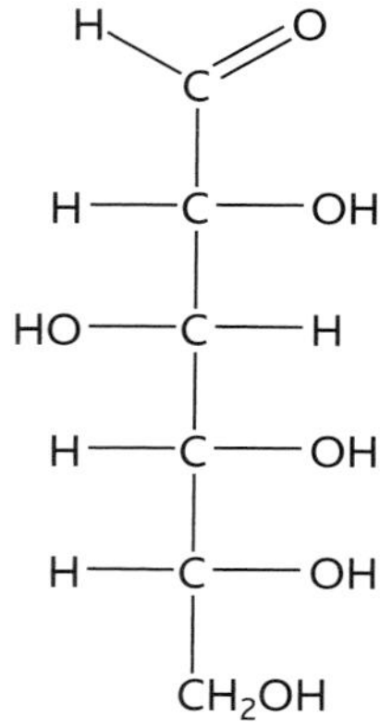

Notice how glucose possesses six carbon atoms, four of which are attached to four different groups of atoms. Each of these carbon atoms is a chiral centre.

As the number of chiral centres in a molecule increases, so the number of enantiomers of the molecule that are possible increases also. For example:

- 1 chiral centre generates 1 pair of enantiomers
- 2 chiral centres generate 2 pairs of enantiomers
- 3 chiral centres generate 4 pairs of enantiomers.

Why is more than one pair of enantiomers possible? Each chiral centre has its mirror image–a group of atoms with the same connectivity, but with a different configuration. So, as the number of chiral centres increases, so the number of possible 'combinations' of mirror images increases also.

Self-check 10.14 Look at this structure of adrenaline. How many chiral centres are there, and where are they?

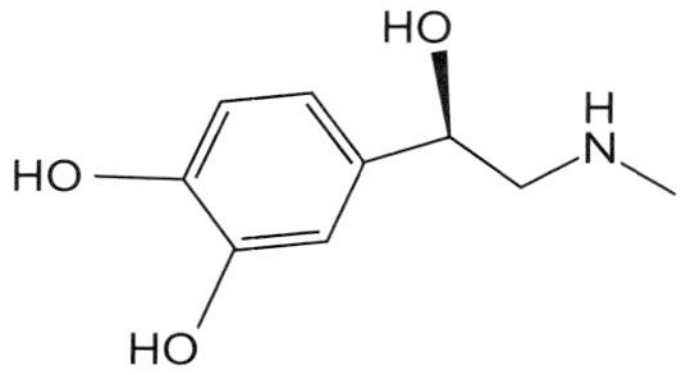

Look at Box 10.4 to find out about molecules that possess *more* than one chiral centre.

How do we distinguish one enantiomer from its mirror image?

Enantiomers are sometimes called **optical isomers**, because the only difference in their physical properties is how they affect the behaviour of **light**. Specifically, enantiomers cause **optical rotation**–the rotation to the left or the right of the plane in which a wave of light is travelling, as depicted in Figure 10.20. Look at this figure, and notice how one enantiomer causes rotation to the left, while the other enantiomer causes rotation to the right.

- ***In physical terms, a pair of enantiomers differ only in the way they affect the rotation of light travelling in a single plane.***

It is this difference that originally enabled pairs of enantiomers to be distinguished from one another. However, several *different* forms of nomenclature have been used

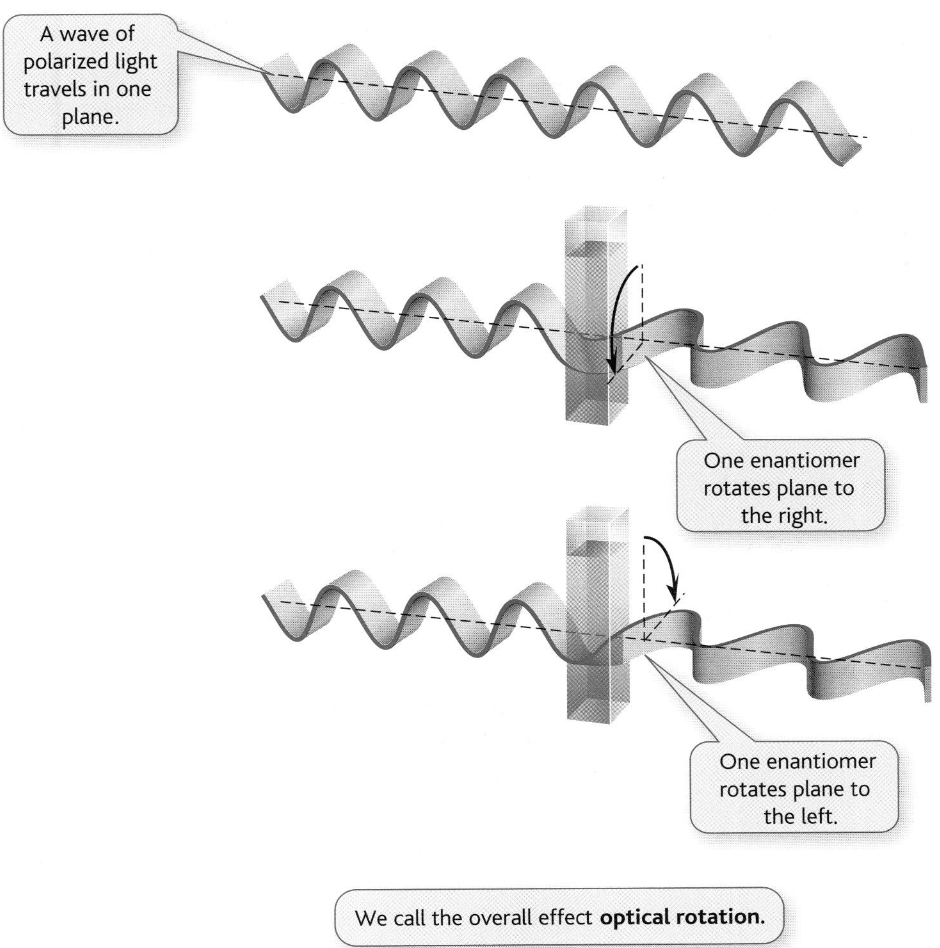
A wave of polarized light travels in one plane.
One enantiomer rotates plane to the right.
One enantiomer rotates plane to the left.
We call the overall effect **optical rotation.**

Figure 10.20 The principle of optical rotation. Optical rotation is the rotation to the left or the right of the plane in which a wave of polarized light is travelling.

historically, and remain in use today: a pair of enantiomers may be labelled (+) and (–), (D) and (L), or *R*- and *S*-.

Let's now consider each of these nomenclatures in turn.

The (+) and (–) notation denotes the opposing effects that a pair of enantiomers has on the behaviour of light, as described above. Look again at Figure 10.20: the enantiomer that rotates plane-polarized light to the right is assigned the label (+); the enantiomer that rotates plane-polarized light to the left is assigned the label (–):

H, NH_2, H_3C, CO_2H — (+)-alanine

H, NH_2, HO_2C, CH_3 — (–)-alanine

The alternative notation (D) and (L) also refers to the rotation of plane-polarized light: (D) represents dextrorotatory (rotating to the right); (L) represents laevorotatory (rotating to the left). We still see the labels D- and L- used commonly today, particularly to distinguish between enantiomers of biological molecules, and especially the amino acids, and sugars. For example, the D- and L- forms of glyceraldehyde are shown in Figure 10.21.

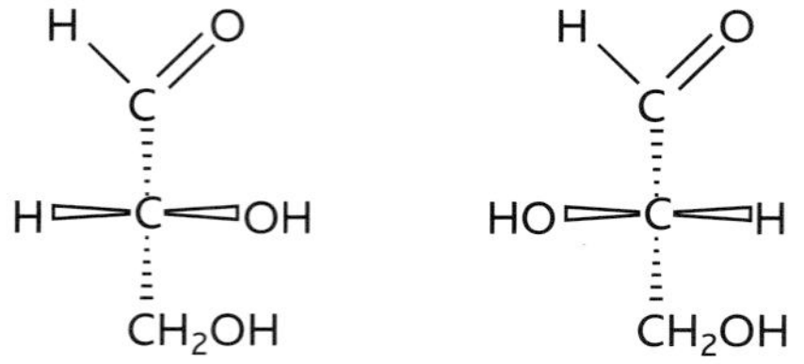

Figure 10.21 The D- and L- forms of glyceraldehyde.

D-glyceraldehyde L-glyceraldehyde

We no longer use optical rotation of light as the property for distinguishing between enantiomers. Instead, we use **X-ray crystallography**, a powerful technique that enables us to deduce the three-dimensional structure of a molecule. Enantiomers differ in their three-dimensional structure and so, by revealing their 3D structures, X-ray crystallography enables us to readily distinguish one enantiomer from another.

We find out more about X-ray crystallography in Chapter 11.

X-ray crystallography tells us the way in which the four different groups attached to the chiral centre are arranged relative to one another. We use the relative arrangement of the four groups to assign each enantiomer a label–either R or S–according to rules based on the Cahn–Ingold–Prelog method, which we encountered in the context of geometric isomers in Box 10.2. Look at Box 10.5 to see how we apply the Cahn–Ingold–Prelog method in the context of enantiomers.

BOX 10.5 How sinister! The *R/S* nomenclature for distinguishing between enantiomers

We see in Box 10.2 how the Cahn–Ingold–Prelog convention is used to assign a priority ranking to four different groups attached to the ends of a carbon–carbon double bond. A similar priority ranking is also used to assign the labels *R* and *S* to a pair of enantiomers.

Let's consider the pair of enantiomers of the amino acid alanine shown in Figure 10.22. Our first step in assigning the *R* and *S* labels to these two molecules is to pick one of them, and to prioritize the four groups that are attached to its chiral centre. Using the Cahn–Ingold–Prelog convention, we consider the atomic numbers of the atoms joined to the chiral carbon, shown here in blue:

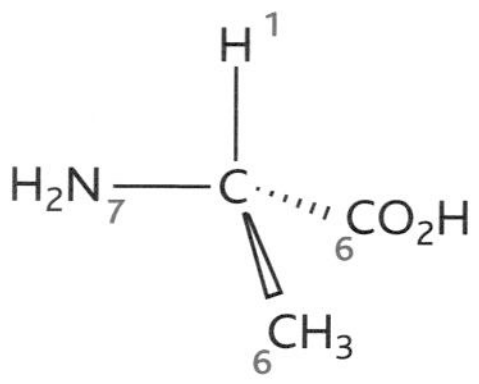

This tells us that the highest priority group is the $-NH_2$ group, and the lowest priority group is the H. However, notice that we find two atoms with equal atomic numbers. Therefore, we must consider the atoms one step further out from the chiral carbon: the carbon forming the $-CH_3$ group has three Hs attached, all of which have an atomic number of 1. The carbon forming the carboxyl group has two oxygen atoms attached, both of which have an atomic number of 8. This higher value means that we give the $-CO_2H$ group a higher priority than the $-CH_3$ group. So, the overall priority ranking is:

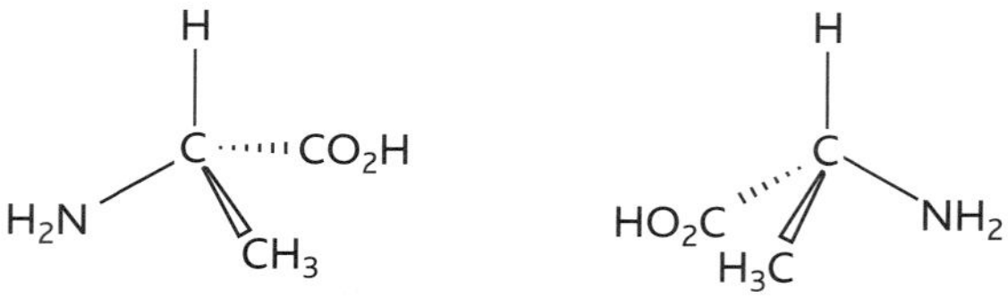

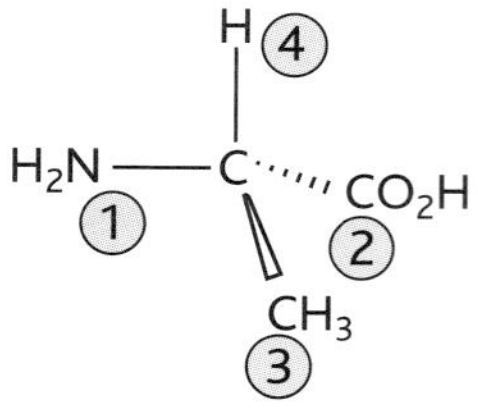

Figure 10.22 The enantiomers of the amino acid alanine.

BOX 10.5 Continued

We now have one more step to assign either the *R* or *S* label. To do this, we need to mentally rotate the molecule so that the lowest-ranking group (in this case, the H atom) is pointing away from us, behind the plane of the page, as illustrated in Figure 10.23(a). (To help visualize this, try to imagine the alanine molecule in Figure 10.23(a) being picked up by the H atom, and the molecule being lifted to face you.)

We now ask: which direction do we have to travel to go from highest-priority group to lowest-priority group? Looking at Figure 10.23(b) we find that we must move in an *anticlockwise* direction. If we were turning a steering wheel, this would be the equivalent of steering to the *left*. We therefore assign this enantiomer the label *S* (from the Latin *sinister*, meaning left).

If we carry out the same exercise with the other enantiomer, as shown in Figure 10.23 (c), we find that we must travel clockwise to move from the highest to lowest priority group–the equivalent of steering to the right. We therefore assign this enantiomer the label *R* (from the Latin *rectus*, meaning right).

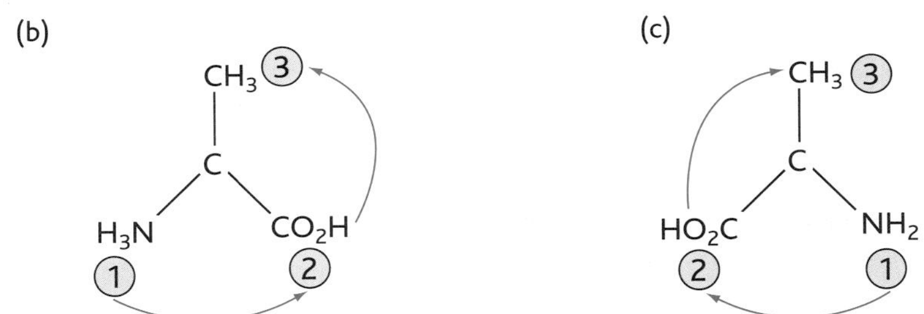

Figure 10.23 The application of the *R*/*S* nomenclature to distinguish between enantiomers. (a) First, we rotate the molecule so that the lowest-ranking group is pointing away from us; (b) We then consider the direction in which we have to move to travel from highest to lowest priority group. In this case, we must travel in the anticlockwise direction, so this is the enantiomer of alanine. (c) When we carry out the same actions for the other enantiomer of alanine, we find we must travel in a clockwise direction; this is the *R* enantiomer.

Self-check 10.15 Which enantiomer of the amino acid serine is shown here?

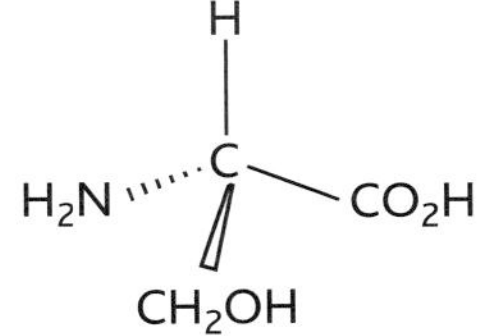

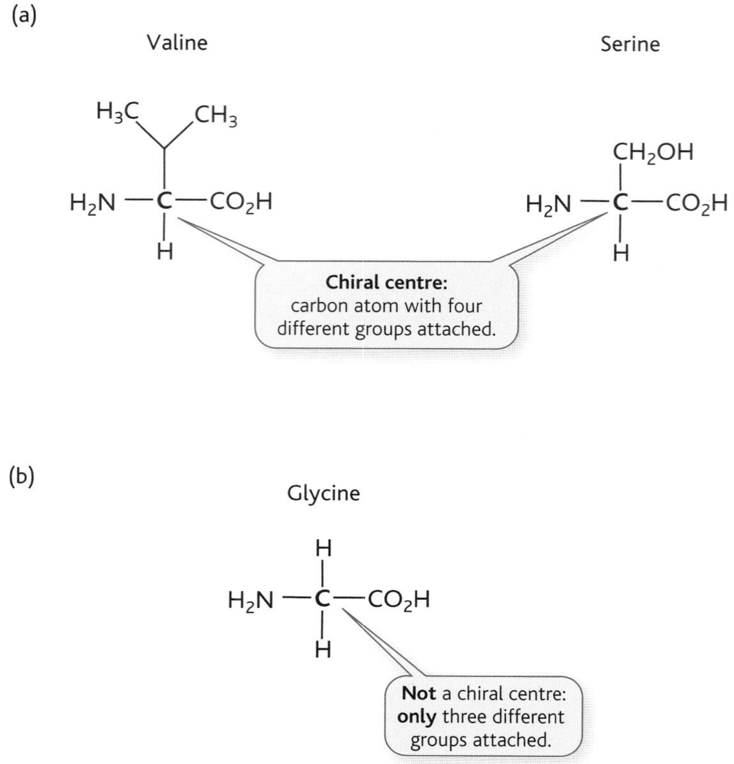

(a)

Valine

H_3C CH_3

$H_2N—C—CO_2H$ with H below

Serine

CH_2OH

$H_2N—C—CO_2H$ with H below

Chiral centre: carbon atom with four different groups attached.

(b)

Glycine

H

$H_2N—C—CO_2H$ with H below

Not a chiral centre: **only** three different groups attached.

Figure 10.24 (a) All but one of the amino acids are chiral: they possess a central carbon atom to which four different groups are attached. (b) The only exception is glycine, which has only three different groups attached, and so is not chiral.

Chirality in biological systems

Chirality is prevalent in biological systems: most biological molecules are chiral. For example, all but one of the twenty naturally occurring amino acids are chiral: the central carbon atom of each is a chiral centre, and has four different groups attached, as illustrated in Figure 10.24(a).

The exception is glycine, which has only three different groups attached to the central carbon atom–COOH, NH_2, and two identical hydrogens–and so lacks a chiral centre, as shown in Figure 10.24(b).

Similarly, the sugars (including ribose and glucose) are all chiral, possessing carbon atoms to which four different groups are attached. (Indeed, we see in Box 10.4 how sugars such as glucose possess *multiple* chiral centres.)

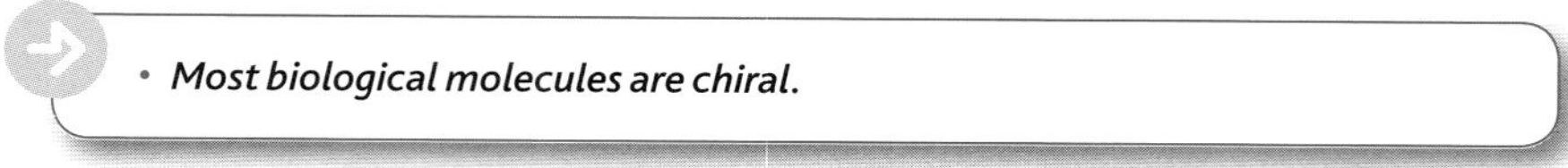
• Most biological molecules are chiral.

Any chiral molecule has an associated non-superimposable mirror image, and represents one half of a pair of enantiomers. Very often, however, only one of a pair of enantiomers is active in a biological system. For example, while the nineteen chiral amino acids all form pairs of enantiomers (the L- and D- forms, as discussed above), only *one* form, the L- form, is generally found in biological systems. (A significant exception is the presence of the unusual D-alanine in bacterial cell walls, as we discover in Box 10.6.) Indeed, certain amino acids that do not fall within the group of 20 used within the human body to produce proteins can be poisonous to us.

BOX 10.6 **Enantiomers and the evolutionary race for survival**

There is a good reason why the incorporation of D-alanine into bacterial cell walls has been selected for during the evolution of bacteria. Animals contain enzymes capable of digesting proteins comprising the 'natural' L-amino acids. However, being chiral (and therefore being able to interact with only one enantiomer of a pair) these enzymes are unable to interact with the unusual D-alanine. Therefore, whereas a bacterial cell wall containing L-alanine would be a target for digestion by animal enzymes, causing destruction of the bacterium, the presence of D-alanine shields the bacterial cell wall from destruction. This protection leaves the bacterium free to multiply, and facilitates infection of the host.

Somewhat ironically, pharmaceutical companies are using the presence of D-alanine in bacterial cell walls as a chink in the bacterial armour. The antibiotic vancomycin has been designed to specifically recognize and bind to D-alanine molecules. While vancomycin does not recognize the 'natural' L-alanine found in animal cells, it selectively binds to the unusual D-alanine in bacterial cell walls, targeting the bacteria for destruction. Therefore, it has no effect on the host cells when administered to humans but destroys any foreign bacterial cells that may have invaded.

For example, the plant-derived amino acid hypoglycine A (9) is found in the plant Ackee (*Blighia sapida*).

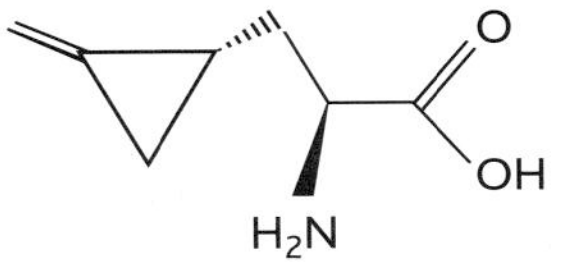

O

OH

H_2N

(9)

The ackee produces a pear-shaped fruit, which is only edible when ripe: the unripened flesh (and even certain parts of the fruit in its ripened state) contains hypoglycine A.

The fruit of the ackee is used widely in Jamaican cuisine and, if properly prepared, is highly nutritious. However, if ingested, hypoglycine A can cause Jamaican vomiting sickness, with symptoms of vomiting, seizures, and hypoglycaemia (abnormally low blood sugar levels).

Similarly, while both ribose and glucose also exist in D- and L-forms, only the D-forms are important in biological systems: D-glucose is the source of much biochemical energy in cells[8], while D-ribose is the sugar component of the backbone of the nucleic acids, DNA and RNA.

We explore the chemical structures of biological molecules such as the nucleic acids in Chapter 7.

There are some instances in which both enantiomers are present in a biological system. In such instances, we find that the two enantiomers fulfil very distinct functions.

Consider the two enantiomers of limonene (**10, 11**).

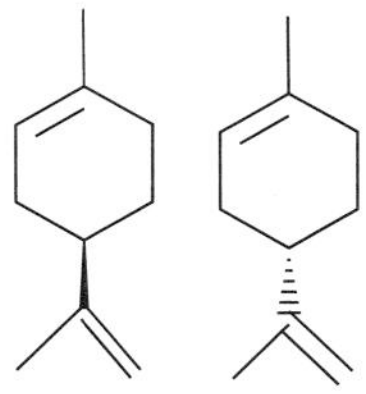

(10) (11)

8 To demonstrate the differences in the way enantiomers interact with biological systems, D-glucose tastes sweet; L-glucose has no taste at all.

While the two molecules look identical (other than being mirror images of each other), (+)-limonene tastes of orange, while (−)-limonene tastes of lemon.

Lactic acid also exists as a pair of enantiomers (**12, 13**) which have contrasting biological functions.

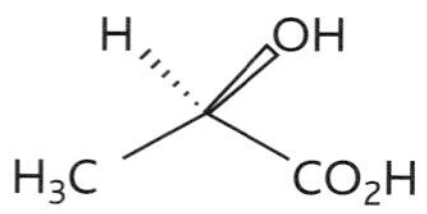

H, OH, H_3C, CO_2H (12) H, OH, HO_2C, CH_3 (13)

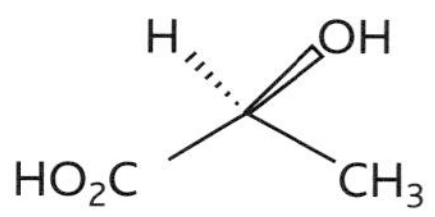

The burning that we experience in our muscles during vigorous exercise is the result of the accumulation of lactic acid in our muscle tissue; the souring of milk is also due to the accumulation of lactic acid as a by-product of the fermentation of milk by certain bacteria. However, these two biological situations involve different enantiomers of lactic acid: the lactic acid in our muscle is (*S*)-lactic acid, whereas the lactic acid produced by certain bacteria is solely (*R*)-lactic acid.

We see on p. 304 that the molecules that comprise a pair of enantiomers have identical chemical and physical properties–they behave in exactly the same way, and can be distinguished only by the effect they have on the behaviour of light. This being the case, how is it that biological systems distinguish so readily between enantiomers? How is it that, despite both molecules having identical chemical and physical properties, we recognize (+)-limonene as the taste of oranges and (–)-limonene as the taste of lemons, for example?

The answer is that biological systems are themselves chiral. As such, they can *discriminate* between the individual molecules that make a pair of enantiomers and, hence, exhibit **enantiomeric specificity** (or **selectivity**). Let's consider an analogy to help us see how the chirality of a biological system might confer upon it the ability to discriminate between enantiomers.

We see on p. 300 how our hands, feet, and shoes are all chiral, and exist as pairs of enantiomers, that is, non-superimposable mirror images. Let's now take the left shoe. Consider what happens when this chiral object interacts with other chiral objects–namely one of our feet. One of the first things we discover when learning to dress ourselves as children is that the left shoe only fits on the left foot, and that the right shoe only fits on the right foot: each shoe can only 'interact' with one enantiomer of a pair–it distinguishes one enantiomer from the other.

This situation is mirrored in biological systems: biological systems are composed of chiral molecules (such as proteins), which are able to distinguish between enantiomers in an analogous fashion to a shoe distinguishing between a left and a right foot.

It is the chirality of proteins that explains how our senses can distinguish between (+)-limonene and (−)-limonene. Our taste buds contain chiral receptor molecules, which only bind specific targets. (Different areas of our tongue are sensitive to different tastes–some areas are sensitive to salt, some to sweet, some to bitter, and so on. The taste to which a particular area of the tongue is sensitive depends on the specific receptor molecules that are present in the taste buds in that region of the tongue.)

One particular taste receptor binds (+)-limonene while a *different* receptor binds (−)-limonene, just as only the left foot fits in a left shoe, and the right foot fits in the right shoe. The selective binding of (+)-limonene by its target receptor triggers a nerve impulse that is perceived by the brain as an 'orange' taste, while the binding of (−)-limonene triggers the perception of a 'lemon' taste.

- ***Biological systems can often discriminate between a pair of enantiomers.***

If only one of a pair of enantiomers is present (as is the case in many biological systems), we say that the system is **enantiomerically pure**.

10.5 The chemistry of isomers

We see in Chapter 6 how different families of organic compounds have contrasting chemical and physical properties. The chemical and physical properties exhibited by a molecule are dictated by its composition, connectivity, and configuration–that is, by the atoms from which the molecule is made, how these atoms are connected, and how they are arranged in three-dimensional space.

But *how* different do compounds have to be to exhibit different chemical and physical properties? Do geometric isomers share the same chemical and physical properties, for example?

The answer is that only pairs of enantiomers share the same physical and chemical properties. Structural isomers and geometric isomers do *not* share the same physical or chemical properties. Let's look at some examples to verify these statements.

Consider the two pairs of structural isomers that we mention on p. 288: propanol and methoxyethane, and butanal and butanone. The atoms of these molecules are connected in distinct ways, giving rise to characteristic functional groups–specific groups of atoms such as the hydroxyl group of propanol, and the carbonyl groups of both butanal and butanone, which confer upon the molecules characteristic (and contrasting) physical and chemical properties. For example, propanol and methoxyethane, and butanal and butanone, exhibit greatly different boiling points, as shown in Table 10.2.

Further, Figure 10.25 shows how butanal's aldehyde group allows it to participate in different chemical reactions from butanone, which possesses a ketone group.

Table 10.2 A comparison of the boiling points of two pairs of structural isomers.

	Boiling point (°C)
Propanol	98
Methoxyethane	8
Butanal	75
Butanone	80

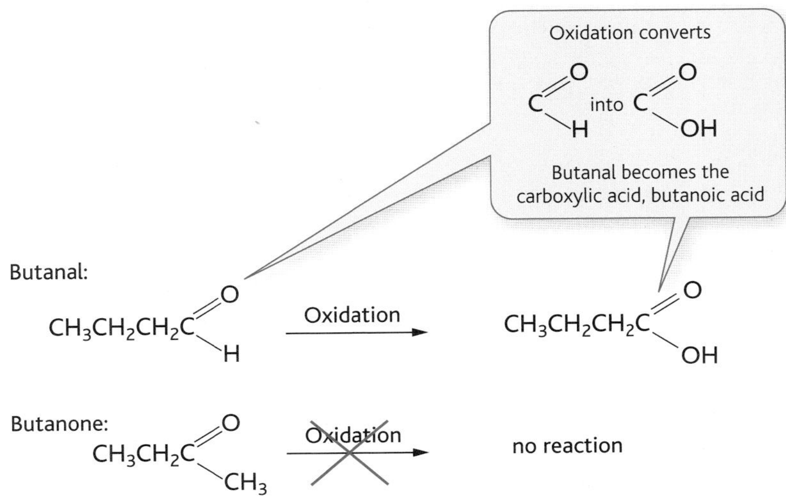

Figure 10.25 The geometric isomers butanal and butanone participate in different chemical reactions because of their different functional groups.

Now let's consider two geometric isomers, fumaric acid (**14**)

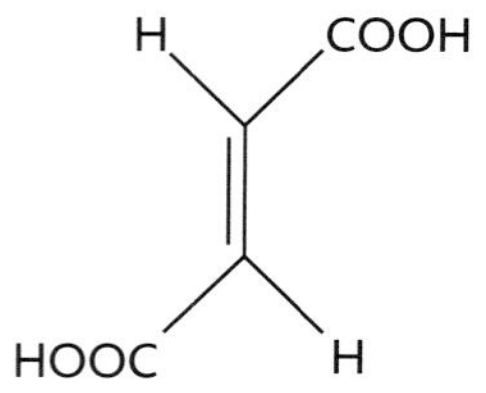

(14)

and maleic acid (**15**).

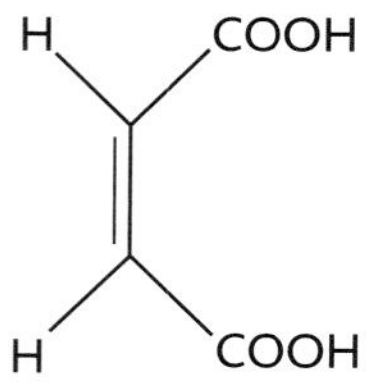

(15)

Remember: geometric isomers have the same connectivity; it is the difference in their configurations that affects how they interact–and subsequently react–with other molecules, or with themselves.

Fumaric acid and maleic acid exhibit different physical properties; for example, fumaric acid has a melting point of 300 °C, while maleic acid has a melting point of 240 °C.

The two molecules also exhibit different chemical properties, due to their contrasting configurations. Figure 10.26 shows how when maleic acid is heated its two carboxyl groups react together to form maleic anhydride. By contrast, fumaric acid does not undergo chemical change when heated: its carboxyl groups are too far apart from one another for a chemical reaction to occur.

Enantiomers are physically and chemically identical. They exhibit the same intermolecular non-covalent forces, and so have the same physical properties (melting point, boiling point, etc.). Pairs of enantiomers also possess the same functional groups, arranged in the same *relative* orientations[9]. Therefore, enantiomers participate in the same chemical reactions.

However, there is one very significant exception to this rule: while enantiomers *do* behave the same way in the laboratory, they behave very differently in biological systems. This is a very important distinction, as we discover in the next section.

---

9 Note, however, that enantiomers have different *absolute* orientations. Look at this pair of enantiomers:

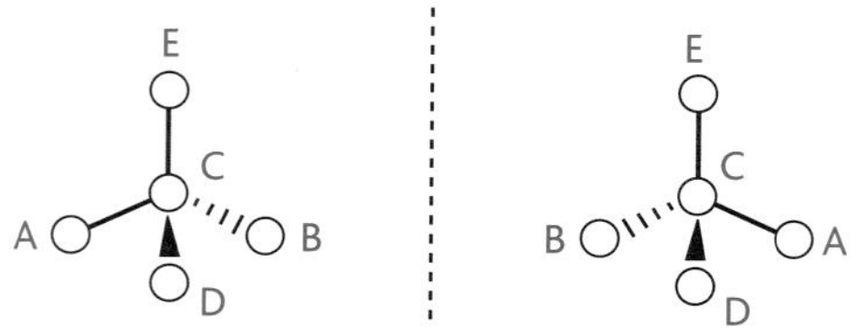

Molecule 1 Molecule 2

Notice how the component atoms of both molecules are oriented in the same way *relative to one another*: for example, atoms A and D are the same distance apart in both molecules; atoms B and D are also the same distance apart in both molecules.

However, notice how atom A is pointing to the left in molecule 1, but is pointing to the *right* in molecule 2. Their **absolute** orientations are different.

We can write this:

H COOH

H COOH

cis

Like this:

H C O OH

H C O OH

If we heat ...

(a)

Carboxyl groups are close together ...

H C O OH

H C O OH

cis

heat

H C O

O

H C O

... so can react with one another.

Water is given out.

$+H_2O$

(b)

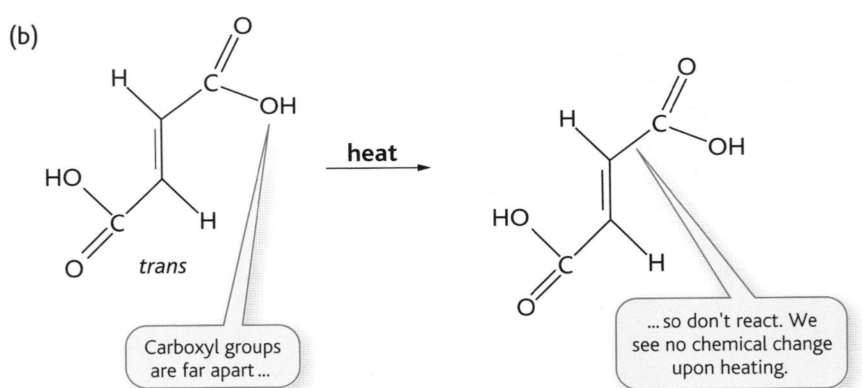

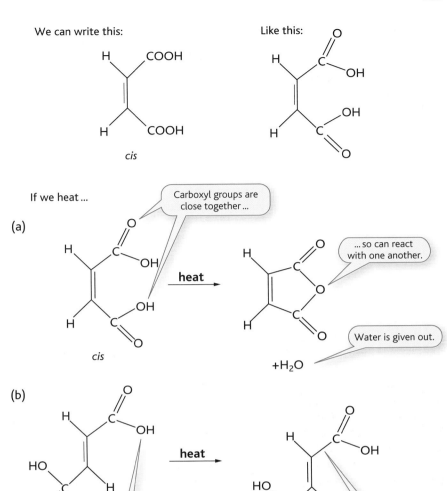

Figure 10.26 Geometric isomers usually exhibit different chemical properties. For example, when maleic acid is heated, its two carboxyl groups react together to form maleic anhydride (a). By contrast, fumaric acid (a geometric isomer of maleic acid) does not undergo chemical change when heated: its carboxyl groups are too far apart from one another for a chemical reaction to occur (b).

- ***Structural isomers and geometric isomers exhibit different physical and chemical properties.***
- ***Enantiomers share the same physical and chemical properties.***

The biological chemistry of enantiomers

Why do enantiomers behave differently in biological systems, compared to the laboratory? We see on p. 308 how one of a pair of enantiomers may be prevalent in a biological system over the other, and that biological systems can often discriminate between pairs of enantiomers because biological systems are, themselves, chiral. The impact of chirality is also felt at the chemical level, and is the key reason for the difference in chemical behaviour exhibited by a pair of enantiomers.

We find out more about how enzymes catalyze reactions in Chapter 14, when we explore the rate at which chemical reactions occur, and how rates can be increased.

All chemical reactions that occur in biological systems are facilitated by enzymes, which make reactions more feasible (or more likely to happen). We say that an enzyme **catalyzes** the reaction. For a given enantiomer to participate in a biochemical reaction, it–like any other molecule in a biological system–must typically interact with an enzyme.

But why do enzymes prevent a pair of enantiomers from exhibiting the same chemical properties in a biological system? Enzymes are chiral molecules, just like any other protein. This chirality gives enzymes exquisite sensitivity when it comes to being selective about which molecules they interact with. Being chiral, enzymes are sensitive enough not only to discriminate between structural isomers or geometric isomers, but to discriminate between enantiomers also. It is this sensitivity and specificity that explains why enantiomers behave differently in biological systems: in order to participate in a chemical reaction in a biological system, an enantiomer must interact with an enzyme (just like any other molecule). However, an enzyme only catalyzes a reaction involving one specific enantiomer, as illustrated in Figure 10.27.

- ***Enantiomers exhibit different biological properties, despite sharing the same chemical and physical properties.***
- ***Enantiomers have different biological properties because of the inherent chirality of biological systems.***

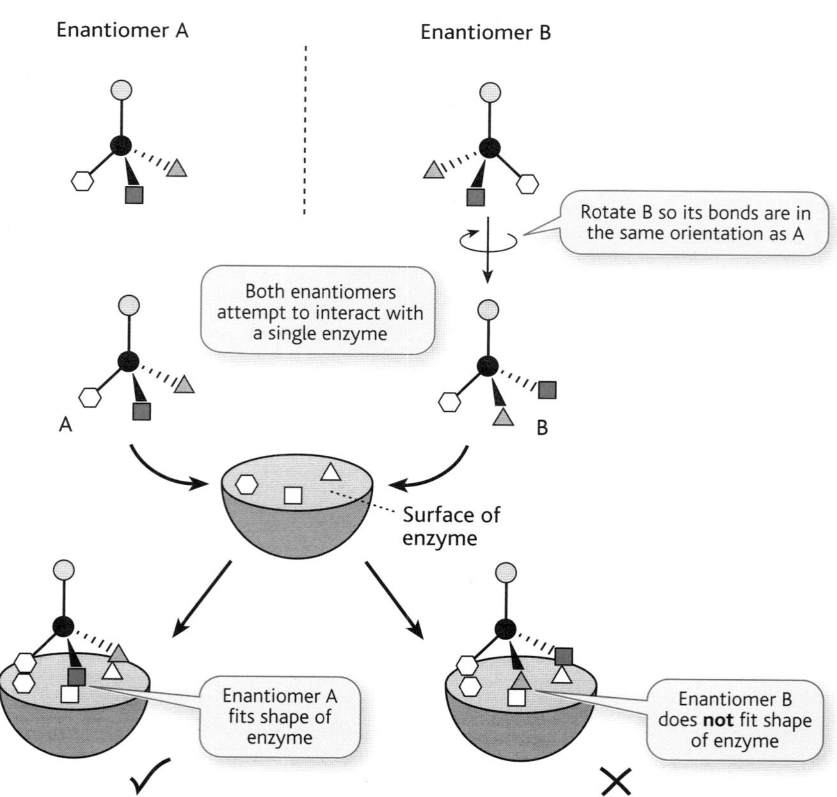

Figure 10.27 Typically, an enzyme only catalyzes a reaction involving *one* specific enantiomer. This ability to discriminate between enantiomers is because enzymes themselves are chiral molecules.

The impact of chirality on medicinal chemistry

The ability of chiral molecules, such as enzymes and receptor molecules, to discriminate between enantiomers creates huge challenges for the field of medicinal chemistry–the design and manufacture of drugs for therapeutic use. Virtually all drug molecules are chiral and so exist as pairs of enantiomers. Just as enantiomers of molecules such as limonene are distinguished between by the body (one being 'interpreted' as the taste of orange, the other as the taste of lemon) so different enantiomers of a particular drug may be 'interpreted' by the body in very different ways too. Let's consider a couple of examples to illustrate this.

Firstly, let's consider the analgesic[10] **methadone**.

Methadone is widely used during the rehabilitation of individuals who have developed a dependency on, or addiction to, the drugs morphine and heroin. Morphine and heroin belong to a family of drugs called opiates, which are components of the biological compound opium. Opium is produced naturally by poppies, from which these compounds can be extracted.

Morphine and heroin have serious side effects: morphine, for example, is associated with nausea, constipation, and respiratory problems. Both drugs also cause debilitating withdrawal symptoms, including chills, sweating, abdominal cramps, and muscle spasms. While methadone also has side effects, these effects are less serious than those associated with either morphine or heroin. Methadone also has the advantage that it can be administered orally (by mouth) rather than intravenously (by direct injection into the bloodstream).

Figure 10.28 shows how methadone has a single chiral centre. Therefore, methadone exists as a single pair of enantiomers. However, the two enantiomers have contrasting activities. While the *R*-enantiomer is a potent analgesic, the *S*-enantiomer has no activity in the body. It is thought that this difference in activity is due to the enantiomeric selectivity of the receptor in the body with which methadone interacts. The receptor, itself a chiral biological molecule, can interact only with the *R*-isomer.

We see earlier in this section how (+)- and (–)-limonene trigger different taste sensations; and how this is due to the taste receptors for limonene, themselves chiral biological molecules, discriminating between the two enantiomers of limonene. While there is one receptor for each of the enantiomers of limonene, such that both enantiomers have biological activity (they both generate taste sensations, albeit *different* taste sensations), it seems that there is a receptor for only *one* of the two enantiomers of methadone, the *R*-isomer. Hence only the *R*-enantiomer is biologically active.

In fact, *R*-methadone is twice as potent as morphine. This characteristic is central to the use of methadone in drug rehabilitation: individuals undergoing rehabilitation

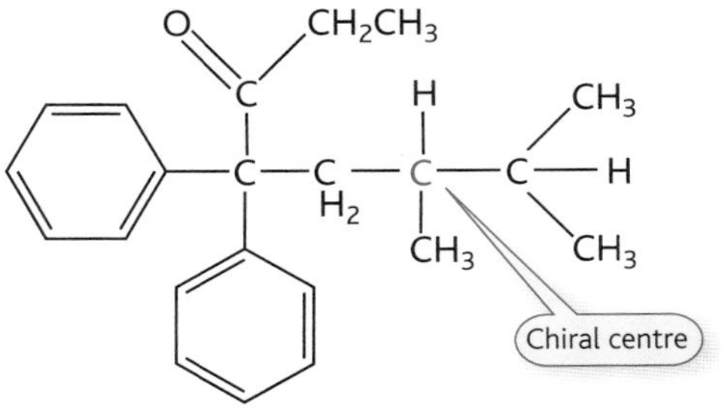

Figure 10.28 The structural formula of methadone. Methadone has a single chiral centre. Therefore, methadone exists as a single pair of enantiomers.

10 An analgesic is more commonly known as a **painkiller**.

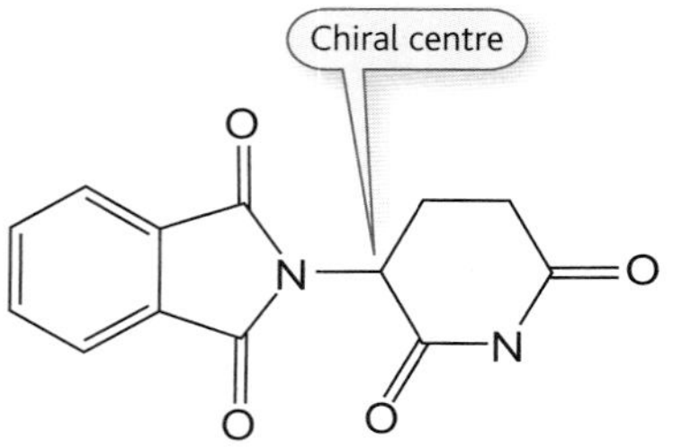

Figure 10.29 The structural formula of thalidomide. Thalidomide has one chiral centre, and so exists as a pair of enantiomers.

may take methadone less regularly than they would the less potent morphine, but still feel the same effect. By reducing the frequency with which the drug is administered, methadone is used to gradually break the 'routine' of drug-taking, and hence eventually wean users away from drug-taking altogether.

Perhaps the most notorious difference in biological activity between enantiomers of a drug was that observed for thalidomide; look at Figure 10.29 to see the structure of this drug. Notice how there is only one chiral centre in thalidomide, despite it being quite a large molecule. Therefore thalidomide exists as only one pair of enantiomers.

Thalidomide was routinely prescribed in the late 1950s and early 1960s as a sedative to offer pregnant women relief from morning sickness. Crucially, thalidomide was administered as a **racemic** mixture of its two enantiomers[11], rather than as an enantiomerically pure preparation of either the *R*-isomer or *S*-isomer.

As women who had taken thalidomide during pregnancy gave birth, however, it became apparent that the drug had devastating side-effects: many babies that had been exposed to thalidomide as foetuses exhibited growth abnormalities–most frequently, truncated or missing limbs. Thalidomide was quickly withdrawn from routine use, and has retained its notoriety ever since.

We call substances that can cause developmental abnormalities **teratogens**, and we say that they have a **teratogenic** effect.

Later studies revealed that the two sharply contrasting properties of thalidomide–that of a sedative and that of a teratogen–were associated with the two different enantiomers of thalidomide. While *R*-thalidomide was found to be an effective sedative, *S*-thalidomide was found to be a potent teratogen. Read Box 10.7 to learn more about *S*-thalidomide's teratogenic effect on the human body.

An instinctive response might be that doctors should only have administered the 'safe' enantiomer of thalidomide (*R*-thalidomide). However, the biochemistry of thalidomide has a sting in the tail: *R*-thalidomide can be converted in the body to the teratogenic *S*-enantiomer. Therefore, even if enantiomerically pure *R*-thalidomide had been administered to patients, a proportion would have been converted to the dangerous *S*-thalidomide, which would have inflicted the same teratogenic damage upon the growing foetus as that witnessed in the 1960s.

Enantiomeric purity and stereoselective synthesis

The different enantiomers of many chiral drugs exhibit contrasting activities in the body, as we have seen for both methadone and thalidomide. This difference in activity is due to biological systems being chiral, and therefore possessing the ability to discriminate between other chiral molecules.

If two enantiomers of a drug exhibit contrasting activities, the efficacy (effectiveness) of the drug will often be increased if only the most active enantiomer is administered.

11 A **racemic mixture** is a mixture of enantiomers, in which the different enantiomers are present in equal proportions.

BOX 10.7 Putting a block on growth: thalidomide's dark side

Thalidomide is believed to have its teratogenic effect by inhibiting angiogenesis–the development of blood vessels. We say that thalidomide has an antiangiogenic effect. While this antiangiogenic effect has negative consequences upon the developing foetus (as evidenced by the birth defects witnessed in the 1960s), it is now at the centre of research into potentially positive applications of the drug. Cancerous tumours require a ready supply of blood to fuel their growth, and so undergo angiogenesis: they develop blood vessels that 'tap' into the blood supply of the tissue that they are encroaching upon. There is some evidence that thalidomide may offer the potential for treating cancerous tumours, by inhibiting angiogenesis and, hence, denying tumours the blood supply that they require for growth. There is also powerful evidence that thalidomide is useful in treating leprosy. It has also been used against HIV.

Historically, the preparation of enantiomerically pure drugs was difficult. As a result, most drugs were prepared as racemic mixtures. However, modern drug synthesis is very different: pharmaceutical companies have now developed methods of drug synthesis that allow them to synthesize only one enantiomer of a pair.

We call the synthesis of just one enantiomer of a pair **stereoselective synthesis**, to reflect the way that we selectively synthesize just one stereoisomer.

Where two enantiomers have contrasting biological activities, stereoselective synthesis allows pharmaceutical companies to manufacture only the most active enantiomer. Superior synthetic routes offer two key advantages: it makes the manufacturing process more economical (drug companies no longer have to waste money synthesizing both enantiomers), and it increases the potency of drugs (the drug comprises solely the active enantiomer, and not a 'watered down' mix of 50% active and 50% inactive enantiomer.)

Check your understanding

To check that you've mastered the key concepts presented in this chapter, review the checklist of key concepts below, and attempt the multiple-choice questions available in the book's Online Resource Centre at **http://www.oxfordtextbooks.co.uk/orc/crowe2e/**.

Checklist of key concepts

Isomers

- Isomers are groups of compounds that comprise exactly the same atoms

Structural isomers

- Structural isomers are molecules that have the same chemical composition, but whose atoms are joined together in different ways
- We call the way in which a molecule's atoms are joined together its connectivity
- Structural isomers share the same molecular formula, but have different structural formulae
- We use particular nomenclature to help us distinguish between structural isomers
- Structural isomers may be members of the same chemical family or different chemical families

Stereoisomers

- Stereoisomers possess the same atoms, which exhibit the same connectivities. However, stereoisomers exhibit different **configurations**
- Stereoisomers fall into two groups: geometric isomers, and enantiomers

Geometric isomers

- Geometric isomers are stereoisomers that possess groups of atoms locked in **specific configurations**
- Geometric isomerism is most common in molecules that possess one or more double bonds
- A *cis*-isomer possesses groups that are positioned on the **same side** of the double bond
- A *trans*-isomer possesses groups that are positioned on **opposite sides** of the double bond
- *cis* and *trans* isomers are distinguished by using the ***E*/*Z* nomenclature**
- Some **cyclic** compounds also exhibit geometric isomerism

Enantiomers

- Enantiomers are non-superimposable mirror images of one another
- Pairs of objects that are non-superimposable mirror images are what we call **chiral**

Chirality

- Chiral objects lack a plane of symmetry–they are **asymmetric**
- We call a central atom that has four different groups attached a **chiral centre**
- A chiral centre is **asymmetric**
- Most biological molecules are chiral
- Some biological molecules possess multiple chiral centres
- Pairs of enantiomers often behave differently in biological systems–we say that the system is showing **enantiomeric specificity**
- Enantiomers behave differently in biological systems because they interact with other chiral molecules, which are able to discriminate between enantiomers
- Enantiomers are also called **optical isomers**
- Enantiomers differ in how they affect the behaviour of light
- A pair of enantiomers may be distinguished using the nomenclatures of (+) and (–), D and L, or (*R*) and (*S*)
- These three systems are broadly complementary

Chemistry of the isomers

- Structural isomers **do not** share the same chemical and physical properties
- Geometric isomers **do not** share the same chemical and physical properties
- Enantiomers **do** share the same chemical and physical properties, but their biological properties differ greatly

The biological chemistry of enantiomers

- If only one of a pair of enantiomers is present, we say that the system is **enantiomerically pure**
- The difference in biological activity between enantiomers is an important consideration in drug design
- A **racemic mixture** contains a pair of enantiomers in equal proportions
- An enantiomerically pure drug can be produced through **stereoselective synthesis** of the drug

Chemical analysis 1: how do we know what is there?

INTRODUCTION

We are surrounded every day by tens of thousands of different chemical compounds, with contrasting characteristics. Sometimes, the presence of a compound is essential for our existence: we need oxygen in the air and glucose from the food we eat to give us the energy required to sustain life. By contrast, the presence of a compound may be detrimental: pollutants in water may make the water unfit for us to drink; too much alcohol in our bloodstream may make us unfit to drive. We need ways of determining what is in our chemical environment–of knowing what is there, and how much.

In this chapter we learn about chemical analysis–the process by which we characterize chemical compounds. We ask how we can separate mixtures to isolate individual compounds; how we can identify the chemical compounds that we have isolated; and how we can determine the structure and composition of these compounds. In the next chapter, we then explore how, having identified a compound, we can determine how *much* of it we have.

11.1 What is chemical analysis?

Chemical analysis is the process of characterizing the chemical nature of a particular system. In this context, the 'system' may represent the contents of a test tube in a chemical laboratory; or it may be a biological sample such as blood. Whatever the system, we can use chemical analysis to answer two general questions:

1. What is the identity of a particular compound?
2. How much of the compound do we have?

We may not realize it, but we carry out chemical analysis on a daily basis to answer precisely these two questions. For example, let's imagine that we're making tea and coffee for a couple of friends. We're ready to hand out the tea and coffee, but can't remember which mug contains tea and which contains coffee (and they look too similar to tell the difference by sight alone). We don't want to risk giving the wrong drink to one of our friends, so we carry out a quick chemical analysis: we take a sip

BOX 11.1 Qualitative and quantitative analysis: the reality

There are two types of test, which are performed by tens of thousands of people every day, that illustrate qualitative and quantitative analysis in action.

The pregnancy testing kit is an example of **qualitative** analysis. A typical pregnancy testing kit detects the presence of the hormone human chorionic gonadotrophin (HCG) in the urine. HCG is secreted by the pre-embryo (the fertilized egg) after it implants in the lining of the uterus; HCG acts as a signal to maintain the lining of the uterus, giving the embryo a stable environment in which to develop. (If HCG is not present, the lining of the uterus is shed as part of the menstrual cycle.) Because HCG is **only** present if a fertilized egg has implanted in the lining of the uterus, its presence is used as a positive indicator of pregnancy. The pregnancy testing kit does not seek to **quantify** HCG, i.e. to determine how much is there, it seeks solely to qualify its presence–to determine whether it's there or not.

By contrast, the blood sugar testing kit is an example of quantitative analysis. People with diabetes must measure their blood sugar levels regularly to ensure these levels do not exceed particular limits. It is not sufficient for diabetics to perform a qualitative analysis–to determine whether sugar is there or not. They must perform a *quantitative* analysis–to determine exactly *how much* of the sugar is present in their blood. We learn more about how people with diabetes measure their blood sugar levels in Box 12.4.

from one of the two mugs and taste the contents. We are asking the question 'What is the identity of the compound?' We are seeking to **qualify** whether the compound is coffee or tea.

We then ask our friends whether we've added enough milk to the tea and coffee, or whether either drink is too strong. To answer this question, our friends also take a sip to see if the flavour of their drink is too strong. If it *is* too strong, they may add more milk to make the flavour less intense. By tasting the drink we are asking the question 'How much of the compound (tea or coffee) do I have?' We are trying to **quantify** the amount of tea or coffee present. Read Box 11.1 to find out about two chemical analyses–one qualitative and one quantitative–that are carried out by thousands of people every day.

In this chapter we explore qualitative analysis ('What is there?') and the tools that are available to answer this question. We explore quantitative analysis ('How much is there?') in the next chapter.

11.2 How do we separate out what is there?

Biological systems such as cells comprise a diverse mixture of compounds. Before we can start to characterize a particular component, we often need to separate it from all the other compounds accompanying it, so that we know what we're looking at. This way, we ensure our data relate to only one compound, and is not a jumble of information from many different compounds. (Think about being in a crowded pub or club: you're trying to talk to a friend, but there are tens of different conversations happening around you, making it really difficult to hear only what your friend is saying. The easiest way to be able to identify what just your friend is saying is to take them to a quiet corner–to separate them from the rest of the crowd. Having done so, you can now clearly identify what your friend is saying.)

Most of the separatory techniques used to isolate different compounds are types of **chromatography** or **centrifugation**, and we explore both of these later in this section. However, we begin by considering the separation of compounds based

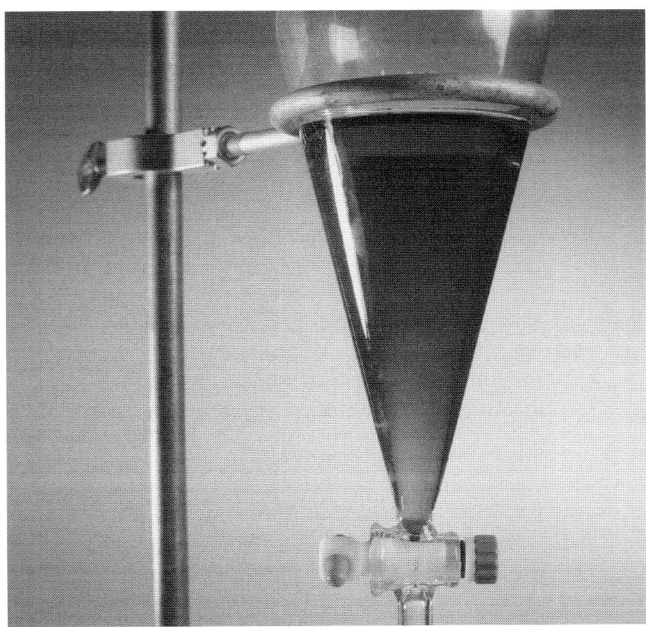

Figure 11.1 A separating funnel. When two immiscible liquids are mixed and allowed to separate, the lower of the two liquids can be drained from the funnel using the tap, leaving the upper liquid behind.

on differences in behaviour when exposed to different solvents, the technique of solvent extraction.

Solvent extraction

A common piece of apparatus in the chemistry laboratory is the separating funnel, as depicted in Figure 11.1. Separating funnels are used to separate two immiscible liquids[1], such as an aqueous phase (a hydrophilic, water-based solution) and an organic phase (a hydrophobic liquid such as chloroform). When the two liquids are mixed, they form two distinct layers, which can be drawn off from the funnel separately by using the tap at the bottom of the funnel.

So how are two immiscible liquids used as a means of separation? If a third component, such as a chemical compound, is introduced into the funnel and the components shaken vigorously, the chemical compound will distribute itself between the two liquids. (We say that it undergoes **partitioning** between the two liquids[2], a process illustrated in Figure 11.2.) Importantly, different types of compound show different affinities for hydrophilic or hydrophobic solvents. Typically, 'like dissolves like' such that hydrophobic compounds (as typified by organic compounds) are more likely to dissolve in hydrophobic (organic) solvents than in hydrophilic, aqueous solvents. By contrast, hydrophilic compounds (such as many inorganic compounds) are more likely to dissolve in hydrophilic, aqueous solvents.

---

1 'Immiscible' means unable to mix. Two immiscible liquids–typically a hydrophobic liquid and a hydrophilic liquid–will separate into two layers after being shaken together. A classic example of two immiscible liquids is the way oil (hydrophobic) and water (hydrophilic) separate into two layers after mixing, as we see in Figure 4.20.

2 Different chemical species prefer to occupy different physical environments (phases), and move to occupy their *preferred* environment if exposed to two different environments simultaneously. This differential movement is called partitioning.

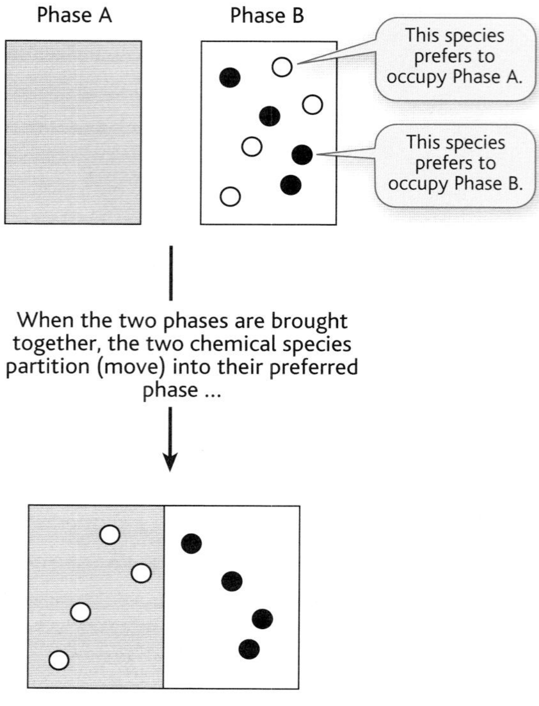

Figure 11.2 The concept of partitioning. Partitioning is the separation of the two species on the basis of the environment they prefer to occupy.

It is this difference in affinity that is used to drive separation: if a hydrophobic compound and a hydrophilic compound are shaken with a mixture of both a hydrophobic, organic solvent and a hydrophilic, aqueous solvent, and the mixture is allowed to separate, the hydrophobic compound will preferentially partition into the organic solvent, and the hydrophilic compound will preferentially partition into the aqueous solvent, as illustrated in Figure 11.3. By partitioning in this way, we achieve separation of the hydrophobic and hydrophilic compounds.

- ***Solvent extraction exploits the relative affinity that a compound shows for an organic (hydrophobic) versus an aqueous (hydrophilic) solvent***

We can determine the relative affinity of a particular compound for a hydrophilic, aqueous solvent versus hydrophobic, organic solvent (and, hence, determine which phase a compound is most likely to end up in when exposed to both solvents in a separating funnel) by examining its **distribution ratio, K**. The distribution ratio (which we also see referred to as the distribution coefficient or partition coefficient) describes the relationship between the concentration of a compound in one solvent and the concentration in another solvent when the system is at equilibrium; the relationship can be represented by the following equation:

$$K = \frac{[X]_{aq}}{[X]_{og}}$$

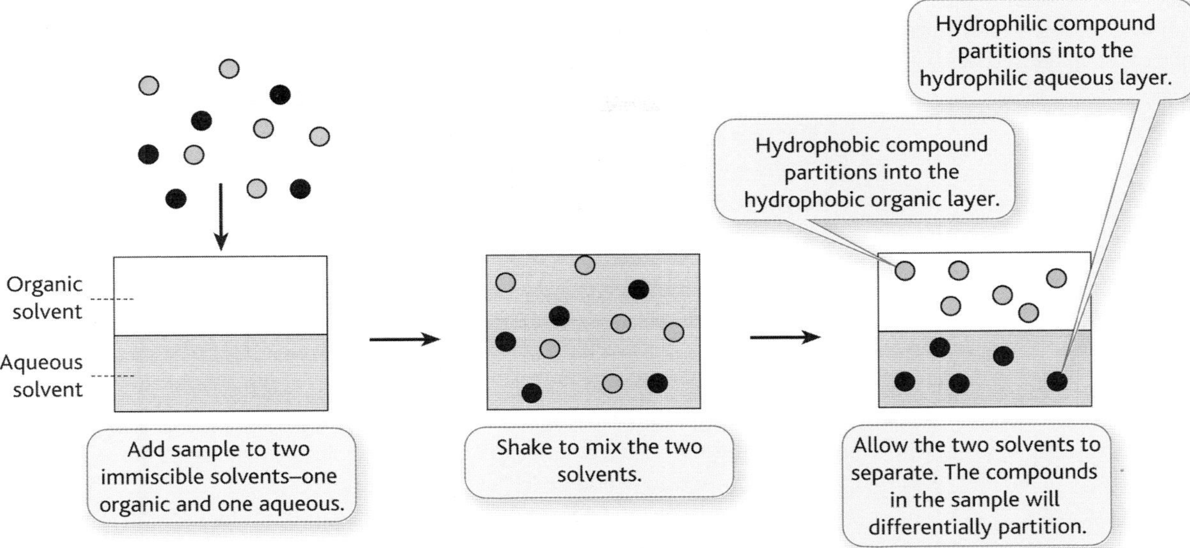

Figure 11.3 If a hydrophobic compound and a hydrophilic compound are shaken with a mixture of both a hydrophobic, organic solvent and a hydrophilic, aqueous solvent, and the mixture is allowed to separate, the hydrophobic compound will preferentially partition into the organic solvent, and the hydrophilic compound will preferentially partition into the aqueous solvent.

where $[X]_{aq}$ is the concentration of compound in the aqueous phase and $[X]_{og}$ is the concentration of the compound in the organic phase.

If a given compound is more soluble in the aqueous phase than in the organic phase (such that it preferentially partitions into the aqueous phase) the value of $[X]_{aq}$ will be greater than the value of $[X]_{og}$, such that the value of K is greater than 1. (If $[X]_{aq}$ is larger than $[X]_{og}$ then $[X]_{aq}/[X]_{og}$ must be greater than 1.)

By contrast, if a compound is more soluble in the organic phase than in the aqueous phase (such that it preferentially partitions into the organic phase) the value of $[X]_{aq}$ will be smaller than the value of $[X]_{og}$, such that the value of K is less than 1. (If $[X]_{aq}$ is smaller than $[X]_{og}$ then $[X]_{aq}/[X]_{og}$ must be less than 1.)

For example, look at Table 11.1, which lists the distribution ratios for a series of fatty acids when mixed with water (an aqueous phase) and sec-butyl alcohol (an organic phase). As we move down the table, the length of the carbon chain increases in steps of one. If we look at the distribution ratio values, we see that they decrease as we descend the table, telling us that the fatty acids are partitioning more into the organic layer. Why is this? As the carbon chain increases in length so the acid becomes increasingly hydrophobic. Consequently, it partitions more readily into the organic layer.

- *A K value > 1 tells us that a compound will preferentially partition into an aqueous phase.*
- *A K value < 1 tells us that a compound will preferentially partition into an organic phase.*

The technique of solvent extraction exploits the different partitioning behaviour of compounds with different distribution ratios. So, if we have a mixture containing

Table 11.1 The distribution ratios for a series of fatty acids of increasing carbon chain length.

Compound (length of carbon chain in parentheses)	Concentration in aqueous layer (mM)	Concentration in organic layer (mM)	Distribution ratio $[X]_{aq}/[X]_{og}$
Ethanoic acid (2C)	151	192	**0.787**
Propionic acid (3C)	72.2	178.5	**0.405**
n-Butyric acid (4C)	31.3	159.6	**0.195**
Valeric [pentanoic] acid (5C)	27.42	311.4	**0.088**
Caproic [hexanoic] acid (6C)	11.65	288.6	**0.040**

a compound with a distribution coefficient that is > 1 and a compound with a distribution coefficient < 1 we would expect one compound to end up in one phase, and the other compound to end up in the other phase.

It is not so easy to separate compounds with very similar values of *K*. In this case we need to repeat the separation several times. Each time, the aqueous phase and organic phase are isolated from the separating funnel, and fresh organic phase is added to the aqueous phase. The separation process is then repeated, as illustrated in Figure 11.4. Gradually, all of the compound that shows a slight preference for the organic phase (that is, the one that is slightly more hydrophobic) will partition into it, leaving the aqueous phase enriched for the more hydrophilic of the two compounds.

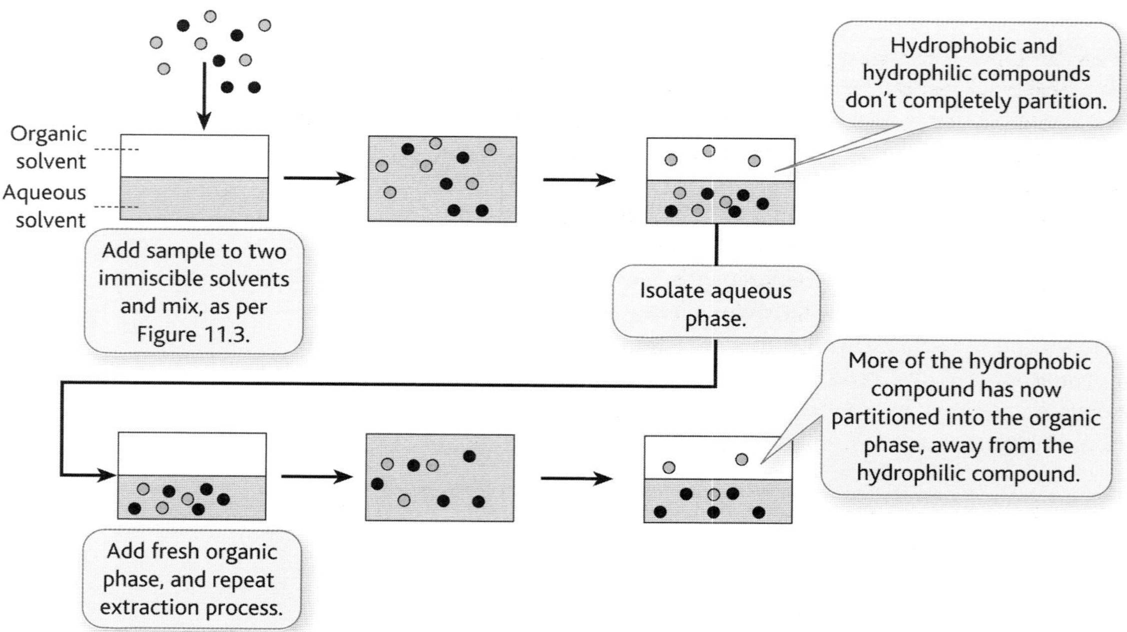

Figure 11.4 Compounds with similar values of *K* are difficult to separate, and solvent extraction has to be repeated several times. With each round of solvent extraction, fresh organic solvent is added. As the cycle is repeated, the compound showing the slight preference for the organic phase will gradually partition away from the aqueous phase.

Self-check 11.1 State whether the following compounds are most likely to partition into an aqueous or an organic phase when shaken in a separating funnel.

(a) O OH

(b) OH

Differential partitioning between hydrophilic and hydrophobic phases occurs within all biological systems and is a hugely important consideration in the design of pharmaceuticals. For example, lipid membranes are strongly hydrophobic, while blood serum is hydrophilic. Consequently, if a pharmaceutical compound is strongly hydrophilic it is most likely to partition into the blood serum and not cross into the lipid membrane. While this isn't a problem if you want to keep the compound in the blood for the purposes of its transport around the body, it *is* an issue if the site of action of the drug is in a particular tissue, such that you need the drug to leave the bloodstream and cross the lipid membrane into a cell. For this, the drug must be at least somewhat hydrophobic.

Having now seen how we can effect the separation of two compounds according to their differential behaviour when exposed to hydrophobic and hydrophilic liquid solvents, let us now move on to explore a family of techniques, which have widespread applications in biological research–chromatographic techniques.

Chromatography

The various types of chromatographic methods available to us utilize the contrasting physical attributes of different compounds to achieve separation. They thereby act as a potent means of distinguishing between and separating the compounds. These physical attributes include solubility, size, electrical charge, and selective binding characteristics.

Chromatography utilizes the differential partitioning of chemical species between two phases, just as we see for solvent extraction above. While solvent extraction uses two liquid phases, chromatography uses different types of phase–solid, liquid, and gas. A chromatographic technique features a **mobile phase** and a **stationary phase**. The mobile phase contains the substances that we seek to separate. The stationary phase has some property that can effect separation based on the physical characteristic that we are seeking to exploit. During chromatography, the mobile phase passes over the stationary phase, during which time separation occurs. However, the two phases do not mix: they are **immiscible**.

During a chromatography experiment, compounds partition between the stationary and mobile phases in a manner analogous to solvent extraction–with compounds separating according to their relative preferences for the stationary and mobile phases. We see above how we may have to carry out solvent extraction a number of times to separate compounds with similar distribution coefficients; during chromatography, partitioning events can happen many times as a sample passes along the chromatography column, helping to ensure that separation of compounds with even quite similar distribution coefficients can occur.

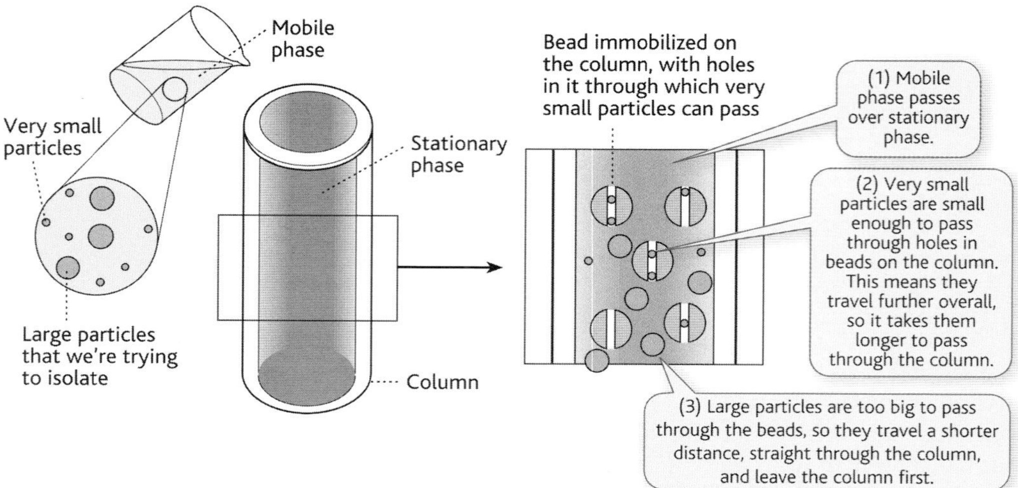

Figure 11.5 The conceptual basis of chromatography. During chromatography, a mobile phase (containing the substance(s) of interest) is passed over a stationary phase. The stationary phase effects separation of substances present in the mobile phase by virtue of a chosen physical attribute. This figure depicts a column that effects separation on the basis of size. We call this molecular exclusion chromatography or gel filtration chromatography.

Most chromatographic techniques use a **column** filled with tiny beads on which the stationary phase is immobilized. Figure 11.5 shows how the mobile phase is then passed through the column (and over the stationary phase). Notice how the mobile phase passes right through the column, and can be collected again once it's passed through the column. We call the fluid that's poured into the column the **eluent**, and the fluid exiting the column the **eluate**. We call the process of passing the mobile phase through the column **elution**.

In Figure 11.5, substances are being separated on the basis of size using a type of chromatography called **molecular exclusion chromatography**. In this case the mobile and stationary phases are the same (that is, they are both liquids) but the stationary phase occupies the beads with which the column is filled. These beads contain tiny holes, which are small enough to allow small molecules to pass through, while precluding the passage of larger particles. Consequently, because smaller molecules spend time travelling through the particles while large molecules do not, the larger molecules pass through the column more rapidly.

By collecting the eluate at different intervals as it leaves the column, we can collect different-sized compounds separately from one another. Larger compounds pass through the column most quickly and can be collected first; smaller compounds pass through much more slowly, and can be collected *after* the larger ones. Because the different molecules are retained within the column to differing extents, we say that the different-sized compounds have different **retention times**[3].

Figure 11.6 shows how the time at which a compound leaves a column can be used to identify that compound. Notice how each compound, A–C, leaves the column at a different time (because they each have different retention times). If we were then to test an unknown sample, X, which showed a retention time of 2 minutes, we could deduce that it must be a sample of compound A.

No matter which type of chromatography we consider, the principle is the same: compounds that partition more strongly into the stationary phase will be retained

3 The retention time is the length of time that a compound takes to pass through the column.

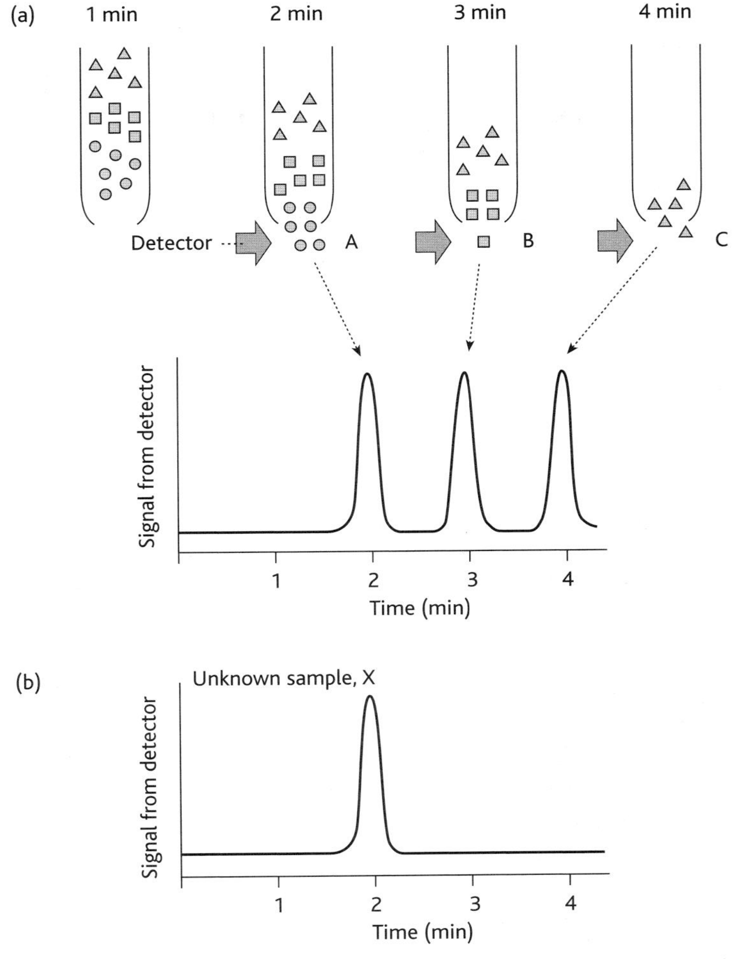

Figure 11.6 (a) The time at which a compound leaves a column can be used to identify that compound. Notice how we can use a detector to monitor when the separated substances leave the column. (b) We can compare the retention time of an unknown sample with the retention times of known substances to deduce the identity of the unknown sample. In this case, we can deduce that the unknown sample, X is compound A.

by the column while those compounds that don't partition into the stationary phase and spend their time in the mobile phase will come off the column first.

- ***Chromatography exploits the differential partitioning of a compound between a mobile and a stationary phase.***
- ***Different compounds in a mixture can be separated on a chromatographic column due to their different retention times on the column.***

We can use different stationary phases to select for compounds based on different criteria. For example, we may wish to separate out compounds on the basis of electrical charge. Figure 11.7 shows how, in this instance, we would use a stationary phase that attracts species of a particular charge, allowing neutral species, or species of opposite charge, to pass through the column more quickly.

Alternatively, we may wish to isolate a compound based on its specific binding properties. Our immune system is mediated by tens of thousands of different

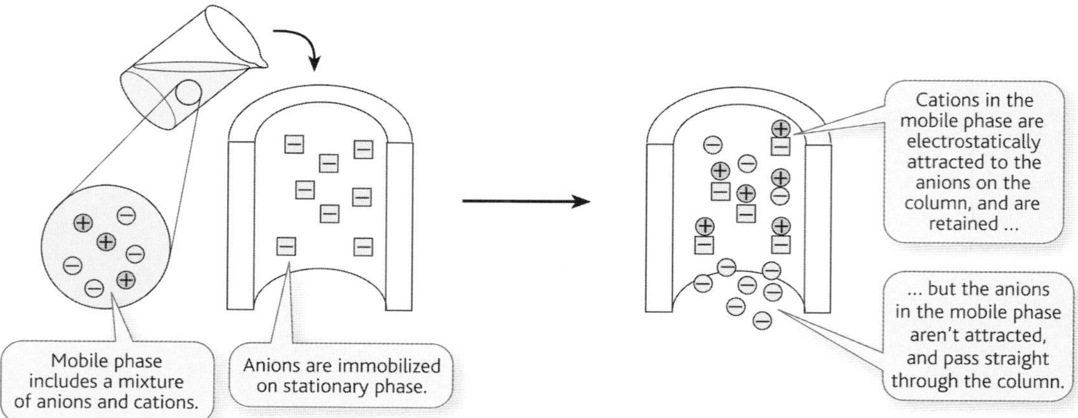

Figure 11.7 To separate out compounds on the basis of electrical charge we would use a stationary phase that attracts species of a particular charge, allowing neutral species, or species of opposite charge, to pass through the column more quickly.

antibodies. Antibodies protect our bodies from infection by specifically recognizing and binding to foreign particles (dust, viruses, bacteria, etc.). The process of binding triggers a cascade of biochemical reactions which result in the destruction of the foreign particle, hopefully before it can cause our body any damage. Antibodies recognize and bind to particles on the basis of **shape**: an antibody can only bind to a particle if the particle possesses a complementary shape.

We can use the selective binding properties of antibodies to separate molecules from the mobile phase. Look at Figure 11.8, which shows how a particular antibody can be immobilized on the stationary phase of a column. Notice how the antibody then binds to, and retains, only the particle to which it has a complementary shape (and, therefore, a binding affinity). We call this process **affinity chromatography**.

Liquid and gas chromatography: changing the mobile phase

When we carry out column-based chromatography we use one of two types of mobile phase. In **liquid chromatography** (**LC**) the mobile phase is a liquid. In **gas chromatography** (**GC**) the mobile phase is a gas. In both instances, the

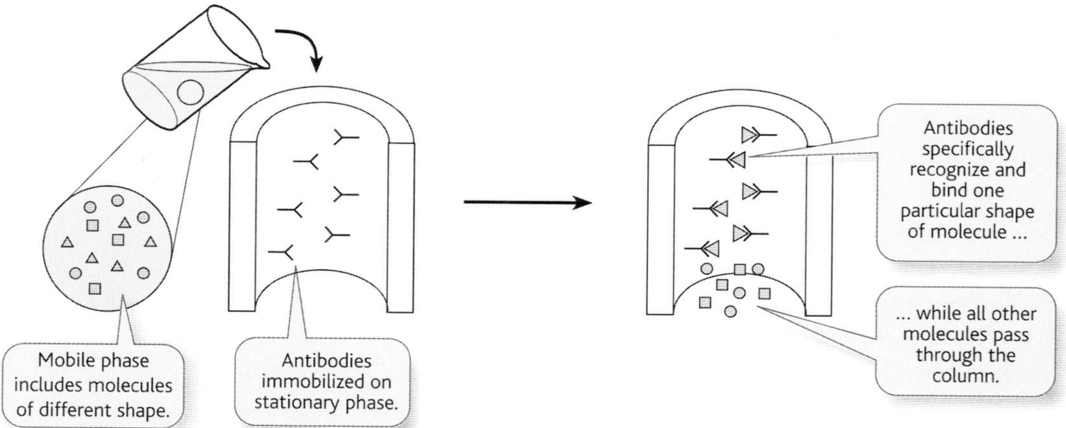

Figure 11.8 We can use the selective binding properties of antibodies to separate molecules from the mobile phase. A particular antibody can be immobilized on the stationary phase of a column. The antibody then binds to, and retains on the column, only the particle to which it has a complementary shape (and, therefore, a binding affinity). We call this process affinity chromatography.

substances that we are trying to separate are dissolved in the mobile phase: in liquid chromatography our samples of interest are carried in a liquid solvent; in gas chromatography our samples are carried in a gas. The *principle* behind chromatographic separation remains the same regardless of the nature of the mobile phase: both liquid and gas chromatography rely on the mobile phase moving over a stationary phase and molecules with different partition coefficients partitioning between the two phases as they travel down the column, leading to their separation. However, the nature of the compounds that can be separated by LC and GC is somewhat different: GC is only appropriate for more volatile compounds (those that can exist in the gas phase), whereas LC is typically more appropriate for a wider range of compounds, including those that are involatile.

We carry out liquid chromatography over a liquid or solid stationary phase. By far the most common chromatography is liquid–liquid chromatography. In this case both mobile and stationary phases are immiscible liquids; one is an aqueous liquid, the other organic. Similarly, gas chromatography can also use either a solid or a liquid stationary phase, as depicted in Figure 11.9. We call gas chromatography in which the stationary phase is a liquid, **gas–liquid chromatography** or **GLC**, and that in which the stationary phase is a solid, **gas–solid chromatography** or **GSC**.

One major advantage of both gas chromatography and liquid chromatography is that they can separate hundreds of different compounds from a single sample. This is particularly useful when studying biological systems, which typically comprise a complicated mixture of different compounds, including proteins, nucleic acids and the many different compounds produced from the biochemical reactions that are continually taking place in our cells. For example, GC is being applied in the study of metabolomics, the analysis of all the different compounds present in a biological sample at a given moment in time. By carrying out such analysis to determine how the biochemical composition of a biological system (a particular cell type or tissue, for example) changes in response to different stimuli, we can begin to elucidate how biological systems change as the result of environmental factors, or through the activity of drugs, for example.

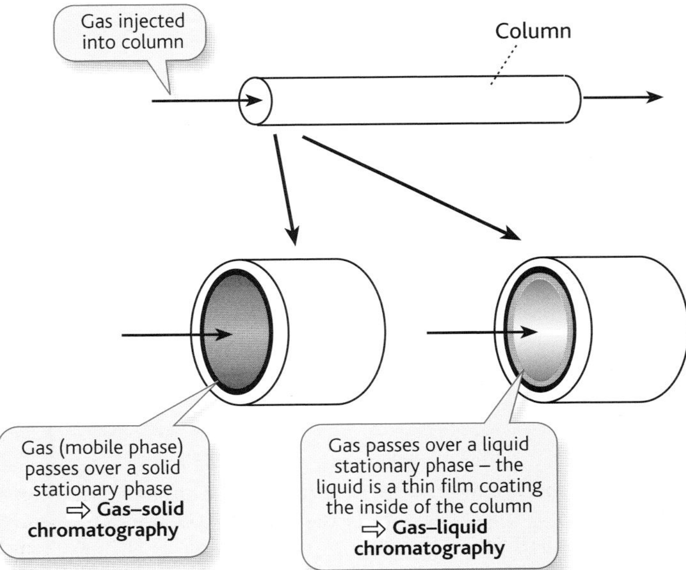

Figure 11.9 Gas chromatography can use either a solid or a liquid stationary phase.

We revisit liquid and gas chromatography again on p. 346, when we see how they can be *coupled* to identification techniques to rapidly identify the components of **mixtures** of compounds.

- ***Liquid and gas chromatography are powerful tools for separating out complicated mixtures of compounds, including those extracted from biological systems.***

Look at Box 11.2 to read about high-performance liquid chromatography, the most widely used type of liquid chromatography.

Electrophoresis

One of the most powerful analytical tools for studying and characterizing biological systems occurs not in a column but along a flat, two-dimensional sheet. **Electrophoresis** uses a combination of electrical charge, size, and shape to effect the separation of molecules as they travel through a porous sheet-like material.

The most commonly used material for studying biological molecules is acrylamide, which polymerises to form a gel. When a gel is formed in the presence of sodium dodecylsulfate, the system can be used to separate proteins on the basis of their molecular mass, a technique called sodium dodecylsulfate polyacrylamide gel electrophoresis. (Mercifully, sodium dodecylsulfate polyacrylamide gel electrophoresis is usually abbreviated to **SDS-PAGE**.)

Look at Figure 11.10, which shows how SDS-PAGE is used to separate out and identify proteins on the basis of their molecular weight. SDS is a detergent which gives proteins a uniform electrical charge. When an electrical current is set up through the gel, the charged sample molecules are pulled through the gel at a rate that is determined by their size: the movement of large molecules is impeded more by the gel than is the movement of smaller molecules, such that smaller molecules travel a greater distance through the gel per unit time.

DNA molecules can also be separated by electrophoresis. However, individual DNA molecules can be much larger than proteins and so require a different type of gel: instead of using polyacrylamide, we use a gel formed from agarose. Agarose solidifies to form a very 'loose' gel whose pores are large enough for the DNA to travel through once a current is applied.

BOX 11.2 High-performance liquid chromatography (HPLC)

The most widespread type of liquid chromatography carried out today is **high-performance liquid chromatography** or **HPLC**. This technique is sometimes called **high-pressure** liquid chromatography, which gives us a clue to the procedure used.

During HPLC, the liquid mobile phase is injected through the chromatographic column at high pressure. The use of high pressure results in much quicker separation (and so greater time efficiency). However, the truly characteristic feature of HPLC is the very small particle size of the stationary phase.

The small particle size generates a huge surface area for the stationary phase (the surface area of some HPLC columns, just a few centimetres long, now exceeds the surface area of the pitch at Wembley Stadium). The large surface area, in turn, improves the **resolution** of the technique (the minimum limit in the difference between two species at which they can be reliably separated), allowing separation to be more **sensitive**. This is because the partitioning between the mobile and stationary phase can occur many hundreds of thousands of times as the eluent moves along the column. Consequently, substances with only subtle differences in a physical characteristic (for example, only a very small difference in size or shape) can be separated. Indeed, HPLC is so powerful that it is even able to separate enantiomers (which, as we see in Chapter 10, differ from each other in only very subtle ways).

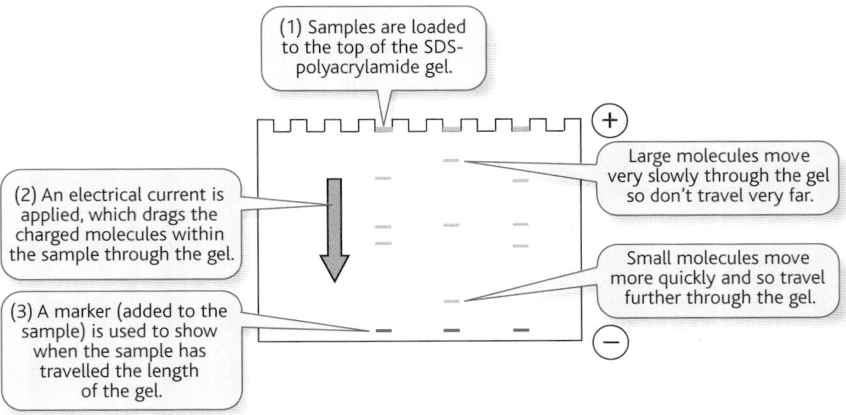
(1) Samples are loaded to the top of the SDS-polyacrylamide gel.
(2) An electrical current is applied, which drags the charged molecules within the sample through the gel.
(3) A marker (added to the sample) is used to show when the sample has travelled the length of the gel.
+
Large molecules move very slowly through the gel so don't travel very far.
Small molecules move more quickly and so travel further through the gel.
−

Figure 11.10 Sodium dodecyl sulfate polyacrylamide gel electrophoresis (SDS-PAGE). SDS-PAGE is used to separate out and identify proteins on the basis of their size. A similar experimental set-up is also used for the separation of nucleic acids, though an agarose gel is typically used rather than polyacrylamide.

- ***The electrophoretic separation of proteins is typically conducted on an SDS-polyacrylamide gel.***
- ***The electrophoretic separation of nucleic acids is conducted on an agarose gel.***

One of the most powerful applications of electrophoresis has been in the development of DNA profiling techniques, as described in Box 11.3.

BOX 11.3 DNA profiling

Everyone (except identical twins) has a very slightly different genome, a set of 23 pairs of chromosomes comprising different DNA sequences. Having now sequenced the human genome, we are getting closer to the goal of being able to sequence the genomes of individuals on a routine basis and then using this information as a way of identifying individuals.

Instead of examining entire genomes, however, DNA profiling exploits regions of great variability that occur at precise locations on the chromosomes of all individuals (just as a fingerprint is a region of great variability at a precise location–the fingertip). Instead of sequencing large portions of the genome, we need only sequence specific regions–a far more manageable job!

Specifically, DNA profiling examines the length of **short tandem repeats** (STRs), areas of high variability that occur at specific locations (or **loci**) on a chromosome. An STR comprises a small number of base pairs (usually around four) that may be duplicated (or **repeated**) several times. It is the variation between individuals in the number of STR repeats at each locus that gives each of us our own unique DNA profile.

Look at Figure 11.11, which illustrates the principle behind DNA profiling. Imagine that we are interested in the number of STR repeats at a particular point on a chromosome, and we're examining the same chromosome from three individuals, a, b, and c. Each individual has a different number of STR repeats at the locus of interest: a has 1 STR; b has 4 STRs; and c has 6 STRs. In step 1, we isolate only the region of each chromosome that contains the STR, and subject these fragments to SDS-PAGE (step 2). As we see above, SDS-PAGE separates molecules on the basis of size: small molecules travel further down the gel than larger molecules.

In step 3, we use a detector to identify the location of each fragment on the gel; the detector converts position on the gel to a peak on a DNA profile: large fragments appear to the left of the profile; smaller fragments appear to the right.

Real DNA profiles show the size of STR repeats from a range of different loci on different chromosomes: they build up a pattern of fragment sizes which varies from person to person. It is this pattern of STR repeat sizes at different loci that is your DNA profile.

DNA profiles from suspects and convicted criminals, together with profiles from unsolved crime evidence, are held in the National DNA Database (NDNAD). By comparing profiles of DNA of unknown origin (i.e. an unidentified offender) with profiles held in the database, the police are able to see if the profile matches a profile of *known* origin–that is, a known suspect or convicted criminal.

BOX 11.3 **Continued**

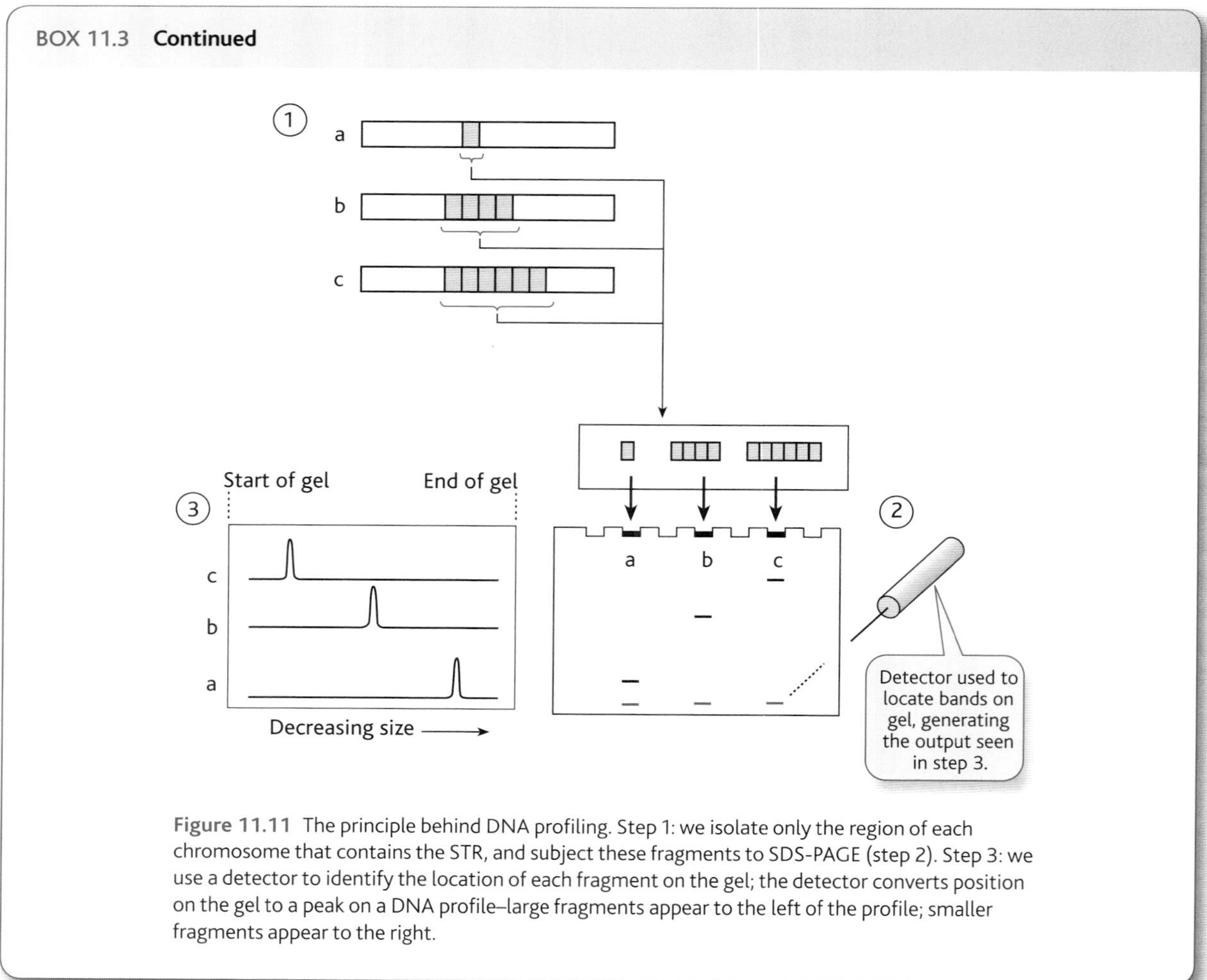

Figure 11.11 The principle behind DNA profiling. Step 1: we isolate only the region of each chromosome that contains the STR, and subject these fragments to SDS-PAGE (step 2). Step 3: we use a detector to identify the location of each fragment on the gel; the detector converts position on the gel to a peak on a DNA profile–large fragments appear to the left of the profile; smaller fragments appear to the right.

We learn more about pH in Chapter 16.

A more powerful method for the separation of polypeptides and proteins couples SDS-PAGE to a gel-based method that separates molecules solely on the basis of charge. All polypeptide molecules carry a slight electrical charge because both the amino and carboxyl groups at the ends of the polypeptide, and some amino acid side chains, are ionized in aqueous solution, as depicted in Figure 11.12. However, the extent to which a polypeptide is ionized (and hence the size of the charge it carries) depends on the pH of the solution in a way that varies from polypeptide to polypeptide, depending on the particular combination of amino acids from which the polypeptide is formed. The pH value at which a given polypeptide carries no electrical charge is called its pI value; the pI value varies from protein to protein. For example, hen albumin has a pI value of 4.6, while human haemoglobin has a pI value of 7.1. This means that hen albumin and human haemoglobin carry a zero electrical charge at pH 4.6 and 7.1 respectively.

Polypeptides can be separated on the basis of their pI values in a process we call **isoelectric focusing,** as illustrated in Figure 11.13(a). This technique requires a gel that features a pH gradient, with a low pH at one end of the gel, and a high pH at the other. If the location of a protein on the gel has a pH that is *different* to the protein's pI value, then the protein will carry an electrical charge. Consequently, when an electrical charge is applied across the gel, the protein will migrate through

Aspartic acid Glutamic acid Histidine Lysine Arginine

Figure 11.12 All polypeptide molecules carry a slight electrical charge, in part because of the way that some amino acid side chains are ionized in aqueous solution, as illustrated here.

the gel: a positively charged protein will be drawn towards the cathode, while a negatively charged protein will be drawn towards the anode. However, when the protein reaches a location on the gel whose pH is the same as the protein's pI value, the protein will no longer carry an electrical charge. Consequently, it will no longer be affected by the charge being applied across the gel, and will no longer move. Instead, the protein will become concentrated (or 'focused') at that location. Because different proteins have different pI values, they will be focused at different locations on the gel–locations whose pH values match the pI values of the proteins in question. An example of an isoelectric focusing gel is shown in Figure 11.13(b).

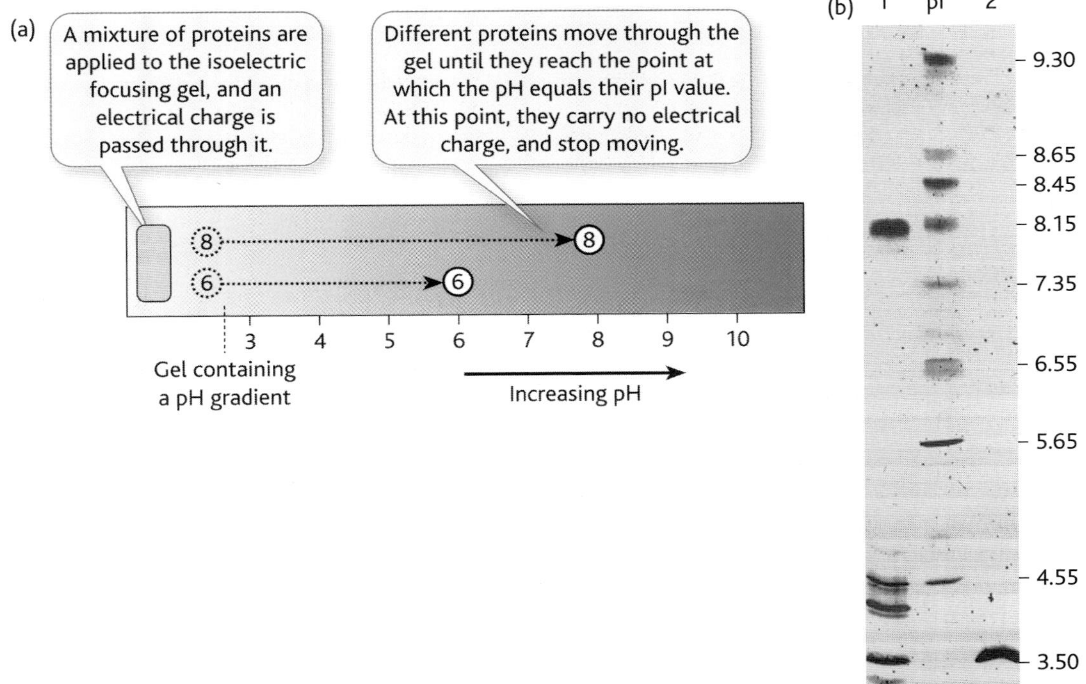

Figure 11.13 Isoelectric focusing. (a) Polypeptides in a sample move through an isoelectric focusing gel until the pH of the gel is equal to their pI value. At this point, they carry no electrical charge, and stop moving. (b) An example of an isoelectric focusing gel. The middle lane (pI) shows markers with known pI values. Lanes 1 and 2 show the separation of proteins of interest, whose pI values can be deduced by comparison with the known markers in the middle lane. © The Biochemical Society, 1996.

- *Isoelectric focusing separates proteins based on their pI values.*
- *The pI value is the pH at which a protein has no net electrical charge.*

Polypeptides and proteins that have been separated on an isoelectric focusing gel can then be separated on the basis of their mass using SDS-PAGE as depicted in Figure 11.14. We call this process **two-dimensional PAGE**. Look at Figure 11.14 and notice how a very complicated mixture of polypeptide molecules can be separated using this powerful technique because polypeptides that travel the same distance on the isoelectric focusing gel (and are hence not separated by this technique) *can* be separated on the subsequent SDS-polyacrylamide gel.

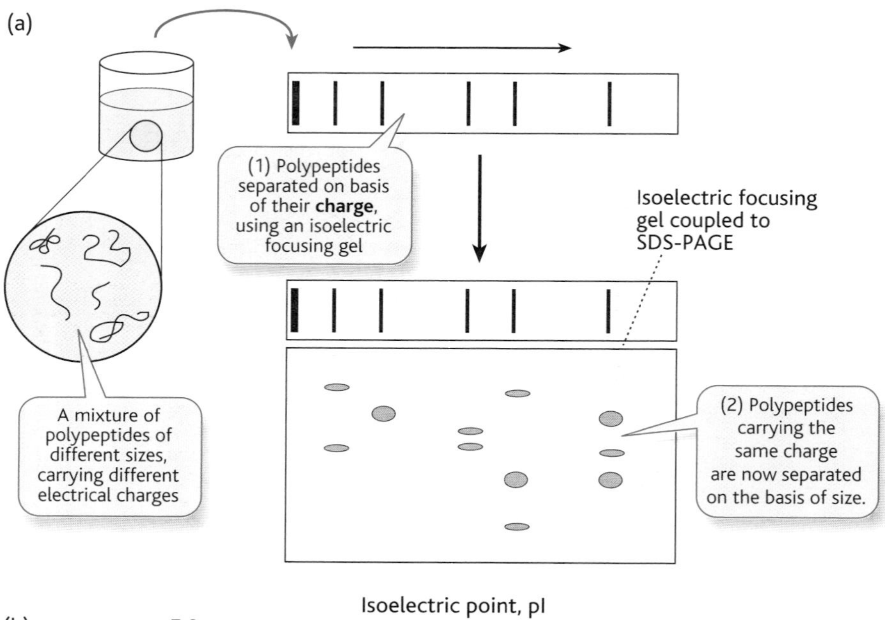

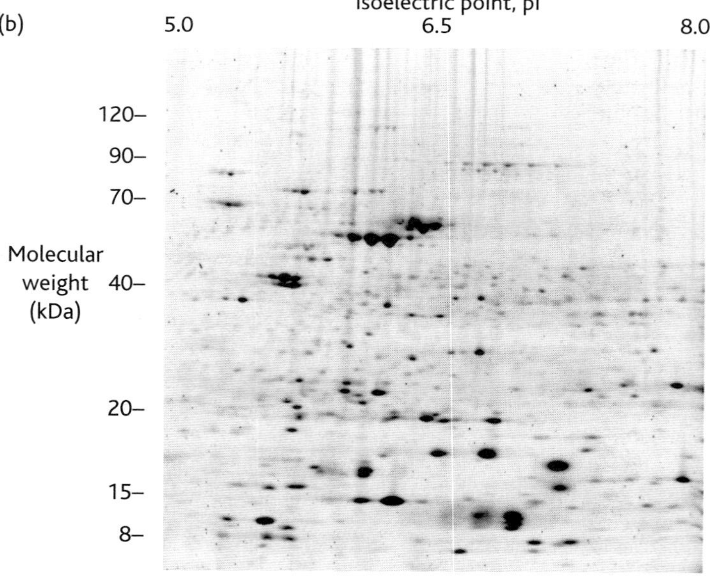

Figure 11.14 (a) Polypeptides and proteins that have been separated on an isoelectric focusing gel can then be separated on the basis of their mass using SDS-PAGE. We call this two-dimensional PAGE. (b) An example of the outcome of running 2D-PAGE on a mixture of proteins. (Image courtesy of Dr S. Cordwell.)

Centrifugation

We now turn to a technique that separates biological molecules and complexes according to the speed at which they sink through a liquid medium when subjected to some kind of force, a phenomenon called **sedimentation**.

Any material that is suspended in a liquid will undergo sedimentation over time–that is, it will sink to the bottom of the vessel in which it is stored. Usually, the force being applied to the suspended material is simply the force of gravity, which acts on the material to drag it down through the liquid. For example, if we look at a bottle of cloudy apple juice or wheat beer, we can often see a sediment at the bottom of the bottle. This sediment has accumulated over time as gravity has acted on material that was originally suspended in the apple juice or beer.

Sedimentation under the force of gravity can be a slow process. However, the rate of sedimentation can be increased by increasing the force being applied to the suspended material. We can do this by applying a centrifugal force–that is, by spinning it.

Many of us may recall a typical bank holiday weekend spent playing on a sandy beach with bucket and spade in hand, and, with it being a bank holiday, a light drizzle in the air, and a stiff breeze whipping in from the sea. If a bucket of wet sand is allowed to sit undisturbed for a while, the sand will settle to the bottom of the bucket. We say the sand sediments under the force of gravity. As it sediments under this force, the sand experiences acceleration, the value of which is equal to g, the Earth's gravitational field; g has a value of approximately 9.81 m s^{-2} at sea level.

If we were to swing the bucket containing the slurry of sand and water, round and round in a circle, the sand would sediment more quickly than under the influence of gravity alone. The circular (or centrifugal) motion of the bucket creates an increased force on the sand, which depends on the speed of rotation. This force is called the **centrifugal force**.

We learn more about determining the value of the centrifugal force in Box 11.4.

How do we apply a centrifugal force to a sample in order to effect sedimentation? We use a technique called **centrifugation**, and an instrument called a **centrifuge**.

Centrifuges vary in size from those that can sit on the laboratory bench to those that need a small room to accommodate them. Like many of the techniques we are discussing, centrifugation can be used both analytically and preparatively. Analytical centrifugation is used to study the properties of purified biological molecules and complexes such as proteins or ribosomes, while preparative centrifugation is used to separate components of biological mixtures such as cells and subcellular organelles. For example, centrifugation can be used to separate blood into plasma and the different types of blood cells, or to isolate viruses or bacteria from a growth medium.

Types of centrifuges

All centrifuges contain an electric motor to drive a rotor around a central shaft. The rotor, which can vary in size, has a sample holder attached; the sample holder may be at a fixed angle or free swinging. A simplified representation of a centrifuge is illustrated in Figure 11.15.

Centrifuges are usually categorized as low-speed, high-speed, or ultracentrifuges according to the speed they can attain. Benchtop, low-speed centrifuges are designed to handle volumes up to 50 mL and generate a centrifugal force of around 7,000 g; such centrifuges are used to sediment bacteria from the medium they are grown in, for example. Benchtop microfuges are used routinely for small volumes and can generate centrifugal forces up to 10,000 g.

BOX 11.4 What connects spinning to *g*?

The centrifugal force exerted on a particle by rotational motion depends both on the speed of rotation and the distance the particle is from the central axis of rotation. The applied centrifugal force, G, is calculated from the following equation:

$$G = \omega^2 r$$

where ω is angular velocity in radians per second (rad s^{-1}) and r is the distance from the axis of rotation in centimetres.

Since the circumference of the circle is equal to $2\pi r$, ω–the speed at which the rotor rotates–can be rewritten in terms of revolutions per minute (rev min^{-1}):

$$\omega = \frac{2\pi\ (\text{rev min}^{-1})}{60}$$

If we substitute this expression for ω into our expression for G we obtain

$$G = \frac{4\pi^2 (\text{rev min}^{-1})^2 r}{60^2}$$

Centrifugal force is usually expressed relative to the Earth's gravitational field, g (981 cm s^{-2}), under which conditions we call it the relative centrifugal force, RCF.

$$\text{RCF} = \frac{4\pi^2 (\text{rev min}^{-1})^2 r}{60^2 \times 981}$$

(Notice how this is the same as our expression for G, but with the bottom part of the expression multiplied by the value of g.)

By calculating the actual values of $4\pi^2$ and ($60^2 \times 981$) we obtain a simplified expression for RCF:

$$\text{RCF (in } g) = 1.12 \times 10^{-5}\ (\text{rev min}^{-1})^2\ r$$

So, if we have a rotor with a radius of 50 cm travelling at 1000 rpm (rev min^{-1}) our sample will experience a relative centrifugal force of

$$(1.12 \times 10^{-5}) \times (1000)^2 \times 50 = 560\ g$$

–that is, 560 times the force of gravity.

High-speed centrifuges are much larger and are usually floor-standing. They may contain a refrigeration unit which allows the samples to be kept cool during long spins (to prevent biological materials such as heat-sensitive proteins becoming denatured during the centrifugation process, for example). These centrifuges can generate up to 70,000 *g*.

Ultracentrifuges are exceedingly sophisticated instruments and must be equipped with a refrigeration unit and a vacuum pump. They are capable of generating centrifugal force up to 800,000 *g*. To achieve this level of centrifugal force the rotor compartment must not only be cooled to remove the heat generated, but must also be evacuated of all air to reduce friction. Ultracentrifuges enable subcellular organelles such as ribosomes, lysosomes, and the endoplasmic reticulum to be isolated and studied.

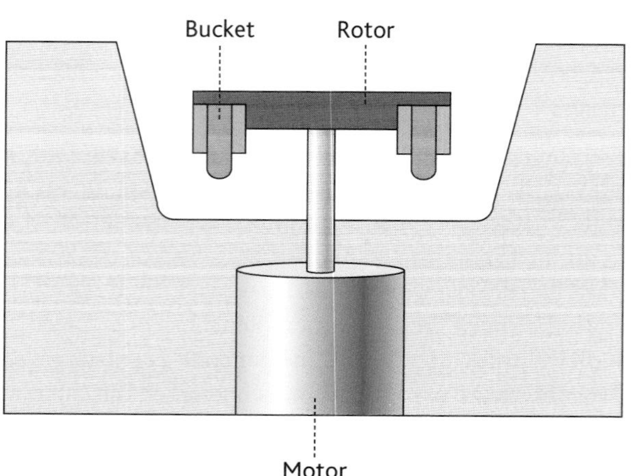

Figure 11.15 A schematic diagram of a centrifuge. Samples are loaded into buckets, which are located on the rotor. The buckets may either be at a fixed angle (whereby the buckets don't move when the centrifuge is in operation) or free-swinging (in which case the buckets swing up when the centrifuge is spinning).

Sample collection from centrifuges

Once a sample has been subjected to centrifugation it separates into two fractions. Figure 11.16 shows how solid matter forms a **pellet** at the bottom of the vessel in which the sample has been centrifuged, while the medium in which the solid matter was originally suspended forms the **supernatant**.

When whole blood is centrifuged, for example, the solid material (the blood cells) forms the pellet, while the blood plasma (in which the cells were originally suspended) forms the supernatant.

Supernatant

Pellet

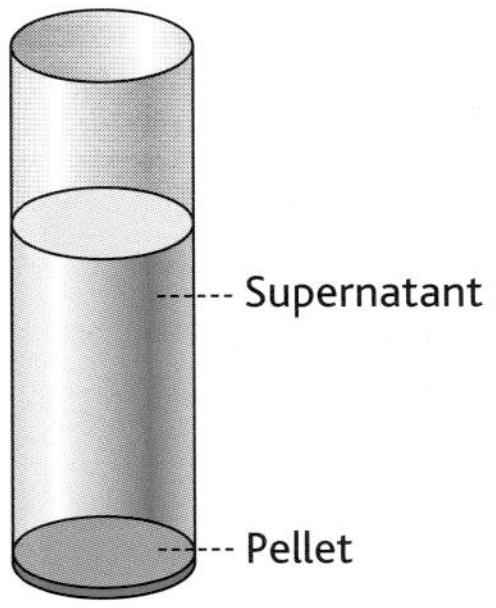

Figure 11.16 Following centrifugation, a sample separates into a pellet, comprising solid matter that was originally suspended in the sample medium, and the supernatant, the fluid in which the solid matter was originally suspended.

- *Centrifugation speeds up sedimentation of a sample by spinning it to apply a centrifugal force.*
- *After centrifugation, solid matter forms a pellet, while the medium in which the solid was originally suspended forms the supernatant.*

Sometimes we may want to do more than just separate solid from liquid. For example, we may wish to separate white blood cells from red blood cells (as well as separating both types of cell from the blood plasma). This higher-resolution separation can be achieved by centrifuging a sample through an appropriate gradient, as illustrated in Figure 11.17. The gradient may comprise layers of liquid possessing different densities, which inhibit the sedimentation of solid matter to different degrees. Large, heavy particles will travel through dense liquid more readily than lighter particles, causing particles of particular size to be sequestered at the interfaces between the different layers of liquid.

Having now explored various ways of separating out different components of a mixture, we now turn to the key theme of this chapter: how do we identify what we've got once we've isolated it?

11.3 Measuring mass: mass spectrometry

The use of chemical analysis to identify the molecule or molecules present in a sample centres on determining some kind of defining characteristics of the molecules, and using these characteristics to deduce what the molecule must be.

In our example involving tea and coffee in Section 11.1 we use what we already know about the two compounds–their contrasting tastes–to distinguish one from

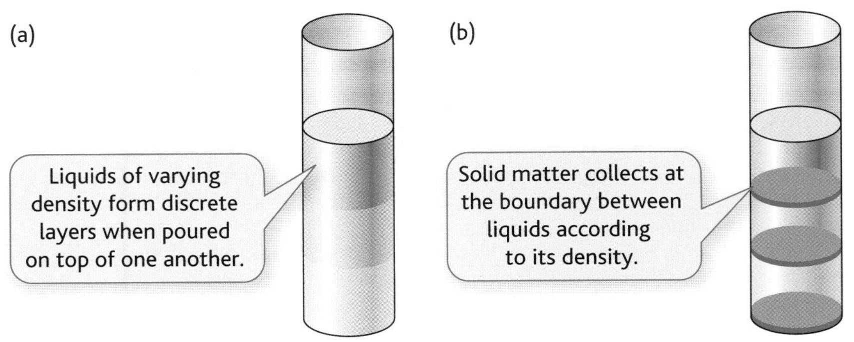
(a)
Liquids of varying density form discrete layers when poured on top of one another.
(b)
Solid matter collects at the boundary between liquids according to its density.

Figure 11.17 Density gradient centrifugation. (a) Liquids of different density can be layered to form a gradient, as depicted by the different shades of blue here. (b) When a sample is added to the gradient, and the tube centrifuged, the different components of the sample separate at the boundaries between layers of different density. The different layers can then be removed one at a time.

the other. However, it would be a spectacularly bad idea to attempt to use taste to distinguish between the different compounds we might encounter in the lab. Instead, we use various other characteristics, including mass, chemical composition, and three-dimensional structure. In this section we learn about a technique that uses the measurement of mass to aid the identification of a sample molecule.

Remember: the atomic masses of the chemical elements are assigned units which we call atomic mass units (amu).

Mass spectrometry measures the **mass** of a compound, and uses this information to identify the compound.

We see in Chapter 2 how every element in the periodic table has a characteristic atomic mass, which tells us the mass of one atom of the element relative to one atom of ^{12}C. The mass of an atom of ^{12}C is the 'gold standard' against which the atomic mass of each of the other chemical elements is calibrated: the mass of an atom of ^{12}C is assigned a value of 12 amu[4].

The atomic mass of hydrogen is assigned a value of 1 amu; on the same scale, the atomic mass of sodium is assigned a value of 23 amu[5].

The mass of a compound is the sum of the masses of all the atoms that comprise the compound. For example, water, H_2O, comprises two atoms of hydrogen (each with a mass of 1 amu), and one atom of oxygen (with a mass of 16 amu). Hence, the mass of one molecule of water is (2 × 1amu) + (1 × 16 amu) = 18 amu.

Self-check 11.2 What is the mass of one molecule of methanol, CH_3OH?

How does mass spectrometry work?

Mass spectrometry exploits the difference in behaviour of particles of different mass and, specifically, the fact that it is more difficult to make a heavy moving object change direction than it is a lighter moving object, when both objects are travelling at the same speed. A moving object has forward **momentum**–a force that continues to drive the object forward temporarily even when no energy is being supplied to propel it. We experience momentum every time the brakes are applied in a car in which we're travelling: as the brakes slow the car down, our momentum continues to drive us forward.

It is the phenomenon of momentum that makes it so important that we wear car seat belts. If we're in a car travelling at 70 miles per hour, and are involved in a sudden collision that brings the car to an almost immediate stop, momentum will continue to propel our bodies forward at 70 mph, even though the car surrounding us is now stationary. Our seat belts will inhibit this forward motion, tethering us to our seats.

---

4 1 amu is defined as 1/12 mass of one ^{12}C atom.

5 Remember: if an element exists as only one isotope, then its atomic mass is equal to its mass number. Sodium exists as a single isotope, ^{23}Na; the atomic mass of sodium is 23 amu–equal in value to its mass number.

If an element exists as a mixture of more than one isotope, its atomic mass is an average of the mass numbers of the different isotopes, with the contribution of each isotope to the atomic mass of the element weighted according to the isotope's relative abundance. For example, chlorine exists as a mixture of 75.78% ^{35}Cl (with mass number 35) and 24.22% ^{37}Cl (with mass number 37). Consequently, the atomic mass of chlorine is 35.5 amu.

By contrast, oxygen exists as a mixture of three isotopes, ^{16}O, ^{17}O, and ^{18}O. However, 99.76% of naturally occurring oxygen is ^{16}O; consequently, the other two isotopes make almost no contribution to the observed, average atomic mass of oxygen, which we take to be 16 amu–equal to the mass number of the predominant isotope, ^{16}O.

If we're not wearing a seat belt, however, momentum will propel us forward until our body meets something stationary that blocks our path. This stationary object may be the seat in front of us (if we're a rear passenger), the steering wheel, or even the front windscreen.

A heavy object has *more* momentum than a light object, even if both objects are moving at the same speed. This is why, if a car and lorry are travelling at the same speed in adjacent carriageways on the motorway and both vehicles encounter a traffic queue, it takes longer for the lorry driver to bring the vehicle to a complete stop than it does the car driver: the lorry driver has more momentum to overcome in order to bring the vehicle to a standstill.

Mass spectrometry separates out different chemical species on the basis of their mass by exploiting the phenomenon of a heavy object having more momentum than a light object. Figure 11.18 shows how, during mass spectrometry, chemical species are accelerated along a straight path, and a force is applied to **deflect** the chemical species from this straight path. The deflected species follows a new **trajectory** (curved path) before striking a detector. Only particles that are deflected by an appropriate amount follow the right trajectory to hit the detector: if a particle is not deflected enough, or is deflected too much, its trajectory will not take it to the detector.

How does mass spectrometry separate chemical species on the basis of mass?

Mass spectrometry measures the **force** that is required to deflect particles so that their new trajectories take them to the detector. The size of the force that needs to be applied tells us about the mass of the particle being studied: a light particle requires a small force to be deflected, while a heavy particle requires a large force to be deflected.

Mass spectrometry uses an electromagnetic force to drive the deflection of the chemical species being studied. (The electromagnetic force is supplied by an electromagnet.) Figure 11.19 shows how, during mass spectrometry, we gradually increase the strength of the electromagnetic force being applied by the electromagnet so that particles of increasing size are deflected by the right amount for them to reach

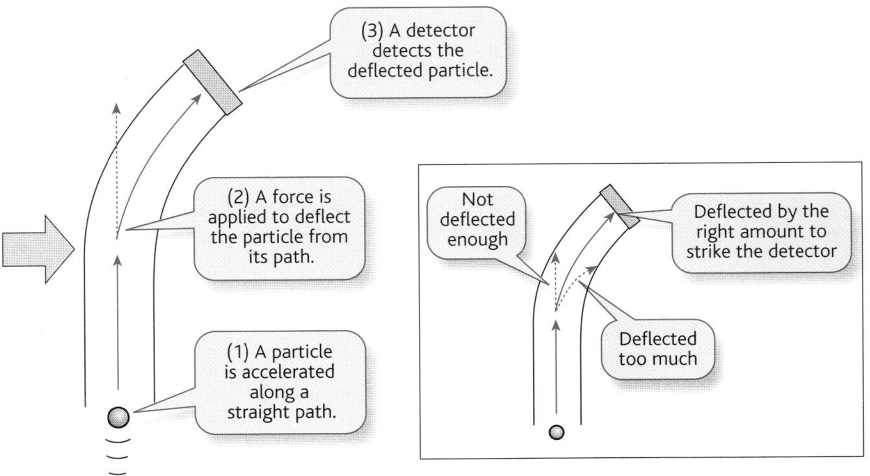

Figure 11.18 The conceptual basis of mass spectrometry. During mass spectrometry, chemical species are accelerated along a straight path, and a force is applied to deflect the chemical species from this straight path. The deflected species follows a new trajectory (curved path) before striking a detector.

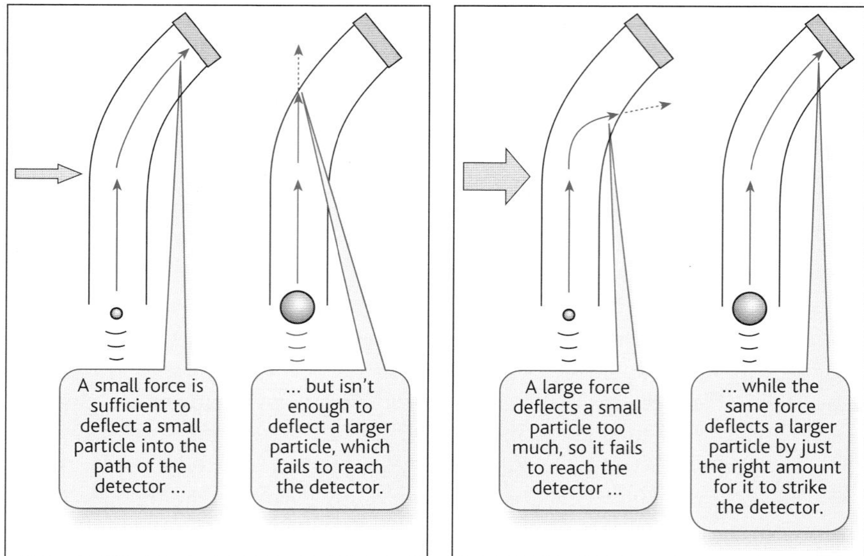
A small force is sufficient to deflect a small particle into the path of the detector ...
... but isn't enough to deflect a larger particle, which fails to reach the detector.
A large force deflects a small particle too much, so it fails to reach the detector ...
... while the same force deflects a larger particle by just the right amount for it to strike the detector.

Figure 11.19 During mass spectrometry, we gradually increase the strength of the electromagnetic force being applied by the electromagnet so that particles of increasing size are deflected by the right amount for them to reach the detector. The precise force required to deflect the particle by the right amount for it to hit the detector depends on the mass of the particle.

the detector. By measuring the electromagnetic force that is required to deflect a particular particle by the right amount for it to strike the detector, we can deduce the **mass** of that particle.

An electromagnetic force can only cause the deflection of a **charged particle** (an ion). Hence, any chemical species being studied must be **ionized** before it is deflected. There are various ways of effecting ionization (which we won't discuss in detail here); each method generates cations (positively charged ions), which can then be deflected and detected in the mass spectrometer.

Look at Figure 11.20, which illustrates the five main stages in mass spectrometry:

1. Vaporization–turning the sample into a gas
2. Ionization–turning uncharged molecules into ions
3. Acceleration–causing the ions to move fast
4. Deflection–causing the ions to follow a circular trajectory (path)
5. Detection–determining which ionic masses are present.

Notice how the sample being tested must be a vapour (i.e. in gaseous form) prior to being ionized, and that an electric field is used to accelerate the sample once it has been ionized.

Ionization of sample molecules can be achieved in a number of different ways, and the approach taken is often dictated by the sample being analysed. For example, electron impact (EI) mass spectrometry ionizes samples by bombarding them with high-energy electrons. By contrast, in chemical ionization (CI) mass spectrometry, the sample is mixed with a gas before the mixture is bombarded by electrons.

One potential drawback of mass spectrometry is that it needs the sample to be vaporized before ionization occurs. While this is readily achievable for volatile compounds, it is more problematic for those samples that don't enter the gas phase so readily (for example, compounds with a high boiling point). Consequently, alternative

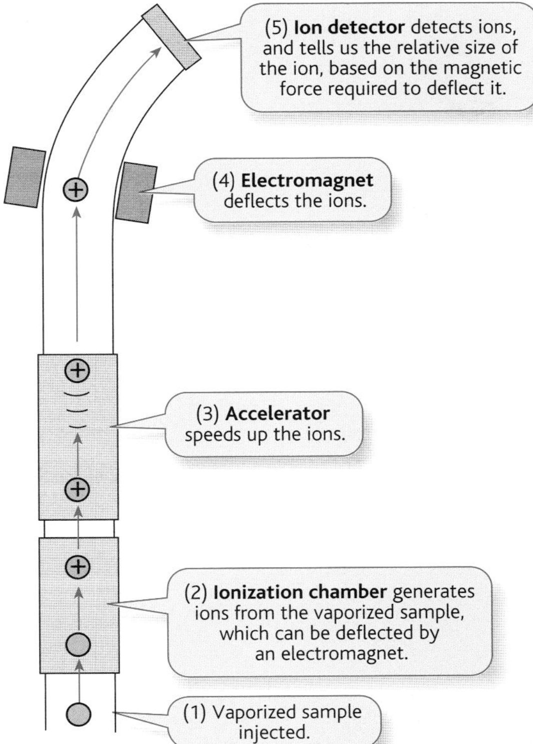
(5) **Ion detector** detects ions, and tells us the relative size of the ion, based on the magnetic force required to deflect it.
(4) **Electromagnet** deflects the ions.
(3) **Accelerator** speeds up the ions.
(2) **Ionization chamber** generates ions from the vaporized sample, which can be deflected by an electromagnet.
(1) Vaporized sample injected.

Figure 11.20 The five main stages in mass spectrometry.

approaches have been developed to achieve vaporization and ionization. These include fast atom bombardment (FAB), electrospray ionization (ESI), and MALDI (matrix-assisted laser desorption/ionization). We won't explore each of these in detail, but it is worth just being aware that, when you see terms such as 'MALDI' or 'FAB', the same basic principles of mass spectrometry–as illustrated in Figure 11.20–are being applied, but with certain modifications to the vaporization and ionization stages depending on the nature of the sample being analyzed.

The mass spectrum: the outcome of mass spectrometry

Look at Figure 11.21, which shows a typical mass spectrum. The mass spectrum shows the **masses** of the particles detected by the mass spectrometer, and the **relative abundance** of each particle.

Notice that the *x*-axis of the mass spectrum has the label 'm/z', which denotes the 'mass to charge ratio' of the ion being detected. The 'mass to charge ratio' is the same as saying the 'mass of the ion divided by charge on a single ion'. The vast majority of ions detected carry a single positive charge, i.e. $z = 1$. Consequently, the mass to charge ratio, m/z, equals the mass/1, which simply equals the mass.

The species with the highest m/z ratio (i.e. largest mass) usually represents the **molecular ion**, i.e. the ionized version of the molecule being studied.

Notice, however, how the mass spectrum includes a *number* of peaks, indicating the detection of a number of different particles from the testing of a single compound. Why is this?

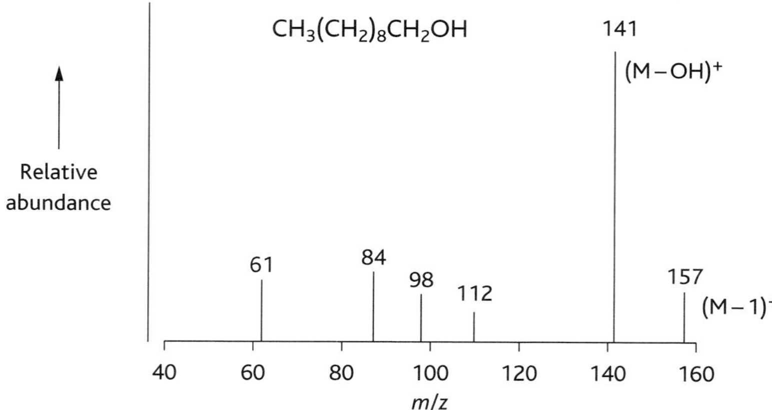

Figure 11.21 A typical mass spectrum. Notice how the mass spectrum shows the masses of the particles detected by the mass spectrometer, and the relative abundance of each particle.

The process of ionization mentioned above rarely produces just the molecular ion:

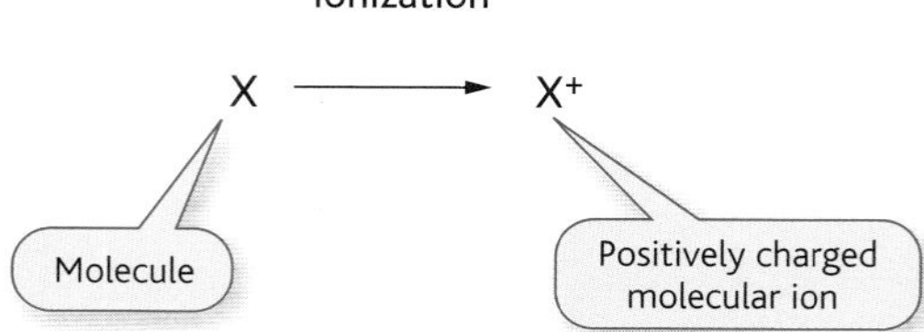

More often, ionization causes **fragmentation** of the molecule, yielding a number of smaller particles. However, only some of these fragments may be ions, and can be detected by the mass spectrometer. Uncharged molecular fragments are 'invisible'.

What can mass spectrometry tell us?

Mass spectrometry can be used to tell us two things:

1. The identity of an unknown molecule
2. Some aspects of the structure of a molecule.

Let's consider each of these in turn.

The identity of an unknown molecule

If we know the mass of a molecule (deduced from the peak in a mass spectrum that corresponds to the molecular ion), we are increasingly able to deduce the chemical identity of the molecule: advances in the sensitivity of mass spectrometry mean that mass spectra are of much higher resolution than in the past. But why does the resolution of a mass spectrum make the difference between being able to identify a molecule and not?

We see in Section 2.2 how many elements exist as a mixture of different isotopes, and each isotope makes a contribution to the atomic mass of a particular element. For convenience, we usually quote atomic masses as integer values (i.e. whole numbers). For example, oxygen has an atomic mass of 16 amu, carbon has an atomic mass of 12 amu, and hydrogen has an atomic mass of 1 amu. In reality, however, atomic masses don't have integer values *exactly*. If we're more precise–and quote atomic masses to more decimal places–we have the kind of values shown in Table 11.2.

Table 11.2 High-precision atomic masses for several biologically important elements.

Element	Atomic mass (amu)
C	12.0107
H	1.0079
O	15.9994
N	14.0067

Using these values, ethanoic acid, CH_3COOH, has a molecular mass of 60.0518, while propanol, C_3H_7OH, has a molecular mass of 60.0947.

Imagine that we generate a mass spectrum that gives only integer values, and it yields a molecular ion peak with an *m*/*z* value of 60. This peak could represent the molecular ion of either ethanoic acid, CH_3COOH, or propanol, C_3H_7OH. It is not of high enough resolution to distinguish between the two compounds; both compounds have a molecular mass of 60 if we consider only integer values.

However, a high-resolution mass spectrum, which detects *m*/*z* values to a greater degree of precision, could distinguish 60.0518 from 60.0947: if we obtained a mass spectrum with a molecular ion peak at 60.052 we could confidently predict that we had a sample of ethanoic acid, and not propanol.

Self-check 11.3 An unidentified sample produces a high-resolution mass spectrum with a molecular ion peak at an *m*/*z* value of 44.0524. The unidentified sample is most likely to consist of which of the two following compounds?

(a) Propane, $CH_3CH_2CH_3$

(b) Ethanal, CH_3CHO

We have to be careful however: there are limits to the degree to which even high-resolution mass spectrometry can help us to identify different compounds. Inspection of the molecular ion peak does *not* enable us to distinguish between structural isomers with the same molecular formula. Imagine that the high-resolution mass spectrum showed a molecular ion peak at 60.095. We could confidently attribute this to the presence of propanol, C_3H_7OH. However, we could not tell whether we were detecting the presence of propan-1-ol (1)

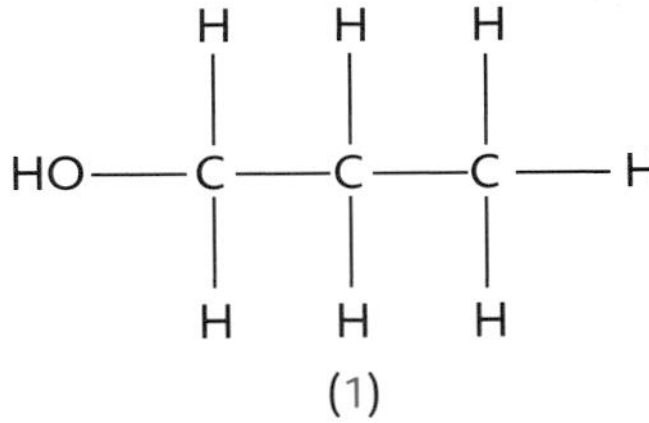

(1)

or propan-2-ol (2)

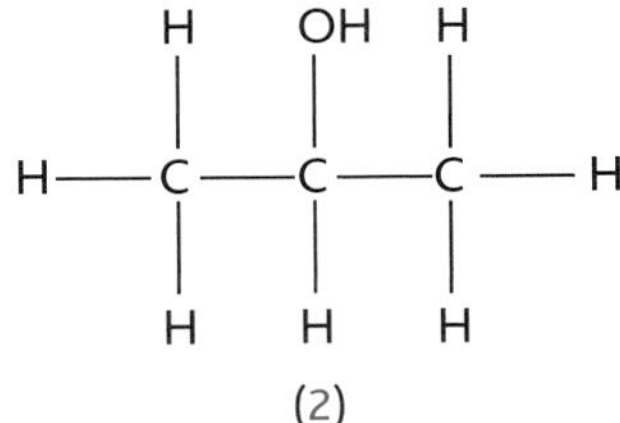

(2)

because both have exactly the same molecular formula and, hence, yield the same molecular ion peak on a mass spectrum.

- ***Mass spectrometry separates and identifies compounds on the basis of their mass.***
- ***Mass spectrometry cannot distinguish between structural isomers.***

We learn more about structural isomers in Chapter 12.

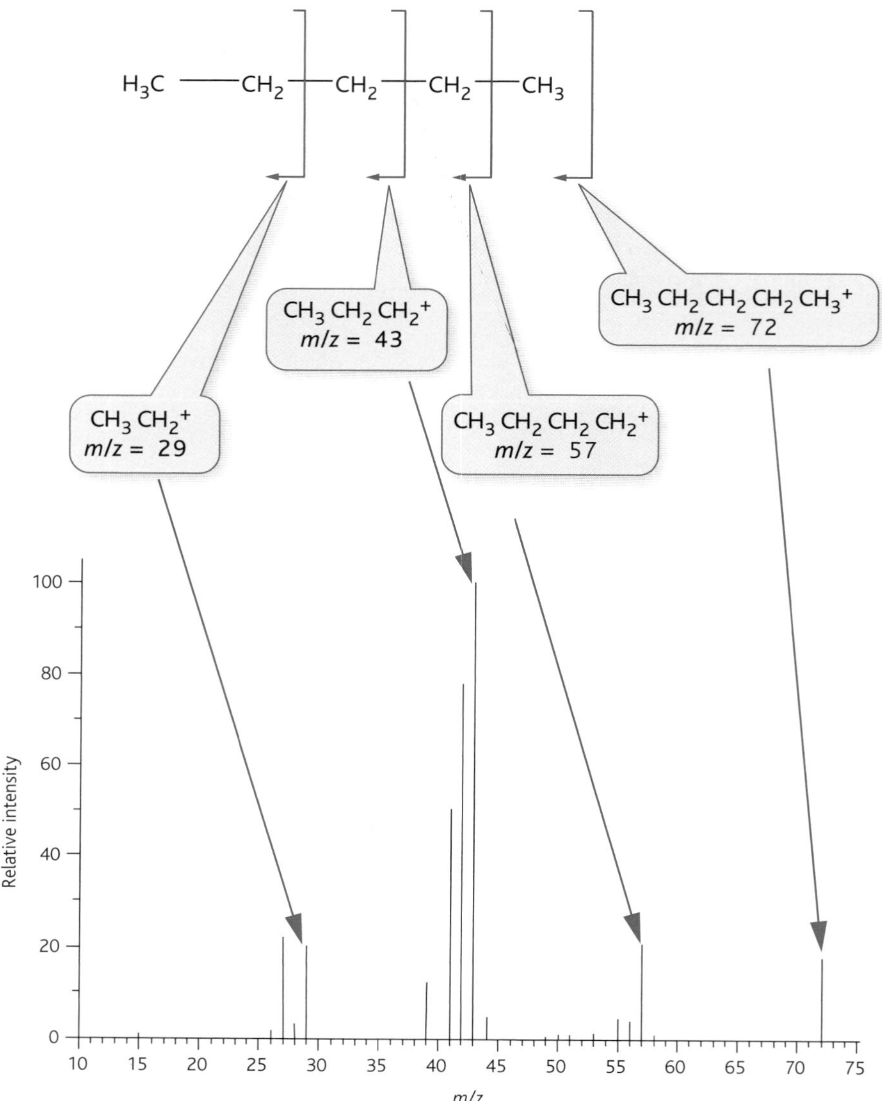

Figure 11.22 Pentane can yield a number of fragments upon ionization, generating a mass spectrum with a **series** of peaks.

The structure of a molecule

The fragmentation of a molecule caused by ionization can help us to deduce some aspects of the structure of a large molecule.

Let's consider the alkane pentane. Look at Figure 11.22, which shows how pentane can yield a number of fragments upon ionization, generating a mass spectrum with a series of peaks. Notice how several of the peaks occur at equal distances apart. These peaks correspond to fragments of pentane that possess fewer and fewer $-CH_2-$ groups.

The information yielded from the fragmentation pattern of molecules is particularly useful for characterizing the structure of complicated molecules, such as polypeptides. We see in Chapter 7 how a polypeptide comprises a number of amino acid residues joined together; different polypeptides have characteristic primary structures–the sequence of amino acid residues from which they are formed. Mass spectrometry can be used to deduce the primary structure of polypeptides identifying the fragments of the polypeptide chain, as illustrated in Figure 11.23.

(a)

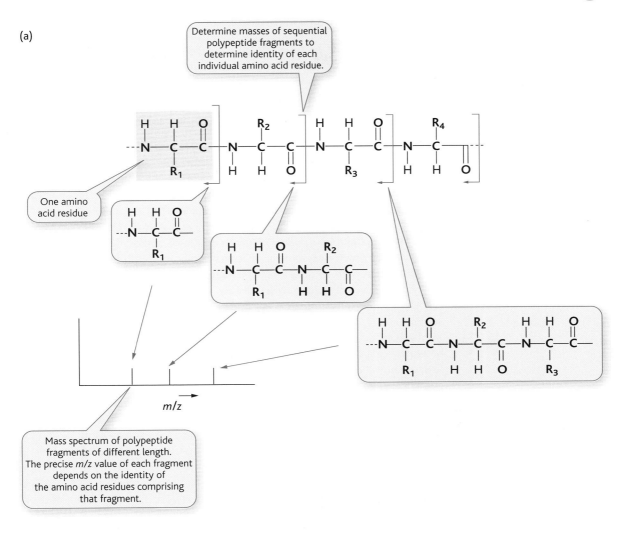

Determine masses of sequential polypeptide fragments to determine identity of each individual amino acid residue.

One amino acid residue

Mass spectrum of polypeptide fragments of different length. The precise m/z value of each fragment depends on the identity of the amino acid residues comprising that fragment.

m/z

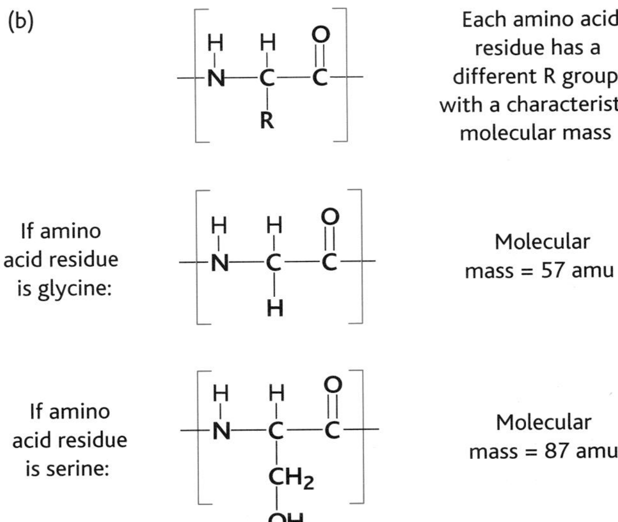

Figure 11.23 Mass spectrometry can be used to deduce the primary structure of polypeptides by leading to the identification of fragments of the polypeptide chain. (a) Different fragments give discrete peaks on the mass spectrum according to the amino acid residues from which the fragments are composed. (b) Different amino acid residues have different molecular masses. By comparing the mass of a fragment of length n with one of length $n + 1$, we can deduce the identity of the amino acid residue that has been added in moving from n to $n + 1$ by comparing the difference in mass of fragment n and fragment $n + 1$.

Coupling separation to identification

We see in Section 11.2 how both liquid and gas chromatography are routinely used as analytical tools throughout the chemical and biological sciences as a potent means of separating components of a mixture. However, rather than simply separating different compounds, devices such as liquid and gas chromatography columns can be **coupled** to instruments that can rapidly *identify* the substances that have been separated as they leave the column. Such coupling allows researchers to move quickly from having a complicated mixture of compounds whose identities are unknown to having separated and identified all the different components in that mixture, making it a powerful experimental tool.

Most often, gas and liquid chromatography are coupled to mass spectrometry. We see above how we can use mass spectrometry to identify compounds on the basis of their mass. In gas chromatography-mass spectrometry (GC-MS) and liquid chromatography-mass spectrometry (LC-MS), the eluate can be diverted to a mass spectrometer as soon as it leaves the column, and the resulting mass spectrum then used to determine the identity of the compound(s) contained in the eluate. These coupled techniques are finding widespread applications in a range of contexts. Read Box 11.5 to find out about one of these contexts: forensic science.

Alternatively, the identification of compounds using a mass spectrometer can be enhanced by an approach called **tandem mass spectrometry** (or MS-MS). We see above how the mass spectrum of just a single compound can yield quite a complicated fragmentation pattern; if a sample contains a mixture of compounds, the fragmentation patterns can be somewhat overwhelming to the point of being indecipherable. Tandem mass spectrometry overcomes this problem by running two mass spectrometers in sequence: the first mass spectrometer ionizes all of the compounds in the sample, and separates them on the basis of their m/z ratio (as per a conventional mass spectrum). However, at this point, only one molecular ion is selected for further analysis. This ion is subjected to fragmentation in a collision chamber (which links the two mass spectrometers). The fragments produced enter the second mass spectrometer for analysis.

By isolating a specific molecular ion from a mixture before fragmentation (and detection of the fragments) occurs, we know that the fragmentation pattern we are looking at is attributable to only one compound from a mixture, rather than a potential assortment of different compounds. As a result, the process of characterizing the compound is greatly facilitated. Look at Box 11.6 (p.348) to learn about the application of tandem mass spectrometry to studies of genomes and proteomes.

Advances in the sensitivity of mass spectrometry mean that this technique often represents a 'one-stop shop' for the identification of some compounds. However, as the example of structural isomers such as propan-1-ol and propan-2-ol illustrates, mass spectrometry can sometimes only *begin* to reveal all of what there is to find out about the structure and composition of a compound. Instead, we can use a number of other techniques to build on the information yielded from mass spectrometry, and eventually produce a more detailed picture of what the compound is like–just as a forensic scientist collects different types of evidence from the crime scene to piece together a picture of what happened when the crime took place. Each technique yields a different piece of the overall jigsaw puzzle.

BOX 11.5 The age of forensic science

GC-MS and LC-MS techniques have become powerful tools in the forensic examination of body fluids. In Germany, for example, a bill amending part of German traffic law related to driving under the influence of drugs, which came into effect in August 1998, made it illegal to drive a vehicle after the consumption of drugs. For this legislation to be enforceable, a reliable way of detecting drug metabolites (the compounds generated by the body from the metabolism of a drug) was needed.

Consequently, scientists in Germany have developed both GC-MS and LC-MS techniques which can quickly and reliably detect whether a sample of body fluid (typically a urine sample) contains chemical compounds indicative of drug use. For example, Figure 11.24(a) shows the mass spectrum of a particular cannabinoid (a metabolite of cannabis), and Figure 11.24(b) shows the result of the LC-MS analysis of a urine sample, which confirms the presence of the cannabinoid in the sample.

Gas chromatography-mass spectrometry has also proved to be a powerful tool in tackling drug abuse in sport. The Olympic Analysis laboratory at the University of California, Los Angeles, uses GC-MS to rapidly screen urine samples for traces of banned drugs, and particularly performance-enhancing steroids. These screens have proved particularly effective at detecting the use of 'designer steroids'–steroids that are synthesized artificially, often with the expressed purpose of being used in a performance-enhancing context, and which are often not detected by standard drug tests.

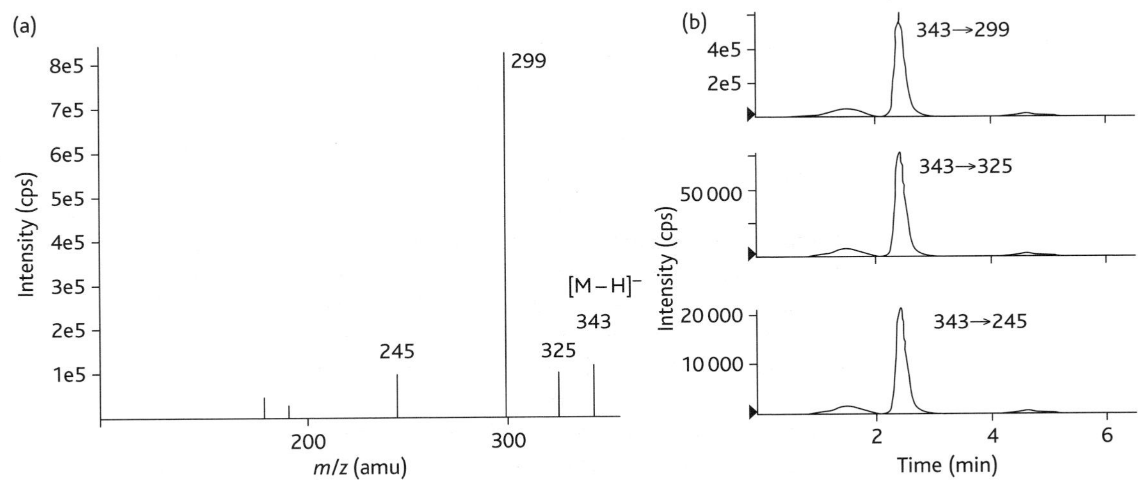

Figure 11.24 (a) The mass spectrum of the cannabinoid THC–COOH (a metabolite of cannabis). (b) The result of the LC-MS analysis of a urine sample, which confirms the presence of the cannabinoid in the sample. Notice how the peaks from the LC-MS analysis correspond to species of the same mass as those shown in the mass spectrum, confirming that the substance being detected in the urine sample is the same substance characterized by the mass spectrum. (Data reproduced by kind permission of Dr Wolfgang Weinmann from Weinmann, W. *et al.* (2001) *Forensic Science International* **121**: 103–107.)

11.4 Building up the picture: spectroscopic techniques

Mass spectrometry often gives us just a thumbnail sketch of the chemical nature of a compound: it can tell us what a compound *is*, but reveals little about the compound's connectivity (how the atoms are joined together) or conformation (how the atoms are arranged in space). In this section, and the sections that follow, we learn about other spectroscopic techniques that can be used to help us determine the structural aspects of molecules.

Spectroscopy and electromagnetic radiation

A number of important analytical tools exploit the way in which chemical compounds interact with **electromagnetic radiation**. We see in Chapter 2 how electromagnetic radiation is characterized by its wavelength and frequency–low-energy electromagnetic radiation has a long wavelength, while high-energy electromagnetic radiation has a short wavelength. The range of different energies (or wavelengths) that electromagnetic radiation may possess is represented by the electromagnetic spectrum, which is divided into different regions, each encompassing a specific range of wavelengths.

Look back at Figure 2.17 for an illustration of the electromagnetic spectrum.

Techniques that use electromagnetic radiation as a tool for probing chemical compounds are what we call **spectroscopic techniques**, and belong to a type of analysis called **spectroscopy**.

Essentially, spectroscopy measures a *change* in the behaviour of a molecule when exposed to electromagnetic radiation (that is, when irradiated). The main changes are:

- A change in the way neighbouring nuclei are arranged relative to each other;
- The rotation of molecules;

BOX 11.6 Genomics gets bitten by the tandem mass spectrometry bug

The mosquito *Anopheles gambiae* (a species complex comprising around five different individual species) is a major carrier (or 'vector') of malaria in sub-Saharan Africa. Malaria is caused by parasites of the *Plasmodium* genus, which are transmitted to humans by the bite of an infected female mosquito. (When biting occurs, the mosquito draws blood from the human, during which time the *Plasmodium* parasite can enter the human bloodstream.)

Malaria kills over 1.5 million people per year, mostly African children. Having a better understanding of the biology of *Anopheles gambiae* is an important element in controlling the spread and impact of this disease: if we understand more about the mosquito, we may discover ways of inhibiting the development of the parasite within the mosquito, or its transmission from the mosquito to humans. Further, the interaction between the female mosquito and human depends on olfactory and taste receptors expressed by the mosquito; the identification of the genes that encode these receptors may lead to the development of new repellents or odour traps (to attract the mosquito, and lure them away from their intended human target).

The genome of *Anopheles gambiae* was first sequenced in 2002, an outcome of the International Anopheles Genome Project. Since then, researchers have explored the transcripts produced from expression of the genome, and the protein products yielded from these transcripts. But what have these studies to do with tandem mass spectrometry?

Tandem mass spectrometry has been an invaluable tool in analyzing the proteins produced by the mosquito, and in mapping the locations within the *Anopheles* genome of the genes encoding these proteins. For example, Dario Kalume and colleagues at Johns Hopkins University in the US used tandem mass spectrometry to study proteins synthesized in the *Anopheles* salivary glands. Peptide sequences identified by tandem mass spectrometry were compared to the *Anopheles* genome database using bioinformatics tools to help understand which proteins are encoded by which parts of the *Anopheles* genome.

Figure 11.25 shows some examples of the MS-MS spectra produced from these studies. Notice how these spectra were used to determine the primary structure of the peptides that were isolated. (The primary structures are displayed above each spectrum, using the one-letter code for the amino acids.) These sequences were then used to verify that a particular region of the *Anopheles* genome was actively expressed to produce a protein product: the peptide fragments were mapped to different regions of a known transcript as illustrated in Figure 11.26.

So, researchers are using approaches such as this to gradually build up a portfolio of 'evidence' of how the *Anopheles* genome functions to direct the way the organism develops and behaves–the way the genome is expressed to produce different proteins, for example. They then hope the knowledge gained can be applied to the development of combative strategies to control the spread of malaria, and ultimately save the lives of millions of children each year.

BOX 11.6 **Continued**

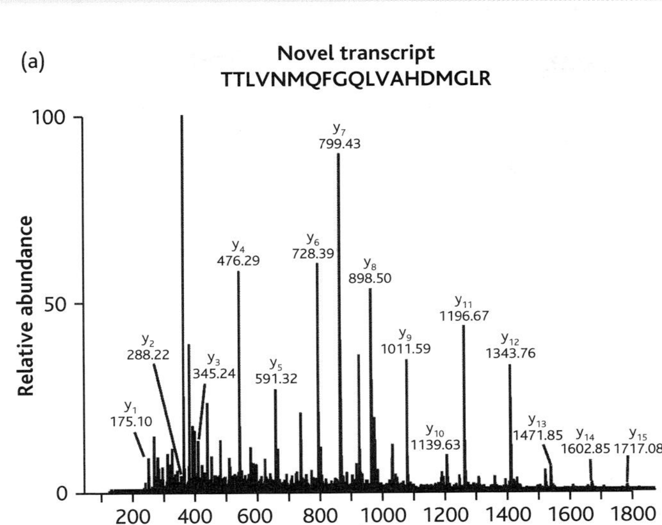

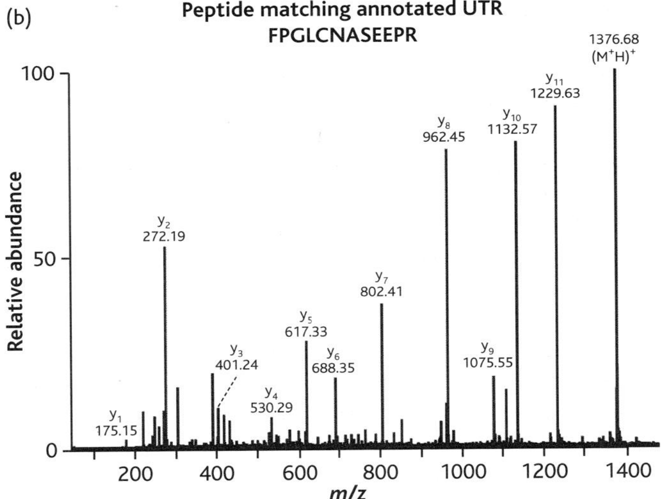

Figure 11.25 Examples of tandem mass spectra that were used to help annotate the *Anopheles gambiae* genome. (a) MS/MS spectrum of a peptide whose sequence was used to validate the identity of an exon of a novel transcript encoding a peroxidase family of proteins. (b) MS/MS spectrum of a peptide that maps to a predicted untranslated region (UTR) of a known transcript encoding Antigen 5-related 1 protein. Reproduced from Kalume *et al.* (2005).

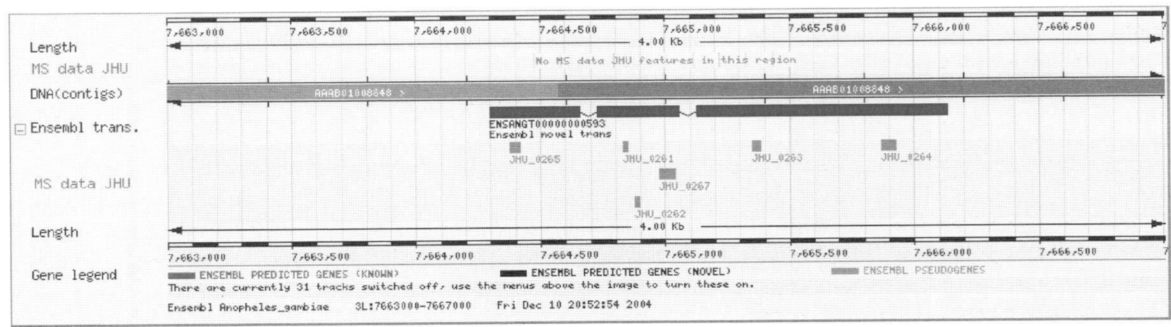

Figure 11.26 A screenshot depicting validation of a novel transcript from the *Anopheles gambiae* genome using mass spectrometry-derived peptide sequences, shown here in blue. Reproduced from Kalume *et al.* (2005).

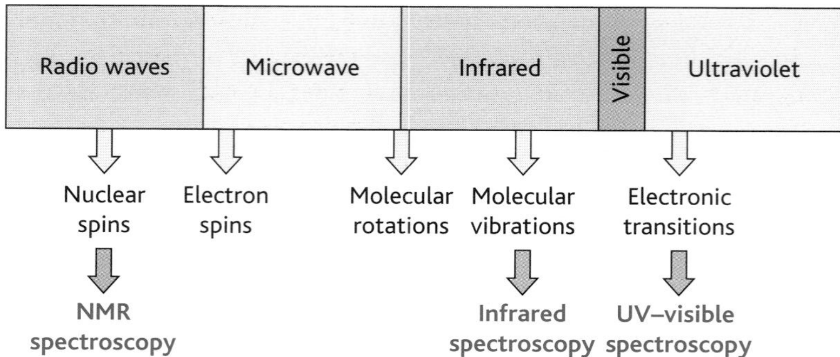

Figure 11.27 Electromagnetic radiation elicits different behaviours in atoms and molecules depending on the wavelength of the radiation. The types of behaviour elicited are shown below the different regions of the electromagnetic spectrum. The techniques that exploit these different behaviours are shown in blue.

- The vibration of molecules;
- The movement of electrons between energy levels.

Importantly, these types of behaviour only happen when the molecules are exposed to electromagnetic radiation of particular energies. We say that the energy required to make these behaviours happen are quantized[6]. The particular energies required to make a given molecule exhibit a certain type of behaviour–to vibrate or rotate, for example–depends on the chemical composition of that molecule. So, a molecule with a certain characteristic chemical composition will exhibit rotations or vibrations when exposed to radiation of a specific energy. By taking this a step further, by studying the energy required to make a molecule behave in a certain way, we can deduce the chemical composition of that molecule.

Different spectroscopic techniques use different parts of the electromagnetic spectrum to ask different questions about chemical structure. They do so by eliciting different types of change at the molecular level. For example, some techniques cause molecular vibrations; some cause the movement of electrons between energy levels. Look at Figure 11.27, which shows the types of behaviour elicited by different parts of the electromagnetic spectrum. Notice how specific segments of the spectrum elicit certain behaviours. Different spectroscopic techniques exploit different regions of the electromagnetic spectrum; the techniques associated with several of these regions are also indicated in Figure 11.27, and we explore these techniques throughout the rest of this chapter.

- ***Different spectroscopic techniques use light from different regions of the electromagnetic spectrum to characterize different aspects of a compound.***

What are we measuring during spectroscopy?

We note above how particular energies are required to make molecules exhibit different types of behaviour. We use spectroscopy to determine the precise energies at which these types of behaviour are happening. But how can we tell that these behaviours are happening–that a molecule is vibrating, or electrons are moving between energy levels? The answer is that, when a molecule is triggered by a certain

6 We encounter the concept of quantization in Section 2.5, when we see how the energy of electrons is quantized: electrons only possess certain, discrete amounts of energy, which correspond to the energy levels they are able to occupy and move between.

frequency or wavelength of energy to behave in a certain way, the molecule **absorbs** energy. So, the process of spectroscopy detects whether or not a sample of interest is absorbing energy or not, and determines the energies at which absorption occurs.

When we are trying to identify a compound using spectroscopy, we use an appropriate instrument to expose the sample to electromagnetic radiation of gradually increasing (or decreasing) energies; the instrument includes some kind of detector to determine whether or not the molecule is absorbing the energy it is being exposed to. Figure 11.28 shows that, if the energy of the radiation does not correspond to a level at which a certain behaviour is triggered, the radiation will simply pass through the sample and will be detected by the detector. If, however, the energy of the radiation *does* correspond to a level at which a certain behaviour is triggered, the radiation will be absorbed by the sample, and will fail to reach the detector.

We generally see two types of spectrum, as illustrated in Figure 11.29. An **absorption spectrum** measures the amount of radiation being absorbed by the sample as the energy of the radiation is varied; a **transmission spectrum** measures the amount of radiation being transmitted through the sample (and reaching the detector) as the energy of the radiation is varied. Notice how the two types of spectrum produce plots of contrasting shape: an absorption spectrum features peaks, while a transmission spectrum features troughs. Different spectroscopic techniques vary according to whether they produce absorption or transmission spectra. We encounter examples of both types of spectrum later in this chapter.

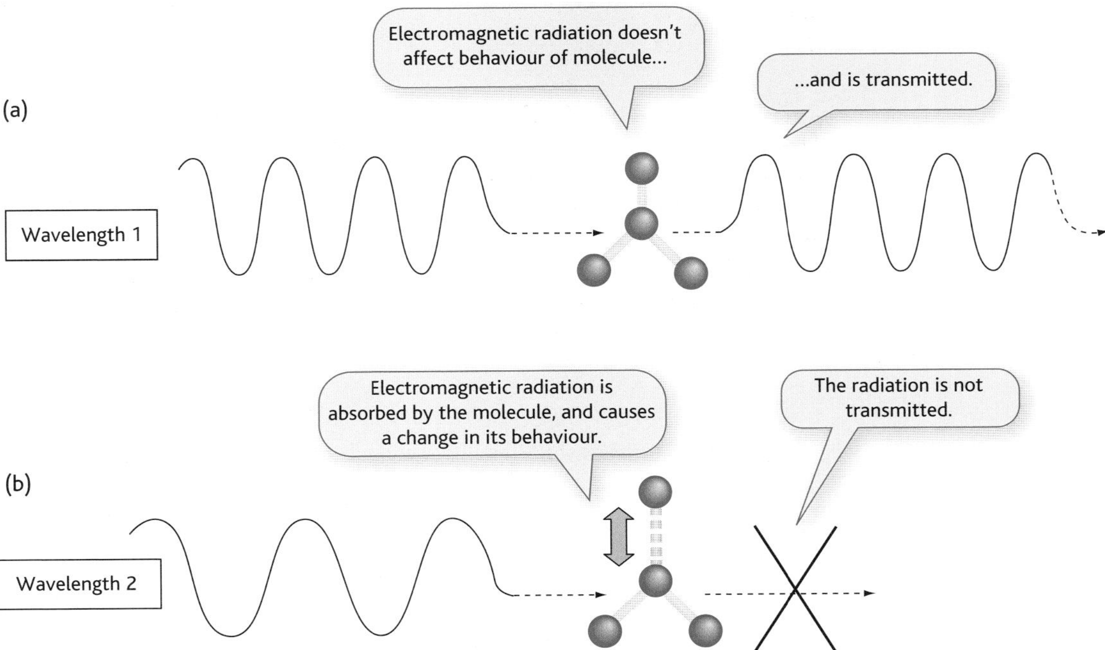

Figure 11.28 Only electromagnetic radiation of specific wavelengths will be absorbed by a given molecule, according to which wavelengths induce a change in the behaviour of that molecule. In (a) we see that the wavelength of the radiation is not one that induces a change in behaviour in the molecule. Consequently, none of the radiation is absorbed but, instead, is transmitted through the sample. By contrast, in (b), a different wavelength does induce a change in behaviour (in this instance, a stretching of a bond) and so we see absorption of the radiation; none of the radiation is transmitted through the sample.

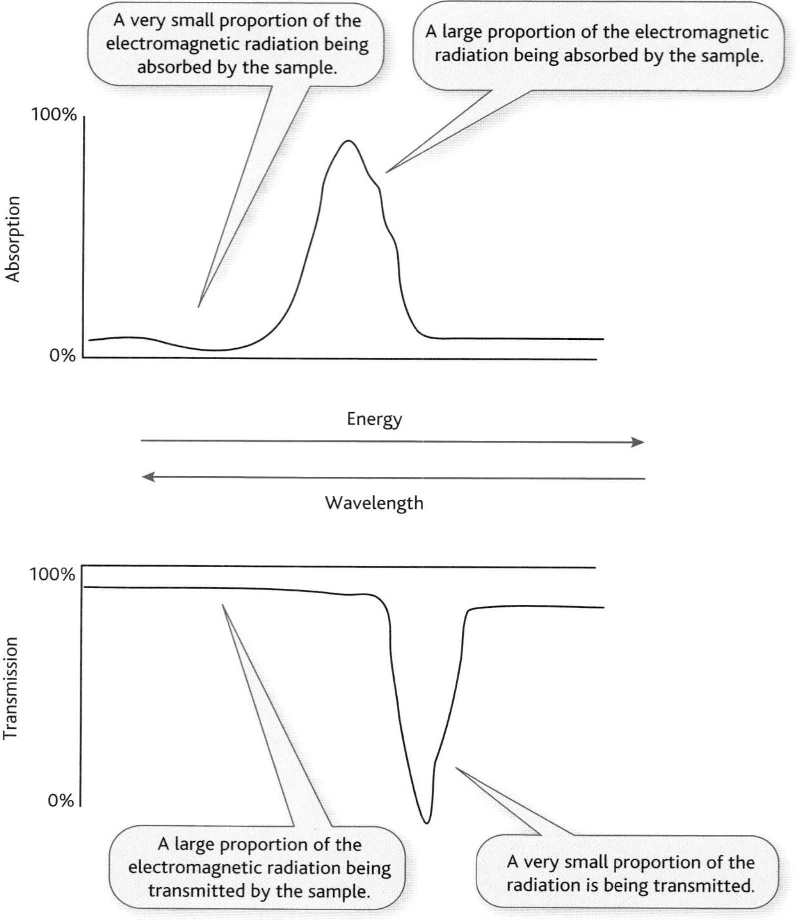

Figure 11.29 Absorption versus transmission spectra. (a) In an absorption spectrum, we measure the relative amount of electromagnetic radiation being absorbed by a sample. Wavelengths at which absorbance occurs are seen as peaks on the spectrum. (b) In a transmission spectrum, we measure the relative amount of electromagnetic radiation being transmitted by a sample. We see wavelengths at which transmission does not occur as troughs on the spectrum.

Sometimes, when a molecule has absorbed energy, it then releases some of it again. We say that it **emits** energy. (We see an example of the emission of energy in Chapter 2: the line emission spectra that are used to determine the electronic structure of atoms are generated when an atom emits some of the energy that it absorbed in the process of an electron being excited to a higher energy level.) The emission of energy from a molecule can be detected in the same way as the absorption of energy, with a suitable detector forming part of the spectroscopic instrument. We encounter types of spectroscopy that exploit the emission of energy by molecules in Chapter 12.

- ***An absorption spectrum measures the absorption of energy by a sample.***
- ***An emission spectrum measures the release of energy from a sample once absorption has occurred.***

Using spectroscopy to characterize chemical compounds

Let's think for a moment about a typical organic compound:

$$H_3C—CH_2—CH(CH_3)—C(=O)—CH_3$$

What characteristics of this compound would help us to build up a really detailed picture of the compound? It would be helpful to know the following:

- **The nature of the carbon framework**–how many carbon atoms comprise the framework? What is the conformation of the framework?
- **The nature of any functional groups** that are attached to the carbon framework–what is the identity of any functional groups, and where on the carbon framework are they attached?

In the remaining sections of this chapter we see how a range of different spectroscopic techniques can be used to probe these characteristics, and help to identify a compound. We begin by considering a technique that helps us to characterize the hydrocarbon framework of organic compounds.

11.5 Characterizing the carbon framework: nuclear magnetic resonance spectroscopy

Nuclear magnetic resonance spectroscopy (NMR) exploits the behaviour of the nuclei of certain atoms when placed in a very strong magnetic field.

The nuclei of certain atoms, including ^{1}H and ^{13}C, behave like tiny bar magnets. When these nuclei are placed in a very strong magnetic field, they **align** with the magnetic field. This is analogous to the behaviour of a compass needle: a compass needle is itself a tiny magnet which aligns with the Earth's magnetic field in a precise way. (It is this precise alignment with the Earth's magnetic field that enables us to use a compass for navigation.)

Remember: the nucleus is the tightly packed cluster of protons and neutrons at the centre of an atom.

Nuclei orient themselves in a magnetic field in one of two ways: we say that the nuclei have two different **spins**. Figure 11.30 shows how these spins can be directly **aligned** with the magnetic field, or **opposed** to the magnetic field. (We can visualize the two spins as 'pointing up' and 'pointing down'.) The two spins are of different energy: one spin is a stable, low-energy spin; the other is a slightly more unstable, higher-energy spin.

Nuclei can change spin, just as an electron can move from a lower-energy orbital to a higher-energy orbital. However, just as an atom must absorb energy of the correct wavelength for an electron to become excited and change orbital, so changing the spin of a nucleus also requires energy. (Remember one of our essential concepts: change–of any sort–involves energy.) Nuclei change spin by absorbing energy corresponding to the **radio frequency** region of the electromagnetic spectrum. When a nucleus absorbs energy to change spin we say that it **resonates**.

Read p. 32 for a reminder of the different regions of the electromagnetic spectrum.

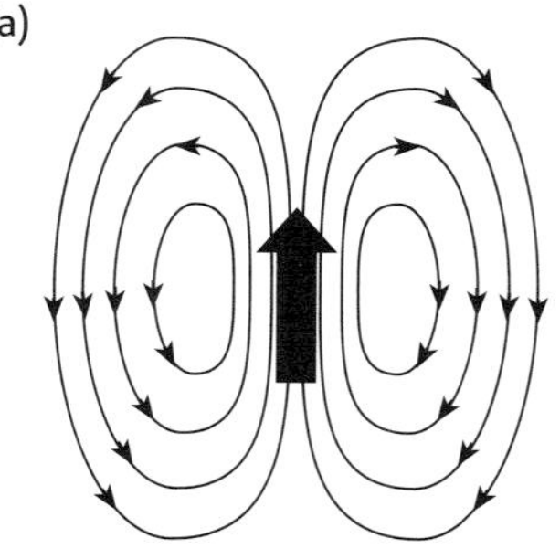

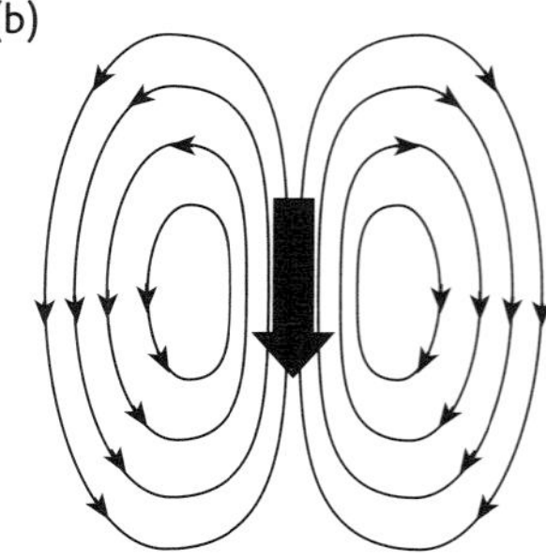

Figure 11.30 Nuclei can orient themselves in a magnetic field in one of two ways: their spins can be directly aligned with the magnetic field (a), or opposed to the magnetic field (b).

The precise amount of energy to effect a change in spin depends on the strength of the magnetic field to which the nucleus is exposed: the stronger the magnetic field, the greater the energy required to change the spin.

We see in Chapter 2 how an atom can absorb electromagnetic radiation of particular wavelengths to drive the movement of an electron from a low-energy orbital to a higher-energy orbital. This excitation is only transitory: the atom is more unstable when the electron occupies the higher-energy orbital, so the atom releases the energy that it absorbed, and the electron falls back to the lower-energy orbital. An equivalent situation occurs when the nuclei of ^{1}H and ^{13}C atoms absorb energy to effect a change in spin, from low-energy to high-energy: the high-energy spin is more unstable than the low-energy spin, so the nuclei release the absorbed energy to return to their former, more stable, spin.

It is this release of energy that we measure when performing NMR spectroscopy[7].

The energy required to effect a change in spin depends not only on the strength of the magnetic field in which the nucleus is placed, but also on the other atoms that surround the nucleus in question. This is because the electrons surrounding other atoms interfere with the magnetic field that the nucleus is exposed to, and so affect the amount of energy required for it to change spin.

- ***Nuclear magnetic resonance spectroscopy uses radiation with energy corresponding to the radio wave region of the electromagnetic spectrum.***
- ***It exploits the way nuclei change their spin when they absorb radiation in the radio wave region of the electromagnetic spectrum.***

For example, look at Figure 11.31, which shows the alcohol propan-1-ol. Propan-1-ol possesses three carbon atoms. Notice how the nucleus of the carbon at one end of the framework has rather different surroundings to the carbon atom at the other end of the framework. Consequently, when placed in a magnetic field, the two carbon atoms experience slightly *different* magnetic fields, and so require different amounts of energy to change their spins. Therefore their nuclei absorb electromagnetic radiation (they resonate) at slightly different frequencies.

7 One thing to note: when we consider energy in the context of NMR spectra we refer to **frequencies** rather than wavelengths, because we are considering electromagnetic radiation in the radio **frequency** region of the electromagnetic spectrum.

H—C—C—C—OH (structure of propan-1-ol)

The nucleus of this carbon is surrounded by three H atoms, and one C atom.

The nucleus of this carbon is surrounded by one C atom, two H atoms, and one O atom.

Figure 11.31 The structure formula of propan-1-ol. Notice how the nucleus of the carbon at one end of the framework has rather different surroundings to the carbon atom at the other end of the framework.

It is this difference in the frequency at which atoms of the same element absorb electromagnetic radiation that can be used to tell us about the structure of an organic compound.

Look at Figure 11.32, which shows the ^{13}C-NMR spectrum of propan-1-ol[8]. Notice how there are three peaks on the spectrum, each occurring at a different frequency. Each peak corresponds to one of the three carbon atoms in propan-1-ol, each of which occupies a different chemical environment.

Look again at Figure 11.32 and notice how the *x*-axis of this spectrum is labelled with units of ppm (parts per million), and is not labelled 'frequency'. In practice, it is normal for the *x*-axis of an NMR spectrum to display the **chemical shift**, δ, at which a peak occurs, rather than the frequency. The chemical shift, which is measured in parts per million (ppm), is the difference between the resonance frequency of the compound being studied and the resonance frequency of a reference compound, usually **tetramethylsilane, TMS** (which, by definition, has a chemical shift of 0 ppm).

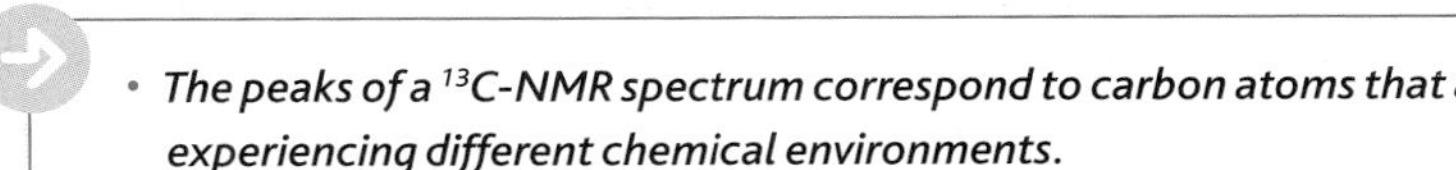

- ***The peaks of a ^{13}C-NMR spectrum correspond to carbon atoms that are experiencing different chemical environments.***

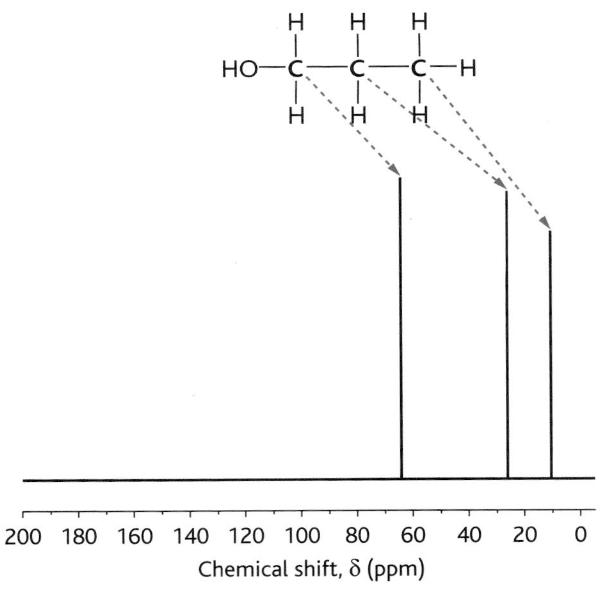

Figure 11.32 The ^{13}C-NMR spectrum of propan-1-ol. Each peak corresponds to one of the three carbon atoms in propan-1-ol as indicated here; notice how each of the carbon atoms occupies a different chemical environment and so generates a separate, distinct peak on the spectrum.

8 A ^{13}C-NMR spectrum tells us specifically about the carbon atoms in a compound.

The *position* of each peak helps us to analyze the chemical environment that each carbon atom is experiencing–whether the carbon atom is in an environment like this:

```
    H
    |
H—C—C·····
    |
    H
```

or in an environment like this:

```
     H
     |
HO—C—C·····
     |
     H
```

- ***The* position *of a peak in an NMR spectrum reveals an atom's chemical environment. Atoms in different chemical environments (that is, attached to atoms whose chemical identities are different) generate peaks at different positions in an NMR spectrum.***

What happens if two atoms share the same chemical environment? If two atoms experience exactly the same chemical environment, their nuclei change spins at the *same* frequency–i.e. both nuclei absorb electromagnetic radiation at the same frequency. Therefore, the **intensity** of absorption at that frequency (the relative amount of electromagnetic radiation that is absorbed) is greater than if just one atom, which experiences a particular chemical environment, is present. Consequently, we see a *larger* peak (a greater area under the curve) on the NMR spectrum.

- ***The area under the curve of a peak in an NMR spectrum (roughly proportional to the* height *of the peak) relates to the number of equivalent atoms present in a compound.***

Look at Figure 11.33, which shows the structure of propane and its ^{13}C-NMR spectrum. Notice how two of the carbon atoms share the *same* chemical environment. Consequently, there are only two peaks on the NMR spectrum, with one peak being approximately double the height of the first (having double the area under the curve). This tells us that there are two atoms that share a common chemical environment, and one atom that experiences a different environment.

Figure 11.34 shows the ^{13}C-NMR spectrum of butan-1-ol (3).

```
    H   H   H   H
    |   |   |   |
H—C—C—C—C—OH
    |   |   |   |
    H   H   H   H
```

(3)

This time, we notice there are four peaks, which tells us that the carbon atoms occupy four different chemical environments.

We mention at the start of this section that only certain nuclei behave as tiny magnets and exhibit specific spins when placed in a strong magnetic field. The two commonest nuclei that do change their spin are those of ^{13}C atoms (as we have seen above)

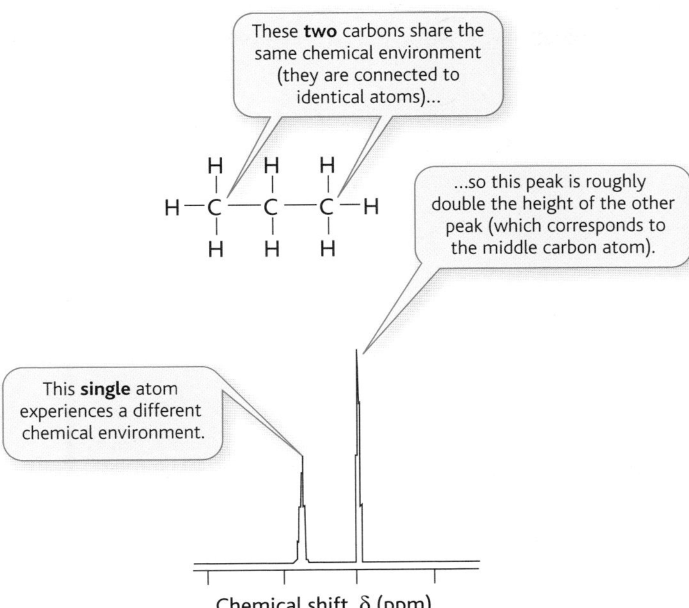

Figure 11.33 The structure of propane, and its ^{13}C-NMR spectrum. There are only two peaks on the NMR spectrum, with one peak being approximately double the height of the first, telling us that there are two atoms that share a common chemical environment, and one atom that experiences a different environment.

and those of ^{1}H atoms. This is perfect for characterizing the *framework* of organic compounds which, as we see in Chapter 5, comprises carbon and hydrogen atoms.

Self-check 11.4 An unidentified sample generates three peaks on a ^{13}C-NMR spectrum, two of which are roughly double the height of the third. Which of the following compounds is most likely to have generated this spectrum? (You may find it helpful to draw out the structure of each compound first.)
(a) propane, $CH_3CH_2CH_3$; (b) butan-1-ol, $CH_3CH_2CH_2CH_2OH$; (c) pentan-3-ol, $CH_3CH_2CHOHCH_2CH_3$

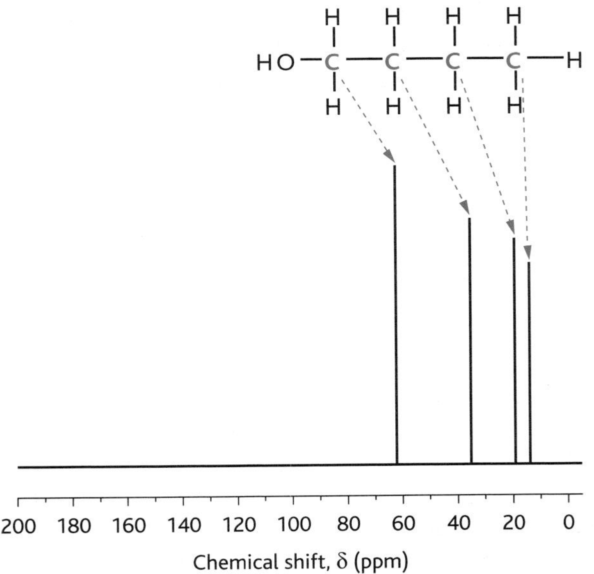

Figure 11.34 The ^{13}C-NMR spectrum of butan-1-ol. This spectrum has four different peaks, because the four carbon atoms in butan-1-ol all experience different chemical environments.

A particular NMR spectrum tells us about only one isotope. We run ^{13}C-NMR spectra to characterize the carbon atoms in a compound; we run ^{1}H-NMR spectra to characterize the hydrogen atoms.

Let's now examine a ^{1}H-NMR spectrum.

Figure 11.35 shows the ^{1}H-NMR spectrum of methanol (**4**).

(4)

^{1}H-NMR spectra tell us the same *kind* of information as ^{13}C-NMR spectra, except they tell us about hydrogen atoms, not carbon atoms.

Remember: when interpreting a spectrum, we need to look for the *position* of the peaks, and the *height* of the peaks. When looking at Figure 11.35, we notice two things:

- There are *two* peaks, which tells us that the hydrogen atoms occupy two chemical environments.
- One peak is significantly *higher* than the other peak, telling us that multiple hydrogen atoms occupy the *same* chemical environment. We see from the structure of methanol that it has *three* equivalent hydrogen atoms (occupying one chemical environment), and one further hydrogen atom (occupying a different chemical environment).

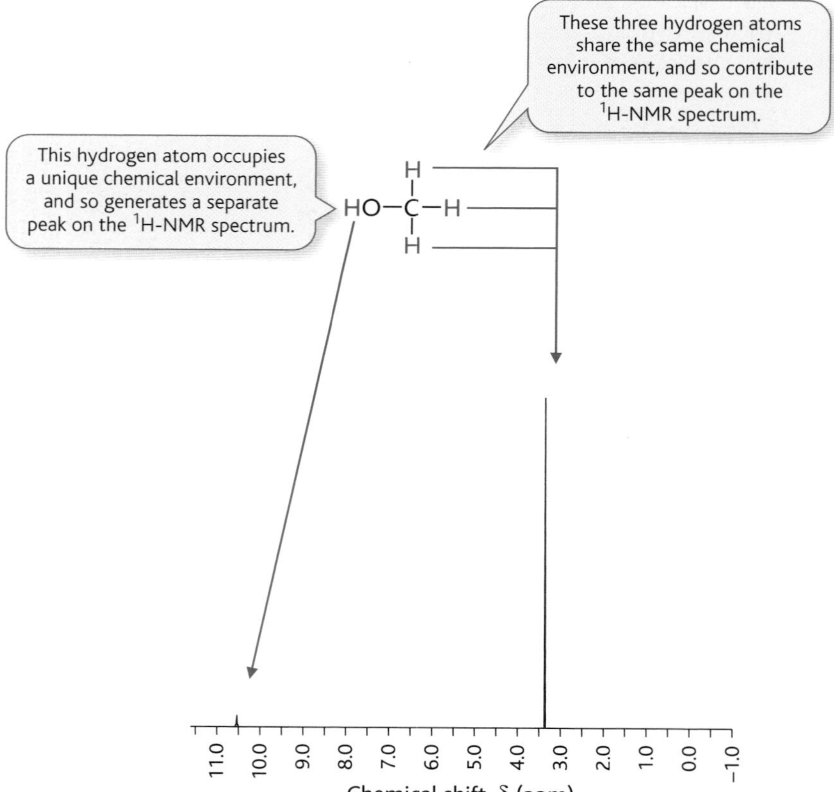

Figure 11.35 The ^{1}H-NMR spectrum of methanol. Notice how there are two peaks, with one peak significantly larger than the other. This taller peak is generated by three hydrogen nuclei that share the same chemical environment. The other smaller peak is generated by a single hydrogen nucleus in a different chemical environment.

The principle of nuclear magnetic resonance is not exploited solely for chemical research. Look at Box 11.7 to see how nuclear magnetic resonance is also used by healthcare professionals to study the human body.

BOX 11.7 A picture of health: magnetic resonance imaging

The principles underlying nuclear magnetic resonance are not only exploited to study the structure of chemical compounds. One of the most valuable applications of nuclear magnetic resonance is **magnetic resonance imaging** (MRI)–the use of nuclear magnetic resonance to generate images of the human body.

During MRI, we place part of the body–or indeed the whole body–in a magnetic field, and detect the resonance of ^1H nuclei in water molecules in the body tissue. ^1H nuclei located in different types of body tissue (bone, muscle, etc.) give different signals, just as ^1H nuclei in different chemical environments give different signals, as we see above.

MRI scans are now routinely used in the healthcare profession to detect diseased or damaged tissues throughout the body to aid diagnosis and treatment. Figure 11.36 shows an MRI scan of a human head, in which we can clearly see detail of the brain, the top of the spinal cord, and even the tongue at the back of the mouth. An MRI scan can reveal much more detail than a typical X-ray. A great advantage of MRI over X-ray imaging is that MRI can image soft tissue, whereas X-ray imaging is limited to bone and other hard tissue.

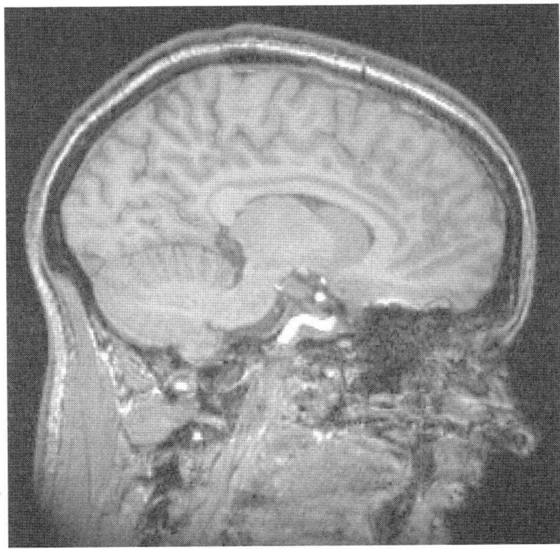

Figure 11.36 An MRI scan of a human head. Notice how an MRI scan can reveal much more detail than a typical X-ray. (Image courtesy Dr G. Williams, Wolfson Brain Imaging Centre, Cambridge.)

It is not just medics who are exploiting the technology, however. Cognitive neuroscientists now use MRI to generate images of the brain. **Functional MRI** (fMRI) images blood flow in the brain, and enables us to see which parts of the brain are active as it carries out particular cognitive processes. fMRI is helping us to characterize the brain activity that underpins processes such as memory, visual perception, and emotion. Look at Figure 11.37 which shows two fMRI scans of the human brain as it performs a specific task. Notice how the scans highlight specific regions of the brain; it is these regions of the brain that are active at the moment the scans are taken.

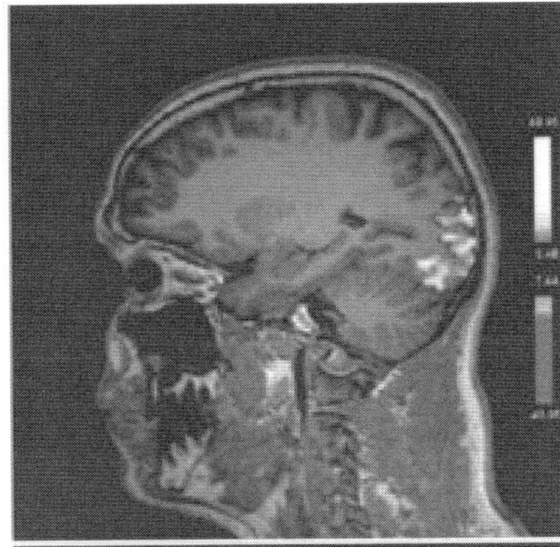

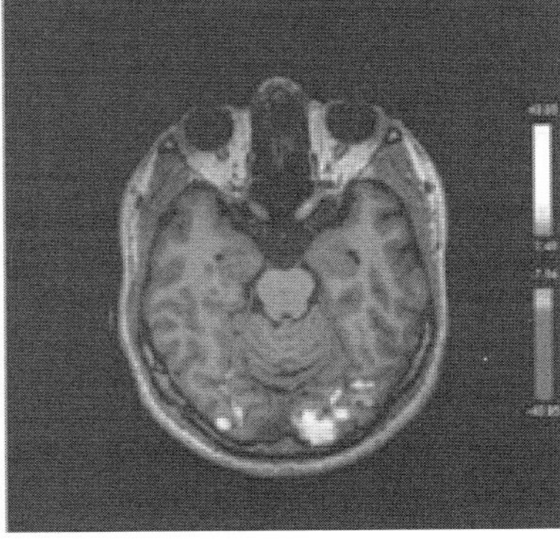

Figure 11.37 fMRI scans of the human brain. The scans highlight specific regions of the brain; it is these regions of the brain that are active at the moment the scans are taken. (© Nature Publishing Group.)

11.6 Identifying functional groups (1): infrared spectroscopy

We see above how a combination of ^{13}C-NMR and ^{1}H-NMR can help us to characterize the carbon framework of an organic compound. The next step in our process of deducing the structure and composition of an organic compound is to identify the functional groups attached to the hydrocarbon framework.

We can use **infrared spectroscopy** to help us identify the functional groups that are present in an organic compound. Infrared (IR) spectroscopy exploits the **vibration** of the covalent bonds that join two atoms to reveal to us how atoms in a molecule are connected to one another[9].

When we say that a covalent bond 'vibrates', we are in fact describing the slight movement of the atoms that are joined by the bond. We can picture two atoms joined by a covalent bond as two snooker balls joined by a tightly coiled spring, as depicted in Figure 11.38. The spring is sufficiently tightly coiled that the two balls (atoms) are held in a relatively straight line, while still having the freedom to move slightly.

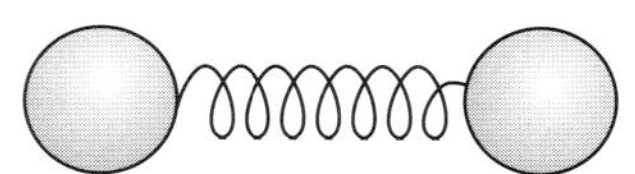

Figure 11.38 We can picture two atoms joined by a covalent bond as two snooker balls joined by a tightly coiled spring.

There are several different types of bond vibration, which are distinguished by the way that the joined atoms move relative to one another in 3D space.

A bond may **stretch**, **scissor**, **rock**, **wag**, or **twist**, as depicted in Figure 11.39. Look at this figure and notice how stretching involves the movement of just two atoms, while the other types of vibration relate to the relative movement of three atoms joined in sequence.

There are two main factors that influence how *fast* a bond vibrates:

1. The strength of the bond joining the atoms

Covalent bonds vibrate with a frequency proportional to their strength: high-energy bonds vibrate very quickly; low-energy bonds vibrate more slowly[10]. We see in Chapter 13 how different covalent bonds possess different energy. C≡C bonds are strong, high-energy bonds; C−C bonds are weaker, lower-energy bonds, for example. Therefore, C≡C bonds vibrate at a higher frequency (a faster speed) than lower-energy C−C bonds.

2. The combined mass of the joined atoms

Covalent bonds vibrate with a frequency **inversely proportional** to their combined mass: as the combined mass of the joined atoms *increases*, the speed at which the bond vibrates *decreases*. We see in Chapter 2 how atoms of different elements have different atomic masses: carbon has an atomic mass of 12; oxygen has an atomic mass of 16; hydrogen has an atomic mass of 1. Therefore, a C−H bond vibrates with higher frequency (greater speed) than a C−C bond: the combined mass of carbon and hydrogen (12 + 1 = 13) is lower than the combined mass of two carbon atoms (12 + 12 = 24). Similarly, an O−H bond vibrates at a higher frequency than a C−O bond, because the combined mass of oxygen and hydrogen (16 + 1 = 17) is lower than the combined mass of oxygen and carbon (16 + 12 = 28).

---

9 We call the speed at which a bond vibrates its frequency.

10 This relationship may seem counter-intuitive. But consider a taut piece of string and a slacker piece of string: if we pluck both pieces of string, the taut string vibrates more quickly than the slacker one.

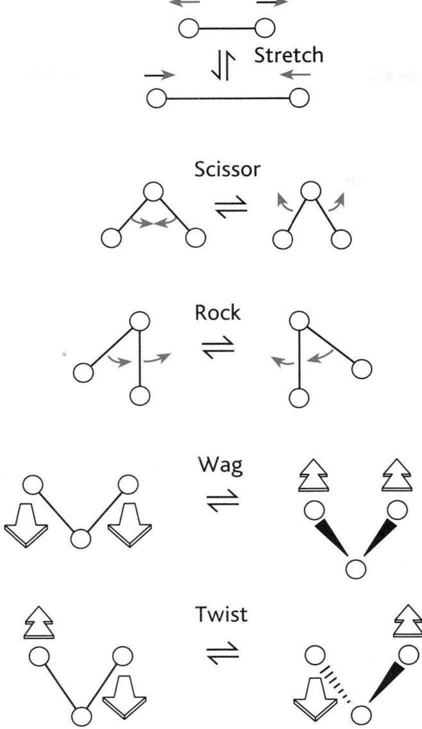

Figure 11.39 The different modes of bond vibration. Different modes of bond vibration are characterized by the way that joined atoms move relative to each other in three-dimensional space, as illustrated here.

As with any type of movement, bond vibration involves energy[11]. As the frequency of vibration increases, so the energy required to power this vibration increases also. (Think of a high-frequency (fast) vibration as a 'high-energy' vibration, and a low-frequency (slow) vibration as a 'low-energy' vibration.)

Infrared spectroscopy uses electromagnetic radiation as the source of energy to power the vibration of covalent bonds in a compound. Specifically, as the name implies, infrared spectroscopy uses electromagnetic radiation with wavelengths in the **infrared region** of the electromagnetic spectrum.

Recall that electromagnetic radiation with a short wavelength has high energy, while radiation with a long wavelength has low energy. Consequently, IR radiation with a relatively long wavelength is sufficient to drive low-frequency (slow) bond vibration; IR radiation with a short wavelength is required to drive high-frequency (fast) bond vibration.

- ***Infrared spectroscopy exploits the way that molecules vibrate when irradiated with radiation whose energy corresponds to that within the infrared region of the electromagnetic spectrum.***
- ***Low-energy infrared radiation with a long wavelength drives low-frequency (slow) bond vibration.***
- ***High-energy infrared radiation with a short wavelength drives high-frequency (fast) bond vibration.***

11 Two of our essential concepts are that change involves both movement and energy; bond vibration illustrates these two concepts nicely.

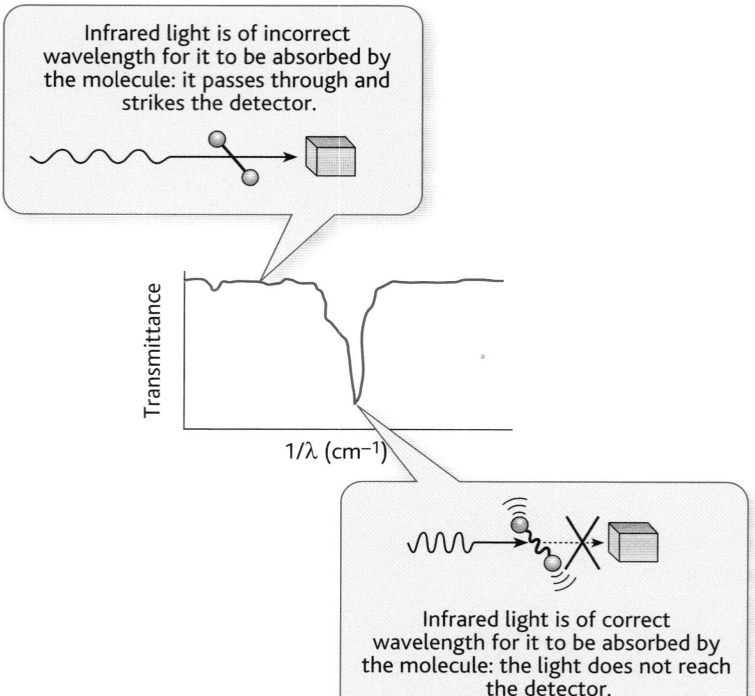

Figure 11.40 The conceptual basis of infrared spectroscopy. When there is no bond vibration, the IR light is not absorbed by the compound under investigation and passes through the sample to hit a detector. When the wavelength of IR light possesses sufficient energy to drive the vibration of a particular bond, the bond absorbs the light. Consequently, the light does not reach the detector, and we see a *peak* in the IR spectrum.

Look at Figure 11.40, which shows what is happening during IR spectroscopy. The compound under investigation is irradiated with infrared light, and the amount of light that passes through the sample of the compound as the wavelength varies is measured to give us an **infrared spectrum**, as depicted in Figure 11.41. When there is no bond vibration, the IR light is not absorbed by the compound under investigation and passes through the sample to hit a detector. By contrast, when the wavelength of IR light possesses sufficient energy to drive vibration (bending or stretching) of a particular bond, the bond absorbs the light and uses it to drive its vibration[12]. Consequently, the light does not reach the detector, and we see a *peak* in the IR spectrum.

A 'peak' in an IR spectrum is really a trough. Look at Figure 11.41 and notice how the 'spikes' point downwards. Each trough occurs because we are seeing a *decrease* in the amount of light that is passing through the sample to be detected by the IR spectrometer, the device used to measure an IR spectrum (that is, the **transmittance** of the sample decreases). Rather than passing through the sample to be detected, the light is being absorbed by the sample to power bond vibration.

Notice that the values on the *x*-axis of the IR spectrum are **wavenumbers** rather than wavelengths. A wavenumber is the reciprocal of its corresponding wavelength, i.e. $1/\lambda$.

This means that, as the wavelength *increases*, the wavenumber *decreases*–as the wavelength increases we are dividing 1 by an increasingly large number to calculate the wavenumber, giving a smaller and smaller value as a result. (Think about

12 We say that, as a sample absorbs, so its transmittance decreases.

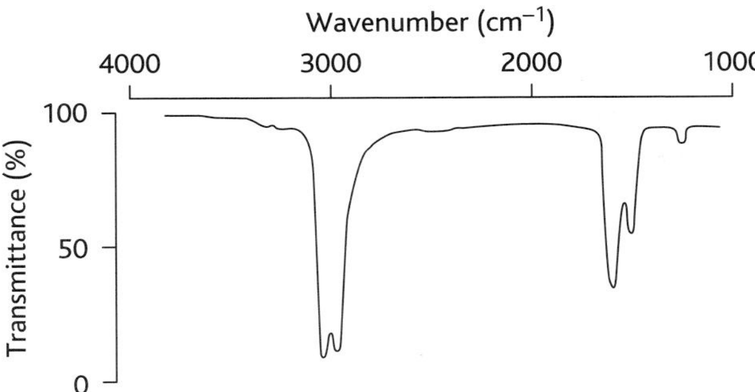

Figure 11.41 A typical infrared spectrum. This is the infrared spectrum of octane, C_8H_{18}.

this: 1/2 = 0.5; 1/4 = 0.25; 1/8 = 0.125–as the number that we're dividing 1 by gets larger, the result is smaller.) We say that wavenumber and wavelength are **inversely proportional**.

As we move across the IR spectrum (from left to right) we scan from short to long wavelength so, because wavelength and wavenumber are inversely proportional, the wavenumber *decreases* from left to right, as illustrated in Figure 11.42. Because wavenumbers are the reciprocal of wavelength (measured in cm), they have the units cm^{-1} (cm^{-1} is the same as writing 1/cm).

The position of the peak on the IR spectrum tells us the nature of the bond that has absorbed the IR light. Peaks to the left of the spectrum correspond to bonds vibrating at a high frequency, which absorb high-energy IR light of shorter wavelength. By contrast, peaks to the right of the spectrum correspond to bonds vibrating at a low frequency, which absorb low-energy IR light of longer wavelength. Consequently, different covalent bonds absorb in different regions of the IR spectrum. In general, the IR spectrum can be divided into regions that correspond to the types of covalent bond that are most likely to absorb energy in each region of the spectrum, as illustrated in Figure 11.43.

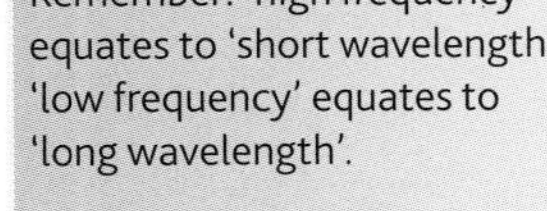
Remember: 'high frequency' equates to 'short wavelength'; 'low frequency' equates to 'long wavelength'.

Now look at Table 11.3, which shows the wavenumbers at which peaks occur on an IR spectrum for various covalent bonds that are prevalent in organic compounds. Notice how a particular covalent bond can give rise to peaks at more than one position in the IR spectrum. For example, the O–H bond may absorb infrared light with a wavenumber that falls within three distinct regions.

There are two main reasons for this variation:

1. 'Bending' and 'stretching' vibrations are of different frequency

A stretching vibration requires more energy than a bending vibration, and so has a higher frequency. Therefore, if a bond exhibits both bending and stretching, we see

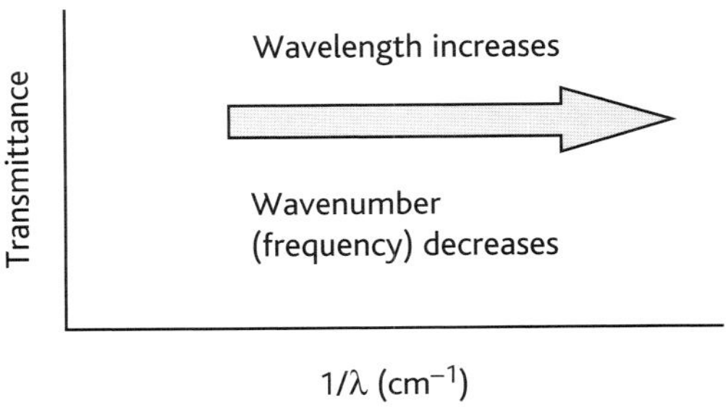

Figure 11.42 Wavelength and wavenumber are inversely proportional: as the wavelength *increases*, the wavenumber *decreases*.

Table 11.3 The wavenumbers of characteristic peaks in the IR spectrum, and the functional groups to which they are attributed.

Bond	Compound it is in	Wavenumber (cm^{-1})
C–H	Alkanes, alkenes, aromatic	2840 to 3095
C=C	Alkenes	1610 to 1680
C=O	Aldehydes, ketones	1680 to 1750
C–O	Alcohols, ethers	1000 to 1300
O–H	'Free' (not hydrogen-bonded)	3580 to 3670
O–H	Hydrogen-bonded in alcohols	3230 to 3550
O–H	Hydrogen-bonded in acids	2500 to 3300
N–H	Primary amines	3100 to 3500

a 'stretching' peak at a slightly shorter wavelength than the corresponding 'bending' peak for the same bond.

2. The wavelength at which a specific bond absorbs IR light depends on its chemical environment

This situation is analogous to the varying positions of peaks that we see in a ^{13}C-NMR spectrum, which are determined by the surroundings of the carbon atoms that are being detected. For example, the O–H bond in an alcohol gives a peak in an IR spectrum at a wavenumber of between 3230–3550 cm^{-1} (typically with a peak at around 3300 cm^{-1}), while an O–H bond in a carboxylic acid gives a peak over a broader range of wavenumbers, from 2500–3300 cm^{-1}.

Let's now take a look at a real infrared spectrum to illustrate the points we've just learned. Look at Figure 11.44, which shows an infrared spectrum of ethanol (**5**).

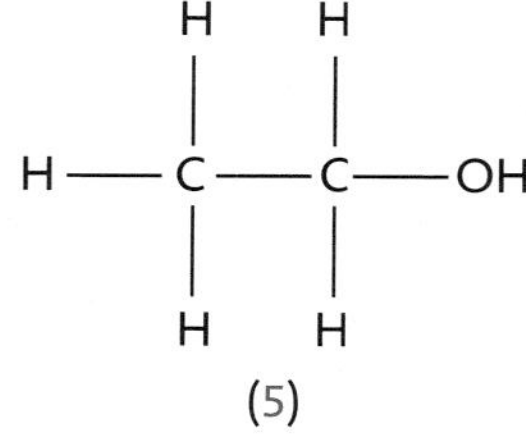

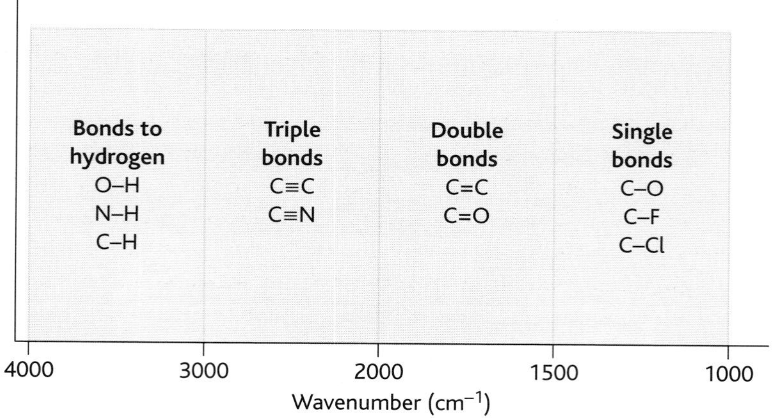

Figure 11.43 The infrared spectrum can be divided into regions according to the types of covalent bond that absorb infrared radiation most strongly in each region.

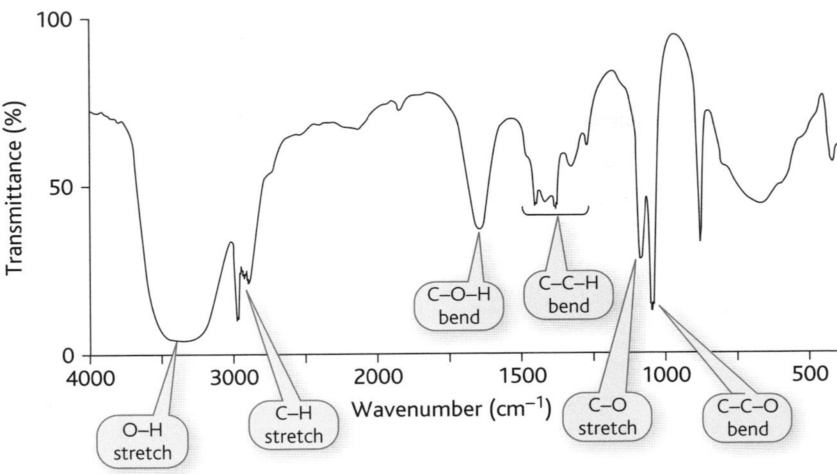

Figure 11.44 The infrared spectrum of ethanol, showing the bonds that give rise to the different peaks on the spectrum.

Notice how the C–H and C–O bonds have distinct 'stretching' and 'bending' peaks, and how each bond type gives rise to a distinct peak at a characteristic wavelength.

Self-check 11.5 Look at Figure 11.44 and notice the frequency of the peak attributed to the O–H stretch. Now look at Table 11.3. What does the position of this peak tell us about whether hydrogen bonds form in ethanol?

11.7 Identifying functional groups (2): UV–visible spectroscopy

We see in Sections 11.5 and 11.6 how NMR spectroscopy exploits the way nuclei behave when exposed to light within the radio wave region of the electromagnetic spectrum, and how infrared spectroscopy exploits the way that molecules vibrate when exposed to certain wavelengths of light within the infrared region of the electromagnetic spectrum. In this section, we consider a type of spectroscopy that employs the ultraviolet (UV) and visible parts of the spectrum.

When an atom or molecule absorbs radiation of an appropriate wavelength from within the UV or visible part of the spectrum it enters what we call an **excited state**. Excitation occurs when photons of light in this part of the spectrum have the right amount of energy to interact with the outer bonding and non-bonding electrons in atoms and molecules, causing them to move to higher energy, unoccupied orbitals. (This excitation is analogous to the excitation of atoms that we discuss in Chapter 2, when considering the way that electrons can move between fixed energy levels, or orbitals.)

Atoms and molecules stay in an excited state for only a fraction of a second before emitting (that is, releasing) the excess energy, and returning once again to the ground state. The energy may be released as thermal energy or sometimes as a photon of light[13]. Ultraviolet and visible spectroscopic techniques can measure both the

13 The release by a compound of photons of light as it relaxes from an excited to a ground state is called luminescence We explore two examples of luminescence–fluorescence and phosphorescence–in Chapter 12.

Table 11.4 λ_{max} values for several organic functional groups.

Functional group	λ_{max} value (nm)
Alkene ($-CH_2=CH_2-$)	175–185
Amino ($-NH_2$)	195–200
Carbonyl	186; 280
Aromatic	200

We learn more about λ_{max} values and what they represent on page 393.

absorption and emission of ultraviolet and visible light by compounds; absorption and emission spectra are employed in different contexts to explore different biological phenomena and processes.

So how do we use UV–visible spectroscopy to characterize the chemical structure of compounds? Some functional groups absorb light at characteristic wavelengths in the UV–visible region of the electromagnetic spectrum, just as they do in the infrared region of the spectrum. Functional groups that absorb strongly in the UV–visible region of the electromagnetic spectrum are what we call **chromophores**. Look at Table 11.4, which shows the λ_{max} values for several functional groups. Notice that most of these functional groups are unsaturated–they contain double bonds. But why do only unsaturated functional groups absorb strongly in the UV–visible region of the electromagnetic spectrum, and why is the absorption restricted to only narrow regions of the UV–visible spectrum?

Chromophores absorb energy from ultraviolet and visible light to drive the excitation of electrons, the movement of electrons from a low-energy to a higher-energy orbital, as noted above. Electrons can only be excited if they have a high-energy orbital to move to, and it happens that UV–visible light is only of sufficient energy to excite electrons to the high-energy anti-bonding orbitals found predominantly in unsaturated compounds, but not other compounds. Therefore, the ability of UV and visible light to cause excitation is limited to unsaturated compounds.

Molecules that absorb light in the ultraviolet and visible part of the spectrum produce a band spectrum–that is, fairly broad peaks rather than the more pronounced, jagged peaks of IR spectroscopy. This is because electronic energy transitions are accompanied by rotational and vibrational excitations, all of which merge to generate one continuous, broad band. However, the characteristic shape of the absorption band is marked by a particular wavelength of maximum intensity, the **absorption maxima**.

- ***Ultraviolet–visible spectroscopy exploits the way that molecules become excited when they absorb radiation with energies corresponding to that found in the UV and visible regions of the electromagnetic spectrum.***
- ***Only unsaturated compounds can absorb energy in the UV–visible region of the electromagnetic spectrum; the parts of these compounds that absorb this energy are called chromophores.***

The UV–visible spectrum

When we scan a range of UV–visible wavelengths, the spectrometer generates a UV–visible spectrum, which is analogous to the IR spectrum that we see in Section 11.6, but simply measured at different wavelengths within the electromagnetic spectrum.

By looking at the peaks on the UV–visible spectrum, we can see if our sample absorbs strongly at any wavelengths that are characteristic of the kind of functional group shown in Table 11.4, and use this data to identify the compound(s) present in our sample. For example, look at Figure 11.45 which shows the UV–visible absorption spectrum of propanone (6).

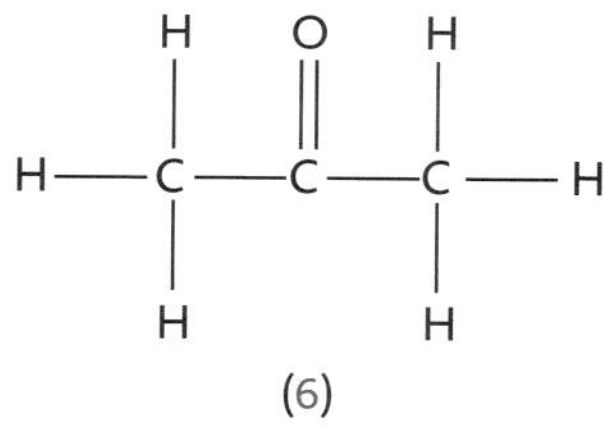

(6)

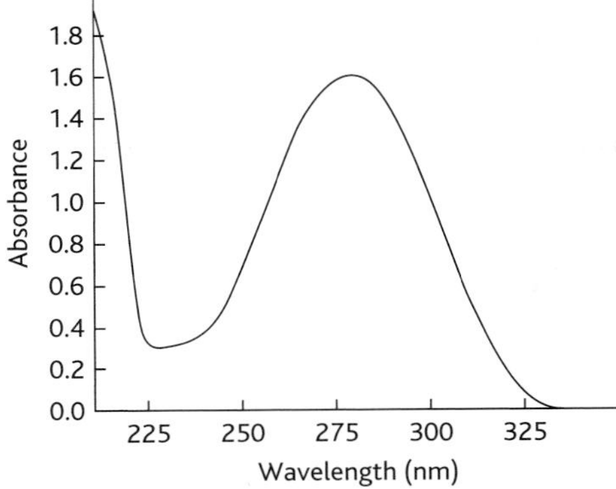

Figure 11.45 The UV–visible spectrum of propanone. Notice how there are two absorption peaks here, one at the very left of the spectrum (around 190 nm), and another at around 280 nm.

Notice how there are two absorption maxima (λ_{max} values) at approximately 190 nm and 280 nm, which correspond with the characteristic λ_{max} values for a carbonyl group, C=O.

Self-check 11.6 A sample containing an unidentified amino acid produces a UV–visible spectrum with a prominent peak at a wavelength of around 200 nm. Which of the following amino acids is most likely to be in the sample? (Look back at Figure 7.2 to remind yourself of the structures of the different amino acid side chains.) (a) Aspartic acid; (b) Phenylalanine; (c) Serine.

Read Box 11.8 to see how UV–visible spectroscopy can not only be used to characterize the chemical composition of a given compound, but can also be used to characterize biological processes.

BOX 11.8 Using UV–visible spectra to monitor biological processes

Many biological molecules such as proteins and nucleic acids can be investigated by monitoring the light absorbed at different wavelengths. Shifts in the absorption maxima of proteins can be used to monitor the interactions between proteins and other molecules and, hence, help to characterize binding activities etc. Moreover the appearance and disappearance of absorption bands can be used to follow the course of a reaction.

Nicotinamide adenine dinucleotide, NAD^+, is found in all living cells. NAD^+ is a coenzyme: it works together with particular enzymes to bring about oxidation reactions in the cell. (For example, we see in Chapter 18 how NAD^+ is required during the process of glycolysis to mediate the oxidation of glyceraldehyde-3-phosphate.) During the course of such reactions, NAD^+ is reduced to NADH:

$$NAD^+ + H^+ + 2\,e^- \rightarrow NADH$$

Both the oxidized and reduced form of NAD absorb in the UV part of the spectrum. However, they have different absorption spectra: NAD^+ has a single absorption maximum at 259 nm, while NADH has absorption peaks at both 259 nm and 339 nm. Consequently, the course of a reaction can be followed by monitoring the change in the absorption spectrum: as NAD^+ is reduced to NADH, a band corresponding to a wavelength of 339 nm will gradually appear. Look at Figure 11.46, which shows the UV–visible spectra for both NAD^+ and NADH. Notice how NADH has two absorption peaks; it is the emergence of the second peak around 339 nm that can be used to monitor the reduction of NAD^+ to NADH.

BOX 11.8 **Continued**

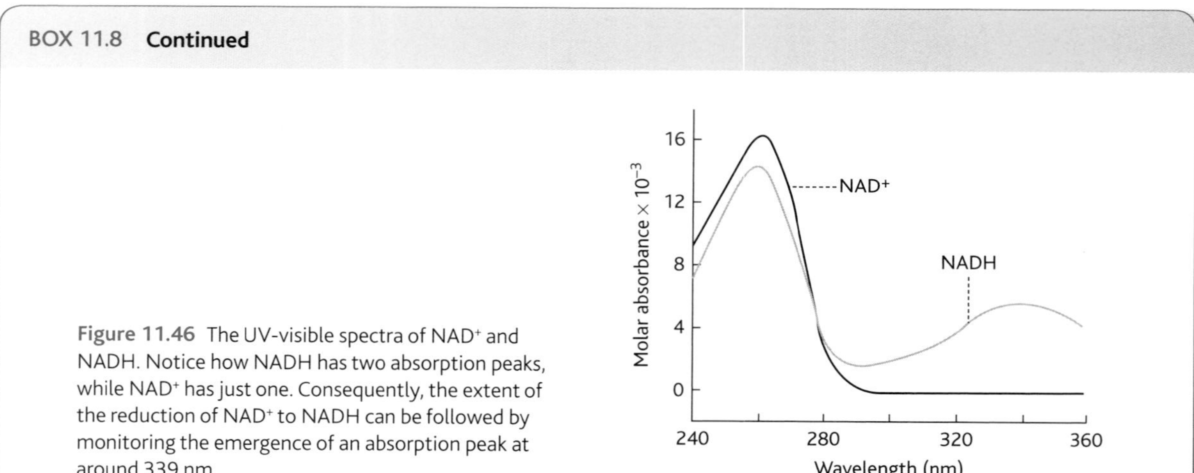

Figure 11.46 The UV-visible spectra of NAD^+ and NADH. Notice how NADH has two absorption peaks, while NAD^+ has just one. Consequently, the extent of the reduction of NAD^+ to NADH can be followed by monitoring the emergence of an absorption peak at around 339 nm.

11.8 Establishing 3D structure: X-ray crystallography

We learn more about how the chemical nature of a compound dictates its structure and shape in Chapters 8 and 9.

Used in tandem on a same sample, mass spectrometry, NMR, and IR spectroscopy can tell us a great deal about the composition of a compound and how its atoms are connected. However, they tell us nothing about how the compound's atoms are arranged three-dimensionally in space.

We can solve this final piece in the puzzle of chemical structure using **X-ray crystallography**. As the name suggests, X-ray crystallography exploits electromagnetic radiation in the X-ray region of the electromagnetic spectrum to characterize the three-dimensional structure of a molecule.

The principle on which X-ray crystallography is based is familiar to anyone who's dropped a bag of rice, frozen peas, or marbles on the floor: if you strike something solid with small particles, the particles will scatter. When a photon of electromagnetic light is deflected (or scattered) by a solid object, we say that it is **diffracted**. The way in which the photons are diffracted tells us about the shape of the object that caused the diffraction.

Look at Figure 11.47, which depicts what is happening during X-ray crystallography:

- X-rays are fired at a (solid) crystal.
- The individual atoms that make up the crystal diffract the X-rays.
- The diffracted X-rays generate a **diffraction pattern**, which we can analyze to determine the conformation of the atoms.

The 3D arrangement of atoms in the crystal dictates the exact nature of the diffraction pattern so analysis of a diffraction pattern allows us to work backwards and determine the conformation.

Now look at Figure 11.48, which shows a typical X-ray diffraction pattern. What is the first thing you notice? A diffraction pattern is far from being a photographic image of a crystal's structure. Rather, a diffraction pattern is a fairly abstract image. To make sense of what a diffraction pattern is telling us, sophisticated mathematical

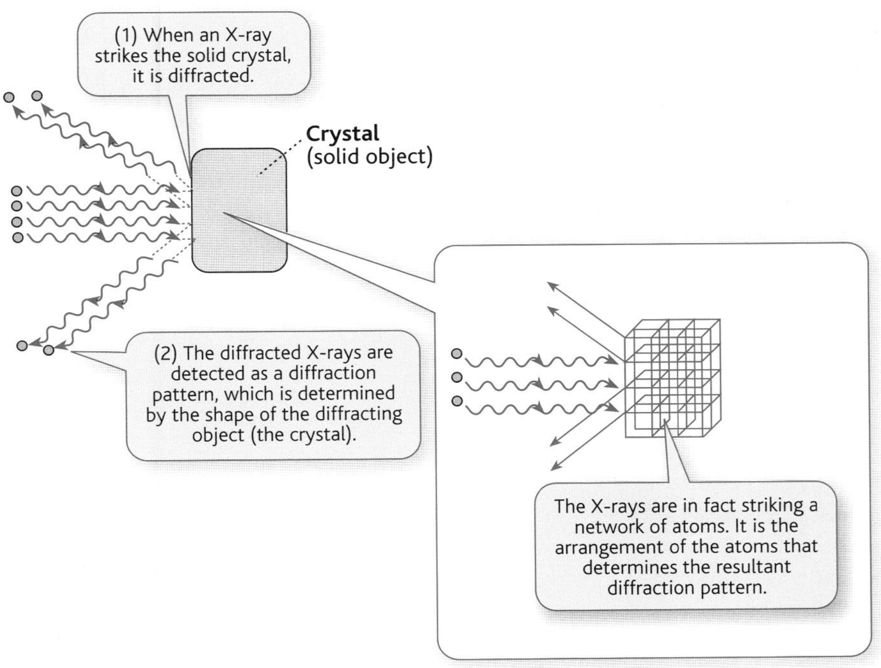
(1) When an X-ray strikes the solid crystal, it is diffracted.
Crystal (solid object)
(2) The diffracted X-rays are detected as a diffraction pattern, which is determined by the shape of the diffracting object (the crystal).
The X-rays are in fact striking a network of atoms. It is the arrangement of the atoms that determines the resultant diffraction pattern.

Figure 11.47 The conceptual basis of X-ray crystallography, in which X-rays are scattered when they strike a solid object (a crystal).

and computational tools are used to 'translate' the diffraction pattern into a structural model, and thereby reveal the 3D structure of the compound we're investigating.

One of the most acclaimed biological applications of X-ray crystallography was to determine the structure of DNA, as discussed in Box 11.9.

The crystal problem–keeping things real

The biggest drawback of X-ray crystallography is its requirement for the sample to be supplied in crystalline form. Biological molecules very rarely exist as crystals in their native state. Consequently, the structure of a biological molecule as revealed by X-ray crystallography is not 100% representative of its 'real' structure *in vivo*. In particular, many biological molecules must possess a degree of structural flexibility in order to function as they should.

We explore molecular flexibility in more detail in Chapter 9, when we learn about the shape and structure of molecules.

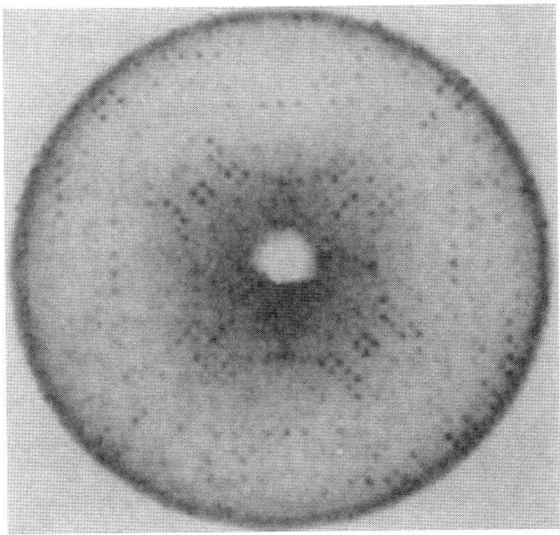

Figure 11.48 A typical X-ray diffraction pattern. (Image courtesy Prof P. Atkins.)

BOX 11.9 X-ray crystallography and the structure of DNA

On 25 April 1953, James Watson and Francis Crick published a paper in the journal *Nature* entitled 'A Structure for Deoxyribose Nucleic Acid', in which they proposed, for the first time, that the structure of DNA comprises 'two helical chains each coiled round the same axis'. Their vision is now a cornerstone of modern molecular biology.

In the same volume of *Nature* (indeed, on the pages immediately following Watson and Crick's paper) appeared two papers entitled 'Molecular Structure of Deoxypentose Nucleic Acids' and 'Molecular Configuration in Sodium Thymonucleate' published by Maurice Wilkins and Rosalind Franklin respectively. The latter two papers featured two X-ray diffraction patterns generated by crystalline DNA fibres, similar to that shown in Figure 11.49. In her paper, Franklin notes that 'The X-ray diagram... shows in a striking manner the features characteristic of helical structures...'.

While Watson and Crick failed to directly attribute any part of their discovery to the X-ray diffraction patterns of Wilkins or Franklin, it later emerged that Watson had seen Franklin's image prior to the publication of his paper with Crick. As Franklin's own observations suggest, the diffraction pattern gave real experimental weight to Watson and Crick's more theoretically based vision.

In 1962, Watson, Crick, and Wilkins received the Nobel Prize for the discovery of the structure of DNA. By this time, Franklin had tragically died of ovarian cancer, just five years after the *Nature* papers had been published, and at the age of just 37. Nobel Prizes are never awarded posthumously–otherwise Rosalind Franklin would have shared this same prize.

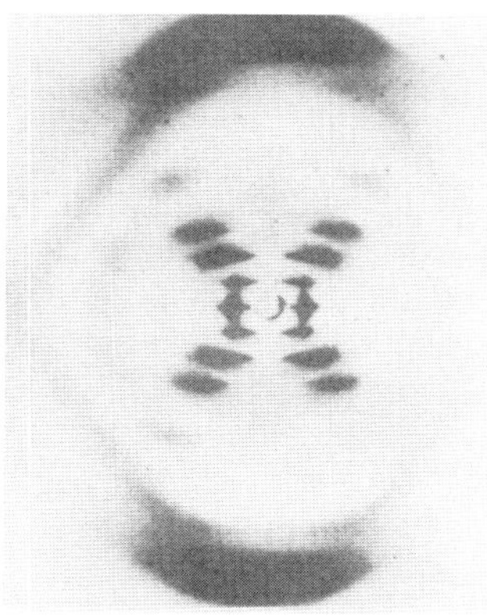

Figure 11.49 The X-ray diffraction pattern generated by crystalline DNA. (Image adapted from an illustration in J. P. Glusker and K. N. Trueblood (1972).)

By contrast, the arrangement of atoms in a crystal is fixed, such that the three-dimensional structure of a biological molecule as determined by X-ray crystallography may fail to reveal to us those variations in structure which may occur *in vivo*.

Throughout this chapter we have considered a number of different tools that are available to help us characterize a chemical compound–to help us identify it. Having learned how we can answer the question 'How do we know what is there?' we now go on to explore an equally important question: 'How do we know how *much* is there?'.

Check your understanding

To check that you've mastered the key concepts presented in this chapter, review the checklist of key concepts below, and attempt the multiple-choice questions available in the book's Online Resource Centre at **http://www.oxfordtextbooks.co.uk/orc/crowe2e/**.

Checklist of key concepts

- Chemical analysis characterizes the chemical nature of a particular system
- Qualitative analysis tells us what is there
- Quantitative analysis tells us how much is there

Separation techniques

Solvent extraction

- Solvent extraction exploits the way that a compound of interest partitions differently between a hydrophobic (non-polar) and a hydrophilic (polar) liquid
- A hydrophobic compound will preferentially partition into a hydrophobic (organic) solvent
- A hydrophilic compound will preferentially partition into a hydrophilic (aqueous) solvent
- The distribution ratio, K, describes how a compound partitions between two different solvents: a value of K greater than 1 tells us that the compound is more soluble in an aqueous phase; a value of K less than 1 tells us the compound is more soluble in an organic phase

Chromatography

- Chromatographic techniques separate out compounds on the basis of differing physical attributes, including size and electrical charge
- A chromatographic separation requires a mobile phase and a stationary phase
- Liquid chromatography uses a liquid as the mobile phase
- Gas chromatography uses a gas as the mobile phase
- Gas and liquid chromatography can separate out hundreds of different compounds from a single sample.
- Different compounds are separated because they have different retention times in a chromatographic column

Electrophoresis

- Electrophoresis uses both electrical charge and size to separate molecules
- SDS-polyacrylamide gel electrophoresis (SDS-PAGE) is widely used to separate proteins on the basis of size
- An agarose gel is typically used to separate nucleic acids
- Polypeptides can also be separated using isoelectric focusing
- Isoelectric focusing separates proteins according to the charge that they carry
- A protein's pI value is the pH at which it carries no overall charge
- Two-dimensional PAGE separates out polypeptides on the basis of both size and charge, to give higher-resolution separation

Centrifugation

- Centrifugation separates molecules according to the speed they sink through a liquid medium when subjected to a force
- When rotated at high speed, particles experience a centrifugal force
- After centrifugation, solid material forms a pellet, while the medium in which the solid was originally suspended forms the supernatant

Mass spectrometry

- Mass spectrometry measures the mass of a compound
- A mass spectrometer separates ions on the basis of their mass/charge ratio
- A mass spectrum tells us the relative abundance of ions of different mass derived from a sample molecule
- The molecular ion is the ion of the molecule being studied; its peak occurs at the highest m/z value on the mass spectrum
- Mass spectroscopic techniques vary according to the way in which the sample is vaporized and ionized
- A high-resolution mass spectrum can identify compounds solely on the basis of their mass
- Liquid and gas chromatography are often coupled to mass spectrometry so that identification can rapidly follow separation
- Tandem mass spectrometry (MS-MS) runs two mass spectrometers in sequence: one effects separation of a mixture of compounds in a sample; the other analyzes each isolated compound separately

Spectroscopy

- Spectroscopic techniques use the interaction of compounds with electromagnetic radiation to help characterize the compounds
- Different spectroscopic techniques use different parts of the electromagnetic spectrum to ask different questions about chemical structure
- An absorption spectrum measures the amount of electromagnetic radiation being absorbed by a sample

- A transmission spectrum measures the amount of radiation being transmitted through the sample

Nuclear magnetic resonance

- Nuclear magnetic resonance spectroscopy examines how compounds interact with radio frequency radiation when placed in a magnetic field
- A ^{13}C-NMR spectrum tells us about the number and nature of carbon atoms in a compound–it can tell us about the carbon framework of an organic compound
- A ^{1}H-NMR spectrum tells us about the number and nature of hydrogen atoms in a compound
- The position of a peak in an NMR spectrum tells us about the chemical environment of a particular nucleus
- The height of the peak in an NMR spectrum tells us about the number of atoms sharing the same chemical environment

Infrared spectroscopy

- Infrared spectroscopy examines how compounds interact with infrared radiation
- Different covalent bonds vibrate at different frequencies
- The frequencies at which bonds vibrate give rise to peaks on an infrared spectrum when the compound is exposed to infrared radiation
- The peaks on an infrared spectrum tell us about how atoms are joined together in a compound
- Different bonds generate peaks at different frequencies. Consequently, different compounds generate IR spectra which vary according to the bonds they contain

UV–visible spectroscopy

- UV–visible spectroscopy exploits the way nuclei behave when exposed to visible or ultraviolet light
- UV–visible techniques can measure both the absorption and emission of UV and visible light
- Functional groups that absorb strongly in the UV–visible region of the electromagnetic spectrum are called chromophores
- Chromophores absorb UV–visible light most strongly at particular wavelengths giving absorption peaks in a UV–visible spectrum
- The wavelength at which absorption peaks occur depends on the functional groups that are present in a compound such that a given compound generates a characteristic UV–visible spectrum

X-ray crystallography

- X-ray crystallography tells us about the three-dimensional structure of a compound in crystalline form
- The atoms in a crystal diffract X-rays to produce a diffraction pattern
- A diffraction pattern can be analyzed to tell us about the three-dimensional conformation of the atoms in the crystal

Chemical analysis 2: how do we know how much is there?

INTRODUCTION

In Chapter 11, we discover how we can separate out mixtures of compounds, and how we can use chemical analysis to identify compounds. However, we often need to do more than simply identify organic compounds: we need to know *how much* of a compound we have. (Think about ordering a drink at a bar: we don't simply ask for 'some beer' or 'some wine'; we typically state *how much* we want–a pint or a half of beer; a small (typically 175 mL) or large (250 mL) glass of wine; a single measure (25 mL) or double measure (50 mL) of gin or vodka.)

In this chapter we explore how we can answer the question 'How **much** of a compound do we have?'–to **quantify** what we have. Knowing how much of a compound we have in a system is often vital: having too much of a compound may be toxic to the body, to the point of it being lethal; too little of a compound may be equally detrimental. We begin by learning about the language of measuring chemical quantities.

12.1 The mole

Consider this chemical equation, which depicts the joining together of carbon and oxygen to generate carbon dioxide:

$$C + O_2 \rightarrow CO_2$$

This equation tells us not about the mass of atoms that are reacting but the *number* of atoms. The equation tells us that one atom of carbon reacts with two atoms of oxygen to generate one molecule of carbon dioxide.

We find out more about describing chemical reactions in Chapter 17.

However, when a chemical reaction occurs in a test tube, or in a cell, it is not just one or two atoms that are reacting, but thousands of millions of atoms. Often we want to quantify how many atoms are present in a given system, but it would be ludicrous to try to write down directly the exact number of atoms that are present. The numbers would be too large to be manageable. Instead, we need a way of describing (in a sensible way) the number of atoms that are present. Rather than talking about the number of atoms in absolute terms, we use a special chemical quantity: the **mole**.

The mole is a convenient way of scaling down large numbers: one mole of atoms represents **6×10^{23}** atoms–that is 600 000 000 000 000 000 000 000 atoms. We call the value 6×10^{23} the **Avogadro constant**. The Avogadro constant is defined as the same number of particles as there are atoms in exactly 12 g of ^{12}C. When we measure numbers of particles in moles, we talk of **molar** quantities, which have the units **mol**. One mole is written as 1 mol rather than 1mole.

There is no hidden sophistication behind using the mole as a unit of measurement. We use one mole to refer to 6×10^{23} separate entities, just as we use one dozen to refer to *twelve* entities. Both 'one mole' and 'one dozen' have a specific numerical amount associated with them; it just happens that the numerical amount associated with a mole is much larger than that associated with a dozen!

There is a good reason why the mole is usually only used in a chemical context. We often encounter twelve of a particular entity in everyday life (twelve eggs; twelve apples), so one dozen can be meaningfully used in an everyday context. But how often would you encounter a mole of eggs, or a mole of apples? A mole of eggs (or apples) is an actual quantity–it's 6×10^{23} eggs, or 6×10^{23} apples. Could you ever imagine needing to refer to a mole of eggs? This example helps to show why the mole, as a numerical amount, only becomes useful when we consider objects on an atomic scale.

One mole of *any* element contains the same number of atoms: 1 mol of Na contains 6×10^{23} atoms of sodium; 1 mol of N contains 6×10^{23} atoms of nitrogen.

A mole is a way of referring to numbers of any type of **particle** (atom, ion, or molecule) not just numbers of atoms. So there are as many molecules in one mole of H_2O as there are atoms in one mole of elemental hydrogen:

$$1 \text{ mol of } H_2O = 6 \times 10^{23} \text{ molecules}$$
$$1 \text{ mol of } H = 6 \times 10^{23} \text{ atoms}$$

- ***One mole is just a number, equal to 6×10^{23}.***
- ***We can have one mole of any substance–be it an atom, a molecule, an ion etc.***

Connecting molar quantities to mass

While one mole of two different elements contains the same number of atoms, each has a different mass.

We see in Section 11.3 how atoms of different elements have different atomic masses, all of which are calibrated against the mass of ^{12}C. For example, carbon has an atomic mass of 12 amu, while hydrogen has an atomic mass of 1 amu. Therefore an atom of ^{12}C is *twelve times* heavier than an atom of hydrogen.

Because one atom of carbon has a mass that is twelve times heavier than one atom of hydrogen, we can deduce that *one mole* of carbon (6×10^{23} atoms) has a mass that is twelve times heavier than *one mole* of hydrogen (also 6×10^{23} atoms). We call the mass of one mole of a substance (atom, ion, or molecule) its **molar mass**, with the unit **g mol^{-1}** (which can also be written **g/mol**).

- ***The mass of one mole of a substance is its molar mass.***

The molar mass of different substances is expressed in relation to the mass of one mole of ^{12}C, just as the mass of an *atom* of each element (atomic mass) is expressed in relation to the mass of an *atom* of ^{12}C. The mass of one mole of ^{12}C is defined as 12 g. Consequently, one mole of hydrogen, which has an atomic mass of 1 amu and is twelve times lighter than ^{12}C, has a mass of just 1 g, and one mole of sodium, with an atomic mass of 23 amu, has a mass of 23 g. So we say that hydrogen has a molar mass of 1 g, and sodium has a molar mass of 23 g.

Notice that the molar mass of an element (the mass of one mole), expressed in grams, has the same value as the element's **atomic mass**. Similarly, the molar mass of a compound has the same value as the compound's molecular mass[1].

- ***The molar mass of an element has the same value as its atomic mass.***

Self-check 12.1 What is the molar mass of (a) nitrogen, N; (b) fluorine, F?

Calculating the number of moles of an element in a sample

Now, suppose that we have a sample of a particular element, and want to know how many moles of the element the sample contains. If we know the mass of the sample, we can calculate the number of moles of the element present.

For example, imagine that we have 24 g of C. We know that carbon has a molar mass of 12 g mol^{-1}; in other words, one mole of carbon has a mass of 12 g.

If 12 g of carbon contains one mole, then 24 g of carbon must contain (24/12) moles, or 2 mol:

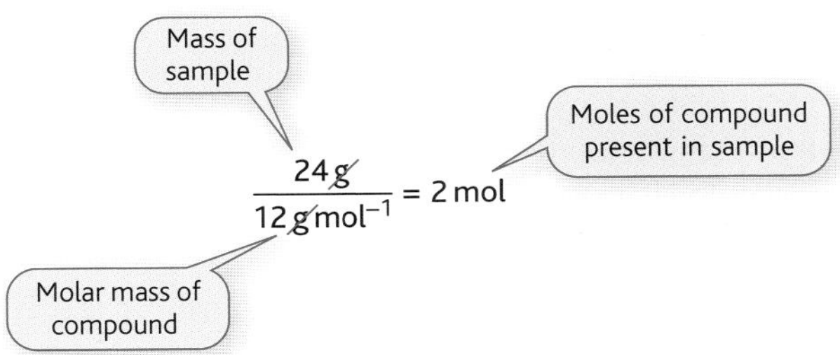

Notice how the units of g on the left-hand side of the equation cancel each other out to give the number of moles the unit **mol**. (The units remaining on the left-hand side of the equation are $\frac{1}{\text{mol}^{-1}}$ which is the same as mol.)

1 In fact, the mass of one mole of *anything* has the same value, in grams, as its molecular mass.

We summarize this relationship by saying:

$$\frac{\text{mass of sample}}{\text{molar mass}} = \text{number of moles present in sample}$$

Let's consider another example. Suppose we have 60 g of potassium, K. The molar mass of potassium is 39.1 g mol^{-1}.

Using the relationship:

$$\frac{\text{mass of sample}}{\text{molar mass}} = \text{number of moles present in sample}$$

we can calculate the number of moles of potassium present as follows:

$$\frac{\text{mass of sample}}{\text{molar mass}} = \frac{60\,\text{g}}{39.1\,\text{g mol}^{-1}} = 1.53\,\text{mol}$$

So, 60 g of potassium contains 1.53 mol.

- ***number of moles present in a sample*** $= \dfrac{\textit{mass of sample}}{\textit{molar mass}}$

Self-check 12.2 How many moles are there in:

(a) 25 g of potassium, K?

(b) 25 g of sodium, Na?

(c) 50 g of nitrogen, N?

The molar mass of a compound

The molar mass of a compound is the sum of the molar masses of each individual atom from which the compound is composed.

Let's consider the compound methane, CH_4. One molecule of methane contains one atom of carbon and four atoms of hydrogen–five atoms in total. The molar mass of methane is the sum of the molar masses of each of these five atoms:

Molar mass of carbon = 12 g mol^{-1}

Molar mass of hydrogen = 1 g mol^{-1}

There is one carbon atom
There are four hydrogen atoms

Molar mass of methane = (1 × (molar mass of C)) + (4 × (molar mass of H))

= (1 × 12 g mol^{-1}) + (4 × 1 g mol^{-1})

= 16 g mol^{-1}

Therefore, the molar mass of methane is **16 g mol^{-1}**.

We can calculate the molar mass of ethanol, CH_3CH_2OH, in the same way:

$$\text{Molecular formula of ethanol} = C_2H_6O$$

$$\text{Molar mass of carbon} = 12\ \text{g mol}^{-1}$$

$$\text{Molar mass of hydrogen} = 1\ \text{g mol}^{-1}$$

$$\text{Molar mass of oxygen} = 16\ \text{g mol}^{-1}$$

The molar mass of ethanol is **46 g mol^{-1}**.

Self-check 12.3 What is the molar mass of:

(a) Propane, $CH_3CH_2CH_3$?

(b) Methanol, CH_3OH?

(c) Ethanoic acid, CH_3COOH?

Calculating the number of moles of a molecule in a sample

We see above how we can calculate the number of moles in a sample of an element if we know both the mass of the sample and the molar mass of the element. This is because the three variables[2] are connected to one another by the following relationship:

$$\frac{\text{mass of sample}}{\text{molar mass}} = \text{number of moles present in sample}$$

Exactly the same relationship links the mass of a sample of compound and the molar mass of that compound, and the number of moles of the compound present in the sample.

For example, let's consider a 100 g sample of sodium chloride, NaCl. The molar mass of NaCl is 58.5 g mol^{-1}: the molar mass of Na = 23 g mol^{-1}; the molar mass of Cl = 35.5 g mol^{-1}. So the molar mass of NaCl = 23 + 35.5 = 58.5 g mol^{-1}.

So the number of moles present in 100 g of NaCl =

$$\frac{\text{mass of sample}}{\text{molar mass}} = \frac{100\ \text{g}}{58.5\ \text{g mol}^{-1}} = 1.71\ \text{mol}$$

Self-check 12.4 How many moles are there in:

(a) 40 g of sodium chloride, NaCl?

(b) 50 g of calcium carbonate, $CaCO_3$?

(c) 100 g of glucose, $C_6H_{12}O_6$?

2 A **variable** is a specific parameter that we can quantify but which can have different values.

12.2 Concentrations

The chemical reactions that occur in biological systems (including those occurring in the cells of our body) take place in water-based solution. If we want to know how many moles of a substance are in a particular sample, we need to know its **concentration**.

The concentration of a solution tells us how much of a substance is present *in a particular volume of the solution*. A solution with a high concentration contains more of the substance per unit volume (i.e. in a particular volume) than does a solution with a lower concentration.

For example, think about preparing a glass of fruit cordial, as illustrated in Figure 12.1. If we want a weak-tasting drink we pour in a small volume of cordial and fill the glass with water. If we want a stronger-tasting drink we start with more cordial, and then fill the glass with water. In both cases, the total volume of drink after adding the water is the same. However, the amount of cordial present in that particular volume (the filled glass) is greater in the strong-tasting drink than in the weak-tasting drink. So, we say that the concentration of the cordial is greater in the strong-tasting drink than in the weak-tasting drink.

When we consider the concentration of substances in chemical terms, we consider the **number of moles** of the substance present in **one litre** of solution[3].

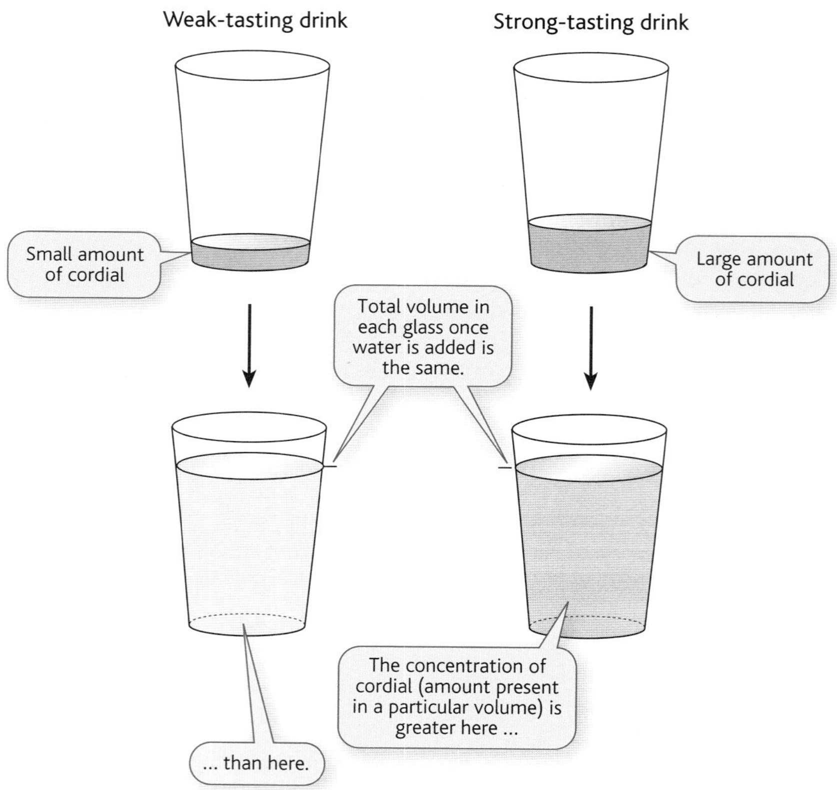

Figure 12.1 We can alter the concentration of a fruit cordial drink by diluting different volumes of fruit cordial to the same final volume by adding water.

3 1 litre (L) is exactly the same volume as 1 dm^3 (decimetre^3). 1 dm = 0.1 metre.

We call the concentration of a chemical compound in solution its **molarity**.

The units of molarity are cited as mol L^{-1} or mol dm^{-3}. However, we often replace these units with the symbol M. (The value of the concentration is the same whichever of these three units we choose to use.)

- *Molarity = concentration of a chemical substance = number of moles present in 1 dm^3 of solution. Molarity is measured in mol dm^{-3}, but is represented by the symbol M.*

For example, a 1M solution of hydrogen chloride, HCl, has a concentration of 1 mol L^{-1}.

So, if we take one litre of this solution, it will contain one mole of hydrogen chloride.

By contrast, a 3M solution of HCl has a concentration of 3 mol L^{-1}; it contains three moles of HCl per litre of solution.

When we write equations involving concentrations of different species, we often write these species within square brackets. For example, [A] denotes 'the concentration of A'. Specifically, square brackets are used to denote the concentration when measured in mol L^{-1}.

Remember: the superscript $^{-1}$ changes the unit L (litre) to 'per L'.

Calculating the number of moles of substance in a sample of solution

If we know the concentration of a solution, we can then calculate the amount of a substance (in moles) present in a *sample* of that solution. We merely need to multiply the concentration of the solution by the volume of the sample (measured in L).

We summarize this relationship as follows:

$$\text{number of moles (mol)} = \text{concentration (mol L}^{-1}) \times \text{volume (L)}$$

EXAMPLE

Suppose we have 100 mL (i.e. 0.1 L) of a NaCl solution, and we know that the solution concentration is 2 M (it has a concentration of 2 mol L^{-1}). How many moles do we have in the 100 mL (0.1 L) sample?

Remember: there are 1000 mL in 1 L. So 100 mL = 0.1 L.

Strategy

Multiply the volume of sample (in L) by the concentration of solution:

Volume of sample

Concentration of solution

$$0.1\,\not{L} \times 2\,\text{mol}\,\not{L}^{-1} = 0.2\,\text{mol}$$

Notice that the L and L^{-1} cancel each other out

So there are **0.2 mol** of NaCl in 100 mL of a 2 M solution.

- *Number of moles (mol) = concentration (mol L^{-1}) × volume (L)*

Self-check 12.5

(a) How many moles of NaCl are there in 50 mL of a 0.5 M solution?
(b) How many moles of HCl are there in 100 mL of a 1 M solution?
(c) How many moles of HCl are there in 2 L of a 0.1 M solution?

Preparing a solution of known concentration

Often, we need to prepare a solution of a particular concentration. How can we go about this? We can also use the relationship:

$$\text{number of moles (mol)} = \text{concentration (mol L}^{-1}) \times \text{volume (L)}$$

to help us determine how many moles of a substance we need to prepare a solution that has a particular concentration.

EXAMPLE

Remember: number of moles = volume of sample × concentration of sample. In this case, number of moles = 1 L × 1 mol L^{-1} = 1 mol.

Let's say that we want to prepare 1 L of a 1 M HCl solution. How many moles of HCl do we need?

Strategy

Number of moles = concentration × volume

= 1 mol L^{-1} × 1 L = 1 mol

So we need one mole of HCl.

By contrast, to prepare 0.5 L of a 1 M HCl solution we need only 0.5 mol of HCl:

$$\text{number of moles} = 0.5\ \text{L} \times 1\ \text{mol L}^{-1} = 0.5\ \text{mol}$$

To prepare 0.5 L of a 2 M HCl solution we need:

$$0.5\ \text{L} \times 2\ \text{mol L}^{-1} = \mathbf{1\ mol\ HCl}$$

Self-check 12.6

(a) How many moles of HCl do we need to prepare 2 L of a 0.5 M solution?
(b) How many moles of HNO_3 do we need to prepare 1 L of a 2 M solution?
(c) How many moles of NaOH do we need to prepare 500 mL of a 1 M solution?

A second step in the process of preparing the solution is to work out the **mass** of HCl that needs to be weighed out to give us the desired number of moles.

To do this, we merely need to remember that:

$$\frac{\text{mass of sample}}{\text{molar mass}} = \text{number of moles present in sample}$$

This time, however, we know the number of moles that we need, and we can work out the molar mass. The **unknown quantity** is the mass of sample. So we need to rearrange the above equation so that the two quantities we know are on one side of the equation, leaving us the unknown quantity on the other side:

$$\text{mass of sample} = \text{molar mass} \times \text{number of moles present}$$

We see how to rearrange equations (including this equation) in Maths Tool 12.1.

EXAMPLE

Returning to our previous example, we need 1 mol of HCl in 1 L of solution to produce a 1 M solution. What mass of HCl contains 1 mol?

Strategy

$$\text{Mass} = \text{molar mass} \times \text{number of moles}$$

Molar mass of HCl = 36.5 g mol^{-1}

So mass = 36.5 g mol^{-1} × 1 mol

= 36.5 g

We need **36.5 g** HCl to make 1 L of 1 M HCl.

Molar mass of HCl = Molar mass of H + Molar mass Cl = 1 g mol^{-1} + 35.5 g mol^{-1} = 36.5 g mol^{-1}

EXAMPLE

Let's consider another example. What mass of glucose ($C_6H_{12}O_6$) do we need to prepare 500 mL of a 4 M glucose solution?

Strategy

Step 1: determine how many moles of glucose we need.

Step 2: determine the mass of glucose that this number of moles equates to.

Step 1:

Number of moles (mol) = concentration of solution (mol L^{-1}) × volume of solution (L)

= 4 mol L^{-1} × 0.5 L

= 2 mol

We need 2 mol of glucose.

Remember: we express the volume of solution in litres, so we convert 500 mL into the equivalent volume in litres, 0.5 L.

Step 2:

Mass of sample = molar mass × number of moles present.

Molar mass of glucose = 180 g mol^{-1}

Mass of sample = 180 g mol^{-1} × 2 mol

= 360 g

We need **360 g** of glucose to prepare 500 mL of a 4 M solution.

MATHS TOOL 12.1 **Rearranging equations**

When we rearrange an equation, we typically want to get an unknown term (the variable we're trying to calculate) by itself on one side of the equation, and any known terms (the variables we already know the value of) on the other side of the equation.

To rearrange an equation, we effectively move the terms around one at a time to separate the known and unknown terms. In order to move a term from one side of an equation to the other, we need to reverse whatever mathematical operation is being applied to the term in the original equation:

- To reverse an addition, we have to subtract
- To reverse a subtraction, we have to add
- To reverse a multiplication, we have to divide
- To reverse a division, we have to multiply.

EXAMPLE

Imagine we have the equation $a + b = c$, and our unknown term is *a*.

Strategy

We need to isolate *a* on one side of the equation. This means we need to move *b* to the other side of the equation. In the original equation, *b* is being added to *a*. So, to reverse this operation, we subtract *b* from both sides of the equation.

Solution

$$a + \cancel{b} - \cancel{b} = c - b$$
$$a = c - b$$

There are two key things to note here:

(1) We're performing the same action on *both* sides of the equation;

(2) The two *b*s on the left-hand side of the equation cancel each other out, isolating *a* as required.

EXAMPLE

Now imagine that our original equation is $\frac{a}{b} = c$. Again, our unknown term is *a*.

Strategy

To isolate *a* we need to remove *b* from the left-hand side of the equation. In the original equation *a* is being divided by *b* so, to reverse this operation, we have to multiply by *b*.

Solution

$$\frac{a}{\cancel{b}} \times \cancel{b} = c \times b$$
$$a = c \times b$$

Again, notice how the two *b* terms on the left-hand side of the equation cancel each other out, while *b* then appears on the right-hand side of the equation instead.

If we need to move more than one term to isolate a particular, unknown term, we just need to move one term at a time, by reversing the operation being applied to it in the original equation.

EXAMPLE

Imagine our original equation is $\frac{a}{b} - c = d$. This time our unknown term is *b*.

Strategy

This time, we need to get *a*, *c*, and *d* together, leaving *b* by itself. We could do this in the following three steps.

Solution

Move *c* to the right-hand side by adding it to both sides (remembering that adding will reverse the effect of *c* being subtracted from the left-hand side of the original equation):

$$\frac{a}{b} - c + c = d + c$$
$$\frac{a}{b} = d + c$$

Next, we multiply both sides of the equation by *b*, to extract *b* from the bottom of the fraction on the left-hand side of the equation:

$$\frac{a}{b} = d + c$$
$$\frac{a}{b} \times b = (d + c) \times b$$
$$a = (d + c) \times b$$

Notice how we write *d* and *c* together in brackets here. This is because we're multiplying both *d* and *c* by *b*, and this saves us writing $(d \times b) + (c \times b)$.

Finally, we need to isolate *b*. At this stage, *b* is the subject of a multiplication operation; to reverse this we divide by $(d + c)$:

$$a = (d + c) \times b$$
$$\frac{a}{(d + c)} = \frac{(d + c) \times b}{(d + c)}$$
$$\frac{a}{(d + c)} = b$$

Let us now return to our original equation.

$$\frac{\text{mass of sample}}{\text{molar mass}} = \text{number of moles present in sample}$$

Our unknown term is the mass of sample. To rearrange the equation we:

(1) Multiply both sides of the equation

$$\frac{\text{mass of sample}}{\text{molar mass}} = \text{number of moles present in sample}$$

by molar mass:

$$\text{molar mass} \times \frac{\text{mass of sample}}{\text{molar mass}} = \text{molar mass} \times \text{number of moles present}$$

(2) Cancel any terms that appear on the same side of the equation:

$$\cancel{\text{molar mass}} \times \frac{\text{mass of sample}}{\cancel{\text{molar mass}}} = \text{molar mass} \times \text{number of moles present}$$

(3) To leave us with:

mass of sample = molar mass × number of moles present

Self-check 12.7 Rearrange $\frac{\text{mass of sample}}{\text{molar mass}} = \text{number of moles present in sample}$ to isolate the molar mass as the unknown term.

Self-check 12.8

(a) What mass of glucose do we need to prepare 250 mL of a 2 M solution?

(b) What mass of NaCl do we need to prepare 1 L of a 0.2 M solution?

(c) What mass of KCl do we need to prepare 100 mL of a 0.05 M solution?

Calculating the concentration of a solution

We can use the relationship:

$$\text{number of moles (mol)} = \text{concentration (mol L}^{-1}\text{)} \times \text{volume (L)}$$

to tell us the concentration of a solution if we know the volume of the solution and the number of moles of a substance present in the solution.

If we rearrange this equation to isolate the unknown term we can see how the concentration is related to the number of moles present, and the volume of solution:

$$\text{concentration (mol L}^{-1}\text{)} = \frac{\text{number of moles (mol)}}{\text{volume (L)}}$$

EXAMPLE

Let's imagine that we have dissolved 2 mol of glucose to make 500 mL of solution. What is the concentration of this solution?

Strategy

$$\text{concentration} = \frac{\text{number of moles}}{\text{volume}}$$

Remember: we quote volume in litres: 500 mL = 0.5 L

$$= \frac{2\text{ mol}}{0.5\text{ L}}$$

$$= 1\text{ mol L}^{-1} = 1\text{ M}$$

So, 2 mol in 500 mL produces a solution of concentration 4 M.

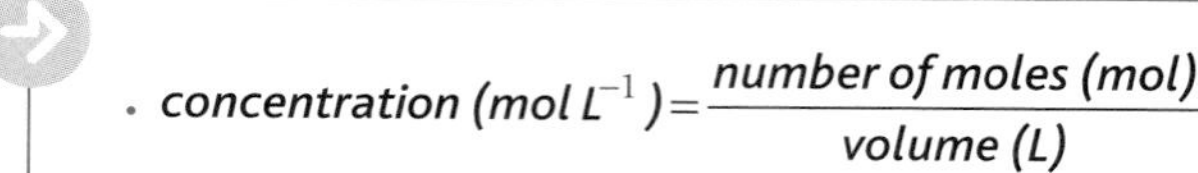

• concentration (mol L⁻¹) = number of moles (mol) / volume (L)

Self-check 12.9

(a) What is the concentration of 250 mL of solution containing 0.5 mol of glucose?

(b) What is the concentration of 1 L of solution containing 3 mol of HCl?

(c) What is the concentration of 500 mL of solution containing 0.2 mol of HNO_3?

(d) What volume of a 0.2M solution would contain 0.5 mol?

12.3 Changing the concentration: solutions and dilutions

In chemistry and biochemistry, we often talk of compounds being 'in solution', or being 'solvents' and 'solutes'. So it is worth us pausing for a moment to be clear about what these terms mean. A **solvent** is the medium in which a substance (called the **solute**) is dissolved. In aqueous solutions, the solvent is **water**.

When we say that something is soluble, it is good practice for us to specify the solvent that we are considering. In day-to-day life, we usually take the word 'soluble' to mean 'water-soluble'. For example, we may take 'soluble aspirin' if we have a headache. The word 'soluble' indicates that the aspirin is water-soluble, and we can take the aspirin having first dissolved it in water. However, when describing chemical systems, we should not assume that the solvent is water.

Why is this distinction important? A particular compound may be insoluble in water, but may be *soluble* in another solvent or vice versa. As a general rule, polar compounds are soluble in polar solvents, while non-polar compounds are soluble in non-polar solvents—we say that 'like dissolves like'. For example, methanol, CH_3OH, is polar, and is soluble in water, which is also polar. By contrast, propane, $CH_3CH_2CH_3$ is non-polar, and is *not* water-soluble. However, propane *is* soluble in the non-polar solvent dichloromethane, CH_2Cl_2. Therefore, we cannot state absolutely that a compound is 'soluble' or 'insoluble' without specifying the solvent that we are considering.

The principles of diluting a solution

Many of the solutions we encounter on a daily basis have to be **diluted** before we can use them as intended. When we dilute an aqueous solution, we reduce

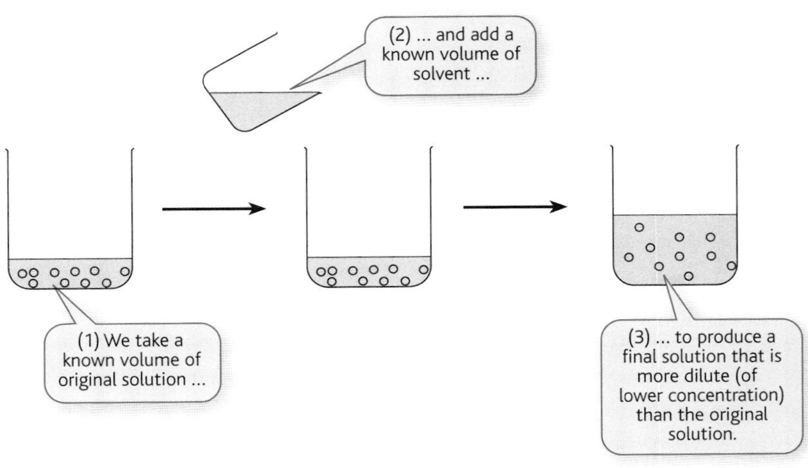

Figure 12.2 The process of diluting a solution. When we dilute a solution, we have the *same* number of solute molecules in a *greater* overall volume. Consequently, the concentration of the solution is *reduced*.

its concentration by adding more water to it. For example, fruit cordials must be diluted before consumption; the coolants that we use in car radiators are diluted to different extents depending on the season: in summer, we are instructed to dilute the coolant more than in winter (i.e. to add more water to the same volume of coolant before pouring the diluted coolant into the radiator).

But how do we know the concentration of a solution once we have diluted it? And how much water do we need to add to dilute a solution down to a desired concentration? We'll consider each of these questions in turn.

As a first step, let's think about what's happening when we dilute a solution, as illustrated in Figure 12.2.

Notice that, while the total volume changes, the number of solute particles present in the solution (represented by the dots) remains the same. This is an important point: when we dilute a solution, the number of moles of the solute (the substance that is dissolved in the solution) remains unchanged.

However, if we take a sample of a certain, fixed volume from a solution before and after dilution, the sample taken after dilution will contain fewer moles than the sample taken before dilution.

- ***When we dilute a solution, the number of moles of the solute remains unchanged.***

We see above how the concentration of a solution is related to the volume of solution, and number of moles of substance as follows:

$$\text{concentration (mol L}^{-1}\text{)} = \frac{\text{number of moles (mol)}}{\text{volume (L)}}$$

So, if we know the total number of moles present (because this number doesn't change when the solution is diluted), and we can measure the volume of the final solution, then we can work out the concentration of the final solution.

EXAMPLE

Let's imagine that we've diluted 100 mL of a 2 M solution of HCl with 900 mL of water to produce 1 L of a diluted HCl solution, as illustrated in Figure 12.3.

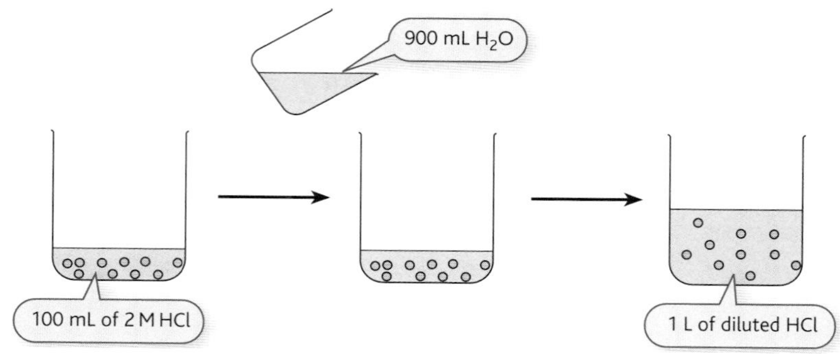

Figure 12.3 The dilution of 100 mL of a 2M solution of HCl with 900 mL of water.

What is the concentration of the diluted HCl solution?

Strategy

Step 1: work out the number of moles of HCl that are present.

Step 2: use this value to calculate the concentration of the diluted solution.

Step 1:

We know the number of moles of HCl that are present because we know the concentration and volume of the original solution:

Remember: 100 mL = 0.1 L.

Number of moles = concentration × volume

$= 2\ \text{mol L}^{-1} \times 0.1\ \text{L}$

$= 0.2\ \text{mol}$

So there are 0.2 mol of HCl in the original solution.

Step 2:

$$\text{Concentration of diluted solution} = \frac{\text{number of moles of HCl}}{\text{volume of diluted solution}}$$

$$= \frac{0.2\ \text{mol}}{1\ \text{L}}$$

Total volume of the final solution

$= 0.2\ \text{mol L}^{-1}$ or 0.2 M

So the concentration of the final solution is **0.2 M**.

Self-check 12.10

(a) What is the concentration of the final solution when 20 mL of a 1 M HCl solution is diluted by adding 80 mL of water?

(b) What is the concentration of the final solution when 50 mL of 0.5 M glucose solution is diluted by adding 150 mL of water?

(c) What is the concentration of the final solution when 100 mL of 2 M HNO_3 solution is diluted by adding 1.9 L of water?

EXAMPLE

Now, let's imagine that we have 200 mL of a 2 M solution of HCl, and we want to dilute it to a concentration of 0.1 M. How much water would we need to add to the 200 mL of original solution to produce a final solution with a concentration of 0.1 M?

In this instance, we need to work out the volume of final solution that is required to give a 0.1 M solution. We can then deduce how much water to add to the original solution to reach this final volume.

Strategy

Step 1: work out the number of moles of HCl present.

Step 2: use this value to calculate the volume of the diluted solution and, hence, the volume of water that needs to be added to the original solution.

Step 1:

We know the number of moles of HCl present because we know the concentration and volume of the original solution:

number of moles = concentration × volume

$= 2\ \text{mol L}^{-1} \times 0.2\ \text{L}$

$= 0.4\ \text{mol}$

So there are 0.4 mol of HCl in the original solution.

Notice how we rearrange the equation to isolate the unknown term on one side of the equation.

Step 2:

$$\text{Volume of diluted solution} = \frac{\text{number of moles of HCl}}{\text{concentration of diluted solution}}$$

$$= \frac{0.4\ \text{mol}}{0.1\ \text{mol L}^{-1}} = 4\ \text{L}$$

So, the diluted solution must have a volume of 4 L to produce a concentration of 0.1 M. We started with a volume of 200 mL of the 2 M solution, so must add **3.80 L** (3800 mL) of water to bring the final volume up to 4 L, and achieve a concentration of 0.1 M.

Take care here: we have said that the diluted solution must have a volume of 4 L to produce a concentration of 0.1 M. However, we have *not* said that we must add 4 L of water to the original solution to achieve the required final volume. Instead, we work out how much water to add by subtracting the volume we start off with (in this case 200 mL) from the final volume required (4 L, or 4000 mL).

Self-check 12.11

(a) How much water would we need to add to 200 mL of a 1 M HCl solution to dilute it to a concentration of 0.2 M?

(b) How much water would we need to add to 100 mL of a 0.5 M HCl solution to dilute it to a concentration of 0.03 M?

(c) How much water would we need to add to 400 mL of a 0.9 M glucose solution to dilute it to a concentration of 0.5 M?

Serial dilutions

If we have a solution that we need to dilute by a significant amount, it can be impractical to dilute it 'in one go'. For example, imagine that we have a growth medium that contains 5×10^6 microorganisms per mL. If we tried to look at a sample of this solution under the microscope, we'll not be able to see anything meaningful because it would be swamped with microbes. Instead, we may need a concentration of no greater than 100 microorganisms per mL to be able to perform our microscopic examination. If we had a 10 mL starting volume, we'd have to add around *50 L* of solvent to reduce the concentration to the required 100 microorganisms per mL. This would clearly be impractical.

Instead, we perform a **serial dilution**. A serial dilution reduces the concentration of a solution in a series of successive steps, such that the volumes being manipulated at each step remain manageable. The strategy employed when carrying out a serial dilution is illustrated in Figure 12.4. Look at this figure, and notice how we carry out a serial dilution by taking a certain volume of our starting solution, and adding a fixed volume of solvent. We then take a sample of this diluted solution–the same volume that was taken from our starting solution–and add the same volume of solvent as before. We can repeat this process of taking a sample of dilute solution and diluting it further as often as we like until we achieve our desired final concentration.

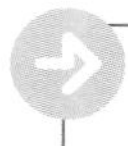

- ***A serial dilution reduces the concentration of a solution in a series of steps, such that the concentration gradually falls from step to step.***

So how does the concentration of our solution vary between each step of our serial dilution? As an example, let's assume that we have 10 mL of starting solution with a concentration of 2 mol L^{-1}. We then carry out a serial dilution whereby we take 1 mL of solution and dilute it back up to a volume of 10 mL by adding 9 mL of solvent; we carry out four serial dilution steps, as illustrated in Figure 12.5. What is the concentration of the solution at each step? Let's work through one step at a time:

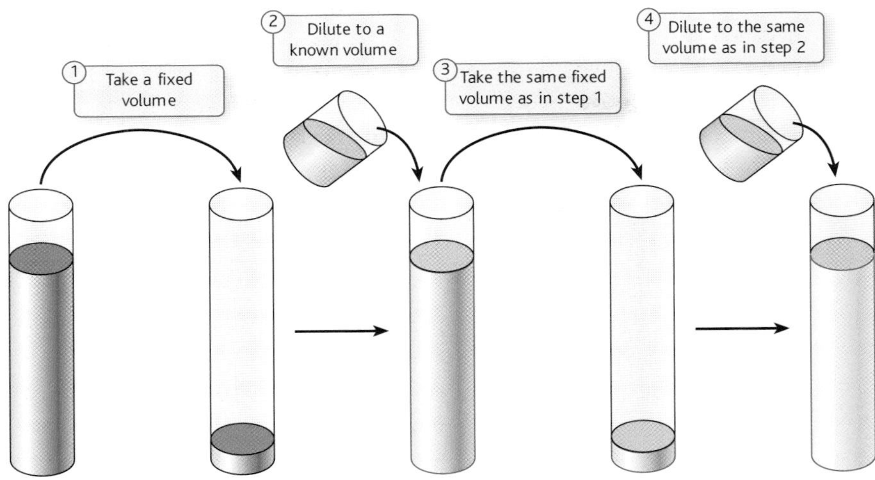

Figure 12.4 The strategy behind a serial dilution. During a serial dilution, a sample goes through repeated rounds of dilution, whereby an aliquot of an original sample is removed and diluted, and the same-sized aliquot of the diluted sample is taken and rediluted, and so on.

Step 1:

We take 1 mL of a 2 mol L^{-1} solution, and dilute it to a total volume of 10 mL. The number of moles in our 1 mL sample is given by the following:

$$\begin{aligned}\text{Number of moles} &= \text{concentration of solution} \times \text{volume of solution}\\ &= 2\ \text{mol L}^{-1} \times 0.001\ \text{L}\\ &= 0.002\ \text{mol}\end{aligned}$$

Remember: 1 mL = 0.001 L.

The concentration of our solution at step 1 is given by:

$$\text{concentration} = \frac{\text{number of moles}}{\text{volume of solution}}$$

We know there are 0.002 mol, and the total volume at step 1 is 10 mL. Therefore:

$$\text{concentration} = \frac{0.002\ \text{mol}}{0.01\ \text{L}}$$

$$\text{concentration} = 0.2\ \text{mol L}^{-1}$$

10 mL = 0.01 L.

Step 2:

We take 1 mL of a 0.2 mol L^{-1} solution, and dilute it to a total volume of 10 mL.

$$\begin{aligned}\text{Number of moles} &= \text{concentration of solution} \times \text{volume of solution}\\ &= 0.2\ \text{mol L}^{-1} \times 0.001\ \text{L}\\ &= 0.0002\ \text{mol}\end{aligned}$$

The concentration of our solution at step 2 is given by:

$$\text{concentration} = \frac{\text{number of moles}}{\text{volume of solution}}$$

$$\text{concentration} = \frac{0.0002\ \text{mol}}{0.01\ \text{L}}$$

$$\text{concentration} = 0.02\ \text{mol L}^{-1}$$

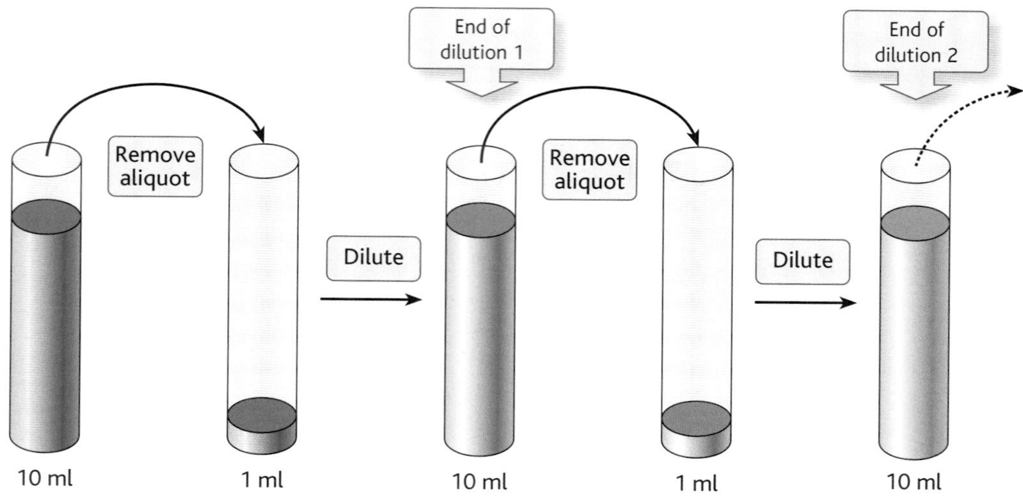

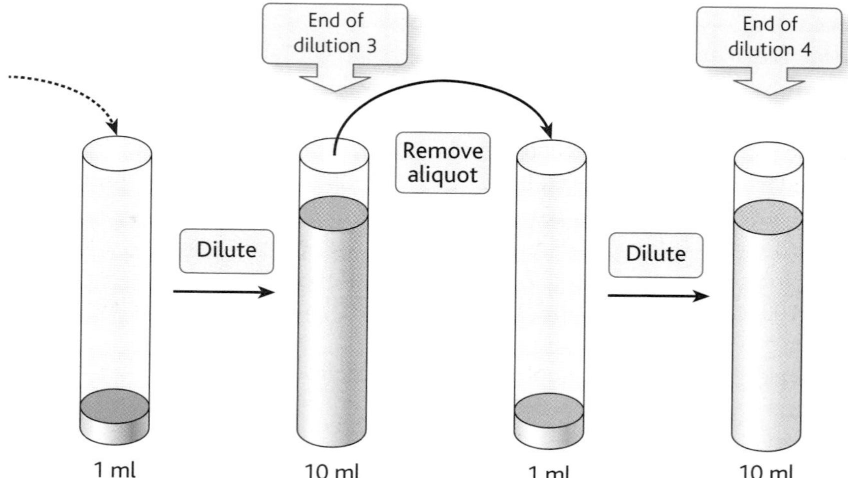

Figure 12.5 A four-step serial dilution, in which we take 1 mL of solution and dilute it back up to a volume of 10 mL by adding 9 mL of solvent.

Step 3:

We take 1 mL of a 0.02 mol L^{-1} solution, and dilute it to a total volume of 10 mL.

$$\text{Number of moles} = \text{concentration of solution} \times \text{volume of solution}$$

$$= 0.02 \text{ mol L}^{-1} \times 0.001 \text{ L}$$

$$= 0.00002 \text{ mol (or } 2 \times 10^{-5} \text{ mol)}$$

The concentration of our solution at step 2 is given by:

$$\text{concentration} = \frac{\text{number of moles}}{\text{volume of solution}}$$

$$\text{concentration} = \frac{2 \times 10^{-5} \text{ mol}}{0.01 \text{ L}}$$

$$\text{concentration} = 0.002 \text{ mol L}^{-1}$$

Step 4:

We take 1 mL of a 0.002 mol L^{-1} solution, and dilute it to a total volume of 10 mL.

$$\text{Number of moles} = \text{concentration of solution} \times \text{volume of solution}$$
$$= 0.002 \text{ mol L}^{-1} \times 0.001 \text{ L}$$
$$= 0.000002 \text{ mol (or } 2 \times 10^{-6} \text{ mol)}$$

The concentration of our solution at step 2 is given by:

$$\text{concentration} = \frac{\text{number of moles}}{\text{volume of solution}}$$

$$\text{concentration} = \frac{2 \times 10^{-6} \text{ mol}}{0.01 \text{ L}}$$

$$\text{concentration} = 0.0002 \text{ mol L}^{-1}$$

So, our concentrations at each step are as follows:

Original	**2 mol L^{-1}**
Step 1	0.2 mol L^{-1}
Step 2	0.02 mol L^{-1}
Step 3	0.002 mol L^{-1}
Step 4	0.0002 mol L^{-1}

What do you notice? The concentration has reduced by a factor of 10 at each dilution step, and by a factor of 10^4 overall. However, the volume of our final solution is the same as our starting volume: 10 mL.

If we'd have tried to carry out this dilution in one step–by diluting the initial sample from a concentration of 2 mol L^{-1} to 0.0002 mol L^{-1}–we would have needed to add 100 L of solvent (less the 10 mL we've started with). This would have left us with a huge volume to deal with!

If we plot the concentration of our serial dilution at each step we obtain a graph like that shown in Figure 12.6. Look at this graph, and notice how the curve becomes more shallow as we move from left to right. The shape of this curve is an example of **exponential decay**; it shows how the concentration doesn't decrease in a linear manner from step to step, but instead decreases by a factor of 10 at each step.

We see another example of a curve whose shape represents exponential decay when we discuss first-order reaction kinetics in Section 14.1.

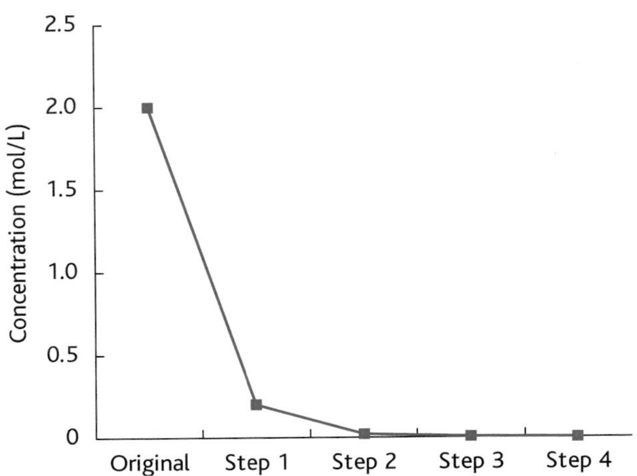

Figure 12.6 The change in concentration of a sample during successive stages of a serial dilution.

So, serial dilutions are a useful tool when we need to decrease the concentration of a compound (or some other entity, from blood cells to microbes) without having to manipulate large volumes of solvent.

Self-check 12.12 A bacterial growth medium contains bacteria at a concentration of 15 000 per mL. If we carry out three dilution steps using aliquots of 1 mL at each dilution step, and dilute to a total volume of 5 mL at each stage, what would be the final concentration of bacteria after the three dilution steps (expressed as number of bacteria per mL)?

12.4 Measuring concentrations

Many of us have heard of the 'litmus test': a strip of red or blue paper, which changes colour to tell us whether a solution is acidic or alkaline. Litmus paper gives us a qualitative measurement of the concentration of hydrogen ions, H^+, in the solution being tested. (We use the **pH scale** to give us a *quantitative* measurement of the concentration of hydrogen ions in a given solution.)

We learn more about pH in Chapter 16.

Being able to measure the concentration of chemical compounds is an important part of everyday life. Our body contains a diverse mixture of chemical compounds, many of which must only be present within certain ranges of concentration for us to remain healthy. For example, while we need sugar to give us energy, it is dangerous for the concentration of sugar in our bloodstream to rise above a concentration of 2 g per litre on a regular basis. (Elevated blood sugar levels may lead to damage to the kidney, eye, and nervous system.) Blood sugar concentration is regulated by the hormone insulin, which triggers the removal of glucose from the bloodstream, and its storage as glycogen in our cells. (When we need a sudden burst of energy, this process is reversed, and the stored glycogen is broken back down to produce glucose for fuel.)

However, individuals with diabetes are unable to regulate their blood sugar in this way: their bodies either fail to produce insulin or cannot respond to the insulin that is produced. Either way, the body lacks this in-built mechanism for converting blood glucose to glycogen. Consequently, people with diabetes must monitor their blood sugar concentrations on a regular basis, to ensure that the concentration of sugar in their blood stays within a safe range[4].

People with diabetes measure their blood sugar concentration with a blood-sugar meter: they prick their finger to produce a small drop of blood, and place the drop of blood on a test strip, which is inserted into the meter to measure the concentration of sugar in the blood sample. We find out more about how blood-sugar meters work in Box 12.4, later in this chapter.

We don't only measure the concentration of chemicals in the body for health reasons. Traffic police are also measuring chemical concentrations when they carry out a breathalyser test: they are determining the concentration of ethanol in the blood.

So how can we determine the concentration of a compound in solution? There are a number of different analytical tools available to us; we explore some of the key tools in the rest of this section.

4 Specifically, people with diabetes measure the concentration of **glucose** in the blood. Glucose is the predominant sugar found in the blood, and the sugar that our body uses directly as a source of energy.

UV–visible spectrophotometry

The intensity of colour of a solution relates to the solution's *concentration*. When we make a mug of tea or coffee, we know that a darker-looking liquid is stronger (more concentrated) than a lighter-looking liquid. Similarly, if we're preparing a glass of fruit cordial, a glass containing a solution whose colour is more intense is more concentrated (tastes 'stronger') than a solution whose colour is weaker.

UV–visible spectrophotometry is a technique for measuring the colour intensity of a solution, from which we can then calculate the *concentration*. Spectrophotometry exploits the difference in absorption of light by solutions of different concentration: a more concentrated solution absorbs more light than a less concentrated solution.

Spectrophotometry is closely related to UV–visible spectroscopy, which we discuss in Chapter 11, and uses electromagnetic radiation in the same portion of the electromagnetic spectrum: light whose wavelengths fall within the **ultraviolet** and **visible regions** of the electromagnetic spectrum, between 200 and 700 nm. However, while UV–visible spectroscopy uses radiation of varying wavelengths, spectrophotometry uses light of a *single, specific* wavelength.

Remember that 1 nm = 10^{-9} m or 0.000 000 001 m.

Self-check 12.13 Does UV–visible spectrophotometry use light of lower or higher energy than infrared spectroscopy?

During spectrophotometry, light is passed through a sample of the solution being tested, and the amount of light absorbed during its passage through the sample is detected, as illustrated in Figure 12.7. Notice from this figure how the amount of light absorbed by the solution is proportional to the concentration of the solution: if the solution is of high concentration, then a lot of light is absorbed, and only a little light passes through the sample to be detected. By contrast, if the solution is of low concentration, only a little light is absorbed by the sample, and much of the light passes straight through it, and is then detected.

Spectrophotometry is performed using an instrument called a **spectrophotometer**. The spectrophotometer compares the intensity of the light shone onto the sample with the amount that has passed through it to calculate the amount of light *absorbed* by the sample. A measure of the light absorbed by a sample is called its **absorbance, A**.

Notice that we say spectrophotometry works by shining light of a *specific wavelength* through a sample. Why is this? Substances of different colour absorb light most strongly at *different* wavelengths. To get the most accurate measurements from the spectrophotometer, we use the wavelength at which the sample absorbs *most* strongly. We call this wavelength λ_{max}.

For example, vitamin A (**1**) absorbs light most strongly at 328 nm, while ATP (**2**) absorbs light most strongly at 259 nm.

(1)

(2)

Read Box 12.1 to find out about the difference between the absorption and reflection of light, and what this difference means in the context of the colour of the objects around us.

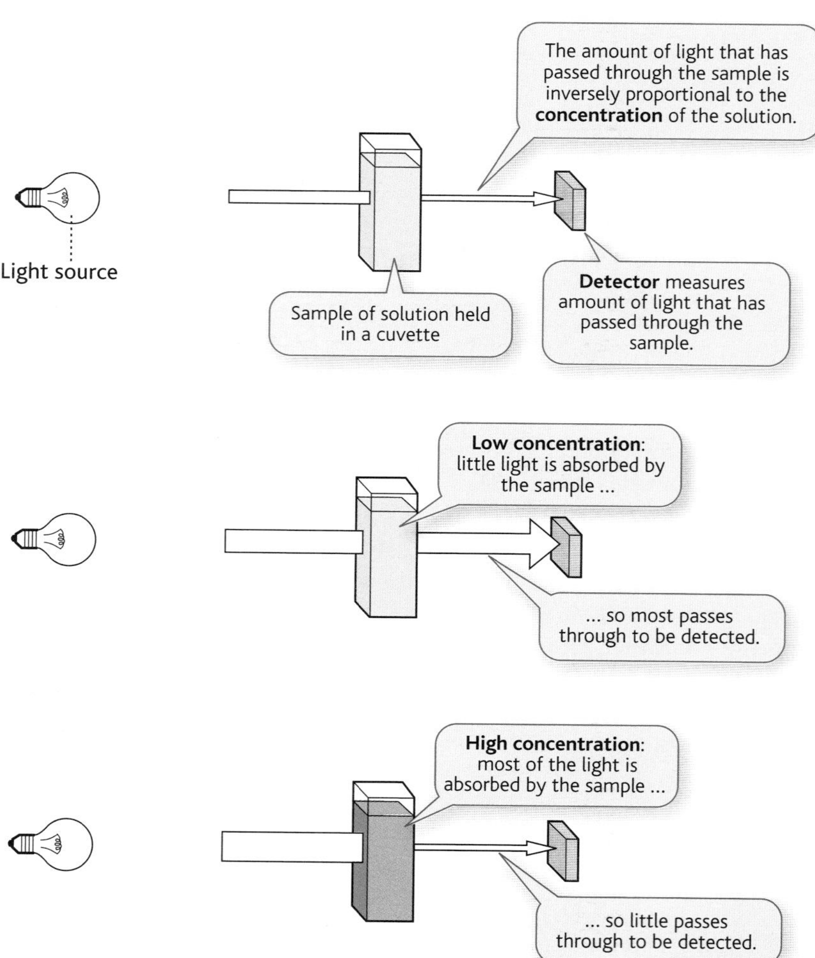

Figure 12.7 Spectrophotometry works by shining light of a specific wavelength through a sample of the solution being tested, and detecting the amount of light that passes through the sample. The amount of light absorbed by the solution is proportional to the concentration of the solution.

BOX 12.1 Absorption versus reflection

The colours of objects we see around us relate to the wavelengths of light that are *not* absorbed by the object. Instead, our eyes detect the light of wavelengths that are **reflected**.

Imagine that we are looking at a blue cube, as illustrated in Figure 12.8. Look at this figure and notice how the cube appears blue because it *reflects* light in the blue region of the visible spectrum (~450–475 nm), which is then detected by our eyes.

Most plants appear to us as being green in colour. Plants contain pigments (coloured compounds) called **chlorophylls** in their cell membranes. Figure 12.9 shows how the chlorophyll in the plant's cell wall absorbs light with wavelengths of approximately 450 nm and 700 nm most strongly. These wavelengths correspond to blue and red light respectively. However, chlorophyll does not absorb light with wavelengths in the region 500–600 nm. It is this light–in the **green** region of the visible spectrum–which is *reflected* by the plant, and which is detected by our eyes.

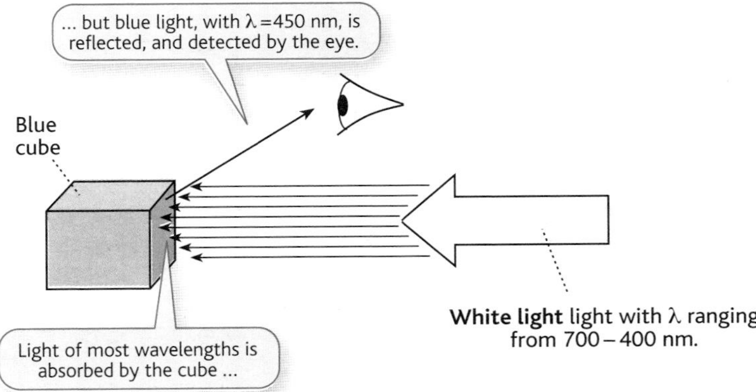

Figure 12.8 A blue cube appears blue because it reflects light in the blue region of the visible spectrum (~450–475 nm), which is then detected by our eyes.

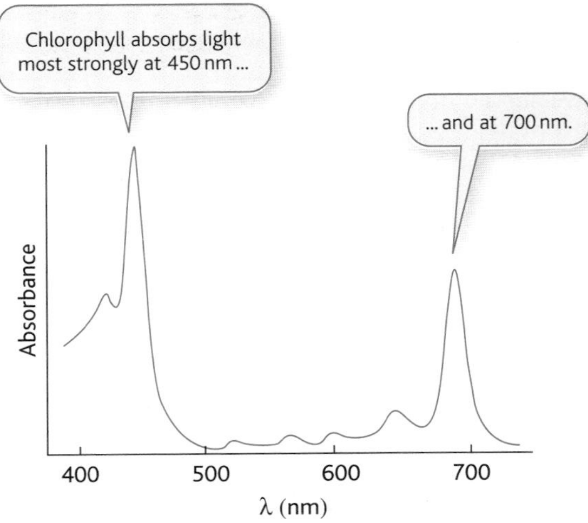

Figure 12.9 The pigment chlorophyll absorbs light with wavelengths of approximately 450 nm and 700 nm most strongly.

So, how can we use the absorbance of a sample to tell us the concentration of a compound in that sample? The absorbance of a sample of a compound in solution is related to the concentration of the compound by the following relationship:

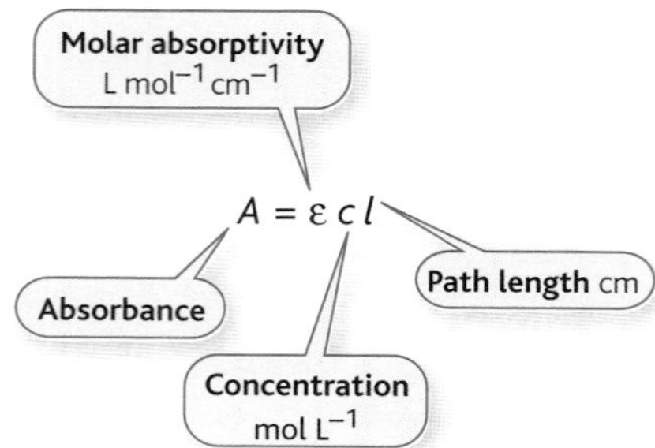

This relationship is the **Beer–Lambert law**, which describes how the concentration of a substance is related to its absorption of light.

- *The absorbance of a sample of a compound in solution is proportional to the concentration of the compound.*
- *The relationship between absorbance and concentration is described by the Beer–Lambert law.*

The **path length**, l, is the distance that the light travels through the sample. Figure 12.10 shows how the path length is dictated by the width of the container–the 'cuvette'. The path length is usually 1 cm.

The **molar absorptivity**, ε, tells us how much light is absorbed by a compound at a particular wavelength and allows the values for absorbance and concentration to tie in correctly with each other. Different compounds have different molar absorptivities.

We should note that the molar absorptivity has to be determined experimentally by taking a solution of *known* concentration and measuring its absorbance at a known wavelength. For our purposes, though, we'll assume that we know the value of ε, having determined it already[5].

So how do we use the Beer–Lambert law to help us calculate the concentration of a compound in solution? We know that absorbance and concentration are related by the following equation:

$$A = \varepsilon c l$$

Light travels this distance through cuvette = width of cuvette

Figure 12.10 The path length is dictated by the width of the container in which the sample is held–the 'cuvette'.

5 In fact, we can find tables of molar absorptivity values in chemical data books.

Our unknown quantity is the concentration, c. So we rearrange the equation to isolate c on one side of the equation:

$$c = \frac{A}{\varepsilon l}$$

This tells us that, to calculate the concentration of our compound, we merely divide the absorbance by ($\varepsilon \times l$). Because the path length, l, is usually 1 cm, $\varepsilon \times l$ becomes $\varepsilon \times 1 = \varepsilon$.

Let's consider an example.

EXAMPLE

Vitamin A has a molar absorptivity of 1550 L mol^{-1} cm^{-1} at 328 nm. Imagine that we have a solution containing an unknown concentration of vitamin A. We place a sample of the solution in a cuvette with a path length of 1 cm, and measure the absorbance of the sample at 328 nm. The spectrophotometer gives us an absorbance, A, of 1.4.

We see above that $c = \frac{A}{\varepsilon l}$. Because $l = 1$ cm, we simplify the equation to $c = \frac{A}{\varepsilon}$

We can then feed in our observed values:

$$c = \frac{1.4}{1550} = 0.000\,903 \text{ mol L}^{-1} = 9.03 \times 10^{-4} \text{ mol L}^{-1}$$

Remember: concentration is measured in mol L^{-1}. We mustn't forget to include the correct units!

Self-check 12.14 ATP has a molar absorptivity of 15 400 L mol^{-1} cm^{-1} at 259 nm. A solution of ATP of unknown concentration gives an absorbance of 1.88 at 259 nm when placed in a cuvette with a path length of 1 cm. What is the concentration of ATP in the solution?

Atomic spectroscopy

Spectroscopic techniques such as UV–visible spectroscopy exploit the way that electromagnetic radiation in the UV–visible region of the electromagnetic spectrum causes the movement of electrons between energy levels in molecules (such that the molecules become excited). However, as we see in Chapter 2 (p. 32), excitation also occurs in atoms–for example, to produce the line emission spectra that are characteristic of the different energy levels present in atoms of a given element.

Beyond its use in characterizing the energy levels present in an atom, atomic spectroscopy can also be used to determine the concentration of atoms of a given element that are present in a sample. So how is this achieved?

When atoms of a given element are atomized (that is, heated until they form a gas), excited by the absorption of energy, and then allowed to return to their original ground state, they emit energy whose wavelength corresponds to electromagnetic radiation in the visible region of the electromagnetic spectrum. (That is, upon heating, the atoms emit light.) The *intensity* of the light emitted is proportional to the concentration of atoms present: the higher the concentration, the greater the intensity of the light emitted. (This is similar to the link between the concentration of a compound in solution and the amount of energy it absorbs when studying concentration using UV–visible spectrophotometry, as described above. However, in UV–visible spectrophotometry, the sample is absorbing energy, not emitting it.)

The use of atomic spectroscopy to determine the concentration of an element in a sample is called **atomic emission spectroscopy** (AES). During AES, the sample is atomized by heating in a flame or other heat source; the intensity of the colour in the flame is determined by the concentration of the element present in the sample. AES can be used to determine the concentration of around 75 different metals and other elements.

Atoms of different elements emit light of characteristic wavelengths, which we see as light of different colours. For example, potassium produces a lilac flame, while sodium emits light of an intense orange colour (the colour associated with sodium street lights). Therefore, the concentration of a particular element present in a sample can be determined by measuring the intensity of light at the wavelength characteristic of the element in question. For example, a wavelength of 589 nm is used to determine the concentration of sodium in a sample, while a wavelength of 760 nm is used for potassium.

A flame is only hot enough to excite atoms of certain elements (including potassium and sodium). For other elements, much higher temperatures are required. One common source of these higher temperatures is plasma, produced from the ionization of argon gas. One significant advantage of atomic emission spectroscopy is its sensitivity: concentrations of elements as low as between 1 $\mu g\ dm^{-3}$ and 5 $\mu g\ dm^{-3}$ can be detected in some instances.

We link the intensity of emission detected by AES at a given wavelength to the concentration of the element present in the sample by using a **calibration curve**. A calibration curve is generated by determining the emission intensity produced from solutions containing known concentrations of the element of interest. We can then compare the emission intensity generated from a sample of unknown concentration to the calibration curve to determine the concentration of the unknown sample, as illustrated in Figure 12.11.

Atomic emission spectroscopy is particularly important in clinical laboratories where blood and urine samples are continually monitored in the diagnosis and treatment of disease. For example, AES can be used to monitor the concentrations of sodium and potassium in the blood: normal concentrations are 135 mmol dm^{-3}–145 mmol dm^{-3} for Na^+ and 3.5 mmol dm^{-3}–5.0 mmol dm^{-3} for K^+. If concentrations vary widely from these normal limits the consequences can be fatal.

- *Atomic emission spectroscopy can be used to determine the concentration of individual metals in a sample, even if a metal is only present in very low concentrations.*

Atomic absorption spectroscopy is similar to atomic emission spectroscopy, except the technique measures the *absorption* by a sample of light of a specific wavelength, rather than the *emission* of light. To irradiate a sample with light of a specific wavelength, a special lamp called a hollow cathode lamp is used. The cathode of this lamp is coated with the same element as that under investigation; when the lamp burns, the element coated onto the cathode becomes excited (in the same

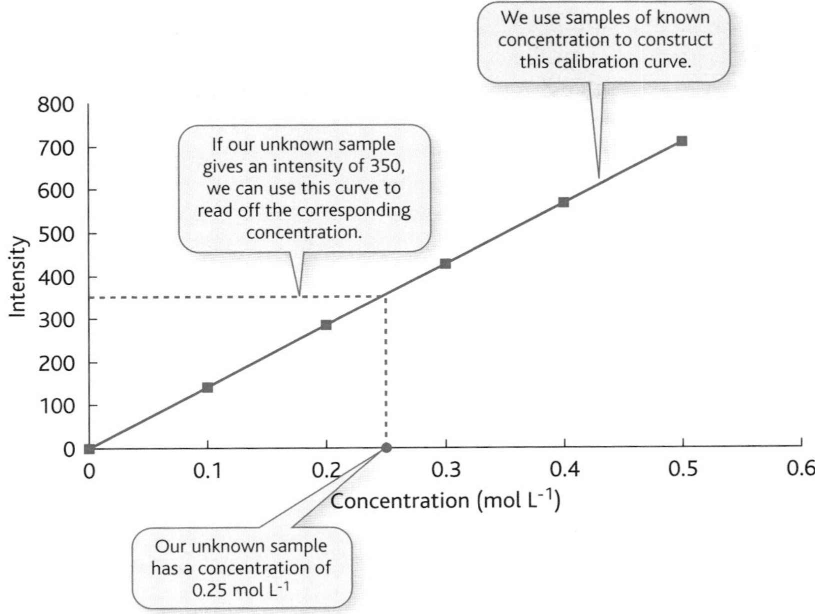

Figure 12.11 Using a calibration curve to calculate the concentration of an unknown sample. We use a series of samples of known concentrations to construct a calibration curve. When we measure the intensity of a sample of unknown concentration, we can use the calibration curve to determine the concentration that corresponds to the intensity measurement that we have obtained. In this case, our sample of interest gives an intensity of 350, which corresponds to a concentration of 0.25 mol L^{-1}.

manner as the sample exposed to atomic emission spectroscopy) and emits light at its characteristic wavelength. This light is then absorbed by the same element in the sample as that coated on the lamp's cathode. The amount of light absorbed is proportional to the concentration of the element present in the sample. Indeed, there is a direct relationship between the amount of light absorbed and the concentration of atoms in the sample, which obeys the Beer–Lambert law, as seen on p. 396.

Fluorescence spectroscopy

When molecules are excited by the absorption of light, they then lose this extra energy and return to the ground state. This excess energy is lost by several processes; for example, the molecules may simply bump into other molecules (losing energy as they do so) or collide with the sides of vessel in which they are contained. Sometimes, however, biological molecules in an excited state relax by emitting light as they return to the ground state. When the wavelength of the emitted light is of a longer wavelength (lower energy) than that used to excite the molecule in the first place, we see this emission as the release of light of a different colour. This process–of the emission of light of a different wavelength from that to which a substance is exposed–is what we call **fluorescence**.

The process of fluorescence is illustrated in Figure 12.12. The energy released during fluorescence is lower than that with which the sample is irradiated; it is therefore of a longer wavelength and falls in a different part the electromagnetic spectrum.

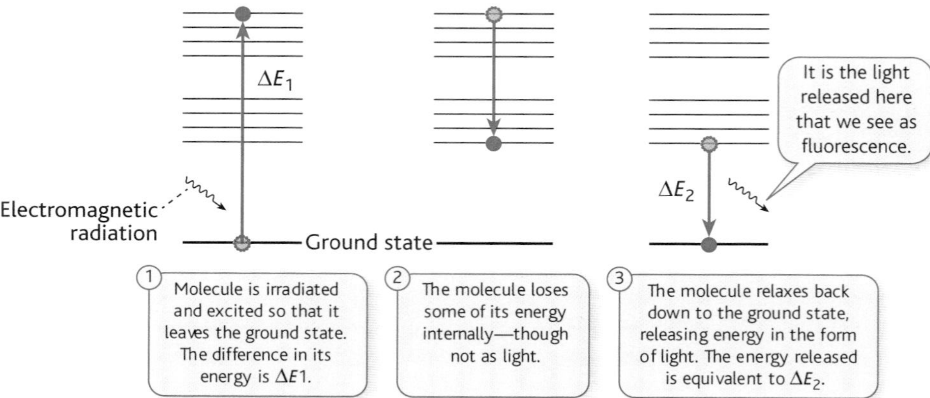

Figure 12.12 The conceptual basis of fluorescence. Fluorescence occurs when an excited molecule relaxes and emits energy of a longer wavelength than that with which it was initially irradiated. We see the emitted energy as fluorescent light.

Fluorescence spectroscopy is far more sensitive than absorption spectroscopy and is a much more accurate way of determining very low concentrations of biological molecules–concentrations that are too low to be detected by absorption techniques. The instruments used to measure fluorescence are called spectrofluorimeters; a spectrofluorimeter is often the detector of choice for HPLC systems where very small amounts of material are eluted.

The measurement of fluorescence can also be used to study the behaviour of biological molecules. A **fluorophore** is a component of a molecule that undergoes fluorescence (that is, it will absorb light of one wavelength and emit light of a different wavelength). Changes in the behaviour of fluorophores can be used to probe what is happening in their immediate surroundings. For example, changes in the pattern of fluorescence of a fluorophore located at the active site of a protein might tell us about the binding of different ligands to that active site: if fluorescence is detected in the presence of one ligand but not another, it could indicate that only ligand is interacting with the active site[6].

- *Fluorescence spectroscopy exploits the way that certain groups of atoms, called fluorophores, absorb light of one wavelength and emit light of another longer wavelength.*
- *The intensity of light emitted is proportional to the concentration of the compound present.*

Fluorescence is also now widely used to study patterns of gene expression, as explained in Box 12.2.

12.5 Using chemical reactions to measure concentration

Having now considered various ways in which we can use different instruments to determine the concentration of a compound of interest, we end the chapter by

6 The suppression of fluorescence is what we call **quenching**.

BOX 12.2 Green fluorescent protein: shining a light on gene expression

Green fluorescent protein, or GFP, is found in the jellyfish *Aequorea victoria*, and has naturally occurring fluorescent properties. GFP is a barrel-shaped protein, as illustrated in Figure 12.13(a), and has a fluorophore *p*-hydroxybenzylidene-imidazoline, which is found at the centre of this barrel structure as depicted in Figure 12.13(b). The fluorophore is formed from the amino acid residues glycine, tyrosine, and either threonine or serine, which react together to form the fluorophore structure as illustrated in Figure 12.13(c).

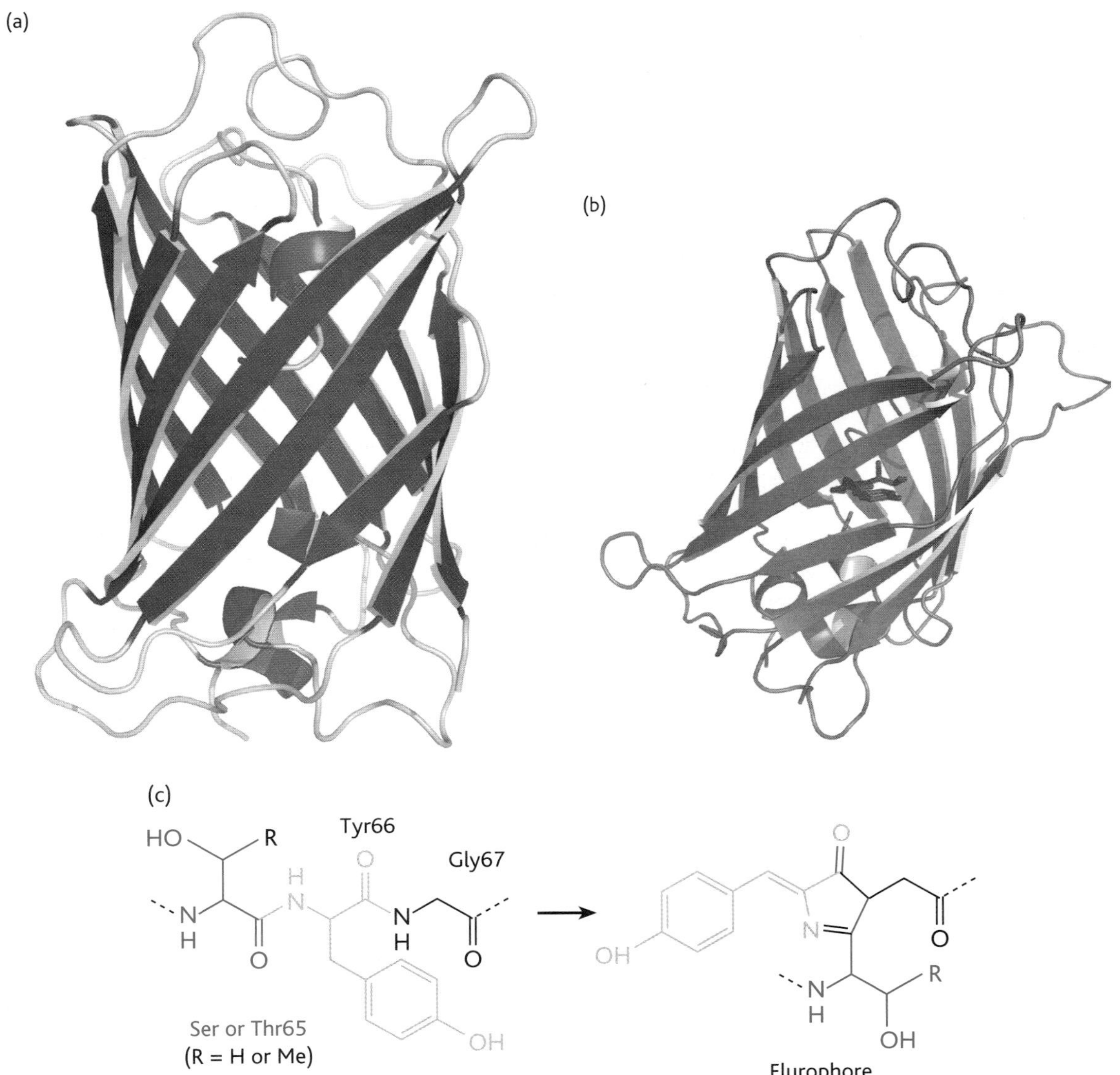

Figure 12.13 Green fluorescent protein (GFP). (a) The three-dimensional structure of GFP. (b) GFP contains a fluorophore at its centre, shown here in blue. (c) The fluorophore is derived from the side chains of three amino acid residues: either serine or threonine 65, tyrosine 66, and glycine 67.

BOX 12.2 **Continued**

GFP's fluorescence stems from the way that it exhibits an absorption maximum at 475 nm when exposed to UV light; when it absorbs light at this wavelength it emits a green fluorescence at the longer wavelength of 510 nm.

But how can this fluorescence be exploited? The GFP gene is commonly used as a **reporter gene**–that is, it is coupled to a gene of interest such that, when the gene of interest is expressed, the GFP protein gene is also expressed. Because expression of the GFP gene results in the formation of a product that can be visualized–the GFP protein itself–it can be used to show where, within an organism, the gene of interest is being expressed.

The use of GFP in this way is having a huge impact on our understanding of the genes underpinning developmental processes, the expression of which are often tightly regulated, both in terms of the time at which genes are expressed, and the precise locations in which expression occurs. It is only through the careful regulation of gene expression in time and space that the proper development of organisms, with the establishment of right three-dimensional distributions of cell types and structures, can occur. GFP reporter gene constructs are revealing both the spatial and temporal patterns of expression of many genes that are central to development.

For example, Figure 12.14 shows the patterning of expression of GFP-tagged *engrailed* in an adult *Drosophila*. *Engrailed* is involved in the control of compartmentalization and segment formation during *Drosophila* growth and development; Figure 12.14 shows how *engrailed* is expressed in very specific regions of the embryo, which demark the individual body segments in the fully developed organism. Notice, for example, how engrailed is only expressed in the posterior (rear) of the wings. This difference delineates the front and rear of the wing in morphological terms.

Figure 12.14 The expression of *engrailed* in *Drosophila*, as visualized using GFP. Notice how *engrailed* is expressed in defined regions; it is this tight delineation of expression that helps to set down the *Drosophila* body plan. (Image courtesy Professor Ansgar Klebes, Free University Berlin.)

discovering some techniques that use chemical reactions to tell us about chemical systems, and the quantities of compounds present in those systems.

Titrations

Titrations use chemical reactions to help us establish how much of a compound is present in a sample. We see in Section 17.1 how a chemical reaction proceeds with a particular stoichiometry–reactants and products are present at precise, unvarying ratios relative to one another.

We can use the known stoichiometry of a chemical reaction to deduce the concentration of a particular compound in solution if we know the concentration of another compound with which it interacts (or reacts).

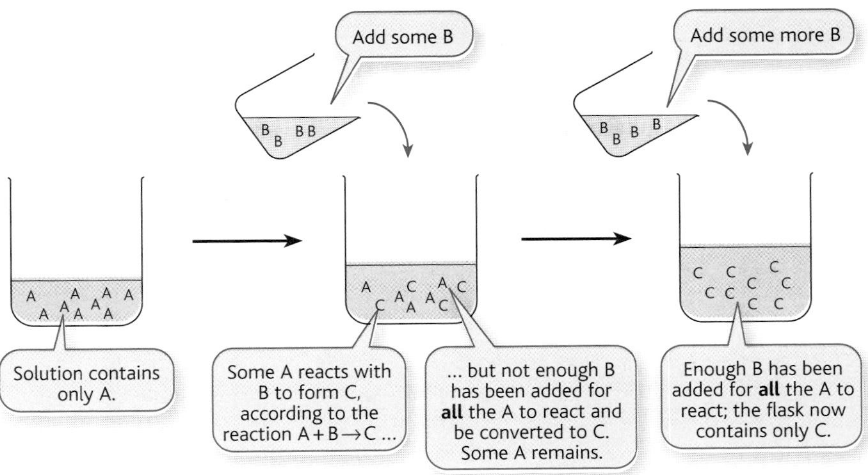

Figure 12.15 The use of the titration technique to determine the concentration of a particular compound in solution.

Consider this general reaction:

$$A + B \rightarrow C$$

This reaction tells us that one mole of A reacts with one mole of B to produce one mole of C. Notice that we are specifying precisely the relative amounts of A and B that react–in this case, equal moles of each.

Now, imagine that we have a flask that contains a solution of A of unknown concentration. Look at Figure 12.15, which shows how we can use the titration technique to deduce the concentration of A. Notice how we are adding a known amount of B to the unknown amount of A until all of the A that is initially present has reacted. How does this help us determine how many moles of A were present in the initial sample? The rationale works like this:

- We use a solution of B of known concentration, so we know how many moles of B we are adding to A.
- For all of the A to be converted into C, the same number of moles of B need to be added as there are moles of A, because the stoichiometry of the reaction is 1 : 1, i.e. $(1 \times A) + (1 \times B) \rightarrow C$.
- So if, for example, we need to add 2 mol of B to convert all the A to C, then there must be 2 mol of A present in the sample initially. Similarly, if we need to add 1 mol of B to convert all the A to C, there must be 1 mol of A present initially.

Now consider a reaction with the stoichiometry 1 : 2, i.e. $A + 2B \rightarrow C$. In this instance, one mole of A reacts with *two* moles of B. So, if we needed to add two moles of B to convert all of the A present into C, then we could deduce that there was one mole of A present in the sample initially.

Self-check 12.15 In a reaction with the stoichiometry 1 : 2, i.e. $A + 2B \rightarrow C$, we need 1 mol of B to convert all of the A into C. How many moles of A are present in the solution initially?

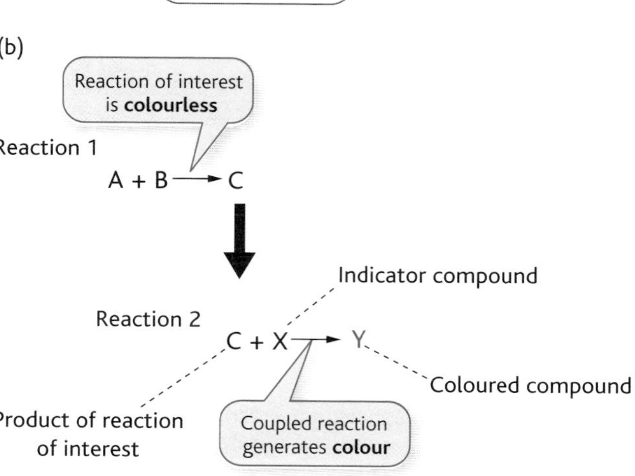

Figure 12.16 (a) Titrations often rely on a *change of colour* in the solution to indicate when a particular point has been reached. (b) If the solution itself is colourless, an indicator can be used. Indicators form coloured solutions when they undergo a particular chemical reaction. The reaction in which the indicator is involved is often coupled to the reaction in which we're interested.

But how do we know when all of our compound of unknown concentration ('A' in our example above) has reacted? Titrations often rely on a **change of colour** in the solution to indicate when a particular point has been reached. For example, look at Figure 12.16(a) to see how we could use colour to indicate the instant at which all of compound A had been converted to C through the addition of compound B.

Not all compounds are coloured when in solution, however. In which case, we use an **indicator** as a source of colour from which to gauge the status of a titration. Indicators are typically compounds that form coloured solutions when they undergo a particular chemical reaction. The reaction in which the indicator is involved is often **coupled** to the reaction in which we're interested, as illustrated in Figure 12.16(b). Notice how, because reactions 1 and 2 are coupled, the colour generated from reaction 2 can be used to tell us about what's happening in reaction 1.

For example, the indicator phenolphthalein is used as an indicator of the pH of a solution: phenolphthalein forms a colourless solution below about pH 8.0. However, as the pH increases beyond pH 9.6 the solution turns from colourless to intense puce pink. The change of colour is linked to a change in the chemical structure of phenolphthalein, as illustrated in Figure 12.17.

- ***Titrations exploit the change in colour that occurs during a certain chemical reaction to help us determine the concentration of a compound of interest.***

Titrations are used as analytical tools in the analysis of water quality. The quality of drinking water is tightly regulated such that the concentrations of certain metal ions in water supplies must stay below specific levels. Similarly, rivers are monitored for signs of pollution, either from industry or from farming, to protect wildlife. For an example, look at Box 12.3, which explains how titrations are used to determine the concentration of chloride ions in river water.

Electrochemical sensors

A final technique for determining concentration exploits the **electrochemical** nature of chemical reactions. We see in Section 17.2 how chemical reactions are the result of the movement and redistribution of electrons between chemical species. **Redox** reactions are coupled reactions in which one reaction uses up electrons released by the other.

We encounter redox reactions again in Chapter 17.

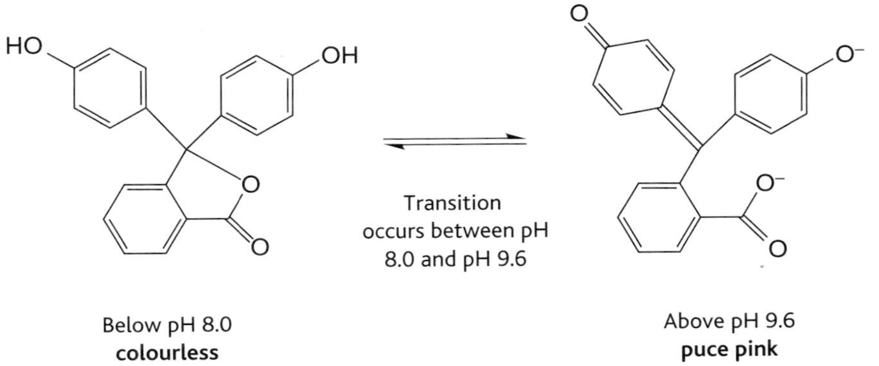

HO … OH ⇌ O … O⁻ … O⁻ … O

Transition occurs between pH 8.0 and pH 9.6

Below pH 8.0 **colourless**

Above pH 9.6 **puce pink**

Figure 12.17 The indicator phenolphthalein forms a colourless solution below pH 8.0, but forms a puce pink solution above pH 9.6. The change in colour comes about as a consequence of a change in the structure of phenolphthalein, as illustrated here.

BOX 12.3 River water analysis

The wildlife in our rivers is sensitive to the concentration of many elements that occur in the environment: elevated concentrations can be toxic. So it is important that we have ways of monitoring the concentrations of key elements such as chlorine in river water.

We can determine the concentration of chloride ions in a sample of river water by titrating the water sample with silver ions, $Ag^+(aq)$, and using chromate ions, $CrO_4^{2-}(aq)$, as an indicator[7]. (Remember: chlorine reacts to form Cl^- ions when it is dissolved in water.)

Look at Figure 12.18 to see how the titration proceeds. The Ag^+ ions are able to react with both the Cl^- and CrO_4^{2-} ions in solution, but react *preferentially* with the Cl^- ions.

We add Ag^+ ions to the water sample until all the Cl^- ions in the water sample have reacted to form AgCl, an insoluble compound that precipitates out of solution. Once all the Cl^- ions have been used up, the Ag^+ ions then react with the CrO_4^{2-} ions to form silver chromate, Ag_2CrO_4.

Silver chromate is a **red-coloured** compound: we can use the transformation of the solution from white to red-coloured (which happens as soon as the Ag^+ ions start to react with the chromate ions) as an indicator that all the Cl^- ions have been used up. By measuring the number of moles of Ag^+ ions that had to be added to the solution to effect the change in colour of solution from colourless to red, we can deduce the number of moles of Cl^- ions that were present in the sample initially.

7 The source of Ag^+ ions used for such a titration is typically silver nitrate solution, $AgNO_3(aq)$, which comprises both Ag^+ and NO_3^- ions. The source of CrO_4^{2-} ions is often potassium chromate solution, $K_2CrO_4(aq)$.

BOX 12.3 **Continued**

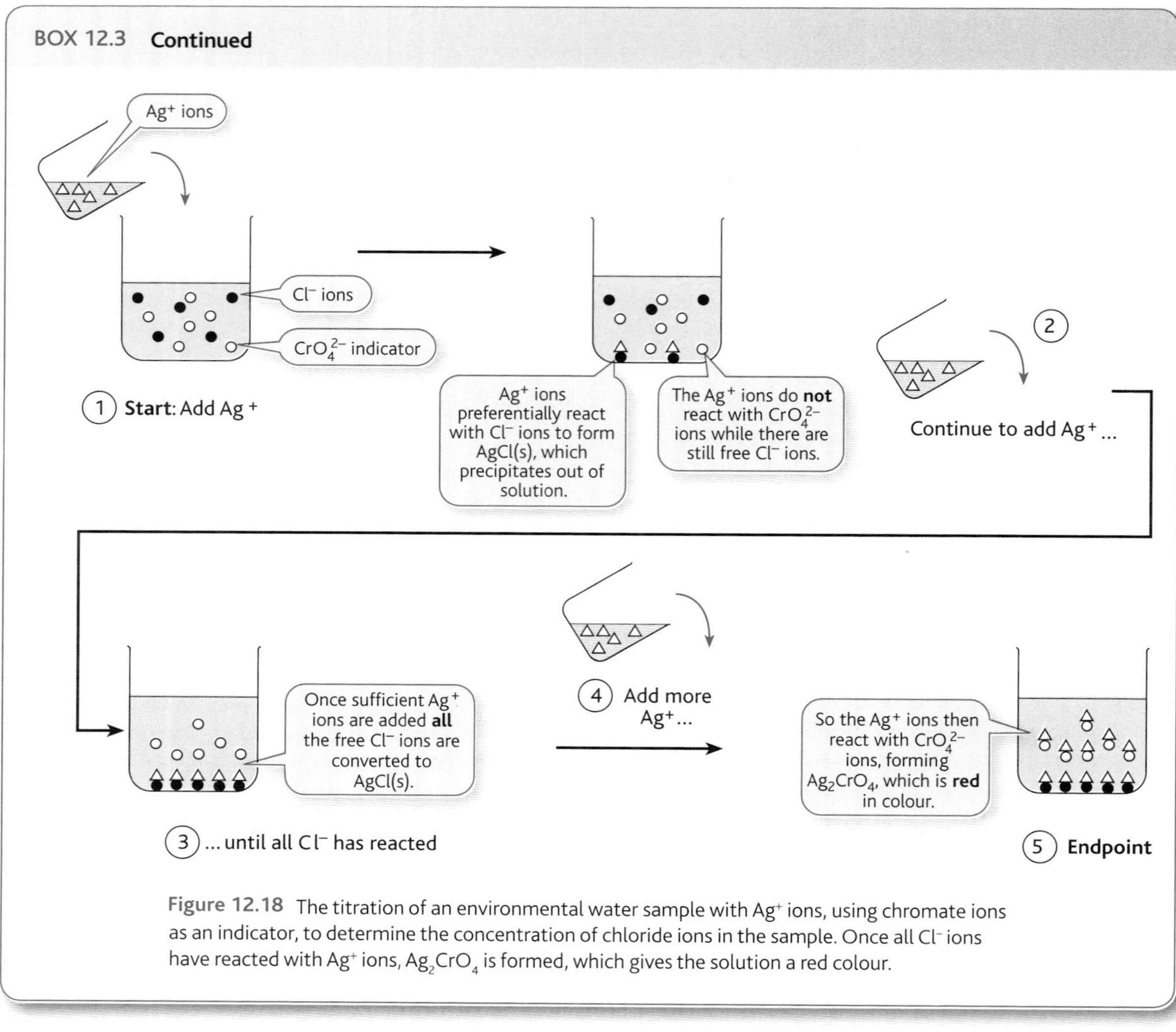

Figure 12.18 The titration of an environmental water sample with Ag^+ ions, using chromate ions as an indicator, to determine the concentration of chloride ions in the sample. Once all Cl^- ions have reacted with Ag^+ ions, Ag_2CrO_4 is formed, which gives the solution a red colour.

Look at Figure 12.19, which shows how, when a redox reaction occurs in solution, it is possible to use a pair of **electrodes** to conduct electrons out of the solution, and round an electrical circuit[8].

The electrical current that flows round such a circuit is proportional to the concentration of the chemical compounds in the solution: the greater the concentration of compound reacting, the greater the number of electrons that are free to be conducted round the circuit.

By measuring the electrical current within the circuit, we can deduce the concentrations of compounds reacting in the solution. The link between concentration of compounds in solution and the electrical current that flows between electrodes placed in the solution has been exploited in the development of **biosensors**. Biosensors are specially designed instruments that measure the electrical current generated

8 An electrode is a conductive device that transports electrons into or out of a solution.

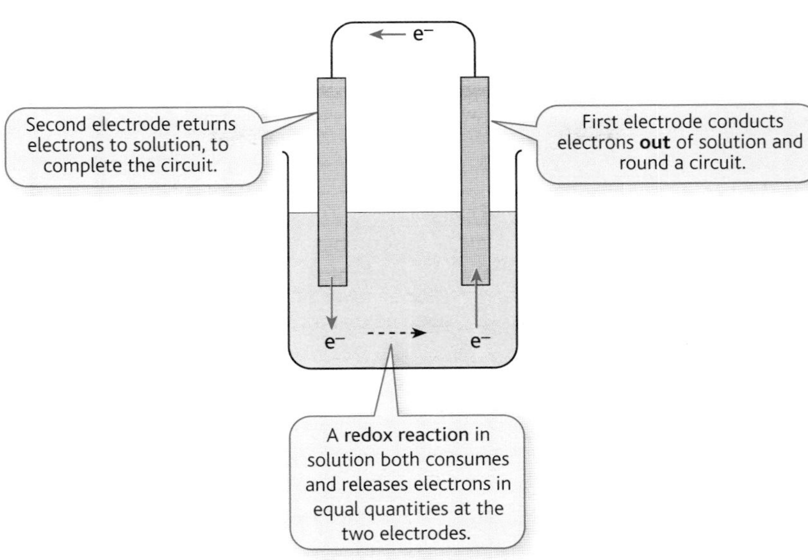
e^-
Second electrode returns electrons to solution, to complete the circuit.
First electrode conducts electrons **out** of solution and round a circuit.
e^-
e^-
A redox reaction in solution both consumes and releases electrons in equal quantities at the two electrodes.

Figure 12.19 When a redox reaction occurs in solution, it is possible to use a pair of electrodes to conduct electrons out of the solution, and round an electrical circuit. The electrical current that flows round such a circuit is proportional to the concentration of the chemical compounds in the solution.

through a redox reaction and relate the current to the concentration of chemical compounds in biological tissues such as blood.

We see on p. 392 how people with diabetes must test their blood sugar levels regularly; Box 12.4 explains how modern blood sugar testing kits comprise a biosensor that uses electrochemistry to measure blood glucose concentrations.

- ***Electrochemical sensors exploit the transfer of electrons during redox reactions to give a measure of the concentration of a particular compound: the greater the flow of electrons, the higher the concentration of the compound of interest.***

BOX 12.4 Biosensors and the measurement of blood glucose concentration

People with diabetes can use a fully portable blood glucose sensor (a type of biosensor) to measure their blood glucose concentration.

The blood glucose sensor has two key components:

1. A strip containing an enzyme that specifically metabolizes (breaks down) glucose
2. An electrode that picks up electrons yielded from the breakdown of glucose, and uses the electrons to generate a current round an electrical circuit.

Look at Figure 12.20, which illustrates how the blood glucose sensor works.

The size of the electrical current (step 6) is proportional to the number of electrons present which, in turn, is proportional to the amount of glucose in the blood sample (step 1): the higher the concentration of glucose, the greater the number of electrons, and the larger the current that flows.

The sensor measures the electrical current, and uses this information to calculate the concentration of glucose present in the blood sample. This value is then displayed on the sensor's display screen.

BOX 12.4 **Continued**

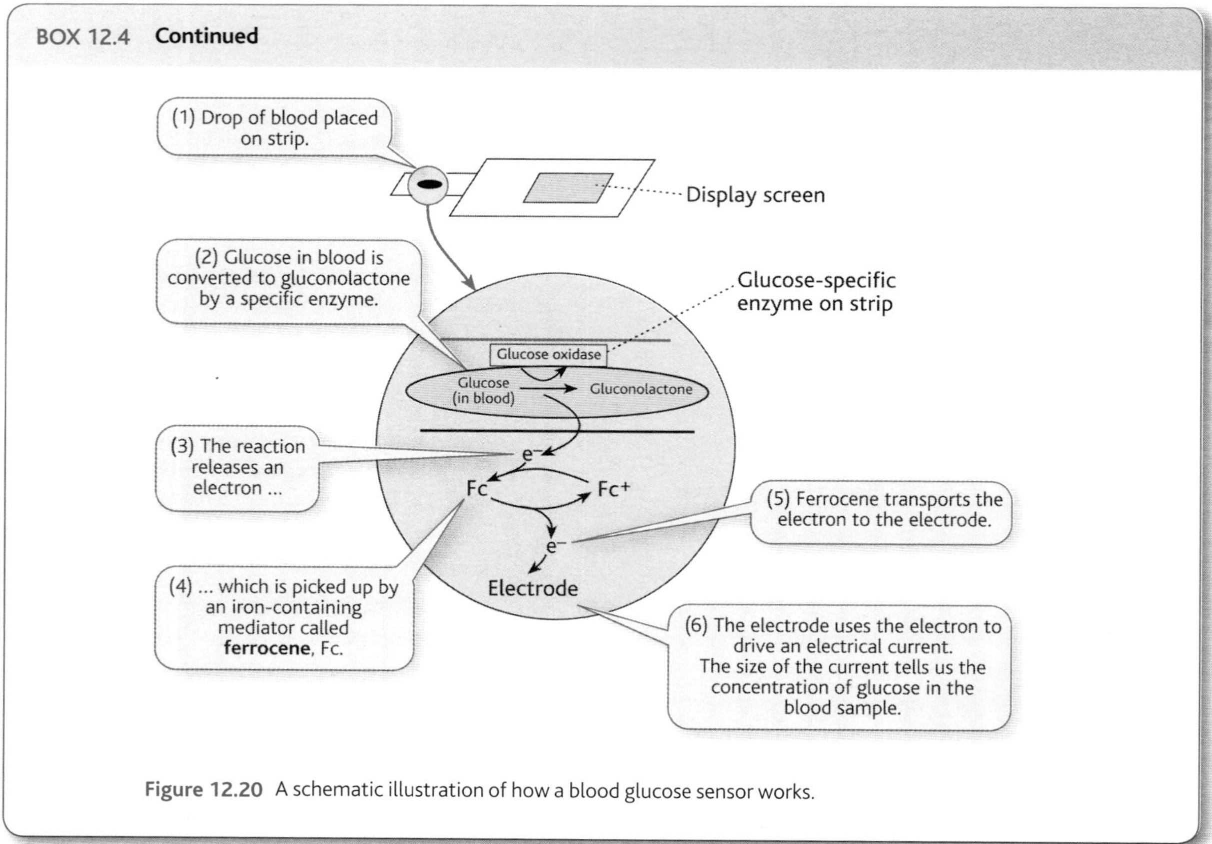

(1) Drop of blood placed on strip.
Display screen
(2) Glucose in blood is converted to gluconolactone by a specific enzyme.
Glucose-specific enzyme on strip
Glucose oxidase
Glucose (in blood)
Gluconolactone
(3) The reaction releases an electron ...
e⁻
Fc
Fc⁺
(5) Ferrocene transports the electron to the electrode.
e⁻
(4) ... which is picked up by an iron-containing mediator called ferrocene, Fc.
Electrode
(6) The electrode uses the electron to drive an electrical current. The size of the current tells us the concentration of glucose in the blood sample.

Figure 12.20 A schematic illustration of how a blood glucose sensor works.

Check your understanding

To check that you've mastered the key concepts presented in this chapter, review the checklist of key concepts below, and attempt the multiple-choice questions available in the book's Online Resource Centre at **http://www.oxfordtextbooks.co.uk/orc/crowe2e/**.

Checklist of key concepts

Amounts and concentrations: how much is there?

- The mole is 6×10^{23} particles
- 6×10^{23} is called the Avogadro constant
- The molar mass is the mass of one mole of a substance
- The number of moles, molar mass, and mass of a sample are related according to the relationship:

$$\frac{\text{mass of sample}}{\text{molar mass}} = \text{number of moles present in sample}$$

- The concentration of a solution tells us how much of a substance is present in a particular volume of the solution

- **Molarity** is the concentration of a compound, expressed in the units mol L^{-1}
- The number of moles of a substance in a given solution is related to the concentration and volume of the solution by the following:

$$\text{number of moles (mol)} = \text{concentration (mol L}^{-1}) \times \text{volume (L)}$$

Solutions and dilutions

- A **solvent** is the medium in which a substance (the **solute**) is dissolved.
- When we **dilute** a solution, the number of moles of the solute remain the same, but the total volume increases, so the concentration decreases
- A **serial dilution** reduces the concentration of a solution in a series of successive steps

Measuring concentrations

Spectroscopic approaches

- **UV–visible spectrophotometry** can measure the concentration of a compound in solution
- UV–visible spectrophotometry measures the **absorbance** of a solution at a given wavelength
- The absorbance of a solution is proportional to its concentration
- The relationship between the absorbance of a solution and its concentration is described by the **Beer–Lambert law**: $A = \varepsilon cl$
- **Atomic spectroscopy** measures the emission of energy from a sample following its irradiation with electromagnetic radiation
- The intensity of the light emitted is proportional to the concentration of atoms present in the sample
- **Atomic emission spectroscopy** is used to determine the concentration of a specific element in a sample
- **Fluorescence** is the emission of light by a substance, following irradiation with electromagnetic radiation, whose wavelength is shorter than the wavelength used to excite the substance in the first place.
- **Fluorescence spectroscopy** is very sensitive, and can be used to determine very low concentrations of biological molecules

Chemical approaches

- A **titration** uses a chemical reaction to establish how much of a compound is present in a solution
- Titrations often rely on a change of colour
- **Indicators** are often used to help track the progress of a titration
- **Electrochemical sensors** exploit the movement of electrons during chemical reactions to determine the concentration of a particular substance in solution

Energy: what makes reactions go?

INTRODUCTION

'Energy' is a word we hear used daily in all sorts of contexts. We might wake up in the morning and feel that we just do not have the energy to get into work on time, or to attend a 9 o'clock lecture. Later in the day we might meet up with a friend and notice they seem in high spirits and comment that they are 'full of energy'. The government talks constantly about its 'energy policy'; people debate whether wind energy is preferable to nuclear energy, or whether or not 'dark energy' really exists. Most crucially, we know that our bodies require food to give us the energy we need to function properly.

But what is 'energy'? How can we describe it scientifically? What have all the examples of energy we see above got in common?

In this chapter we look at the nature of energy. We explore the different forms of energy; we look at the way in which it is transformed from one form to another, and see how energy transfer drives the biochemical processes on which organisms depend for life.

13.1 What is energy?

On reflection, we can name several forms of energy used every day. In the morning we may use electrical energy to heat the water for a shower or a cup of coffee; we use chemical energy (in our muscles) to run for the bus; the bus uses chemical energy in its fuel to power its engine.

However, it is really difficult for us to describe what energy actually 'is'. Energy does not have a physical form that we can easily visualize and describe. Instead, energy is an inherent attribute of a substance. Trying to describe energy is a bit like trying to describe the human mind: we all have a 'mind' but the mind doesn't really have a physical form (despite it being an intrinsic part of who we are, how we behave, etc.). And, like a mind, which is bounded and confined by the physical boundaries of the brain, so energy is contained within a chemical bond, or body. But energy has no physical form, as such; it is merely an intrinsic part of any substance–be that an atom, a molecule, or a biological system.

More formally, energy is the capacity to do work. The more energy something has, the more able it is to bring about some kind of change. If we're feeling 'more energetic', for example, we typically get more done than at times when we're feeling lethargic, and feel like we have 'no energy'.

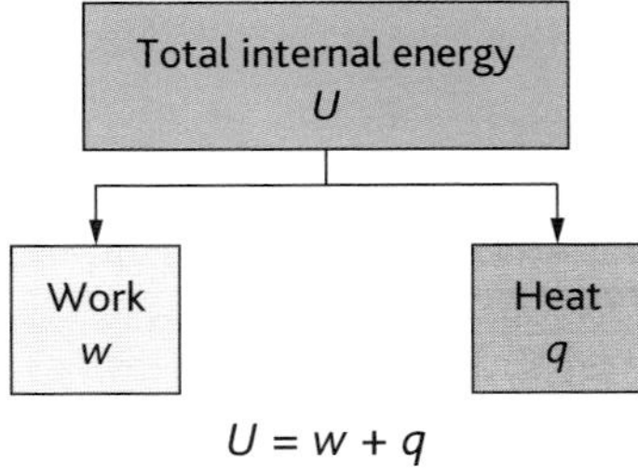

Figure 13.1 The total internal energy of a system can be used either to do useful work, w, or to warm cooler surroundings through the transfer of heat, q.

So what are we really referring to when we talk about something's energy? Everything has what is referred to as '**internal energy**', which we represent with the symbol U. We measure the internal energy in joules.

The total internal energy of a system, U, can be used in two ways: as **work** and as **heat**, as depicted in Figure 13.1. The energy associated with work, represented by the symbol w, is the energy that can be channelled in an organized way to do something useful–to drive an ion across a membrane, or power the contraction of a muscle. The energy associated with heat, represented by the symbol q, is the energy that is lost from a hot entity to a neighbouring cooler entity–from a warm human body to its cooler surroundings, for example. We will explore the concepts of work and heat, and their biological significance, much further in Section 13.2.

- ***The total energy of a system is its internal energy, U.***
- ***Internal energy has two components–the energy associated with work, w, and the energy associated with heat, q.***

The conservation of energy

Let's return to the processes we mentioned at the start of the chapter–the use of electrical energy to heat water in a kettle to make a coffee, or the use of chemical energy in fuel to power a bus. All the processes mentioned involve the **transformation** of energy: electrical energy is *transformed* into thermal energy to heat the water for a coffee, and chemical energy is *transformed* into mechanical energy to power the engine in the bus.

Why do we say that energy is 'transformed' rather than 'used'? The answer is that energy cannot be **created** or **destroyed**. Instead, energy is **conserved.** When electrical energy heats water in a kettle, the energy is not 'used up' (or destroyed), it is transformed into thermal energy. Similarly, when we light a match to produce a flame, we are not creating energy, we are merely transforming energy that is already 'stored' chemically in the match into different types of energy, including the thermal energy of the flame, which we can feel as heat, and light.

This important principle is captured by the **first law of thermodynamics**[1]: 'The internal energy of an isolated system is constant'. Another way of expressing this

1 Thermodynamics is the field of physical science that studies the conversion of energy from one form to another, the relationship between energy and heat, and the availability of energy that is available to do work. We explore the concepts of both heat and work later in this chapter.

is to say that energy cannot be created or destroyed, but only converted from one form to another.

- ***Energy cannot be created or destroyed, but only converted from one form to another.***

There are many different 'types' of energy that contribute to the overall internal energy of a substance. However, just three key types have the *most* influence. These are kinetic energy, potential gravitational energy, and potential chemical energy.

Kinetic energy

Kinetic energy is the energy a system has due to its **motion**[2]. A fast-moving object has more kinetic energy than a slow-moving object. (For example, a car travelling at 70 miles per hour has about three times more kinetic energy than a car travelling at 40 miles per hour.) Equally, the water molecules in a pan of boiling water, in which the water molecules are moving around vigorously, have more kinetic energy than those molecules in a pan of cold water, in which the molecules move less quickly.

The kinetic energy of an object also depends on its mass: a heavy object travelling at a particular speed has more kinetic energy than a lighter object travelling at the same speed. (Consider a train and a car, both travelling at 70 miles per hour: it takes longer (both in terms of time and distance) to bring the train to a standstill than it does the car. This is because the train has more kinetic energy than the car–energy which must be overcome in order to bring the train to a stop.)

The kinetic energy of an entity (e.g. an atom, molecule, or car) is related to its mass and velocity as follows:

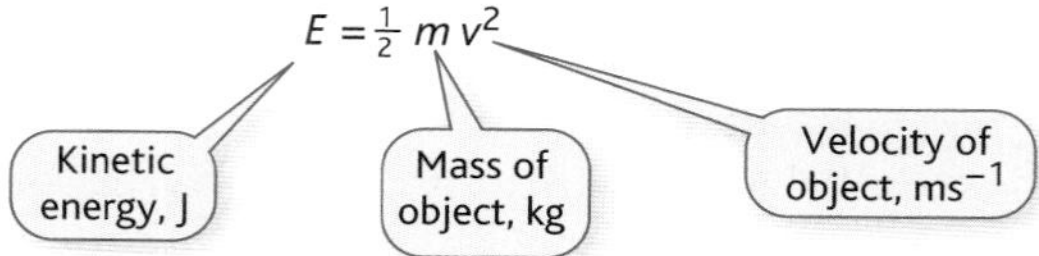

EXAMPLE

Imagine a 55 kg cheetah running at a velocity of 110 km per hour. What kinetic energy would the cheetah have?

First, we need to convert the velocity into units of metres per second. Let's do this in two stages. First, we convert km into m. There are 1000 m in 1 km, so the cheetah runs at a velocity of (110 × 1000) m–that is, 110 000 m–per hour.

Next, we convert 'per hour' into 'per second'. There are 3600 seconds in 1 hour. If the cheetah runs 110 000 m in one hour (3600 s) then the cheetah runs (110,000/3600) m or 30.56 m in just one second (that is, 'per second').

So, the cheetah runs at a velocity of 30.56 m s^{-1}.

2 We refer to the speed and direction of motion of an object as its velocity. By contrast, 'speed' refers only to the distance travelled per unit time.

We can now return to our original question. If the cheetah has a mass of 55 kg, then its kinetic energy when running at 30.56 m s^{-1} is given by:

$$E = \tfrac{1}{2} mv^2$$
$$= \tfrac{1}{2} \times 55 \text{ kg} \times (30.56 \text{ m s}^{-1})^2$$
$$= \tfrac{1}{2} \times 55 \text{ kg} \times 933.91 \text{ m}^2 \text{ s}^{-2}$$
$$= 25\,682.53 \text{ kg m}^2 \text{ s}^{-2}$$

Notice how both the number *and* the units are squared.

Now, 1 kg m^2 s^{-2} is equivalent to 1 J (joule). Therefore, the kinetic energy of the cheetah is **25 682.53 J** (or 25.68 kJ).

Self-check 13.1 What is the kinetic energy of a 5000 kg elephant walking at a velocity of 7.2 km per hour?

Look at Box 13.1 to find out more about the relationship between energy, mass, and velocity.

Potential energy

Potential energy is energy that is 'stored' in some way, ready to be exploited when the entity with which the energy is associated changes in some way–that is, it is energy that is waiting to be stored to have some kind of effect. There are two key types of potential energy: potential gravitational energy, and potential chemical energy. Let us now consider each of these in turn.

BOX 13.1 The link between energy and speed

The amount of energy possessed by a moving car depends upon both its mass and its velocity. However, the relationship is such that dependence on velocity is far more significant.

A car travelling at 10 m s^{-1} requires 5 m to stop when brakes are applied, while the same car travelling at 20 m s^{-1} requires 20 m. Notice how the same car takes *four times* longer to stop despite its speed having only *doubled*.

Why is this? If we look at the relationship $E = \tfrac{1}{2} mv^2$ we see how energy, E, is proportional to the square of the velocity, v^2. So, if the velocity doubles, say from 1 m s^{-1} to 2 m s^{-1}, the value of v^2 changes from $(1 \times 1) = 1$ to $(2 \times 2) = 4$. Notice how the value of v^2 has increased four times while the value of v has just doubled.

So, when a car doubles its velocity, its kinetic energy quadruples, and it takes four times the energy to bring the car to a stop. This relationship explains why police authorities often stress the importance of 30 mph speed limits in residential areas: a relatively modest increase in speed over the speed limit equates to a *substantial* increase in kinetic energy, making it take longer to bring the car to a stop in an emergency.

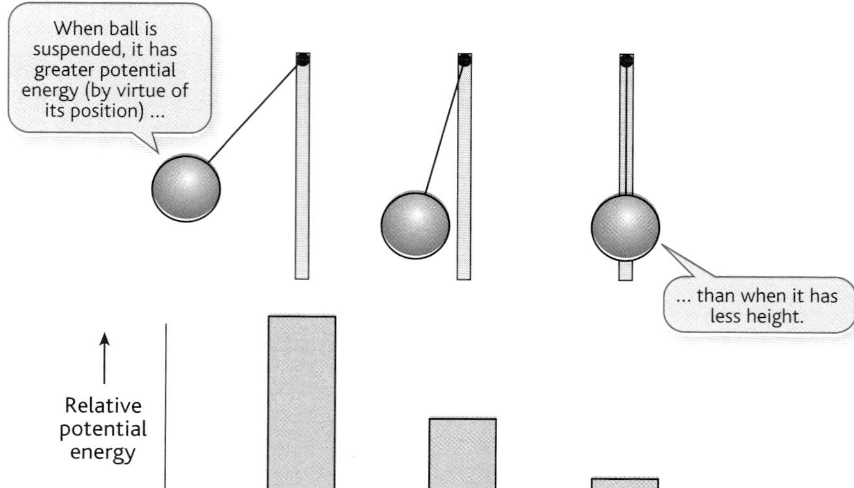

Figure 13.2 The potential gravitational energy of a demolition ball changes as its position is altered. The ball has *more* potential energy when it is raised than when it is lowered.

Potential gravitational energy

Potential gravitational energy is the energy an entity has due to its relative **position**–energy that can be used to do work when an entity *changes* its position.

How does an object's position determine the amount of potential energy it possesses? Let's consider an example. Old buildings are often demolished with the aid of a demolition ball swung by a crane. The ball is raised to a position high above the intended point of impact. As it waits to be released, the large, heavy metal ball possesses a large amount of potential energy waiting to be released. When the ball is released, the potential energy is transformed into kinetic energy (energy of motion) as the ball gathers speed under the force of gravity. It is the ball's motion that is harnessed to demolish the wall. By contrast, when the demolition ball is not suspended in the air but is lying on the ground, it has less potential energy. Look at Figure 13.2, which compares the demolition ball's potential gravitational energy at different positions. Again, notice how the potential energy of the demolition ball changes when its position is altered.

The transformation of energy from potential to kinetic is not an irreversible one. Consider a pendulum swinging from side to side, as depicted in Figure 13.3. As the pendulum swings through 90° from left to right its total energy changes from potential energy to kinetic energy as its position lowers, and back from kinetic to potential energy as its position rises again. The pendulum's total energy remains the same regardless of its position; it is merely the *nature* of the energy that changes.

Potential chemical energy

Potential chemical energy (or, simply 'chemical energy') is the energy stored in molecules by virtue of their having bonds. More specifically, we can think of chemical energy as the energy that holds together atoms joined by covalent or ionic bonds–and the energy that must be overcome to break these bonds. Chemical energy is a type of potential energy because it is energy that is stored in the chemical bonds of a compound, only to be harnessed when a compound undergoes chemical change–that is, when bonds are formed or broken[3].

We see in Section 3.1 how compounds form as a result of the redistribution of valence electrons into more stable (lower-energy) arrangements. The amount of

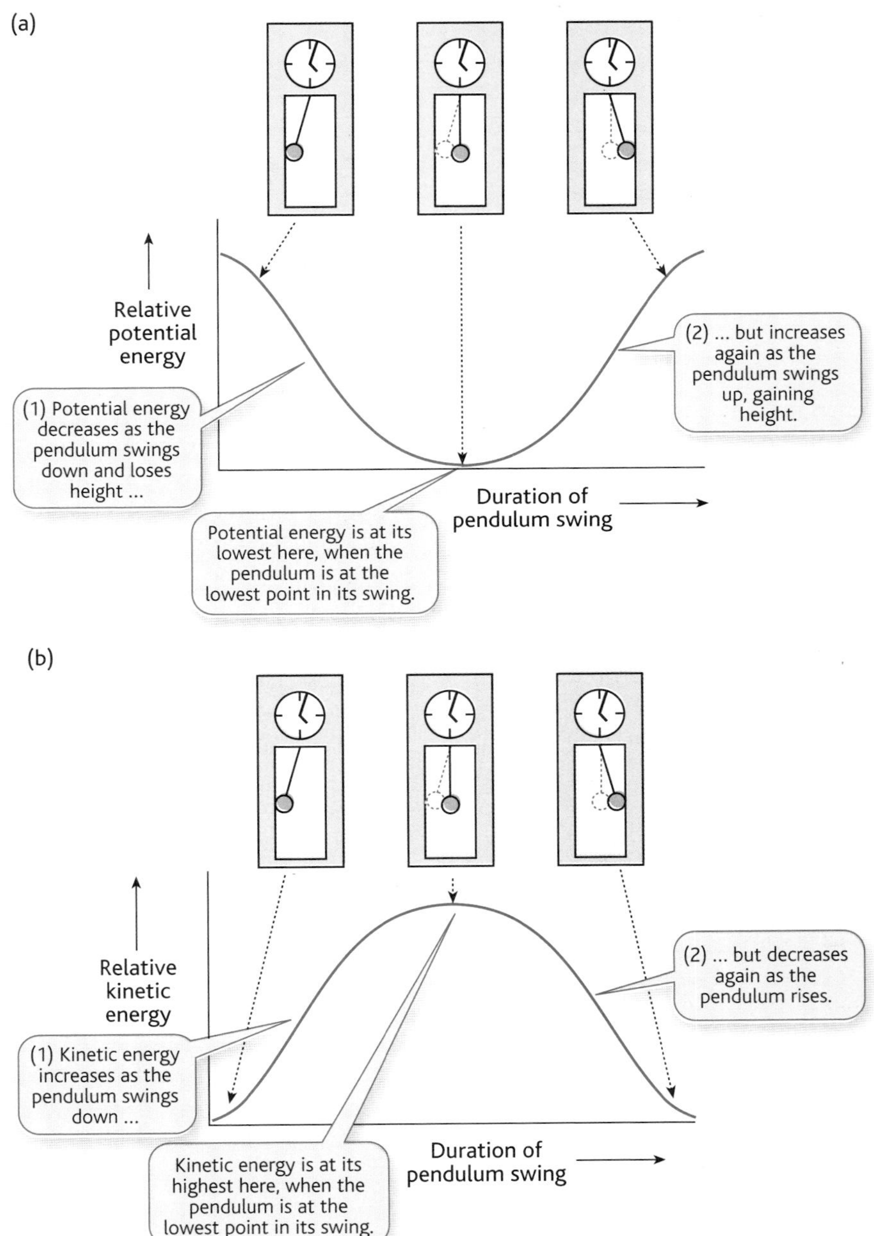
(a)
Relative potential energy
(1) Potential energy decreases as the pendulum swings down and loses height ...
(2) ... but increases again as the pendulum swings up, gaining height.
Potential energy is at its lowest here, when the pendulum is at the lowest point in its swing.
Duration of pendulum swing
(b)
Relative kinetic energy
(1) Kinetic energy increases as the pendulum swings down ...
(2) ... but decreases again as the pendulum rises.
Kinetic energy is at its highest here, when the pendulum is at the lowest point in its swing.
Duration of pendulum swing

Figure 13.3 As a pendulum swings through 90° from left to right its total energy changes from potential energy to kinetic energy as its position lowers, and back from kinetic to potential energy as its position rises again. Compare (a) and (b), and notice how, as the potential energy of the pendulum increases the kinetic energy decreases and vice versa.

3 Notice the similarity between potential gravitational energy and potential chemical energy: both rely on change occurring to the entity in which the energy is stored for the energy to be liberated and be used to do something useful.

(c)

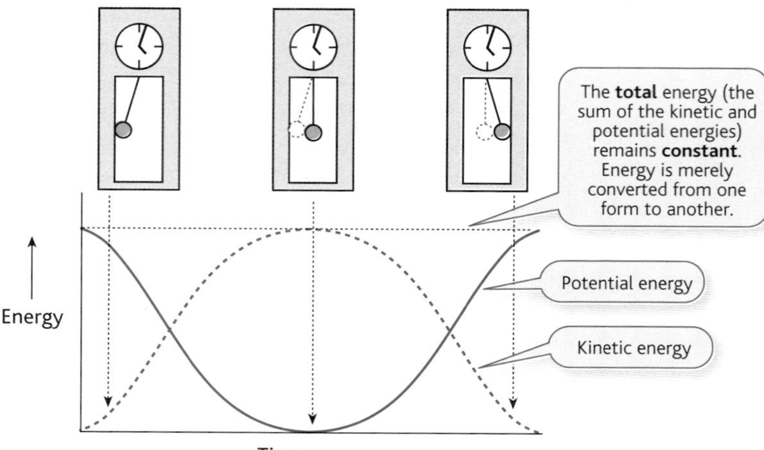
The **total** energy (the sum of the kinetic and potential energies) remains **constant**. Energy is merely converted from one form to another.
Potential energy
Kinetic energy
Energy
Time

Figure 13.3 (*continued*)
(c) shows how the total energy is conserved: it is merely converted from one form to another.

energy required to separate two atoms that are joined by a covalent bond depends on just how stable the bond is. More energy is required to separate two atoms joined by a very stable bond (one in which the valence electrons are shared in a low-energy arrangement) than is required to separate two atoms joined by a less-stable bond (one in which the valence electrons are shared in a higher-energy arrangement).

We call the amount of energy 'stored' in a bond–and the amount of energy required to break an ionic or covalent bond to separate two joined atoms–the **bond energy**.

For example, a typical C–H bond has an energy of 412 kJ mol^{-1}, while a typical N–H bond has an energy of 390 kJ mol^{-1}.
(Notice that we measure the value of bond energies in joules per mole (J mol^{-1}), though we often cite this amount in a slightly different way, as 'kJ mol^{-1}'. To obtain kJ mol^{-1} from J mol^{-1}, divide the original value by a factor of 10^3.)

The energy of a bond joining a particular pair of atoms depends on the identity and environment of the other atoms, or groups of atoms, to which the pair of atoms are attached. For example, the O–H bond in water has a slightly different bond energy from the O–H bond in methanol:

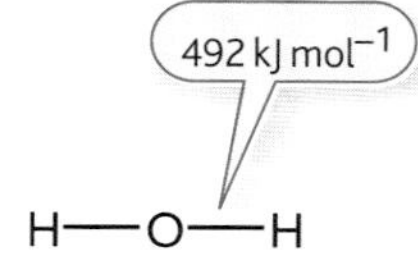
492 kJ mol^{-1}
H—O—H

H_3C—O—H

437 kJ mol^{-1}

(Notice how the –OH group in methanol is attached to a different atom compared to the –OH group in water.)

BOX 13.2 A disastrous transformation of energy

On 26 December 2004 there was a major movement of the Earth's crust. Just off the coast of Indonesia, deep under the Indian Ocean, two tectonic plates moved about 10 m relative to each other, creating a massive earthquake registering greater than magnitude 9.0 on the Richter scale.

The position of the seawater relative to the Earth's crust was altered by the crust's movement: Figure 13.4 shows how the seawater consequently had more potential gravitational energy than it did before the earthquake. However, the potential gravitational energy of the seawater was immediately transformed into kinetic energy–energy of motion–as the force of gravity pulled the seawater down, causing a forward-moving wave.

This forward-moving wave–a tsunami–spread out across the Indian Ocean, causing widespread destruction as far away as the coast of Africa.

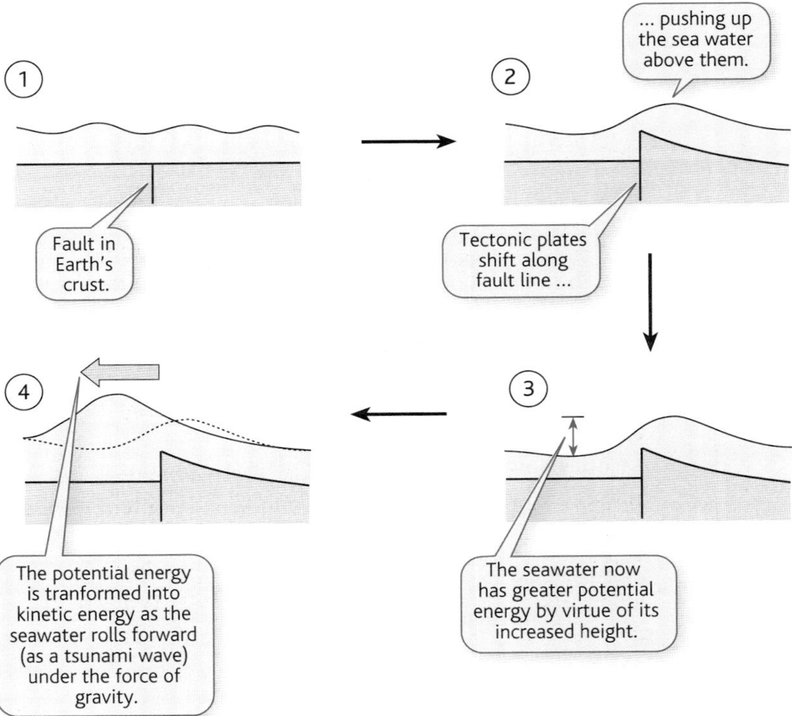

Figure 13.4 The movement of the Earth's crust during the earthquake off the coast of Indonesia increased the potential energy of the seawater above it. This potential energy was then converted to kinetic energy as a tsunami wave, triggered by the earthquake, rolled forward.

Similarly, the C–C bond energy in ethane is different from the C–C bond in cyclobutane:

347 kJ mol^{-1}

$H_3C—CH_3$

417 kJ mol^{-1}

(cyclobutane structure: four C atoms in a ring, each bearing two H atoms)

However, the bond energy for a particular bond varies only slightly between different compounds. (Notice how the bond energy for the O–H bond in water is not vastly different from the bond energy for the O–H bond in methanol.) So we can calculate an **average** bond energy for a particular bond, and use it as the 'typical' bond energy for that bond. Look at Table 13.1; this lists the average bond energies for a range of chemical bonds.

Bond formation **generates** (or releases) energy, while bond cleavage **consumes** energy. The bond energy for a particular bond tells us two things:

1. The amount of energy released when the bond is formed.
2. The amount of energy required to break the bond once it is formed.

For example, the average C–H bond energy is 412 kJ mol^{-1}. This tells us that 412 kJ of energy is *released* for every mole of C–H bonds that form. It also tells us that 412 kJ of energy is *consumed* for every mole of C–H bonds that are broken.

Notice how we talk about energy release or consumption per *mole* of bonds. We are *not* considering the energy consumed or released per single bond.

- ***Bond formation generates energy.***
- ***Bond cleavage requires energy.***

Table 13.1 Average bond energies for a range of chemical bonds. All bond energy values are quoted in kJ mol^{-1}.

Bond	Bond energy	Bond	Bond energy	Bond	Bond energy
C–H	412	N–H	390	H–H	436
C–C	348	N–N	163	H–F	565
C=C	610	N≡N	945	H–Br	365
C≡C	835	N–O	201	H–Cl	431
C–N	305	N–F	272		
C–O	358	N–Cl	200		
C=O	740	N–Br	243		
C–F	495				
C–Br	280	O–H	463		
C–S	259	O=O	497		

Bond energies can also help us to quantify the energy changes associated with chemical reactions. We learn more about the energy changes associated with chemical reactions in Section 13.3.

13.2 Energy transfer

We see in Section 13.1 how energy cannot be created or destroyed, but how it is transformed from one type to another. For example, chemical energy in fuel is transformed into kinetic energy in a car engine, to give the car motion. Muscle tissue transforms chemical energy stored in food into kinetic energy, which we witness as the contraction of the muscle. In both of these cases, the transformation of energy is associated with the **transfer** of that energy between two entities: the energy from the combustion of fuel in a car engine (chemical energy) is *transferred* to the engine cylinder, which moves to power the car (kinetic energy).

We call the two entities between which energy transfer occurs the **system** and the **surroundings**.

A **system** can be defined as the particular thing in which we are interested, which is contained within a boundary. It may be a bacterial cell in a growth medium, or part of a cell such as a mitochondrion; it may be a reaction flask in a water-bath, or a large industrial vat in a brewery.

The **surroundings** are everything else in contact with the system. For example, in the case of the bacterial cell, the surroundings may be the growth medium with which the cell is surrounded. We define the surroundings in such a way that nothing outside them changes what is within the 'boundary', nor is affected by changes within the 'boundary'.

Figure 13.5 shows that there are three different types of system.

An **open** system is one in which both energy and matter[4] can be transferred across the boundary between the system and its surroundings. For example, a kettle is an open system: thermal energy can be transferred from the kettle (the system) to the air around it (the surroundings). Similarly, water vapour (matter) can also be freely transferred: we see this happening when droplets of water and vapour leave the spout of a kettle as it boils.

A **closed system** is one in which it is possible for energy to be transferred both ways across the boundary between the system and its surroundings, but in which matter cannot be exchanged. For example, any container with a lid is a closed system. Energy can transfer from the container (the system) to its surroundings, but the lid prevents the transfer of matter between the system and surroundings. (This is, in fact, the whole purpose of the lid: to prevent the contents of the container spilling out into its surroundings!)

An **isolated system** is one in which neither energy nor matter can be transferred across the boundary with its surroundings. A Thermos™ flask is designed to be an isolated system. In theory, no energy should be able to be transferred from the system (the flask interior) to the surroundings. Instead, the system retains its energy: if perfectly isolated, the contents of the flask stay hot (or cold) indefinitely.

4 'Matter' is any kind of physical entity. When we consider matter in chemical terms, we are usually thinking about atoms and molecules.

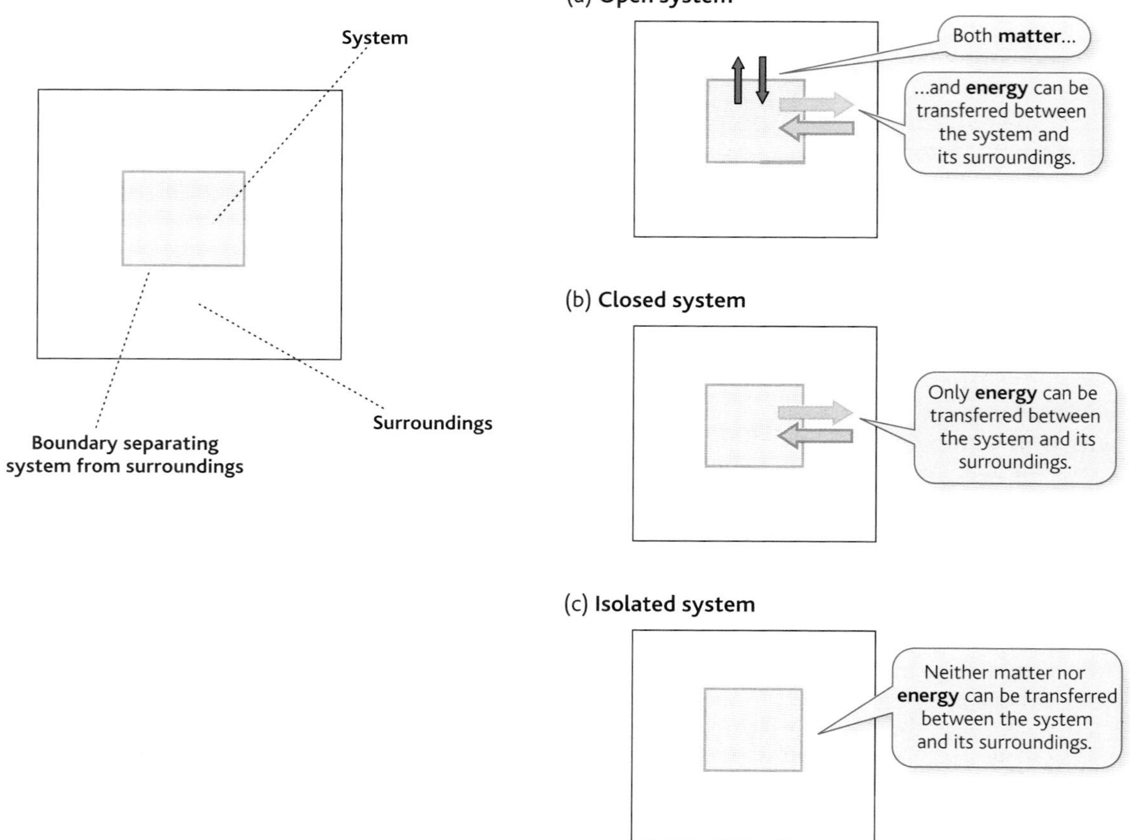

Figure 13.5 The three different types of system. (a) An open system is one in which both energy and matter can be transferred across the boundary with its surroundings. (b) A closed system is one in which it is possible for energy to be transferred both ways across the boundary between the system and its surroundings, but in which matter cannot be exchanged. (c) An isolated system is one in which neither energy nor matter can be transferred across the boundary with its surroundings.

While energy is transferred between a system and its surroundings (and vice versa), the total energy of the system and its surroundings always stays *the same*. This is a crucial point to remember, and relates directly to our interpretation of the first law of thermodynamics: energy cannot be created or destroyed, but just converted from one form to another.

Essentially, we can consider a system and its surroundings to constitute a sealed environment: energy can flow between the system and surroundings, but it cannot escape. If it's not in one part of the overall sealed environment, then it must be in the other.

But how is the energy transferred from a system to its surroundings (and vice versa)? Energy transfer occurs in two ways: as the process of **work**, and as the process of **heat**. We first encountered work and heat on p. 411, where we saw them to be the two processes through which internal energy can be used. Let's now consider each of these in turn.

The transfer of energy as work

'Work' is a bit of a vague term. Many of us may think of work as a task that we perform in order to earn some kind of reward (be it money, or a qualification such as a university degree). Scientifically, though, 'work' is defined as any process that can be used to **lift a weight**. When we lift a book from the floor to the table, we operate against the force of gravity and are doing work on the book. This process is a transfer of energy from a system to its surroundings[5]: chemical energy in our muscles is converted to kinetic energy and counteracts the potential energy of the book as the muscles contract during the lifting process. Our body is the system, and the book (to which the energy is transferred) is the surroundings. (The air may also warm slightly, if the lifting is difficult.) The amount of energy transferred between the system and the surroundings is proportional to the amount of work done by the system on the surroundings. So, the greater the amount of energy transferred between system and surroundings, the greater the amount of work done by the system on its surroundings.

In biological systems, we see various types of work happening: our muscles perform mechanical work; our nerve cells perform electrical work. Whatever type of work is occurring, however, the underlying principle is the same: energy is being transferred.

Look at Box 13.3 to learn how we can *quantify* the amount of work done during energy transfer.

The transfer of energy as heat

It is most natural for us to think of heat as a 'thing'–a physical characteristic. Late into the nineteenth century scientists thought that heat was a substance possessed by systems which could be transferred from one system to another; they even gave it a name: 'caloric' (from which we get our everyday word 'calorie').

We now know that heat is merely one form of energy transfer, just like work. Specifically, heat is the transfer of energy from hot to cold: from a region of high temperature to a region of lower temperature. For example, if a system is at a higher temperature than its surroundings, energy flows (that is, transfers) as heat from the system to its surroundings. By contrast, if a system is at a lower temperature than its surroundings, energy flows from the surroundings into that system–again, from hot to cold.

When energy is transferred as heat, we say the entity from which the energy has transferred undergoes cooling.

- ***Heat is the transfer of energy from hot to cold: from a region of high temperature to a region of lower temperature.***
- ***When something loses energy in this way, it undergoes cooling.***

Our everyday experience tells us much about the transfer of energy as heat. From everyday life, we know that when we place a hot object next to a cooler object thermal energy flows from the hot object to the cooler object until they both reach the

5 We often see energy defined as the 'capacity to do work': the greater the energy something possesses, the greater its capacity to do work. We often feel more able to do work if we're feeling energetic: even the smallest task can often feel like a chore if we feel we have no energy, and are tired.

BOX 13.3 How much work has been done?

When an object is moved against a resistance we say work has been done on it. We can calculate the amount of work done by multiplying the force exerted (that is, the force needed to overcome the inertia that keeps an object stationary) by the distance travelled by the object it moves:

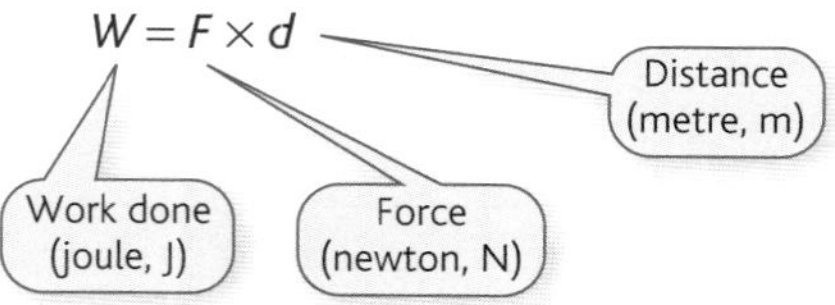

Let's consider the energy required to lift a tonne of bricks (1000 kg) up the side of a house by a distance of 10 m.

The force exerted by the bricks is determined by multiplying the mass, m, by the gravitational constant, g, which quantifies the gravitational attraction of the Earth:

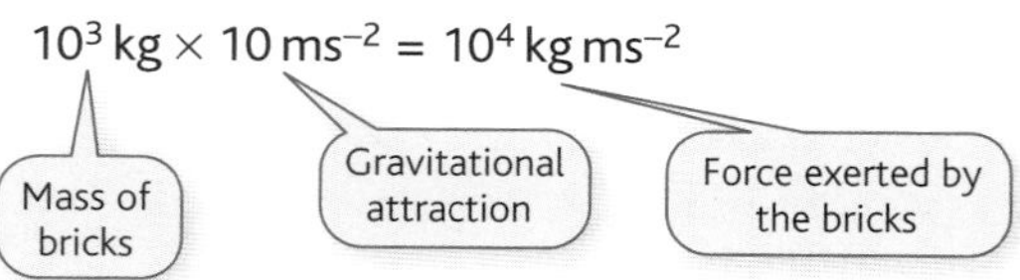

Notice how this calculation assigns units of kg m s^{-2} to the value for the force, even though the previous equation says that force is measured in newtons, N. In fact, 1 N is the same as 1 kg m s^{-2}–that is, the two units are equivalent. Rather than quoting force in N here, though, we'll keep units of kg m s^{-2} as we work through the rest of the calculation.

We can now calculate the energy required to raise the bricks 10 m:

$$W = F \times d$$
$$= 10^4 \text{ kg m s}^{-2} \times 10 \text{ m}$$
$$= 10^5 \text{ kg m}^2 \text{ s}^{-2}$$
$$1 \text{ kg m}^2 \text{ s}^{-2} = 1 \text{ J, so}$$
$$10^5 \text{ kg m}^2 \text{ s}^{-2} = 10^5 \text{ J}$$
$$= 100 \text{ kJ}$$

We may think that lifting a tonne of bricks a distance of ten metres may require a great deal of energy. To put the 100 kJ required to lift the bricks into some kind of perspective, however, the energy released by the oxidation of just one mole (a mere 180 g) of glucose is nearly thirty times larger, at 2870 kJ.

Self-check 13.2 How much energy would you expend carrying two 8 kg bags of shopping to a first floor apartment, up a flight of steps 4 m high?

same temperature. For example, Figure 13.6 shows how, if we prepare a hot drink and leave it at room temperature, the drink cools down until it reaches the same temperature as its surroundings: thermal energy from the drink (the system) is transferred to the air in contact with it (the surroundings).

The transfer of energy as heat–from high temperature to lower temperature–is a **spontaneous** process: it happens without any effort or work being required to bring it about.

The transfer of energy does *not* flow spontaneously from cold to hot. Let's consider a hot drink again. If we stand a hot drink in a room which is at a cooler temperature, the drink does not get hotter–we do not see energy transferring from the surroundings (the air around the drink) to the system (the drink) to give it more thermal and kinetic energy (which we would observe as an increase in temperature of the system).

We learn more about spontaneity in Section 13.4.

- ***The transfer of energy as heat is a spontaneous process.***

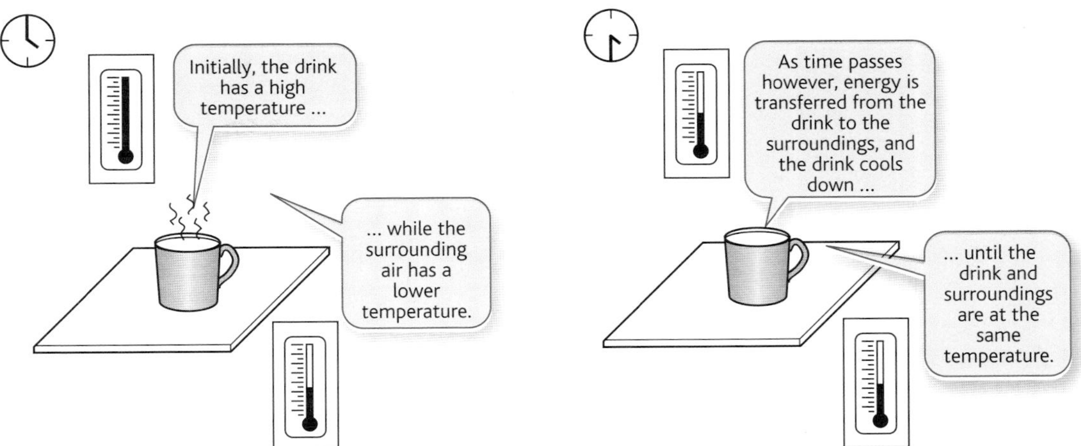

Figure 13.6 If we prepare a hot drink and leave it at room temperature, the drink cools down and the temperature of the surrounding air increases, until they reach the same temperature.

Remember: energy cannot be created or destroyed. Consequently, the total energy of a system and its surroundings remains unchanged as a result of energy transfer; we merely see a redistribution of energy between the two.

Heat versus temperature

We often use the words 'heat' and 'temperature' as if they mean the same thing. Food packaging may tell us to cook the contents 'on a high heat', or 'at a temperature of 180 °C'. However, temperature and heat are *not* the same thing.

We see above how heat is the *transfer* of energy from a system with a high temperature to surroundings with a lower temperature. The temperature itself gives us a measure of what we call the **thermal energy** of a system. The main component of thermal energy is the *kinetic energy* of molecules, the energy associated with motion.

The atoms and molecules making up a system are in constant, rapid motion; they collide with each other and with the walls of the vessel in which they are contained. A system with a large amount of thermal energy comprises atoms and molecules that are moving around a great deal (they have a lot of kinetic energy). By contrast, a system with little thermal energy comprises atoms and molecules that are moving around less (they have less kinetic energy).

We measure the thermal energy of a system as its **temperature**. A system with a large amount of thermal energy has a **high** temperature; a system with a small amount of thermal energy has a **low** temperature.

It is not useful to talk of the temperature of a microscopic entity such as an atom, molecule, or cell. Even if it were, we could not measure it. Rather, at a microscopic level we talk of 'energy'. At a macroscopic level, we can correlate energy with temperature, which *is*, then, measurable.

- ***The temperature of a system is proportional to its energy. A system with a high temperature has a large amount of energy; a system with a low temperature has a smaller amount of energy.***

When we measure the temperature of a system we are obtaining one measure of the **average** energy of all the atoms or molecules in the system. Figure 13.7 shows

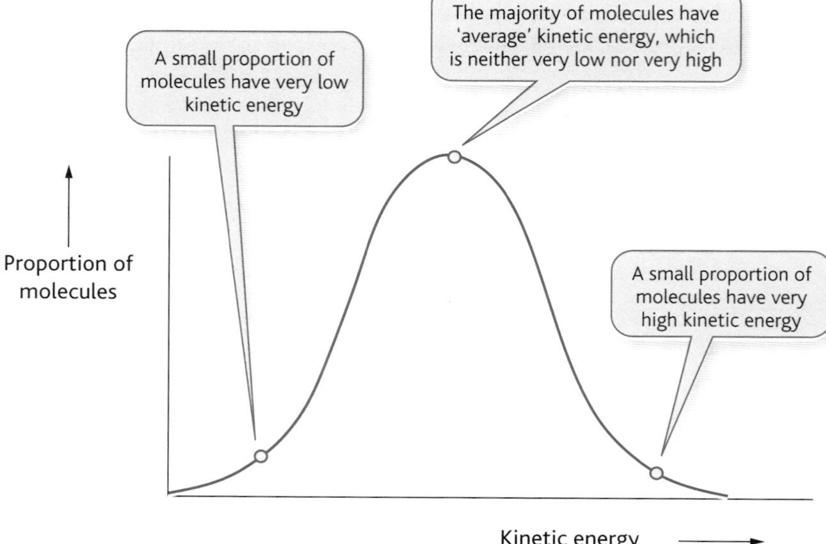

Figure 13.7 The atoms and molecules in a system do not all possess the same thermal or kinetic energy; some atoms and molecules possess a great deal of energy, some possess very little energy; the other atoms and molecules in the system have energies that fall somewhere between the two extremes.

that the atoms and molecules in a system do not all possess the same thermal or kinetic energy: some atoms and molecules possess a great deal of energy, some possess very little energy; the other atoms and molecules in the system have energies that fall somewhere between the two extremes[6].

13.3 Enthalpy

We see in Section 13.1 how chemical energy is the energy stored in the chemical bonds holding a compound together. When a chemical bond is formed, energy is given out; when a chemical bond is broken, energy is consumed.

These two concepts help us to answer the question: what is happening at the level of energy when one or more reactants undergo a chemical reaction to form products?

When a chemical reaction occurs, certain bonds are broken and others are formed. It is most unlikely that the same amount of energy is liberated by the breaking of bonds as is consumed by the forming of new ones. If there is more energy associated with the formation of new bonds than with the breaking of bonds, the reaction gives out energy. If there is more energy associated with the breaking of bonds than with the formation of bonds, the reaction consumes energy.

The overall **energy change** for the reaction is the difference between the energy consumed when bonds break and when new bonds form. We call this change in energy the **enthalpy change of reaction**, and give it the symbol ΔH[7].

---

6 The plot in Figure 13.6, with its characteristic bell shape, represents what we call a normal, or Gaussian distribution. In such a distribution, a minority of the population have a characteristic (in this case, kinetic energy) which lies at two extremes–very high, or very low–whereas the majority of a population have a characteristic whose value (in this case, kinetic energy) falls somewhere between these two extremes. We see normal distributions throughout biology.

7 Remember: we use the Greek letter delta (Δ) to denote 'change in'.

But the enthalpy is more than just a label given to the energy change associated with chemical reactions. The enthalpy is a distinct thermodynamic property that is very closely related to the internal energy of a system, as explained in Box 13.4.

- ***The enthalpy change of a reaction is the difference between the energy consumed to break bonds, and the energy liberated when bonds are formed.***

Energy transfer during chemical reactions

Biochemical reactions happen at a constant pressure. Under such conditions, the enthalpy change happening in a system is equal to the heat transferred to the system during the reaction[8]. We can write that:

$$\Delta H = q$$

BOX 13.4 Enthalpy and internal energy

The enthalpy change of reaction helps us to describe the way in which the energy of a system changes when a reaction happens. But what is enthalpy, in thermodynamic terms?

Enthalpy is closely related to the internal energy of a system according to the following relationship:

$$H = U + pV$$

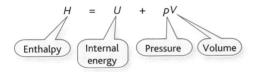

This tells us that the enthalpy is simply the internal energy plus the product of the pressure of the system and the volume of the system (that is, $p \times V$).

So, the change in enthalpy is given by the relationship:

$$\Delta H = \Delta U + \Delta(pV)$$

At constant pressure, however (which applies to biochemical systems), we write:

$$\Delta H = \Delta U + p\Delta V$$

(There is no change in pressure, so only the volume changes, as denoted by ΔV.)

But what does the term $p\Delta V$ denote? This term accounts for the energy used when the system either contracts or expands during the course of a chemical reaction, depending on the number of molecules involved, as illustrated in Figure 13.8. If a chemical reaction results in the generation of a smaller number of molecules than at the start of the reaction, the system will occupy a smaller volume: the pressure being exerted on the system by the surroundings (in the case of a beaker of water, for example, the pressure of the surrounding air) will squeeze the smaller number of molecules so that they occupy a smaller space than the larger number of molecules at the start of the reaction.

By contrast, if the reaction generates a larger number of molecules, these molecules will occupy a larger volume than the starting molecules. To occupy this larger volume, the molecules must do work on the surroundings–pushing out against the surroundings to expand the volume that the system occupies.

So $p\Delta V$ represents the work done on the system by the surroundings if the volume of the system contracts during a reaction, or the work done by the system on the surroundings if the volume of the system expands during a reaction.

---

8 There is one caveat: any work being performed on or by the system for this relationship to hold true must only be that of expansion–that of the system pushing against the surroundings as the volume of the system increases (or the surroundings pushing against the system if the volume of the system decreases), as mentioned in Box 13.4.

BOX 13.4 **Continued**

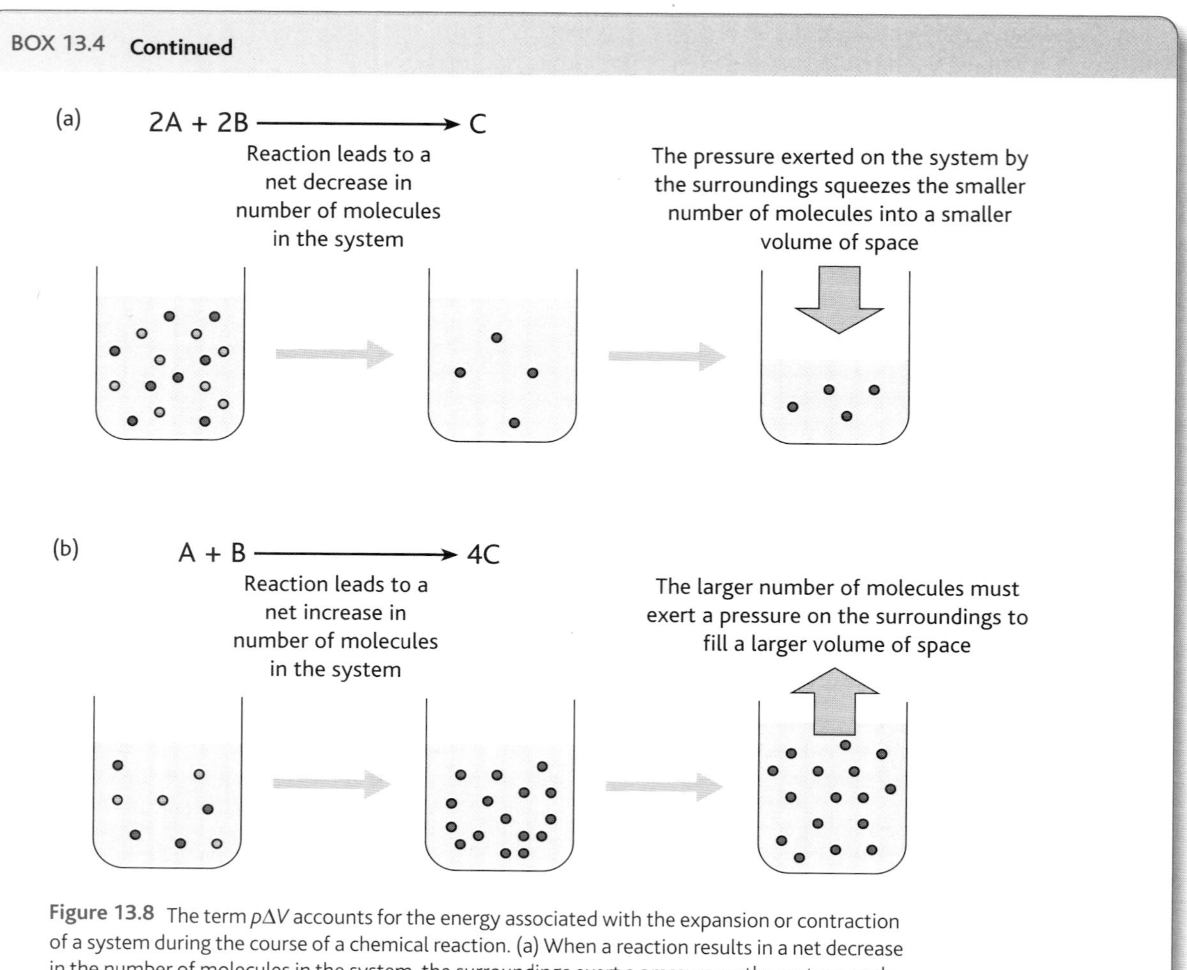

Figure 13.8 The term $p\Delta V$ accounts for the energy associated with the expansion or contraction of a system during the course of a chemical reaction. (a) When a reaction results in a net decrease in the number of molecules in the system, the surroundings exert a pressure on the system, such that it fills a smaller volume of space. (b) When a reaction results in a net increase in the number of molecules in the system, the system must exert a pressure on the surroundings, such that the larger number of molecules can expand to fill a larger volume of space.

With this in mind, we can consider the transfer of energy between a system and its surroundings that is happening during a chemical reaction. Look at Figure 13.9. This figure shows how, if the energy required to break bonds during a chemical reaction is *greater* than energy released from the formation of bonds, the enthalpy change for the reaction is **positive**. If the enthalpy change is positive, the system needs a supply of energy for the reaction to occur. To satisfy this requirement, energy is **absorbed** into the system from the surroundings to provide the net **increase** in energy required for the reaction to happen. That is, heat is transferred to the system from the surroundings.

By contrast, if the energy required to break bonds is *less* than the energy released during the formation of bonds, the enthalpy change for the reaction is **negative**. If the enthalpy change is negative, there is a surplus of energy when the reactants form

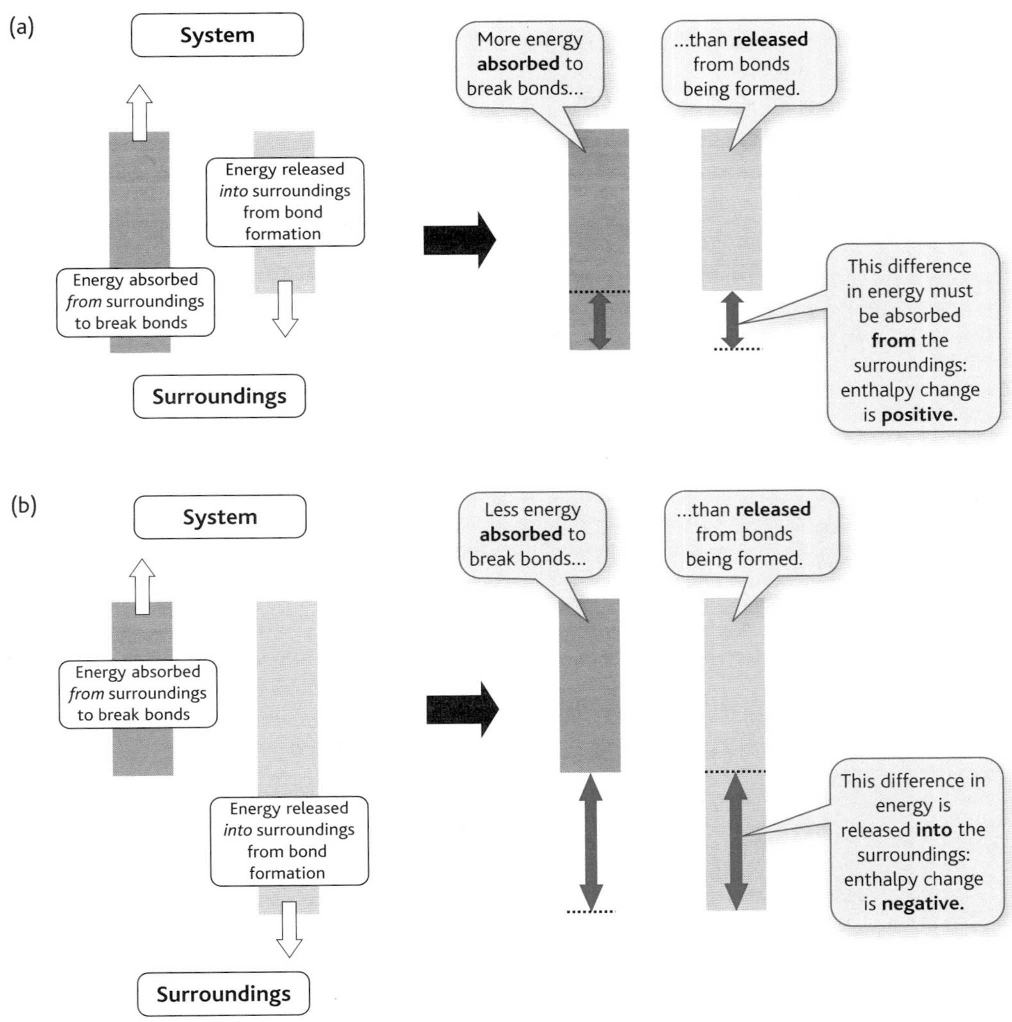

Figure 13.9 (a) If the energy required to be absorbed from the surroundings to break bonds during a chemical reaction is *greater* than energy released to the surroundings from the formation of bonds, the enthalpy change for the reaction is *positive*. (b) If the energy required to be absorbed from the surroundings to break bonds is *less* than the energy released to the surroundings from the formation of bonds, the enthalpy change for the reaction is *negative*.

products, which can be transferred from the system to the surroundings. We see energy **released** by the system as the reaction proceeds. That is, heat is transferred from the system to the surroundings.

Reactions for which the enthalpy of reaction, ΔH, is negative (i.e. for which energy is transferred from the system to its surroundings) are called **exothermic** reactions[9].

9 'Exo-' = 'outwards'; exothermic = heat flows outwards (from the system to the surroundings).

Reactions for which the enthalpy of reaction, ΔH, is positive (i.e. for which energy is transferred from the surroundings to the system) are called **endothermic** reactions[10].

- *An exothermic reaction is one that releases energy from the system into the surroundings.*
- *During an exothermic reaction, heat is transferred from the system to the surroundings.*
- *An endothermic reaction is one that absorbs energy from the surroundings into the system.*
- *During an endothermic reaction heat is transferred from the surroundings to the system.*

Self-check 13.3 State whether the following reactions are endothermic or exothermic:

(a) $CaCO_3(s) \rightarrow CaO(s) + CO_2(g)$ $\Delta H = +178$ kJ mol^{-1}
(b) $H_2O(g) \rightarrow H_2O(l)$ $\Delta H = -44.1$ kJ mol^{-1}
(c) $NH_4Cl(s) \rightarrow NH_4^+(aq) + Cl^-(aq)$ $\Delta H = +16$ kJ mol^{-1}

How can we determine the enthalpy change for a reaction?

We can determine whether a reaction has a positive or negative enthalpy change by calculating the difference between the total bond energies of the reactants and products for a given reaction.

We can write this calculation as follows:

$$\Delta H = \Sigma E_{(reactants)} - \Sigma E_{(products)}$$

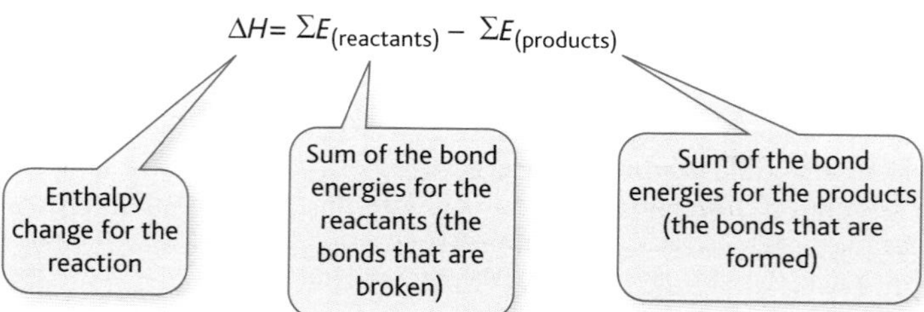

If $\Sigma E_{(reactants)}$ is *greater* than $\Sigma E_{(products)}$ (that is, more energy is required to break bonds than is liberated from the formation of bonds) the value of ΔH is positive.

If $\Sigma E_{(reactants)}$ is *less* than $\Sigma E_{(products)}$ (that is, less energy is required to break bonds than is liberated from the formation of bonds) the value of ΔH is negative.

Let's consider an example, in which we ask whether the reaction is exothermic or endothermic.

10 'Endo-' = 'inwards'; endothermic = heat flows inwards (from the surroundings to the system).

EXAMPLE

$CH_4(g) + 2\ O_2(g) \rightarrow CO_2(g) + 2\ H_2O(l)$[11]

$\Delta H = \Sigma E_{(reactants)} - \Sigma E_{(products)}$

Bonds broken — Bonds formed

$\Sigma E_{(reactants)}$:	CH_4 +	$2\ O_2$
Molecules:	H–C–H with H above and H below (methane)	$2 \times O{=}O$
Bonds:	$4 \times (C–H)$	$2 \times (O{=}O)$
	$4 \times (412\ kJ\ mol^{-1})$	$2 \times (497\ kJ\ mol^{-1})$
	$1648\ kJ\ mol^{-1}$	$994\ kJ\ mol^{-1}$
$\Sigma E_{(reactants)}$:	$1648 + 994$	
	$= 2642\ kJ\ mol^{-1}$	
$\Sigma E_{(products)}$:	CO_2 +	$2\ H_2O$
Molecules:	$O{=}C{=}O$	$2 \times H–O–H$
Bonds:	$2 \times (C{=}O)$	$4 \times (O–H)$
	$2 \times (740\ kJ\ mol^{-1})$	$4 \times (463\ kJ\ mol^{-1})$
	$1480\ kJ\ mol^{-1}$	$1852\ kJ\ mol^{-1}$
$\Sigma E_{(reactants)}$:	$1480 + 1852$	
	$= 3332\ kJ\ mol^{-1}$	
ΔH	$= \Sigma E_{(reactants)}$	$- \Sigma E_{(products)}$
	$= 2642\ kJ\ mol^{-1}$	$- 3332\ kJ\ mol^{-1}$
	$= \mathbf{-690\ kJ\ mol^{-1}}$	

The combustion of methane is an **exothermic** process, which has a negative enthalpy change of reaction. We can predict that energy is liberated into the surroundings during the course of the reaction. Our experience tells us that this is indeed the case: we associate all combustion reactions with the release of energy as heat.

Self-check 13.4 In both of the following cases, use the bond energy values shown in Table 13.1 to determine the value of the enthalpy change for the reaction presented. Is each reaction exothermic or endothermic?

(a) When we consume alcohol, the body metabolizes it according to the following reaction:

$$CH_3CH_2OH + 3\ O_2 \rightarrow 2\ CO_2 + 3\ H_2O$$

11 This reaction is the combustion (burning) of methane in oxygen which powers the gas cooker and central heating in many homes.

Use this as the structural formula of the reactant, ethanol:

```
   H   H
   |   |
H—C—C—OH
   |   |
   H   H
```

(b) Consider the following reaction:

$$6\,CO_2 + 6\,H_2O \rightarrow C_6H_{12}O_6 + 6\,O_2$$

This reaction represents the process of photosynthesis: the net chemical reaction from which plants evolve oxygen. Use this as the structural formula of glucose:

```
     H   H   H   OH  H
     |   |   |   |   |      H
HO—C—C—C—C—C—C<
     |   |   |   |   |      O (double bond)
     H   OH  OH  H   OH
```

Importantly, the value of a thermodynamic quantity such as a change in enthalpy depends on the conditions under which the quantity is measured. Read Box 13.5 to learn more about quoting thermodynamic quantities, and the reference points called standard states from which energy changes are measured.

Depicting enthalpy changes: the energy diagram

The enthalpy changes associated with chemical reactions are best illustrated pictorially on an **energy diagram**. An energy diagram contains a wealth of information about what is happening at the level of the energy of the system during the course of a reaction.

Look at Figure 13.10, which shows a generic energy diagram. The reactants and the different species that subsequently form during the course of the chemical reaction are plotted on an energy diagram at vertical positions which reflect their relative energies: species with greater enthalpies are plotted at positions higher on the energy diagram than species with smaller enthalpies.

Notice how the reactants are plotted on the left-hand side of the energy diagram; the products are plotted on the right-hand side. By using a line to link the different species represented on the energy diagram, we are able to visualize the overall energy changes that occur during the course of a chemical reaction.

Let's consider the reaction of glucose with oxygen:

$$C_6H_{12}O_6 + 6\,O_2 \rightarrow 6\,CO_2 + 6\,H_2O$$

We can represent the energy changes occurring during the course of this reaction with the energy diagram shown in Figure 13.11. Notice how the energy diagram shows how the reaction is exothermic: the enthalpy change for the reaction is negative, so heat is evolved during the course of the reaction.

BOX 13.5 Standard states: making sense of measurements

As we see in Chapter 1, measurements only make sense if we know the units that apply to them. Equally, the things we might measure during the course of scientific investigation–the rate at which a certain process proceeds, for example–are often influenced by the environment they are exposed to. For example, the rate at which food decomposes through the activity of bacteria depends on the temperature of the surroundings. (Our experience tells us that milk left at room temperature on a warm summer's day will turn sour more quickly than milk left out during winter.) This is why we store many foods in a refrigerator–to slow down the process of decomposition, and help the food remain edible for longer.

Equally, the energy changes associated with many chemical and biochemical reactions are influenced by the surrounding environment. Therefore, to make real sense of any kind of measurement, we need to know the conditions under which the measurement was taken.

To address this issue, it is often useful to benchmark our measurements against some common reference points. These reference points are values of thermodynamic properties such as enthalpy and free energy, which are measured when substances are in their **standard states**. Different substances have different standard states as listed in Table 13.2.

Table 13.2 The standard states of different substances.

Substance	Standard state
Solid	Pure solid
Liquid	Pure liquid
Solute	Concentration of 1 mol L^{-1}
Gas	Pure gas at a pressure of 1 atmosphere

The biochemical standard state is a particular standard state when the concentration of all solutes is 1 M with the exception of protons, which are present at a concentration of 10^{-7} M. (If protons were also present at a standard state concentration of 1 M the pH of the solution would be pH 0, which is far too acidic for the biological processes occurring in the cell to operate. Instead, a proton concentration of 10^{-7} M equates to a pH of 7.0, which is a much better approximation of the pH of the solutions in which many biological processes occur.)

A measurement taken when a substance is in its standard state is denoted with the superscript $^{\circ}$ (though the plimsoll symbol, $^{\ominus}$ is often used, too.) By contrast, a measurement when a substance is in the biochemical standard state is denoted with the superscript $^{\circ\prime}$.

Measurements taken when compounds are in their standard state can happen at different temperatures. Therefore, it is good practice to quote the temperature at which a particular measurement was taken. Most often, however, chemical measurements are taken at room temperature, which is taken to be 25 °C (or 298 K). By contrast, biochemical measurements are often taken at body temperature, around 37 °C (or 310 K).

Look out for the standard state and biochemical standard state symbols throughout this chapter (and the rest of the book), and bear in mind that they are telling you something specific about the thermodynamic quantity being quoted. Don't be surprised if you see seemingly different values quoted for the same thing (for example, the enthalpy change associated with the oxidation of glucose). If the values are different, ask yourself whether the conditions under which the quantities are quoted are the same or different–are they referring to the reaction in the standard state, or not?

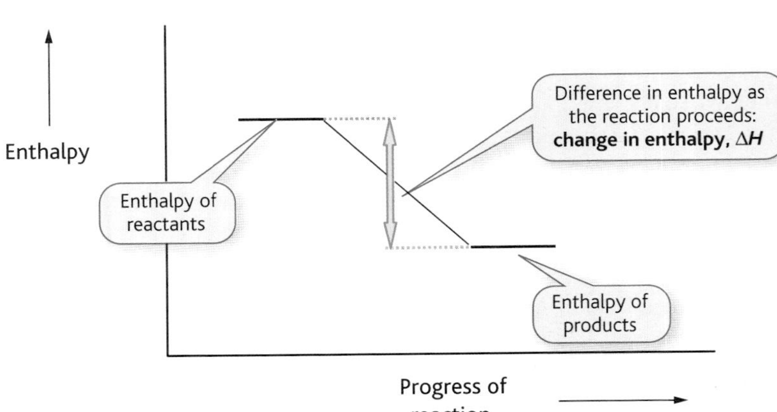

Figure 13.10 A generic energy diagram. Notice how different species that are present during the course of the chemical reaction are plotted at vertical positions which reflect their relative energies.

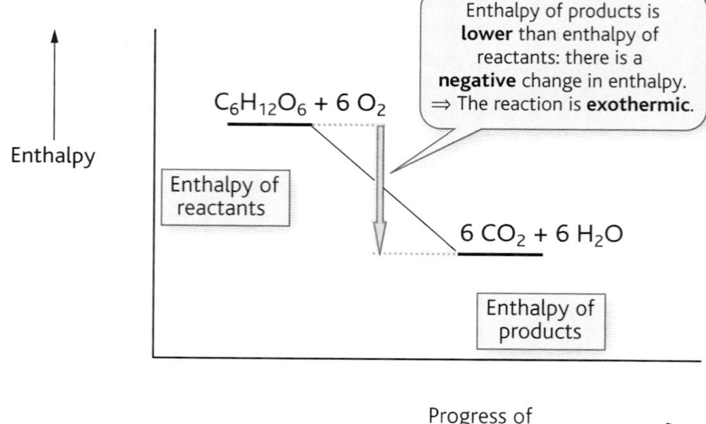

Figure 13.11 The energy change for the reaction of glucose with oxygen. ΔH for this reaction is *negative*, so we say the reaction is exothermic.

In fact, for every mole of glucose consumed under standard conditions[12], 2808 kJ of energy is released. We say $\Delta H^{\circ} = -2808\ \text{kJ mol}^{-1}$, where the '–' sign says that energy is released, rather than consumed. 2808 kJ mol^{-1} is a great deal of energy. This energy is harnessed by the body to do work: to power the contraction of muscles; to be transformed into energy to power other reactions, for example.

We can use energy diagrams to depict changes in different types of energy. Here we see energy diagrams used to depict changes in *enthalpy*. By contrast, in Chapter 17 we see energy diagrams used to depict the overall change in energy associated with a chemical reaction in a more qualitative way. We encounter energy diagrams again in Section 13.6, later in this chapter, in the context of depicting the change in another type of energy, the Gibbs free energy.

Measuring the enthalpy of a reaction

It is important to realize that we can't measure *directly* the enthalpy of a particular compound. We can only measure the *change* in enthalpy when a particular reaction occurs.

The enthalpy change of a reaction can be measured using **calorimetry**, using what we call a **bomb calorimeter** (depicted in Figure 13.12).

A bomb calorimeter is a thermally insulated container inside which a chemical reaction is performed. The enthalpy of reaction is measured as a change in the temperature of the environment immediately surrounding the reaction.

The reactants inside the calorimeter (the system) are in contact only with their surroundings, so any energy gained by the surroundings must have come from its respective system (the chemical reaction).

We use a bomb calorimeter to study combustion reactions. In consequence, the bomb calorimeter only measures the enthalpy change associated with an *exothermic* reaction (i.e. when energy is transferred from the system to the surroundings). The magnitude of the enthalpy change for an exothermic reaction is proportional to the increase in temperature of the surroundings that we are measuring–generally

12 Read Box 13.5 if you're not sure what this term means.

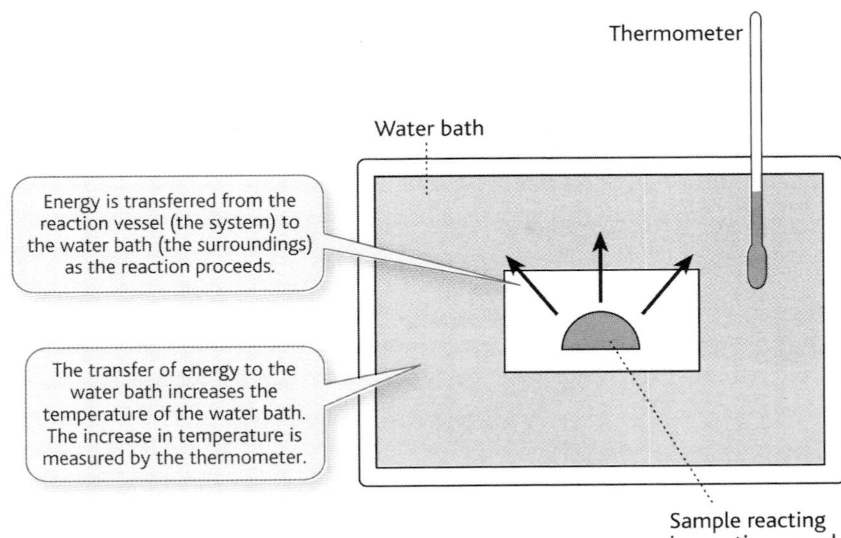

Figure 13.12 A simplified representation of a bomb calorimeter.

a reservoir of water. We can measure the extent of cooling of the surroundings of other calorimeters to allow us to study endothermic reactions.

- ***We can't measure the enthalpy of a compound directly. We can only measure the change in a compound's enthalpy when some kind of chemical reaction occurs.***

Enthalpy changes for different processes

Any chemical reaction has associated with it a particular enthalpy change. However, we can give enthalpy changes particular names, depending on the type of reaction they relate to.

For example, the enthalpy change of **formation**, ΔH°_f, is the enthalpy change associated with the formation of a compound from its constituent elements in their standard states[13].

For example, the enthalpy change of formation of glucose is the enthalpy change for the following reaction:

$$6\,C(graphite) + 6\,H_2(g) + 3\,O_2(g) \rightleftharpoons C_6H_{12}O_6(s)$$

The enthalpy change for this reaction is -1250.1 kJ mol^{-1}.

By contrast, the enthalpy change of **combustion**, ΔH°_c, is the enthalpy change when one mole of a substance reacts completely with excess oxygen.

For example, the enthalpy change of combustion of ethanol is the enthalpy change for the following reaction:

$$C_2H_5OH\,(l) + 3\,O_2(g) \rightarrow 2\,CO_2(g) + 3\,H_2O(l)$$

13 The superscript ° denotes that the enthalpy of formation relates to the components in their standard states. For a reminder of standard states, look at Box 13.5.

The enthalpy change for this reaction is also large at −1368 kJ mol^{-1}. This large value of ΔH explains why alcohols are often used as fuels.

But what can enthalpies of reaction tell us about biological systems? For one thing, they can give us an insight into the relative value of different biochemical reactions as potential sources of energy. Let us return to glucose, which is oxidized by the cells of our body as a 'fuel' to drive other biochemical reactions.

When there is plenty of oxygen in our cells, glucose is oxidized completely:

$$C_6H_{12}O_6(s) + 6\,O_2(g) \rightarrow 6\,CO_2(g) + 6\,H_2O(l)$$

The enthalpy change for this reaction, ΔH°_c, = −2808 kJ mol^{-1}.

When there are lower levels of oxygen available–for example, in our muscle cells during vigorous exercise–glucose is only partially oxidized to lactic acid:

$$C_6H_{12}O_6(s) \rightarrow 2\,CH_3CH(OH)CO_2H(s)$$

(It is the accumulation of lactic acid that gives the burning sensation in our muscles during exercise.) As we breathe deeply at the end of a period of exercise to replenish the supply of oxygen to our muscle tissue, this lactic acid is then completely oxidized:

$$2\,CH_3CH(OH)CO_2H(s) + 6\,O_2(g) \rightarrow 6\,CO_2(g) + 6\,H_2O(l)$$

The standard enthalpy change for the conversion of glucose to lactic acid during the first of these two processes is just −120 kJ mol^{-1}. So we see that the metabolism of glucose under anaerobic conditions (low levels of oxygen) generates just 4% of the energy of the metabolism of glucose when oxygen is plentiful, telling us that the metabolism of glucose in plentiful oxygen is a much more useful source of energy for our cells.

Enthalpy changes and the stability of chemical compounds

An exothermic reaction, with a negative ΔH, equates to a system losing energy to its surroundings–that is, the system becoming more stable[14].

By contrast, an endothermic reaction, with a positive ΔH, equates to a system *gaining* energy from its surroundings–that is, the system becoming *less* stable.

What does the enthalpy change for a reaction tell us about the relative stability of the reactants and products? An exothermic reaction leads to products that are *more stable* than their reactants, because there is a net decrease in energy in the system. By contrast, an endothermic reaction leads to products that are *less* stable than their reactants; there is a net *increase* in energy of the system.

Let's consider the stability of glucose, the formation of which is mentioned above. We see above that the enthalpy of formation of glucose has a negative value, −1250.1 kJ mol^{-1} (and so we say it is **energetically favourable**). So glucose is more stable than its component elements in their standard states.

However, we also see above how the standard enthalpy of combustion of glucose is also a negative value, −2808 kJ mol^{-1}. Because this value is *more* negative than the enthalpy of formation of glucose, it tells us that the products of the combustion of glucose (its complete reaction with oxygen) are more stable than glucose itself (even though glucose itself is more stable than the elements from which it is formed).

---

14 Remember: lower energy equates to greater stability.

By contrast, the enthalpy of formation of adenine (1), one of the two purines found in DNA, is +96.9 kJ mol^{-1}.

(1)

This is an endothermic reaction for which energy must be gained from the surroundings. Consequently, adenine is *less* stable than the elements from which it is formed.

- ***For a reaction whose enthalpy change is negative, the products are more stable than the reactants.***
- ***For a reaction whose enthalpy change is positive, the reactants are more stable than the products.***

13.4 Entropy: the distribution of energy as the engine of change

Let us pause for a moment to recall the two important things we have learned about a system and its surroundings:

- The total energy of a system and its surroundings (collectively called a **universe**) is constant (unchanging), because energy cannot be created or destroyed.
- Energy flows (transfers) spontaneously as heat from a hot system to cooler surroundings, or from hot surroundings to a cooler system.

Now, with these two points in mind, let us consider the situations shown in Figure 13.13(a) and (b). Figure 13.13(a) shows a mug of hot drink (system) standing in a cool room (the surroundings); Figure 13.13(b) shows the same system and surroundings, but after energy has transferred from the hot system to the cooler surroundings.

We know that the total energy of the system and surroundings in Figure 13.13(a) and (b) is the same (because energy is conserved), yet they must be different to some degree because energy transfer has occurred between the system and surroundings. So we need to ask: why does the energy transfer occur, and how can we describe the difference between the two situations shown in Figure 13.13?

In order to answer these questions we need to explore a new concept, essentially one that describes a property of all matter. We say that during the process of heat transfer, as energy spreads from the hot system to cooler surroundings, the **entropy** of the system and surroundings changes.

But what is 'entropy'? Entropy, like energy, is an inherent property of something, rather than a physical characteristic we can easily visualize.

A system is typically made up of millions and millions of atoms or molecules. The atoms and molecules typically possess a wide range of internal energies, which,

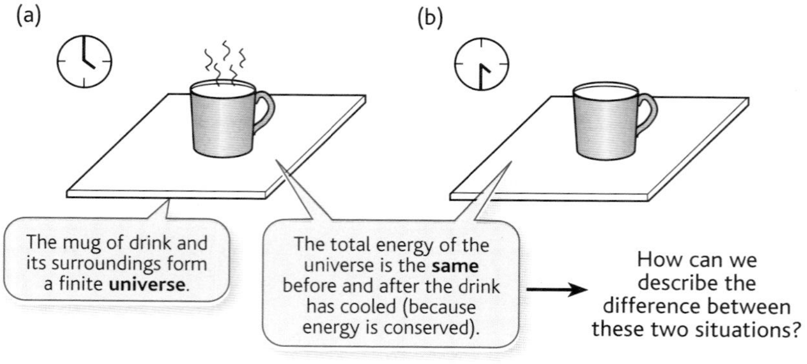

(a)

(b)

The mug of drink and its surroundings form a finite **universe**.

The total energy of the universe is the **same** before and after the drink has cooled (because energy is conserved).

How can we describe the difference between these two situations?

Figure 13.13 A mug of drink (the system) standing in a cool room (the surroundings) when the drink is hot (a), and when it has cooled (b). We are interested in how we can describe the difference between (a) and (b).

together, make up the total energy of the system. This total energy reflects the kinetic and potential energy of each species in the system. Some atoms and molecules possess relatively little energy, while others possess a great deal: it is the overall distribution of energy across the entire population of atoms and molecules that gives the total energy of the system.

There are many different ways in which the total energy of a system can be distributed amongst its composite atoms and molecules. Let's consider an analogy. Imagine we have two litres of water (the equivalent of the total energy of the system), which we need to share between several one-litre buckets (the equivalent of the atoms and molecules in the system) in 'units' of one litre. Look at Figure 13.14, which illustrates how there are various ways in which we can distribute the water: if we have four one-litre buckets, we can distribute the water in six different ways.

Entropy gives us a measure of all the different ways in which energy can be distributed amongst the atoms and molecules in a system: it gives us a measure of the **distribution** of energy through a system. If the entropy of a system is low, there is

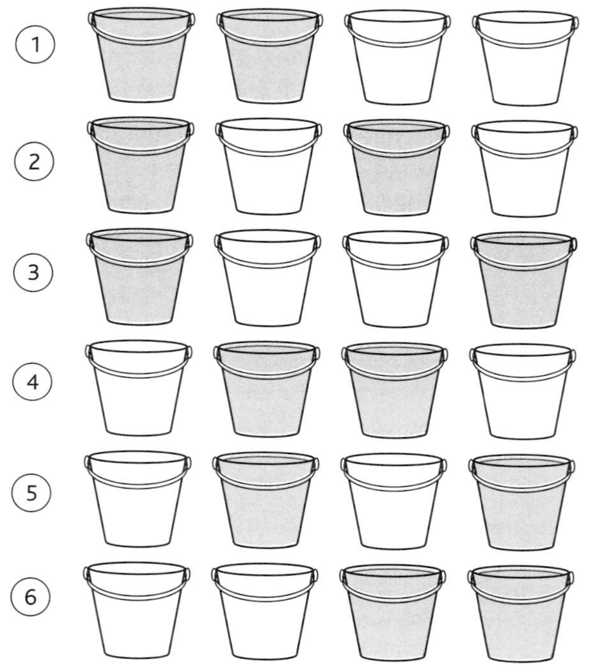

Figure 13.14 Two litres of water can be distributed between four one-litre buckets in a total of six ways (assuming the water is decanted in one-litre measures).

little spread of energy through the system–there is little spread of kinetic energy, for example, so components of the system interact in a fairly ordered, low-key way. By contrast, if the entropy of the system is high, energy is spread widely throughout the system–there is a greater spread of kinetic energy through the system, for example, so components of the system jostle around in a more disordered way.

More formally, we can describe entropy as a measure of the **energetic disorder** of a system. The greater the entropy (the spread of energy through a system or its surroundings), the greater the level of the disorder of the energy the system comprises.

- *Entropy gives us a measure of the distribution of energy in a system.*
- *The greater the spread of energy (and greater the entropy) the greater the disorder in the system.*

Entropy in chemical and biological systems

Taking the above concepts into account, what can we predict about the entropies of different chemical and biological systems? Let's consider three examples, which illustrate how entropy varies within chemical and biological systems, but is always an inherent property.

First, consider water in its three different states–solid (ice), liquid, and gas (water vapour). We see in Chapter 4 how the different physical states have different energies:

- Solids have the lowest energy: their component atoms are arranged in an ordered way, and show little movement (kinetic energy);
- Liquids possess a higher energy than solids: their component molecules have more kinetic energy than solids, and are arranged in a less ordered way.
- Gases possess more energy than liquids: their component molecules are high-energy, and move around freely, in a disorganized way.

So, given that low entropy equates to low energy, and high entropy equates to high energy, we can place the different states of water in the following sequence, in terms of their entropy, with solids having the lowest entropy, and gases having the highest:

$$\text{ice} < \text{liquid} < \text{water vapour}$$

Let us now consider a protein comprising a single polypeptide chain. When first synthesized, we can visualize a polypeptide as a long chain that is free to twist and bend in an unconstrained, disorganized way, as depicted in Figure 13.15. By contrast, most proteins have to fold into a distinct, carefully organized three-dimensional shape to become active. What is the difference in entropy between the unfolded and folded protein? When folded, the protein adopts a more organized, lower-energy, stable arrangement; this lower-energy state therefore has lower entropy than an unfolded protein (which has higher energy–for example, it is freer to move, and so has more kinetic energy).

The low entropy of a folded protein compared to its unfolded form is a critically important issue in biological systems, as we explore further in Section 13.5.

Finally, let's consider a biochemical reaction that involves the breakdown of large molecules into numerous smaller ones, as illustrated in Figure 13.16. How can we

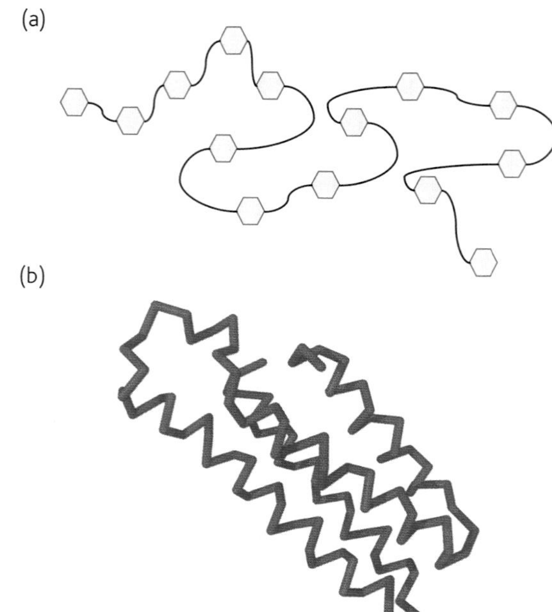

Figure 13.15 (a) When first synthesized, a polypeptide chain may adopt a disorganized structure, in which the chain can twist and bend in an unconstrained way. (b) To perform its correct function, however, a protein must typically fold into an organized three-dimensional structure.

describe the entropy of the system at the start and end of this reaction? At the end of the reaction the energy of the system is spread out amongst more entities than at the start of the reaction, and the system is now inherently more disorganized because it contains more components. Therefore, the entropy of the system has *increased*.

These examples help us to draw the following general conclusions:

- The different physical states possess increasing entropy, in the order solid < liquid < gas;
- Molecules that are more organized and more stable (like folded proteins) have lower entropy than molecules that are less organized and less stable;
- Reactions that increase the number of entities in a system increase the entropy of that system.

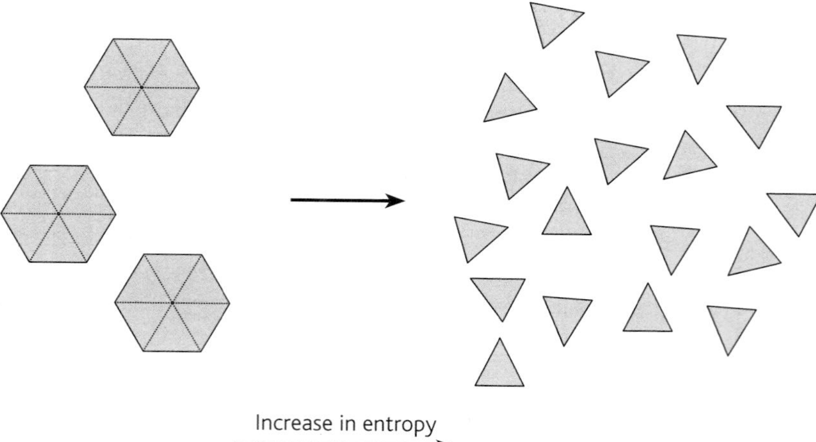

Figure 13.16 When large molecules break down into more numerous smaller ones, the entropy of the system increases because it contains more components, which are inherently more disorganized.

Self-check 13.5 For each of the following processes, predict whether the entropy of the system has increased or decreased, giving a reason for your answer.

(a) The oxidation of glucose: $C_6H_{12}O_6 + 6\ O_2 \rightarrow 6\ CO_2 + 6\ H_2O$

(b) The denaturing (unfolding) of an enzyme on heating

(c) The synthesis of a polypeptide chain from free amino acids

(d) The formation of solid gallstones from compounds in solution.

The link between entropy and energy

What happens to the entropy of a system and its surroundings when a chemical reaction occurs? The entropy of a system is proportional to the energy transferred to or from a system (as heat) when a reaction occurs: when energy is transferred to the system (during an endothermic reaction), entropy increases. When energy is transferred *from* the system (during an exothermic reaction), entropy decreases.

However, we need to be a bit more specific here, and think about what is happening to both the system *and* the surroundings. In essence, whatever happens to the system is the opposite of what is happening to the surroundings, as energy is transferred *from* one *to* the other.

Consider an exothermic reaction, in which energy transfers from a system to its surroundings. If the enthalpy of the system decreases, so the *entropy* of the system must also decrease. However, as the enthalpy of the system decreases, the enthalpy of the *surroundings* increases (because it has absorbed energy from the system), so the entropy of the surroundings also increases. Notice how the system and surroundings are being affected in opposite ways.

During an *endothermic* reaction, the enthalpy of the system *increases*. Consequently, the entropy of the system also increases. However, as the enthalpy of the system increases, the enthalpy of the surroundings decreases (because energy has flowed *from* the surroundings *to* the system), so the entropy of the surroundings also decreases.

Look at Figure 13.17, which summarizes how enthalpy and entropy differ during an exothermic and an endothermic reaction.

- ***During an exothermic reaction, the entropy of the system decreases but the entropy of the surroundings increases as energy transfers from the system to the surroundings.***
- ***During an endothermic reaction, the entropy of the system increases but the entropy of the surroundings decreases as energy transfers from the surroundings to the system.***

We see in the previous pages how changes in entropy are directly related to the changes in energy of a system: as the energy increases, entropy decreases, and vice versa. We can describe the relationship between entropy and the transfer of heat energy, q, as follows[15]:

15 As is often the case, such a relationship only holds true under certain conditions–in this case, that the system is not undergoing significant change–that is, it is at equilibrium, such that there are only minor fluctuations in the system as it acts to maintain its steady state. We learn more about chemical equilibria in Chapter 15.

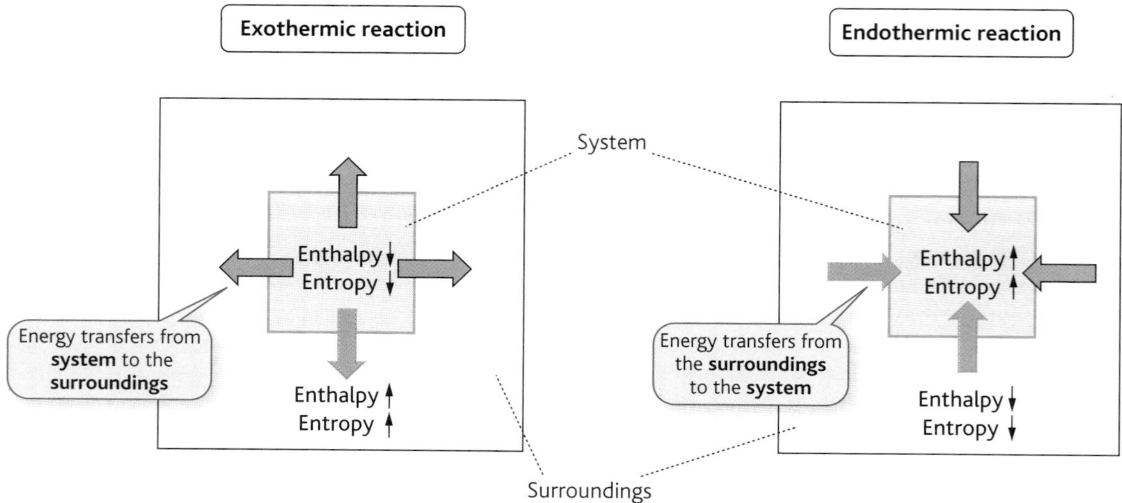

Figure 13.17 The variation in enthalpy and entropy of a system and its surroundings during an exothermic and an endothermic reaction. The black arrows indicate whether the enthalpy and entropy are increasing (↑) or decreasing (↓).

Notice that temperature is always measured in the SI unit of kelvin (K).

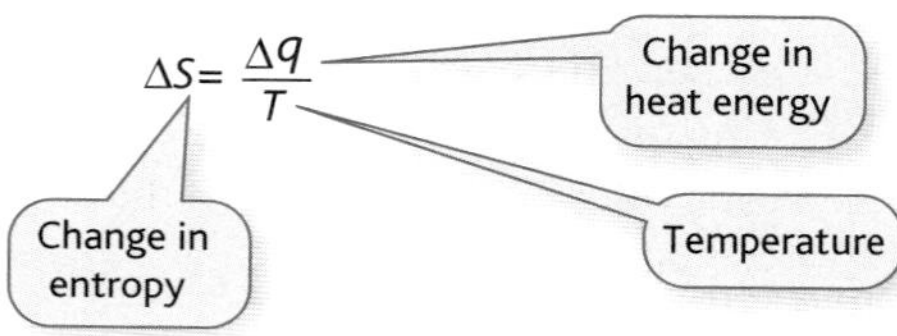

So, for example, if 50 kJ of heat energy is transferred to a system at 298 K, the entropy of the system will increase as follows:

$$\Delta S = \frac{\Delta q}{T}$$
$$= \frac{+50 \text{ kJ}}{298 \text{ K}}$$
$$= +0.1678 \text{ kJ K}^{-1}$$

Note that it is standard practice to quote entropy values in J K^{-1} and not kJ K^{-1}. Therefore, we need to multiply our result by 10^3 (that is, 1000) to change from kJ to J. So, the change in entropy is (0.1678×10^3) or +167.8 J K^{-1}.

Self-check 13.6 A gecko, basking in the sun at 30 °C, absorbs 4.932 kJ of energy in an hour. What entropy change does the gecko experience in this time? (Hint: Temperature (K) = Temperature (°C) + 273)

The influence of temperature on entropy

The magnitude (size) of the change in entropy of a system depends on the **temperature** of the system, as represented by the presence of the term T in the relationship $\Delta S = \frac{\Delta q}{T}$.

What this relationship tells us is that, at low temperatures, the change in entropy, ΔS, is *greater* than at higher temperatures. (As the temperature, T, increases, the overall term $\Delta q/T$ gets smaller.)

For example, if energy transferred to the system, Δq, is 40 kJ at 278 K (5 °C), the change in entropy is:

$$\frac{\Delta q}{T} = \frac{40 \text{ kJ}}{278 \text{ K}} = +0.144 \text{ kJ K}^{-1} = +144 \text{ J K}^{-1}$$

By contrast, if the temperature is higher (say, 298 K (25°C)) the change in entropy is:

$$\frac{\Delta q}{T} = \frac{40 \text{ kJ}}{298 \text{ K}} = +0.134 \text{ kJ K}^{-1} = +134 \text{ J K}^{-1}$$

Notice how the increase in S is relatively smaller at a high temperature compared to a lower temperature. This is because the transfer of energy to the system has *less* impact on the entropy of the system when the temperature of the system is high, compared to when the temperature is lower.

In essence, it's like shouting out to a friend across a crowded bar, and shouting out across a quiet library: your voice barely registers in the already noisy bar, but is highly conspicuous in the virtually silent library.

- ***The entropy change associated with a reaction depends upon the temperature at which the reaction is occurring.***

Self-check 13.7 What entropy change occurs when 50 kJ of energy is transferred to a large vat of water at (a) 0 °C and (b) 100 °C?

So far, we have focused on the entropy of the *system* at different temperatures. However, the entropy of the *surroundings* varies in a similar way, specifically according to the relationship[16]:

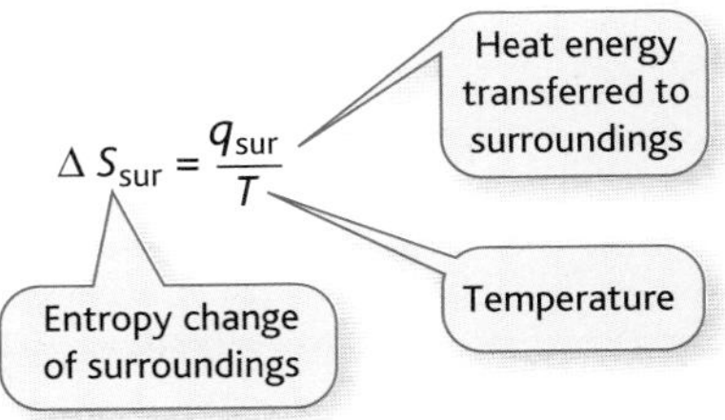

16 The subscript 'sur' denotes 'surroundings'

EXAMPLE

A human loses heat to their surroundings at a rate of about 6 kJ per minute at a temperature of 20°C. What is the entropy change of the surroundings in one hour as a result of this transfer of energy?

First, we calculate the energy transferred to the surroundings in an hour. There are 60 minutes in an hour, so (6 kJ × 60) or 360 kJ are transferred to the surroundings in an hour.

Entropy is quoted in units of J rather than kJ, so we multiply by 10^3 to obtain the transfer of energy in J: $360 \times 10^3 = 3.6 \times 10^5$ J per hour.

Remember to convert temperature from °C to K: 20 °C = 293 K.

Now, $\Delta S = \frac{q_{sur}}{T} = \frac{3.6 \times 10^5 \text{ J}}{293 \text{ K}} = +3871 \text{ J K}^{-1}$

So, a human increases the entropy of their surroundings by about 3871 J K^{-1} in an hour.

We see on p. 425 how $q = H$ (the enthalpy) at constant pressure. So, when the surroundings are at constant pressure (which are the normal conditions for chemical and biochemical reactions), we can write that:

$$\Delta S = \frac{\Delta H_{sur}}{T}.$$

When using this relationship, we use the units J K^{-1} mol^{-1}, rather than just J K^{-1}, as explained in Box 13.6.

Spontaneous reactions and entropy

The overall entropy change accompanying a chemical reaction is the sum of the entropy change of the system, and the entropy change of the surroundings:

$$\Delta S_{total} = \Delta S_{system} + \Delta S_{surroundings}$$

For a chemical reaction to be spontaneous, the **total** change in entropy, ΔS_{total}, must be **positive**: there *must* be an overall **increase** in entropy.

This principle is stated by what is called the second law of thermodynamics: **'for a process to occur spontaneously, the entropy of the thermodynamic universe increases'**[17]. Let's consider this statement for a moment. If the entropy of

BOX 13.6 A note on units

The change in enthalpy, ΔH, is measured in units of joules per mole, J mol^{-1}

Temperature is measured in units of kelvin, K.

$$\Delta S = \frac{\Delta H_{sur}}{T}$$

So the units of entropy when using this relationship are:

$$\Delta S = \frac{\Delta H_{sur}}{T} = \frac{\text{J mol}^{-1}}{\text{K}} = \text{J K}^{-1}\text{mol}^{-1}$$

17 The 'entropy of the universe' is a different way of saying S_{total}.

the universe (S_{total}) must increase for a reaction to be spontaneous, then the *change* in entropy, ΔS_{total}, must be greater than 0. (If ΔS is < 0, then the entropy would *decrease*; for it to increase, the entropy change must be positive–that is, greater than 0.) So, we can say:

For a reaction to be spontaneous, $\Delta S_{total} > 0$.

We can express this in a slightly different way by saying that a process that leads to an overall increase in entropy is favoured over one that does not.

- ***For a reaction to be spontaneous, the overall entropy change–ΔS_{total}–must be greater than zero. That is, there must be a net increase in entropy.***

EXAMPLE

Let us consider the reaction of oxygen and hydrogen to form water:

$$\tfrac{1}{2}O_2 + H_2 \rightarrow H_2O$$

ΔS_{system} for this reaction is negative: $-89\ \text{J K}^{-1}\ \text{mol}^{-1}$. Considering just the entropy change for the system, we might predict that the reaction would not be spontaneous. However, if we witness the reaction occurring we find that the reaction *is* spontaneous. Why is this? The reaction of oxygen and hydrogen is explosive, such that a huge amount of energy is transferred from the system to the surroundings. Consequently, we see a large, positive value for $\Delta S_{surroundings}$, $+1623\ \text{J K}^{-1}\ \text{mol}^{-1}$.

The $\Delta S_{surroundings}$ term is so large that it compensates for the negative value of ΔS_{system}. Therefore, when we consider ΔS_{total} we find that it is positive:

$$\begin{aligned}\Delta S_{total} &= \Delta S_{system} + \Delta S_{surroundings}\\ &= (-89\ \text{J K}^{-1}\text{mol}^{-1}) + (1623\ \text{J K}^{-1}\ \text{mol}^{-1})\\ &= +1534\ \text{J K}^{-1}\ \text{mol}^{-1}\end{aligned}$$

and the reaction is therefore spontaneous.

So we see how there are two key issues that influence the entropy of a reaction: the temperature at which the reaction happens, and the overall entropy change across both the system and its surroundings. Only by considering both of these issues can we gain a complete picture of the nature of the entropy change associated with a given reaction.

We can use the ideas explored above to explain why water spontaneously freezes at temperatures below 0 °C but *doesn't* spontaneously freeze at temperatures above 0 °C. Look at Box 13.7 to find out how.

These ideas can also help us to rationalize the behaviour of biological systems, which can sometimes seem to contradict the principle that an overall increase in entropy is favoured. For example, biological systems are inherently organized, and

BOX 13.7 Water and ice

Why does water freeze spontaneously below 0 °C but not at higher temperatures? The answer lies in the entropy change associated with the change water → ice.

We need to draw on two relationships to help with our explanation:

1 $\Delta S_{total} = \Delta S_{system} + \Delta S_{surroundings}$

and

2 $\Delta S_{surroundings} = \dfrac{\Delta H_{sur}}{T}$

We also need to remember that, for a reaction to be spontaneous, the total entropy change needs to be positive (i.e. greater than 0).

Let's consider the entropy change associated with the change water → ice at two different temperatures, 10 °C and −10 °C (283 and 263 kelvin).

We're interested in the total entropy change, ΔS_{total}, and we know that $\Delta S_{total} = \Delta S_{system} + \Delta S_{surroundings}$. So we need to determine the values of ΔS_{system} and $\Delta S_{surroundings}$.

ΔS_{system}

We can use tables to look up entropy changes in the system for different chemical reactions. The entropy change in the system, ΔS_{system}, for the change water → ice is −22.0 J K^{-1} mol^{-1}.

$\Delta S_{surroundings}$

The value of $\Delta S_{surroundings}$ depends on temperature: we need to find its value at 10 °C and −10 °C.

We know that $\Delta S_{surroundings} = \dfrac{\Delta H_{sur}}{T}$

The enthalpy change, ΔH, for the reaction water → ice is −6010 J mol^{-1}. Notice that this tells us about the enthalpy from the point of view of the system: the negative sign tells us it is an exothermic reaction during which heat is transferred *from* the system (the water) *to* the surroundings. If 6010 J mol^{-1} is transferred *from the system*, then the enthalpy change for the surroundings, ΔH_{sur} must be + 6010 J mol^{-1}.

So, at 10°C:

$$\Delta S_{surroundings} = \frac{\Delta H_{sur}}{T} = \frac{+6010\ \text{J mol}^{-1}}{283\ \text{K}} = +21.2\ \text{J K}^{-1}\ \text{mol}^{-1}$$

At −10°C:

$$\Delta S_{surroundings} = \frac{\Delta H_{sur}}{T} = \frac{+6010\ \text{J mol}^{-1}}{263\ \text{K}} = +22.9\ \text{J K}^{-1}\ \text{mol}^{-1}$$

We can now calculate ΔS_{total} at both temperatures.

At 10°C:

$$\Delta S_{total} = \Delta S_{system} + \Delta S_{surroundings}$$
$$= -22.0\ \text{J K}^{-1}\ \text{mol}^{-1} + 21.2\ \text{J K}^{-1}\ \text{mol}^{-1}$$
$$= \mathbf{-0.8\ J\ K^{-1}\ mol^{-1}}$$

At − 10°C:

$$\Delta S_{total} = \Delta S_{system} + \Delta S_{surroundings}$$
$$= -22.0\ \text{J K}^{-1}\ \text{mol}^{-1} + 22.9\ \text{J K}^{-1}\ \text{mol}^{-1}$$
$$= \mathbf{+0.9\ J\ K^{-1}\ mol^{-1}}$$

Next, we look at these numbers and notice that the total entropy change at the two temperatures has a different sign.

At 10 °C the total entropy change is negative (−0.8 J K^{-1} mol^{-1}), so the reaction water → ice cannot happen spontaneously at 10 °C. ΔS must be positive for a spontaneous change.

However, at −10 °C the total entropy change *is* positive (+0.9 J K^{-1} mol^{-1}); the reaction water → ice *can* happen spontaneously at −10 °C because ΔS is positive.

So entropy, and the second law of thermodynamics, is sufficient to explain why water freezes spontaneously at some temperatures, and not at others.

tend towards a state of stability. If they weren't, we would be unable to develop in the carefully controlled and tightly regulated way that we do. But surely this tendency towards organization violates the principle that life needs to tend towards a state of disorganization to ensure that entropy increases, as required by the second law of thermodynamics?

There are several ways in which biological systems operate to ensure that they *do* obey the second law of thermodynamics to the greatest extent possible. First, biological systems typically transfer energy (as heat) to their surroundings. (For example, all warm-blooded mammals are continually radiating heat into their surroundings.) This radiation of heat leads to an increase in the entropy of the surroundings (as we see above, where a change in entropy is related to the heat transferred to the surroundings). This transfer of energy–and the resulting increase in energy of the surroundings–helps to offset the low entropy of the organized biological system that has radiated the heat in the first place. (Remember: we only really care about the overall entropy of the 'universe'–the system *and* the surroundings. It doesn't matter if the entropy of either the system or surroundings is low, as long as, when taken together, the overall entropy tends to increase–that is, the change is greater than zero.)

But what about the folding of proteins–something that is central to proteins operating as they should, despite requiring a lowering of the entropy of the system in the process? Again, the key is to look at the bigger picture–the overall entropy change across both the system and surroundings.

In the cell, proteins are surrounded by water molecules, which form fluid hydrogen-bonded networks, as described in Chapter 4. Hydrogen-bonded water molecules are inherently more organized (and therefore possess lower entropy) than water molecules that are free, and not hydrogen-bonded. (This is because hydrogen bonding acts to tether water molecules to one another, restricting their movement.) When a polypeptide chain folds, the water molecules that surround it become more disorganized, as illustrated in Figure 13.18. As a result, the entropy of the water molecules surrounding the folded polypeptide *increases* during the folding process; this increase in entropy acts to offset the decrease in entropy associated with the folding of the polypeptide itself.

If a change in entropy of the surroundings is still insufficient to make the overall change in entropy–across both the system and its surroundings–positive, the process will be non-spontaneous. We explore the implication of this statement in Section 13.5.

13.5 Spontaneous versus non-spontaneous processes: how much energy do we need?

We discuss spontaneous processes several times throughout this chapter: we see how the transfer of energy as heat, from a hot entity to a cold one, happens spontaneously; we see how there must be an overall increase in entropy for a reaction to be spontaneous at all. But why is the notion of spontaneity important? Why should we care whether a process is happening or not? The answer lies in energy and, specifically, the question 'Just how much energy do we need to make things happen as they should?'

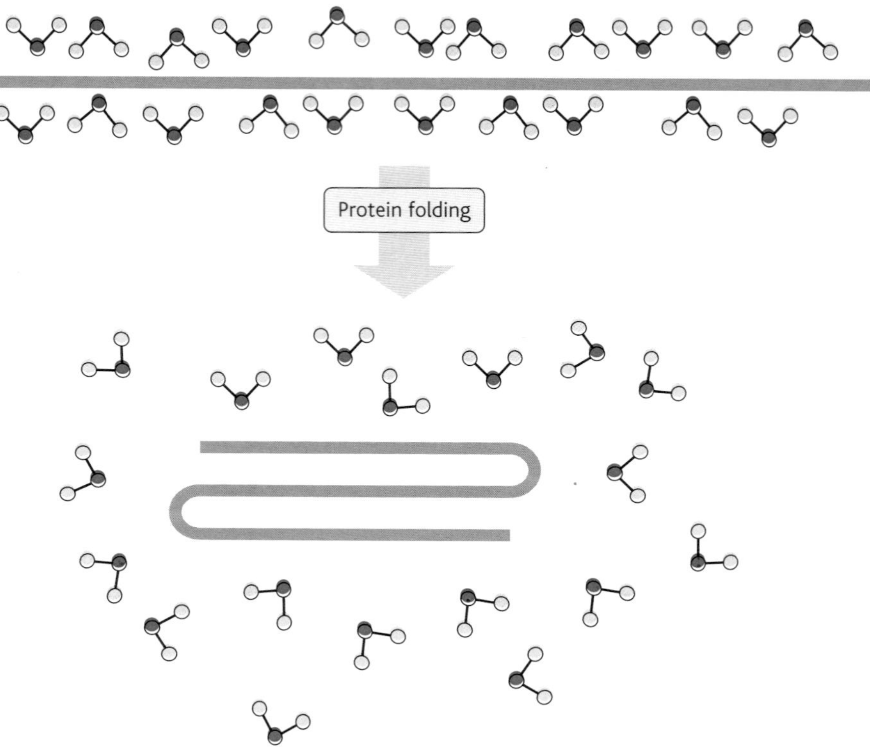

Figure 13.18 When a protein folds, the water molecules that were previously arranged around the unfolded chain in an organized manner become relatively disorganized, leading to an increase in entropy of the surroundings.

A spontaneous process is one that occurs without an input of energy to keep it going once started. We see in everyday life examples of things that happen spontaneously. If we leave a car at the top of a slope with its handbrake off, the car will spontaneously roll to the bottom of the slope. If we store liquid water in a freezer below 0 °C it will spontaneously freeze. Both processes are spontaneous because they occur without any additional help from us, or other forms of intervention. That is, they do not require an input of energy.

Some spontaneous events, however, occur so slowly without intervention that they *appear* not to occur spontaneously. The rate at which glucose is converted to carbon dioxide and water at room temperature is so slow we might wrongly conclude that it is not a spontaneous process. This is an important point that bears stressing: whether a reaction happens spontaneously or not tells us nothing about the *rate* at which the reaction occurs–whether it happens quickly or slowly. Once started, however, a spontaneous reaction proceeds to completion without any external intervention (i.e. a transfer of energy from surroundings to the system) even if the reaction itself occurs extremely slowly.

We learn more about the rate of chemical reactions in Chapter 14.

Spontaneous processes happen in *one direction* only. If we leave a car at the *bottom* of a slope with its handbrake off, it will *not* spontaneously roll to the top of the hill. Ice in a freezer won't spontaneously melt to form liquid water. Such reverse processes are non-spontaneous: they require some kind of external intervention to make it occur. (In the example of the car at the bottom of the slope, we would need to start the engine and drive it to the top of the slope, for example, which requires the input of much energy via the liquid fuel in its tank.)

So how do we reverse a spontaneous process, or simply make a non-spontaneous process happen at all? As we see in the example of moving a car from the bottom to the top of a slope, reversing a spontaneous process–or making a non-spontaneous process happen–requires an input of energy.

- *A spontaneous process proceeds to completion without an input of energy.*
- *A non-spontaneous process requires energy to drive it forward.*

We know from looking at a waterfall, or dishwater going down the plughole, that water spontaneously flows downhill, from a high elevation to a lower elevation, without any work having to be done on it (i.e. without us providing energy to drive its movement). If we want to move water from a low elevation to a higher elevation, however, we must do **work** on the water: we need to provide energy to pump the water from the low elevation up to the higher elevation.

Many everyday processes are non-spontaneous and require an input of energy. Consider a refrigerator. We use a refrigerator to keep food cool. When we place food inside the fridge, heat spontaneously transfers from the relatively warm object to the colder air of the fridge that surrounds it. This transfer of energy from hot to cold is a spontaneous process. In time, however, the air inside the fridge warms up as energy is transferred from the food to the air surrounding it. But we don't want the air in the fridge to warm up to the extent that thermal energy starts to flow back from the air in the fridge to the food; we need to find a way of getting thermal energy to flow out of the fridge. This requires energy, as explained in Box 13.8.

Similarly, if a biochemical process is not spontaneous, it requires the input of energy to drive it forward. Without the supply of energy, such reactions simply won't happen. But how can we tell which reactions are spontaneous–and can happen without an external helping hand–and which ones are not? We discover the answer in the final section of this chapter.

While each of the instances of heat transfer we see in Box 13.8, from food to fridge interior, fridge interior to coolant, and coolant to fridge exterior, are spontaneous and occur from hot to cold, they all depend on a non-spontaneous process–the compression of the coolant, which requires energy from an electric motor–to make them happen.

BOX 13.8 How does a refrigerator keep its cool?

The proper function of a fridge relies on the way that a gas cools as it expands, and warms up as it is compressed.

Outside a fridge, at its rear, is a system of pipes that contain a 'coolant'. The coolant is a low-boiling point liquid. Most of the time, the coolant is a gas. This gas goes through cycles of being compressed (having work done on it) by an electric motor and then expanding. When the coolant is compressed, it heats up; when the motor stops, the coolant expands again, and cools. But how does this help the inside of a fridge to keep cool?

Figure 13.19 shows how. The expanded coolant is cooler than the air inside the fridge. Consequently, energy spontaneously flows from the warm air in the fridge to the colder coolant, enabling the air inside the fridge to become cooler as it loses energy.

The motor then moves the coolant away from the fridge interior to the rear, where it is compressed so that it warms up. The hot coolant passes through coils on the exterior of the fridge; these pipes are in contact with the outside air (e.g. the air in the kitchen). The coolant is compressed to the extent that it is hotter than the air surrounding the fridge, so energy now flows spontaneously from the hot coolant to the cooler air around the fridge. (We can feel this heat transfer for ourselves if we touch the coils at the back of a fridge: they usually feel quite hot.)

BOX 13.8 **Continued**

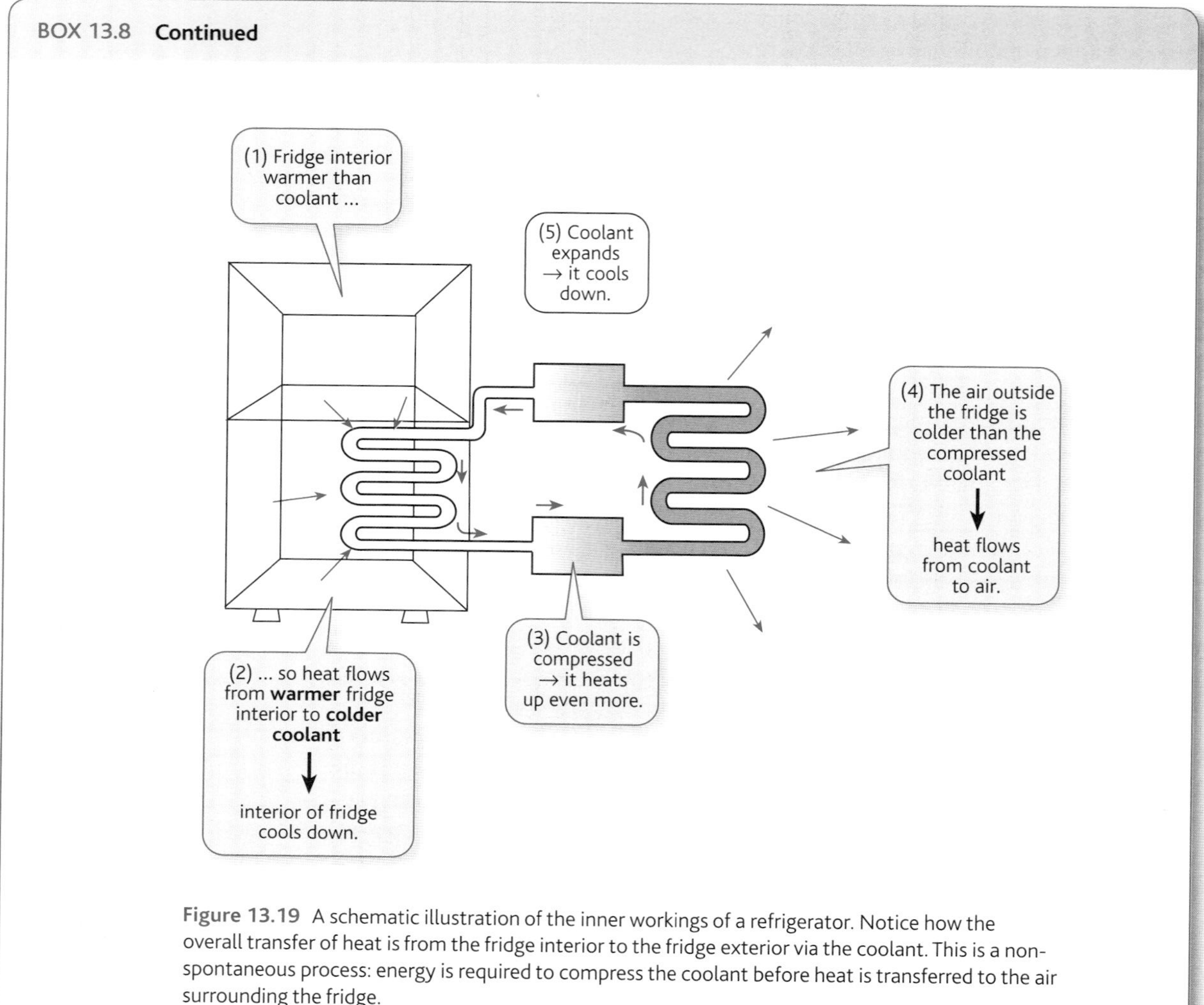

Figure 13.19 A schematic illustration of the inner workings of a refrigerator. Notice how the overall transfer of heat is from the fridge interior to the fridge exterior via the coolant. This is a non-spontaneous process: energy is required to compress the coolant before heat is transferred to the air surrounding the fridge.

13.6 Gibbs free energy: the driving force of chemical reactions

We are now in a position to return to the question posed in the title of this chapter, 'What makes reactions go?'. This is the question that is of most importance when it comes to maintaining biological systems in a fully functioning state.

We see in Section 13.2 how energy transfer happens as work and as heat. What we're really interested in, however, is just the energy that is free to do work. This is the energy that a reaction, a process, or the cells in our body can harness to carry out all the other processes required to maintain life–those processes that aren't spontaneous, but which require a supply of energy to make them happen. However, not all the chemical energy generated by a reaction (equal to the change in enthalpy, ΔH)

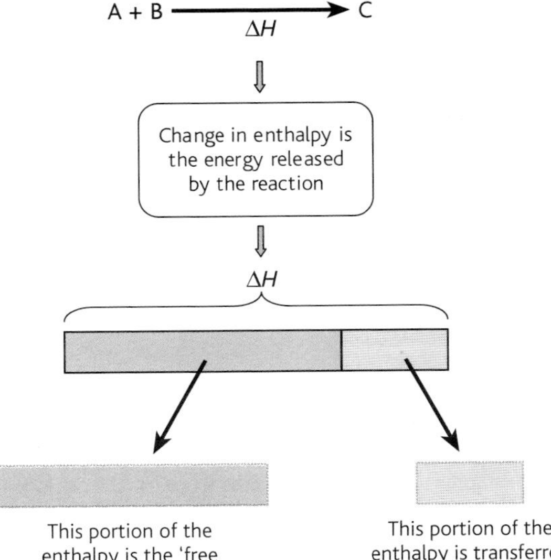

Figure 13.20 The Gibbs free energy is the proportion of energy from a reaction that can be harnessed to do useful work. The other portion of the energy is used to increase the entropy. Therefore, not all the energy released from a reaction is free to do useful work.

is free to do work. We have to take into account the proportion of energy transferred during the reaction as heat, which is used to alter the overall entropy[18].

The energy that is truly free to do work is what is called the **Gibbs free energy**, represented by the symbol **G**. ΔG is the *change* in Gibbs free energy associated with a particular reaction; the value of ΔG represents the energy from the reaction of reactants $\rightarrow$ products that is free to do work. Formally, ΔG is 'the energy available for reaction having adjusted for the entropy changes of the surroundings'. Look at Figure 13.20, which illustrates this point. Notice how, for the reaction denoted in the figure, not all of the energy released by the reaction is free to be harnessed to do something useful: part of the energy has to be set aside for altering the entropy of the surroundings[19].

For example, the Gibbs free energy change for the combustion of carbon in oxygen to form carbon dioxide:

$$C(s) + O_2(g) \rightarrow CO_2(g)$$

under standard conditions is -394.4 kJ mol^{-1}. For every mole of carbon that is burnt, 394.4 kJ of useful energy is available, after adjusting the entropy of the surroundings. This energy can then be used to do work, e.g. to drive other chemical reactions.

- ***The change in Gibbs free energy for a reaction, ΔG, is the energy that is free to be harnessed to do something useful—that is, to do work.***

18 Remember: the total (internal) energy of a system comprises work energy, w, and heat energy, q. Only work energy can be harnessed to do something 'useful'; heat energy can only be used to increase entropy.

19 Remember: entropy is related directly to the energy transferred to the surroundings as heat, as we see on p. 439.

We've just said that the energy from a reaction that is free to do work is the portion of the energy that is *not* required to drive a change in entropy. The change in Gibbs free energy is related to the change in enthalpy and change in entropy as follows:

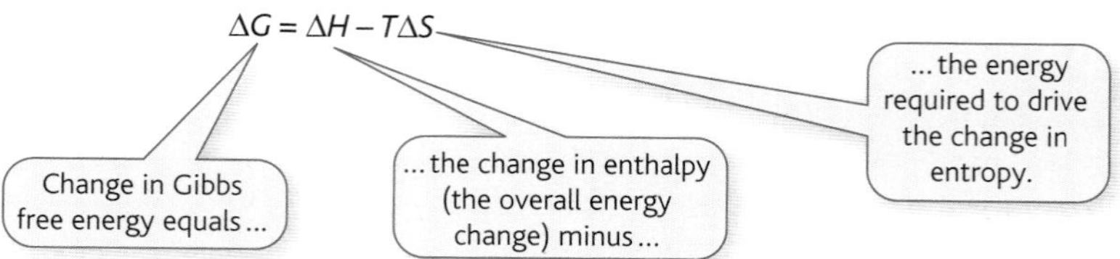
ΔG = ΔH – TΔS
Change in Gibbs free energy equals...
... the change in enthalpy (the overall energy change) minus...
... the energy required to drive the change in entropy.

So, ΔG is the energy available for a reaction, once some of the energy has adjusted the entropy of the surroundings.

We measure the value of ΔG in units of **joules per mole, J mol^{-1}** (though we typically divide values by a factor of 10^3 and quote in **kJ mol^{-1}**).

EXAMPLE

Let us consider the combustion of hydrogen:

$$2\,H_2(g) + O_2(g) \rightarrow 2\,H_2O(l)$$

The enthalpy change for this reaction (under standard conditions) is −286 kJ per mole of H_2. By contrast, the Gibbs free energy change for the reaction, $\Delta G°$, is slightly smaller at −237 kJ mol^{-1}. What energy is used to drive the change in entropy associated with the reaction?

We use the relationship:

$$\Delta G° = \Delta H° - T\Delta S°$$
$$\Delta G° = -237 \text{ kJ mol}^{-1} \text{ and } \Delta H° = -286 \text{ kJ mol}^{-1}$$

So substituting in these values:

$$-237 \text{ kJ mol}^{-1} = -286 \text{ kJ mol}^{-1} - T\Delta S°$$

Rearrange terms to isolate the term we're interested in: $-T\Delta S°$.

$$-237 \text{ kJ mol}^{-1} + 286 \text{ kJ mol}^{-1} = -T\Delta S°$$
$$+49 \text{ kJ mol}^{-1} = -T\Delta S°$$

So $T\Delta S° = -49$ kJ mol^{-1}

This is the 'unfree' energy required to drive the change in entropy associated with the reaction. Notice how this value is negative, so the entropy change for the reaction must also be negative–that is, there is a *reduction* in entropy during this reaction. Look again at the reaction scheme:

$$2\,H_2(g) + O_2(g) \rightarrow 2\,H_2O(l)$$

Notice how there are two ways in which this reduction occurs:

1. A change from higher-entropy gas molecules to lower-entropy liquid;
2. A reduction in the relative number of species present, from three reactants to two products.

We see earlier how the standard enthalpy change, $\Delta H°$, for the combustion of glucose–the reaction from which our bodies derive energy to drive other reactions:

$$C_6H_{12}O_6(s) + 6\,O_2(g) \rightarrow 6\,CO_2(g) + 6\,H_2O(l)$$

is –2808 kJ mol^{-1}. By contrast, the standard Gibbs free energy change, $\Delta G°$, is –2878 kJ mol^{-1}. Again, let's consider the energy that is tied up with the change in entropy for the reaction, using the same strategy as above:

$$\Delta G° = \Delta H° - T\Delta S°$$
$$-2878 \text{ kJ mol}^{-1} = -2808 \text{ kJ mol}^{-1} - T\Delta S°$$
$$-70 \text{ kJ mol}^{-1} = -T\Delta S°$$
$$T\Delta S° = +70 \text{ kJ mol}^{-1}$$

This time we see that the term $T\Delta S°$ is positive, such that the entropy change associated with the reaction is a positive one. Again, by inspecting the reaction scheme, we see a couple of reasons why the entropy might be positive:

1. There is a net change from a low entropy solid to a higher-entropy liquid. (The six gas molecules appear on both sides of the scheme, so can be ignored.)
2. There is an increase in the relative number of species present, from seven reactants to twelve products.

The Gibbs free energy of spontaneous reactions

We mention above how spontaneous reactions occur without an input of energy, whereas non-spontaneous reactions need an input of energy to drive them forward. What does this mean for the Gibbs free energy change for a reaction? For a reaction to be spontaneous, ΔG must be **negative**: the products must have **lower** Gibbs free energy than the reactants, as depicted in Figure 13.21. Look at this figure, and notice how the difference in Gibbs free energy between the products and reactants is essentially 'free' energy, that is, energy that can be released by the system to do work. (The '$T \times \Delta S$' component is not available; ΔG is therefore a net energy.)

Spontaneous reactions that release excess Gibbs free energy are called **exergonic** ('outward energy') reactions. Any reactions that require an input of energy are called **endergonic** ('inward energy') reactions[20].

If the value of ΔG is positive, however, energy must be **supplied** to the system for the reaction to happen: a reaction with a positive ΔG does *not* occur spontaneously.

For example, during the process of photosynthesis, plants convert carbon dioxide and water into glucose:

$$6\,CO_2 + 6\,H_2O \rightarrow C_6H_{12}O_6 + 6\,O_2 \qquad \Delta G° = +\,2870 \text{ kJ mol}^{-1}$$

This process has a positive ΔG, and so does not occur spontaneously. As with all non-spontaneous processes, energy must be continually provided for the reaction

20 You may also see exergonic spelt 'exogenic', and endergonic spelt 'endogenic'.

(a)

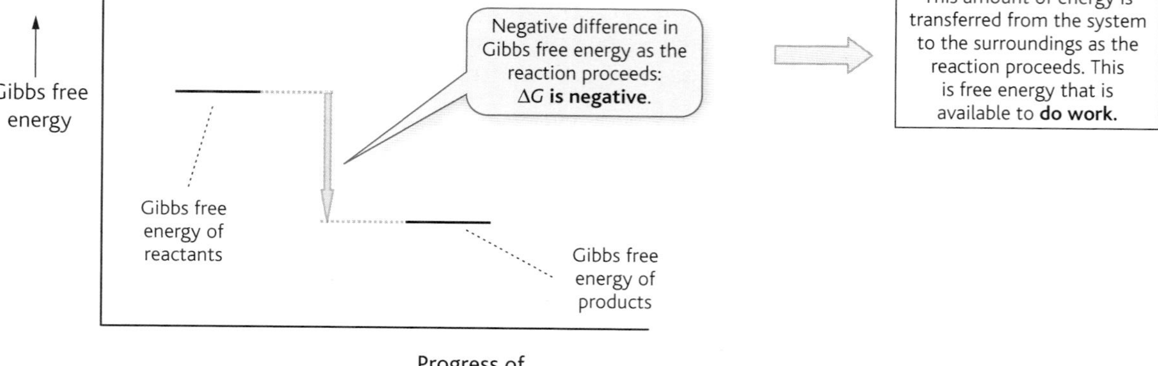

(b)

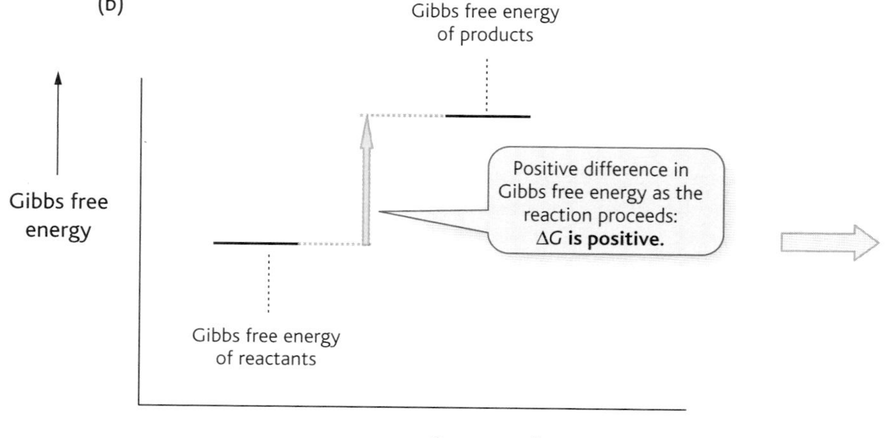

Figure 13.21 (a) When ΔG for a reaction is negative, free energy transfers from the system to its surroundings, and is available to do work. (b) When ΔG for a reaction is positive, free energy must be supplied by the surroundings to the system. There is no free energy available to do work.

to occur. However, plants have evolved a ready solution to this problem: they get the energy they require by absorbing photons of sunlight. (We see in Chapter 2 how sunlight comprises electromagnetic radiation of various wavelengths; plants are able to utilize the energy that sunlight possesses to drive the endergonic process of photosynthesis.)

The Gibbs free energy, $\Delta G°$, for a reaction also tells us useful information about the direction in which an equilibrium reaction occurs, as we see in Chapter 16.

By contrast, we see above how the metabolism of glucose has a standard Gibbs free energy change of -2878 kJ mol^{-1}. This is a strongly exergonic spontaneous reaction, which releases a large amount of free energy—energy that can then be used to do work in the cell (for example, powering important reactions that are not spontaneous and need an input of energy to make them happen), as we see below.

Let us see how we can use the Gibbs free energy change for a reaction to predict whether the reaction will happen spontaneously or not.

EXAMPLE

Ribonuclease is an enzyme that digests RNA into small fragments. The standard enthalpy change for the folding of ribonuclease at pH 6.0 is –209 kJ mol^{-1}, and the standard entropy change is –554 J K^{-1} mol^{-1}. Will this enzyme fold spontaneously at a temperature of 298 K?

We use the relationship $\Delta G° = \Delta H° - T\Delta S°$ to determine the value of $\Delta G°$ for the folding process.

$$\Delta G° = \Delta H° - T\Delta S°$$

We substitute in the values for $\Delta H°$, $\Delta S°$, and T given above:

$$\Delta G° = -209 \text{ kJ mol}^{-1} - (298 \text{ K} \times -554 \text{ J K}^{-1} \text{ mol}^{-1})$$
$$= -209 \text{ kJ mol}^{-1} - (-165\,092 \text{ J mol}^{-1})$$

Notice that the K and K^{-1} units cancel each other.

We write the second term as kJ mol^{-1} by dividing by 1000 to then complete the calculation:

$$\Delta G° = -209 \text{ kJ mol}^{-1} - (-165 \text{ kJ mol}^{-1})$$
$$= -209 \text{ kJ mol}^{-1} + 165 \text{ kJ mol}^{-1}$$
$$= -44 \text{ kJ mol}^{-1}$$

Remember: x – (–y) is the same as writing x + y.

Therefore, the standard Gibbs free energy change for this reaction is –44 kJ mol^{-1}. This is a negative value, telling us that this process occurs spontaneously.

Self-check 13.8 The breakdown of glucose in muscle tissue during exercise (in the absence of excess oxygen) is called anaerobic glycolysis. The second half of this process can be represented by the following reaction scheme:

$$\text{Glyceraldehyde-3-phosphate} + \text{Pi} + 2\text{ ADP} \rightleftharpoons \text{Lactate} + 2\text{ ATP} + H_2O$$

The standard enthalpy change for this reaction, $\Delta H°$, is –63.0 kJ mol^{-1} and the standard entropy change, $\Delta S°$, is +220 J K^{-1} mol^{-1}. Assuming that the reaction is taking place at 310 K, what is the standard Gibbs free energy change for this reaction?

What is the ultimate driver of spontaneity?

In Section 13.4 we see how a spontaneous reaction requires the *entropy change* for the reaction to be positive. In this section we see how a spontaneous reaction requires the *Gibbs free energy change* for the reaction to be negative. Which of these is the ultimate driver of spontaneity? The answer is that both are important:

- The criterion for spontaneity at the level of the **system** is that ΔG for the reaction must be negative.
- The criterion for spontaneity at the level of the chemical **universe** (the system *and* its surroundings) is that ΔS must be positive.

A system needs a reaction to have a negative ΔG in order for it to be able to do something useful as a consequence of the reaction–that is, to be able to do work.

The universe–the system and its surroundings–must show an overall increase in entropy (a positive value for ΔS_{total}) to satisfy the second law of thermodynamics: 'for a spontaneous process, the entropy of the thermodynamic universe increases'.

Self-check 13.9 State whether the following reactions are spontaneous or non-spontaneous. If the reactions are spontaneous, are they spontaneous at the level of the thermodynamic universe or at the level of the thermodynamic system?

(a) $H_2(g) + \frac{1}{2} O_2(g) \rightarrow H_2O(g)$ $\Delta G° = -457.2$ kJ mol^{-1} at 298 K

(b) $H_2O(l) \rightarrow H_2O(s)$ $\Delta G° = +0.6$ kJ mol^{-1} at 298 K

Gibbs free energy and cell metabolism

In the physical world, energy utilized as work can be used to drive a turbine in a power station, or a piston in a car engine. In the biological and chemical worlds, the energy from exergonic (spontaneous, energy-releasing) reactions can be used to drive endergonic (non-spontaneous, energy-consuming) reactions. The work that the free energy is used to do is to enable non-spontaneous reactions to occur.

Animals, including humans, get the energy they require to drive the endergonic reactions of the cell by oxidizing high-energy organic compounds such as carbohydrates and fats. The most important source of energy for humans is **glucose**. As we see above, a great deal of energy is released when glucose is oxidized completely to form water and carbon dioxide:

$$C_6H_{12}O_6 + 6\,O_2 \rightarrow 6\,CO_2 + 6\,H_2O \qquad \Delta G° = -\,2878 \text{ kJ mol}^{-1}$$

The reaction as written here does not occur in one step. Instead, about thirty separate reactions are involved when glucose is oxidized to carbon dioxide and water, in the biochemical process called **cellular respiration**. The 2878 kJ of energy released per mole of glucose consumed is not released in one go, but rather a little at a time–spread out over these thirty reactions that make up the process as a whole.

Coupling biochemical reactions

The array of reactions occurring in a cell constitutes the overall process known as **metabolism**. Metabolic reactions fall into two groups. The first group are called **catabolic** reactions and break down high-energy compounds into simpler molecules. Most of these reactions are **exergonic** and so occur *spontaneously* without additional energy.

A second group, called **anabolic** reactions, build up more complicated molecules (such as proteins) from simpler subunits. Most of these reactions are non-spontaneous endergonic reactions, and require a constant supply of energy to drive them forward.

So, for metabolism to occur, there must be the right balance between exergonic catabolic reactions and endergonic anabolic reactions. There must be enough energy yielded by exergonic reactions to drive the non-spontaneous endergonic reactions. If the balance is wrong, and not enough energy is put in, the endergonic reactions will not happen, and the metabolic processes will grind to a halt.

One of the most important coupled reactions in biological systems is the formation of adenosine triphosphate (ATP), coupled to the oxidation of glucose.

We see above how the oxidation of glucose is highly exergonic (with a value of $\Delta G°$ of -2878 kJ mol^{-1}). While some of the energy released during the oxidation of glucose is transferred in the form of heat (which the body needs to maintain its constant temperature), a proportion of the remainder is used to drive the synthesis of ATP:

$$ADP(aq) + Pi(aq) + H^+(aq) \rightarrow ATP(aq) + H_2O(l)\ \Delta G^{\circ\prime} = +\ 30.5\ kJ\ mol^{-1}$$

'Pi' denotes the phosphate group, PO_4^{3-}. You may also see it represented by 'P' or Ⓟ.

Notice that the synthesis of ATP is endergonic; energy must be supplied to make this non-spontaneous reaction happen. Because the Gibbs free energy yielded by the oxidation of glucose is much greater than the Gibbs free energy *consumed* during the synthesis of ATP, the cellular oxidation of one mole of glucose molecules can be used to drive the synthesis of ATP. In practice, around 32–36 moles of ATP are synthesized per mole of glucose oxidized.

We learn more about the synthesis of ATP in Chapter 18.

Once synthesized, ATP is a valuable store of free energy, which is exploited in biological systems to drive other endergonic reactions, as explained in Box 13.9.

So we see throughout this chapter how carefully balanced the energy requirements of a biological system are: with thermodynamic rules determining whether reactions happen spontaneously or not, and the cell coupling energy-yielding spontaneous reactions with energy-requiring non-spontaneous ones in an intricate–but ultimately efficient and impressive–way.

BOX 13.9 ATP: the cell's energy currency

One of the most important compounds in any cell is adenosine triphosphate (ATP). ATP is the universal energy currency in all biological systems on earth.

Cells need a constant supply of energy to drive processes that feature endergonic reactions. For example, muscle cells need energy to contract; nerve cells need energy to generate nerve impulses.

A molecule of ATP comprises the purine base adenine, the sugar ribose, and three phosphate groups:

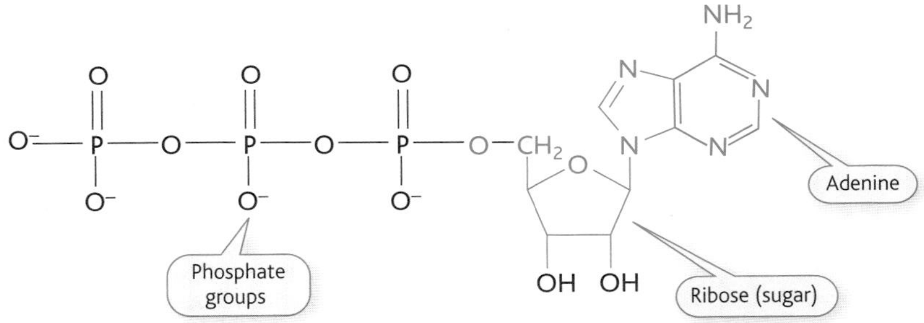

Because 30.5 kJ is required to *synthesize* one mole of ATP according to the reaction:

$$ADP + Pi + H^+ \rightarrow ATP + H_2O$$

30.5 kJ of energy is *released* when one mole is hydrolyzed:

$$ATP + H_2O \rightarrow ADP + Pi + H^+$$

That is, the Gibbs free energy change for the hydrolysis of ATP, $\Delta G^{\circ\prime}$, is $-30.5\ kJ\ mol^{-1}$.

This amount of energy is roughly the amount needed to drive most of the reactions in cells that require energy to be supplied. So ATP acts as an invaluable source of Gibbs free energy, which is harnessed to drive many essential biochemical reactions.

But what is it about ATP that makes it such a good energy reserve? Part of the answer lies in the net effect of cleaving a high-energy phosphate ester (O–P) bond during the process of ATP hydrolysis, as illustrated in Figure 13.22.

BOX 13.9 **Continued**

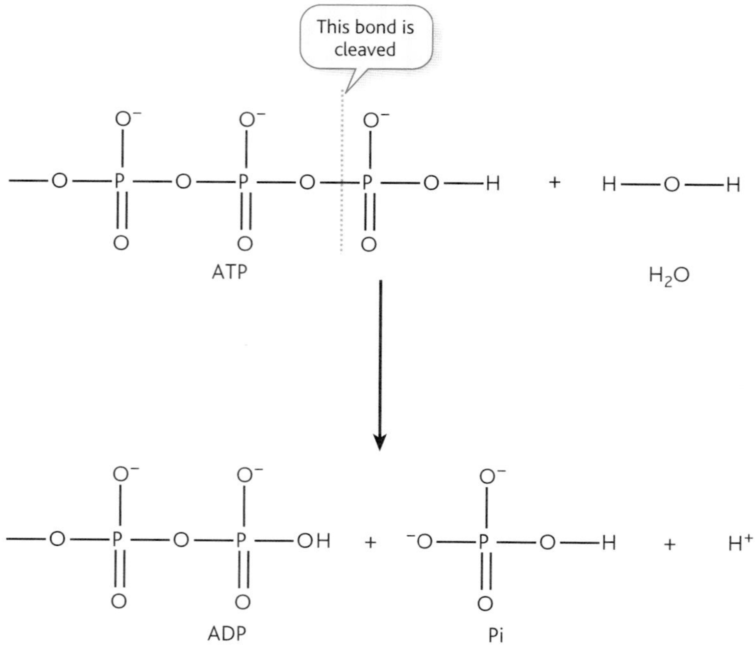

Figure 13.22 During ATP hydrolysis, a high-energy phosphate ester bond is cleaved, as denoted here in blue.

This process is exothermic, with a value of $\Delta H^{\circ\prime}$ of -20.1 kJ mol^{-1}, and has a positive entropy; $\Delta S^{\circ\prime} = +33.5$ J K^{-1} mol^{-1}. Both of these values help to ensure that the value of ΔG is negative, such that there is free energy available to perform work in the cell.

The large negative free energy of hydrolysis tells us that products of the reaction are much more stable than the reactants. So why should hydrolysis of these phosphate ester bonds release so much energy? ATP is structurally unstable partly because of the repulsion between its ionized oxygen atoms, which are all in close proximity. Hydrolysis of the terminal phosphate group produces a free phosphoric acid molecule, which is stabilized by resonance and relieves some of the repulsion forces in ATP. Both of these effects contribute to the free energy.

A similar reduction of repulsive forces within ATP is achieved if the middle phosphate ester bond is hydrolyzed; in this case, ATP forms AMP and diphosphoric acid. By contrast, AMP is much more stable than ATP; hydrolysis of the final phosphate group yields little in the way of free energy.

Another major advantage of ATP as an energy store is that the normal hydrolysis of ATP (in the absence of an enzyme) is very slow, even though it happens spontaneously. This behaviour means that the hydrolysis of ATP can be tightly controlled: it will only happen to an appreciable extent if a suitable enzyme is present. This allows ATP to be available in a kinetically stable form (that is, one that won't disintegrate when we don't want it to), only to release its energy when an enzyme is present to catalyze the process of hydrolysis.

In fact, it is wrong to think that the body has huge stockpiles of ATP ready to be hydrolyzed. In practice, a cell has enough ATP to last it for only a few minutes at any one time. (The 'stockpile' of ATP in the human body at any one time is about 80 g, whereas, even at rest, the daily turnover is huge, at around 80 kg.) Anything that might inhibit its production is fatal to the cell and, potentially, the entire organism. Sodium cyanide is a deadly poison because it inhibits the production of ATP; a sufficient amount will kill a person within a few minutes. Also, cyanide, CN^-, binds irreversibly to the iron of haemoglobin. The binding of cyanide to haemoglobin inhibits the transport of oxygen, leading to a state of hypoxia (that is, oxygen deficiency).

Check your understanding

To check that you've mastered the key concepts presented in this chapter, review the checklist of key concepts below, and attempt the multiple-choice questions available in the book's Online Resource Centre at **http://www.oxfordtextbooks.co.uk/orc/crowe2e/**.

Checklist of key concepts

Energy

- Energy is the capacity something has to do work
- The total energy that a substance has is its **internal energy**, U
- The internal energy has two components: work, w, and heat, q
- Energy cannot be created or destroyed. It can only be **transformed** from one form to another
- Many different types of energy contribute to the internal energy–the main contributors are **kinetic** energy, **potential** gravitational energy, and potential **chemical** energy
- **Kinetic** energy is the energy an object has due to its **motion**
- An object's kinetic energy depends on its mass and velocity
- **Potential gravitational** energy is the energy an object has due to its **position**
- **Potential chemical** energy is the energy a chemical compound has **stored** in its chemical bonds
- The amount of energy stored in a bond is called the **bond energy**
- When a bond is **broken**, an amount of energy equal to the bond energy is **consumed**
- When a bond is **formed**, an amount of energy equal to the bond energy is **liberated**

Energy transfer

- Energy transfer occurs between a **system** and its **surroundings**
- A **system** is a particular thing we're interested in, contained within a boundary
- The **surroundings** are everything else in contact with the system
- There are three types of system: a **closed** system, an **open** system, and an **isolated** system
- Energy is transferred between a system and its surroundings in the form of work and heat

Work and heat

- **Work** is any process that can lift a weight
- **Heat** is the transfer of energy from hot to cold
- Heat and temperature are not the same thing
- A substance's temperature gives a measure of its **energy**: a system with a high temperature has a large amount of energy; a system with a low temperature has a smaller amount of energy
- The transfer of energy as heat–from hot to cold–is a **spontaneous** process

Enthalpy

- The energy change associated with a chemical reaction is called the enthalpy change of reaction, ΔH
- ΔH is the difference between the energy consumed to break bonds and the energy liberated when bonds are formed
- An **exothermic** reaction has a **negative** value for ΔH: heat is transferred from the system to its surroundings (**heat is released**)
- An **endothermic** reaction has a **positive** value for ΔH: heat is transferred from the surroundings to a system (**heat is absorbed**)
- We can determine the enthalpy change for a reaction by using the following relationship: $\Delta H = \Sigma E_{(\text{reactants})} - \Sigma E_{(\text{products})}$
- We can depict the change in energy associated with a reaction by using an **energy diagram**
- We can measure the enthalpy change for an exothermic reaction using a **bomb calorimeter**
- The **enthalpy change of formation** is the enthalpy change when one mole of a compound forms from its constituent elements in their standard states

- The enthalpy change of combustion is the enthalpy change when one mole of a compound burns completely in excess oxygen
- For a reaction whose enthalpy change is negative, the products are more stable than the reactants
- For a reaction whose enthalpy change is positive, the products are less stable than the reactants

Entropy

- Entropy is a measure of the energetic disorder of a system: the greater the entropy, the greater the degree of disorder
- Gases have a larger entropy than liquids, which have a larger entropy than solids
- More disorganized systems have a larger entropy than organized ones: unfolded proteins have a larger entropy than folded ones
- Systems with more components have an inherently larger entropy than those with fewer components
- The entropy of a substance increases as its energy increases
- The extent of the increase in entropy of the surroundings depends on the temperature of the surroundings: the higher the temperature, the smaller the increase in entropy

Spontaneous processes

- For a process to happen spontaneously, there must be an overall increase in entropy
- A spontaneous process only happens in one direction
- If we want to reverse a spontaneous process we must provide a source of energy to drive the process

Gibbs free energy

- The change in Gibbs free energy, ΔG, is the amount of energy yielded by a reaction that is actually free to do work
- The change in Gibbs free energy equals the change in enthalpy minus the energy required to drive a change in entropy: $\Delta G = \Delta H - T\Delta S$
- For a reaction to be spontaneous, ΔG must be negative
- If ΔG is positive, energy must be supplied to the system for the reaction to happen
- Reactions that release Gibbs free energy are called exergonic
- Reactions that require an input of energy are called endergonic
- Catabolic reactions are typically exergonic: they involve the breakdown of large molecules into smaller components
- Anabolic reactions are typically endergonic: they involve the building up of large molecules from simpler subunits
- Anabolic and catabolic reactions can be coupled such that the energy yielded by exergonic reactions can be used to drive endergonic reactions

Kinetics: what affects the speed of a reaction?

INTRODUCTION

Objects made of iron, when left in a damp atmosphere, will rust if unprotected by paint. The process of rusting is a slow chemical reaction between iron, oxygen, and water; it often takes years for a particular piece of iron to rust completely. By contrast, the compound nitroglycerine reacts explosively: it undergoes an extremely fast reaction that occurs in less than one second.

The tens of thousands of chemical reactions known to us have characteristic reaction rates–some very slow, some very fast. But what influences how fast a reaction happens: whether we face an explosion or a reaction that's likely to proceed so slowly that its progress is virtually unnoticeable? And how do we ensure that reactions that *need* to happen can happen fast enough?

In this chapter we explore the factors that control the rate of a chemical reaction. We ask how we can influence the rate of reaction, how different reaction rates impact on biological systems, and how biological systems control the reactions that enable life.

14.1 The rate of reaction

We see in Chapter 13 how the Gibbs free energy of a particular reaction can tell us whether the reaction happens spontaneously or not. However, the Gibbs free energy of a reaction tells us nothing about how *fast* the reaction happens: the **rate** of a reaction[1].

A chemical reaction involving carbon monoxide and oxygen, such as:

$$2\,CO + O_2 \rightarrow 2\,CO_2$$

tells us only part of the story about the chemical changes involved in this particular reaction. It reveals the chemical identity of both the reactants and products, and tells us the stoichiometry of the overall reaction: the relative quantities of each species present in the equation, which are present during the course of the reaction.

We learn more about stoichiometry in Chapter 17.

However, such an equation does not tell us about the rate at which the reaction proceeds.

1 The study of the speed (or rate) of chemical reactions is called **reaction kinetics**.

Knowing the rate of a reaction can be an invaluable tool in helping us to understand how chemical compounds behave when they interact. For example, carbon monoxide is toxic to humans and causes numerous fatalities each year amongst those who live in poorly maintained accommodation. Carbon monoxide is also a chronic poison when inhaled in small amounts, for example in cigarette smoke. This colourless, tasteless gas is formed when fuels such as gas, oil, and wood are burned, but the supply of oxygen is limited:

$$C_3H_8 + 4\,O_2 \rightarrow CO_2 + 2\,CO + 4\,H_2O$$

Faulty household heating appliances can produce sufficient amounts of CO to kill any of a home's occupants. Why is this? The reaction depicted in Figure 14.1(a) is slow enough for the carbon monoxide to be inhaled *before* it can react with oxygen in the air. However, if the reaction were to proceed more quickly, as depicted in Figure 14.1(b), the reaction between the CO produced by the heating appliance and oxygen in the air might proceed quickly enough for most of the CO to be consumed before any occupant was able to inhale the gas in toxic quantities.

An understanding of reaction rates is also an important part of biology. For example, most metabolic reactions happen at a suitable rate because of the presence of suitable catalysts (enzymes). However, problems can arise if the enzyme activity is not present at a suitable level, due perhaps to an inherited condition that impairs enzyme activity. Further, when pharmaceutical companies develop new drugs that will be taken into the human body, they need to understand how long the drug will take to be absorbed by the target tissue, and how long it will remain in the body before being excreted or metabolized. A failure to understand such drug behaviour could result in a drug accumulating in the body to toxic levels, with potentially harmful consequences.

Just as reaction kinetics are an important part of understanding the stability of drugs in the human body, the rate of a reaction is also an important factor in determining the stability of some compounds that are a ubiquitous part of everyday life, as discussed in Box 14.1.

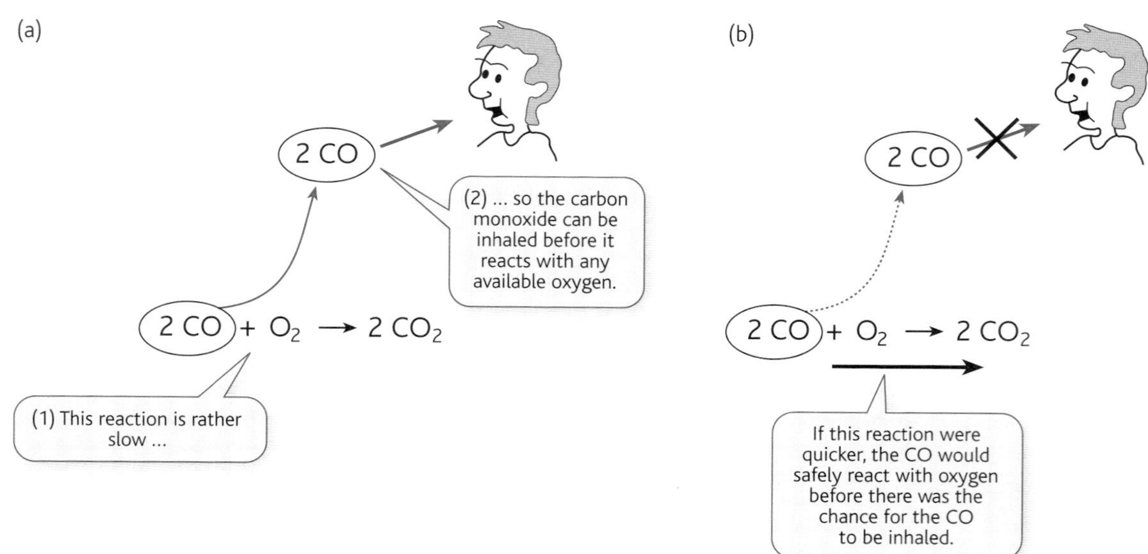

Figure 14.1 The consequences of fast and slow reaction rates. (a) The reaction between CO and O_2 is rather slow, such that CO can be inhaled before it reacts with the available oxygen. (b) If the reaction between CO and O_2 were quicker, the CO could safely react with the oxygen before being inhaled.

BOX 14.1 Diamonds are forever. Or are they?

We see in Box 10.1 how diamond and graphite are two allotropes of carbon: they both comprise only atoms of carbon, but the way in which the carbon atoms bond to one another is different in the two compounds. Figure 14.2(a) shows how graphite comprises sheets of atoms, in which each carbon atom is covalently bonded to three other atoms. By contrast, diamond comprises an extensive network of atoms, in which each carbon atom is covalently bonded to *four* other atoms, as depicted in Figure 14.2(b).

Over time, diamond undergoes a slow chemical change to form graphite:

$$\text{C(diamond)} \rightarrow \text{C(graphite)}$$

This requires the breaking of a covalent bond joining a carbon atom to one of its four neighbours.

This reaction happens spontaneously; in other words, if left at room temperature, a diamond will change into graphite without requiring the input of any energy from its surroundings.

However, despite being spontaneous, this reaction is extremely slow–in fact, immeasurably slow at room temperature. Technically, then, diamonds aren't forever: if we wait long enough, a diamond will change into graphite. Fortunately for jewellers, however, we would have to wait an extraordinarily long time–about a million years for the diamond even to turn black!

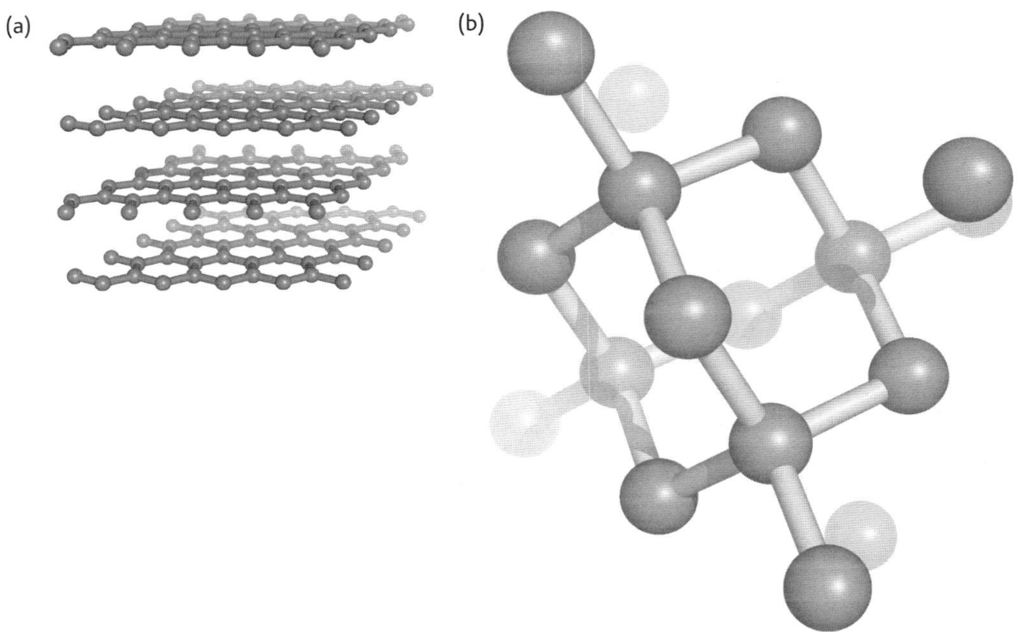

Figure 14.2 (a) Graphite comprises sheets of atoms, in which each atom is covalently bonded to three other atoms. (b) Diamond comprises an extensive network of atoms, in which each carbon atom is covalently bonded to *four* other atoms.

What is the rate of a reaction?

The word 'rate' refers to changes occurring over a fixed timescale. For example, an interest rate tells us how our bank savings, or the amount on our credit card account, changes over time.

A 'rate' can be expressed by the following general equation:

$$\text{Rate} = \frac{\text{change measured}}{\text{change in time}}$$

Consider the speed of a moving car. 'Speed' is a type of rate. The speed tells us the change in distance travelled by the car over a certain period of time. If the car is travelling at a speed of 70 kilometres per hour, we can write the speed in the form of the following rate equation:

$$\text{Speed (rate)} = \frac{\text{distance travelled (change measured)}}{\text{change in time}} = \frac{\text{70 kilometres}}{\text{one hour}}$$
$$= \text{70 kilometres per hour}$$

Biological processes also exhibit characteristic rates. For example, the kidneys filter blood plasma at a rate of around 125 mL min^{-1}. In this instance, we are looking at the volume of plasma passing through the kidney over a certain time period (in this case, one minute):

$$\text{Rate} = \frac{\text{change measured (change in volume)}}{\text{change in time}} = \frac{\text{125 mL}}{\text{1 min}} (\text{or 125 mL min}^{-1})$$

Self-check 14.1 Identify which of the following are rates:
(a) 15 cm (b) 30 m s^{-1} (c) 25 °C (d) 5 °C min^{-1}

- $\textit{Rate} = \dfrac{\textit{change measured}}{\textit{change in time}}$

Measuring the rate of a reaction

We see above how we can quantify the rate of a process (be it a chemical reaction, or the filtering of blood plasma by the kidneys, for example) by monitoring the change in some variable over a specific period of time. Most often, the variable we monitor when determining the rate of a chemical reaction is the **concentration** of reactants or products per unit time (e.g. per minute or per second). That is, we measure the *change* in concentration of reactants or products over time, as the reaction proceeds.

During the course of a chemical reaction, reactant molecules are consumed and products are formed. Consider a simple reaction between hydrogen and iodine to form hydrogen iodide:

$$H_2 + I_2 \rightarrow 2\,HI$$

During the course of this reaction, molecular hydrogen and iodine are consumed, and hydrogen iodide is formed. Consequently, as the reaction proceeds, the concentration of hydrogen and iodine **decreases** and the concentration of hydrogen iodide **increases**.

We can determine the *rate* of a reaction such as this by measuring how fast the concentration of hydrogen or iodine *decreases*, or how fast the concentration of hydrogen iodide *increases*. Firstly, we would need to measure the concentration of one of the species present in the reaction mixture (the hydrogen, iodine, or hydrogen iodide) at different time intervals. We could then use the data gathered from

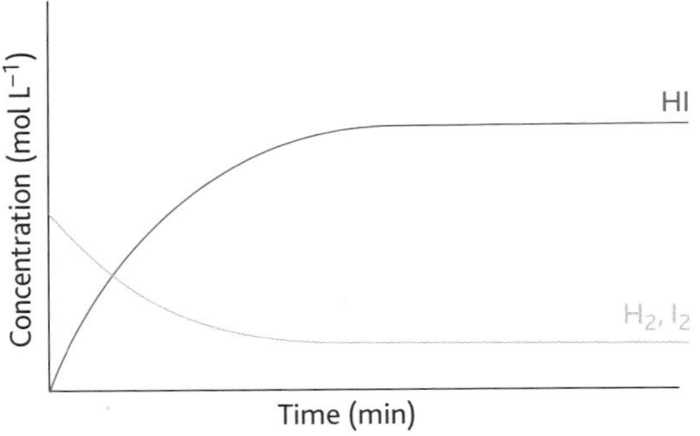

Figure 14.3 A plot of the concentration of hydrogen iodine, hydrogen, and iodine at different time intervals during the course of a reaction. Notice how the plots are curves rather than being linear, telling us that the rate of the reaction does not remain constant throughout the reaction.

such an experiment to plot a graph showing the concentration measured at each time point, as illustrated in Figure 14.3.

The shape of a plot of concentration of reactant (or product) versus time, as shown in Figure 14.3, can be used to tell us about the rate of the reaction occurring, and how that rate varies with time. For example, Figure 14.4 shows how the concentration of alcohol in the blood varies with time following an intravenous infusion of alcohol. What do you notice about the shape of the graph in Figure 14.4? The plot is a straight line. This tells us that the concentration of blood alcohol is falling by a constant amount–that is, the *rate* of removal of alcohol from the blood is constant.

By contrast, Figure 14.5 shows how the concentration of warfarin, a widely used anticoagulant (an agent that inhibits blood clotting) in the blood plasma varies with time. Notice how this plot is not linear, but is a curve. This tells us that the rate of the reaction is *not* constant, but varies during the course of the reaction.

We can measure the rate of reaction at a *particular time* during a chemical reaction from a plot of the change in concentration with time (such as that shown in Figure 14.5). We do this by determining the **gradient** (slope) of the curve at the time point we're interested in, as explained in Maths Tool 14.1.

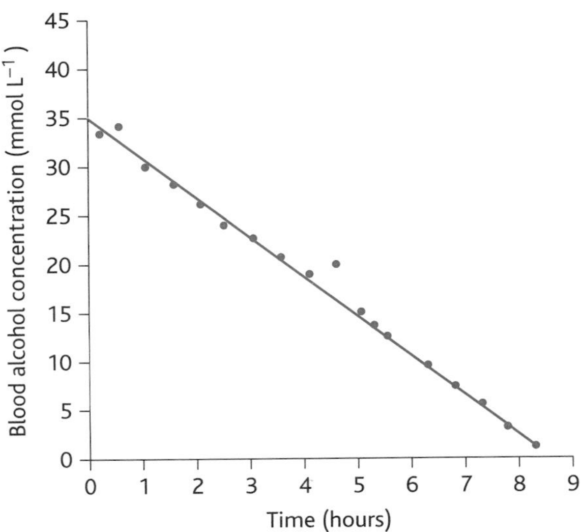

Figure 14.4 A plot showing the concentration of alcohol in the blood at different time intervals after intravenous infusion of alcohol. Notice how this plot is a straight line: the concentration of blood alcohol is falling at a constant rate throughout.

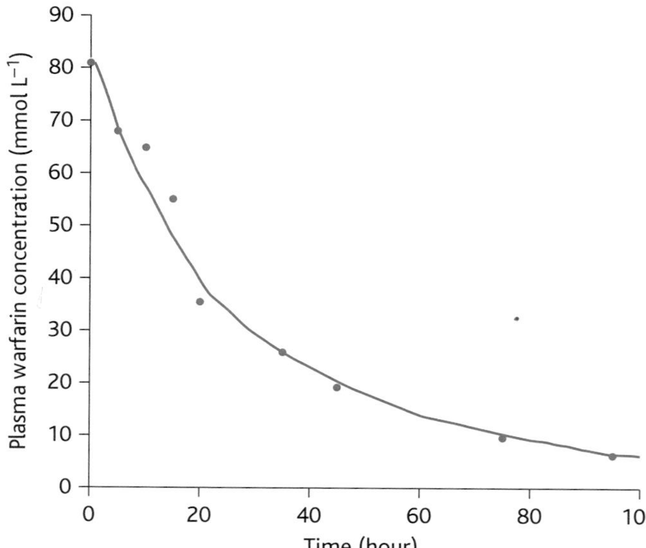

Figure 14.5 A plot showing the concentration of warfarin in the blood plasma at different time intervals following its administration. Notice how this plot is a curve, telling us that the rate of removal of warfarin varies with time.

But why do reaction rates change with time? Why don't they just remain constant? Read Box 14.2 to find the answer.

When we see a reaction rate quoted in scientific literature (e.g. journal articles), we are typically being told the **initial** reaction rate–the rate observed when the reactants are *first* mixed.

We often refer to a reaction with a fast reaction rate as **kinetically favourable**, and a reaction with a slow reaction rate as **kinetically unfavourable**. (Contrast these phrases with the terms used to describe the energy of reactions: if a reaction is spontaneous, we say that it is energetically favourable; if it is not spontaneous, we say that it is energetically unfavourable.)

The dependence of reaction rate on concentration

As we see from the examples above, the rate of a reaction can vary with the concentration of the reactant(s) in characteristic ways. For example, the rate of removal of alcohol from the blood is completely independent of the blood alcohol concentration: the rate at which the alcohol is removed stays the same (as indicated by the linear plot in Figure 14.4) regardless of the concentration of alcohol. By contrast, the rate of removal of warfarin is dependent on its concentration: as the concentration of warfarin in the plasma decreases, the rate of the reaction decreases. This is indicated by the gradual shallowing of the curve in Figure 14.5.

In fact, we can represent the relationship between reaction rate and concentration of reactants(s) mathematically. Let's consider the following hypothetical reaction:

$$\text{A} \rightarrow \text{Products}$$

We can represent the rate of this reaction by the following equation:

$$\text{Rate of reaction} = k[\text{A}]^x$$

where k is what we call the **rate constant**.

MATHS TOOL 14.1 **Measuring the gradient of a curve**

We measure the gradient (or slope) of a curve by drawing a **tangent** to the curve at the time point of interest, as depicted in Figure 14.6. A tangent is a line drawn perpendicular (at right angles) to the curve. A tangent helps us to answer the question 'If the rate were to be constant at this particular time point, what would that rate be?'

Once we have drawn the tangent, we then measure the difference between the coordinates (the *x*- and *y*-values) of any two points on the tangent. (We usually choose points at opposite ends of the tangent because the gradient will be calculated more accurately when the points are further apart.) The difference between the values of the coordinates at each end of the tangent tells us how the concentration of the substance being measured differs between two time points. This difference is the rate of the reaction between the two points, which we take to be the rate of reaction at the point on the graph at which the tangent is drawn.

For example, the tangent in Figure 14.6 gives us an iodine concentration of 0.33 mol L^{-1} after 0.5 minutes, and a concentration of 0.18 mol L^{-1} after 1.5 minutes. So, using the general rate equation shown earlier, the rate of the reaction of iodine (the rate of its consumption) at 1 minute is:

$$\text{Rate} = \frac{\text{change measured}}{\text{change in time}}$$

$$= \frac{\text{change in concentration}}{\text{change in time}} = \frac{(0.33 - 0.18)\ \text{mol L}^{-1}}{(1.5 - 0.5)\ \text{min}}$$

$$= \frac{0.15\ \text{mol L}^{-1}}{1\ \text{min}} = 0.15\ \text{mol L}^{-1}\ \text{min}^{-1}$$

The rate of consumption of iodine at 1 minute is 0.15 mol L^{-1} min^{-1} (or 0.0025 mol L^{-1} s^{-1}).

(Remember: the symbol 's' denotes units of seconds.)

We have to remember that a tangent takes a 'snapshot' of the rate of a reaction at a particular time point: it provides us with an approximation of the rate of reaction at a given time. Because the rate of a chemical reaction typically changes as the reaction proceeds, the gradients of tangents to the curve also change.

Look at Figure 14.7, which shows tangents to the curve at different time points. Notice how the gradients (slopes) of the tangents vary, telling us that the rate *changes* as the reaction proceeds.

We can measure directly the gradient of a linear curve—a straight line denoting a constant rate. We merely pick two

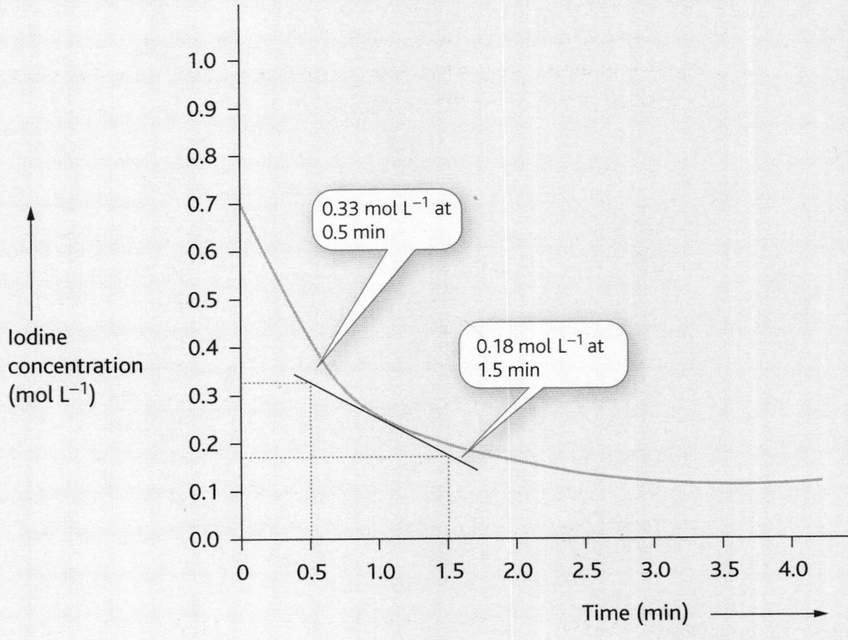

Figure 14.6 The gradient (slope) of the tangent to one of the curves shown in Figure 14.3 at a time of one minute tells us the rate of the reaction at a time of one minute. The rate is equal to the difference between the coordinates (the *x*- and *y*-values) of any two points on the tangent.

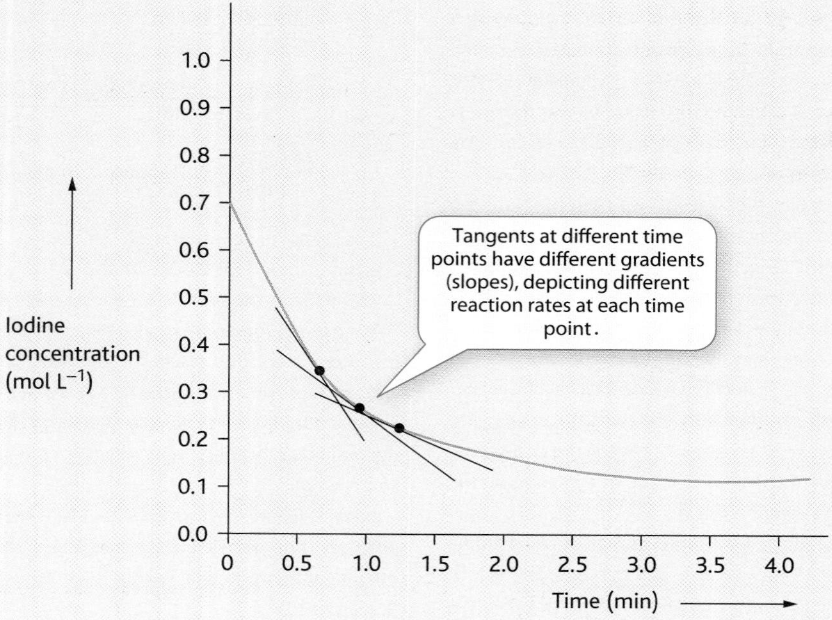

Figure 14.7 Tangents to the curve shown in Figure 14.6 at different time points. Notice how the gradient (slope) of each tangent is different, telling us that the rate of reaction varies at each time point.

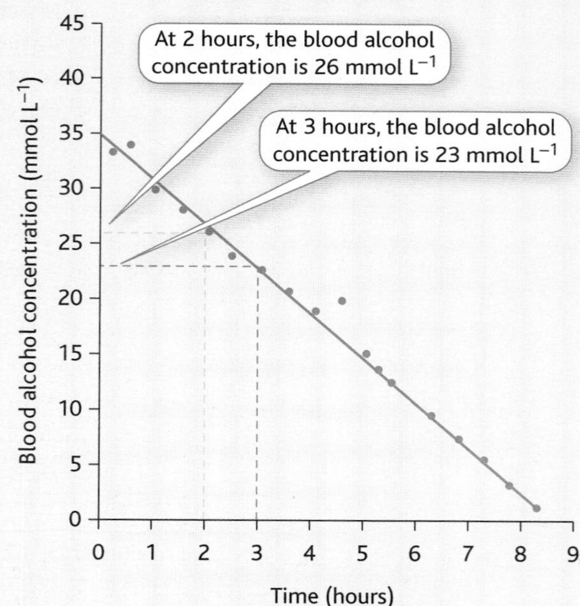

Figure 14.8 To determine the gradient of a linear curve, we merely pick two pairs of x and y values at any point along the curve, and calculate the difference between the values.

points on the line, and measure the difference between their two values. For example, to determine the rate of the decrease in blood alcohol depicted in Figure 14.4, we merely pick two points on the line, and use the *x* and *y* values of each point to determine the gradient of the line, as depicted in Figure 14.8. We use the same method as shown above to calculate the rate from the gradient of this line.

$$\text{Rate} = \frac{\text{change measured}}{\text{change in time}}$$

$$= \frac{\text{change in concentration}}{\text{change in time}} = \frac{(26-23)\ \text{mmol L}^{-1}}{(3-2)\ \text{hr}}$$

$$= \frac{3\ \text{mmol L}^{-1}}{1\ \text{hr}} = 3\ \text{mmol L}^{-1}\ \text{hr}^{-1}$$

Self-check 14.2 The rate of hydrolysis of sucrose to form glucose and fructose was monitored by measuring the concentration of sucrose at different time points, giving the following data.

Time (min)	0	14	39	60	80	110	140	170	210
[sucrose] (mol L^{-1})	0.316	0.300	0.274	0.256	0.238	0.211	0.190	0.170	0.146

What was the rate of the reaction at 80 minutes?

Notice how the rate of reaction depends on the concentration of the reactant *raised to a particular power* (denoted here by *x*). The values of *x* are usually 0, 1, or 2. We say that this value of *x* represents the **order** of the reaction with respect to the reactant.

Let's pause for a moment to digest what we're saying here. The value of *x* tells us how the rate of the reaction varies in proportion to the concentration of the reactant, A. If the order is zero (what we call a zero-order reaction), the rate of the reaction is given by $k[A]^0$. Now, because anything to the power of 0 is equal to 1, the rate of reaction must be completely independent of the concentration of A. Instead, the rate of the reaction is simply equal to the value of the rate constant, *k*. Let's show how this is the case:

$$\text{Rate of reaction} = k[A]^0 = k \times 1 = k$$

So we can say that, for a zero-order reaction, the rate of reaction is completely independent of the concentration of the reactant.

A biological reaction often exhibits zero order kinetics if there is so much substrate present that the enzyme responsible for catalyzing the reaction is saturated. When saturated (such that all available enzyme is bound to substrate) the enzyme simply can't work any faster: even if more substrate is added, the reaction will proceed at the same, constant rate.

By contrast, if the order is one (what we call a 'first-order' reaction), the rate of reaction is given by $k[A]^1$. In this instance, we use the rule that A^1 simply equals A. Therefore, the rate of the reaction $= k[A]^1 = k[A]$. In this case, the rate of the reaction is directly proportional to [A]: as A increases or decreases by a particular magnitude, so the rate will increase or decrease by an equivalent magnitude.

BOX 14.2 **Changing rates: why don't things stay constant?**

As a reaction proceeds, reactant molecules are gradually consumed as they react to form the products. Consequently we see a decrease in the concentration of reactant molecules during the course of the reaction.

Fewer reactant molecules equates to fewer opportunities for reactants to interact, and hence for new product to be formed. So we see a **slowdown** in the rate at which the reaction–the formation of products from reactants–can occur.

Look at Figure 14.9. Notice how the concentration of iodine decreases as the reaction proceeds. This decrease correlates with a decrease in the rate of reaction, as indicated by a decrease in the gradient of the tangent to the curve.

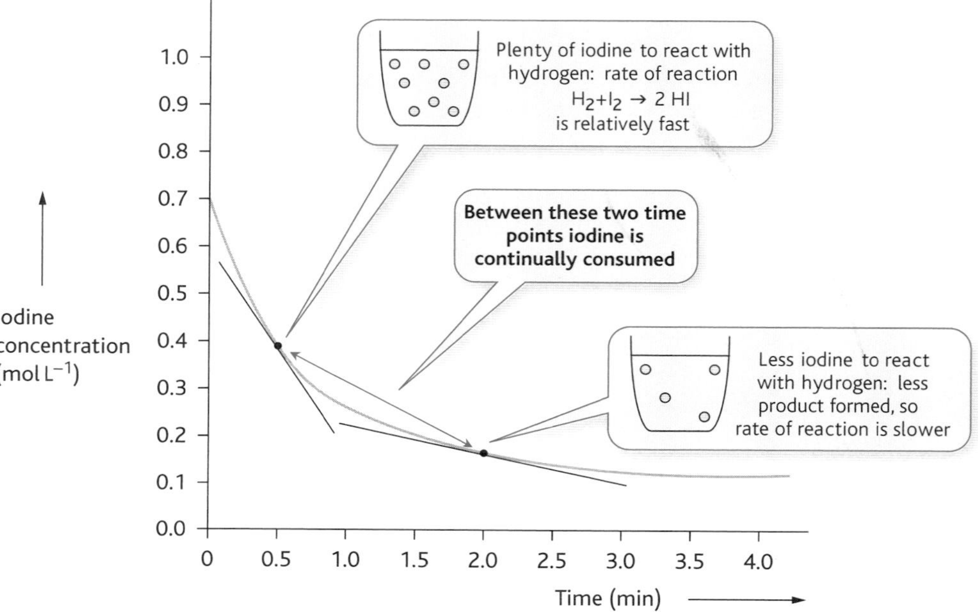

Figure 14.9 The concentration of iodine decreases during the progress of the reaction $H_2 + I_2 \rightleftharpoons 2HI$. This decrease correlates with a decrease in the rate of reaction, indicated by the shallower gradients of the tangents to the curve as time passes.

If we now look back at Figures 14.4 and 14.5 we can see how the elimination of alcohol from the body is an example of a zero-order reaction: it doesn't matter what the concentration of the alcohol is: the reaction rate stays the same. By contrast, the elimination of warfarin is a first-order reaction: the rate of reaction changes in proportion to the concentration.

- *For a zero-order reaction, the rate of reaction is independent of the concentration of reactants: the reaction rate is constant, regardless of the concentration of reactants present.*
- *For a first-order reaction, the rate of reaction is directly proportional to the concentration of reactants: as the concentration increases, so the reaction rate also increases.*

The value of k, the rate constant, is specific to a given reaction: it acts as a 'correction factor' to make sure the concentration and equivalent reaction rate equate as they should.

If a reaction involves more than one reactant–for example, $A + B \rightarrow$ Products–we write the relationship between the reaction rate and the concentration of each reactant as follows:

$$\text{Rate of reaction} = k[A]^x[B]^y$$

So, each individual reactant has its own order (power), and the overall order of the reaction is given by the sum of the orders of the reaction with respect to each individual reactant. In this case, for example, the overall order is equal to $x+y$. So, if the reaction were to be zero order in respect of A (so the rate of the reaction is completely independent of the concentration of A) but first order in respect of B, the rate of reaction is given by $k[A]^0[B]^1$, and the overall order $= 0 + 1 = 1$.

Self-check 14.3 Here is the graph produced from the data in Self-test 14.2.

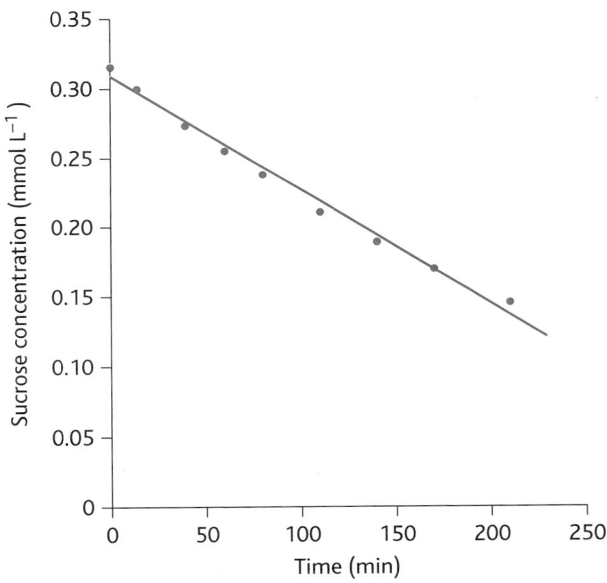

Based on the shape of the graph, would you predict this to be a zero-order or first-order reaction?

The half life

The rate of a reaction tells us how fast that reaction is occurring at a certain point in time. However, as we see in our examples above, this rate can vary depending on the particular point in time at which we measure it. Another valuable way of describing how quickly a process happens is to consider the **half life** of the reaction.

The half life, $t_{1/2}$, is the time it takes for the concentration of a given entity to fall to half of its current, observed value. So, for example, if an enzyme and its substrate bind to form an enzyme–substrate complex, and that complex has a half life of 10 s, it means that, after ten seconds, the concentration of the enzyme–substrate complex will have reduced to 50% of its initial value (that is, the concentration will have halved). (This reduction may be because the enzyme and substrate have

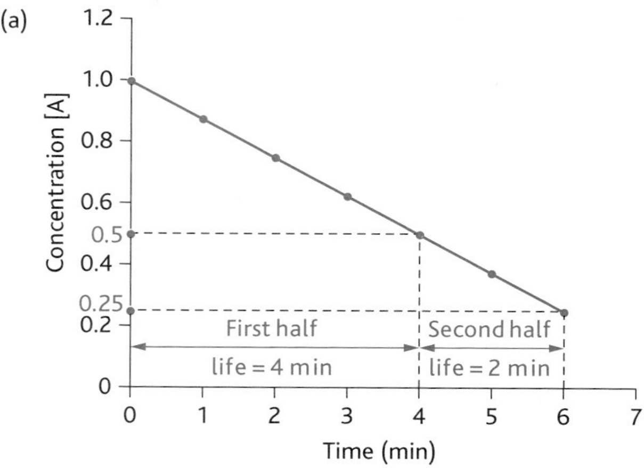

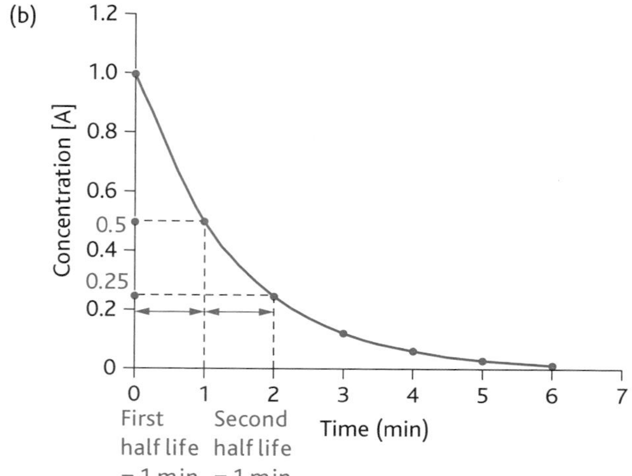

Figure 14.10 (a) For a zero-order reaction, the half life reduces as the concentration of reactants decreases. Notice how the first half life–the time it takes for the concentration to fall by half (from 1 to 0.5)–is four minutes, but the second half life–the time it takes for the concentration to fall by half again (from 0.5 to 0.25)–is just two minutes. (b) For a first-order reaction, the half life remains constant. Notice how the first half life and the second half life in this instance are both one minute.

simply dissociated again, or because the substrate has undergone a reaction to form a distinct product.)

The half life exhibited by a given process will depend on the order of the reaction: for a zero-order reaction, the half life decreases as the concentration of reactant decreases. By contrast, the half life of a first-order reaction remains constant with time–that is, it takes as long for the concentration of A to reduce by half from 0.5 mol L^{-1} to 0.25 mol L^{-1} as it does for it to fall from 1.0 mol L^{-1} to 0.5 mol L^{-1}. Look at Figure 14.10, which illustrates this difference; notice how the half life for the zero-order process gets shorter as the concentration of reactant decreases, while the half life of the first-order process stays the same.

Half lives are an important consideration in understanding the time course of drug excretion by the human body: they help us to establish how long an administered drug will remain in the body, and whether it will be excreted before it can start to have potentially harmful effects. For example, the removal of warfarin, the anticoagulant mentioned above, from the body has a half-life of 25.5 hours (based on an initial concentration of 80 μmol L^{-1}): it takes 25.5 hours for the concentration of warfarin to drop from an initial concentration of 80 μmol L^{-1} to half of this value, 40 μmol L^{-1}, and a further 25.5 hours to drop to a quarter of the original value, 20 μmol L^{-1}.

Self-check 14.4 We see in Self-check 14.2 how sucrose is hydrolyzed to glucose and fructose. Self-check 14.3 asks you to predict the order of this reaction.

(a) Given your answer to this question, explain how the half life of sucrose varies as its concentration decreases. (b) Assuming that the half life of sucrose behaves as depicted in Figure 14.10(a), how long will it take for the concentration of sucrose to fall from 8 mmol L^{-1} to 1 mmol L^{-1}, assuming that the first half life, from a starting concentration of 8 mmol L^{-1}, is 90 minutes?

- *The half life tells us how long it takes for the concentration of a reactant to fall to half of its initial value.*
- *The half life for a zero-order reaction decreases as the concentration of reactant decreases.*
- *The half life for a first-order reaction stays constant, regardless of the concentration of reactant.*

Having now seen how the rate of a reaction is related to the concentration of reactants that are present at a given time point, let's now move on to consider *why* the concentrations of reactants–and other factors too–affect reaction rates.

14.2 The collision theory of reaction rates

We see in Chapter 17 how many chemical reactions happen when two or more molecules interact: by 'interact', we mean that existing covalent bonds are broken, and new covalent bonds are formed. The most important aspect of a chemical reaction is the coming together of reactants at the start of the reaction: if two reactants fail to come together, no reaction can occur. Reactant molecules do not merely 'come together', however; in order for a chemical reaction to happen, the reacting molecules must **collide** energetically.

In order for a collision to be successful (to result in a net chemical change), the colliding molecules must meet with sufficient energy, and in the right orientation, for bonds to break and new bonds to form. Think of a game of snooker: the white cue ball must strike a coloured ball with just the right amount of energy, and in the right orientation in order for a coloured ball to move towards a pocket on the table, and be 'potted'. Too little energy, and the coloured ball won't reach the pocket; too much energy, and the coloured ball may ricochet off the back of the pocket and jump back onto the table. If the cue ball strikes with the wrong orientation, the coloured ball may be sent off in completely the wrong direction.

We see in Chapter 13 how the molecules comprising a particular substance all have different energies. Consider a flask containing a mixture of two reactants. Figure 14.11 shows how there is a wide distribution of energy between the molecules in the flask: some have a great deal of energy; some have very little energy. Importantly, of all the trillions of molecules present in the flask, statistically, only a *very few* have sufficient kinetic energy for their collision with one another to result in a chemical reaction. Look at the graph, and notice how only a few molecules, on the extreme right of the graph, have sufficient energy to react.

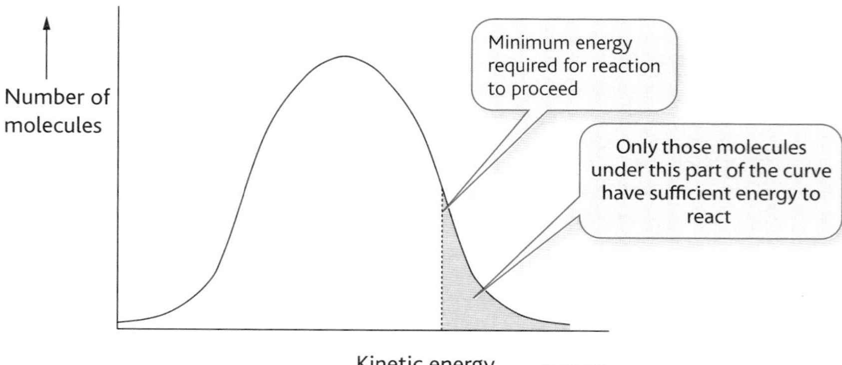

Figure 14.11 The distribution of energy within a population of molecules. There is a wide distribution of energy within a population of molecules: some have a great deal of energy; some have very little energy.

By contrast, collisions between the majority of molecules (those with energies that fall further to the left on the graph in Figure 14.11) do not result in a successful chemical reaction: the colliding molecules simply bounce off each other.

If the 'default' state is that only a minority of molecules involved in a chemical reaction have enough energy to react, how can any reaction happen at an appreciable rate? In order to increase the rate of a reaction we need to *increase* the number of molecules with sufficient energy to react. There are two key ways in which we can do this:

1. Increase the concentrations of one or more of the reactants
2. Increase the temperature of the system[2].

Let us consider each of these in turn.

Increasing the concentration

When we increase the concentration of a reactant, we increase the **overall number** of molecules occupying the same space. (Remember: the concentration equals the number of molecules per unit volume, i.e. in a *particular* volume.) If we increase the number of molecules, there will be a greater number of molecules with enough energy to react: if 10% of a population of molecules always has enough energy to react, then a concentration of 100 molecules in a litre will yield 10 molecules with enough energy; by contrast, a concentration of 200 molecules in a litre will yield 20 molecules with enough energy.

In addition to this, Figure 14.12 shows how, if there are more molecules in a given space, they will collide *more often*. So the chance of two molecules, which have enough energy to react, colliding (and in the right orientation) is increased.

Increasing the temperature

An increase in the temperature of a system has two effects:

1. It increases the movement of molecules in the system, so the number of collisions increases.
2. It increases the overall energy of the molecules in the system, so a greater number of molecules have sufficient energy to react successfully.

---

2 A third possibility for increasing the rate of the reaction is to add a catalyst, as we discuss further in Section 14.4.

Low concentration
Only a few molecules have enough energy to react ...
... and they don't collide with each other very often.

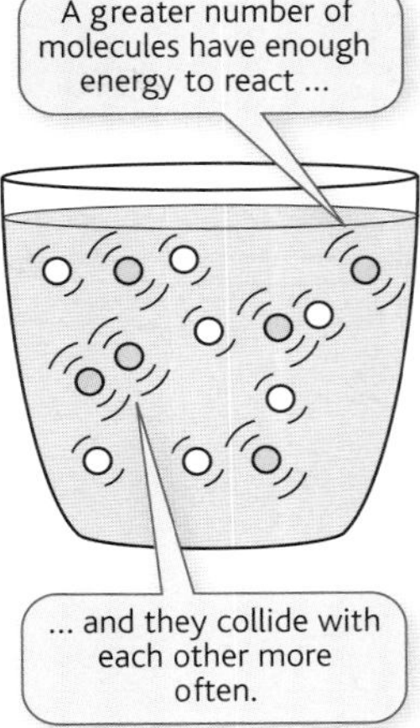
Higher concentration
A greater number of molecules have enough energy to react ...
... and they collide with each other more often.

Figure 14.12 The effect of increasing the concentration of reactants on the rate of a reaction. As the concentration is increased, the chance of two molecules which have enough energy to react colliding (and in the right orientation) is increased. This results in an increase in the rate of the reaction.

Both of these effects increase the overall rate of reaction, as depicted in Figure 14.13. For example, increasing the temperature of a system by 10 °C approximately *doubles* the reaction rate. However, it is the second of these–an increase in the overall energy of molecules in the system–that is numerically much more important.

We use temperature as a way of controlling reaction rates on a daily basis, in the context of storing and cooking food. Read Box 14.3 to find out how.

When we increase the temperature of a system we:

- ***increase the average kinetic energy of the molecules in the system.***
- ***increase the rate of reaction.***

BOX 14.3 **Ready, steady . . . cooking against the clock**

The link between temperature and reaction rates has a huge influence upon how we prepare and store food. We use heat to speed up the chemical reactions involved in the transformation of food from raw to cooked: the breakdown of starch in pasta and potatoes, the chemical change in proteins that underpins the difference in texture between raw and cooked meat, or simply the melting of cheese.

Conversely, we also use temperature to slow down chemical reactions and prolong the life of foods. Refrigerators and freezers use lower temperatures so that chemical reactions that cause food to spoil happen more slowly, e.g. the growth of harmful bacteria.

Many chemical reactions associated with the spoilage of food are carried out by bacteria living in or on the surface of the food. (For example, the souring of milk is the result of the accumulation of acid released by bacteria living in the milk.) Lower temperatures slow down the chemical reactions carried out by the bacteria, keeping food edible for longer.

The heating of foods during cooking has two important benefits. Not only does it speed up chemical reactions that make food more palatable (such as the breakdown of starch in potatoes), it also kills the potentially harmful bacteria that may be present in or on the food. For example, the process of **pasteurization** uses heat to kill harmful bacteria in dairy products such as milk.

Low temperature:
Only a few molecules have enough energy to react ...
... and they have little kinetic energy (energy of motion) so they don't collide very often.
Relatively **low** reaction rate
Higher temperature:
A greater number of molecules have enough energy to react (because the increase in temperature increases the average energy of molecules in the system) ...
... and they have greater kinetic energy, so collide more often.
Relatively **high** reaction rate

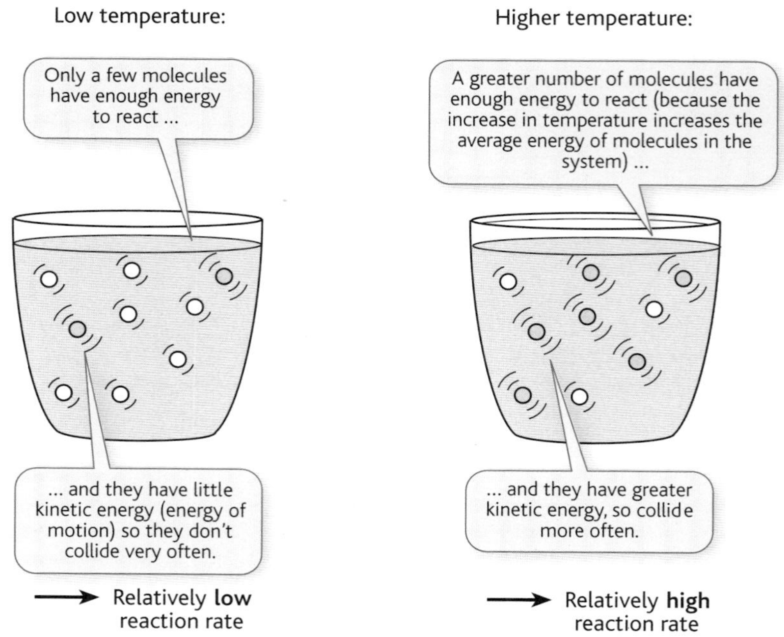

Figure 14.13 The effect of increasing the temperature on the rate of reaction. An increase in temperature increases the movement of molecules in the system, so the number of collisions increases. It also increases the overall energy of the molecules in the system, so a greater number of molecules have enough energy to react. Both of these factors lead to an increase in the reaction rate.

14.3 The activation energy: getting reactions started

Many of us have experienced the frustration of a car breakdown; if we're really unfortunate we find ourselves having to push the broken-down car off the road to a safer location. Consider the task of pushing the broken-down car: it takes a real effort to get the car moving initially, from standstill. However, once the car is moving, we find it easier to sustain motion. It has momentum.

A similar notion applies to chemical reactions: there is an energy threshold to be overcome before a reaction gets started, just as there is a force threshold above which we have to push the car before it starts to move.

The energy threshold that must be reached for reaction to occur is called the **activation energy, E_a**. The activation energy represents the minimum energy a reactant molecule must possess to initiate the bond cleavage and formation reactions associated with 'the reaction'. *All* reactions have activation energies; some activation energies are large, others are smaller[3].

Consider a high jump athlete. We can consider the height of the high jump bar to represent the magnitude of the activation energy. When the bar is relatively low, the high jumper needs relatively little energy to thrust themselves over the bar. This is analogous to a reaction with a low activation energy. When the bar is higher, however, the jumper needs to exert more effort to push themselves up and over the bar. This is analogous to a reaction with a higher activation energy.

Look at p. 430 to learn more about energy diagrams, and what they can tell us about the relative energies of reactants and products.

It is the size of the activation energy that determines, in part, the rate of the reaction. The larger the energy barrier is, the slower the rate of the reaction.

3 The activation energy illustrates one of our essential concepts: change requires energy. Chemical reactions (chemical *change*) require an input of energy equal to the activation energy before they occur.

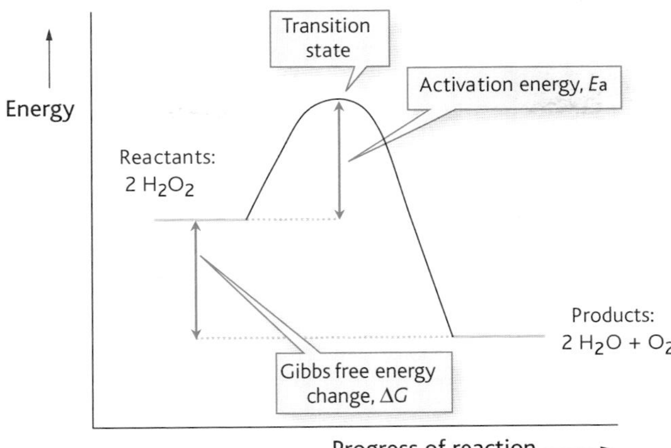

Figure 14.14 An energy diagram representing the reaction of hydrogen peroxide to give water and oxygen.

Figure 14.14 is an energy diagram representing the decomposition of two moles of hydrogen peroxide to give two moles of water and one mole of oxygen.

$$2\,H_2O_2 \rightarrow 2\,H_2O + O_2$$

This is an energetically favourable reaction because hydrogen peroxide (the reactant) is less stable than the products. (Look at Figure 14.14 and notice how the line representing the energy of the reactant is at a higher position vertically than the line representing the energy of the products.) The distance on the vertical axis between the two lines representing reactant and products shows the Gibbs free energy change (ΔG) for the reaction. ΔG for this reaction is negative. Therefore the reaction happens spontaneously.

However, the reaction has a relatively high activation energy. Consequently, this particular reaction is very slow at room temperature[4]. We can watch the reaction occurring over several weeks and months as the bubbles of oxygen form and rise to the surface. Our observations tell us that very few hydrogen peroxide molecules have sufficient energy at room temperature to get over the barrier set by the activation energy. So we join reactant and products with a curve with a high peak, representing a large activation energy.

- ***The activation energy is the minimum energy a reactant molecule must possess for the reaction it is involved with to proceed.***

Breaking the energy barrier: the transition state

Let's consider what happens at the molecular level as hydrogen peroxide decomposes to yield water and oxygen. As the reaction proceeds, bonds in hydrogen peroxide break while, simultaneously, new bonds form. The hydrogen peroxide molecule passes through a **transition state**–a high-energy species whose chemical character falls part way between being a reactant and a product.

Look at Chapter 17 to learn more about transition states.

4 Remember: whether or not a reaction is energetically favourable–whether it happens spontaneously or not–tells us nothing about the **rate** at which the reaction occurs.

The transition state is the highest-energy species that exists during the course of a chemical reaction. Therefore, it occurs at the point in the reaction where the energy of the reaction is at its highest. Look again at Figure 14.14: we see that the point at which energy reaches its highest level in a reaction is equivalent to the activation energy.

So, the transition state occurs at the point in the reaction when the activation energy is achieved: when the reacting species break through the energy barrier.

We need to remember that *every* reaction has its own activation energy. A reactant is more stable (i.e. has lower energy) than the transition state it passes through on the way to forming products. So the reactant needs a 'nudge' to get the reaction started (just as we have to give a broken-down car a hefty shove to get it rolling). Even exergonic reactions (where the products are more stable than the reactants, and energy is evolved during the reaction) require an input of energy initially.

- ***The transition state is the highest-energy species that exists during the course of a chemical reaction.***

Given that every reaction has its own activation energy, we need to ask ourselves an important question: how can we ensure that the activation energy is overcome, so that the reaction can proceed?

14.4 Catalysis: lowering the activation energy

We see in Section 14.2 how there are two important ways in which the rate of reaction can be increased: we can increase the concentration of the reactant(s), and/or we can raise the temperature of the system. Increasing the concentration of reactant and raising the temperature both increase the number of collisions between reactant molecules occurring per second. Increasing the temperature also increases the proportion of those molecules with at least the energy of activation.

Returning to our analogy of a high jumper at an athletics event, both increasing the concentration of reactants and increasing the temperature of the system are equivalent to giving the high jumper an extra push up over the bar: they increase the likelihood of the high jumper clearing the bar successfully. However, we have not altered the height of the bar itself.

Similarly, if we increase the proportion of molecules in a system that possess energies equal to the activation energy (either by increasing their concentration or by heating to increase their temperature) it is important to appreciate that the activation energy itself *remains the same*. We've merely increased the likelihood of molecules being able to overcome the barrier presented by the activation energy.

However, there is another powerful way in which we can increase the rate of reaction: we can *lower* the activation energy. Let's go back to our high jumper at the athletics championships. There are two ways in which we can make it easier for the high jumper to clear the bar: we can give them a boost to clear the bar (as we've just seen) or we can lower the bar so that the high jumper has to put in less effort to jump yet still clear the bar.

Lowering the activation energy for a reaction has a similar effect on the rate of reaction as lowering the bar for a high jumper: less energy is required to overcome

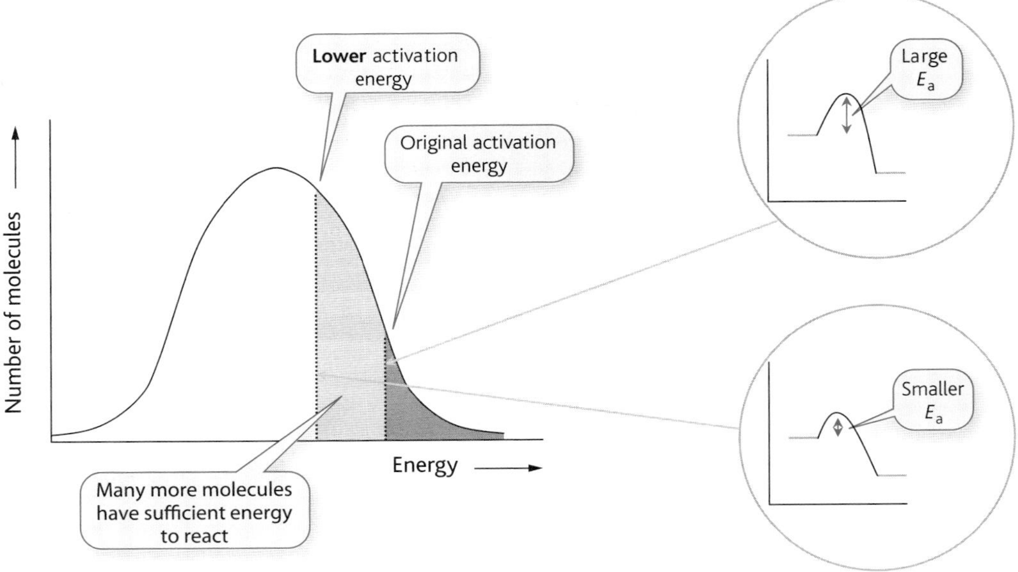

Figure 14.15 The effect of lowering the activation energy of a reaction. When the activation energy is relatively high, only a small proportion of molecules (to the right of the graph) have energies equal to or above the activation energy 'threshold'. As we decrease the activation energy, however, molecules with lower energies (moving from the right towards the left of the graph) now cross this threshold; the *overall* number of molecules with energies at least as high as the activation energy increases.

the barrier presented by the activation energy, so it's more likely that a molecular collision will be successful. Overall, the rate of reaction **increases**.

Why do we say that the reaction is 'more likely' to happen if we lower the activation energy? If the activation energy is reduced, a **larger proportion** of molecules now have sufficient energy to pass over the barrier of the activation energy. Look at Figure 14.15, which shows how this is achieved. When the activation energy is relatively high, only a small proportion of molecules (to the right of the graph) have energies equivalent to, or greater than, the activation energy. By contrast, as we decrease the activation energy, molecules with lower energies (moving from the right towards the left of the graph) now possess energies that are equivalent to, or greater than the activation energy; the *overall* proportion of molecules with energies at least as high as the activation energy has increased.

So how do we lower the activation energy? To answer this question, we must consider a particular class of chemical species: **catalysts**.

The role of catalysts in chemical reactions

Many reactions have high activation energies, even if the reaction is energetically favourable and occurs spontaneously. Consequently, the reaction rates are very slow.

In some cases, it is imperative that reaction rates are increased beyond their 'default' level. For example, some spontaneous reactions in the cell would be so slow at body temperature that life could not exist. Since increasing the temperature is not an option in a living cell we must seek to lower the activation energy[5]. But how can we achieve this? The answer is to use a catalyst.

A catalyst is defined as a substance that alters the rate of a chemical reaction but is itself chemically unchanged at the end of the reaction. A catalyst itself does not undergo any permanent change during the reaction and can be recovered and used again.

The role of a catalyst is to facilitate bond cleavage and formation by providing an alternative mechanism or reaction path from reactants to product–one with a much lower activation energy. In essence, a catalyst serves the same function as a pedestrian subway under a busy road: the subway provides the pedestrian with a quicker, easier route to get from one side of the road to the other (without having to wait for gaps in the traffic, for example); the catalyst provides a quicker route from reactants to products, bypassing a high-energy transition state in favour of a lower-energy transition state[6].

- ***A catalyst is a substance that alters the rate of a chemical reaction but is itself chemically unchanged at the end of the reaction.***
- ***A catalyst alters the rate of reaction by lowering the activation energy for the reaction.***

Look at Figure 14.16, which depicts the energy profiles for a reaction with and without a catalyst. Notice how the reactants and products possess the same energies in both instances. However, notice how the 'peak' in the energy profile representing the activation energy is lower in the presence of a catalyst than for the uncatalyzed reaction.

The presence of a catalyst is analogous to lowering the bar on the high jump: the high jumper starts at the same place, and ends in the same place regardless of the height of the bar, but expends less energy if the bar is lower because they have to jump less high. Remember: the lowering of the activation energy by a catalyst speeds up a reaction by increasing the proportion of molecules that possess energies equal to or greater than the activation energy.

Importantly, catalysts alter no other aspect of a chemical reaction than the rate. For example, a catalyst does not change ΔG, ΔH, or ΔS: if chemicals do not react at all because the reaction is energetically unfavourable, adding a catalyst has no effect. A reaction must be energetically possible if a catalyst is to have any effect.

- ***A catalyst does not affect the energetics of a reaction–whether or not it will happen spontaneously. It merely influences the rate at which a spontaneous reaction occurs.***

5 The adverse effect of heat on the human body was seen during the heatwaves that affected Europe in 2003 and 2006. In 2003, more than 30,000 people were killed. As *The Guardian* reported in 2006, 'When the human body gets to 42 °C it starts to cook. The heat causes the proteins in each cell to change irreversibly, like an egg white when it boils. Even before that, the brain shuts down because of a lack of blood coming from the overworked, overheated heart. Muscles stop working, the stomach cramps, and the mind becomes delirious. Death is inevitable'. In 2006, temperatures in southern England reached 36.5 °C; in Bosnia, 41 °C.

6 Remember: the transition state represents the highest-energy point of a reaction; the energy at this point of the reaction corresponds to the activation energy.

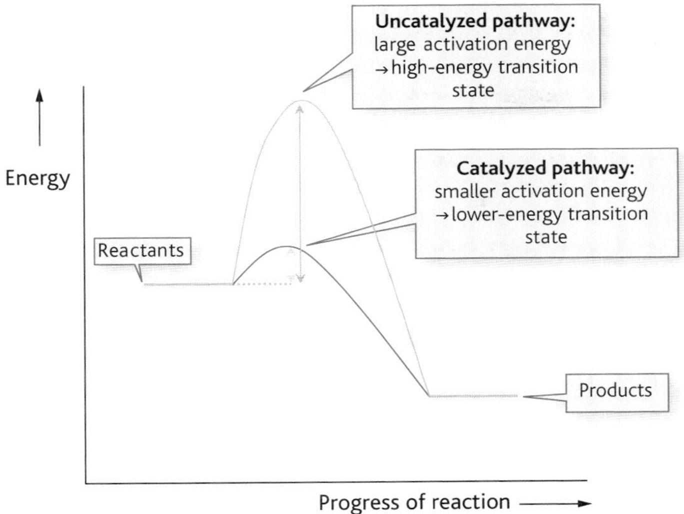

Figure 14.16 The energy profiles for a reaction with and without a catalyst. Notice how the catalyst provides an alternative reaction pathway, which has a lower activation energy.

Because a catalyst is not used up (consumed) during a catalyzed reaction, generally only a very small amount of the substance is required. Catalysts are used to speed up reaction rates that are so slow they would otherwise take a very long time, perhaps millions of years to reach equilibrium. Environmental waste removal and industrial processes employ catalysts to clean up waste water and produce valuable chemicals more quickly, for example. Catalysts are also used in the exhaust systems of many modern cars to increase the breakdown of carbon monoxide in exhaust fumes, as explained in Box 14.4.

BOX 14.4 Catalytic converters

When an internal combustion engine (i.e. a car engine) is functioning optimally, carbon monoxide and several nitrogen oxides are produced. These pollutants are toxic, and must be removed from the exhaust before emission into the atmosphere. Though the degradation of carbon monoxide is a thermodynamically spontaneous process, the reaction rate is very slow. Consequently, car manufacturers have sought ways of increasing the rate at which carbon monoxide is broken down to reduce the amounts of carbon monoxide emitted in car exhausts.

Modern car exhausts feature catalysts built into the exhaust system, which speed up the oxidation of carbon monoxide to carbon dioxide, and the reduction of nitrogen oxides to nitrogen, such that smaller quantities of these two pollutants escape into the atmosphere:

$$2\,CO + O_2 \rightarrow 2\,CO_2$$

$$NO_X \rightarrow \frac{X}{2} O_2 + \frac{1}{2} N_2$$

However, although these so-called catalytic converters prevent the accumulation of carbon monoxide in the atmosphere (by speeding up its reaction with oxygen) they actually encourage the build up of carbon dioxide, a gas with its own devastating effects on the environment. They also cause the chemical reduction of sulfur-containing oxides (SO_2 and SO_3) to form hydrogen sulfide, H_2S, which is far more toxic than the carbon monoxide for which the catalytic converter was developed. Sulfur is found in variable (but low) amounts in different crude oils. Because of the problems with the poisoning of the catalytic converter by sulfur compounds and the production of toxic H_2S, there is a great premium on reducing the sulfur content of fuels during the refining process.

Catalysts come in many guises. Some catalysts are pure elements such as platinum or nickel, usually in the form of a very fine powder (which provides a large surface area). Compounds, including ferric oxide and manganese dioxide, and metal ions in solution, such as copper ions, can also serve as catalysts. Perhaps the most important catalysts from the point of view of humans, however, are enzymes: the family of proteins which act as catalysts in biological systems.

14.5 Enzymes: the biological catalysts

As we note above, we cannot simply increase the temperature of cells in order to overcome high activation energies. Neither can we sprinkle cells with platinum or nickel–the metal catalysts used in the catalysis of numerous industrial processes. Instead, the most important family of catalysts from a biological perspective are **enzymes**.

Of the thousands of chemical reactions that occur in living cells, the vast majority have activation energies of such a high magnitude that the reactions could not occur at a fast enough rate to sustain life without the intervention of some kind of catalyst. Fortunately, enzymes can increase reaction rates by many orders of magnitude, from a million (10^6) to 10^{17} times. Reactions that would take millions of years to complete in the absence of enzymes can occur in seconds.

Most, but not all, of the known enzymes are proteins. However, look at Box 14.5 to find out about a class of enzyme that comprises the nucleic acid, RNA.

BOX 14.5 Ribozymes: a new type of biological catalyst

RNA, ribonucleic acid, has several important roles in the cell. As messenger RNA (mRNA) it conveys genetic information from the nucleus to the cytosol, and it is also part of the machinery (the ribosome) that translates the genetic information into proteins.

More recently, however, it has been discovered that RNA molecules can function as catalysts. Catalytic RNA molecules are referred to as **ribozymes** to distinguish them from protein-based catalysts: enzymes.

Ribozymes were originally discovered to have a catalytic role in their own maturation–they were found to catalyze their own cleavage (cutting the RNA chain to generate a mature, active molecule from a precursor). However, ribozymes are now believed to catalyze the cleavage of other RNAs, and to be responsible for the catalytic activity (the formation of peptide bonds between amino acid residues) of the ribosome–the assembly of polypeptides and RNA molecules that mediates the synthesis of polypeptide chains using mRNA as a template. Ribozymes are also thought to be involved in the regulation of gene expression. For example, in bacteria, the control of a specific gene is mediated by a ribozyme that is embedded in the 5′ untranslated region (UTR) of the gene's mRNA, and which has self-cleaving activity (that is, it catalyzes its own excision from the 5′ UTR).

It is even thought that RNA might once have been both the genetic material through which genetic information is passed from generation to generation (a function now fulfilled by DNA) *and* the key structural and catalytic molecule (a function now fulfilled by proteins), placing RNA at the heart of the evolution of life.

The first ribozyme was discovered in the 1980s by Thomas R. Cech (who won the 1989 Nobel Prize in chemistry, together with Sydney Altman, for his discoveries). Today, a number of ribozymes are known, including the 'hairpin', 'hammerhead', and 'hepatitis delta virus' ribozymes. Recent research has shown that a self-cleaving hammerhead ribozyme can reduce protein expression in mouse cells, just as the ribozyme noted above had been found to control gene regulation in bacteria, suggesting ribozymes may play just as important a range of roles in eukaryotes as they appear to in prokaryotes.

Researchers are even trying to artificially synthesize ribozymes with particular catalytic activities. Ribozyme research is still in its infancy, but much has already been discovered, and doubtless much remains to be explored.

The central importance of enzymes in the chemistry of life is emphasized by the role of the International Union of Biochemistry and Molecular Biology (IUBMB) whose **Enzyme Commission (EC)** defined strict rules for classifying and naming enzymes. The thousands of enzymes identified to date are placed into one of six classes (depending on the particular nature of the reaction) and given a unique four-figure code. This four-figure code identifies exactly the reaction the enzyme catalyzes.

Each enzyme may also be given a trivial name, which usually ends with the suffix '**-ase**', and describes the reaction the enzyme performs. For example, alcohol dehydrogenase (EC number 1.1.1.1) catalyzes the oxidation of an alcohol to an aldehyde or ketone (by 'removal' of hydrogen atoms). Similarly, DNA polymerase (EC 2.7.7.7) does exactly what the name implies: it catalyzes the addition of nucleotide bases to a DNA polymer.

The six enzyme families defined by the Enzyme Commission are listed in Table 14.1. Notice how these different families catalyze quite contrasting processes, including those that join molecules together (the ligases) and those that break them down by hydrolysis (the hydrolases).

The specificity of enzymes

Enzymes and inorganic (metallic) catalysts have a common role: they increase the rate of chemical reactions by lowering the activation energy for the reaction. Indeed, some enzymes and inorganic catalysts even have in common the chemical reactions they catalyze. (For example, look at Box 14.6, which compares the decomposition of hydrogen peroxide when catalyzed by enzymes versus inorganic catalysts.)

However, while many solid state inorganic catalysts can catalyze many different types of reaction because they operate in a non-specific way, enzymes are totally **specific**. A particular enzyme catalyzes only one type of reaction. Consequently, the human body needs upward of 3000 different enzymes to catalyze all the reactions throughout the body that are required for the maintenance of life.

Table 14.1 Some common enzyme families and the types of reaction they catalyze. (Adapted from Price *et al*, 2001.)

EC Subclass number	Name	Type of reaction
EC 1	Oxidoreductases	Catalyze redox reactions, in which one substrate is reduced while another is simultaneously oxidized
EC 2	Transferases	Catalyze the transfer of a chemical group from one substrate to another
EC 3	Hydrolases	Catalyze the hydrolysis of a substrate
EC 4	Lyases	Catalyze reactions in which a group is eliminated from a substrate, forming a double bond in the substrate molecule
EC 5	Isomerases	Catalyze the isomerization of a substrate
EC 6	Ligases	Catalyze the joining together of two molecules at the expense of an energy source (usually ATP)

BOX 14.6 The decomposition of hydrogen peroxide

Hydrogen peroxide is a very toxic chemical, which is produced in the body by several metabolic reactions involving oxygen.

We see in Section 14.4 how the decomposition of hydrogen peroxide:

$$2\,H_2O_2 \rightarrow 2\,H_2O + O_2$$

occurs very slowly, despite the reaction being spontaneous. However, our bodies cannot wait for the hydrogen peroxide to decompose spontaneously. If our bodies were to rely on the spontaneous decomposition of hydrogen peroxide as the means of its removal, the hydrogen peroxide would inflict serious damage on components of our cells. It would oxidize anything on sight, causing nasty chemical burns.

Outside of the body, we can increase the rate of reaction by heating the solution or by adding an inorganic catalyst. However, sprinkling a handful of metal catalyst around inside the body is not an option. It would cause toxic damage to our cells.

Instead, our bodies synthesize the enzyme catalase, which increases the rate of decomposition of hydrogen peroxide a staggering 10 000 000 times at 37 °C (equivalent to lowering the activation energy by around 30%). Consequently, any hydrogen peroxide generated in the body is removed as rapidly as it is formed, keeping cellular damage to a minimum.

Some enzymes accept several different reactants, but others are so specific they accept only one particular reactant[7].

- *Enzymes catalyze only specific reactions.*

How is enzyme specificity achieved?

Most enzymes are globular proteins–they comprise long polypeptide chains that are folded to adopt distinctive three-dimensional shapes. On the surface of the enzyme is a pocket, sometimes called a **cleft**. Look at Figure 14.17(a), which depicts the structure of the enzyme yeast hexokinase. The location of the enzyme's cleft is indicated by the arrow.

The cleft is the site at which the substrate binds to the enzyme. The shape of the cleft is complementary to that of the substrate such that only the substrate is able to fit into the cleft. The complementary shapes of the enzyme cleft and the substrate molecule allow specific non-covalent interactions (such as ionic bonds, hydrogen bonds, and hydrophobic interactions) to operate between the cleft and substrate, stabilizing their association. It is not only the shape of the cleft, but also the charges that operate within the cleft that determine enzyme–substrate specificity. For example, charged amino acid residues at specific locations in an enzyme's cleft may experience ionic interactions with charged groups within a substrate molecule, acting to stabilize the association of enzyme and substrate.

We learn more about molecular interactions, such as hydrogen bonds, hydrophobic interactions, and ionic bonding in Chapter 4.
Look at Section 7.1 to remind yourself about those amino acid residue side chains that are charged.

The most important part of the cleft is the **active site**–a region that exhibits the highest degree of complementarity with the shape of the substrate, and participates in the strongest interactions with the substrate, binding it firmly in place. Very few of the amino acids in an enzyme are involved in this specific binding, which may involve hydrogen bonds, hydrophobic interactions, and ionic bonding. Look at Figure 14.17(b); this shows the location of a molecule of glucose associated with the

7 When considering enzyme-catalyzed reactions, we refer to the reactant as the **substrate**.

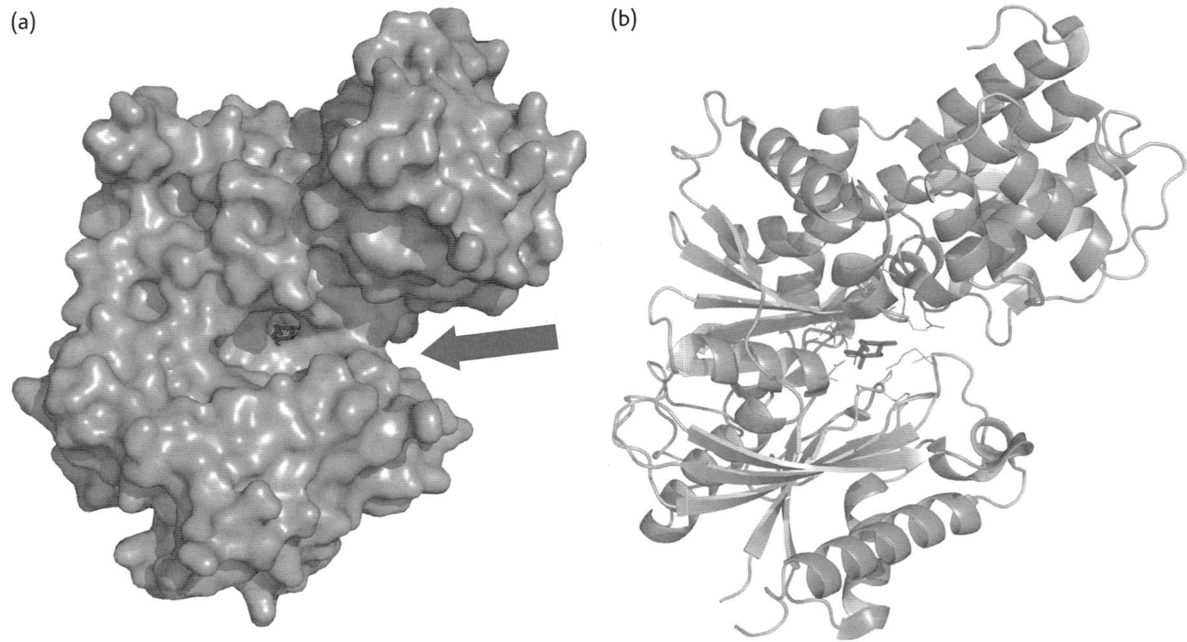

Figure 14.17 (a) The structure of the enzyme yeast hexokinase. The location of the enzyme's cleft is indicated by the arrow. (b) The location of a molecule of glucose (shown here in blue) within the enzyme's active site. Notice how this molecule is bound deep within the enzyme's cleft, as emphasized by its location in (a).

active site of yeast hexokinase. Notice how when you look back at Figure 14.17(a) this binding site is deep within the enzyme's cleft.

While other molecules may have shapes that are *roughly* complementary to the enzyme cleft, the exquisite selectivity that arises from an enzyme's specific shape means that these molecules do not bind anything like as strongly to the active site as the substrate; their association with the enzyme is only transient.

The specificity of enzyme–substrate interaction can also have its downsides, however. The interaction between active site and substrate is so specific that a mutation in just one of the amino acids at the active site can lead to complete loss of specificity and enzyme activity. We see in Box 7.2 (p. 190) how the condition phenylketonuria can arise from the loss of activity of the enzyme phenyalanine hydroxylase (PAH).

There are a number of possible causes for the loss of enzyme activity. These include a *single* base change in the gene encoding PAH (from CGG to TGG), leading to the substitution of the amino acid tryptophan for arginine at residue 408 of the PAH polypeptide chain. This change in just a single amino acid is sufficient to affect the enzyme to the extent that it loses its activity, and is no longer able to bind its substrate at the active site.

What happens during enzyme catalysis?

Chemical reactions occur at the active site of an enzyme, and proceed via a number of distinct steps, as illustrated in Figure 14.18:

1. **The substrate (S) binds to the active site of the enzyme, forming an enzyme–substrate complex (ES).** The binding of substrate molecules to the

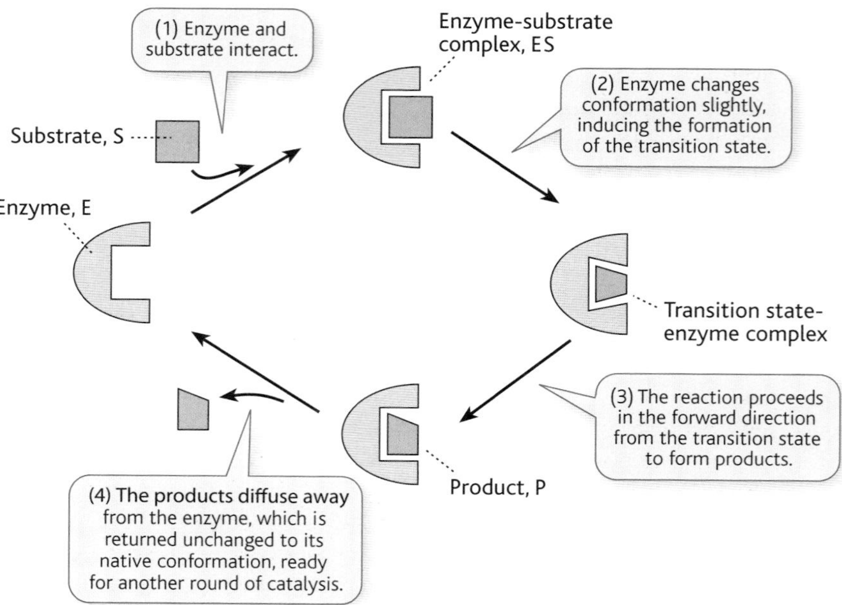

Figure 14.18 The stages of a typical enzyme-catalyzed reaction. (1) The substrate (S) binds to the active site of the enzyme, forming an enzyme-substrate complex (ES). (2) After the substrate binds to the active site, the enzyme adjusts its conformation slightly so that the active site and substrate together take on the shape of the transition state. (3) The transition state-enzyme complex gives rise to the product (P). (4) The product is released from the active site, and the enzyme is returned unchanged to its native conformation, ready to bind another substrate.

active site necessitates the reacting molecules adopting the orientation that most favours the reaction being catalyzed.

2. **After the substrate binds to the active site, the enzyme adjusts its conformation slightly so that the active site and substrate together take on the shape of the transition state.** It is important to appreciate that biological macromolecules are flexible and not structurally rigid. When the substrate binds to the active site, slight movements of the protein backbone induce strain in the bonds between the atoms of the substrate, pushing the complex towards the shape of the transition state.

We learn more about the shape of biological molecules, and the importance of structural flexibility, in Chapter 9.

The transition state binds to the active site more strongly than does the initial substrate. (The change in conformation of the enzyme during this step allows more molecular forces to operate between the active site and the bound molecule.) Because it is more tightly bound, the transition state is more **stable** than the substrate. Applying our essential concept that stability is preferable to instability, we see how the enhanced stability of the transition state, relative to the substrate, drives forward the formation of the transition state from the substrate.

3. **The transition state–enzyme complex gives rise to the product (P).**
4. **The product is released from the active site, and the enzyme is returned unchanged to its native conformation, ready to bind another substrate.**

Why is the product released from the active site of the enzyme? The product of the reaction does not bind specifically to the active site because its shape is not complementary to that of the active site (in contrast to the shape of both the substrate and the transition state). Because its shape does not allow close association between the product and the active site, fewer intermolecular forces hold the product in place. In effect, the product, P, is free to diffuse out of the enzyme after the reaction is complete.

The enzyme now returns to its native shape and is ready for another round of catalysis.

The overall process is summarized by the following equation:

$$E + S \rightleftharpoons ES \rightarrow E + P$$

We call this process the **induced-fit mechanism** of enzyme activity. Notice how this name highlights the subtle change in the enzyme's conformation in step 2 of the process.

The key function of an enzyme is to facilitate a smooth, low-energy pathway from substrate to transition state. As we note above, the stable binding of the transition state to the active site means the transition state is of **lower energy** than if it were not bound to the active site. Consequently, the activation energy for the reaction–the energy required to form the transition state from substrate–is also decreased. The enzyme has successfully 'lowered the high jump bar', making it quicker and easier to make the jump from reactant to transition state, thereby speeding up the reaction.

- ***The key function of an enzyme is to facilitate a smooth, low-energy pathway from substrate to transition state.***

In addition to enhancing reaction rates through the binding of a substrate and the subsequent stabilization of the transition state for a given reaction, enzymes can also participate chemically in reactions. That is, amino acid residue side chains from the enzyme, or other groups that are associated with the enzyme but do not form part of its integral structure[8], participate directly in the chemical reaction through which the substrate undergoes change.

For example, Figure 14.19 shows how the side chain of the amino acid residue Glu35 in lysozyme, the enzyme that catalyzes the hydrolysis of polysaccharides into

Figure 14.19 The side chain of the amino acid glutamine 35 in lysozyme participates directly in the hydrolysis of a glycosidic linkage to effect catalysis of this process. The curly arrows show the movement of electrons during this reaction, as explained further in Chapters 17 and 18.

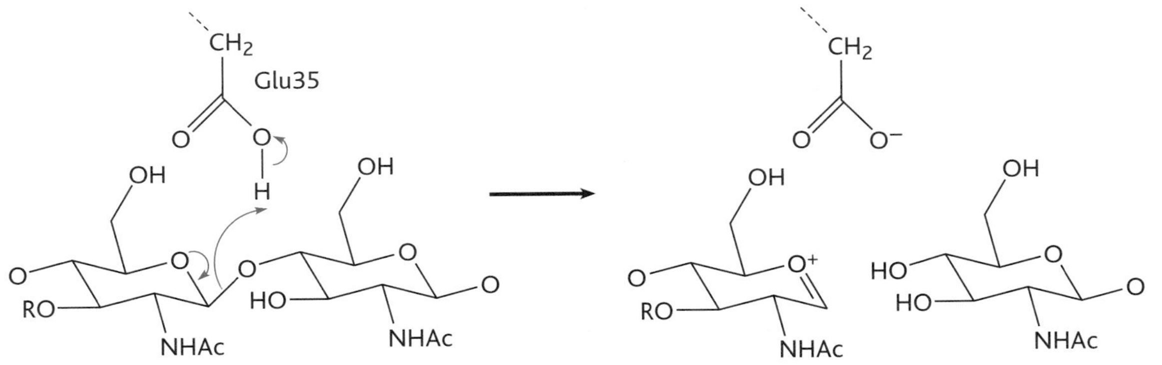

8 These 'other groups' are called **cofactors**. They may include organic groups, such as vitamins and nucleotide electron carriers such as NAD and FAD, or they may include inorganic groups, such as metal ions.

smaller sugar groups, participates in the chemical reaction through which hydrolysis occurs. (In this instance, Glu35 is acting as an acid catalyst by donating a proton during the course of the reaction. We learn more about acids and bases–and acid/base catalysis–in Chapter 16.)

14.6 Enzyme kinetics

We see above how a simple enzyme-catalyzed reaction can be represented by the following reaction scheme:

$$E + S \rightleftharpoons ES \rightarrow E + P$$

This is a two-step process, in which each step operates at a characteristic rate:

$$E + S \underset{k_{-1}}{\overset{k_1}{\rightleftharpoons}} ES \xrightarrow{k_2} E + \text{Products}$$

(Notice how the first step is an equilibrium reaction, with a forward reaction rate constant of k_1 and a back reaction rate constant of k_{-1}; the second step has a rate constant of k_2.)

Just as reaction rates can be increased independently of the presence of a catalyst by increasing the concentration of the reactant, and by increasing the temperature of the system, making these changes also increases the rate of catalyzed reactions. Not only is the bar lowered by having the catalyst present, but we also give the molecules an extra push to clear the already lowered bar. We experience the best of both worlds.

Let us now explore how substrate concentration and temperature can both influence the rate of an enzyme-catalyzed reaction. We begin by considering substrate concentration.

Increasing substrate concentration: the limitation of the enzyme's active site

Look at Figure 14.20, which shows how the rate of an enzyme-catalyzed reaction varies as the substrate concentration is increased. There are several things we should note:

1. Initially, the rate steadily increases as we increase the substrate concentration.
2. There comes a point where there is no further substantive increase in rate, regardless of any further increase in substrate concentration.

The maximum rate (or velocity) of an enzyme-catalyzed reaction is called $\mathbf{V_{max}}$. Once the reaction rate approaches V_{max}, any further increase in substrate concentration does not produce any increase in reaction rate.

Why do we see a levelling-off of reaction rate when V_{max} is approached? At V_{max} the enzyme is working at full capacity: every active site has a substrate bound to it. We say the enzyme is **saturated**: it cannot work any faster. Each enzyme has a characteristic V_{max} at which the enzyme is processing substrate at the fastest possible rate. The number of substrate molecules processed by an enzyme molecule per second is called the **turnover number**. Look at Table 14.2, which shows the turnover numbers of a number of common enzymes, that is, when they've reached V_{max}.

The concentration of substrate that is required for an enzyme to operate at **one-half** maximum velocity is called the **Michaelis constant** (K_M), as illustrated in

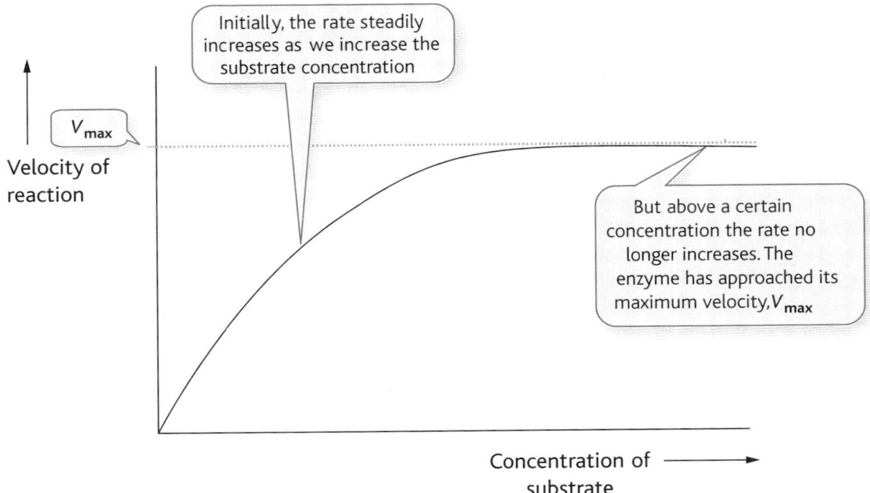

Figure 14.20 The rate of an enzyme-catalyzed reaction varies as the substrate concentration is increased. The maximum rate of an enzyme-catalyzed reaction is called V_{max}.

Figure 14.21. The Michaelis constant for different enzymes varies from as low as 10^{-8} M to around 1.0 M: in other words, some enzymes require a low concentration of substrate to operate at half their maximum velocity, whereas some require a much higher concentration. For example, the enzyme hexokinase has a K_M value for glucose of about 40 μM; glucose must be present at a concentration of 40 μM (0.00004 M) for hexokinase to operate at ½ V_{max}. By contrast, the enzyme catalase has a K_M value of 25 mM (0.025 M); its substrate must be present at a higher concentration of 0.025 M for the enzyme to operate at ½ V_{max}.

- ***V_{max} is the maximum velocity at which an enzyme-catalyzed reaction can proceed.***
- ***The K_M tells us the concentration of substrate at which the velocity of the reaction is ½ V_{max}.***

The value of K_M tells us about the strength of the interaction between the substrate and the enzyme at the active site: when K_M is small, the substrate binds strongly to the active site, and the substrate is converted readily to product at low concentration. By contrast, when K_M is large, binding is weaker and there is little enzyme activity

Enzyme	Turnover number (s^{-1})
Catalase	40 000 000
Carbonic anhydrase	1 000 000
Lactate dehydrogenase	1000
Chymotrypsin	100
DNA polymerase	15
Lysozyme	0.5

Table 14.2 The turnover numbers for some common enzymes.

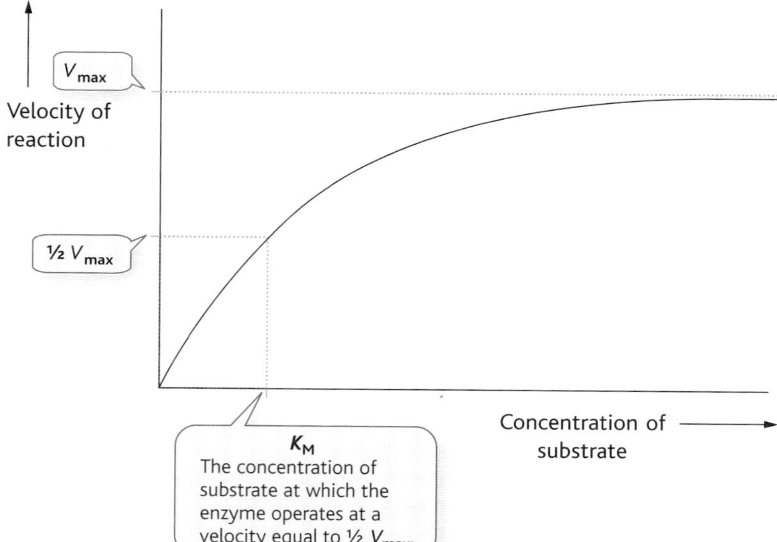

Figure 14.21 The concentration of substrate that is required for an enzyme to operate at one-half maximum velocity (that is, one-half of V_{max}) is called the Michaelis constant (K_M).

at low concentrations: the product is not formed until the concentration of the substrate is much higher.

Consequently, the value of K_M also reflects the *specificity* that an enzyme shows for a substrate: an enzyme will preferentially bind a substrate that exhibits a low K_M than it will a substrate with a high K_M because its association with the substrate with low K_M will be more stable (the association is stronger). And, as we might expect, if an enzyme preferentially binds a substrate, it will catalyze the reaction involving that substrate more efficiently–hence the way we see a substrate with low K_M being converted readily to product even at low concentrations.

- ***A small K_M value tells us that an enzyme binds strongly to its substrate, and needs only a low concentration of substrate to catalyze effectively the formation of product.***
- ***A large K_M value tells us that the enzyme binds weakly to its substrate, and needs a high concentration of substrate to catalyze effectively the formation of product.***

The values of both K_M and V_{max} may be determined experimentally by measuring the rate of an enzyme-catalyzed reaction at different substrate concentrations. The rate at each substrate concentration is related to K_M and V_{max} by the following equation:

$$v = \frac{V_{max}[S]}{K_M + [S]}$$

where v is the velocity (or rate). This is called the **Michaelis–Menten equation.** Notice how the velocity given by this equation is dependent on the concentration of the substrate (hence the presence of the [S] in both the numerator and denominator of the fraction on the right-hand side of the equation), mirroring the presence of the reactant concentration in the rate equations we explore on pp. 467–469.

The relationship between substrate concentration and rate of reaction (velocity) as described by the Michaelis–Menten equation is as depicted in Figure 14.20[9]. At low substrate concentrations, the velocity is directly proportional to the substrate concentration: as the substrate concentration increases, so does the rate of reaction. However, at higher substrate concentrations, the rate starts to level off (as denoted by the levelling off of the plot in Figure 14.20), and approaches V_{max}. As V_{max} is approached the reaction rate becomes independent of substrate concentration. This levelling-off of rate reflects the saturation with substrate of the enzyme's active site, as noted above.

Self-check 14.5 The enzyme urease catalyzes the breakdown of urea according to the following reaction:

$$CO(NH_2)_2 + H_2O \rightleftharpoons CO_2 + 2\,NH_3$$

K_M for this reaction = 105 mM, and V_{max} = 15.2 mmol urea consumed $mg^{-1}\,min^{-1}$. Using the Michaelis–Menten equation, what is the velocity of the reaction when the concentration of urea is 60 mM?

Determining the values of K_M and V_{max}

Look again at Figure 14.20, which shows a graph of the rate of an enzyme reaction plotted at different substrate concentrations. As we see above, the graph for an enzyme-catalyzed reaction is a curve rather than the straight line we might expect with an uncatalyzed reaction. The rate of an enzyme reaction increases as the substrate concentration increases, progressively approaching the maximum velocity, V_{max}. Importantly, however, the curve never actually reaches V_{max}. We say the curve approaches the maximum value *asymptotically*. The line representing the maximum velocity would only be reached by the curve at an infinitely high substrate concentration.

Because the maximum velocity is never actually reached, the use of such a curve to determine V_{max} of an enzyme is unreliable (because the data collected when plotting such a graph only ever *approaches* V_{max}, but doesn't actually reach it). Further, the value of K_M calculated from such a plot can only be approximate too: given that K_M is the substrate concentration at half the value of V_{max}, if the value used for V_{max} is approximate, then the value for K_M derived from it must also be approximate.

It is often important to obtain more accurate values of K_M and V_{max} for an enzyme than can be obtained directly from plotting the experimental data directly as in Figure 14.20. For example, when designing new pharmaceutical drugs which may alter these parameters, more accurate values may be needed. It is difficult to get accurate values from a curve without the use of software. So, for rapid analysis, scientists will often mathematically transform their data so that instead of describing a curve, the data describes a straight line, from which more accurate values of K_M and V_{max} can be determined.

Lineweaver–Burk plot

There are several ways of transforming the experimental data obtained by measuring reaction rate against substrate concentration from a curve into a straight line. One such method involves rearranging the Michaelis–Menton equation by taking

9 Remember: we are considering a simple enzyme-catalyzed reaction of the form $E + S \rightleftharpoons ES \rightarrow E + P$.

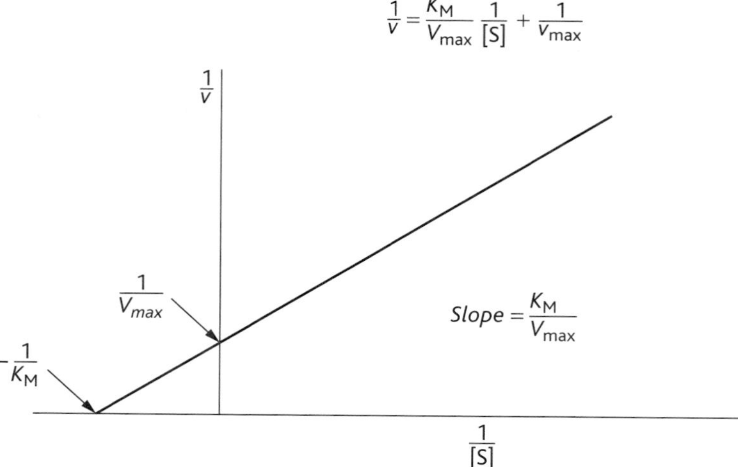

Figure 14.22 An example of a Lineweaver–Burk plot. Notice how the x-axis intercept gives a value of $-(1/K_M)$ and the y-axis intercept gives a value of $1/V_{max}$.

reciprocals of both sides of the equation to give the Lineweaver–Burk or double reciprocal plot:

$$\frac{1}{v} = \frac{K_M}{V_{max}}\left(\frac{1}{[S]}\right) + \frac{1}{V_{max}}$$

Notice how this is now the equation of a straight line, which can be represented as $y = mx + b$

$$\underset{y}{\frac{1}{v}} = \underset{m}{\frac{K_M}{V_{max}}} \underset{x}{\left(\frac{1}{[S]}\right)} + \underset{b}{\frac{1}{V_{max}}}$$

Look at Figure 14.22, which illustrates a Lineweaver–Burk plot. When you plot $1/v$ against $1/[S]$ the graph is a straight line with the intercept on the x-axis (the point at which the line crosses the x-axis) giving a value of $-(1/K_M)$ from which K_M can be calculated, and the intercept on the y-axis giving a value of $1/V_{max}$, from which V_{max} can be calculated.

EXAMPLE

The rate of hydrolysis of a penicillin analogue at different substrate concentrations in a reaction catalyzed by the enzyme β-lactamase gives the following data. What are the values of K_M and V_{max} for this reaction?

[Substrate] (μM)	Rate (nmol min^{-1})
1	0.22
2	0.38
3	0.50
5	0.68
10	0.90
30	1.16

Strategy

In order to use a Lineweaver–Burk plot we need to plot values of 1/[S] versus $1/v$. We can then read off the x-axis and y-axis intercepts and use these to calculate our two parameters.

Solution

First, we calculate these values:

[Substrate] (μM)	1/[S]	Rate (nmol min^{-1})	$1/v$
1	1	0.22	4.55
2	0.5	0.38	2.63
3	0.333	0.50	2
5	0.2	0.68	1.47
10	0.1	0.90	1.11
30	0.033	1.16	0.86

We then plot the values on a graph, as illustrated in Figure 14.23. Notice that we don't have data for values of $y < 0$. Instead, we *extrapolate* the line down to the x-axis–following exactly the same slope as that generated by our actual data–to get an x-axis intercept value.

This graph gives the following intercepts:

$$x\text{-axis} = -0.19$$

$$y\text{-axis} = 0.74$$

We know that the x-axis intercept $= -(1/K_M)$, so

$$-\left(\frac{1}{K_M}\right) = -0.19$$

$$\left(\frac{1}{K_M}\right) = 0.19$$

$$K_M = \frac{1}{0.19}$$

$$= 5.26\ \mu\text{M}$$

And we know that the y-axis intercept $= 1/V_{max}$, so

$$\frac{1}{V_{max}} = 0.74$$

$$V_{max} = \frac{1}{0.74}$$

$$= 1.36\ \text{nmol min}^{-1}$$

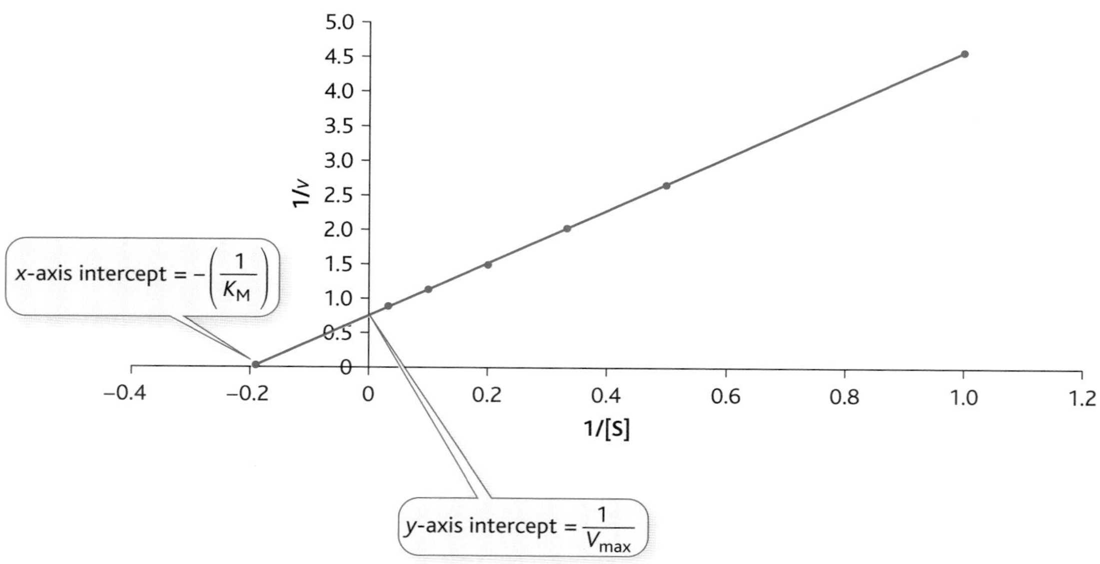

Figure 14.23 A Lineweaver–Burk plot generated from data gathered from the β-lactamase-catalyzed hydrolysis of a penicillin analogue at different substrate concentrations.

Hanes plot

Another very useful transformation gives rise to the Hanes (or Hanes–Woolf) plot. In this case the Michaelis–Menton equation is rearranged to give the following equation:

$$\frac{[S]}{v} = \frac{[S]}{V_{max}} + \frac{K_M}{V_{max}}$$

This equation allows the kinetic parameter to be determined from the straight line plot shown in Figure 14.24. In this instance, we plot [S] on the *x*-axis, and [S]/*v* on the *y*-axis. Reading values from the plot, the *x*-axis intercept gives a value of $-K_M$, and the *y*-axis intercept gives a value of K_M/V_{max}.

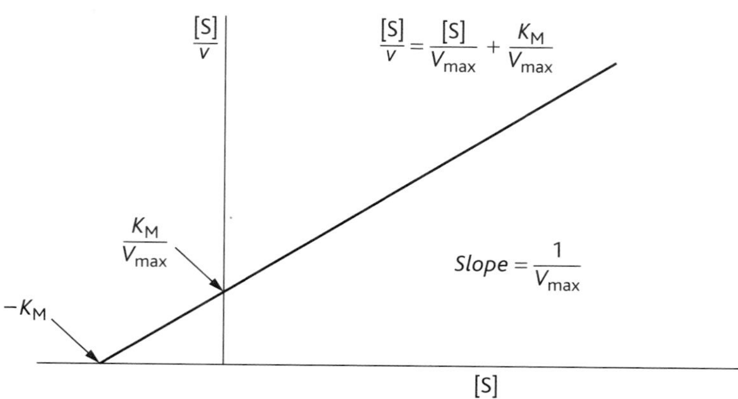

Figure 14.24 An example of a Hanes plot. Notice how the *x*-axis intercept gives a value of $-K_M$, while the *y*-axis intercept gives a value of K_M/V_{max}.

EXAMPLE

Let's return to our earlier example of the hydrolysis of a penicillin analogue. This time we need to plot [S] versus [S]/v so we need a slightly modified table of data:

[Substrate] (μM)	Rate (nmol min^{-1})	[S]/v
1	0.22	4.55
2	0.38	5.26
3	0.50	6
5	0.68	7.35
10	0.90	11.11
30	1.16	25.86

This data gives the plot shown in Figure 14.25, which gives the following intercepts:

$$x\text{-axis} = -5.12$$
$$y\text{-axis} = 3.76$$

We note above that the x-axis intercept $= -K_M$. Therefore $K_M = 5.12$ μM.
We also note that the y-axis intercept $= K_M/V_{max}$. So $K_M/V_{max} = 3.76$.
We know that $K_M = 5.12$.

$$\frac{5.12}{V_{max}} = 3.76$$
$$5.12 = 3.76 \times V_{max}$$
$$V_{max} = \frac{5.12}{3.76}$$

So $V_{max} = 1.36$ nmol min^{-1}.

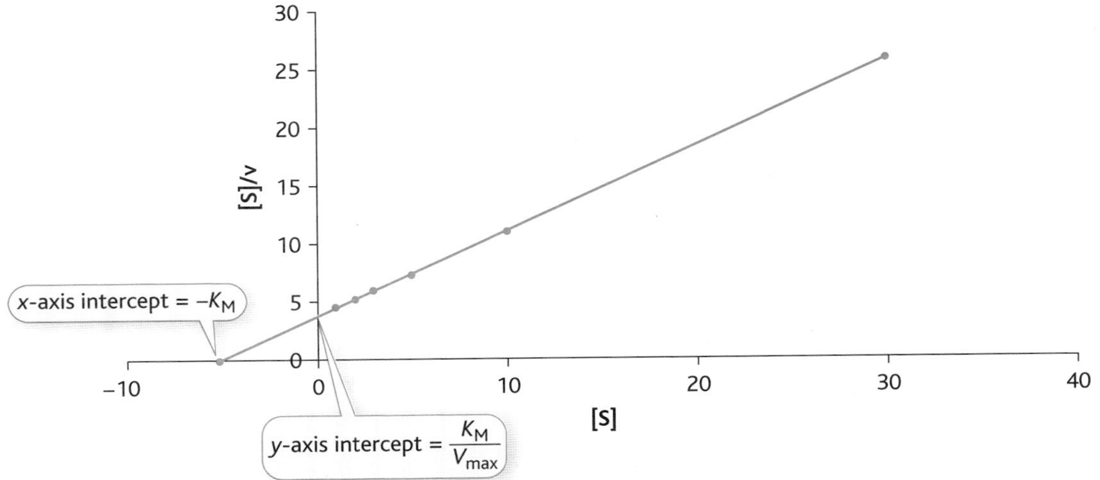

Figure 14.25 A Hanes plot generated from data gathered from the β-lactamase-catalyzed hydrolysis of a penicillin analogue at different substrate concentrations.

Notice how the values of K_M and V_{max} calculated using the Hanes plot are in close agreement–though not exact agreement–with those obtained from the Lineweaver–Burk plot above.

The Hanes plot is used more frequently than the Lineweaver–Burk plot as it gives more reliable values of K_M and V_{max} when the experimental method is prone to errors.

Self-check 14.6

(a) Another experiment measuring the rate of hydrolysis of a penicillin analogue in the presence of β-lactamase generated the following data. What are the values of K_M and V_{max} when determined using (a) a Lineweaver–Burk plot, and (b) a Hanes plot?

[Substrate] (μM)	Rate (nmol min^{-1})
1	0.24
2	0.34
3	0.56
5	0.76
10	0.90
30	0.95

(b) Using the table of substrate concentrations below, plot an appropriate graph to determine K_M and V_{max} for the enzyme aspartate kinase in the presence of its substrate aspartate.

[aspartate], mM	Rate, μmol min^{-1} mg^{-1}
0.10	0.071
0.25	0.132
0.50	0.183
1.00	0.227
2.00	0.259

Having now considered how substrate concentration can influence the rate of an enzyme-catalyzed reaction, we end this chapter by considering how temperature can also influence the reaction rate.

Increasing temperature: the limitation of being a protein

Look at Figure 14.26, which shows how the rate of an enzyme-catalyzed reaction varies as temperature is increased. Notice the following:

1. Initially, the reaction rate increases as the temperature increases.
2. As temperature increases further, the reaction rate slows down.
3. There comes a temperature above which the reaction rate drops towards zero.

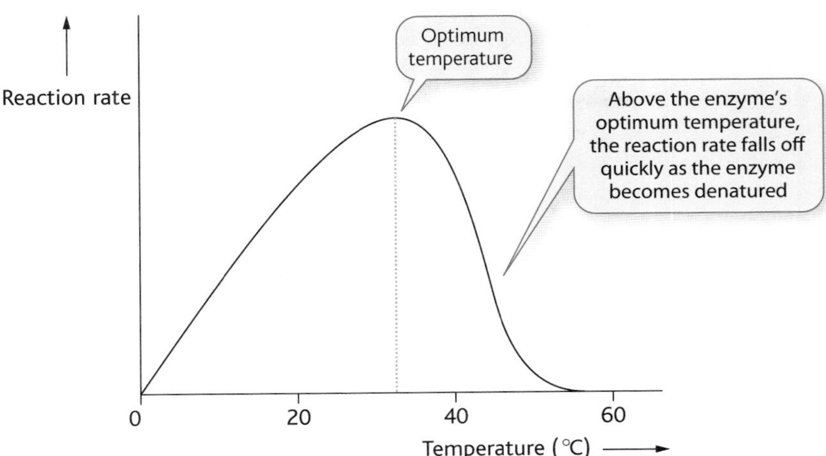

Figure 14.26 The rate of an enzyme-catalyzed reaction varies as temperature is increased. Once the temperature exceeds the enzyme's optimum temperature, the enzyme starts to denature, and its activity is lost. The rate of the reaction rapidly falls off towards zero.

Initially this result seems odd: how can increasing the temperature beyond a certain point actually slow down a reaction to the point that it almost stops? The answer is obvious as soon as we remember that enzymes are proteins. We see in Chapter 9 how biological molecules (such as proteins) adopt specific conformations, which are stabilized by intramolecular forces (such as hydrogen bonds). Any protein (including enzymes) must have the correct structure if it is to function properly. However, an increase in temperature disrupts the molecular forces that hold proteins in their correct conformations. Without molecular forces to hold them in their correct shape, proteins collapse into incorrect conformations: we say that they become **denatured**.

We see protein denaturation in action every time we cook an egg. Egg white comprises the protein albumin. When albumin is heated it denatures to form coagulated clumps of unravelled polypeptide; we see this denatured protein as the cooked egg white–which is quite different from the translucent, fluid egg white seen prior to cooking.

When enzymes denature, their active sites change shape, so that they no longer complement their substrate. Consequently, the substrate can no longer bind, and the enzyme loses its catalytic activity.

The optimum temperature for enzyme activity varies, but for most human enzymes it is around 37 °C, which is the temperature of a healthy adult's body. Above this temperature, we see a gradual decrease in activity, until the enzyme denatures fully, at which point all catalytic activity is lost.

Not all enzymes cease to operate at high temperatures, however. We see in Box 14.7 how some thermophilic bacteria have evolved to thrive in conditions under which most proteins would cease to function.

Check your understanding

To check that you've mastered the key concepts presented in this chapter, review the checklist of key concepts below, and attempt the multiple-choice questions available in the book's Online Resource Centre at **http://www.oxfordtextbooks.co.uk/orc/crowe2e/**.

BOX 14.7 Can't stand the heat? Meet the thermophilic bacteria

Few organisms have evolved to be able to survive at temperatures above 60 °C. Of these, arguably the most remarkable are a group of Archaebacteria called **hyperthermophiles**, which can thrive even in boiling water. Thermophiles have been studied most extensively in hot springs, such as those found at the Yellowstone National Park, Wyoming. Methane-producing bacteria have been found to survive at temperatures as high as 110 °C, while sulfur-dependent bacteria cope with temperatures up to 115 °C.

But how do thermophiles survive? Why don't the proteins they contain simply denature, like our own enzymes would at such temperatures?

We see in Chapter 9 how the function of all molecules (including proteins) depends on their structure. When enzymes denature (and lose their preferred three-dimensional structure) they also lose their function. It is thought that thermophiles have evolved proteins that are unusually good at retaining their three-dimensional structure, even at high temperatures. It appears that the thermophiles employ a number of strategies to stabilize their proteins, including more extensive intramolecular bonds to hold the polypeptide together particularly tightly; lower amounts of the amino acid glycine (which would normally give polypeptides extra flexibility); and the use of 'chaperonin' proteins, which help the polypeptide adopt its correct structure.

Thermophiles also require heat-stable DNA, and employ enzymes such as reverse DNA gyrase (which helps the DNA to pack tightly) and DNA binding proteins (which do for DNA what chaperonins do for polypeptides–helping to maintain the correct folded structure).

It is thought that thermophiles may give us a glimpse into our own evolution. Many scientists believe that life might first have arisen three billion years ago in environments hotter than those we experience today. Thermophiles would have been ideally suited to such conditions, making them potential ancestors of all the forms of life that subsequently evolved as the Earth cooled down in the intervening three billion years.

Checklist of key concepts

- The **rate of a chemical reaction** is the speed of change in concentration of reactants or products per unit time (e.g. per minute or per second) as the reaction proceeds
- We can determine the rate of a reaction by measuring the concentrations of reactants or products at different times during the course of a reaction, and plotting on a graph the change of concentration with time
- The rate of reaction at a particular time is equal to the **gradient** (slope) of the graph at that time
- We determine the gradient of a curve by plotting a **tangent** to the curve, and measuring the slope of the tangent
- The rate of a reaction is dependent upon the concentration of the reactants in a way that is determined by the **order** of the reaction
- For a **zero-order reaction**, the rate of reaction is independent of the concentration of reactants, and the rate is constant, regardless of the concentration of reactants present
- For a **first-order reaction**, the rate of reaction is directly proportional to the concentration of reactants
- The **half life**, $t_{1/2,}$ is the time it takes for the concentration of a given entity to fall to half of its current, observed value

The collision theory of reaction rates

- In order for a reaction to happen, the reacting molecules must **collide**
- The collision between reacting molecules must happen with sufficient **energy**, and in the correct **orientation**
- Typically, only a **small** number of reactant molecules have enough energy to be able to react
- We can increase the number of molecules with sufficient energy to react by:
  1. increasing the concentration of the reactants, or
  2. increasing the temperature of the system

The activation energy

- Every reaction has an energy threshold that must be reached before the reaction can occur

- The energy threshold is called the **activation energy**
- The **transition state** is the highest-energy species that exists during the course of a chemical reaction
- The transition state occurs at the point in the reaction when the activation energy is achieved

Catalysts

- We can increase the rate of a reaction by **lowering** its activation energy
- We can lower the activation energy of a reaction by using a **catalyst**
- A catalyst is a substance that alters the rate of a chemical reaction but is itself chemically **unchanged** at the end of the reaction

Enzymes

- Enzymes are **protein based** catalysts, which operate in biological systems
- Enzymes are totally **specific**: a particular enzyme catalyzes only one type of reaction
- When considering enzyme-catalyzed reactions, we refer to the reactant as the **substrate**
- The **cleft** is the site on the enzyme at which the substrate binds to the enzyme
- An enzyme only binds a **specific** substrate
- The **active site** is the region of the cleft that exhibits the highest degree of complementarity with the shape of the substrate and participates in the strongest interactions with the substrate, binding it firmly in place
- Both the shape of the cleft and the charges that operate within the cleft determine enzyme-substrate specificity
- Enzymes operate via the so-called **induced-fit** mechanism
- The key function of an enzyme is to facilitate a smooth, low-energy pathway from substrate to transition state

Enzyme kinetics

- The maximum rate of an enzyme-catalyzed reaction is called V_{max}
- At V_{max} the enzyme is working at **full capacity**: every active site has a substrate bound to it
- As V_{max} is approached, any further increase in substrate concentration does **not** produce any increase in reaction rate
- The number of substrate molecules processed by an enzyme per second is called the **turnover number**
- The concentration of substrate that is required for an enzyme to operate at **one-half** maximum velocity is called the **Michaelis constant** (K_M)
- When K_M is **small**, the substrate binds **strongly** to the active site, and the substrate is converted readily to product at low concentration
- When K_M is **large**, binding is **weak** and there is little enzyme activity at low concentrations; the product is not formed until the concentration of the substrate is much higher
- The values of K_M and V_{max} can be determined experimentally, drawing on the relationship described by the **Michaelis–Menten equation:**

$$v = \frac{V_{max}[S]}{K_M + [S]}$$

- Values for K_M and V_{max} can be determined most accurately by transforming experimental data into straight line plots such as the **Lineweaver–Burk plot** or **Hanes plot**.
- A Lineweaver–Burk plot plots $1/v$ against $1/[S]$. The x-axis intercept is $-(1/K_M)$ and the y-axis intercept is $1/V_{max}$
- A Hanes plot plots [S] against $[S]/v$. The x-axis intercept is $-K_M$, and the y-axis intercept is K_M/V_{max}
- Enzymes operate at optimum temperatures, above which enzymes **denature** and lose their catalytic activity

Equilibria: how far do reactions go?

INTRODUCTION

Anyone who's participated in a tug-of-war will know what it's like to be battling against an opposing force. If your team pulls with less force than the opposing team, you'll be dragged forwards. If you pull with the same force as the opposing team you'll find yourself going nowhere–all your efforts at pulling backwards counteracted by the opposing 'tug' of the other team. You'll have reached a state of equilibrium, in which there's no overall change in the position of the rope, or either team. To make any headway, your team needs to pull with a greater force than your opponents, pulling the rope in your direction to achieve victory.

Similarly, chemical reactions are rarely a matter of a one-way process in which reactants quickly form products. The change from reactant to product may be like a tug-of-war: products may form reactants just as quickly as reactants become products, acting against the forward 'flow' of the reaction to limit how far the reaction gets 'tugged' towards products. Indeed, in biological systems, this two-way flow is essential for certain key biological processes to happen as they should.

In this chapter, we explore equilibrium reactions: the opposing reactions that comprise many chemical processes, and the state of dynamic equilibrium, when opposing reactions simply cancel each other out. We see what effect altering reaction conditions can have on a reaction at equilibrium, and we tie together the concepts of energy and equilibrium to see how the free energy of a reaction reflects how far a reaction goes–and whether we can get the product that we want.

15.1 Equilibrium reactions

When we write a chemical equation, we commonly assume that the reaction proceeds 'to completion'–that the reactants on the left-hand side of the equation are totally consumed to form products. Unfortunately, chemical systems aren't that straightforward. Almost always, reactions won't go all the way to completion–only a proportion of the reactants may form products, or some of the product may decompose to re-form reactants. Sometimes the proportion that reacts is really quite small.

The usual way of representing a chemical reaction is with a balanced equation, using a single arrow to represent the chemical change from reactants to products:

$$A + B \rightarrow C + D$$

This equation tells us that one mole of A reacts with one mole of B to produce one mole each of C and D. However, if A and B can react to form C and D, very often the reverse can be true–C and D can react to form A and B:

$$C + D \rightarrow A + B$$

We have what is called an **equilibrium reaction.**

We represent an equilibrium reaction by using two half arrows $\rightleftharpoons$ to join the two sides of the equation. So this chemical equation:

$$A + B \rightleftharpoons C + D$$

tells us that the reaction is an equilibrium reaction, and can happen in both directions

- from left to right: $A + B \rightarrow C + D$ (the 'forward reaction'), or
- from right to left: $C + D \rightarrow A + B$ (the 'back reaction').

Equilibrium reactions are chemical examples of our essential concept that change can often be reversed[1]. Imagine the surface of a melting ice cube, which is at a temperature of 0 °C. We might think that the water molecules on the surface of the ice cube undergo only a change from solid to liquid. However, in reality, the water molecules participate in the following equilibrium reaction:

$$H_2O(s) \rightleftharpoons H_2O(l)$$

That is, on the surface of an ice cube, molecules from solid water (ice) enter the liquid phase while, simultaneously, molecules from the liquid phase enter the solid phase. (However, our experience tells us that, if the air surrounding the ice cube is at a temperature higher than 0 °C, more molecules will participate in the forward reaction, $H_2O(s) \rightarrow H_2O(l)$, than in the back reaction, $H_2O(l) \rightarrow H_2O(s)$: over time, the ice cube will melt.)

Biological processes also often operate in an analogous way, where a forward reaction is opposed by a back reaction. One of the most striking examples is the binding of oxygen to haemoglobin, the protein complex in our red blood cells that transports oxygen in our blood, from the lungs to our tissues. This binding is a two-way process, as represented by the following reaction scheme (where 'Hb' is a general representation of haemoglobin):

$$Hb + O_2 \rightleftharpoons Hb\text{–}O_2$$

During the forward reaction, $Hb + O_2 \rightarrow Hb\text{–}O_2$, oxygen binds to haemoglobin. In the back reaction, $Hb\text{–}O_2 \rightarrow Hb + O_2$, oxygen is released from haemoglobin. Our survival depends on these two reactions–oxygen binding and release–happening to the right extents, relative to each other, at the right time. We need oxygen to bind to haemoglobin in our lungs (that is, for the forward reaction to outweigh the back reaction), so that oxygen gets absorbed into the blood for transport to our tissues. By contrast, we need haemoglobin to release oxygen (that is, for the back reaction

1 It is tempting to use the term **reversible** to describe a process that comprises both forward and back reactions. However, we should be aware that the term **reversible reaction** has a precise thermodynamic meaning, which isn't always appropriate when considering reactions that go both forwards and backwards. We will use the term **equilibrium reaction** here.

to outweigh the forward one) in our tissues, so that oxygen gets released into our tissues.

We learn later in this chapter about how the correct balance between oxygen binding and release is achieved, and how we can determine whether a forward or back reaction dominates at a particular point in time.

Equilibrium reactions and change

For much of the time, the forward and back reactions that comprise equilibrium reactions in biological systems happen at different rates–that is, one reaction dominates over the other. As a result, we see some kind of *change* occurring in the system. (In the case of the binding of oxygen to haemoglobin, for example, we see oxygen either taken up by haemoglobin, or we see it released, depending on whether the forward or back reaction is dominating at a particular time or in a particular location.)

When the forward and back reactions happen at the same rate, however, there is *no* overall change in the system–one reaction cancels out the other, as illustrated in Figure 15.1. We say that the system is at **equilibrium**.

But how does a system reach such a state of equilibrium, and how can we determine that a state of equilibrium has been reached?

Let's consider a flask which contains just A and B with no C and D present. Only one reaction is possible:

$$A + B \rightarrow C + D$$

There is no C and D present at the start of the reaction, so the back reaction, $C + D \rightarrow A + B$ cannot occur. As soon as A and B start to react, however, C and D are formed. Once C + D are formed, the back reaction, $C + D \rightarrow A + B$, *can* start to happen.

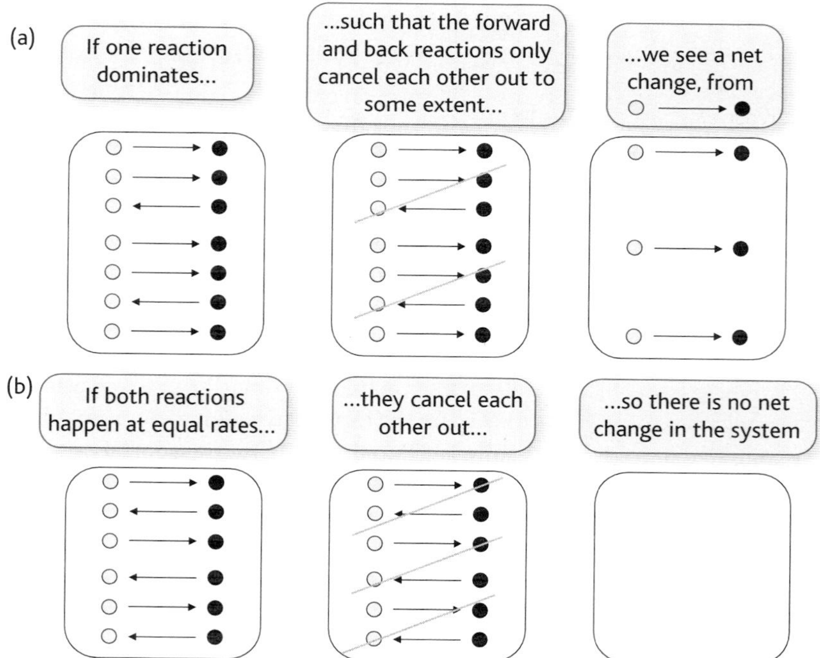

Figure 15.1 (a) When either a forward or back reaction predominates over the other, we see a net change in the system. (b) When both reactions proceed at the same rate, however, there is no overall change in the system.

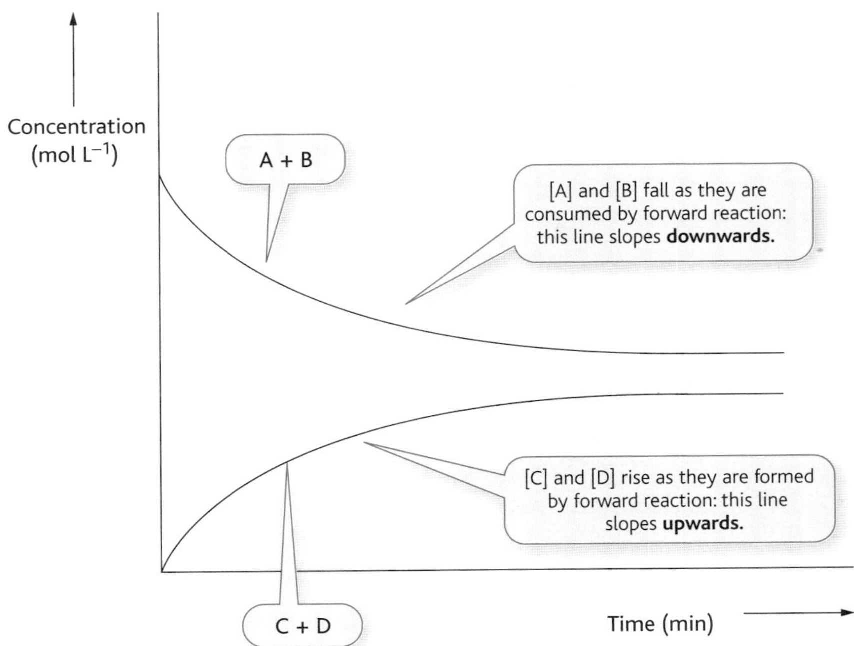

Figure 15.2 A graph showing the changes in concentration of species A, B, C, and D with time as the equilibrium reaction $A + B \rightleftharpoons C + D$ proceeds. Notice how [A] and [B] decrease, while [C] and [D] increase, before the concentrations of all four species attain stable, unchanging values.

We learn in Section 14.1 that the **rate** of a chemical reaction depends on several factors; one of the most important is the **concentration** of the reactants[2]. What does this mean in terms of our equilibrium reaction, $A + B \rightleftharpoons C + D$? When we mix A and B together, they start to react (and are consumed), so the concentrations of A and B begin to fall as shown in Figure 15.2. This decrease in concentration leads to a *decrease* in the rate of the forward reaction $A + B \rightarrow C + D$.

By contrast, the concentrations of C and D *increase* as both species are formed from the reaction of A and B, so the rate of the back reaction $C + D \rightarrow A + B$ also increases. Look again at Figure 15.2 and notice how the concentrations of C and D increase as the concentrations of A and B decrease. Notice how we're seeing the effect of the two competing reactions, the forward reaction **slowing** while the back reaction **accelerates**. These changes are a direct consequence of the changes in concentration.

What happens next? Our instinct might be that the reaction will proceed until all the reactant has formed product and the forward reaction simply comes to a stop. However, this isn't the case. As the extent of the reaction $C + D \rightarrow A + B$ increases, so A and B are formed, **replenishing** some of the A and B. In consequence, the forward reaction can carry on occurring.

So, instead of the concentrations of A and B tailing off to zero, the rate of the forward reaction stabilizes at a non-zero value as A and B are replenished by the back reaction. Look at Figure 15.3: we see the slope of the line in Figure 15.2 that

2 We see in Chapter 15 how the rate of a reaction is directly proportional to the concentration of the reactants: as the concentration of reactants increases, so the rate of reaction increases. As the concentration of reactants decreases, so the rate of reaction decreases.

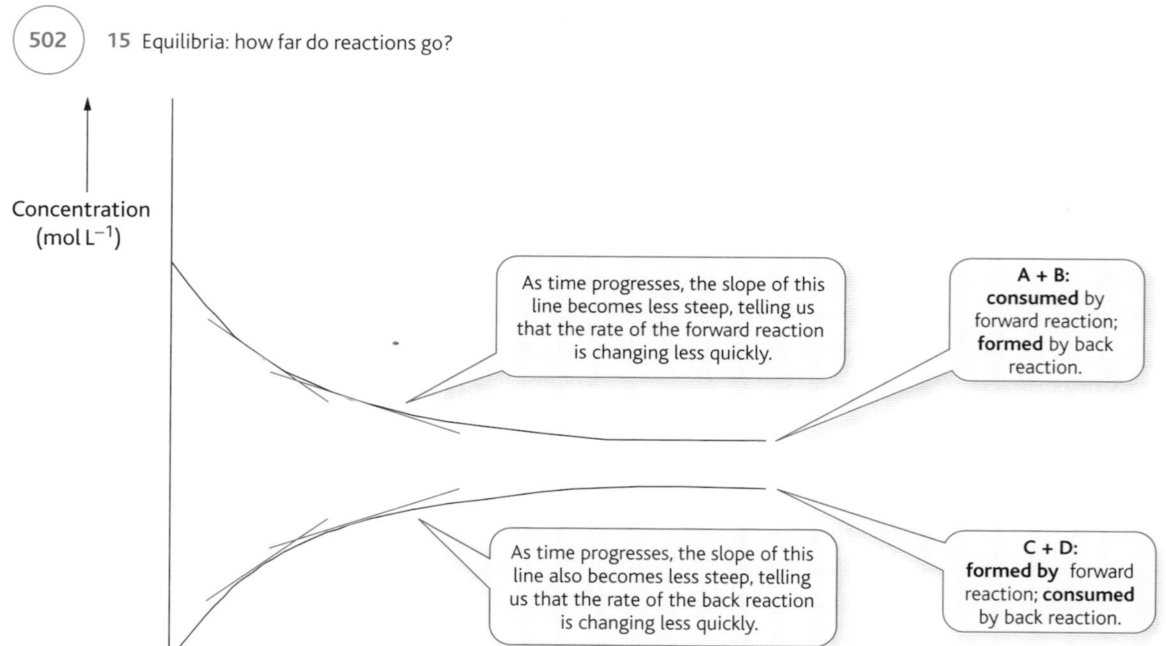
Concentration (mol L^{-1})
As time progresses, the slope of this line becomes less steep, telling us that the rate of the forward reaction is changing less quickly.
A + B: **consumed** by forward reaction; **formed** by back reaction.
As time progresses, the slope of this line also becomes less steep, telling us that the rate of the back reaction is changing less quickly.
C + D: **formed by** forward reaction; **consumed** by back reaction.
Time (min)

Figure 15.3 The gradients (slopes) of the lines depicting the concentrations of A and B, and C and D become shallower over time, telling us that the rate of change in concentrations is decelerating.

represents the concentrations of A and B becoming shallower, telling us the forward reaction isn't decelerating as fast as it was[3].

By contrast, as C and D are consumed to form A and B, their concentrations *decrease*, so the back reaction, $C + D \rightarrow A + B$, slows down. We see the slope of the line in Figure 15.2 that represents the concentrations of C and D also becoming shallower, telling us the back reaction isn't *accelerating* as fast as it was.

This chemical 'tug-of-war' continues until, eventually, the rate of the forward reaction and the rate of the back reaction are **identical**. At this point in the reaction the composition of the chemical system stops changing. Figure 15.4 shows how the rate of the forward reaction $A + B \rightarrow C + D$ exactly balances the rate of the back reaction $C + D \rightarrow A + B$, so that A and B are consumed by the forward reaction just as fast as they are re-formed by the back reaction (and vice versa for species C and D). Consequently, the concentrations of A and B (and C and D) cease changing, and remain **constant**. At this stage we say the reaction is at **equilibrium**.

It appears to the observer that, at equilibrium, the reaction has come to a stop. Indeed, if we use the plots in Figure 15.4 to measure the rates of the forward and back reactions at the point at which equilibrium is reached we would obtain a slope of zero, as depicted in Figure 15.5, suggesting that both reactions had, indeed, stopped. However, it is important to realize that this is not the case. The two complementary (forward and back) reactions are still happening, but, because their rates are the same, we do not observe any **overall** change in the system: the matching rates of the forward and back reactions cancel each other out so that, even though they *are* happening, we can't actually detect that they are. The concentrations of A, B, C, and D remain unchanged as long as the equilibrium conditions are left undisturbed.

3 Remember: the slope of a plot of the time-dependent concentration of a compound tells us the **rate** at which the compound is being consumed or formed. We find out more about the determining the rates of chemical reactions in Chapter 14.

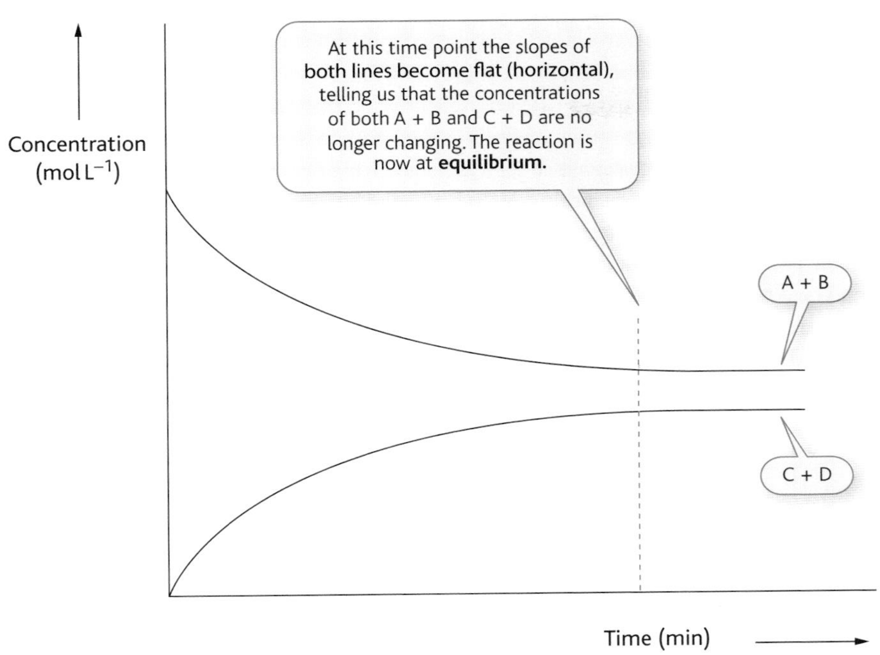

Figure 15.4 The rates of the forward and back reactions converge over time. The system attains a state of equilibrium when the rates of the forward and back reactions are *identical*.

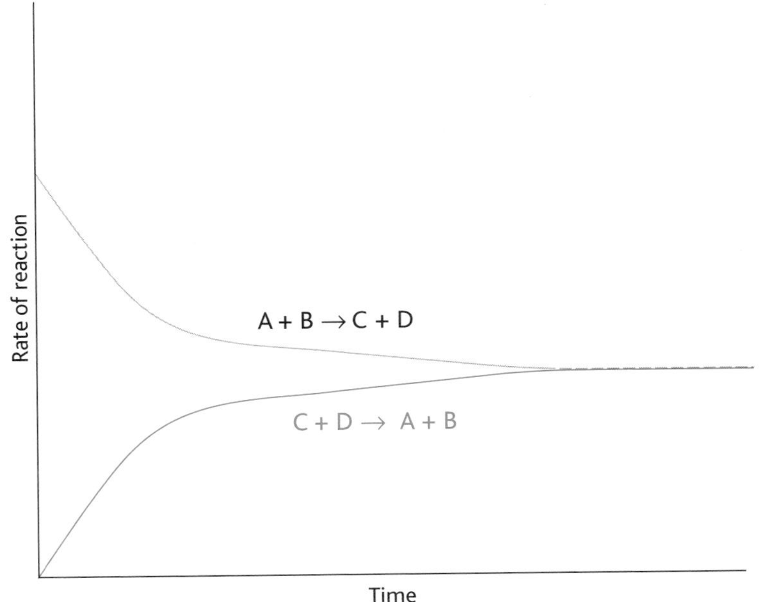

Figure 15.5 At equilibrium, the *change* in rate of the forward and back reactions falls to zero, so we see no overall change in the system. However, the forward and back reactions are still occurring, just at equal rates–so effectively masking each other.

- ***At equilibrium, the rates of the forward and back reactions are equal. There is no overall change in the system, even though the forward and back reactions continue to occur.***

We call a state of equilibrium in which reactions are still taking place, but there is no overall change in the system, a **dynamic equilibrium**. Think about walking on a treadmill–with the treadmill moving backwards as we walk forwards. If the treadmill moves backwards at the same speed we're walking forwards, we'll be in a state of dynamic equilibrium: both of us, the walker and the treadmill, are moving, but there is no overall change–the two sets of movement cancel each other out.

Note that it's only the rates of the forward and back reactions that are equal at equilibrium. The *concentrations* of the reacting species are most unlikely to be equal. Even though the concentrations of the different species may differ, however, the concentrations themselves also do not change when equilibrium is reached.

For example, the metabolism of glucose during glycolysis features the following equilibrium reaction, in which glucose 6-phosphate undergoes isomerization to form fructose 6-phosphate in the presence of the enzyme glucose 6-phosphate isomerase:

Glucose 6-phosphate $\rightleftharpoons$ (*Glucose 6-phosphate isomerase*) Fructose 6-phosphate

At equilibrium, the rate of the forward reaction:

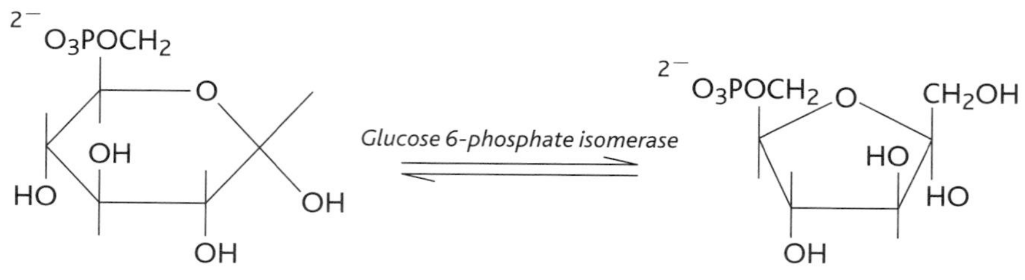

$$\text{glucose 6-phosphate} \rightarrow \text{fructose 6-phosphate}$$

is equal to the rate of the back reaction:

$$\text{fructose 6-phosphate} \rightarrow \text{glucose 6-phosphate}$$

However, the concentrations of the two species present are *not* equal. Instead, 33% of the reaction mixture is fructose 6-phosphate, and 67% is glucose 6-phosphate.

Despite being different, these concentrations do not vary once equilibrium is reached.

- ***At equilibrium, the concentrations of the chemical species present will usually be different from each other, but the actual value of that concentration does not change.***

Does it matter which reaction is 'forward' and which is 'back'?

The extent of reaction, that is, the point at which a state of equilibrium is reached, is the same regardless of the direction from which we approach it. By analogy, Figure 15.6 shows how, if we take a mug of ice-cold water, and a mug of boiling water and leave them both at room temperature, the temperature of the water in each mug will reach the *same* equilibrium temperature, regardless of whether the temperature of the water was initially very hot (and cooled down) or was initially very cold (and warmed up).

In terms of equilibrium reactions, it doesn't matter which side of the chemical equation we start from, the equilibrium point will be the same.

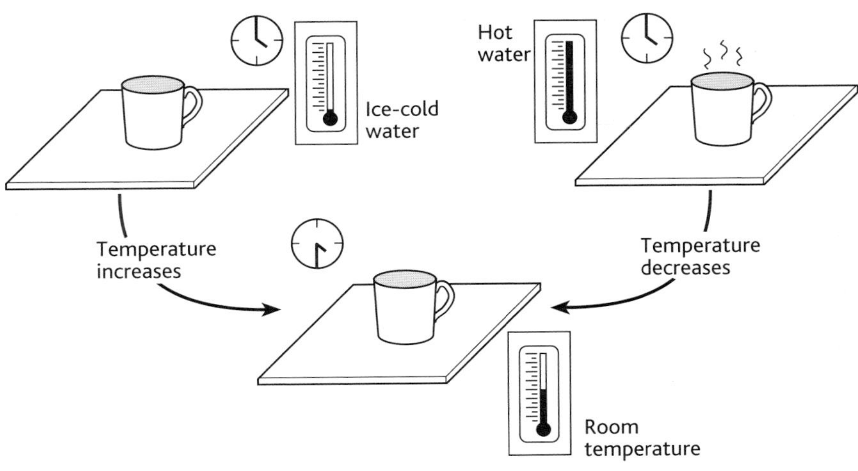

Figure 15.6 If we take a mug of ice-cold water, and a mug of boiling water and leave them both at room temperature, the temperature of the water in each mug will reach the *same* equilibrium point, regardless of their starting temperatures.

For example, let's return to the isomerization of glucose 6-phosphate, catalyzed by glucose 6-phosphate isomerase:

$$\text{glucose 6-phosphate} \rightleftharpoons \text{fructose 6-phosphate}$$

There are two ways in which equilibrium can be reached:

1. We start with a solution of glucose 6-phosphate (with enzyme), which isomerizes to form fructose 6-phosphate. We are approaching the reaction in the direction of the forward reaction, reading the above equation from left to right:

$$\text{glucose 6-phosphate} \rightarrow \text{fructose 6-phosphate}$$

2. We start with a solution of fructose 6-phosphate (with enzyme), which isomerizes to form glucose 6-phosphate. We are approaching the reaction in the direction of the back reaction, reading the above equation from right to left:

$$\text{fructose 6-phosphate} \rightarrow \text{glucose 6-phosphate}$$

It doesn't matter whether we start with glucose 6-phosphate or fructose 6-phosphate (or, indeed, any mixture of the two compounds), the equilibrium position–the rates of the forward and back reactions, and the concentration of each species–is the same.

- ***Equilibrium reactions reach a state of dynamic equilibrium irrespective of the direction from which the equilibrium is approached. In a dynamic equilibrium, both the forward and back reactions occur, so the concentrations of the components do not change.***

15.2 Forward and back reactions: where is the balance struck?

In principle, *all* solution-phase chemical reactions carried out in a closed system are equilibrium reactions. (By definition, this includes the biochemical reactions happening in the cells of our bodies every minute.)

We learn about the distinction between closed, open, and isolated systems on p. 419, Chapter 13.

Forward reaction occurs to large extent

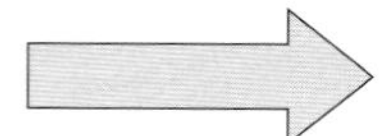

2 H2 + O2 ⇌ 2 H2O
Back reaction occurs to small extent
At equilibrium

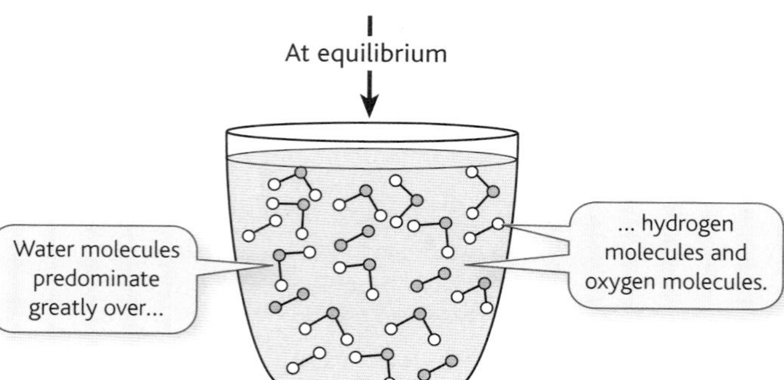
Water molecules predominate greatly over...
... hydrogen molecules and oxygen molecules.

Figure 15.7 The equilibrium reaction $2H_2 + O_2 \rightleftharpoons 2H_2O$ lies heavily to the right. Consequently, at equilibrium, the reaction mixture consists predominantly of H_2O molecules. Very few reactant molecules (H_2 and O_2) are present.

In practice, however, it may be difficult to tell that two separate reactions–the forward and back reactions–are both occurring, because one reaction may be strongly favoured over the other.

For example, hydrogen and oxygen react to form water according to the following equilibrium reaction:

$$2H_2 + O_2 \rightleftharpoons 2H_2O$$

However, the forward reaction, $2H_2 + O_2 \rightarrow 2H_2O$, occurs to a far greater extent than the back reaction, $2H_2O \rightarrow 2H_2 + O_2$. Consequently, at equilibrium, the relative composition of the reaction mixture is heavily biased towards the forward reaction: most of the H_2 and O_2 have been consumed by the forward reaction (to form H_2O), but very little H_2O has been consumed by the back reaction (to form H_2 and O_2), so the reaction mixture predominantly consists of H_2O, as depicted in Figure 15.7.

> Remember: the concentrations of the different components of a reaction need not be *equal* at equilibrium; rather, the concentrations merely do not *change*.

We say that this reaction '**lies to the right**', implying that the forward reaction predominates over the back reaction so that, at equilibrium, there is very much more H_2O present than there is either H_2 or O_2.

By contrast, let's consider the reaction of nitrogen with oxygen to form nitric oxide, an important signalling molecule in many organisms, including humans[4]:

$$N_2 + O_2 \rightleftharpoons 2NO$$

In this instance, the back reaction predominates over the forward reaction: NO decomposes to give N_2 and O_2 more readily than N_2 and O_2 react to form NO. Consequently, at equilibrium, the composition of the reaction mixture is heavily biased towards the back reaction having happened: there is very much more N_2 and O_2 present than there is NO. We say that the reaction '**lies to the left**'.

The descriptions 'lying to the left' and 'lying to the right' are rather hazy terms. Fortunately, however, there is a more precise way of determining whether a reaction lies to the left or to the right at equilibrium. We use what is called an **equilibrium constant**, as we discover in the next section.

---

4 Note that this isn't the actual reaction through which NO is formed in cells.

The equilibrium constant

The equilibrium constant is a useful tool in allowing us to describe the relative equilibrium position of an equilibrium reaction.

We can represent an equilibrium reaction by the following general equation:

$$aA + bB \rightleftharpoons cC + dD$$

Reactants | Products

where a, b, c, and d are the stoichiometric coefficients[5] of the balanced reaction between reagents A, B, C, and D.

We see above how, at equilibrium, the concentrations of the chemical species participating in a reaction remain **constant**. The equilibrium constant is related to these concentrations according to the following relationship:

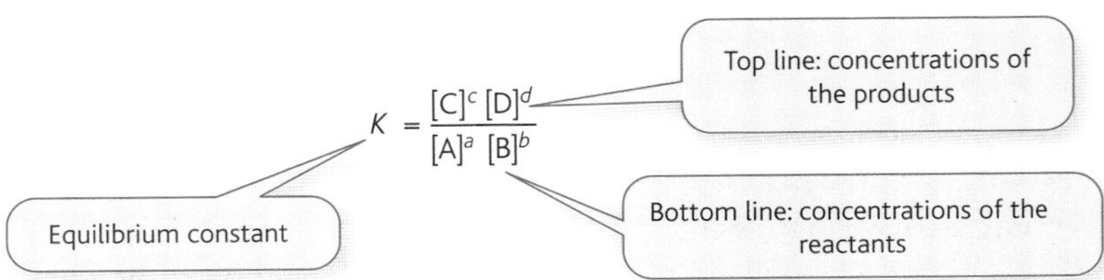

The equilibrium constant is represented by the symbol K, and tells us about the **ratio of products to reactants** when a reaction is at a state of equilibrium.

Notice how the stoichiometric coefficients from the chemical formula become the powers to which the concentrations of the chemical species are raised. For example, if the chemical formula is:

$$A + 2B \rightleftharpoons 3C + D$$

$a = 1, b = 2, c = 3,$ and $d = 1.$

Then the equilibrium constant is given by:

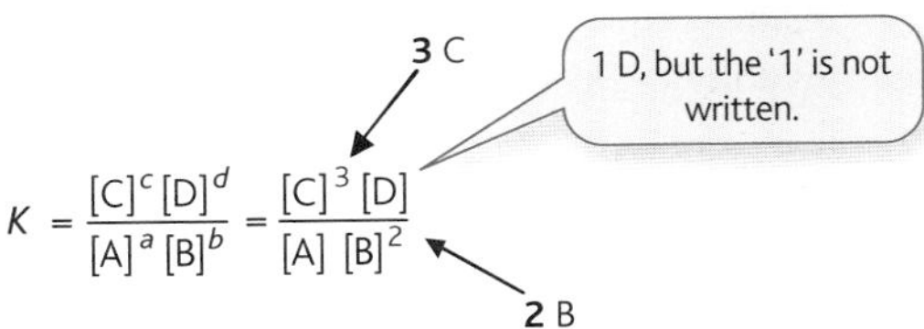

Remember: we don't usually write in the stoichiometric coefficient if its value is 1. Similarly, we don't bother writing 1 in the equation for the equilibrium constant[6]; $[A] \equiv [A]^1$; and $[D] \equiv [D]^1$.

Let's now run through an example, using real data.

Remember: the square brackets around a term denote 'the concentration of'. So [A] denotes 'the concentration of the chemical species A'.

5 Remember: the stoichiometric coefficient tells us the relative amount of a chemical species that is present in the fully balanced reaction equation. We learn more about stoichiometry in Chapter 17.

6 Anything to the power of '1' is the thing itself: $4^1 = 4$, $16^1 = 16$, $[D]^1 = [D]$.

EXAMPLE

Let us consider the reaction of hydrogen and iodine to give hydrogen iodide:

$H_2 + I_2 \rightleftharpoons 2HI$

First, we relate this equation to the general equation for an equilibrium reaction:

Notice that there's no dD as there is only one species on the right-hand side of the equation.

$$\begin{array}{ccccc} H_2 & + & I_2 & \rightleftharpoons & 2HI \\ \uparrow & & \uparrow & & \uparrow \\ aA & + & bB & \rightleftharpoons & cC \end{array}$$

So, $a = 1$; $b = 1$; $c = 2$.

The equilibrium constant for this equilibrium reaction is defined as:

$$K = \frac{[C]^c}{[A]^a[B]^b} = \frac{[HI]^2}{[H_2][I_2]}$$

We observe the following concentrations at equilibrium when two moles of H_2 and one mole of I_2 are heated at 400 °C:

$[HI] = 1.87$ mol L^{-1}

$[H_2] = 1.065$ mol L^{-1}

$[I_2] = 0.065$ mol L^{-1}

Feeding these values into the equation for the equilibrium constant:

$$K = \frac{[HI]^2}{[H_2][I_2]}$$

$$= \frac{(1.87)^2}{(1.065) \times (0.065)}$$

$$= \frac{3.4969}{0.069225}$$

$$= 50.5$$

There are a couple of important things to note here. First, the value of K is dimensionless: it does not have units as they cancel each other. Second, the concentrations used when calculating the value of K must be expressed in units of mol L^{-1} (that is, M), rather than mmol L^{-1} or µmol^{-1} etc.

But what does this value of 50.5 actually tell us about the position of the equilibrium? We discover the answer in the next section.

Self-check 15.1

(a) Write down the equilibrium constant expression for the following reaction:

$$N_2 + 3H_2 \rightleftharpoons 2NH_3$$

(b) At equilibrium, the three components of the reaction mixture are present at the following concentrations: $[N_2] = 0.185$ mol L^{-1}; $[H_2] = 0.098$ mol L^{-1}; $[NH_3] = 0.34$ mol L^{-1}.

What is the equilibrium constant for this equilibrium reaction?

The magnitude of equilibrium constants

The magnitude ('size') of the equilibrium constant, *K*, tells us the relative position of a reaction at equilibrium–whether it lies to the left, to the right, or in between–by indicating the relative ratio of products to reactants (whether it is the products or reactants that predominate). The correlation between the magnitude of *K* and the position of a reaction at equilibrium is illustrated in Figure 15.8.

Large values of *K*

A very large value for *K* tells us the reaction proceeds almost completely to the *right*. For example, the equilibrium constant for the reaction between oxygen and hydrogen (which we explore above):

$$2\,H_2 + O_2 \rightleftharpoons 2\,H_2O$$

is absolutely huge at **3.3 × 10**[81] (at 25 °C).

Why does a very large value for *K* tell us the reaction lies to the right? For *K* to be so large, the numerator (the top line of the fraction) in the equilibrium constant expression must be very much larger than the denominator (the bottom line of the fraction):

$$K = 3.3 \times 10^{81} = \frac{[H_2O]^2}{[H_2]^2[O_2]}$$

A large numerator equates to a *high* concentration of H_2O. By contrast, a small denominator equates to *low* concentrations of H_2 and O_2. If the concentration of

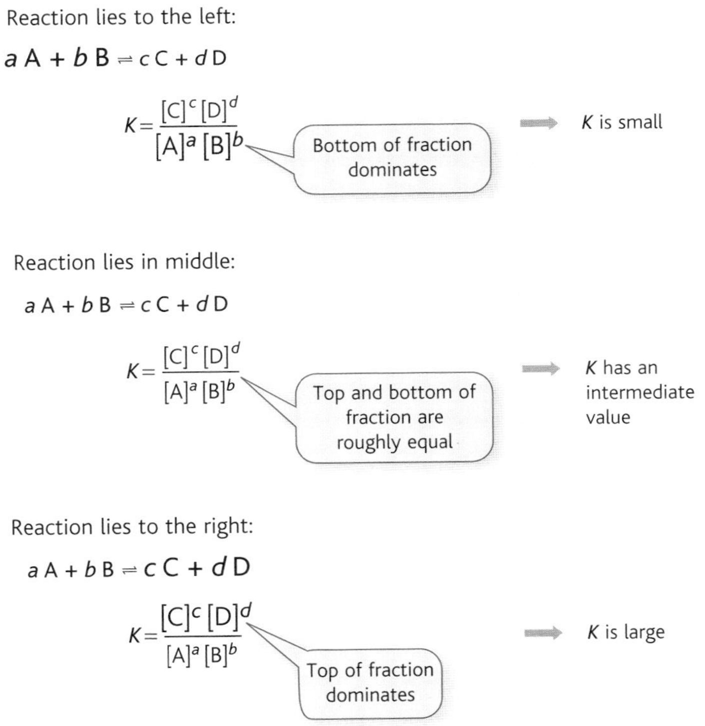

Reaction lies to the left:

$$a\,A + b\,B \rightleftharpoons c\,C + d\,D$$

$$K = \frac{[C]^c[D]^d}{[A]^a[B]^b}$$

→ *K* is small

Reaction lies in middle:

$$a\,A + b\,B \rightleftharpoons c\,C + d\,D$$

$$K = \frac{[C]^c[D]^d}{[A]^a[B]^b}$$

→ *K* has an intermediate value

Reaction lies to the right:

$$a\,A + b\,B \rightleftharpoons c\,C + d\,D$$

$$K = \frac{[C]^c[D]^d}{[A]^a[B]^b}$$

→ *K* is large

Figure 15.8 The correlation between the magnitude of *K* and the position of a reaction at equilibrium. When the reaction lies to the left, *K* is small; when the reaction lies to the right, *K* is large.

H_2O is very much larger than the concentrations of H_2 and O_2, the equilibrium reaction $2\,H_2 + O_2 \rightleftharpoons 2\,H_2O$ must lie to the right:

Forward reaction is favoured ⟶

$$2\,H_2 + O_2 \rightleftharpoons 2\,H_2O$$

Products predominate over reactants

In fact, when this reaction is at equilibrium, the amount of hydrogen and oxygen present is so small as to be virtually undetectable. In this case we would normally not use the equilibrium symbol; it is a rare case in which to use a single arrow between the reactants and products:

$$2\,H_2 + O_2 \rightarrow 2\,H_2O$$

The use of a single arrow in this way is a rare exception: few equilibrium reactions lie so far to the right.

Small values of *K*

A very small value for *K* tells us the reaction lies almost completely to the *left*. For example, let's reconsider the reaction of nitrogen and oxygen:

$$N_2 + O_2 \rightleftharpoons 2\,NO$$

In this case, at 25 °C the equilibrium constant is **4.7×10^{-31}**.

For the value of *K* to be so small, the numerator must be very much smaller than the denominator:

$$K = \frac{[NO]^2}{[N_2][O_2]}$$

Numerator

Denominator

A small numerator equates to a *low* concentration of NO. By contrast, a larger denominator equates to *high* concentrations of N_2 and O_2. If the concentration of NO is very much smaller than the concentrations of N_2 and O_2, the reaction $N_2 + O_2 \rightleftharpoons 2\,NO$ must lie to the *left*:

⟵ Back reaction is favoured

$$N_2 + O_2 \rightleftharpoons 2\,NO$$

Reactants predominate over products

Intermediate values of *K*

When *K* has a value between approximately 10^{-2} and 10^2 (i.e. neither very large nor very small), this tells us that at equilibrium, *all* components of the reaction are present in significant amounts, and that the equilibrium reaction is fairly evenly balanced–neither lying to the left nor to the right.

For example, we see above how, at 400 °C, the equilibrium constant for the reaction between hydrogen and iodine is approximately 50. For the equilibrium constant to have this kind of value, all components of the reaction must be present at concentrations which are of approximately similar magnitudes, so neither the numerator nor the denominator predominates:

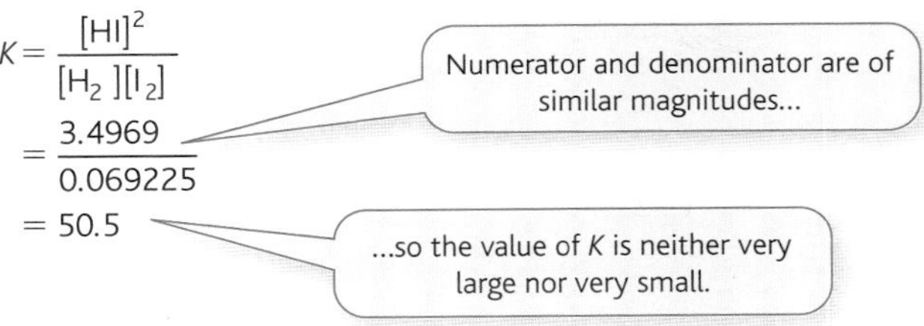
$$K = \frac{[HI]^2}{[H_2][I_2]}$$
$$= \frac{3.4969}{0.069225}$$
$$= 50.5$$
Numerator and denominator are of similar magnitudes...
...so the value of K is neither very large nor very small.

In practice, the reaction is evenly balanced at equilibrium.

- ***If the value of K is large, the equilibrium lies to the right and products dominate.***
- ***If K is small, the equilibrium lies to the left and reactants dominate.***
- ***If K falls between 10^{-2} and 10^2, the equilibrium lies in the middle, and reactants and products are both present in similar amounts.***

Self-check 15.2 Look at the following equilibrium constants. Do these equilibrium reactions lie to the left, to the right, or in the middle?

(a) $H_2CO_3 + H_2O \rightleftharpoons HCO_3^- + H_3O^+$ $\quad K = 4.2 \times 10^{-7}$

(b) $2\,NO + O_2 \rightleftharpoons 2\,NO_2$ $\quad K = 2.25 \times 10^{12}$

(c) $CH_4 + H_2O \rightleftharpoons CO + 3\,H_2$ $\quad K = 9.4 \times 10^{-1}$

(d) $2\,H_2 + O_2 \rightleftharpoons 2\,H_2O$ $\quad K = 3.3 \times 10^{81}$

It is important to note that the equilibrium constant for a particular equilibrium reaction depends on the **temperature** at which the reaction is performed. The value of an equilibrium constant for a reaction at 25 °C is different from the equilibrium constant for the *same* reaction at 50 °C: *K* depends strongly on temperature.

For example, at 400 °C, the value of *K* for the reaction:

$$H_2 + I_2 \rightleftharpoons 2\,HI$$

is 50.5. By contrast, at 425 °C, the value of *K* is **54.5**.

However, when carrying out a reaction *at a particular temperature*, it doesn't matter what concentrations we start with; the final value of *K* is always the same. In other words, changing the initial concentration of the reactants does not affect the equilibrium position.

Indeed, *K* stays the same even if the concentrations of the reactants and products at equilibrium differ (which will happen if different *initial* concentrations are used).

For example, if we start with one particular set of concentrations for H_2 and I_2, we may obtain the following concentrations for the reactants and products when a state of equilibrium is reached:

$$[H_2] = 0.22\ \text{mol L}^{-1}$$

$$[I_2] = 0.22 \text{ mol L}^{-1}$$

$$[HI] = 1.557 \text{ mol L}^{-1}$$

The equilibrium constant under these conditions would be:

$$K = \frac{[HI]^2}{[H_2][I_2]} = \frac{[1.557]^2}{[0.22][0.22]} = 50.1$$

However, a different set of starting concentrations for H_2 and I_2 would give a *different* set of concentrations when a state of equilibrium is reached. For example:

$$[H_2] = 0.38 \text{ mol L}^{-1}$$

$$[I_2] = 0.15 \text{ mol L}^{-1}$$

$$[HI] = 1.69 \text{ mol L}^{-1}$$

$$K = \frac{[HI]^2}{[H_2][I_2]} = \frac{[1.69]^2}{[0.38][0.15]} = 50.1$$

Notice how the equilibrium constant remains exactly the *same*, despite there being different concentrations of the reactants and products at equilibrium.

Look at Box 15.1 to find out how we can use the equilibrium constant, *K*, to decide which reactions will occur and which will not.

BOX 15.1 Using the equilibrium constant, *K*

The value of an equilibrium constant is very useful when deciding which reactions will occur and which will not. Let us look at two situations.

Situation 1

Suppose we wanted to go into the laboratory to make some phosgene (or 'carbonyl chloride') as the first step in synthesizing a new drug. One way of doing this is to react carbon monoxide and chlorine in the following reaction, in the gas phase:

$$CO + Cl_2 \rightleftharpoons COCl_2$$

We look up a table of equilibrium constants and discover the value of *K* at 100 °C is 4.57×10^9. Would we consider this a useful approach to synthesizing this compound or should we try and find a different way of making it?

The equilibrium constant under these conditions is a very large number, which tells us that the reaction proceeds far to the right: virtually all the reactants are converted to product when equilibrium is reached. So we can deduce that this would be an excellent approach to take to make phosgene. We would get a very high yield of our desired product, assuming that the reaction took place at a reasonable rate.

Situation 2

Imagine that we're trying to find a way to 'fix' nitrogen–that is, to turn nitrogen gas from the atmosphere into a form that living organisms can use during metabolic processes, namely NO. One option might be to react nitrogen with oxygen:

$$N_2 + O_2 \rightleftharpoons 2\,NO$$

However, we see above that, at 25 °C, the equilibrium constant for this reaction is 4.7×10^{-31}. Therefore, this reaction lies heavily to the left; we are forced to conclude that this is *not* a practical way to fix nitrogen.

It is worth noting that $N_2 + O_2 \rightleftharpoons 2\,NO$ is still an important *natural* reaction. During electrical storms, lightning generates extremely high atmospheric temperatures of 3000 K, which changes the value of the equilibrium constant, making the reaction more favourable (i.e. *K* is less small). Consequently, the fixing of atmospheric nitrogen, through its reaction with atmospheric oxygen, *does* occur naturally during electrical storms. Such NO then reacts with H_2O in the air to make HNO_3, which acts as a natural fertilizer.

Self-check 15.3 Ethanoic (acetic) acid reacts with water according to the following equation:

$$CH_3COOH + H_2O \rightleftharpoons CH_3COO^- + H_3O^+$$

The equilibrium constant for the reaction at 25 °C is 1.8×10^{-5}. Which of the reaction components predominate in the solution at equilibrium?

15.3 The reaction quotient

We see in Section 15.2 how the equilibrium constant, *K*, relates the concentrations of products to the concentrations of reactants *at the point at which a reaction is at equilibrium*. However, we can, in fact, relate the concentrations of reactants and products to one another even when the reaction is not at equilibrium.

The concentrations of reactants and products *at any point* in an equilibrium reaction are related by what we call the **reaction quotient, Q**[7]. We calculate the reaction quotient from the following expression:

$$Q = \frac{[C]^c[D]^d}{[A]^a[B]^b}$$

Notice how the terms on the right-hand side of this equation appear identical to those on the right-hand side of the equation for the equilibrium constant, *K*. Figure 15.9 shows how the reaction quotient, Q, applies when the concentrations of the components are at *anything other* than the equilibrium concentrations. When Q = *K*, that is, when the reaction quotient is equal to the equilibrium constant, the reaction is at equilibrium.

- ***The reaction quotient relates the concentrations of reactants and products at any point in an equilibrium reaction.***
- ***When a reaction has reached a state of dynamic equilibrium, Q = K.***

Self-check 15.4 The hydrolysis of ATP to ADP and a phosphate group can be represented by the following simplified equation:

$$ATP \rightleftharpoons ADP + P_i$$

In resting muscle, the concentrations of the three species are as follows: [ATP] = 0.01 M; [ADP] = 0.003 M; $[P_i]$ = 0.001M. What is the value of the reaction quotient, Q?

7 The reaction quotient is often referred to by biologists as 'the mass action ratio', represented by the symbol Γ.

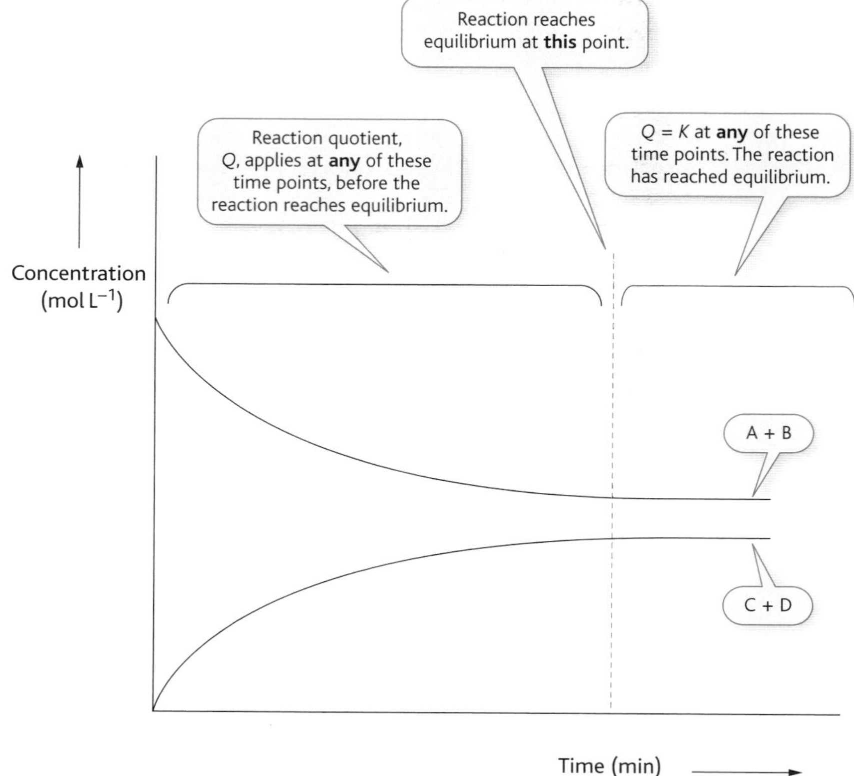
Reaction reaches equilibrium at **this** point.
Reaction quotient, Q, applies at **any** of these time points, before the reaction reaches equilibrium.
Q = K at **any** of these time points. The reaction has reached equilibrium.
Concentration (mol L−1)
A + B
C + D
Time (min)

Figure 15.9 The reaction quotient, Q, applies when the concentrations of the components are at *anything other* than the equilibrium concentrations. When $Q = K$, the reaction is at equilibrium.

Predicting the direction of a reaction

The reaction quotient can be used to help us predict whether the forward or back reaction needs to predominate for a reaction mixture to reach a state of equilibrium.

If the value of the reaction quotient is *less* than the equilibrium constant ($Q < K$), the concentrations of the products are *lower* than they would be at equilibrium. As a result, the reactants dominate the expression for the reaction quotient, making it smaller than it would be at equilibrium:

$$\text{if } Q < K \qquad Q = \frac{[C]^c[D]^d}{[A]^a[B]^b}$$

Reactants dominate: the denominator is larger than at equilibrium.

Therefore, the reaction must proceed from left to right (in the direction of the forward reaction) for the concentration of products to be increased, and for equilibrium to be reached:

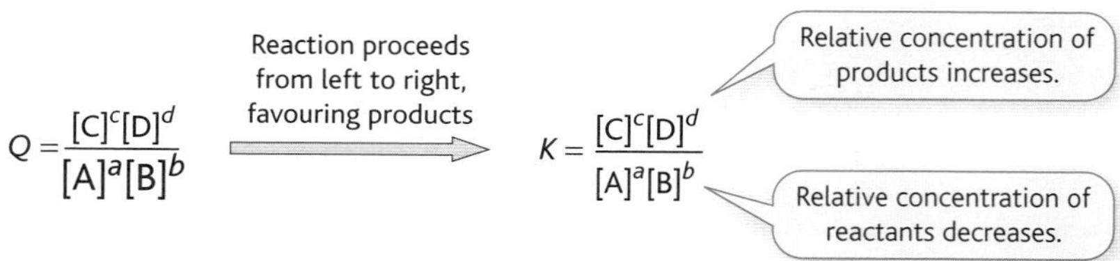
Q = [C]c[D]d / [A]a[B]b
Reaction proceeds from left to right, favouring products
K = [C]c[D]d / [A]a[B]b
Relative concentration of products increases.
Relative concentration of reactants decreases.

By contrast, if the value of the reaction quotient is *greater* than the equilibrium constant ($Q > K$), the concentrations of the products are higher than they would be at equilibrium. As a result, the products dominate the expression for the reaction quotient, making it larger than it would be at equilibrium:

if $Q > K$ $$Q = \frac{[C]^c[D]^d}{[A]^a[B]^b}$$

Products dominate: the numerator is larger than at equilibrium.

Consequently, the reaction must proceed from right to left (in the direction of the back reaction) for the concentration of the products to be reduced, and for equilibrium to be reached:

$$Q = \frac{[C]^c[D]^d}{[A]^a[B]^b} \xrightarrow{\text{Reaction proceeds from right to left, favouring reactants}} K = \frac{[C]^c[D]^d}{[A]^a[B]^b}$$

Relative concentration of products increases.

Relative concentration of reactants decreases.

When the value of the reaction quotient is the same as the equilibrium constant, $Q = K$, the reaction mixture is already at equilibrium, so there is no change in the balance between the forward and back reactions.

EXAMPLE

Let's return to the equilibrium reaction through which glucose 6-phosphate undergoes isomerization to form fructose 6-phosphate in the presence of glucose 6-phosphate isomerase:

glucose 6-phosphate ⇌ fructose 6-phosphate

The equilibrium constant for this reaction, K, is 0.43 at 298 K. Suppose we were to measure the concentrations of the two substances in the reaction mixture at a particular point in time, and obtained the following values:

[glucose 6-phosphate] = 0.08 mol L^{-1}

[fructose 6-phosphate] = 0.02 mol L^{-1}

In which direction would the reaction need to proceed for equilibrium to be reached?

We calculate the reaction quotient from the following expression:

$$Q = \frac{[C]^c[D]^d}{[A]^a[B]^b} = \frac{[\text{fructose 6-phosphate}]}{[\text{glucose 6-phosphate}]}$$

As there is only one term on either side of the equilibrium arrow, the expression for the reaction quotient in this instance simplifies to $Q = [C]^c/[A]^a$

$$Q = \frac{0.02}{0.08}$$

$$Q = 0.25$$

From our measurements, the reaction quotient is *smaller* than the equilibrium constant (0.43), so the reaction must proceed from left to right (in the direction of the forward reaction) for equilibrium to be reached:

glucose 6-phosphate $\rightarrow$ fructose 6-phosphate

Now suppose we were to measure the concentrations of the two substances in a different reaction mixture, and obtained the following, different, values:

[glucose 6-phosphate] = 0.05 mol L^{-1}

[fructose 6-phosphate] = 0.05 mol L^{-1}

In which direction would the reaction need to proceed for equilibrium to be reached?

$$\text{Again, } Q = \frac{[\text{fructose 6-phosphate}]}{[\text{glucose 6-phosphate}]}$$

$$Q = \frac{0.05}{0.05}$$

$$Q = 1$$

This time, we find the reaction quotient is *larger* than the equilibrium constant ($K = 0.43$), so the reaction must proceed from right to left (in the direction of the back reaction) for equilibrium to be reached:

fructose 6-phosphate $\rightarrow$ glucose 6-phosphate

- ***If $Q < K$ the reaction proceeds from left to right (in the direction of the forward reaction).***
- ***If $Q > K$ the reaction proceeds from right to left (in the direction of the back reaction).***
- ***If $Q = K$ the reaction is already at equilibrium concentrations. There is no further change.***

Self-check 15.5 When we measure the concentrations of the components of the equilibrium reaction $H_2 + I_2 \rightleftharpoons 2\ HI$, we obtain the following values:

$[H_2]$ = 1.3 mol L^{-1}

$[I_2]$ = 0.7 mol L^{-1}

$[HI]$ = 0.3 mol L^{-1}

In which direction must the reaction proceed for equilibrium to be reached? The equilibrium constant for this reaction, K, at 400 °C is 50.5.

15.4 Binding reactions in biological systems

Many of the dynamic equilibrium reactions we explore earlier in this chapter are examples of chemical change: one or more species (reactants) interacting to form one or more products, in a manner that can be reversed. However, biological systems feature other more general processes that we wouldn't normally consider to be chemical reactions, but which still take the form of dynamic equilibria. We encounter such a process at the start of this chapter, when we consider the binding of oxygen to haemoglobin:

$$\mathrm{Hb} + \mathrm{O_2} \rightleftharpoons \mathrm{Hb{-}O_2}$$

We can represent this process as an equilibrium reaction, but simply denote the haemoglobin complex with the shorthand 'Hb' where we might usually write a chemical species.

Such binding processes are essential features of biology–whether it be the binding of a signal molecule to its receptor, or an enzyme to its substrate. The specificity exhibited by one molecule when it binds to another is often reflected by how strongly the two entities bind: strong binding is usually taken to indicate a high degree of specificity, whereas two entities that associate only weakly fail to show much specificity for one another.

But what has the strength of binding to do with dynamic equilibria? Let us now explore the answer to this question.

Consider a protein (P) binding to its ligand (L) (perhaps an enzyme binding to its substrate, or a receptor binding to its signalling molecule), a process we can represent by the following general scheme:

$$\mathrm{P} + \mathrm{L} \rightleftharpoons \mathrm{PL}$$

During the forward reaction, the protein and ligand associate to form the protein–ligand complex, PL. During the back reaction, the protein–ligand complex dissociates to form separate protein and ligand molecules, P and L. Where might we expect the equilibrium position to lie if the protein and ligand showed strong binding, and how would the equilibrium position vary if the binding was weak?

If a ligand binds strongly to a given protein then we might expect little in the way of dissociation. That is, once associated, the protein and ligand would be quite likely to remain associated, rather than drift apart. Therefore, we could expect the equilibrium reaction $\mathrm{P} + \mathrm{L} \rightleftharpoons \mathrm{PL}$ to lie to the right, favouring the formation of the PL complex.

By contrast, if a protein and ligand associate only weakly, we might expect the protein–ligand complex to dissociate readily–that is, to proceed in the direction of the back reaction, $\mathrm{PL} \rightarrow \mathrm{P} + \mathrm{L}$. Under such conditions, the equilibrium reaction lies to the left, favouring the formation of the reactants, P and L.

- ***The binding of a protein and its ligand can be represented by the general equilibrium reaction*** $\mathrm{P} + \mathrm{L} \rightleftharpoons \mathrm{PL}$.
- ***If a protein and its ligand bind weakly, the dissociation reaction,*** $\mathrm{PL} \rightarrow \mathrm{P} + \mathrm{L}$, ***is favoured.***
- ***If a protein and its ligand bind strongly, the association reaction,*** $\mathrm{P} + \mathrm{L} \rightarrow \mathrm{PL}$, ***is favoured.***

We see in Section 15.2 how we can estimate the position of an equilibrium reaction using the equilibrium constant, *K*, where

$$K = \frac{[C]^c[D]^d}{[A]^a[B]^b}$$

for the general equilibrium reaction $a\,A + b\,B \rightleftharpoons c\,C + d\,D$. We can use this relationship to write an analogous expression for the equilibrium constant for the reaction $P + L \rightleftharpoons PL$:

$$K = \frac{[PL]}{[P][L]}$$

Because this equilibrium constant relates to a binding reaction–that is, the association of P and L–we give a special name: the association constant, and denote it K_a.

However, in biology, we are usually most interested in the reverse process, that of **dissociation**. (It is the process of dissociation–how likely it is that two entities will come apart once associated–that tells us something useful about the strength of binding.) Focusing on our protein–ligand interaction once again, we can represent the dissociation reaction as follows:

$$PL \rightleftharpoons P + L$$

And so can write down the following equilibrium constant:

$$K = \frac{[P][L]}{[PL]}$$

We call this equilibrium constant the dissociation constant, K_d. Notice how the association constant, K_a, and dissociation constant, K_d, relate to complementary–but opposite–processes[8].

What does the dissociation constant tell us?

We now turn to the question 'How can we use the dissociation constant to tell us something useful about the binding behaviour of two entities and, particularly, whether they bind strongly or weakly?'

We note above that the position of equilibrium for a binding reaction is related to the strength of binding. For the dissociation reaction, $PL \rightleftharpoons P + L$, we would expect the equilibrium position for a protein and ligand that showed strong binding to lie to the *left*. (That is, we would expect little of the PL complex to dissociate to form P and L.) By contrast, we would expect the reaction between a *weakly* bound protein and ligand to lie to the right, favouring the formation of separate protein and ligand (P + L) rather than the bound complex.

How would this be reflected in the value of the dissociation constant, K_d?

We see in Section 15.2 that an equilibrium reaction that lies to the right–where products are favoured–has a value of *K* that is large. By contrast, an equilibrium reaction that lies to the left (favouring the reactants) has a value of *K* that is small. We can use exactly the same ideas to relate the value of K_d to the position of a dissociation reaction, as depicted in Figure 15.10. For the dissociation reaction $PL \rightleftharpoons P + L$, a large value of K_d tells us that the reaction lies to the right. This means that the reaction favours the products of the forward reaction, P and L, so the

8 Further, in numerical terms, K_d is the reciprocal of K_a–that is, $K_d = 1/K_a$

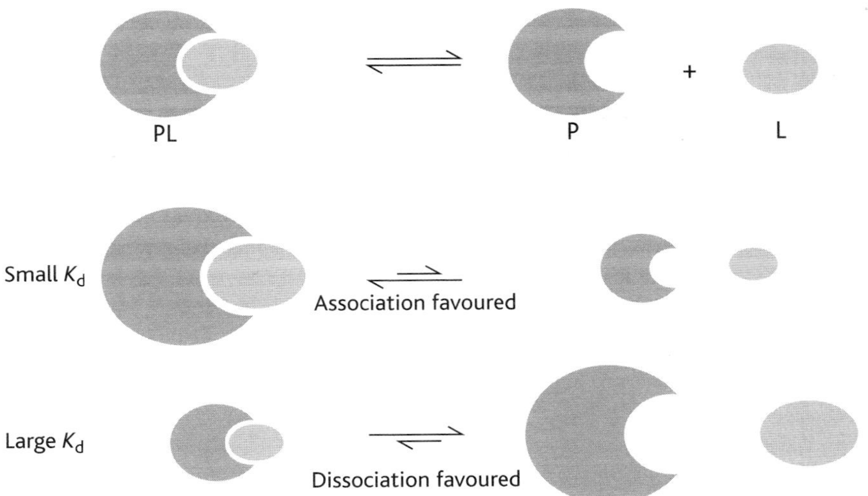

Figure 15.10 The relationship between the value of K_d and the extent of dissociation of a protein–ligand complex. When K_d is small, association (the continuation of binding) is favoured over dissociation. When K_d is large, dissociation of the protein–ligand complex to give separate protein and ligand is favoured.

complex dissociates readily. This being the case, the protein and ligand must bind only weakly, and show a low degree of specificity for each other.

By contrast, if the value of K_d is small, the dissociation reaction lies to left such that the reactant, PL, is favoured. Therefore, dissociation happens only to a small extent; the protein and ligand bind strongly, and so show a high degree of specificity for each other.

- ***The dissociation constant, K_d, gives a measure of the strength of binding between a protein and its ligand.***
- ***If binding is weak, dissociation is favoured, the dissociation reaction lies to the right, and the value of K_d is large.***
- ***If binding is strong, dissociation is not favoured, the dissociation reaction lies to the left, and the value of K_d is small.***

Let us consider a couple of examples to illustrate these concepts.

The enzyme lactate dehydrogenase (LDH) catalyzes the conversion of pyruvate, the end-product of glycolysis, into lactate according to the following reaction scheme.

$$H_3C-C(=O)-CO_2H \xrightleftharpoons[\text{NADH} \rightarrow \text{NAD}^+]{\text{LDH}} H_3C-C(OH)(CH_3)-CO_2H$$

This reaction requires the coenzyme NADH, which binds to LDH and is converted to NAD^+ during the course of the reaction[9]. We can determine the relative strength

9 A coenzyme is a small, non-protein molecule which is required by an enzyme for its catalytic activity. The coenzyme isn't a permanent part of the enzyme but, instead, is one of the enzyme's substrates–that is, it must bind to the enzyme to have its effect.

of binding of NADH and NAD^+ to LDH by considering the dissociation constants associated with the following binding equilibria:

$$\text{LDH-NADH} \rightleftharpoons \text{LDH} + \text{NADH}$$

$$\text{LDH-NAD}^+ \rightleftharpoons \text{LDH} + \text{NAD}^+$$

When LDH is present at a concentration of 0.145 μg mL^{-1}, at a pH of 10.0, the values of K_d are as follows:

$$\text{LDH–NADH} \rightleftharpoons \text{LDH} + \text{NADH} \qquad K_d = 8\ \mu\text{M}$$

$$\text{LDH–NAD}^+ \rightleftharpoons \text{LDH} + \text{NAD}^+ \qquad K_d = 400\ \mu\text{M}$$

What does this tell us about the relative strength of binding of NADH and NAD^+ to LDH? The small K_d value for NADH tells us that this dissociation reaction lies to the *left* and NADH binds strongly to LDH. By contrast, the relatively large K_d value for NAD^+ tells us that this dissociation reaction lies to the right, and NAD^+ binds relatively weakly to LDH. This means that, if we were to mix LDH with equal concentrations of NADH and NAD^+, the LDH–NADH complex would greatly predominate over the LDH–NAD^+ complex.

As another example, the compound naloxone is used to treat opiate overdoses. (Opiates include the drugs heroin and morphine.) But how does it work? Heroin binds to opioid receptors in the body with a K_d of 2 nM. By contrast, naloxone binds much more tightly: it binds with a K_d of 7.3 pM[10]. Therefore, when both heroin and naloxone are present, naloxone displaces heroin from opioid receptors: it binds more tightly (and so preferentially), and so effectively reduces the concentration of heroin present, overcoming the effect of the overdose.

Self-check 15.6 The hormone adrenaline binds to beta-adrenergic receptors in the body. The drug alprenolol is used as a beta-blocker: it binds to beta-adrenergic receptors more tightly than adrenaline, and so acts to reduce its biological activity. When we measure the K_d values for the binding of alprenolol and adrenaline we get two contrasting values: 500 nM and 3.1 nM. But which is the K_d value for alprenolol, and which the K_d value for adrenaline?

Many biological interactions are more complicated than that of a single ligand binding to a protein. (For example, haemoglobin binds four different oxygen molecules to four distinct binding sites.) However, the general principles of binding being an example of an equilibrium reaction–and on the position of equilibrium being a measure of the strength of binding–are ones that remain valid (and instructive) however complicated the particular example we are considering.

Having now considered the notion of a system attaining the steady state of equilibrium let us now consider what happens when this steady state is disrupted, and an equilibrium is perturbed.

10 Remember: 1 pM = 10^{-12} M, while 1 nM = 10^{-9} M. Therefore, 7.3 pM is a much smaller value than 2 nM–around 300 times smaller, in fact.

15.5 Perturbing an equilibrium

When a reaction is at equilibrium, both the forward and back reactions occur to the same extent. However, if the equilibrium is perturbed (that is, disturbed in some way), the concentrations of the components will readjust until equilibrium is re-established.

The response of a reaction mixture at equilibrium to perturbation (to the equilibrium being disrupted) is described by the principle of Henri Le Chatelier. He stated that '**when a chemical system at equilibrium is disturbed, it responds by shifting the equilibrium composition in such a way to counteract the change**'.

Picture two people on a see-saw, as depicted in Figure 15.11. When the weight of the two people is evenly balanced, the see-saw is at equilibrium, and lies horizontally. However, if we add a second person to one side of the see-saw, we

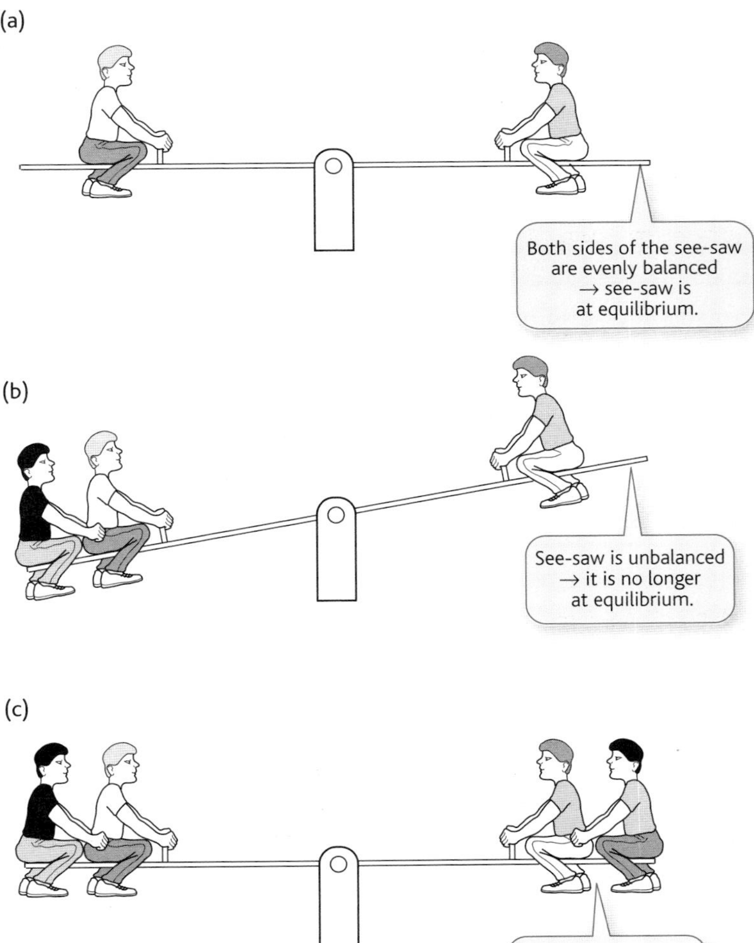
(a)
Both sides of the see-saw are evenly balanced → see-saw is at equilibrium.
(b)
See-saw is unbalanced → it is no longer at equilibrium.
(c)
Equilibrium can only be restored by rebalancing both sides of the see-saw.

Figure 15.11 (a) When the force applied on two sides of a see-saw is equal, the see-saw is at equilibrium, and lies flat, horizontally. (b) If the force applied is unequal, the see-saw is no longer at equilibrium, and tips towards the side on which the greatest force is being applied. (c) Equilibrium is restored by re-balancing both sides of the see-saw.

perturb the equilibrium (the weight on both sides of the see-saw is no longer balanced): the see-saw shifts away from its horizontal position, and tilts. In order to *counteract* this change, we must add more weight to the other side of the see-saw–that is, we shift the balance of weight so that the see-saw tilts back towards the horizontal.

When a chemical equilibrium is perturbed, and shifts away from its balanced position, the reaction mixture carries out the equivalent of rebalancing the see-saw–it undergoes change–to bring conditions back to the balanced state of equilibrium. The 'change' is in the form of a change in the rates of the forward and back reactions so that they are no longer equal–either the forward reaction predominates, or the back reaction predominates. But the disruption is only ever temporary: the predominance of the forward or back reaction occurs until the initial disruption, which perturbed the equilibrium, is overcome, and a state of equilibrium is reached once more.

- ***When an equilibrium is disturbed, the system changes to counteract the disturbance until equilibrium is re-established.***

Before we look in more detail at how a reaction mixture responds to the perturbation of equilibrium, however, we need to think about how a state of equilibrium is perturbed in the first place.

There are three key ways of perturbing a reaction at equilibrium:

1. Actively changing the concentrations of one or more of the components of the reaction, by adding a chemical
2. Changing the pressure of the system (if the reaction proceeds in the gas phase)
3. Changing the temperature of the system.

Let's consider each of these in turn.

Changing the concentration of the system

Let us look again the equilibrium reaction represented by the isomerization of glucose 6-phosphate:

$$\text{glucose 6-phosphate} \rightleftharpoons \text{fructose 6-phosphate} \qquad K = 0.43 \text{ at } 298\text{ K } (25\text{ °C})$$

Let's assume that, at equilibrium, the two species are present in the following concentrations:

$$[\text{glucose 6-phosphate}] = 0.07 \text{ mol L}^{-1}$$

$$[\text{fructose 6-phosphate}] = 0.03 \text{ mol L}^{-1}$$

$$K = \frac{[\text{fructose 6-phosphate}]}{[\text{glucose 6-phosphate}]} = \frac{[0.03]}{[0.07]} = 0.43$$

We now investigate what happens if we *double* the concentration of glucose 6-phosphate present to 0.14 mol L^{-1}. Glucose 6-phosphate is no longer present at its equilibrium concentration (0.07 mol L^{-1}): the state of equilibrium has been perturbed.

The system must now readjust its concentrations to re-establish an equilibrium situation. It does this by changing the relative extents of the forward and back reactions so that more glucose 6-phosphate is converted to fructose 6-phospate (to lower its concentration from 0.14 mol L^{-1} back down towards a new equilibrium value).

Logic tells us that this requires an increase in the rate of the forward reaction, glucose 6-phosphate $\rightarrow$ fructose 6-phosphate. However, we can use the reaction quotient, Q, to verify this.

When we double the concentration of glucose 6-phosphate, the reaction quotient is:

Remember: the reaction is no longer at equilibrium, so we must consider the reaction quotient and *not* the equilibrium constant.

$$Q = \frac{[\text{fructose 6-phosphate}]}{[\text{glucose 6-phosphate}]} = \frac{[0.03]}{[0.14]} = 0.21$$

Notice how the reaction quotient is *smaller* than the equilibrium constant. We see above how, if $Q < K$, the reaction must proceed from left to right (in the direction of the forward reaction) to bring the system to equilibrium. So, in this case, the rate of the forward reaction, glucose 6-phosphate $\rightarrow$ fructose 6-phosphate, is increased to 'consume' some of the added glucose 6-phosphate. The concentration of glucose 6-phosphate is reduced by its conversion to fructose 6-phosphate, such that the concentration of fructose 6-phosphate *increases*. Therefore, if we measure the concentrations again when equilibrium is re-established, we find *less* glucose 6-phosphate and *more* fructose 6-phosphate present:

$$[\text{glucose 6-phosphate}] = 0.119 \text{ mol L}^{-1}$$

$$[\text{fructose 6-phosphate}] = 0.051 \text{ mol L}^{-1}$$

However, even though the concentration of each reaction component at equilibrium is quite different, *the equilibrium constant remains the same*–the overall ratio of products to reactants within the expression for K is unchanged, even though the actual quantities present have changed:

$$K = \frac{[\text{fructose 6-phosphate}]}{[\text{glucose 6-phosphate}]} = \frac{[0.051]}{[0.119]} = 0.43$$

Manipulating concentrations to affect the equilibrium position of a particular reaction can be a useful tool in making things happen in the way we want. Consider an enzyme binding to its substrate–an example of a binding reaction discussed in Section 15.5. Imagine that the enzyme and substrate showed only moderate specificity such that binding is only weak. We could represent the process of dissociation as follows: $ES \rightleftharpoons E + S$. Under conditions of fairly weak binding we might expect the dissociation reaction, $ES \rightarrow E + S$, to be favoured, such that the reaction catalyzed by the enzyme when it associates with its substrate happens to only a modest extent. How might we improve the yield of the desired product?

Imagine that we increased the concentration of free substrate, S, present in the system after a state of equilibrium had been achieved. How would the system respond? According to Le Chatelier's principle, the system would compensate by shifting the reaction to the left:

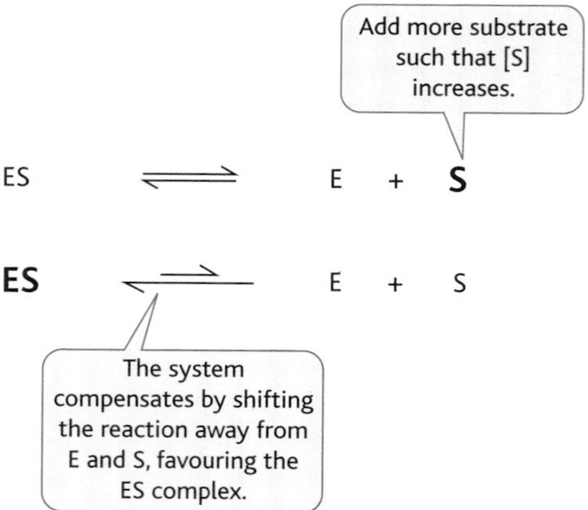

This change would favour the formation of the enzyme–substrate complex, effectively increasing the strength of binding, and reducing the degree of dissociation of the enzyme–substrate complex.

We may think this is a somewhat artificial example. Surely biological systems can't manipulate concentrations of cellular components quite like this? In fact, they can. Such manipulation can happen by virtue of the way that the cell is divided into different compartments (or organelles) by a membrane. The membrane acts as a physical barrier, which can be used to delineate different biochemical conditions. For example, one side of a cell membrane may maintain concentrations of certain molecules at a level higher than on the other side–and the presence of the membrane itself can help to ensure that this difference is maintained. We see such differences across the membrane of a nerve cell, where Na^+ and K^+ ions are present at different concentrations. It is this difference in concentration that lies at the heart of the passage of nerve impulses.

Look at Box 15.2 to see how the notion of counteracting a change in concentration away from the equilibrium position is central to **osmosis**–one of the processes through which the cells of our body regulate the concentrations of the chemicals they contain.

BOX 15.2 Osmosis: balancing concentrations in the cell

When isolated red blood cells are placed in pure water they expand until they eventually burst: water molecules surrounding the red blood cells rush across the cell membrane and fill the cell to the point that the membrane can no longer withstand the internal pressure; it ruptures, destroying the cell.

But why does this happen? Water molecules move across cell membranes in such a way as to balance the concentration of the solutions on either side of the membrane: they move until the concentrations of the solutions on both sides of the membrane have reached equilibrium. This movement is called osmosis, and the force driving the movement of the water molecules is called osmotic pressure.

Red blood cells effectively contain a solution of salt. Pure water contains no salt. In consequence, when the red blood cells are placed in pure water, osmosis drives the movement of water molecules across the cell membrane from the pure water and *into* the cell interior, in an attempt to even up the concentrations of the solutions on either side of the membrane. Adding H_2O in the cell dilutes the salt, so the concentration goes down. This process is depicted in Figure 15.12.

In order to preserve red blood cells whole and intact, they must be placed in an aqueous solution of 0.9% sodium chloride. This solution has the *same* concentration as the interior of the red blood cell, so the aqueous solution and the cell interior

BOX 15.2 **Continued**

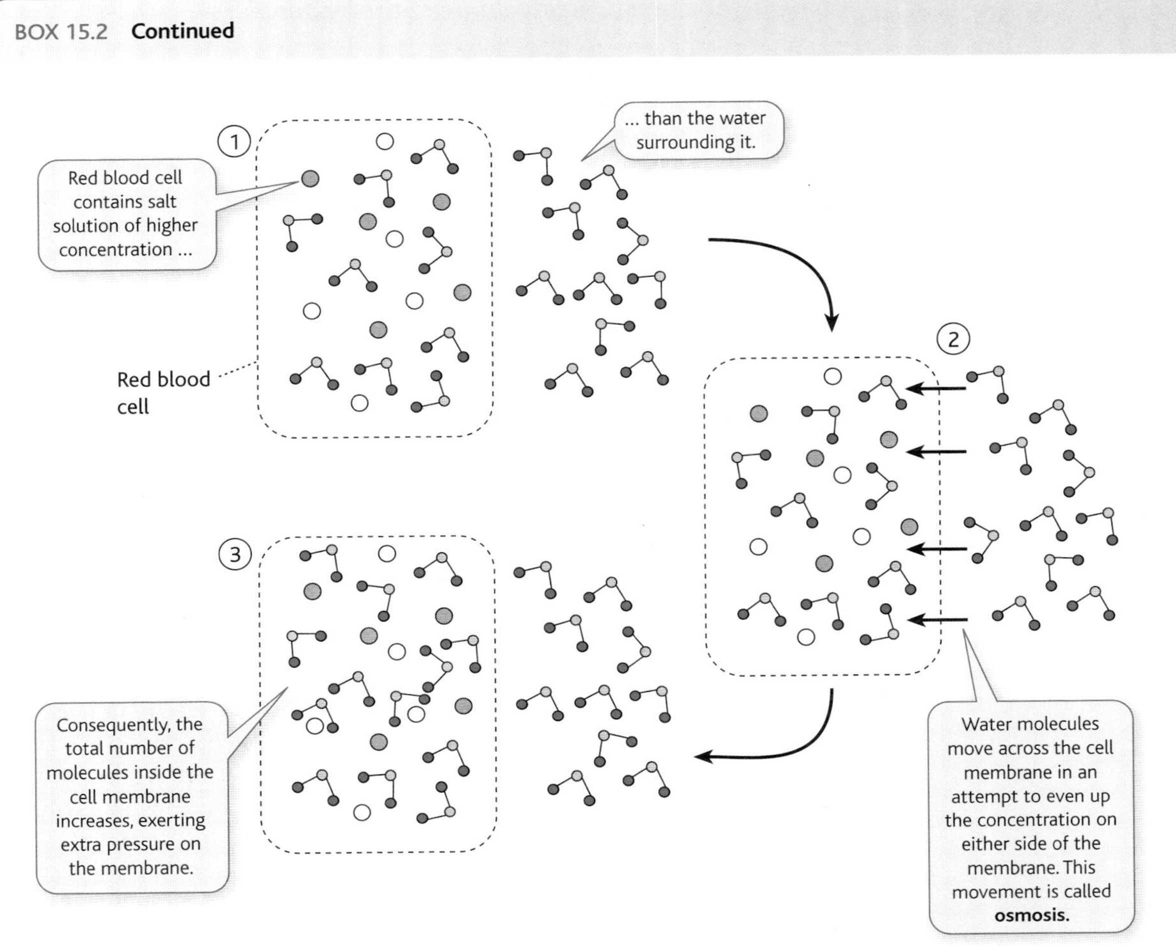
1
Red blood cell contains salt solution of higher concentration ...
... than the water surrounding it.
Red blood cell
2
Water molecules move across the cell membrane in an attempt to even up the concentration on either side of the membrane. This movement is called **osmosis.**
3
Consequently, the total number of molecules inside the cell membrane increases, exerting extra pressure on the membrane.

Figure 15.12 Osmosis in action. Water molecules move across cell membranes in such a way as to balance up the concentration of the solutions on either side of the membrane. The interior of a red blood cell comprises a more concentrated salt solution than water. When a red blood cell is placed in pure water, water molecules move under the force of osmosis into the interior of the red blood cell in an attempt to even up the salt solution concentrations on either side of the membrane.

are already at equilibrium, and no subsequent movement of water needs to occur.

The 0.9% sodium chloride solution is said to be isotonic (iso- is Greek for 'same'): the concentrations of its component molecules are identical to that of the cell. A solution which is less concentrated than an isotonic solution is called hypotonic (hypo = lower); a solution which is more concentrated than an isotonic solution is called hypertonic (hyper = higher).

What happens if red blood cells are placed into a hypertonic solution containing a 9% solution of sodium chloride, ten times more concentrated than the isotonic solution? In this case, water molecules flow *from* the more dilute cell interior *into* the more concentrated cell exterior, causing the cell to shrivel as its volume decreases in accompaniment with water loss.

In some circumstances, however, an imbalance in the concentrations of certain ions across cell membranes is necessary for the cell to function correctly. For example, nerve cells transmit electrical impulses in direct consequence of differences in concentrations of Na^+ and K^+ ions across the nerve cell membrane. However, maintaining different concentrations requires energy: the cell must actively counter the effect of osmosis, which automatically tries to balance concentrations.

Changing the pressure or volume of the system

Changes in pressure only affect the equilibrium position of reactions when gases are involved. Even then, changing pressure only affects those reactions in which the total number of gaseous molecules on either side of the equation is different. Consider the reaction between nitrogen and hydrogen to produce ammonia:

$$N_2 + 3\,H_2 \rightleftharpoons 2\,NH_3$$

The total pressure in the reaction vessel depends upon the number of molecules present–the greater the number of molecules present, the greater the pressure being exerted on the system.

Imagine standing in a phone box, and being joined by a friend. As soon as the second person enters the box, it starts to feel like a tight squeeze. If a third person joins you, you become even more tightly packed: the volume you're occupying is staying the same; you're merely cramming more people in. The more tightly you cram objects into a fixed volume, the greater the pressure that is exerted on that volume. So a sealed jar containing 1 mol of a gas is under greater pressure than the same sealed jar containing just 0.1 mol.

Notice how, for this reaction, there are four reactant molecules for every two product molecules–so the product molecules exert a lower pressure than the reactants (in fact, they exert half the pressure because there are half the number of molecules).

What would be the effect of increasing the pressure on the reaction when it is at equilibrium? The Le Chatelier principle tells us that a system re-establishes equilibrium in such a way as to counteract the stress imposed upon it. If we increase the pressure, the system responds by *decreasing* its own pressure to counteract the change we've imposed artificially.

But how can the system change to decrease the pressure? We have just said that the product (NH_3) exerts half as much pressure as the reactants because there are fewer molecules occupying the same volume. Figure 15.13 shows how, if we increase the pressure, the system responds by converting some of the reactants into products–decreasing the total number of molecules present and, therefore,

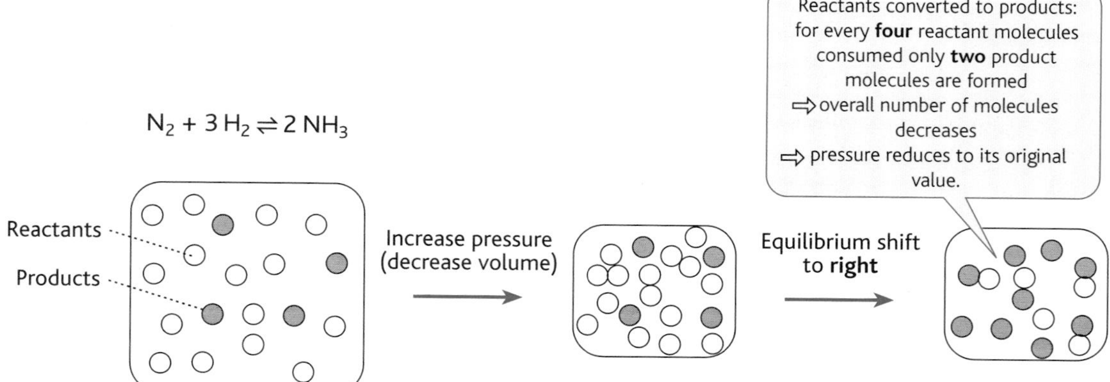

Figure 15.13 If we increase the pressure exerted on a system at equilibrium, the system responds by counteracting the pressure increase. In this case, the system converts some of the reactants into products–decreasing the total number of molecules present and, therefore, decreasing the overall pressure.

decreasing the overall pressure. So, when we increase the pressure, the system responds by shifting the reaction towards the right–increasing the extent of the forward reaction, $N_2 + 3\ H_2 \rightarrow 2\ NH_3$–to convert more N_2 and H_2 to NH_3, and thereby decreasing the total number of molecules present, and causing an overall decrease in the system's pressure.

By a similar argument, the system responds to a *decrease* in pressure (relative to the pressure at equilibrium) by shifting the reaction to the *left*–increasing the rate of the back reaction so that the total number of molecules *increases*, and the overall pressure is elevated back towards its equilibrium value.

In a reaction where the total number of molecules is the same on both sides of the equation, however, a change in pressure has no effect on the equilibrium because changing the external pressure affects both the forward and back reactions to the same extent, i.e. no *net* change. For example, consider the reaction between gaseous nitrogen and oxygen:

$$N_2 + O_2 \rightleftharpoons 2\,NO$$

One molecule of nitrogen reacts with one molecule of oxygen to form two molecules of nitrogen dioxide, so the pressure of reactants and products is the same both sides. Increasing or decreasing pressure has no effect on the position of equilibrium.

The effect of pressure on the position of equilibrium is central to our survival. Read Box 15.3 to find out how.

BOX 15.3 Every breath we take: pressure, equilibrium, and the binding properties of haemoglobin

Every moment of our lives, we depend on changes in pressure on an equilibrium reaction, even though we may not realize it.

The complex protein haemoglobin facilitates the transport of oxygen from our lungs to our body tissues. The reaction by which oxygen binds to the haemoglobin must be capable of reversing: the oxygen must first bind to the haemoglobin but subsequently be released when it reaches the appropriate body tissues (where it is utilized to generate energy).

We can represent the binding of oxygen to haemoglobin ('Hb') by the following equilibrium reaction:

$$Hb + O_2 \rightleftharpoons Hb - O_2$$

Both reactions occur in the lungs, but the forward reaction occurs to the greatest extent:

$$Hb + O_2 \rightarrow Hb - O_2$$

So oxygen binds to the haemoglobin.

By contrast, in the body tissue, it is the back reaction which occurs to the greatest extent

$$Hb - O_2 \rightarrow Hb + O_2$$

i.e. oxygen is released from the haemoglobin complex and becomes available for biochemical reactions in the body.

But what drives this shift in the position of equilibrium–from the reaction initially lying to the right (with the forward reaction favoured), to the position of equilibrium lying to the left (when the back reaction is favoured)? The answer lies in the respective pressures of oxygen in the two systems, lungs and body tissue.

(More precisely, we need to consider the so-called partial pressure of oxygen–that is, the proportion of the overall pressure of the system which is attributed to oxygen, rather than to any other compound in the air we breathe, such as nitrogen, N_2, or argon, Ar.)

The pressure of oxygen in the lungs is about 14 kPa, whereas the pressure of oxygen in the body tissue is much lower at about 5 kPa. We see in this section how *lowering* the pressure of oxygen favours the reaction whereby the greatest number of molecules is formed, while *increasing* the pressure favours the reaction in which the smallest number of molecules is formed. In consequence, we see how the relatively *high* pressure of oxygen in the lungs favours the reaction generating the smallest number of molecules: $Hb + O_2 \rightarrow Hb - O_2$. By contrast, the relatively *low* pressure of oxygen in the body's tissues favours the reaction generating the largest number of molecules: $Hb - O_2 \rightarrow Hb + O_2$.

As a result, oxygen is taken up and released by haemoglobin at the right time, in the right place. Without such a shift in the position of equilibrium, our tissues would fail to receive the supply of oxygen needed for survival–with fatal consequences.

Self-check 15.7 When we increase the pressure exerted on the following reaction systems at equilibrium, do the reactions shift to the left, to the right, or remain unchanged?

(a) $2\ H_2(g) + O_2(g) \rightleftharpoons 2\ H_2O(g)$

(b) $H_2(g) + I_2(g) \rightleftharpoons 2\ HI(g)$

(c) $CH_4(g) + H_2O(g) \rightleftharpoons CO(g) + 3\ H_2(g)$

Changing the temperature

We see in Section 15.2 how the equilibrium constant (a measure of the ratio of products to reactants for a reaction at equilibrium) is **temperature dependent**. If we change the temperature, the existing equilibrium constant changes, causing the reaction mixture to change to a new composition.

Once again, a reaction mixture responds according to Le Chatelier's principle: it **counteracts** the change imposed upon it. If we increase the temperature of the system, the system responds by *decreasing* the temperature again. By contrast, if we decrease the temperature of the system, the system responds by *increasing* the temperature.

How does a system effect a change in temperature? It does so by shifting the position of the equilibrium so energy is either transferred to or released from the system.

Look at p. 427, Chapter 13, to find out more about the terms exothermic and endothermic.

We see in Chapter 13 how some reactions are exothermic (they transfer energy to the surroundings as heat) and others are endothermic (they absorb energy from the surroundings). In the case of an equilibrium reaction, if the forward reaction is exothermic, the back reaction is endothermic, and vice versa.

At equilibrium, the forward and back reactions occur at the same rate, so the overall temperature change in the system is zero–there is no net gain or loss of energy[11]. However, if the forward and back reactions are occurring to *different* extents, we see an overall *change* in temperature:

- If the exothermic reaction predominates, we see an increase in temperature.
- If the endothermic reaction predominates, we see a decrease in temperature.

Let's look again at the reaction between nitrogen and hydrogen to form ammonia:

$$N_2 + 3\ H_2 \rightleftharpoons 2\ NH_3$$

How might we expect an increase in temperature to affect the equilibrium? This reaction is exothermic in the forward direction (it *loses* energy as heat):

$$N_2 + 3\ H_2 \rightarrow 2\ NH_3\ (+\text{heat})$$

and endothermic in the back direction (it *absorbs* energy as heat):

$$2\ NH_3\ (+\text{heat}) \rightarrow N_2 + 3\ H_2$$

11 Remember: the temperature of a system gives us only an indirect measure of the energy of the system, as explained on p. 423. As energy transfers to a system as heat, the kinetic energy of the molecules comprising the system increases–that is, they move around more. This extra movement translates into increased thermal energy–which we measure as the temperature of the system.

If we increase the temperature of the system (such that more energy is transferred to it as heat), the system responds by shifting to the left (favouring the endothermic back reaction) so that the heat we've added is *used* to counteract the temperature increase:

$$N_2 + 3\,H_2 \rightleftharpoons 2\,NH_3$$

By contrast, if we cool the system, the system responds by shifting to the right (favouring the exothermic forward reaction) to produce more heat to compensate for the effect of cooling.

Self-check 15.8 Predict the direction in which each of the following equilibrium reactions will shift following an increase in temperature, given the enthalpy change for each reaction indicated.

(a) glucose + 6 O_2 $\rightleftharpoons$ 6 CO_2 + 6 H_2O; $\Delta H° = -2821.5$ kJ mol^{-1}

(b) glutamine(aq) + NH_4^+(aq) $\rightleftharpoons$ glutamate(aq) + H_2O(l); $\Delta H° = +21.8$ kJ mol^{-1}

We need to bear in mind the important point we encounter on p. 523: when a system shifts its equilibrium position in response to something like a change in temperature the composition of the system when the new equilibrium is reached is *different* from the composition prior to the shift. That is, there may be relatively more product, or relatively more reactant.

This could have potentially serious implications for biological systems, where the molecular composition of cells and tissues needs to be tightly controlled. Imagine a tissue in which a biochemical reaction occurs that increases the acidity (decreases the pH) of the tissue by releasing protons. For example, the conversion of glucose to lactic acid during anaerobic respiration in muscle tissues is an exothermic reaction with a standard enthalpy change of -120 kJ mol^{-1}. If we were to reduce the temperature, this reaction would shift towards producing *more* lactic acid (because the reduced temperature would favour the exothermic forward reaction), increasing the concentration of protons in the cell and hence lowering the pH (which the body needs to guard against).

Fortunately for us, our body has sophisticated regulatory mechanisms to guard against the uncontrolled fluctuations that might come about as the result of a shift in equilibrium. (These mechanisms mediate the process of homeostasis, through which our bodies are maintained in a stable condition.) For example, homeostatic mechanisms keep our body temperature at around 37 °C. In addition, specific components of our body fluids act as buffers, acting to maintain the pH of these fluids (including our blood) within narrow ranges.

We learn more about buffers and buffer solutions in Chapter 16.

Catalysts and chemical equilibria

We see in Section 15.4 that a catalyst is a substance that alters the **rate** of a reaction without itself being consumed by the reaction. A catalyst alters the reaction path to one with lower activation energy; it is this lowering of the activation energy that increases the rate of reaction.

It's important to note that when a catalyst is used with an equilibrium reaction, the catalyst alters the rate of both the forward *and* back reactions by exactly the *same amount*. Therefore, a catalyst does not alter the **position** of the equilibrium. It merely speeds up the rate at which the reaction *reaches* equilibrium.

Table 15.1 The effect of changing reaction conditions on a reaction at equilibrium

Change	Effect
Concentration	**Increasing** [reactants] favours the **forward** reaction, but **decreasing** [reactants] favours the **back** reaction **Increasing** [products] favours the **back** reaction, but **decreasing** [products] favours the **forward** reaction
Pressure	**Increasing** the pressure favours the reaction which yields the **smallest** number of molecules **Decreasing** the pressure favours the reaction which yields the **largest** number of molecules
Temperature	**Increasing** the temperature favours **endothermic** reactions **Decreasing** the temperature favours **exothermic** reactions
Addition of catalyst	Increases the rate of both the forward **and** back reactions, but equilibrium constant remains **unchanged**

The effects of changing different reaction conditions at equilibrium are summarized in Table 15.1.

15.6 Free energy and chemical equilibria

We see in Chapter 13 how the change in Gibbs free energy, ΔG, is the energy available for reaction that is free to 'do work', e.g. to drive other chemical reactions.

A reaction only happens spontaneously if the change in Gibbs free energy during the course of the reaction (ΔG) is *negative* (i.e. less than zero). When the Gibbs free energy change is positive, a reaction does not occur spontaneously. We say it 'is not feasible'. The only way to make such non-favourable reactions happen is to supply energy, usually in the form of heat.

When a reaction is at equilibrium, there is no overall change in the reaction mixture–the concentrations of the reactants and products remain constant–so ΔG, the change in Gibbs free energy, is *zero*.

- ***When a reaction mixture is at equilibrium,*** $\Delta G = 0$.

If the free energy change for a reaction is positive or negative, but the free energy at equilibrium is always zero, we can deduce that the value of the free energy must change as a reaction proceeds–that is, as it *approaches* equilibrium.

We see in Chapter 13 how the change in free energy is negative if the products of a reaction have a lower free energy than the reactants–that is, if the free energy of the system falls when the reaction occurs. By contrast, the change in free energy is positive if the products of the reaction have a higher free energy than the reactants: the free energy of the system *increases* as the reaction proceeds.

We can therefore represent the change in free energy during the progress of an equilibrium reaction as illustrated in Figure 15.14, which depicts the value of the free energy during the progress of an equilibrium reaction where the forward reaction is spontaneous. Notice that the *change* in free energy, ΔG, is given by the slope of the graph. Also, notice the following features:

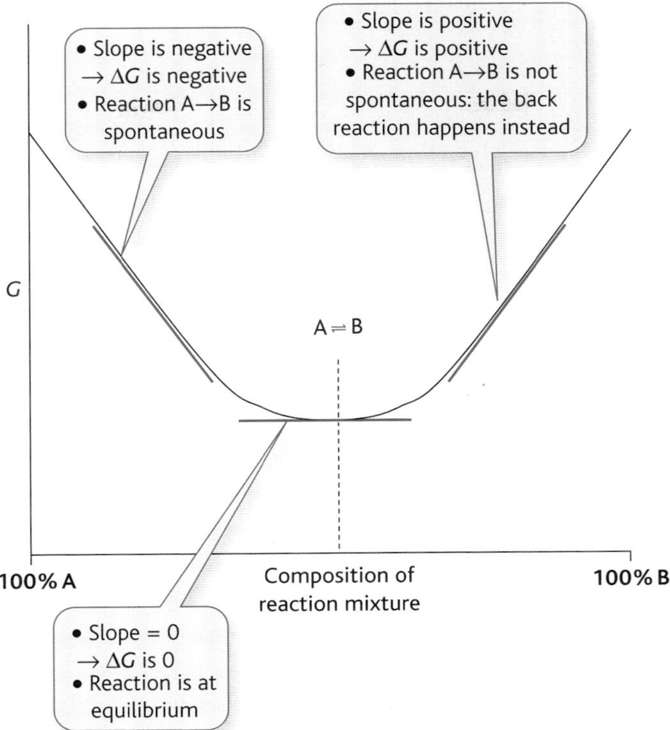

Figure 15.14 The change in free energy during an equilibrium reaction, A → B, in which the forward reaction is spontaneous. Notice how the slope of the graph is negative (denoting a negative value of ΔG) until the reaction mixture reaches equilibrium, at which point ΔG falls to zero.

Equilibrium is reached at the minimum of the curve, where the *change* in free energy is zero.

The forward reaction is no longer spontaneous beyond the equilibrium point. (Notice how the slope of the graph changes from negative to positive.) Beyond this point, it is the back reaction (from products to reactants) that becomes spontaneous.

The observation that ΔG is zero at equilibrium is not the only link between ΔG and equilibrium, however. The value of $\Delta G°$ (the free energy change under standard conditions) is related to the *equilibrium constant*, *K*, by the following expression, which is known as the **van't Hoff isotherm**:

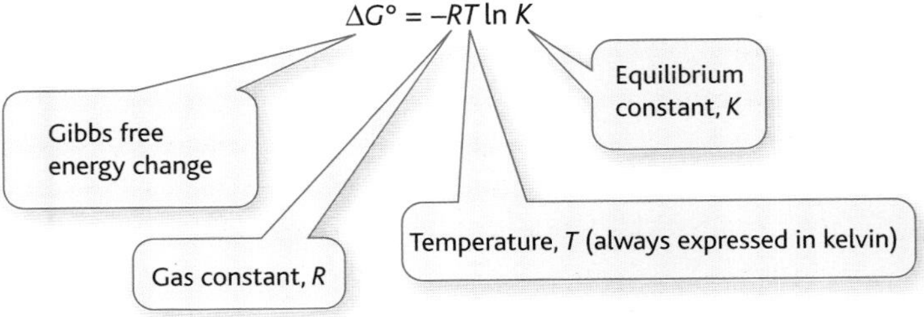

The gas constant has a value of 8.31 J K^{-1} mol^{-1}. It is a fixed number that helps to tie together the value of $\Delta G°$ with the value of *K*.

'ln *K*' means 'the natural logarithm of *K*'. The 'natural log' (ln) is a mathematical function that is used to scale down numbers varying greatly in magnitude so that they become a similar size. We find the natural log of a number by pressing the 'ln' button on a calculator.

We learn more about logarithms in Maths Tool 16.1, p. 552.

Why does this relationship interest us? This relationship enables us to do two things:

1. It gives us a way of predicting whether a particular reaction happens spontaneously or not if we know the equilibrium constant, *K*.
2. It gives us a way of estimating how far a reaction is likely to proceed–that is, whether it lies to the left, or to the right–if we know the value of ΔG°.

Let us consider each of these in turn.

The van't Hoff isotherm and predicting spontaneity

The van't Hoff isotherm gives us a way of predicting whether a reaction happens spontaneously by enabling us to calculate the value of ΔG° if we know the equilibrium constant, *K*. We can then use the value of ΔG° we obtain to predict whether the reaction will be spontaneous or not.

Let's revisit two equilibrium reactions from earlier in the chapter:

1. $CO + Cl_2 \rightleftharpoons COCl_2$ $\quad K = 4.57 \times 10^9$ at 100 °C
2. $N_2 + O_2 \rightleftharpoons 2NO$ $\quad K = 4.7 \times 10^{-31}$ at 25 °C

What are the values of ΔG° for these two reactions? We use the van't Hoff isotherm to calculate a value as shown below.

Reaction 1:

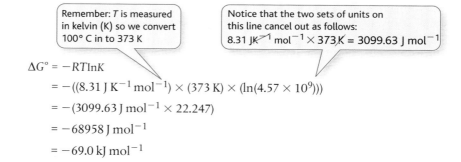

Remember: *T* is measured in kelvin (K) so we convert 100° C in to 373 K

Notice that the two sets of units on this line cancel out as follows: $8.31\ \text{J K}^{-1}\ \text{mol}^{-1} \times 373\ \text{K} = 3099.63\ \text{J mol}^{-1}$

$$
\begin{aligned}
\Delta G^\circ &= -RT\ln K \\
&= -((8.31\ \text{J K}^{-1}\ \text{mol}^{-1}) \times (373\ \text{K}) \times (\ln(4.57 \times 10^9))) \\
&= -(3099.63\ \text{J mol}^{-1} \times 22.247) \\
&= -68958\ \text{J mol}^{-1} \\
&= -69.0\ \text{kJ mol}^{-1}
\end{aligned}
$$

Look at Maths Tool 15.1 to see how we handle multiple sets of brackets in equations.

So, the Gibbs free energy change for the reaction as written ($CO + Cl_2 \rightarrow COCl_2$) is negative, and the reaction occurs **spontaneously**.

Reaction 2:

$$
\begin{aligned}
\Delta G^\circ &= -RT\ln K \\
&= -((8.31\ \text{J K}^{-1}\ \text{mol}^{-1}) \times (298\ \text{K}) \times (\ln(4.7 \times 10^{-31}))) \\
&= -((2476.38\ \text{J K}^{-1}\ \text{mol}^{-1}) \times (-69.8332)) \\
&= +172\,932\ \text{J mol}^{-} \\
&= +173\ \text{kJ mol}^{-1}
\end{aligned}
$$

Notice that we're multiplying together two negative numbers in the preceding line, so the result has a positive value.

This time, the Gibbs free energy change for the reaction as written ($N_2 + O_2 \rightarrow 2NO$) is positive, and the reaction does *not* occur spontaneously.

Self-check 15.9

(a) What are the standard state Gibbs free energy changes (ΔG°) for the following reactions at 298 K?

(i) $H_2CO_3 + H_2O \rightleftharpoons HCO_3^- + H_3O^+$ $K = 4.2 \times 10^{-7}$

(ii) $2NO + O_2 \rightleftharpoons 2NO_2$ $K = 2.25 \times 10^{12}$

(iii) $CH_4 + H_2O \rightleftharpoons CO + 3H_2$ $K = 9.4 \times 10^{-1}$

(iv) $2H_2 + O_2 \rightleftharpoons 2H_2O$ $K = 3.3 \times 10^{81}$

(b) Based on your answers to (a), is each reaction spontaneous or non-spontaneous?

The van't Hoff isotherm and the position of equilibrium

We can also use the van't Hoff isotherm to help us estimate the position of equilibrium–whether a reaction happens almost completely (the position of equilibrium lies to the right, favouring the products), or whether it barely occurs at all (the equilibrium lies to the left, favouring the reactants).

We see in Section 15.2 how a reaction that lies to the left (the reactants predominate over the products at equilibrium) has a very small equilibrium constant, K. If the value of K is < 1, the term '$\ln K$' (in the expression $\Delta G^\circ = -RT \ln K$) is *negative*, so the overall term $-RT \ln K$ is *positive*. (Remember: if we multiply two negative numbers we end up with a positive number.) So ΔG° for the reaction is > 0, and the reaction does not proceed spontaneously. This explains why the reaction lies to the left: it is energetically **unfavourable**.

In reaction 2 above, the value of K is < 1, and we see that the value of ΔG° is positive, so the reaction $N_2 + O_2 \rightleftharpoons 2NO$ is not spontaneous and lies to the *left*: not much of the product forms.

By contrast, an equilibrium reaction that lies to the right (the products predominate over the reactants at equilibrium) has a very large equilibrium constant, K. If the value of K is > 1, the term '$\ln K$' (in the expression $\Delta G^\circ = -RT \ln K$) is *positive*, so the overall term $-RT \ln K$ is *negative*. (If we multiply a positive and a negative number, we end up with a negative number.) So ΔG° for the reaction is < 0, and the reaction *does* proceed spontaneously. This explains why the reaction lies to the right: it is energetically **favourable**.

In reaction 1 above, the value of K is > 1, and we see that the value of ΔG° is negative, so the reaction $CO + Cl_2 \rightleftharpoons COCl_2$ is spontaneous and lies to the *right*: virtually all of the reactant is converted into product.

- ***If the equilibrium constant for a reaction is very large, ΔG° is negative: the reaction is spontaneous.***
- ***If the equilibrium constant for a reaction is very small, ΔG° is positive: the reaction is not spontaneous.***

MATHS TOOL 15.1 Handling brackets

On p. 532 we see this sequence of terms:

$$-((8.31\ \text{J K}^{-1}\ \text{mol}^{-1}) \times (373\ \text{K}) \times (\ln(4.59 \times 10^9)))$$

If we add some colour, we can see the different pairs of brackets in operation here:

$$-((8.31\ \text{J K}^{-1}\ \text{mol}^{-1}) \times (373\ \text{K}) \times (\ln(4.59 \times 10^9)))$$

The use of brackets helps to clarify how we should proceed with the calculation. For example, if we were to write out the right-hand term without brackets–ln 4.59×10^9–we couldn't be sure whether we should be calculating the natural log (ln) of 4.59 and then multiplying this term by 10^9, or calculating the natural log of 4.59×10^9. (The outcome would be quite different.) Brackets help to remove this ambiguity: in this case the light blue brackets tell us that (4.59×10^9) should be treated as a single entity. Consequently, our calculation should be $\ln(4.59 \times 10^9) = 22.247$.

We can now feed this value into our overall equation:

$$-((8.31\ \text{J K}^{-1}\ \text{mol}^{-1}) \times (373\ \text{K}) \times (22.247))$$

We have now dispensed with the innermost set of brackets (the light blue ones), and are left with three terms each within dark blue brackets. We can now simply multiply these terms in sequence: $8.31\ \text{J K}^{-1}\ \text{mol}^{-1} \times 373\ \text{K} \times 22.247 = 68958\ \text{J mol}^{-1}$.

But why do we need the outer pair of black brackets? If we were to write this expression without these brackets:

$$-(8.31\ \text{J K}^{-1}\ \text{mol}^{-1}) \times (373\ \text{K}) \times (22.247)$$

we would take the minus sign to apply only to the first term, which we would read as $-8.31\ \text{J K mol}^{-1}$. Having the minus sign outside of the black brackets, however, tells us that the minus sign applies to the value we obtain when we evaluate *everything* within these brackets. We have seen that

$$(8.31\ \text{J K}^{-1}\ \text{mol}^{-1}) \times (373\ \text{K}) \times (22.247) = 68958\ \text{J mol}^{-1}$$

So our equation becomes $-(68958\ \text{J mol}^{-1})$. The minus sign applies to whatever is in the brackets, so our final value is $-68958\ \text{J mol}^{-1}$.

In fact, we can make several useful predictions about the position of equilibrium based on the value of $\Delta G°$ without having to remember the van't Hoff isotherm at all, but by simply applying some logic. Consider the following reaction:

$$A + B \rightleftharpoons C + D$$

We know that a positive value of $\Delta G°$ (that is, $\Delta G° > 0$) tells us that a reaction is not spontaneous. So, if the above reaction had a value of $\Delta G° > 0$, and we had a vessel containing A and B, we would expect very little (if any) A and B to react to form C and D (because the reaction isn't spontaneous). Therefore, the position of equilibrium would lie very heavily to the left.

By contrast, we know that a negative value of $\Delta G°$ (that is, $\Delta G° < 0$) tells us that a reaction *is* spontaneous. So, if the above reaction had a value of $\Delta G° < 0$, and we had a vessel containing A and B, we would expect a substantial proportion of A and B to react to form C and D. Therefore, the position of equilibrium would lie more heavily to the right.

Let's consider the complete metabolism of glucose during respiration:

$$C_6H_{12}O_6(s) + 6\ O_2(g) \rightarrow 6\ CO_2(g) + 6\ H_2O(l)$$

We see in Chapter 13 how this reaction has a value of $\Delta G°$ of $-2878\ \text{kJ mol}^{-1}$. This large, negative value tells us that the reaction lies heavily to the right, such that the products predominate. (This is why the oxidation of glucose is such a good source of energy: it makes very good use of the starting 'fuel', glucose, by converting a vast majority of it into products, releasing free energy in the process.)

BOX 15.4 Calculating *K* from the van't Hoff isotherm

To calculate the value of *K* from the value of ΔG° using the van't Hoff isotherm we first have to rearrange the expression for the van't Hoff isotherm, $\Delta G^\circ = -RT\ln K$, so that *K* is isolated. We do this in three stages.

First, we multiply through by –1:

$$-\Delta G^\circ = RT\ln K$$

We then divide both sides by *RT*:

$$\frac{-\Delta G^\circ}{RT} = \frac{\cancel{RT}\ln K}{\cancel{RT}}$$

$$\frac{-\Delta G^\circ}{RT} = \ln K$$

Finally, we take the exponent of both sides[12]:

$$\exp\left(\frac{-\Delta G^\circ}{RT}\right) = K$$

Let's now use this new expression to calculate *K*. The value of $\Delta G^{\circ\prime}$ for the hydrolysis of ATP ($ATP + H_2O \rightleftharpoons ADP + P_i + H^+$) at 310K is –30.5 kJ mol⁻¹.

What is the value of *K* for this reaction?

$$K = \exp\left(\frac{-\Delta G^{\circ\prime}}{RT}\right)$$

$$= \exp\left(\frac{-(-30.5\ \text{kJ mol}^{-1})}{8.31\ \text{J K}^{-1}\text{mol}^{-1} \times 310\ \text{K}}\right)$$

$$= \exp\frac{+30.5\ \text{kJ mol}^{-1}}{2576.1\ \text{J mol}^{-1}}$$

$$= \exp\frac{30.5 \times 10^3\ \text{J mol}^{-1}}{2576.1\ \text{J mol}^{-1}}$$

$$= \exp(11.84)$$

$$K = 1.39 \times 10^5$$

This large value of *K* tells us that this reaction lies to the right, and that products predominate when the reaction is at equilibrium (as we might expect for a spontaneous reaction with a negative value for ΔG°).

This form of the van't Hoff isotherm has its limitations, most notably that you cannot calculate the exponent of a negative value. Therefore, you cannot calculate a value of *K* from a value of $\Delta G^\circ > 0$ (because the expression converts this positive value into a negative one).

- ***A reaction with $\Delta G^\circ > 0$ is not spontaneous, and lies to the left–very few products form.***
- ***A reaction with $\Delta G^\circ < 0$ is spontaneous, and lies to the right–virtually all the reactants change into products.***

We can also use the van't Hoff isotherm to calculate a value of *K* if we know the value of ΔG°. Look at Box 15.4 to find out how.

Self-check 15.10 The folding of the enzyme ribonuclease has a standard Gibbs free energy change of –44 kJ mol⁻¹. What is the value of *K* for this process, assuming a temperature of 298 K?

12 The exponent is the inverse function of natural log, ln. That is, it reverses the effect of the function ln. So, exp (ln *K*) = *K*

Check your understanding

To check that you've mastered the key concepts presented in this chapter, review the checklist of key concepts below, and attempt the multiple-choice questions available in the book's Online Resource Centre at **http://www.oxfordtextbooks.co.uk/orc/crowe2e/**.

Checklist of key concepts

Equilibrium reactions

- An equilibrium reaction is a reaction that can proceed in both the **forward** and **reverse** directions simultaneously
- An equilibrium reaction comprises a **forward reaction** and a **back reaction**
- An equilibrium reaction can be generally represented by: $A + B \rightleftharpoons C + D$
- The forward reaction is $A + B \rightarrow C + D$
- The back reaction is $C + D \rightarrow A + B$
- At equilibrium, the rates of the forward and back reactions are **equal**
- At equilibrium, the forward and back reactions continue to proceed, but there is **no overall change** in the system–we call this a **dynamic equilibrium**
- The point at which a state of equilibrium is reached is the same regardless of the direction from which we approach it
- If the forward reaction predominates over the back reaction we say the equilibrium reaction **lies to the right**, and the products are favoured
- If the back reaction predominates over the forward reaction we say the equilibrium reaction **lies to the left**, and the reactants are favoured

The equilibrium constant

- The position of equilibrium is represented by an **equilibrium constant, *K***
- For an equilibrium reaction, $a\,A + b\,B \rightarrow c\,C + d\,D$

$$K = \frac{[C]^c[D]^d}{[A]^a[B]^b}$$

whereby the concentration terms must be expressed in mol L^{-1} (that is, M)

- *K* tells us about the ratio of products to reactants when a reaction is at a state of equilibrium
- The magnitude ('size') of the equilibrium constant, *K*, tells us the relative position of a reaction at equilibrium
- The value of *K* is dimensionless
- A **very large** value for *K* tells us the reaction proceeds almost completely to the **right**
- A **very small** value for *K* tells us the reaction lies almost completely to the **left**
- When *K* has a value between approximately 10^{-2} and 10^2, the equilibrium reaction is **fairly evenly** balanced
- The equilibrium constant for a particular reaction depends on the **temperature** at which the reaction is performed

The reaction quotient

- The **reaction quotient, *Q***, tells us the relative ratio of products to reactants at any point in a reaction
- When $Q = K$ the reaction is at equilibrium
- If $Q < K$ the reaction proceeds from left to right (in the direction of the forward reaction)
- If $Q > K$ the reaction proceeds from right to left (in the direction of the back reaction)
- If $Q = K$ the reaction is already at equilibrium concentrations–there is no further change

Binding reactions

- Binding processes are examples of **dynamic** equilibrium reactions
- If a protein and ligand bind **strongly**, the binding reaction, $P + L \rightleftharpoons PL$, lies to the **right**
- If a protein and ligand bind only **weakly**, the binding reaction lies to the **left**
- The dissociation constant is given by the expression

$$K_d = \frac{[P][L]}{[PL]}$$

- If binding between a protein and ligand is weak, the value of K_d is large, and the dissociation reaction lies to the right
- If binding between a protein and ligand is strong, the value of K_d is small, and the dissociation reaction lies to the left

Perturbing the equilibrium position

- When an equilibrium is perturbed the system acts to counteract the change so that a state of equilibrium is re-established
- This principle is called the Le Chatelier principle
- There are three key ways to perturb an equilibrium:
  - change the concentration of species present
  - change the temperature
  - change the pressure
- Increasing the concentration of the reactants favours the forward reaction, but decreasing the concentration of the reactants favours the back reaction
- Increasing the concentration of the products favours the back reaction, but decreasing the concentration of the products favours the forward reaction
- Increasing the pressure favours the reaction which yields the smallest number of molecules
- Decreasing the pressure favours the reaction which yields the largest number of molecules
- Increasing the temperature favours endothermic reactions
- Decreasing the temperature favours exothermic reactions
- Using a catalyst has no effect on the equilibrium position–it merely increases the rate of both the forward and back reactions by the same amount
- When a system shifts its equilibrium position in response to something like a change in temperature, the composition of the system when the new equilibrium is reached is different from the composition prior to the shift

Equilibria and Gibbs free energy

- When a reaction is at equilibrium the change in Gibbs free energy is zero
- The value of ΔG° is related to the value of K by the van't Hoff isotherm:

$$\Delta G^\circ = -RT \ln K$$

- If the equilibrium constant for a reaction is very large, ΔG° is negative: the reaction is spontaneous
- If the equilibrium constant for a reaction is very small, ΔG° is positive: the reaction is not spontaneous
- A reaction with $\Delta G^\circ > 0$ lies to the left–very few products form
- A reaction with $\Delta G^\circ < 0$ lies to the right–virtually all the reactants change into products

16 Acids, bases, and the aqueous environment: the medium of life

INTRODUCTION

The chemistry of life–the collection of biochemical reactions that keep us alive–occurs not in a test tube in a laboratory, but within the cells of our body. And our cells are filled with cytoplasm, the water-based (aqueous) environment in which the biochemical reactions of life occur.

Our cells are dynamic, active entities: they grow and replicate, consuming certain compounds and generating others. Yet the chemical environment of our cells is carefully regulated. Our bodies maintain a constant temperature of 37 °C, and the concentrations of the many different compounds that we need for survival are kept within limits that we can tolerate. Central among the compounds that comprise the aqueous environment of our cells are two groups of compounds, the acids and the bases.

In this chapter we discover some key properties of aqueous environments that are pivotal to biochemical reactions happening correctly. We learn about the behaviour of acids and bases–why they are important to the chemistry of life, what distinguishes them from other compounds, and how their behaviour is kept in check. We go on to explore how acids and bases feature in the chemical reactions of life in the final two chapters of this book.

16.1 Acids and bases: making life happen

The proton, H^+, and hydroxide ion, OH^-, seem almost too simple to be of significance. However, our bodies depend on these ions to carry out the biological processes that maintain life. But where do these ions come from? How do our bodies maintain a constant supply of ions such as these?

We see various examples of the involvement of protons and hydroxide ions in the chemical reactions that underpin life processes in Chapters 17 and 18.

H^+ and OH^- ions are generated as a result of the way certain groups of compounds behave in aqueous systems. These 'certain groups' of compounds are **acids** and **bases**.

Let's begin by seeing what is meant by the words 'acid' and 'base', and how they relate to the generation of H^+ and OH^- ions.

Defining acids and bases

An acid is a proton donor: a compound that can undergo chemical change to produce a proton, H^+. We can represent this 'chemical change' as follows:

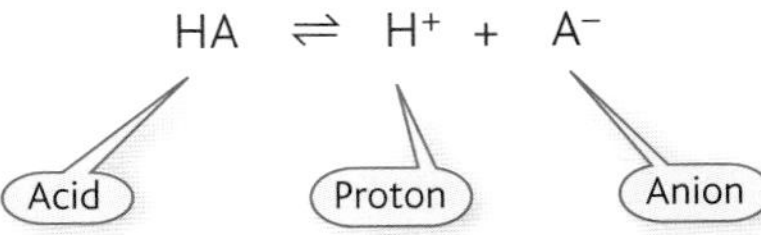

We say that the acid undergoes dissociation, or that it '**dissociates**'. Notice how this dissociation process is an equilibrium reaction, as denoted by the ⇌ symbol.

For example, hydrogen chloride dissociates in water to yield a proton and a chloride ion:

$$HCl \rightleftharpoons H^+ + Cl^-$$

while ethanoic acid dissociates to yield a proton and an ethanoate ion[1]:

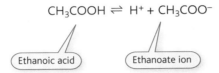

$$CH_3COOH \rightleftharpoons H^+ + CH_3COO^-$$

Ethanoic acid — Ethanoate ion

We explore equilibrium reactions, and the chemical equilibria arising from them, in Chapter 15.

Biological molecules can also undergo similar dissociation. For example, the side chain of the amino acid aspartic acid dissociates as follows:

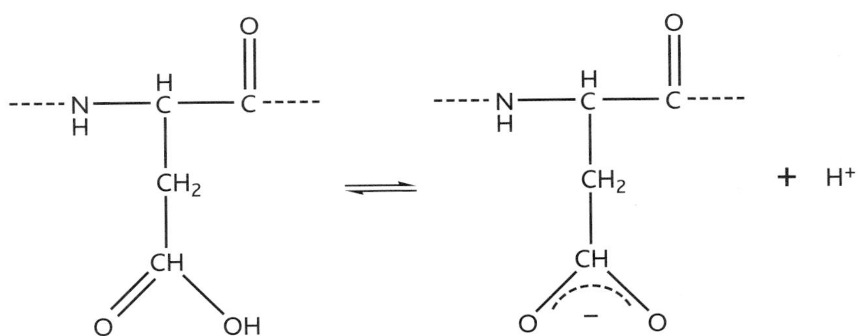

The statement that 'an acid is a proton donor' is called the Brønsted–Lowry definition of an acid. It is named after Johannes Brønsted and Thomas Lowry.

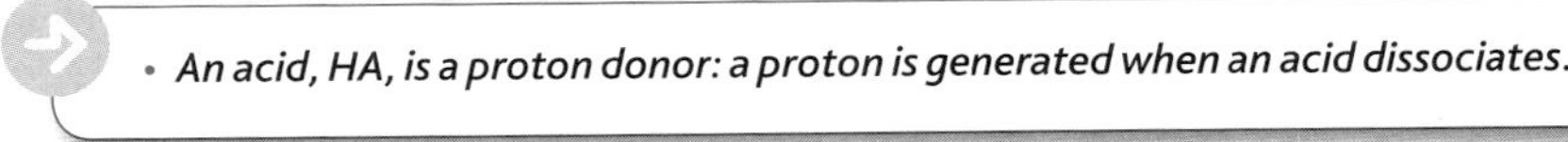

- *An acid, HA, is a proton donor: a proton is generated when an acid dissociates.*

By contrast, a base is a proton **acceptor**; that is, a compound that can undergo chemical change to *combine* with a proton:

$$B + H^+ \rightleftharpoons BH^+$$

The statement that 'a base is a proton acceptor' is called the Brønsted–Lowry definition of a base.

1 Remember: you may see ethanoic acid and ethanoate written as acetic acid and acetate respectively.

For example, when dissolved in water, ammonia can accept a proton to form the ammonium ion:

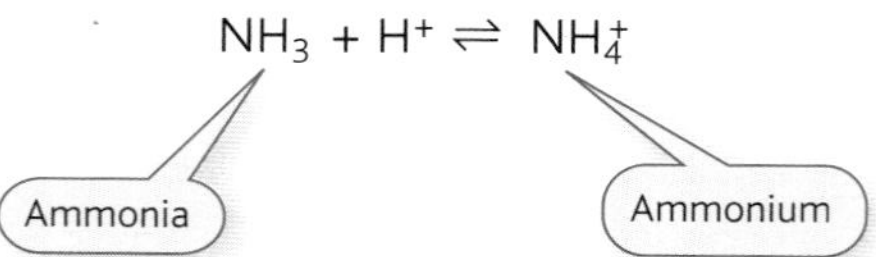

- ***A base is a proton acceptor: it removes a free proton from solution.***

Acids and bases in aqueous solution

The biochemical reactions associated with life processes occur in aqueous (water-based) solution. Water is more than merely a solvent: it plays an important part in how acids and bases behave.

Let's go back to the dissociation of the acid, hydrogen chloride:

$$HCl \rightleftharpoons H^+ + Cl^-$$

While this process should happen as written in theory, protons are too unstable to exist independently. Instead, the proton yielded by an acid when it dissociates is **stabilized** by reacting with a water molecule:

$$H^+ + H_2O \rightleftharpoons H_3O^+$$

The species H_3O^+ is responsible for all acidic properties. It has various names, including 'hydroxonium ion' and 'hydronium ion'. We shall call it a 'solvated proton'.

So the overall dissociation reaction is:

$$HCl + H_2O \rightleftharpoons H_3O^+ + Cl^-$$

When we are considering the behaviour of acids in aqueous solution we usually don't bother including the water molecule in the equation. We merely focus on the acid:

$$HCl \rightleftharpoons H^+ + Cl^-$$

The behaviour of bases in water

By contrast, when a base dissolves in water it *accepts* a proton from a water molecule:

$$B + H_2O \rightleftharpoons BH^+ + OH^-$$

Base has accepted a proton from a water molecule ...

... which becomes a hydroxide ion.

After its reaction with a base, the water molecule becomes a hydroxide ion, OH^-. This reaction is an example of hydrolysis: the splitting of water. (In Greek, hydro = water, and lysis = splitting.)

The behaviour of acids and bases in aqueous solution led to the creation of a second definition for acids and bases by the Swedish scientist Svante Arrhenius, as explained in Box 16.1.

BOX 16.1 The Arrhenius definition of acids and bases

The Arrhenius definition of an acid states that: 'An acid is a compound which gives rise to a proton when dissolved in water.'

Notice how this definition specifies that an acid dissociates to give rise to a proton when it's *dissolved in water*. This definition reflects the role that water plays in stabilizing the formation of a proton when the acid dissociates.

The Arrhenius definition of a base states that: 'A base is a compound which produces hydroxide ions when dissolved in water.'

For example:

$$NaOH \rightarrow Na^+ + OH^-$$

Notice how the Arrhenius definitions of acids and bases are a little restrictive: they refer only to the behaviour of acids and bases in aqueous (water-based) solution. However, not all acids dissociate to yield a proton *only* when dissolved in water (though all acids *do* dissociate in water, if in no other solvent). The Brønsted–Lowry definitions of acids and bases are broader, and refer not only to the behaviour of acids and bases in aqueous solutions, but in other solvents and in the gas phase.

Self-check 16.1 State whether the compound in each of the following reactions is behaving as an acid or a base.

(a) $HBr \rightleftharpoons H^+ + Br^-$

(b) $CN^- + H^+ \rightleftharpoons HCN$

(c) $H_2O \rightleftharpoons H^+ + OH^-$

(d) $NH_3 + H^+ \rightleftharpoons NH_4^+$

Pairing up acids and bases: the conjugate acid–base pair

Let's go back to the dissociation of hydrogen chloride in water:

$$HCl + H_2O \rightleftharpoons H_3O^+ + Cl^-$$

We see above how hydrogen chloride acts as an acid; it donates a proton:

$$HCl \rightarrow H^+ + Cl^-$$

But what about the water molecule? The water molecule **accepts a proton** to become the solvated proton:

$$H_2O + H^+ \rightarrow H_3O^+$$

So, water itself is **acting as a base.**

The acid and base operate as a pair, as depicted in Figure 16.1: the acid donates a proton to the base, which duly accepts it. An acid and a base cannot operate in isolation: an acid *must* also have a base to which it donates a proton, just as a nucleophile must have an electrophile to which it donates electrons (as we see in Chapter 17).

We can write a **general** equation for an acid–base reaction as follows:

$$HA + B \rightleftharpoons A^- + BH^+$$

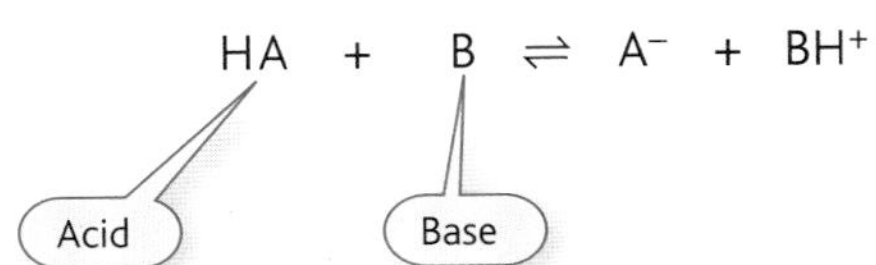

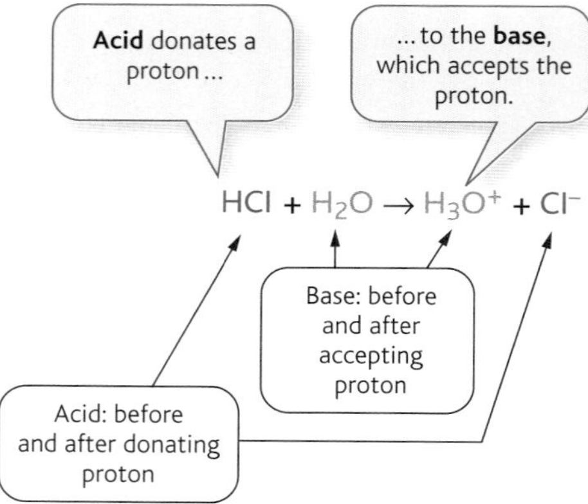

Figure 16.1 An acid and a base act as a pair. An acid must have a base to which it donates a proton.

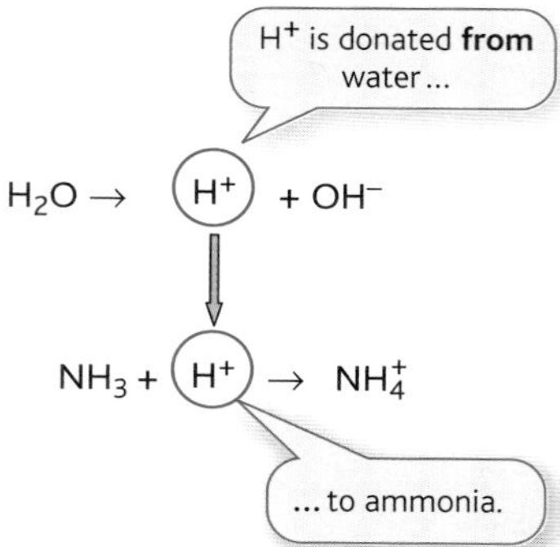

Figure 16.2 During the reaction of ammonia and water, water acts as an acid and donates a proton to ammonia, which acts as a base (and accepts the proton).

- ***Acids and bases operate in pairs: an acid must have a base to which to donate a proton.***

Let's look at another acid–base reaction, the reaction of ammonia with water:

$$NH_3 + H_2O \rightleftharpoons NH_4^+ + OH^-$$

Figure 16.2 shows how:

- Water, H_2O, acts as an **acid**, and donates a proton.
- Ammonia, NH_3, acts as a **base**, and accepts a proton.

Look at this reaction, and notice how ammonia fulfils the Arrhenius definition of a base: OH^- ions are generated when ammonia dissolves in water.

The reaction of ammonia with water is an equilibrium reaction. We've just looked at the forward reaction. But what happens during the back reaction? We see the following:

$$NH_4^+ + OH^- \rightarrow NH_3 + H_2O$$

Figure 16.3 shows how, in this instance:

- NH_4^+ acts as an **acid**, and donates a proton.
- OH^- acts as a **base**, and accepts a proton.

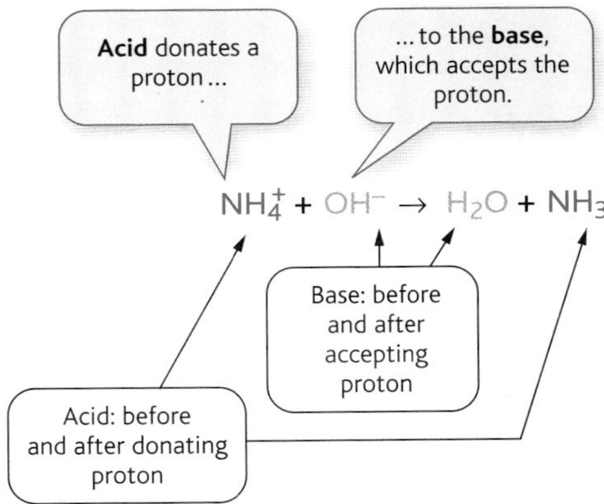

Figure 16.3 When the reaction shown in Figure 16.2 is reversed, the ammonium ion acts as an acid and donates a proton to the hydroxide ion, which acts as a base.

Taking the forward and back reactions together, notice how:

- Water acts as an **acid** in the forward direction, but as a **base** in the reverse direction.
- Ammonia acts as a **base** in the forward direction, but as an **acid** in the reverse direction.

This leads us to one of the most important ideas in the chemistry of life–the concept of the **conjugate acid–base pairs**. In this instance:

- H_2O (an acid) and OH^- (a base) form one conjugate acid–base pair.
- NH_3 (a base) and NH_4^+ (an acid) form a second conjugate acid–base pair.

Figure 16.4 brings together the forward and back reactions, and shows how the two conjugate acid–base pairs are formed. Notice how the reaction features *two* conjugate acid–base pairs.

So acids and bases are not different, independent chemical entities: a conjugate acid–base pair comprises the acid and base versions of the *same* chemical species. When a Brønsted–Lowry acid donates a proton, it becomes a base; and the base which *accepts* the proton becomes an acid.

In a reaction between an acid and a base there are always *two* conjugate acid–base pairs. We can represent this in a general way as follows:

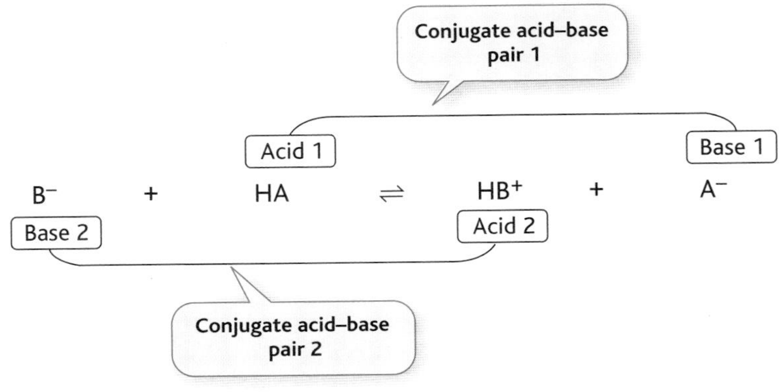

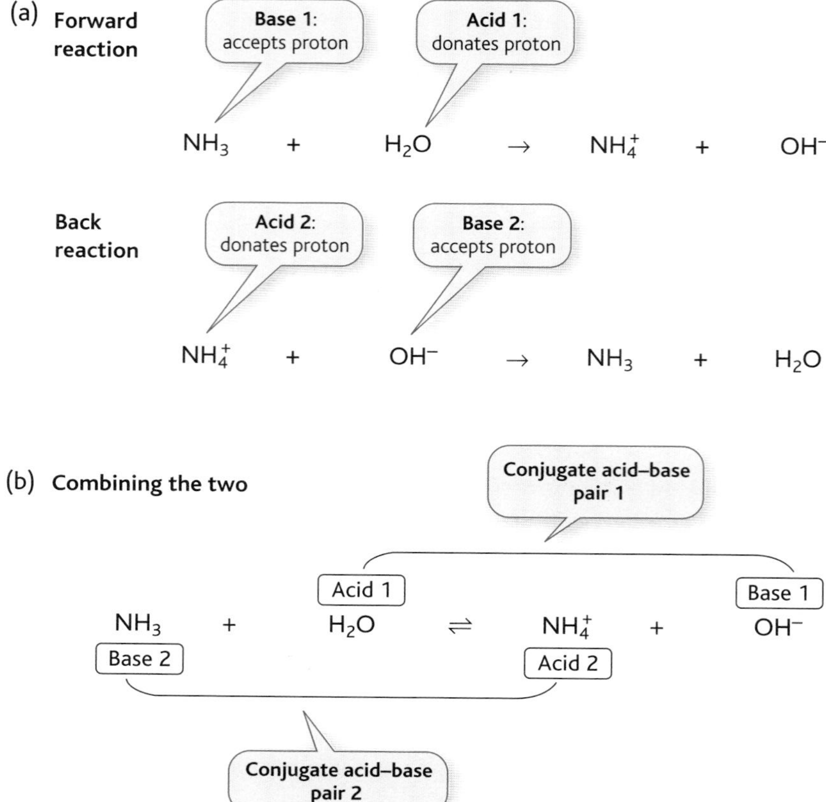

Figure 16.4 (a) Water acts as an acid during the forward reaction, and as a base during the back reaction. By contrast, ammonia acts as a base during the forward reaction, and as an acid during the back reaction.
(b) An acid–base reaction features two conjugate acid-base pairs.

Let's consider another example of conjugate acid–base pairs. Carbonic acid, which plays a central role in regulating the acidity of our blood, reacts with water as follows:

$$H_2CO_3 + H_2O \rightleftharpoons HCO_3^- + H_3O^+$$

We see from this equation that during the forward reaction, carbonic acid (H_2CO_3) acts as an acid and donates a proton to water (H_2O), which acts as a base. During the back reaction, HCO_3^- acts as a *base* and accepts a proton from the solvated proton, H_3O^+. So the two conjugate acid–base pairs in this reaction are:

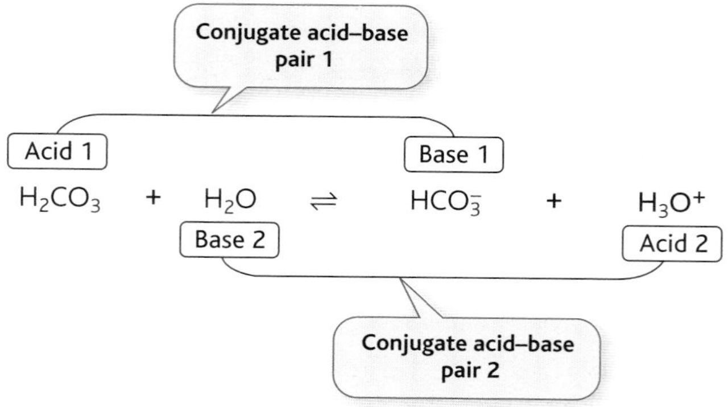

Self-check 16.2

(a) The reaction of hydrogen chloride with water can be represented by the following:

$$HCl + H_2O \rightleftharpoons H_3O^+ + Cl^-$$

Label this equation to show the two conjugate acid–base pairs.

(b) Identify the acid and base for the reactants in the following equation:

$$H_2SO_4 + H_2O \rightleftharpoons HSO_4^- + H_3O^+$$

(c) Label the following equation to show the two conjugate acid–base pairs:

$$CO_3^{2-} + H_2O \rightleftharpoons HCO_3^- + OH^-$$

Water: a split personality

Take a moment to see how water behaves when it reacts with both ammonia or with ethanoic acid. What do you notice? When it reacts with ammonia, water acts as an **acid**. However, when it reacts with ethanoic acid, water acts as a **base**. A compound that can act as either an acid *or* a base is what we call an **amphoteric** compound.

- *An amphoteric compound is one that can act as either an acid or as a base.*

16.2 The strength of acids and bases: to what extent does the dissociation reaction occur?

We see at the start of Section 16.1 how the dissociation of an acid (or a base) is an equilibrium reaction. For an acid:

$$HA \rightleftharpoons A^- + H^+$$

Like any equilibrium reaction, the dissociation of acids and bases reaches a position of equilibrium when the forward and back reactions occur to the same extent, and we see no overall change in the composition of the system.

Look at Section 15.1 to find out more about equilibrium reactions, and the state of chemical equilibrium.

The extent to which an acid dissociates in aqueous solution to generate a proton and its conjugate base–or the readiness with which it donates a proton–is determined by what we call the **strength** of the acid. Similarly, the ease with which a base accepts a proton from water to generate hydroxide ions in aqueous solution reflects the **strength** of the base[2].

Acids and bases fall into two general groups: those that are 'strong' and those that are 'weak'. If an acid readily donates a proton (and so readily dissociates to form its conjugate base in aqueous solution) we say it is a **strong acid**. By contrast, if an acid is poor at donating a proton, and only partially dissociates to give its conjugate base in aqueous solution, we say it is a **weak acid**.

2 It's a common mistake to use 'strength' when we really mean 'concentration'. Thus we can have a low concentration of a strong acid and a high concentration of a weak acid.

For example, nitric acid is a strong acid. When it is dissolved in water, virtually all the nitric acid dissociates to form NO_3^- ions and protons. The reaction lies heavily to the right:

$$HNO_3 + H_2O \rightleftharpoons NO_3^- + H_3O^+$$

Equilibrium position lies almost entirely to the **right:** virtually all the HNO_3 dissociates.

Remember: the proton released by the HNO_3 upon dissociation isn't 'free', but is stabilized by associating with a water molecule to form a H_3O^+ ion. So we don't see a free H^+ in the equation for the reaction.

HNO_3 is such a strong acid that we often write its dissociation reaction as a one-way reaction, because the back reaction happens to such a small degree:

$$HNO_3 + H_2O \rightarrow NO_3^- + H_3O^-$$

There are six key strong acids. Look at Table 16.1, which lists them.

By contrast, ethanoic acid is a *weak* acid. When it is dissolved in water, ethanoic acid only partially dissociates. The reaction lies to the left:

$$CH_3COOH + H_2O \rightleftharpoons CH_3COO^- + H_3O^+$$

Equilibrium position lies far to the **left:** little of the ethanoic acid dissociates.

Consequently, ethanoic acid isn't a particularly good source of protons.

- ***A strong acid readily dissociates to form its conjugate base, releasing a proton in the process.***
- ***A weak acid does not dissociate readily.***

Table 16.1 The six strong acids.

Acid name	Chemical formula
Hydrochloric acid	HCl
Hydrobromic acid	HBr
Hydroiodic acid	HI
Perchloric acid	$HClO_4$
Nitric acid	HNO_3
Sulfuric acid	H_2SO_4

The body contain hundreds of organic acids involved in life processes. However, none of them appear in Table 16.1: they are all weak acids. Table 16.2 shows a range of weak acids, a number of which participate in biochemical reactions in cells. The table lists these acids in order of strength, strongest first.

Table 16.2 A range of weak acids.

Acid name	Chemical formula
Phosphoric acid	H_3PO_4
Pyruvic acid	$CH_3COCOOH$
Lactic acid	$CH_3CHOHCOOH$
Ethanoic acid	CH_3COOH
Carbonic acid	H_2CO_3

We distinguish between strong and weak bases in a similar way. A strong base is one that readily accepts a proton from an acid (or one that readily dissociates in aqueous solution to yield OH^- ions). By contrast, a weak base is one that is poor at accepting a proton, and only partially dissociates in aqueous solution.

For example, the hydroxide ion, OH^-, acts as a strong base–it readily accepts a proton from an acid:

$$OH^- + H_2O \rightleftharpoons H_2O + OH^-$$

Water acting as an acid

Water donates a proton to form an OH^- ion

This reaction lies heavily to the right.

By contrast, the chloride ion, Cl^-, acts as a very weak base. The equilibrium reaction:

$$Cl^- + H_2O \rightleftharpoons HCl + OH^-$$

lies heavily to the *left*.

- ***A strong base readily accepts a proton from solution; a weak base does not.***

Juggling protons: the tug-of-war between conjugate acid–base pairs

Let's consider the dissociation of hydrogen chloride in water:

$$HCl + H_2O \rightleftharpoons Cl^- + H_3O^+$$

Hydrogen chloride is a relatively strong acid, and the equilibrium position for the dissociation of hydrogen chloride lies mainly to the right. What does this tell us about the strength of hydrogen chloride's conjugate base, the Cl^- ion? If the equilibrium reaction lies to the right, not much of the conjugate base, Cl^-, must be able to accept a proton to form HCl. In other words, it must be a **weak** base. So we see how a strong acid has a weak conjugate base.

By contrast, let's consider the dissociation of ethanoic acid in water:

$$CH_3COOH + H_2O \rightleftharpoons CH_3COO^- + H_3O^+$$

Ethanoic acid is a weak acid, so the reaction lies to the left (the back reaction outweighs the forward reaction). This tells us that ethanoic acid's conjugate base, CH_3COO^-, is very good at accepting a proton to form CH_3COOH: it is a **strong** base. Therefore, a weak acid has a strong conjugate base.

- ***Strong acids have weak conjugate bases.***
- ***Weak acids have strong conjugate bases.***

When we see the equation describing the dissociation of an acid or base, we have no way of telling whether the acid or base is strong or weak. So how can we distinguish between different acids on the basis of their strength? How do we know, for example, whether ethanoic acid should be above or below lactic acid in Table 16.2?

We need some way of *quantifying* the position of the equilibrium reaction (the extent to which the acid or base is dissociated at equilibrium). We do this by considering the **acid** and **base dissociation constants**.

The acid dissociation constant: to what extent does an acid dissociate?

The acid dissociation constant gives us a measure of the extent to which an acid dissociates in aqueous solution and, hence, how readily it can donate a proton.

We see in Chapter 15 how the equilibrium position for an equilibrium reaction:

Remember: *a*, *b*, *c*, and *d* are the stoichiometric coefficients. They indicate the *relative* proportions in which the reactants and products react with one another.

$$aA + bB \rightleftharpoons cC + dD$$

is indicated by the value of its equilibrium constant, K:

$$K = \frac{[C]^c[D]^d}{[A]^a[B]^b}$$

We can apply exactly the same principle to quantify the equilibrium position of the dissociation of an acid. We write the reaction as:

$$HA \rightleftharpoons H^+ + A^-$$

So the equilibrium constant is expressed as follows:

$$K = \frac{[H^+][A^-]}{[HA]}$$

When applied to the dissociation of an acid, we call the equilibrium constant the **acid dissociation constant**, and represent it by the symbol $\mathbf{K_a}$. Remember: the acid dissociation constant is exactly the same as the equilibrium constant. It merely tells us that we are specifically considering a reaction that represents the dissociation of an acid.

What does the magnitude (size) of the acid dissociation constant, K_a, tell us? It tells us exactly the same as any other equilibrium constant: the relative position of the reaction at equilibrium. A large value for K_a tells us that the reaction lies to the **right** at equilibrium–that the products outweigh the reactants. In the case of the dissociation of an acid, a large K_a value tells us that it is a **strong** acid and that relatively large amounts of solvated protons are yielded when the acid dissociates.

By contrast, a small value for K_a tells us that the reaction lies to the **left** at equilibrium–that the reactants outweigh the products. So a small K_a value tells us that it is a **weak** acid, and a low concentration of solvated protons are yielded when the acid dissociates.

- $K_a = \frac{[H^+][A^-]}{[HA]}$
- *A strong acid has a large K_a value.*
- *A weak acid has a small K_a value.*

For example, HNO_3 is a strong acid. The acid dissociation constant for the reaction:

$$HNO_3 \rightleftharpoons H^+ + NO_3$$

is so large that it is very difficult to measure accurately. In the expression:

$$K_a = \frac{[H^+][NO_3^-]}{[HNO_3]}$$

the term $[H^+][NO_3^-]$ is so much greater than $[HNO_3]$ that the value of K_a is huge; the products significantly outweigh the reactants.

By contrast, hydrocyanic acid, HCN, is a weak acid. The acid dissociation constant for the reaction:

$$HCN \rightleftharpoons H^+ + CN^-$$

is 4.0×10^{-10}. In other words, the value of [HCN] is very much greater than the value of $([H^+][CN^-])$ in the expression for K_a:

$$K_a = \frac{[H^+][CN^-]}{[HCN]}$$

such that the value of K_a is very small. The reaction lies to the left, and the reactants outweigh the products.

The base dissociation constant: to what extent does a base dissociate?

The base dissociation constant gives us a measure of the extent to which a base reacts with water to yield hydroxide ions or, looking at it another way, how readily it accepts a proton from an acid.

Let's reconsider the equilibrium reaction for the reaction of a base with water:

$$B + H_2O \rightleftharpoons BH^+ + OH^-$$

Again, we can use the general expression for the equilibrium constant for an equilibrium reaction:

$$K = \frac{[C]^c[D]^d}{[A]^a[B]^b}$$

to write down the expression for the equilibrium constant for the reaction of a base with water:

$$K = \frac{[BH^+][OH^-]}{[B][H_2O]}$$

We can actually ignore the concentration of water, $[H_2O]$, because it remains constant, and doesn't affect the concentrations of any of the other species in the reaction mixture[3]. So we write a slightly amended equilibrium constant:

$$K = \frac{[BH^+][OH^-]}{[B]}$$

3 More formally, $[H_2O]$ is numerically equal to 1, and so doesn't affect the value of K obtained.

When applied to the dissociation of a base, we call the equilibrium constant the **base dissociation constant** and represent it by the symbol $\mathbf{K_b}$.

A large value for K_b tells us that the reaction:

$$B + H_2O \rightleftharpoons BH^+ + OH^-$$

lies to the right at equilibrium: the base reacts readily with water to accept a proton, and OH^- ions are readily generated. It is a **strong** base.

By contrast, a small value for K_b tells us that the reaction lies to the left: the base is poor at reacting with water to accept a proton, and so few OH^- ions are generated. It is a **weak** base.

- $K_b = \dfrac{[BH^+][A^-]}{[HA]}$
- ***A strong base has a large K_b value.***
- ***A weak base has a small K_b value.***

For example, the nitrate ion, NO_3^-, is a very weak base. It has a value of K_b of 5×10^{-16}, telling us that, in the expression for the equilibrium constant:

$$K_b = \frac{[HNO_3][OH^-]}{[NO_3^-]}$$

the value of $[NO_3^-]$ is very much larger than the value of $([HNO_3][OH^-])$, and so the reaction:

$$NO_3^- + H_2O \rightleftharpoons HNO_3 + OH^-$$

lies to the left.

Self-check 16.3

(a) Arrange the following acids in order of strength, starting with the strongest acid:

ethanoic acid, $K_a = 1.8 \times 10^{-5}$

formic acid, $K_a = 1.8 \times 10^{-4}$

boric acid, $K_a = 7.3 \times 10^{-10}$

sulfurous acid, $K_a = 1.2 \times 10^{-2}$

benzoic acid, $K_a = 6.3 \times 10^{-5}$

(b) Arrange the following bases in order of strength, starting with the weakest base:

phosphate ion, $K_b = 2.8 \times 10^{-2}$

sulfite ion, $K_b = 1.6 \times 10^{-7}$

ammonia, $K_b = 1.8 \times 10^{-5}$

ethanoate ion, $K_b = 5.6 \times 10^{-10}$

cyanide ion, $K_b = 2.5 \times 10^{-5}$

pK_a and pK_b

The values of K_a can vary over many orders of magnitude (that is, over a large range of powers of ten). A strong acid, such as HNO_3 in water, is entirely dissociated, so the value of K_a is very large (too large in fact to be measured easily for concentrated solutions). By contrast, most of a weak acid, such as ethanoic acid in water, is present as the undissociated molecule, so the value of K_a is small.

Because the values of K_a can vary over such a large range there is an advantage in expressing them on a logarithmic scale. (We see in Maths Tool 16.1 how using a logarithmic ('log') scale gives us a way of scaling down widely varying numbers to bring them within a similar order of magnitude.)

Having taken the logarithm of a K_a value, we obtain the pK_a, using the expression:

$$pK_a = -\log(K_a)$$

Notice how this expression mirrors the expression for pH ($-\log [H^+]$), which we encounter on p. 559.

For example, a weak acid with a K_a value of 3×10^{-6} has a pK_a of:

$$-\log(3 \times 10^{-6})$$
$$= -(-5.5)$$
$$= +5.5$$

We can treat the K_b value for a weak base in a similar way:

$$pK_b = -\log(K_b)$$

So, a weak base with a K_b of 3.8×10^{-4} has a pK_b of:

$$-\log(3.8 \times 10^{-4})$$
$$= -(-3.4)$$
$$= 3.4$$

Look at Table 16.5 (p. 553), which shows the K_a and pK_a values for several weak acids. Notice how the K_a values vary by a factor of one million (1 000 000 or 10^6), while the pK_a values vary over a much smaller range.

Many biological molecules exhibit acidic and basic properties and, hence, have characteristic pK_a values. Look at Box 16.2 to learn about the pK_as of amino acid side chains, and how they can be instrumental in the catalytic function of many enzymes.

- $pK_a = -\log [K_a]$
- $pK_b = -\log [K_b]$

Self-check 16.4

(a) For ethanoic acid, $K_a = 2 \times 10^{-5}$; what is its pK_a value?

(b) For ammonia, $K_b = 1.8 \times 10^{-5}$; what is its pK_b value?

MATHS TOOL 16.1 **The exponential and logarithmic functions**

When we use the term 'exponential', we are usually considering a process that is happening very rapidly. For example, exponential growth is typically used to describe the way that the size of a bacterial population increases in size over time given an adequate supply of nutrients, as illustrated in Figure 16.5. Look at this figure and notice that, as we travel in equally-sized increments along the *x*-axis (denoting time), the population size (measured along the *y*-axis) increases in a very rapid, non-linear way. The plot in Figure 16.5 curves steeply upwards.

We can describe the relationship between variables such as bacterial population size and time by the equation $y = a^x$. This relationship states that y takes the value of some constant (denoted here by *a*) raised to the power of *x*, and is called an exponential function.

In principle, *a* can take any value. In practice, though, we usually encounter two particular values of *a*: base e (where $a = e$), and base 10 (where $a = 10$).

e is what we call a natural constant (like π) and, like π, is an irrational number. This means that we can't evaluate e precisely as a decimal; the string of numbers after the decimal point is infinitely long. However, as an approximation, we can assign e the value of 2.718.

So, we write the exponential function to the base e as $y = e^x$. Similarly, we write the exponential function to the base 10 as $y = 10^x$.

Look at Table 16.3, which shows how *y* varies as we increase *x* by increments of 1 for the exponential functions $y = e^x$ and $y = 10^x$. Notice how *y* quickly reaches very large values, even while *x* is still quite small.

If a variable exhibits exponential *decay*, however, then we see its value decrease at a rapid rate, as illustrated in Figure 16.6. If something exhibits exponential decay (such as we see with the decay of radioactive isotopes, for example) we describe its behaviour by the following function: $y = e^{-x}$.

Table 16.3 The exponential functions $y = e^x$ and $y = 10x$.

x	Value of y when $y = e^x$	Value of y when $y = 10^x$
1	2.718	10
2	7.389	100
3	20.09	1000
4	54.60	10000
5	148.4	100000
6	403.4	1000000
7	1097	10000000
8	2981	100000000
9	8103	1000000000
10	22030	10000000000

Reversing the exponential function

When two variables–such as bacterial population growth and time–can be described by an exponential function of the type we see above, we find ourselves having to deal with very large numbers. It would make life much easier for us if we could manipulate these numbers in some consistent way, to make them smaller. We can do this by using the inverse of the exponential function, called the logarithm. (The inverse of a mathematical function simply reverses it. Therefore, division is the inverse of multiplication, and vice versa: $3/2 = 1.5$; to reverse this function, we carry out multiplication: $1.5 \times 2 = 3$.)

Logarithmic functions are used widely to encompass a very wide range of values. For example, the decibel (dB) system

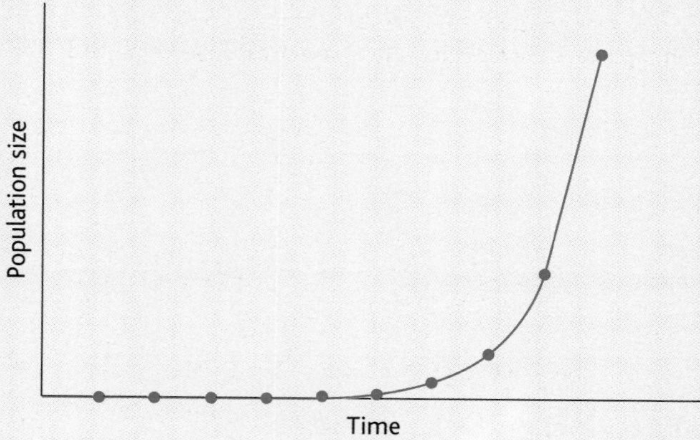

Figure 16.5 A graph denoting exponential growth. Notice how the value of *y* (in this case, population size) increases rapidly as the value of *x* increases.

that measures the intensity of sound uses a logarithmic scale: a value of 0 dB corresponds to the smallest audible sound, and each increase of 10 dB corresponds to a 10-fold increase in sound. (So a heavy lorry at 83 dB would give a 20-fold more intense sound than a car at 70 dB.) The Richter scale to measure the intensity of an earthquake is also logarithmic, where each unit on the scale corresponds to a 10-fold change in ground movement.

We note above how there are two widely used exponential functions: $y = e^x$ and $y = 10^x$. These two functions have different inverse functions.

The inverse of $y = e^x$ is called the natural logarithm, ln. In this instance, if $y = e^x$, then $\ln y = x$. So, for example, when $x = 3$, $y = e^3 = 20.09$, and ln 20.09 (the inverse function) = 3.

We can calculate the natural logarithm by using the 'ln' button on a calculator.

The inverse of $y = 10^x$ is called the logarithm to the base 10, written log. In this instance, if $y = 10^x$, then $\log y = x$. So, for example, when $x = 5$, $y = 10^5$, and $\log 10^5 = 5$.

We can calculate the logarithm to the base 10 by using the 'log' button on a calculator.

Table 16.4 The use of natural logarithms (ln) and logarithms to the base 10 (log) to scale down large numbers

y	Value of x, where $\ln y = x$	Value of x, where $\log y = x$
4.56×10^4	10.73	4.66
1.5×10^6	14.22	6.716
3×10^8	19.52	8.477
2.4×10^9	21.60	9.380
7×10^{13}	31.88	13.85

We can use natural logs or logarithms to the base 10 to scale down any large numbers, as illustrated in Table 16.4.

Notice how, regardless of whether we use ln or log functions, the values of y, which span several orders of magnitude (that is, from values in the thousands through to values of the magnitude 10^{13}) are reduced to be within the same order of magnitude: they are reduced to span a much narrower range of values.

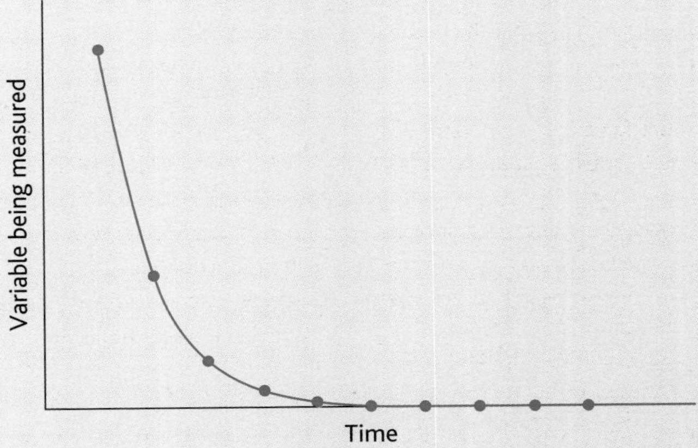

Figure 16.6 A graph denoting exponential decay. Notice how the value of y shows a rapid decrease, in contrast to the rapid increase seen with exponential growth.

Table 16.5 K_a and pK_a values for a range of weak acids.

Weak acid	K_a	pK_a
Lactic acid $CH_3CHOHCOOH$	1.4×10^{-4}	3.85
Ethanoic acid CH_3COOH	1.8×10^{-5}	4.74
Carbonic acid H_2CO_3	4.2×10^{-7}	6.38
Ammonium ion NH_4^+	5.6×10^{-10}	9.25

BOX 16.2 The pK_as of amino acids

A number of the amino acids we find in proteins have side chains that exhibit acidic or basic properties. Consequently, each of these amino acids has a characteristic pK_a value, just like any other acid or base.

Look at Table 16.6, which shows the pK_a values for several different amino acids, and also illustrates the chemical change each side chain undergoes when it acts as an acid or as a base, as applicable. Notice how these amino acids show quite contrasting pK_a values, demonstrating that the different amino acids can act as acids and bases to different extents.

It is not just the side chains of amino acids that undergo acid/base dissociation, however. The carboxyl and amino groups of an amino acid also exhibit characteristic pK_a values, as indicated in this table.

The pK_a values of the different amino acid side chains can have a marked effect on the function of the proteins of which they are part. Nowhere is this more important than in the active sites of enzymes.

We see in Chapter 14 how an enzyme's active site is the location within the enzyme molecule at which the substrate binds. Consequently, it is the site of any chemical changes that occur within the substrate molecule during the course of the enzyme-catalyzed reaction. Very often, an enzyme provides more than a physical structure in which

Table 16.6 pK_a values for a range of amino acids.

Amino acid	pK_a	
Arginine	12.5	$-(CH_2)_3-NH-C(=NH_2^+)-NH_2$
Aspartic acid	3.9	$-CH_2-C(=O)O^-$
Cysteine	8.4	$-CH_2-SH$
Glutamic acid	4.1	$-CH_2-CH_2-C(=O)O^-$
Histidine	6.0	$-CH_2-$ imidazole ring (HN+, NH)
Lysine	10.5	$-(CH_2)_4-NH_3^+$
Tyrosine	10.5	$-CH_2-C_6H_4-OH$
Carboxyl group	~2.2	$-C(=O)OH$
Amino group	~9.5	H_2N-

BOX 16.2 **Continued**

the substrate can 'dock' while the reaction takes place; amino acid side chains located at the enzyme's active site can participate actively in the chemical reaction involving the substrate.

The environment of an enzyme's active site can often be in stark contrast to that of the aqueous medium surrounding it, which can serve to alter the pK_a values of the amino acid side chains present in the active site. This alteration in pK_a values can be of central importance to the enzyme's catalytic function: a shift in pK_a value may make a particular side chain more or less acidic (that is, make it behave as a stronger or weaker acid) than it would otherwise be; this change in acidity then alters the availability of protons within the environment of the active site, which has an impact on the chemical reaction that can occur there.

For example, the pK_a of glutamic acid, when free in aqueous solution, is 4.1. Therefore, under normal conditions, glutamic acid is a rather weak acid. However, in the active site of the enzyme lysozyme, the pK_a of glutamic acid is raised to above 6, making it an even weaker acid. As a result, the side chain of glutamic acid remains protonated to a greater extent; this proton can be used to catalyze the cleavage of the glycosidic bond between monomer units in certain carbohydrate polymers, as illustrated in Figure 16.7. (By contrast, if the pK_a of the glutamic acid remained at 4.1, the side chain would be deprotonated and glutamic acid would no longer have a proton to donate in this fashion.)

Such catalysis is called **acid-base catalysis**; many enzymes feature chemical groups (such as amino acid side chains) in their active sites to participate in such catalysis.

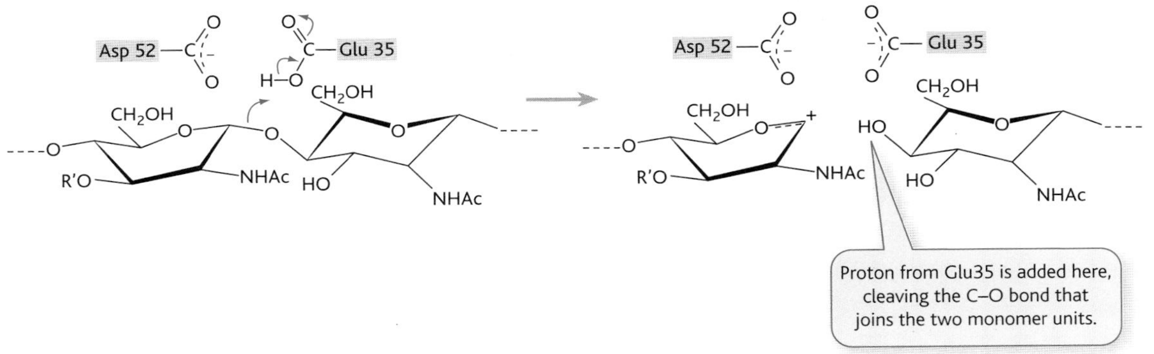

Figure 16.7 The side chain of Glu35 in the active site of the enzyme lysozyme is involved directly in the hydrolysis of carbohydrate polymers. Specifically, the side chain of Glu35 acts as an acid, and donates a proton to a C–O–R group, such that the C–O bond is cleaved. (Adapted from Petsko and Ringe, 2004.)

16.3 Keeping things balanced: the ion product of water

We see in Section 16.1 how water is amphoteric–that is, water behaves as both an acid *and* a base.

$$H_2O \rightleftharpoons H^+ + OH^-$$

But is water a strong acid or a strong base? A weak acid or a weak base? To answer these questions we need to consider the equilibrium constant for the dissociation of water.

Remembering that the equilibrium constant for a reaction is given by:

$$K = \frac{[C]^c[D]^d}{[A]^a[B]^b}$$

We can write the following expression for the dissociation of water:

$$K = \frac{[H^+][OH^-]}{[H_2O]}$$

In practice, the equilibrium constant for water has a value of 1×10^{-14} at 25 °C. This means that only *1 in every 550 000 000* (1 in 5.5×10^8) molecules of water dissociates into a proton and a hydroxyl ion. In other words, the equilibrium reaction:

$$H_2O \rightleftharpoons H^+ + OH^-$$

lies almost totally to the left: the back reaction predominates hugely over the forward reaction.

So few water molecules dissociate, in fact, that we can ignore the contribution of $[H_2O]$ to the value of the equilibrium constant. (Essentially, we could consider the value of $[H_2O]$ to be so invariant that we can simply discount it altogether. In practice, we can set $[H_2O]$ equal to 1.) Look at Box 16.3 to find out exactly what the concentration of pure water is.

Consequently, the expression for the equilibrium constant of water becomes:

$$K = \frac{[H^+][OH^-]}{1} = [H^+][OH^-]$$

We call this special value of the equilibrium constant the **ion product** of water, which we represent by the symbol $\mathbf{K_w}$.

- $K_W = [H^+][OH^-] = 1 \times 10^{-14}$

What does this mean in practice? It means that if a reaction occurring in aqueous solution requires a source of H^+ or OH^- ions to make it happen, we can't rely solely on the dissociation of water to supply them. So few water molecules dissociate that the concentration of H^+ or OH^- ions in the solution is effectively nil. Instead, we have to provide an *alternative* source of an acid or base: we have to add another acid or a base (other than water) to the aqueous solution.

Making use of the ion product of water

The ion product of water is a surprisingly useful expression: it shows a relationship between the concentration of H^+ or OH^- in *any* **aqueous solution**.

In words, the expression '$K_w = [H^+][OH^-] = 1 \times 10^{-14}$' is saying 'the concentration of protons multiplied by the concentration of hydroxide ions always has a value of 1×10^{-14}'. Therefore, if we know either the concentration of protons *or* the concentration of hydroxyl ions in a particular solution we can calculate the concentration of the other species.

BOX 16.3 What is the concentration of water molecules in pure water?

We see in Chapter 12 how we can calculate the number of moles of a compound if we know the mass and molar mass of the compound:

$$\frac{\text{mass of sample}}{\text{molar mass}} = \text{number of moles present in sample}$$

Let's imagine that we have 1 L of pure water. A litre of pure water has a mass of 1000 g, and water has a molar mass of 18 (atomic mass of O = 16; atomic mass of H = 1).

So the number of moles of water in 1 L is:

$$\frac{\text{mass of sample}}{\text{molar mass}} = \frac{1000}{18} = 55.56 \text{ mol}$$

We also see in Chapter 12 how the *concentration* of a solution is the number of moles of a compound in a *particular volume* of solution.

$$\text{concentration (mol L}^{-1}) = \frac{\text{number of moles (mol)}}{\text{volume (L)}}$$

Now, we've just said that there are 55.56 mol of water molecules in 1 L of water, so the concentration of water is:

$$\frac{\text{number of moles (mol)}}{\text{volume (L)}} = \frac{55.56 \text{ mol}}{1 \text{ L}} = 55.56 \text{ mol L}^{-1}$$

So, the concentration of water molecules in pure water is a huge 55.56 M.

EXAMPLE

Let's imagine that we have a solution that contains protons at a concentration of 10^{-3} M.

The ion product of water, K_w, tells us that:

$[H^+][OH^-] = 10^{-14}$

We know that $[H^+] = 10^{-3}$, so:

$(10^{-3}) \times [OH^-] = 10^{-14}$

We then rearrange the equation to get the unknown quantity, $[OH^-]$, by itself:

$$\frac{(\cancel{10^{-3}}) \times [OH^-]}{\cancel{10^{-3}}} = \frac{10^{-14}}{10^{-3}}$$

$$[OH^-] = \frac{10^{-14}}{10^{-3}}$$

We divide both sides of the equation by the term we want to remove from the left-hand side (10^{-3}).

$$= 10^{-11} \text{ M}$$

Self-check 16.5 An aqueous solution contains hydroxyl ions at a concentration of 10^{-7} M. What is the concentration of H^+ ions in the same solution?

Linking K_w, K_a, and K_b

There is a very important relationship between the values of the acid dissociation constant, K_a, and the base dissociation constant, K_b, for a conjugate acid–base pair. The two values multiplied together always equal 1×10^{-14}, the value of K_w.

$$K_w = K_a \times K_b = 1 \times 10^{-14}$$

This relationship tells us that, if we know the value of K_b for a weak base, we can calculate K_a, the acid dissociation constant for its conjugate acid. Similarly, if we know the value of K_a for a weak acid, we can calculate the value of K_b for its conjugate base.

To show how this relationship arises, let us consider the reaction of ammonia with water:

$$NH_3 + H_2O \rightleftharpoons NH_4^+ + OH^-$$

The base in this reaction is ammonia, NH_3, while its conjugate acid is NH_4^+.

We write the dissociation of the base, NH_3, as follows:

$$NH_3 + H_2O \rightleftharpoons NH_4^+ + OH^-$$

So:

$$K_b = \frac{[NH_4^+][OH^-]}{[NH_3]}$$

Remember: the general form of the expression for

$$K_b = \frac{[BH^+][OH^-]}{[B]}$$

where B = the base, and BH^+ = its conjugate acid. We can ignore the contribution of H_2O to the equation.

By contrast, we write the dissociation of ammonia's conjugate acid, NH_4^+, as:

$$NH_4^+ \rightleftharpoons NH_3 + H^+$$

So:

$$K_a = \frac{[NH_3][H^+]}{[NH_4^+]}$$

If we multiply K_b and K_a, we get:

$$\frac{[NH_4^+][OH^-]}{[NH_3]} \times \frac{[NH_3][H^+]}{[NH_4^+]}$$

$$\frac{\cancel{[NH_4^+]}\,[OH^-]}{\cancel{[NH_3]}} \times \frac{\cancel{[NH_3]}\,[H^+]}{\cancel{[NH_4^+]}}$$

We can simplify the equation by cancelling out common terms

$$= [OH^-] \times [H^+]$$

$$= K_W$$

$$= 1 \times 10^{-14}$$

EXAMPLE

Suppose we find that the value of K_b for ammonia is 2×10^{-5}. What is the value of K_a for its conjugate acid, NH_4^+?

We know that $K_b \times K_a = K_w = 1 \times 10^{-14}$. So:

$$(2\times10^{-5})\times K_a = K_W = 1\times10^{-14}$$

If we rearrange to isolate the unknown term, K_a:

$$K_a = \frac{1\times10^{-14}}{2\times10^{-5}}$$

$$= 5 \times 10^{-10}$$

Self-check 16.6

(a) The value of K_b for the hydrogen carbonate ion, HCO_3^-, is 2.4×10^{-8}. What is the value of K_a for its conjugate acid, H_2CO_3?

(b) The value of K_a for methanoic acid, HCOOH, is 2×10^{-4}. What is the value of K_b for its conjugate base, $HCOO^-$?

16.4 Measuring concentrations: the pH scale

We see in the previous section how the acid and base dissociation constants give a measure of the extent to which an acid or a base dissociate to form protons or hydroxide ions respectively in solution. But how can we give an easy indication of the concentrations of these species in solution? The answer lies in the use of the pH scale.

The pH of a solution gives us a measure of the concentration of hydrogen ions in the solution.

We see in Section 12.2 how we usually measure concentrations in units of moles per litre (mol L^{-1}). However, the concentration of protons in an aqueous solution can vary by many orders of magnitude: from as low as 1×10^{-15} M to greater than 10 M, depending on whether the solution contains a strong or a weak acid.

To make it easier for us to compare values of the H^+ concentrations of different solutions, we convert concentrations so they fit the **pH scale**. The pH scale is related to the concentration of protons in solution as follows:

We learn more about logarithms in Maths Tool 16.1 on p. 552.

$$pH = -\log [H^+]$$

Notice that pH is the **negative** of the log of the proton concentration. The log of any number with a value less than 1 is negative, so we use the rule that a negative value multiplied by a negative value gives a positive value to convert the negative value of log $[H^+]$ into a positive value.

EXAMPLE

Let us suppose we have a solution with a H^+ concentration of 3×10^{-4} M. What is the pH of this solution?

$$\begin{aligned} pH &= -\log (3 \times 10^{-4}) \\ &= -(-3.5) \\ &= 3.5 \end{aligned}$$

Self-check 16.7

(a) What is the pH of a solution with a $[H^+]$ of 4.5×10^{-5} M?

(b) What is the pH of a solution with a $[H^+]$ of 3.2×10^{-7} M?

We often group different solutions into families depending on their pH:

- If the pH of a solution is less than 7.0 we say the solution is **acidic**.
- If the pH of a solution is greater than 7.0 we say the solution is **alkaline**.
- If the pH of a solution is equal to 7.0 we say the solution is **neutral**.

For example, we see in Section 16.3 how pure water has a H^+ concentration of 1×10^{-7} M. So the pH of pure water is $-\log(1 \times 10^{-7}) = 7.0$. Pure water is a **neutral** solution.

- $pH = -\log[H^+]$
- $pH < 7 \Rightarrow$ *acidic solution*
- $pH > 7 \Rightarrow$ *alkaline solution*
- $pH = 7 \Rightarrow$ *neutral solution*

Look at Table 16.7, which shows the pH values of a range of different solutions. Notice how some solutions have a pH < 7 and so are acidic, while other solutions have a pH > 7 and are alkaline.

Now look at Table 16.8, which shows the pH values of a variety of components of biological systems. Notice how different biological systems have very different pHs. For example, blood has a pH of 7.4, while the stomach can have a pH as low as 1.5. The differences in pH between different parts of the body present a major challenge when it comes to the formulation of therapeutic agents (that is, drugs) that must enter and interact with the body in a carefully targeted way. We learn more about the challenges to drug design and formulation presented by differing pHs within biological systems in Section 16.6.

The pH of strong and weak acids

Let's imagine that we dissolve 1 mol (63 g) of HNO_3 in 1 L of water, to produce a 1M HNO_3 solution. HNO_3 is a strong acid, which means that it dissociates almost *completely* when it is dissolved in water. In other words, if we dissolve 1 mol of HNO_3, we generate 1 mol of H^+ ions: all the HNO_3 dissociates into protons (which form the more stable solvated proton) and nitrate ions, so the concentration of protons generated is equal to the concentration of the acid added to the solution initially.

Consequently, it is very easy to determine the pH of a strong acid solution if we know the concentration of the acid itself. In the case of HNO_3 at a concentration of

Table 16.7 pH values for a range of solutions

Solution	pH
Gastric juice	1.6
Lemon juice	2.2
Vinegar	2.8
Orange juice	3.5
Apple juice	3.8
Coffee	5.0
Urine	6.0
Milk	6.4
Pure water	7.0
Drinking water	7.2
Bile	8.0
Seawater	8.5
Ammonia	11.0

Table 16.8 pH values for various components of biological systems. (Data from Price *et al.*, 2001).

System	Typical pH
Blood	7.4
Stomach	1.5
Lungs	7.6
Active muscle	6.0
Mitochondrial matrix	8.0
Endoplasmic reticulum	7.2
Golgi	6.5
Lysosomes	5.0

1M, the solution contains protons at a concentration of 1 mol L^{-1} (because all of the HNO_3 has dissociated), so the pH of the solution is:

$$-\log 1 = 0$$

By contrast, a solution of H_2SO_4 at a concentration of 0.5 mol L^{-1} contains protons at a concentration of 1 mol L^{-1}–that is, twice the concentration of the acid. Why is this? H_2SO_4 dissociates as follows:

$$H_2SO_4 \rightarrow 2H^+ + SO_4^-$$

Notice from the stoichiometry of this equation how two moles of protons are yielded for every one mole of H_2SO_4 that dissociates. Therefore, the pH of a 0.5 M solution of H_2SO_4 is $-\log [H^+] = -\log 1 = 0$.

Self-check 16.8 (a) What is the pH of a solution containing the strong acid HBr at a concentration of 3.2×10^{-3} mol L^{-1}? (b) What is the pH of a solution containing the strong acid H_2SO_4 at a concentration of 1.4×10^{-2} mol L^{-1}?

However, things aren't quite so straightforward for weak acids. Weak acids only dissociate **partially**. This means that, if we dissolve 1 mol of a weak acid in a litre of water, we *don't* get 1 mol of protons because not *all* of the 1 mol of weak acid molecules dissociate to form protons. Look at Figure 16.8, which contrasts the formation of protons from strong and weak acids. Notice how a strong acid such as HNO_3 yields as many moles of protons as there are moles of acid added initially. By contrast, a weak acid yields fewer protons than there are moles of acid added initially.

So how can we calculate the pH of a solution containing a weak acid?

The answer lies in a new relationship defined by the **Henderson–Hasselbalch equation**.

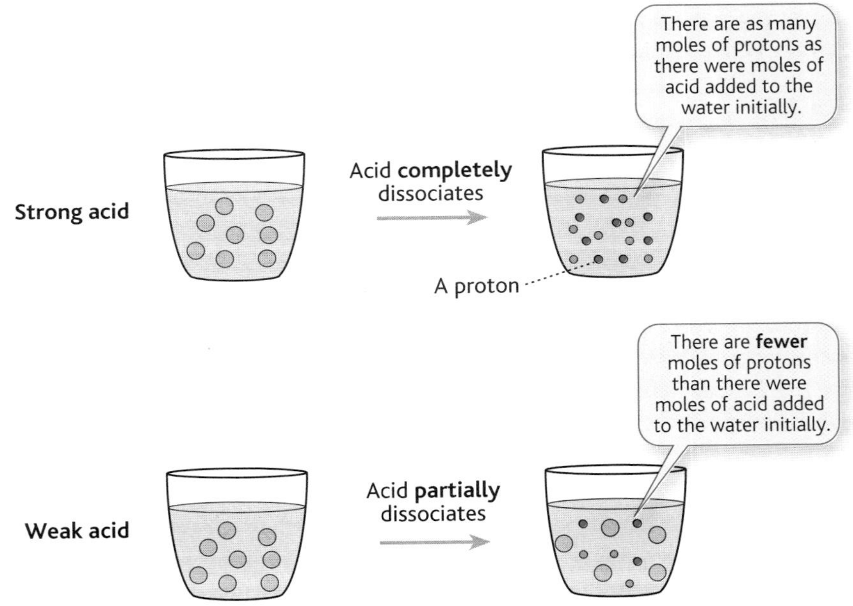

Figure 16.8 The formation of protons from strong and weak acids. A strong acid like HNO_3 yields as many moles of protons as there are moles of acid added initially. By contrast, a weak acid yields *fewer* protons than there are moles of acid added initially.

Linking weak acid strength and pH: using the Henderson–Hasselbalch equation

The Henderson–Hasselbalch equation is used in different forms to relate pH to acidity and basicity in the form of pK_a or pK_b values. In the case of the relationship between the pH and pK_a of a weak acid, the following form of the Henderson–Hasselbalch equation applies:

We most often see a different form of the Henderson–Hasselbalch equation being used to determine the pH of buffer solutions. We learn more about the use of the Henderson–Hasselbalch equation in this context on p. 575.

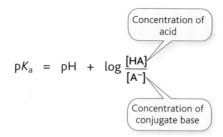

$$pK_a = pH + \log \frac{[HA]}{[A^-]}$$

Concentration of acid

Concentration of conjugate base

Let's imagine that we have a solution containing 0.0025 M ethanoic acid and 0.0005 M ethanoate ion (the conjugate base), and we know that the pK_a value for ethanoic acid is 4.74. What would be the pH of the solution? We merely need to feed these values into the appropriate form of the Henderson–Hasselbalch equation:

$$pK_a = pH + \log \frac{[HA]}{[A^-]}$$

$$4.74 = pH + \log \frac{[0.0025]}{[0.0005]}$$

$$4.74 = pH + \log 5$$

$$4.74 = pH + 0.70$$

By subtracting 0.70 from both sides, we find that

$$4.04 = pH$$

So the pH of the solution is **4.04**.

Self-check 16.9 What is the pH of a solution containing carbonic acid at a concentration of 0.003 M and hydrogen carbonate ion (its conjugate base) at a concentration of 0.0007 M? The pK_a value for carbonic acid is 6.38.

The contribution of water to the pH

The protons present in an aqueous solution don't form solely from the dissociation of acids (and, likewise, the hydroxide ions don't come only from the dissociation of bases). Instead, a proportion of protons and hydroxide ions originate from the dissociation of water:

$$H_2O \rightleftharpoons H^+ + OH^-$$

However, as we see in Section 16.3, water dissociates to such a small extent that the number of protons and hydroxide ions contributed to an aqueous solution by the dissociation of water is negligible. So we can ignore the contribution of the dissociation of water to the concentration of protons or hydroxide ions, and assume that all protons and hydroxide ions originate from the dissociation of compounds other than water.

pH and the ion product of water: balancing $[H^+]$ and $[OH^-]$

The ion product of water tells us that the concentration of protons multiplied by the concentration of hydroxide ions in aqueous solution is always equal to 1×10^{-14}:

$$K_w = [H^+][OH^-] = 1 \times 10^{-14}$$

What does this relationship tell us about the concentration of hydroxide ions in acidic, alkaline, and neutral solutions? The concentration of protons and hydroxide ions in an aqueous solution must be inversely proportional to one another: as the concentration of one ion increases, the concentration of the other ion must *decrease* (and vice versa) so the product of the two concentrations always remains the same.

- In an acidic solution, $[H^+]$ is relatively high, so $[OH^-]$ must be relatively *low*.
- In an alkaline solution, $[H^+]$ is relatively low, so $[OH^-]$ must be relatively *high*.

Look at Table 16.9, which shows how the H^+ and OH^- ion concentrations compare in solutions of different pH. Notice how as the pH increases, so the proton concentration decreases, while hydroxide ion concentration *increases*.

We've already seen how the concentrations of H^+ and OH^- ions compare in a neutral solution (though we may not have realized it). Look again at the calculation of the pH of water above. Pure water is neutral, and has a pH of 7.0. The ion product of water shows how the concentration of both H^+ and OH^- ions is *equal* at 1×10^{-7} M. This is true of any neutral solution at 298 K:

- In a neutral solution, the concentrations of H^+ and OH^- ions are *equal*.

Notice that a solution being neutral does *not* mean that it doesn't contain charged species, i.e. ions. The use of the term 'neutral' here is quite different from the use of 'neutral' in the context of electrical charge.

- ***The concentration of protons multiplied by the concentration of hydroxide ions in solution is always equal to 1×10^{-14} at 25 °C.***

Table 16.9 A comparison of $[H^+]$ and $[OH^-]$ in solutions of different pH. All concentrations are mol L^{-1}.

pH	$[H^+]$	$[OH^-]$
0	1	10^{-14}
1	10^{-1}	10^{-13}
2	10^{-2}	10^{-12}
3	10^{-3}	10^{-11}
4	10^{-4}	10^{-10}
5	10^{-5}	10^{-9}
6	10^{-6}	10^{-8}
7	10^{-7}	10^{-7}
8	10^{-8}	10^{-6}
9	10^{-9}	10^{-5}
10	10^{-10}	10^{-4}
11	10^{-11}	10^{-3}
12	10^{-12}	10^{-2}
13	10^{-13}	10^{-1}
14	10^{-14}	1

Changing pH: neutralization reactions

We see in Section 16.1 how acids and bases must react in tandem: an acid must have a base to which to donate a proton and, likewise, a base must have an acid from which to *accept* a proton. When the base is a hydroxide[4] the reaction between the acid and base is called **neutralization**.

Neutralization involves the following general chemical change:

$$\text{acid} + \text{base} \rightarrow \text{salt} + \text{water}$$

4 A **hydroxide** is an ionic base comprising a cation and the OH^- anion.

When hydrogen chloride reacts with sodium hydroxide, the following neutralization reaction occurs:

$$\underset{\text{Acid}}{HCl} + \underset{\text{Base}}{NaOH} \rightarrow \underset{\text{Salt}}{NaCl} + \underset{\text{Water}}{H_2O}$$

A salt is formed by removing H^+ ions from an acid and replacing them with a metal ion. For example, the salt of hydrochloric acid, HCl, is sodium chloride, NaCl: we remove the H^+ ion from the acid, HCl, and replace it with the sodium ion Na^+ (a metal ion) to form the salt NaCl. We could equally well use the metal ion K^+ (potassium) to form the salt potassium chloride (KCl).

Why is the reaction called a 'neutralization'? When an acid and hydroxide base react, the concentration of protons and hydroxide ions in solution changes, leading to an accompanying change in pH.

For example, let's consider what happens to the pH of the solution when equal quantities of HCl and NaOH undergo neutralization. 1M HCl has a pH of 0, while 1M NaOH has a pH of 14.0. HCl is a strong acid and NaOH is a strong base. Consequently, both compounds undergo virtually complete dissociation when dissolved in water:

$$HCl \rightarrow H^+ + Cl^-$$

and

$$NaOH \rightarrow Na^+ + OH^-$$

However, when the acid and base react together to undergo neutralization, the protons and hydroxide ions yielded from the dissociation come together to form water, as illustrated in Figure 16.9. This means that there are no 'free' H^+ or OH^- ions from the dissociation of HCl or NaOH. The only free H^+ and OH^- ions arise from the dissociation of the water molecules that are the basis of the aqueous solution.

Remember: water always dissociates to form equal numbers of H^+ and OH^- ions, as represented by the ion product of water, K_w.

So what is the pH of the solution? We know that water dissociates to give H^+ and OH^- ions at concentrations of 1×10^{-7}M, which correspond to pH 7.0. The solution is now at a **neutral pH**. Notice how the neutralization reaction has 'cancelled out' the acidity of the HCl (pH 0) and the alkalinity of the NaOH (pH 14.0), to leave a **neutral** solution.

Neutralization reactions do not always yield final solutions with a pH of 7.0, however. The pH of the final solution depends on the strength of the acid and the base that are reacting.

These combine to form H_2O — These combine to form the salt, NaCl

Dissociation of acid:	HCl	$\longrightarrow$	H^+	+	Cl^-
Dissociation of base:	NaOH	$\longrightarrow$	OH^-	+	Na^+
Overall:	**HCl + NaOH**	$\longrightarrow$	**H_2O**	**+**	**NaCl**

Figure 16.9 When HCl and NaOH undergo neutralization the protons and hydroxide ions generated come together to form water.

For example, let's consider the reaction of a weak acid with a strong base:

$$CH_3COOH + NaOH \rightarrow H_2O + CH_3COONa$$

In this instance, the salt yielded from this reaction, CH_3COONa participates in a separate neutralization reaction with water, generating more OH^- ions:

$$CH_3COO^- + H_2O \rightleftharpoons CH_3COOH + OH^-$$

(CH_3COO^- is a strong enough base that it can extract protons from water, disrupting the ion product of water as it does so.)

As a result, at the end of the neutralization reaction, there are relatively more hydroxide ions than protons, so the solution has a slightly **alkaline** pH, and the pH is greater than 7.

We can summarize the pH of the solution resulting from the neutralization of acids and bases of different strengths as follows:

- The salt of a strong acid and a strong base yields a **neutral** solution.
- The salt of a strong acid and a weak base yields an **acidic** solution.
- The salt of a weak acid and a strong base yields an **alkaline** solution.
- The salt of a weak acid and a weak base yields an acidic, alkaline, or neutral solution depending on the pK_a of the ions involved.

Neutralization reactions provide a useful way of altering the pH of biological solutions. Look at Box 16.4 to discover how we use neutralization reactions to treat conditions such as heartburn and indigestion.

pOH: the basic equivalent of pH

The pOH of a solution gives us a measure of the concentration of hydroxide ions in the solution, just as the pH gives us a measure of the concentration of protons.

pOH is related to the concentration of hydroxide ions according to the following relationship:

$$pOH = -\log[OH^-]$$

Notice how this relationship mirrors the relationship between pH and $[H^+]$.

BOX 16.4 Quelling the flames of heartburn

Hydrochloric acid (HCl) is an important constituent of the gastric juices secreted by the stomach lining. Hydrochloric acid is required by pepsin to catalyze the breakdown of proteins in food. However, if the stomach is irritated or overstretched (by eating too much) some of the hydrochloric acid can flow back up the oesophagus (the tube connecting the stomach to the throat), and produces the burning sensation we call heartburn.

The symptoms of heartburn can be treated with anti-acid (i.e. 'antacid') tablets, which are made from bases such as sodium bicarbonate and aluminium hydroxide. The bases in the antacid tablet undergo a neutralization reaction with the hydrochloric acid in the stomach. For example, sodium bicarbonate reacts with HCl as follows:

$$NaHCO_3 + HCl \rightarrow NaCl + H_2O + CO_2$$

Aluminium hydroxide reacts in a similar way:

$$Al(OH)_3 + 3\,HCl \rightarrow AlCl_3 + 3\,H_2O$$

Such neutralization reactions consume excess HCl, making the contents of the stomach less acidic, and so preventing any continued burning sensation should any more of the stomach contents flow back up the oesophagus.

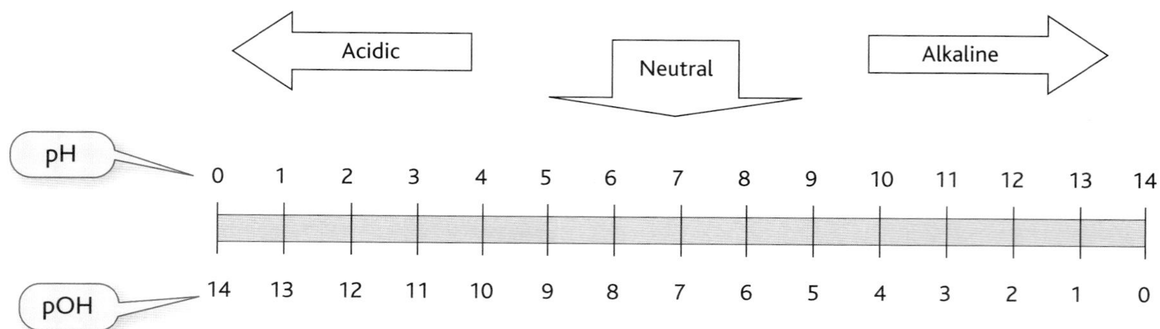

Figure 16.10 A comparison of pOH with pH. Notice how the sum of the pH and pOH of a particular solution always equals 14.0.

EXAMPLE

Let's imagine that we have a solution containing hydroxide ions at a concentration of 1.5×10^{-4} M. What is the pOH of the solution?

$$\begin{aligned} pOH &= -\log [OH^-] \\ &= -\log (1.5 \times 10^{-4}) \\ &= -(-3.8) \\ &= 3.8 \end{aligned}$$

We see on p. 563 how the concentrations of protons and hydroxide ions in solution are inversely proportional: as the hydroxide ion concentration increases, so the proton concentration *decreases*. Consequently, a solution with a high pH (high $[H^+]$) has a *low* pOH (low $[OH^-]$). By contrast, a solution with a low pH (low $[H^+]$) has a *high* pOH (high $[OH^-]$).

Look at Figure 16.10, which shows how pOH varies in solutions of different pH. Notice how the sum of the pH and pOH of a particular solution always equals 14.0. (For example, if a solution has a pH value of 3.0, its pOH value is 11.0.)

- *pOH gives a measure of the concentration of hydroxide ions in solution.*
- *$pOH = -\log[OH^-]$*
- *$pH + pOH = 14$*

Self-check 16.10 What is the pOH of a solution containing hydroxide ions at a concentration of:

(a) 0.5 M (b) 3×10^{-3} M?

16.5 The behaviour of acids and bases in biological systems

In the preceding sections of this chapter, we see how different acids and bases undergo dissociation to different extents in aqueous solution. We can determine the extent of this dissociation by determining the pK_a value of an acid, or the pK_b value for a base. But how does the acidic or basic nature of a particular molecule affect its behaviour in a biological system, and how can the biological environment to which an acidic or basic compound is exposed influence its behaviour? We explore the answers to these two questions in this section.

The effect of acidity and basicity on partitioning between aqueous and hydrophobic systems

We see in Section 16.1 how we can represent the dissociation of an acid by the following general equation:

$$HA \rightleftharpoons A^- + H^+$$

Notice how, before dissociation occurs, the acid (HA) is uncharged, but becomes ionized–to form a charged species–once dissociation has occurred.

This transition–from un-ionized to ionized–has significant implications for the behaviour of a compound in a biological system. But why is this? Un-ionized and ionized compounds differ in the degree to which they interact favourably with hydrophilic and hydrophobic entities. Ionized compounds are typically polar (by virtue of their charge) and so interact favourably with an aqueous environment, such as the cytoplasm of the cell. By contrast, they do not interact so favourably with hydrophobic species, such as the lipid membranes which surround the cell, and the organelles into which the cell is compartmentalized. Therefore, when exposed to both a hydrophilic, aqueous environment, and a hydrophobic, lipid-based environment, an ionized species will partition more readily into the aqueous environment.

We learn more about the concept of partitioning in Chapter 11.

Un-ionized compounds are typically more non-polar in character than ionized compounds, and so typically exhibit opposing characteristics: they favour hydrophobic environments (such as lipid membranes) over hydrophilic ones.

As a consequence, the extent to which a compound dissociates in a particular environment can influence whether it is most likely to partition into an aqueous system (and remain in the cytoplasm, for example) or partition into a hydrophobic system (and associate with a lipid membrane). A strongly acidic or basic compound–one with a low pK_a or low pK_b–is more likely to dissociate than a weak acid or base. Therefore, in general terms, a strong acid or base is more likely to be ionized than a weak acid or base, and is more likely to exhibit hydrophilic tendencies. By contrast, a weak acid or base is less likely to be ionized, and so is more likely to exhibit hydrophobic tendencies.

These differences in behaviour can have important implications for the design of compounds, such as therapeutic drugs, which must interact with biological systems in certain, defined ways. If a drug needs to travel to a certain tissue it must exhibit some degree of solubility in aqueous solution if it is to be stable in the bloodstream and be transported to its target location. However, to enter a target tissue, a drug will need to leave the aqueous environment of the bloodstream and cross at least one membrane. Such partitioning requires the drug to exhibit some degree of hydrophobicity (and, hence, lipophilicity). Such considerations immediately begin to reveal the challenges associated with the design and development of therapeutic agents.

- ***A strong acid is most likely to be ionized in aqueous solution, and therefore to partition into an aqueous phase.***
- ***A weak acid is more likely not to be ionized, and therefore to partition into a hydrophobic/lipophilic phase.***

Remember: a compound's pK_a or pK_b gives a measure of the extent to which a compound dissociates and so gives a measure of the likely degree of ionization of that compound.

Remember: a low pH equates to there being a high concentration of protons; a high pH equates to there being a low concentration of protons. Therefore, a decrease in pH means an increase in $[H^+]$; an increase in pH means a decrease in $[H^+]$.

The effect of pH on acidity and basicity

A compound's pK_a (or pK_b) value can be affected by its surrounding environment–specifically, the pH of its environment. Consequently, while a compound may behave as a strong acid at one pH (and so exhibit a high degree of ionization) it may be a weaker acid at a different pH (and so be far less ionized, giving it more hydrophobic character).

We can sum up the impact of pH on the extent of dissociation of an acid as follows:

- A decrease in pH leads to a decrease in the extent of dissociation of an acid
- An increase in pH leads to an increase in the extent of dissociation of an acid.

Increasing or decreasing pH has the opposite effect on a base, however:

- A decrease in pH leads to an increase in the extent of dissociation of a base
- An increase in pH leads to a decrease in the extent of dissociation of a base.

Why do acids and bases behave in this way when exposed to environments of different pH? The answer lies in Le Chatelier's principle, and the way that a system at equilibrium responds when the equilibrium is disturbed. We see on page 521 how Henri Le Chatelier stated that 'when a chemical system at equilibrium is disturbed, it responds by shifting the equilibrium composition in such a way to counteract the change'. But how does this help to explain the behaviour of acids and bases at different pH values?

pK_a and pK_b values provide a measure of the extent of dissociation of an acid and base respectively when a system is at equilibrium:

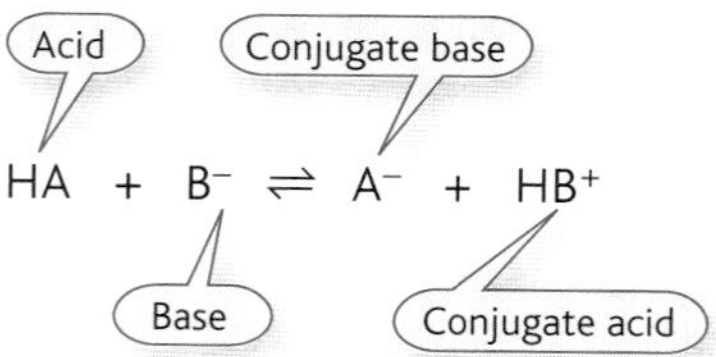

Let us just focus on the dissociation of the acid:

$$HA \rightleftharpoons A^- + H^+$$

At equilibrium, there will be no net change in the system. What happens if we alter the pH? If we decrease the pH, we increase the concentration of protons in solution. The system will respond by countering the increase in $[H^+]$ by shifting the acid dissociation equilibrium to the left–that is, towards the *formation* of HA. If relatively more HA is being formed, then it is behaving as a weaker acid: less of the acid is in dissociated form. So, when equilibrium is re-established, its position will be relatively further to the left than before the pH was decreased. This equates to a smaller pK_a value, characteristic of a weaker acid.

What happens if we *increase* the pH, such that we decrease the concentration of protons in solution? In this case, the system responds by shifting the acid dissociation equilibrium to the right–that is, towards the dissociation of HA into A^- and H^+. (The extra protons yielded by this increase in dissociation act to counter the decrease in $[H^+]$ in the system as a whole as a result of the increase in pH.) If relatively more HA is dissociating, it is acting as a stronger acid. In this instance, when equilibrium is re-established, its position will be relatively further to the right than

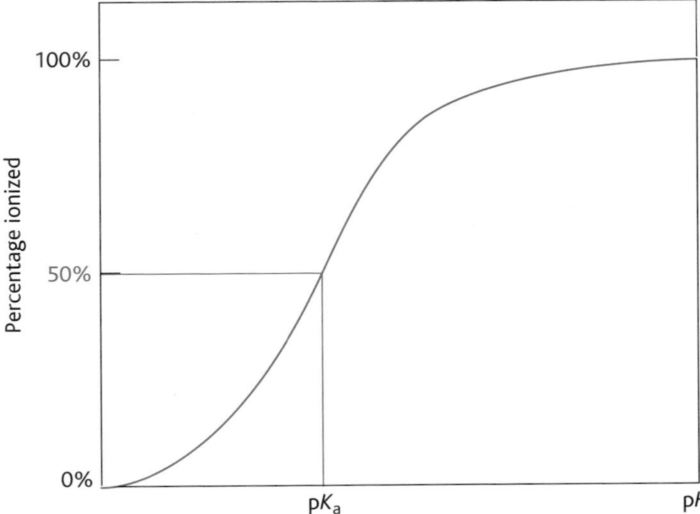

Figure 16.11 The variation in extent of dissociation of an acid with pH. Notice how the pK_a is the pH at which the acid is 50% dissociated.

before the pH was increased. This equates to a larger pK_a value, characteristic of a stronger acid.

Look at Figure 16.11, which illustrates how the extent of dissociation of an acid varies with pH. Notice how, at very low pH, little of the acid is dissociated. As the pH increases, however, (such that the concentration of H^+ in the system falls) more of the acid will dissociate.

Figure 16.11 also illustrates an important general concept: the pK_a is equal to the pH at which 50% of an acid is ionized (that is, dissociated). If an acid has a pK_a of 4.0, it tells us that, when the acid is present in a solution of pH 4.0, 50% of the acid will be dissociated. (That is, of we take the general acid, HA, 50% will be HA, and 50% will be dissociated to form A^- and H^+.) If the pH is *higher* than the pK_a value for the acid, then *more* than 50% of the acid will be dissociated; if the pH is *lower* than the pK_a for the acid, then *less* than 50% of the acid will be dissociated.

- ***The pK_a value of a molecule represents the* pH *at which it is 50% ionized in solution***

Let's just reiterate the key concepts here:

- A decrease in pH reduces the extent of dissociation of an acid
- An increase in pH increases the extent of dissociation of an acid
- If the pH is lower than an acid's pK_a value, the acid will be less than 50% dissociated
- If the pH is higher than an acid's pK_a value, the acid will be more than 50% dissociated.

These concepts have important implications for pharmaceutically active compounds. Such compounds may be exposed to conditions of different pH as they travel through the body. Consequently, they may be ionized to different extents depending on their specific location.

We see on p. 567 how the extent of ionization of a pharmaceutical compound can affect its uptake into tissues or cellular fluids (with a greater degree of ionization favouring partitioning into aqueous media, such as the cell cytoplasm, and a lesser degree of ionization favouring partitioning into hydrophobic environments, such as the lipid bilayer of cell membranes). Therefore, as a pharmaceutical compound travels through the body and encounters environments of different pH, its degree of ionization may favour its partitioning into lipids (enabling it to enter cells) in some locations, but favour its partitioning into an aqueous environment (such as the blood) elsewhere. These differences in behaviour under conditions of different pH can therefore help target pharmaceuticals to their correct sites of action. We learn more about the impact of pH on the behaviour of pharmaceuticals in Box 16.5.

The biochemical behaviour of amino acids can also be modulated by the pH of the environment to which they are exposed. If we look back at Table 16.6, we see how different amino acids exhibit different pK_a values. If an amino acid such as aspartic acid or glutamic acid, with pK_a values around 4.0, are exposed to a physiological pH of around 7.0, they will be highly dissociated, and will carry a negative charge, as illustrated in Figure 16.12. This means that, at pH 7.0, the side chains of these amino acids can participate in ionic interactions with neighbouring positively charged side-chains, which could serve to stabilize the three-dimensional structure of the protein of which they are part. At a lower pH, the same side chains would be less extensively ionized, making them less able to enter into such stabilizing ionic interactions.

We learn more about molecular interactions and their importance in biological systems in Chapter 4.

Self-check 16.11 Are the following amino acid side chains likely to be more than 50% ionized, less than 50% ionized, or roughly 50% ionized at pH 6.0? (a) Lysine (pK_a 10.5) (b) Glutamic acid (pK_a 4.1); (c) Histidine (pK_a 6.0)?

BOX 16.5 Hitting the spot: drug dosage during feast and famine

Indomethacin (1) is an anti-inflammatory drug, which is used to relieve mild to moderately severe pain due to inflammation in conditions such as rheumatoid arthritis and osteoarthritis, whose symptoms include pain, swelling and inflammation in the joints. The pK_a of indomethacin is 4.5. Indomethacin can encounter different pHs in the stomach depending on whether the stomach is full or empty: after fasting the pH of the stomach is 2.0, and after a full meal is closer to 6.0. So what effect do these varying pHs have on the drug?

When the stomach is at pH 2.0, the pH is lower than the pK_a of indomethacin, and pushes the acid dissociation equilibrium to the left, away from the dissociated acid. Therefore, the majority of a dose of indomethacin is un-ionized at pH 2.0. By contrast, a full stomach, with a pH of 6.0, is higher than the pK_a of indomethacin, so we would expect to see more than 50% of the drug in its ionized form under these conditions.

But what does this difference mean in terms of how the drug interacts with the body? We would expect the un-ionized form to be better absorbed across the gut wall into the bloodstream, so greater absorption of the dose from the stomach would be expected at pH 2.0 than at pH 6.0. Therefore, we could expect to be told only to take indomethacin on an empty stomach, in order to maximise the uptake of the drug (and, hence, the dose that we receive).

O, N, Cl, CH_3, OH_3C, CO_2H

(1)

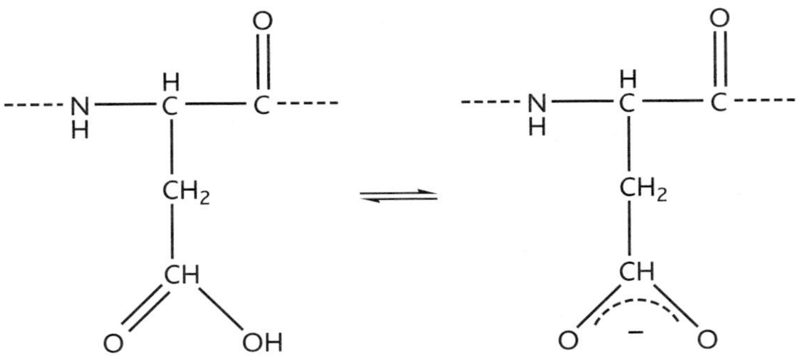

Figure 16.12 When an amino acid such as aspartic acid (illustrated here) is exposed to a physiological pH of around 7.0 its side chain becomes highly dissociated, such that the equilibrium shown here lies heavily to the right. Consequently, at pH 7.0, this side chain carries a negative charge.

16.6 Buffer solutions: keeping pH the same

A large number of reactions in the cell either release protons into solution, or remove them. Despite this influx and efflux of protons, the pH of cellular solutions does not change significantly. How is this so? The answer is that the aqueous solutions that fill cells act as **buffer solutions**.

A buffer solution is a solution that **resists changes** in its pH when small quantities of acid or alkali are added. A buffer solution acts in this way in order to **control** the pH so it remains within certain limits.

Buffer solutions are central to life processes. Normally, the pH of the blood, for example, is buffered to remain at pH 7.4. A deviation from this pH may be potentially lethal, as explained in Box 16.6.

A buffer solution typically has one of the following two compositions:

1. A weak acid and the salt of its conjugate base
2. A weak base and the salt of the conjugate acid.

We see why these compositions are important later in this section.

How does a buffer solution work?

A buffer solution works because it contains appropriate quantities of an acid–base conjugate pair, which can suppress the addition of protons *or* hydroxide ions to prevent the pH from changing. (A buffer solution must be able to counteract either an increase in the concentration of H^+ ions *or* an increase in the concentration of OH^- ions. So a buffer solution must contain an acid to react with additional OH^- ions, and a base to react with additional H^+ ions.) A buffer solution typically contains roughly equal concentrations of an acid and its conjugate base.

First, let's consider what happens when we add an acid to pure water, i.e. to a solution which doesn't act as a buffer. Look at Figure 16.13 to see what happens to the concentration of protons in solution. The concentration of hydrogen ions in a solution of pure water is 1×10^{-7} M, so the pH is 7. If we add just a few drops of 1 M hydrochloric acid to the water, the concentration of protons increases to something of the order of 10^{-3} M and the pH falls to a value of 3.0. Notice how this change corresponds to a **ten thousand-fold** increase in hydrogen ion concentration. A shift in concentration of such magnitude would be disastrous if it were to happen in the cell.

Now let's consider the composition of a buffer solution containing ethanoic acid and the salt of its conjugate base, sodium ethanoate, as illustrated in Figure 16.14.

BOX 16.6 Acidosis

Blood has a pH of 7.4, so it is very slightly alkaline. If the concentration of hydrogen ions in our body rises above its normal limits, we experience a state of acidosis.

Acidosis is a disturbance of the body acid–base balance resulting in elevated H^+ concentrations and, consequently, unusually low pH levels. Metabolic acidosis can occur as a result of many different conditions, such as kidney failure, alcoholism, and hyperthyroidism (an overactive thyroid gland). Any of these conditions are life-threatening, i.e. severe metabolic acidosis can be fatal.

Respiratory acidosis can occur if CO_2 levels in the blood become elevated (often as a result of CO_2 failing to diffuse from the lungs due to lung disease). CO_2 reacts with water to form hydrogen carbonate which, in turn, dissociates to form solvated protons (H_3O^+):

$$CO_2 + H_2O \rightleftharpoons H_2CO_3 \rightleftharpoons H_3O^+ + HCO_3^-$$

It is the extra H_3O^+ ions that lead to a lowering of pH.

By contrast, an abnormally *low* concentration of hydrogen ions (and *elevated* pH) is termed alkalosis.

The body contains a number of buffers, which are involved in maintaining the pH of blood at, or around, a value of 7.4, and help to ensure that conditions such as acidosis and alkalosis are avoided. Some of the most important biochemical buffers are shown in Table 16.10.

Not all parts of the body have the same pH as blood. The stomach, for example, is highly acidic, with a pH value which can be as low as 1.5. (The acidity is required to help our stomach to properly digest food.) By contrast, bile (which is produced by the liver) has a pH of 8.0.

Table 16.10 A selection of biochemically important buffers

Weak acid	Conjugate base
Carbonic acid: H_2CO_3	Hydrogen carbonate ion: HCO_3^-
Dihydrogen phosphate ion: $H_2PO_4^-$	Hydrogen phosphate ion: HPO_4^{2-}
Pyruvic acid: $CH_3COCOOH$	Pyruvate ion: CH_3COCOO^-
Lactic acid: $CH_3CHOHCOOH$	Lactate ion: $CH_3CHOHCOO^-$

In pure water...

Back reaction is favoured: very small K_a

$$H_2O \rightleftharpoons H^+ + OH^-$$

High concentration of water; low concentration of H^+

There are few H^+ ions in pure water because water does not dissociate readily...

When HCl is added...

Forward reaction is favoured: large K_a

$$HCl \rightleftharpoons H^+ + Cl^-$$

High concentration of H^+; low concentration of water

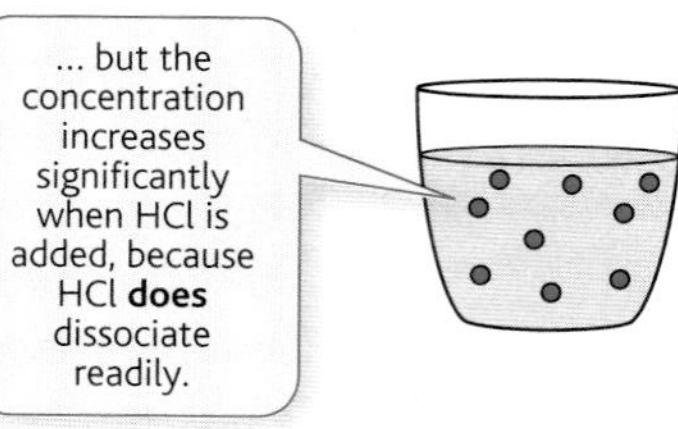

... but the concentration increases significantly when HCl is added, because HCl **does** dissociate readily.

Figure 16.13 The effect on the pH of pure water when a solution of an acid is added. Water is unable to act as a buffer. Consequently, the addition of an acid greatly increases the concentration of protons in solution. We see a marked rise in pH.

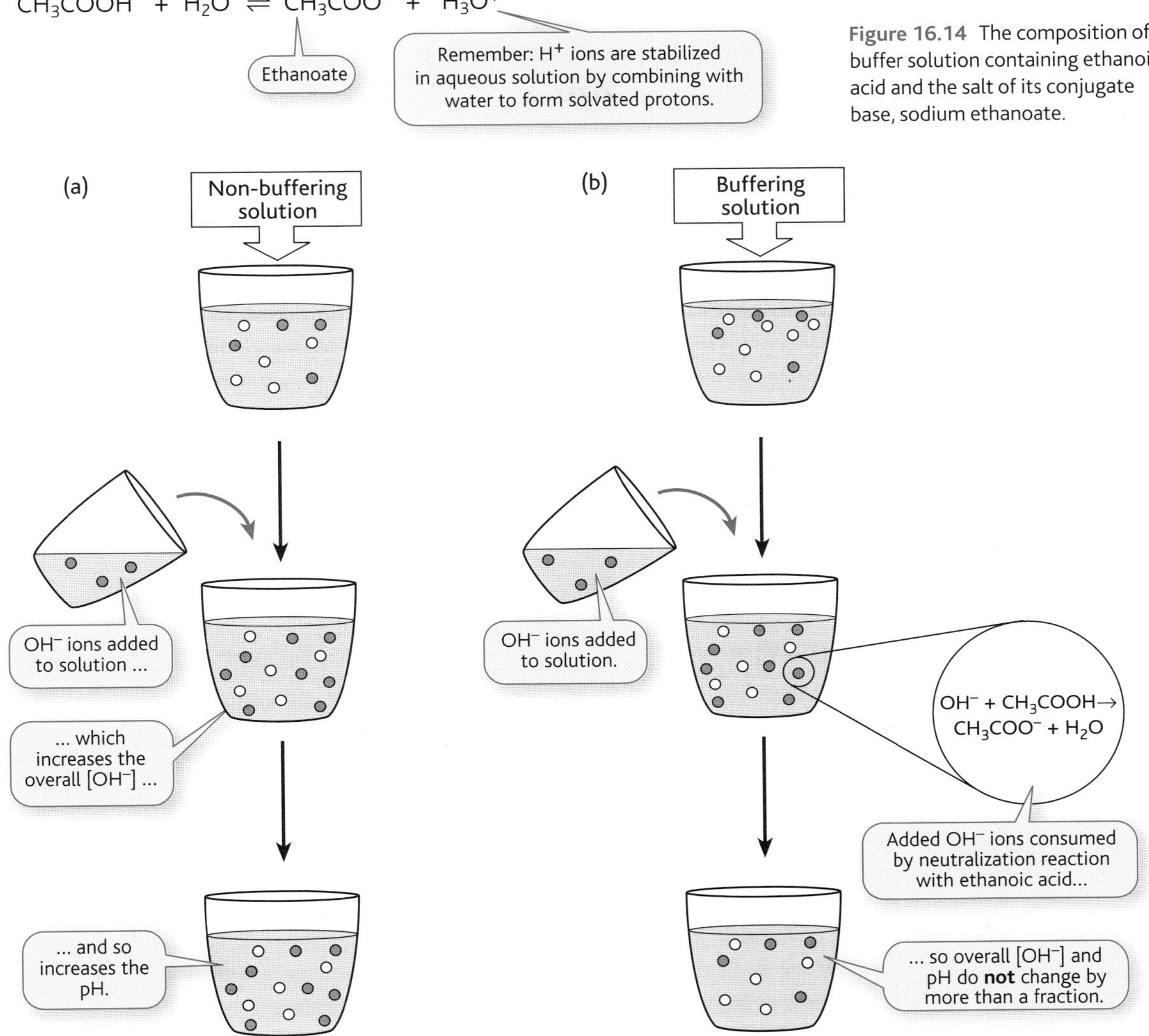

$CH_3COOH + H_2O \rightleftharpoons CH_3COO^- + H_3O^+$
Ethanoate
Remember: H+ ions are stabilized in aqueous solution by combining with water to form solvated protons.
(a)
Non-buffering solution
OH− ions added to solution ...
... which increases the overall [OH−] ...
... and so increases the pH.
(b)
Buffering solution
OH− ions added to solution.
$OH^- + CH_3COOH \rightarrow CH_3COO^- + H_2O$
Added OH− ions consumed by neutralization reaction with ethanoic acid...
... so overall [OH−] and pH do **not** change by more than a fraction.

Figure 16.14 The composition of a buffer solution containing ethanoic acid and the salt of its conjugate base, sodium ethanoate.

Figure 16.15 A comparison of the addition of a base to a non-buffering solution (a), and to a buffer solution (b).

In this solution the main constituents are water, ethanoic acid (0.1M), and ethanoate (0.1 M) all in dynamic equilibrium. For this buffer, the pH is 4.7.

Let's explore the consequences of disrupting this equilibrium, firstly by adding a solution of acid. We see in Chapter 15 how a solution at equilibrium counteracts a change in concentration by shifting its equilibrium position. In this instance, the equilibrium shifts to the left: the back reaction happens at a faster rate, so that H^+ are **consumed** by combining with CH_3COO^- ions to form ethanoic acid according to the back reaction:

$$CH_3COO^- + H_3O^+ \rightarrow CH_3COOH + H_2O$$

Because the added H^+ ions are consumed by the back reaction, the pH of the solution is virtually undisturbed.

What about the addition of a strong base such as OH^- ions to the solution? In this case, Figure 16.15 shows how the added OH^- ions are removed by a **neutralization** reaction with the ethanoic acid to produce ethanoate ions and water molecules:

$$CH_3COOH + OH^- \rightarrow CH_3COO^- + H_2O$$

We need to remember that, during a neutralization reaction, an acid reacts with a base to generate a salt and water. However, in solution, this salt may dissociate into its composite ions–in this case, the anion CH_3COO^- and the cation Na^+. We don't show the Na^+ ion in the equation in this instance because it is passive, and is only needed for completion.

So, once again, we see how the added ions, which would alter the pH of the solution if not dealt with, are consumed by reacting with a component of the buffer solution (in this case, the ethanoic acid, CH_3COOH). As a result, the overall concentration of H^+ and OH^- ions remains virtually unchanged, and so we see *very little change* in the pH of the solution.

There are several important points to note here:

1. We must use an **acid**–base conjugate pair.
2. We must use a **weak** acid or base.
3. When we use a weak acid we include the **salt** of its conjugate base (and when we use a weak base we include the salt of its conjugate acid).
4. Both compounds must be present at sufficiently high concentrations.

Why are these points important? Let's consider each in turn:

1. We must use an acid–base conjugate pair

A buffer solution must contain an acid and a base. However, the acid and base must not be able to react with each other. (Remember: an acid and a hydroxide base can undergo a neutralization reaction in solution to form a salt plus water.) We use an acid and its *conjugate* base because these species do not undergo neutralization with one another.

2. We must use a weak acid or base

We use a **weak** acid and its conjugate base because the position of equilibrium between the two species can shift–either in favour of the weak acid or in favour of the conjugate base. It is this flexibility that allows buffer solutions to counteract the changes in concentration caused by the addition of an acid or a base to the solution.

By contrast, strong acids and bases almost totally dissociate in solution: the equilibrium position for their dissociation (e.g. $HA \rightleftharpoons H^+ + A^-$) lies so far to the right that the back reaction barely occurs at all. In other words, the forward reaction is so heavily favoured that it is very difficult for the forward reaction to be overcome, and for any protons that are added to the solution to be consumed by an increase in the rate of the back reaction.

Consequently, strong acids and bases cannot compensate for fluctuations in H^+ or OH^- concentrations, and therefore they cannot act as buffers.

3. When we use a weak acid we include the salt of its conjugate base

The salt of a conjugate base provides us with a **ready source** of the conjugate base. Why is a 'ready source' of the conjugate base necessary? A buffer solution typically contains roughly equal concentrations of a weak acid and a base. However, we see in Section 16.2 how a weak acid only dissociates to a small extent. It is not feasible for a weak acid to dissociate to the extent that its concentration matches that of its conjugate base. So we need an alternative source of the conjugate base to boost its overall concentration.

Let us consider the buffer solution described above: ethanoic acid and the salt of its conjugate base, sodium ethanoate. A salt dissociates into its composite ions in solution. For example, sodium ethanoate, $NaCH_3COO$, dissociates into its component cation, Na^+, and anion, CH_3COO^-. The anion is the **conjugate base** of the acid, ethanoic acid.

By dissociating into its composite ions in solution, the salt of the conjugate base provides us with the extra source of conjugate base required for the weak acid and conjugate base to be present at roughly equal concentrations.

Some buffers need to operate at a relatively high pH. In such cases we use solutions containing a weak base and the salt of its conjugate acid. For example, ammonia (a weak base) can form a buffer solution when combined with one of its salts–for example, $NH_4^+Cl^-$. The salt dissociates into its component ions, NH_4^+ and Cl^-. NH_4^+ is the conjugate acid of NH_3.

4. Both compounds must be present at sufficiently high concentrations

The acid and base that comprise a buffer solution must be present at sufficiently high concentrations in order to mask any minor fluctuations in H^+ or OH^- concentration that may arise when additional acid or base is added to the solution. Consider an examination hall (silent except for the frantic scribbling of pens) and a busy pub. Imagine that someone enters both of these environments and starts talking. The effect in the exam hall would be one of significant disruption: their voice would really stand out in the silence. By contrast, there are so many conversations already going on in a busy pub we would hardly notice an additional voice.

The addition of an acid or base to a solution is the equivalent of someone walking into a room and starting to speak. A non-buffering solution (such as water) mirrors the situation in the examination hall: there are so few H^+ or OH^- ions in the solution that the addition of just a small amount of acid or base makes a significant difference. A buffering solution mirrors the situation in the busy pub: there are so many acid and base molecules already present in the buffer solution that the minor fluctuations in H^+ or OH^- concentration that may occur when additional acid or base is added are barely noticeable.

Look at Box 16.7 to find out about the compounds that give blood its buffering capability.

The pH of buffer solutions

Different buffer solutions operate at a characteristic pH value. Some buffer solutions maintain a relatively acidic pH, while other buffer solutions maintain a relatively basic pH. But how can we determine the pH of a specific buffer solution?

The pH of a buffer solution depends on the concentration of the weak acid and its conjugate base that are present in the solution (or the concentration of the weak base and the salt of its conjugate acid). The relationship between pH and these two concentrations is described by the following form of the Henderson–Hasselbalch equation:

$$\text{pH} = \text{p}K_a + \log\left(\frac{[\text{A}^-]}{[\text{HA}]}\right) \quad \textbf{or} \quad \text{pH} = \text{p}K_a + \log\left(\frac{[\text{A}^-]}{[\text{HA}]}\right)$$

(Left: [A⁻] = Conjugate base; [HA] = Weak acid. Right: [A⁻] = Weak base; [HA] = Salt of weak acid)

BOX 16.7 Buffering the blood

One of the most important buffers in the blood is carbonic acid, and its conjugate base, bicarbonate:

$$H_2CO_3 + H_2O \rightleftharpoons H_3O^+ + HCO_3^-$$

Carbonic acid — Bicarbonate ion

Carbonic acid is formed by the reaction of carbon dioxide (a product of human respiration) and water:

$$CO_2 + H_2O \rightleftharpoons H_2CO_3$$

Carbonic acid is a weak acid, with a K_a value of 4.2×10^{-7}. Most buffer solutions comprise a weak acid and conjugate base at roughly equal concentrations. However, to keep the blood at the right pH, carbonic acid is present at a concentration of about 0.0024 M, while the concentration of its conjugate base, bicarbonate, is ten times higher, at about 0.024 M.

Like all buffer solutions, the H_2CO_3/HCO_3^- buffer can counter the increases in both acid and base concentrations. An increase in acid concentration shifts the reaction to the left-hand side:

$$H_3O^+ + HCO_3^- \rightarrow H_2CO_3 + H_2O$$

Acid consumed

while an increase in base concentrations shifts the reaction to the right-hand side:

$$H_2CO_3 + H_2O \rightarrow H_3O^+ + HCO_3^-$$

Acid generated to offset increase in base concentration.

For a buffer solution consisting of a weak acid and its conjugate base, the pK_a value cited in the Henderson–Hasselbalch equation is the pK_a of the weak acid; for a buffer solution comprising a weak base and the salt of the weak base, the pK_a cited is the pK_a of the conjugate acid of the weak base.

EXAMPLE

Let's imagine that a buffer solution contains ethanoic acid at a concentration of 0.045 mol L^{-1} and its conjugate base, ethanoate ion, at a concentration of 0.051 mol L^{-1}. What is the pH of this solution, given that the pK_a of ethanoic acid is 4.74?

Strategy

We merely need to feed our known concentrations into the Henderson–Hasselbalch equation, along with the pK_a for ethanoic acid.

Solution

$$\begin{aligned} pH &= pK_a + \log\frac{[A^-]}{[HA]} \\ &= 4.74 + \log\frac{(0.051)}{(0.045)} \\ &= 4.74 + \log 1.133 \\ &= 4.74 + 0.0544 \\ &= 4.79 \end{aligned}$$

EXAMPLE

Now imagine that the buffer solution contains ammonia, NH_3, at a concentration of 0.025 mol L^{-1} and ammonium chloride (NH_4Cl), a salt of its conjugate acid, at a concentration of 0.021 mol L^{-1}. What is the pH of this solution?

Strategy

In this instance, we need to remember that the relevant pK_a is that of the conjugate acid whose salt we are using. The salt is ammonium chloride, and the corresponding weak acid is the ammonium ion, NH_4^+. The pK_a of NH_4^+ is 9.25. We can now feed our known concentrations into the Henderson–Hasselbalch equation as before, along with the pK_a for the ammonium ion.

Solution

$$\begin{aligned}\text{pH} &= \text{p}K_a + \log\frac{[\text{A}^-]}{[\text{HA}]}\\ &= 9.25 + \log\frac{(0.025)}{(0.021)}\\ &= 9.25 + \log 1.190\\ &= 9.25 + 0.0757\\ &= 9.33\end{aligned}$$

Self-check 16.12 A solution contains lactic acid at a concentration of 0.03 mol L^{-1} and lactate ion at a concentration of 0.034 mol L^{-1}. What is the pH of this solution, given that the pK_a of lactic acid is 3.85?

Maximising the buffering effect

Weak acids and their conjugate base (or weak bases and their conjugate acid) provide a maximum buffering effect when their concentrations are approximately equal. If we consider the general equation for the dissociation of an acid:

$$\text{HA} \rightleftharpoons \text{H}^+ + \text{A}^-$$

Weak acid (HA); Conjugate base (A^-)

this means that buffering occurs when [HA] = [A$^-$].

What does this tell us about the pH of the buffer solution under these conditions? We know from the Henderson–Hasselbalch equation that:

$$\text{pH} = \text{p}K_a + \log\frac{[\text{A}^-]}{[\text{HA}]}$$

Now, if [HA] and [A$^-$] are equal, then $\frac{[\text{A}^-]}{[\text{HA}]}$ must be 1. (Think about this for a moment: one number divided by itself is always equal to one.)

So, for a buffer solution in which the weak acid and its conjugate base are present in equal quantities, the Henderson–Hasselbalch equation becomes:

$$\text{pH} = \text{p}K_a + \log 1$$

Because log 1 = 0, this expression becomes pH = pK_a

Table 16.11 Composition, pH, and pK_a values for a range of buffer solutions.

Weak acid	Conjugate base	pK_a of weak acid	Approximate pH at which solution buffers
Lactic acid $CH_3CHOHCOOH$	Lactate ion $CH_3CHOHCOO^-$	3.85	4.0
Ethanoic acid CH_3COOH	Ethanoate ion CH_3COO^-	4.74	5.0
Carbonic acid H_2CO_3	Hydrogen carbonate ion HCO_3^-	6.38	6.0
Ammonium ion NH_4^+	Ammonia NH_3	9.25	9.0

For any weak acid in solution with its conjugate base, when their concentrations are equal, the pH of the solution is equal to the pK_a of the acid.

- *The pH value around which a solution buffers is equal to the pK_a of the acid.*

Look at Table 16.11, which shows the composition of various buffer solutions, and cites the pK_a value for the weak acid, and the pH around which the solution buffers. Notice the close agreement between the value of pK_a and the pH at which each solution buffers. In practice, it is recommended that buffer solutions (of sufficiently high concentration) are used within 1 pH unit of the pK_a. Thus, an ethanoic acid/ethanoate buffer should be used over the range of pH values from 3.7 to 5.7.

Preparing buffer solutions to a desired pH

Many important buffer solutions occur naturally in biological systems. But what happens if we need to prepare a buffer solution ourselves, perhaps to enable us to carry out a biochemical reaction *in vitro*? How do we know which compounds to use as the basis of the buffer solution, and how much of the chosen weak acid and its conjugate base (or weak base and salt of its conjugate acid) to use to produce a buffer solution of the desired pH?

Our choice of weak acid (or weak base) is driven by the pH at which we want to buffer. We see above how a buffer solution buffers most effectively close to the pK_a of its weak acid (or conjugate acid, if we're using a weak base). Therefore, when selecting the components of a buffer solution, we choose a weak acid (or weak base) whose pK_a is as close as possible to the pH we're trying to buffer around.

Notice again how, when we're considering a weak base as the basis of a buffer solution, the pK_a of interest is that of its conjugate acid.

Look again at Table 16.11. If we needed a buffer solution that buffers around pH 5.0, our best choice would be to use ethanoic acid (whose pK_a is 4.74) and its conjugate base, ethanoate ion. By contrast, if we were looking to buffer at a more alkaline pH, say 9.0, we should use ammonia and a salt of its conjugate acid, ammonium ion, whose pK_a is 9.25.

Having chosen the components of our buffer solution, how do we then determine how much of each component to add? The answer is that we use the Henderson–Hasselbalch equation to tell us the *ratio* of weak acid to conjugate base (or weak base to salt of conjugate acid) to use.

Let's imagine that we wish to prepare a buffer solution that buffers around pH 4.5. Our best choice of component would be ethanoic acid (with its pK_a of 4.74) and its conjugate base, ethanoate ion. We determine the ratio of ethanoic acid to ethanoate ion to use by using the Henderson–Hasselbalch equation as follows:

$$pH = pK_a + \log\frac{[A^-]}{[HA]}$$

We are trying to achieve a pH of 4.5, and the pK_a of ethanoic acid is 4.74. Therefore

$$4.5 = 4.74 + \log\frac{[A^-]}{[HA]}$$

Subtracting 4.74 from both sides to isolate the logarithmic term, we get:

$$-0.24 = \log\frac{[A^-]}{[HA]}$$

Now, we see in Maths Tool 16.1 how the logarithm to the base 10 is the inverse of the exponential function, such that if $\log y = x, y = 10^x$. Therefore, if

$$\log\frac{[A^-]}{[HA]} = -0.24, \quad \frac{[A^-]}{[HA]} = 10^{-0.24} = 0.575$$

This tells us that the ratio of ethanoate ion, the conjugate base (A^-), to ethanoic acid, the weak acid (HA), in the buffer solution must be 0.575:1–that is, for every 1 mol of ethanoic acid present, there must be 0.575 mol of ethanoate ion.

This tells us the *relative* quantities of weak acid and conjugate base that must be present in our buffer solution. But how do we determine the *actual* amounts? The easiest approach is to work with solutions of the weak acid and conjugate base that have equal concentrations.

Let us assume we have a 0.05 M solution of ethanoic acid, and a 0.05 M solution of ethanoate, and need 100 mL of buffer solution. We know that the two solutions must be present in the ratio of 0.575 ethanoate: 1 ethanoic acid. If our final buffer solution has a volume of 100 mL, and our two components are present in the ratio 0.575:1, we need 36.51 mL of the 0.05 M ethanoate solution, and 63.49 mL of the 0.05 M ethanoic acid solution. Look at Maths Tool 16.2 to see how we obtain these volumes.

Self-check 16.13 (a) What ratio of ammonia to ammonium ion would you use when preparing a buffer solution that buffers at around pH 9.1, given that the pK_a of the ammonium ion is 9.25? (b) If you have a 0.02 M solution of ammonium ion, and a 0.02 M solution of ammonia, what volume of each solution would you require to make 200 mL of the buffer solution?

Biochemical reactions occurring in the aqueous environment of the cell need to be buffered because enzyme function is highly sensitive to pH. Look at Box 16.8 to learn more about how the pH of an aqueous solution affects the activity of an enzyme.

MATHS TOOL 16.2 From ratios to percentages

Let's imagine that we have a solution comprising 0.575 mol of ethanoate ion and 1 mol of ethanoic acid. These amounts would satisfy our 0.575:1 ratio. Notice, however, that there are 1.575 moles altogether.

We can work out the actual quantities of both components by expressing the number of moles of ethanoate ion and ethanoic acid as *percentages* of the total number of moles, and using the percentages to determine the quantities required.

$$\text{Percentage of ethanoate ion} = \left(\frac{0.575}{1.575}\right) \times 100 = 36.51\%$$

$$\text{Percentage of ethanoic acid} = \left(\frac{1}{1.575}\right) \times 100 = 63.49\%$$

We need a final volume of 100 mL, and now know that 36.51% of this volume must be of the ethanoate solution, and that 63.49% of the volume must be of the ethanoic acid solution. 36.51% of 100 mL = 36.51 mL, and 63.49% of 100 mL = 63.49 mL, giving us our desired volumes.

Note that this strategy only works when we start with solutions of our weak acid and conjugate base (or weak base and salt of conjugate acid) whose concentrations are the same.

BOX 16.8 pH and enzyme activity

All enzymes have a pH optimum at which they have the greatest activity. An enzyme's optimum pH usually reflects the environment in which the enzyme usually functions. For example, the enzymes in blood and saliva have a pH optimum at around pH 7, reflecting the pH of the cytosol of 7.4. By contrast, enzymes in the stomach have a pH optimum at around pH 2, reflecting the extreme acidity of the stomach.

A deviation from the optimum pH can subtly alter the ability of an enzyme to bind its substrate (the molecule or molecules that it specifically binds), either because it disrupts the intermolecular forces that occur at the enzyme's active site (thereby preventing the enzyme from binding its substrate as strongly), or because it alters the molecular shape of the active site, such that the shape of the enzyme and its substrate are no longer complementary.

As a result, as the pH deviates from an enzyme's pH optimum, we see a change in the activity of the enzyme. Look at Figure 16.16, which illustrates the efficiency of an enzyme at different pH values. Notice how the enzyme's activity peaks at its optimum pH, but decreases, sometimes quite rapidly, as the pH moves away from the optimum pH–to either higher or lower values.

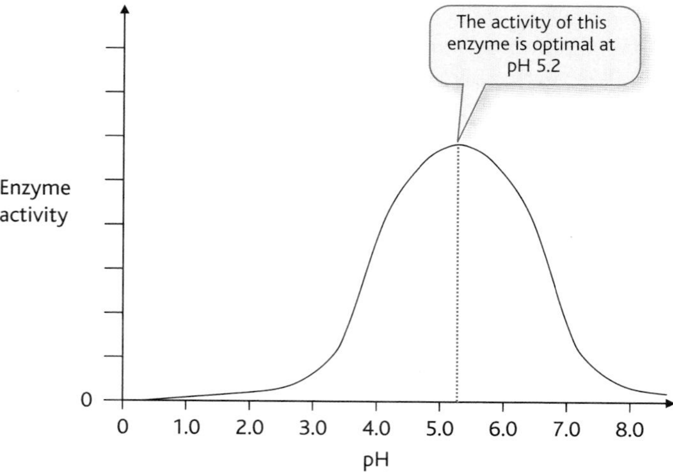

Figure 16.16 The efficiency of an enzyme at different pH values. An enzyme's activity peaks at its optimum pH, but decreases as the pH moves away from the optimum pH–to either higher or lower values.

Check your understanding

To check that you've mastered the key concepts presented in this chapter, review the checklist of key concepts below, and attempt the multiple-choice questions available in the book's Online Resource Centre at **http://www.oxfordtextbooks.co.uk/orc/crowe2e/**.

Checklist of key concepts

Acids and bases

- An acid is a proton donor
- A base is a proton acceptor
- In water, an acid dissociates to generate a proton
- In water, a base accepts a proton from a water molecule
- A proton is stabilized in aqueous solution by reacting with a water molecule to form a solvated proton
- Acids and bases must operate in pairs: an acid must have a base to which to donate a proton

Conjugate acid–base pairs

- A conjugate acid–base pair comprises the acid and base versions of the same chemical species
- When an acid donates a proton it becomes its conjugate base
- When a base accepts a proton it becomes its conjugate acid
- An acid–base reaction involves two acid–base pairs
- Water is amphoteric: it can act as an acid and as a base

The strengths of acids and bases

- The strength of an acid tells us the extent to which the acid dissociates when dissolved in water
- A strong acid dissociates almost completely in water
- A weak acid only partially dissociates in water
- The strength of a base tells us the extent to which it accepts a proton from water to generate hydroxide ions
- A strong base accepts a proton from water very readily
- A weak base is poor at accepting a proton from water
- Strong acids have weak conjugate bases
- Weak acids have strong conjugate bases

The acid and base dissociation constants

- The acid dissociation constant gives us a measure of the extent to which an acid dissociates in aqueous solution
- We represent the acid dissociation constant with the symbol K_a
- A strong acid has a large K_a value
- A weak acid has a small K_a value
- The base dissociation constant gives us a measure of the extent to which a base reacts with water to yield hydroxide ions (or how readily it accepts protons from an acid)
- We represent the base dissociation constant with the symbol K_b
- A strong base has a large K_b value
- A weak base has a small K_b value
- pK_a and pK_b values are scaled-down representations of K_a and K_b
- $pK_a = -\log (K_a)$
- $pK_b = -\log (K_b)$

The ion product of water

- The ion product of water is the equilibrium constant for the dissociation of water into H^+ and OH^- ions
- We represent the ion product of water with the symbol K_w
- K_w has a value of 1×10^{-14} at 25°C
- The product of the proton concentration and hydroxide concentration for any aqueous solution is equal to the value of K_w
- For any acid/conjugate base system, $K_w = K_a \times K_b$

The pH scale

- The pH of a solution gives us a measure of the concentration of protons (H^+ ions) in solution
- $pH = -\log [H^+]$

- A solution with a high $[H^+]$ has a low pH
- A solution with a low $[H^+]$ has a high pH
- If the pH of a solution is less than 7.0 we say the solution is **acidic**
- If the pH of a solution is greater than 7.0 we say the solution is **alkaline**
- If the pH of a solution is equal to 7.0 we say the solution is **neutral**
- The concentration of protons in a strong acid solution is equal to the concentration of the acid itself
- The pH of a weak acid solution is related to the pK_a value for the acid by the **Henderson–Hasselbalch equation**:

$$pH = pK_a + \log\frac{[A^-]}{[HA]}$$

Neutralization reactions

- **Neutralization** is the reaction between an acid and a hydroxide base
- Neutralization involves the following chemical change: acid + base $\rightarrow$ salt + water
- The pH of the final solution depends on the strength of the acid and base that are reacting
- The salt of a strong acid and a strong base gives a **neutral** solution
- The salt of a strong acid and a weak base gives an **acidic** solution
- The salt of a weak acid and a strong base gives an **alkaline** solution
- The salt of a weak acid and a weak base gives an acidic, alkaline, or neutral solution depending on the pK_a values of the ions involved

The pOH scale

- The **pOH** of a solution gives us a measure of the concentration of hydroxide ions in the solution
- $pOH = -\log[OH^-]$
- A solution with a low pOH has a high pH, and vice versa

Acids and bases in biology

- The acidity (or basicity) of a compound affects its **partitioning** between aqueous and hydrophobic systems
- A **strong** acid is most likely to partition into an **aqueous** phase
- A **weak** acid is most likely to partition into a **hydrophobic/lipophilic** phase
- The pH of a compound's environment can affect its acidity (or basicity)
- A decrease in pH leads to a decrease in the extent of dissociation of an acid
- An increase in pH leads to an increase in the extent of dissociation of a acid
- A decrease in pH leads to an increase in the extent of dissociation of a base
- An increase in pH leads to a decrease in the extent of dissociation of a base

Buffer solutions

- A **buffer solution** is a solution that resists changes in pH
- A buffer solution contains large quantities of an acid–base conjugate pair, which can suppress the addition of protons or hydroxide ions to a solution and prevent the pH from changing
- A buffer solution must use a **weak** acid or base
- When we use a weak acid we include the **salt** of its conjugate base
- When we use a weak base we include the **salt** of its conjugate acid
- The pH around which a solution buffers is **equal to the pK_a** of the acid

Chemical reactions 1: bringing molecules to life

INTRODUCTION

When we consume any food, be it a chocolate bar or burger, a salad or a piece of fruit, chemical reactions in our body break down the complex organic compounds the foods contain into simple products. These products in turn undergo further reactions to produce energy (to fuel the processes that keep us alive) and form new macromolecules (to keep our bodies strong and healthy). Life is the result of thousands of different biochemical reactions occurring in every cell of our body. At first glance, many of these reactions seem very complicated; beneath the facade of complexity, however, lie simple principles that underpin all chemical reactions.

In this chapter and the next, we discover the principles that unify even the most complicated chemical reactions. We explore what is happening at the molecular level during a chemical reaction, asking what it is that distinguishes different chemical reactions, and what it is that unifies all reactions, however simple or complicated.

17.1 What is a chemical reaction?

A chemical reaction is a chemical change occurring within one compound, or an interaction between two or more starting compounds, to form one or more end compounds, which are *different* from the compound(s) we started with.

Glycolysis, the first stage in the catabolism of glucose in the body, results in the formation of two molecules of pyruvate for every molecule of glucose that is broken down:

$$C_6H_{12}O_6 + O_2 \rightarrow 2\,CH_3COCO_2^- + 2\,H_2O + 2\,H^+$$

We call the starting compounds the **reactants**, and compounds generated by the chemical reaction the **products.**

A written **reaction scheme**, such as the one above, tells us, at a glance, what we have at the start of the reaction (the reactants) and what we have at the end of the reaction (the products): it depicts the *overall* chemical change that occurs when reactants form products. However, reaction schemes fail to tell us exactly *how* the reactants made the transition from reactants to products. How does glucose undergo chemical change to yield pyruvate, for example?

Consider a train journey from London to Glasgow. If we were to write this journey in the form of a reaction scheme, we could write:

London → Glasgow

However, this 'reaction scheme' tells us nothing about the route we took to get from London to Glasgow, merely that the overall result of the journey was our movement from London to Glasgow. It might have been that we chose to take a direct train on the West Coast line. Alternatively, we might have travelled up the east coast to Edinburgh, and taken a second connecting train to Glasgow:

London → Glasgow (direct route)

or

London → Edinburgh → Glasgow (indirect route; one change)

Similarly, chemical reactions proceed via direct or indirect routes: a reaction may proceed direct from reactants to products (the equivalent of taking a direct train), or the change from reactants to products may occur via two or more **steps**.

Look at Figure 17.1. Notice how the first reaction–the change from reactants to products–proceeds in just one step. By contrast, the second reaction requires two separate steps to complete the transition from reactants to products. Don't worry about what's actually being shown during these reactions: we will find out later in this chapter. Just notice how the first reaction proceeds in only one step, while the other reaction proceeds in two separate steps.

Returning to our example of glycolysis above, we see in Chapter 18 how the overall process of a molecule of glucose being broken down into two molecules of pyruvate occurs through a series of *ten* enzyme-catalyzed steps.

The stoichiometry of chemical reactions

Let's just pause for a moment to consider what else a reaction scheme can tell us. Not only does a reaction scheme tell us the **identity** of reactants and products

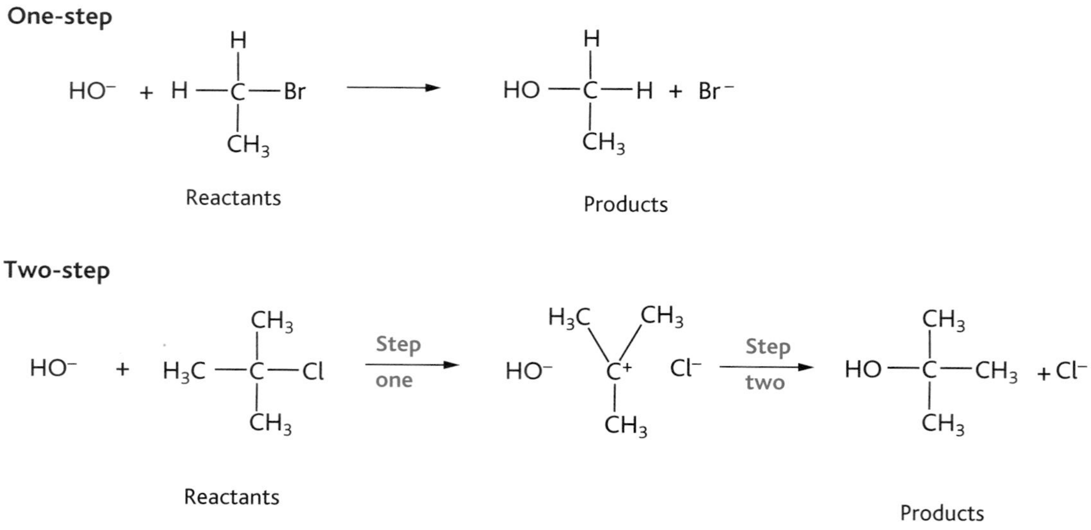

One-step

$HO^- + H{-}C(H)(CH_3){-}Br \longrightarrow HO{-}C(H)(CH_3){-}H + Br^-$

Reactants — Products

Two-step

$HO^- + H_3C{-}C(CH_3)_2{-}Cl \xrightarrow[\text{one}]{\text{Step}} HO^- \quad (H_3C)_2C^+{-}CH_3 \quad Cl^- \xrightarrow[\text{two}]{\text{Step}} HO{-}C(CH_3)_2{-}CH_3 + Cl^-$

Reactants — Products

Figure 17.1 In some reactions, the change from reactants to products proceeds in a single step. By contrast, other reactions proceed in two or more steps.

involved in the reaction, it also tells us the **relative quantities** of reactants that interact, and the relative quantities of products that are formed.

We call the relative quantities of reactants and products associated with a particular reaction the **stoichiometry** of the reaction.

For example, look at this reaction scheme:

CH_3CH_2Br	+	OH^-	→	CH_3CH_2OH	+	Br^-
Bromoethane		Hydroxide ion		Ethanol		Bromide ion

This scheme tells us that one molecule of bromoethane reacts with one hydroxide ion to form one molecule of ethanol and one bromide ion.

These relative quantities are *precise*: for every bromoethane molecule that is consumed during the reaction, exactly one hydroxide ion is also consumed. Similarly, for every molecule of bromoethane and every OH^- ion consumed, one molecule of ethanol and one bromide ion is produced[1]. The stoichiometry of the reaction is:

$$\begin{array}{c} CH_3CH_2Br + OH^- \rightarrow CH_3CH_2OH + Br^- \\ 1 \quad : \quad 1 \qquad\qquad 1 \quad : \quad 1 \end{array}$$

Now look at this reaction scheme, which describes the overall process of photosynthesis:

$$6\,CO_2 + 6\,H_2O \rightarrow C_6H_{12}O_6 + 6\,O_2$$

The stoichiometry of this reaction is:

$$\begin{array}{c} 6\,CO_2 + 6\,H_2O \rightarrow C_6H_{12}O_6 + 6\,O_2 \\ 6 \quad : \quad 6 \qquad\qquad 1 \quad : \quad 6 \end{array}$$

So this reaction scheme tells us that *six* moles of carbon dioxide react with *six* moles of water to produce *one* mole of glucose and *six* moles of molecular oxygen.

The relative number of species participating in the reaction is indicated by the number in front of the chemical formulae. We call this number the **stoichiometric coefficient**.

Notice that we omit the number '1' for chemical species that have a stoichiometry of 1. So we simply write '$C_6H_{12}O_6$' rather than '1 $C_6H_{12}O_6$'.

- ***During a chemical reaction, reactants and products react in precise relative amounts.***
- ***The relative amount of reactants and products associated with a particular chemical reaction is called the stoichiometry of the reaction, and is indicated by coefficients in front of the chemical formulae in the reaction scheme.***

But what is actually happening to the substances involved during a chemical reaction? What processes are required to convert reactants into products?

1 We could equally say that one mole of bromoethane reacts with one mole of hydroxide ions to produce one mole of ethanol and one mole of bromide ions. Remember: 'one mole' is a particular number: 6×10^{23}. The mole is the most convenient way of quantifying the number of particles present in chemical systems. We learn more about the mole on p. 373.

17.2 The molecular basis of chemical reactions

Chemical reactions involve the **breaking** of bonds between atoms in molecules, and the **formation** of new bonds between different groups of atoms to form products. We see in Chapter 3 how chemical bonds comprise pairs of valence electrons, and how other pairs of valence electrons are not involved directly in bonding; these are the non-bonding pairs[2]. The breaking of existing bonds and formation of new bonds involves the **redistribution** of valence electrons–both those involved in bonding prior to the redistribution, and those that were non-bonding–so they become shared between new groups of atoms.

Notice how chemical reactions reflect one of our essential concepts: that change involves movement. The change that occurs during a chemical reaction requires the *movement* of valence electrons. Throughout this chapter, when we consider the movement of valence electrons during chemical reactions, we need to remember that the electrons may have been part of a chemical bond or may have been non-bonding before the reaction commenced.

We explore the sharing of valence electrons during the formation of compounds from atoms in Chapter 3.

Valence electrons are shared between atoms to form molecules in such a way that all participating atoms have full valence shells (comprising four pairs of valence electrons). The movement of valence electrons during chemical reactions always results in full valence shells being maintained.

How do valence electrons move during chemical reactions?

There are two ways in which valence electrons become redistributed when a chemical bond is broken:

1. A pair of valence electrons moves together, remaining as a pair.
2. A pair of valence electrons splits up and they then move separately, as single entities.

Look at Figure 17.2, which shows the two possible ways in which the valence electrons forming the bond between atoms X and Y are redistributed when the bond is broken. Notice how one atom ends up with both electrons and the other atom ends up with none when the valence pair moves together. By contrast, both atoms end up with one electron each when the valence pair splits and each electron moves separately.

We call these two processes **heterolytic** and **homolytic** bond cleavage (breaking)[3].

Heterolytic cleavage occurs when the valence electrons are distributed *differently* (unequally) between two atoms once a bond joining the atoms has been broken.

Homolytic cleavage occurs when the valence electrons are distributed in the *same way* (equally) between the two atoms when the bond is broken.

Hetero- and homolytic bond cleavage, and the movement of valence electrons in intact pairs or as separate single entities, lies at the heart of all chemical reactions.

---

2 Oxygen and nitrogen are particularly good examples of elements that exhibit non-bonding pairs of valence electrons.

3 '-lytic' is a derivative of 'lysis', from the Greek 'to split'. 'Hetero-' = 'different'; 'homo-' = 'the same'. Therefore, 'Heterolytic' = 'different splitting' and 'Homolytic' = 'the same (way of) splitting'.

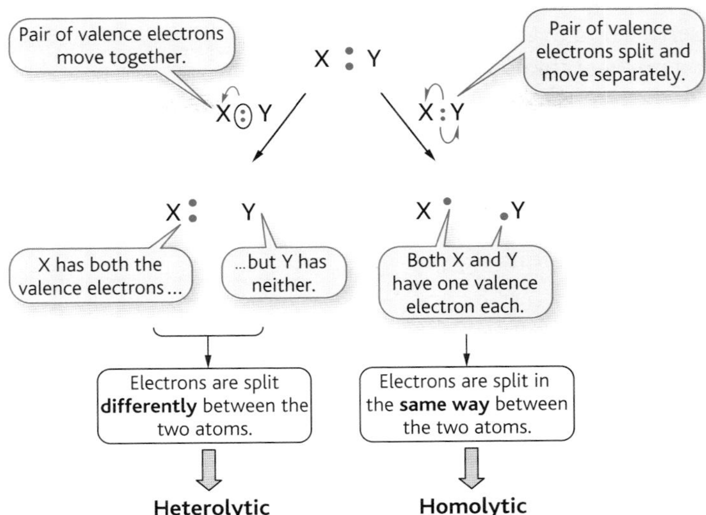

Figure 17.2 There are two possible ways in which the valence electrons forming the bond between the generic atoms X and Y can be redistributed when the bond is broken: the pair of valence electrons move together, and we see **heterolytic** bond cleavage; or the electrons split and move separately, and we see **homolytic** bond cleavage.

When we think about chemical reactions, we first need to consider how electrons are moving: are they moving as pairs, or are they moving singly[4]?

We explore these two types of bond cleavage, and electron movement, in turn, during the rest of this chapter.

- ***Heterolytic bond cleavage results in the unequal sharing of a pair of valence electrons between two atoms when a bond joining them is broken.***
- ***Homolytic bond cleavage results in the equal sharing of a pair of valence electrons between two atoms when a bond joining them is broken.***

Depicting the movement of electrons

We use particular conventions to communicate to others how electrons are moving, just as we use chemical symbols, and molecular and structural formulae to communicate the structure and composition of chemical substances.

The movement of valence electrons is denoted using '**curly arrows**'. We use different curly arrows to depict the movement of a single electron, and a pair of electrons. When used precisely, the *start* of a curly arrow or curly fish hook denotes the location of the valence electron(s) at the beginning of the reaction, and the *end* of the arrow denotes the location of the electron(s) at the end of the reaction.

We use a curly 'fish hook' ⌒ to denote the movement of a **single** electron, and a full curly arrow ⌒ to denote the movement of a **pair** of electrons.

During chemical reactions, valence electrons become redistributed in particular ways: the movement of valence electrons is far from random, but follows

4 Note that we refer to valence electrons (which may include non-bonding electrons) simply as 'electrons' during much of the rest of this chapter. Inner-shell electrons are not involved in chemical reactions, so the only electrons we need to consider are valence electrons.

certain patterns. The way in which valence electrons move during a chemical reaction is described by the **reaction mechanism**.

Reaction mechanisms use a combination of curly arrows, curly fish hooks, and structural formulae to illustrate what is happening, at the molecular level, during a chemical reaction, as reactants interact to form products. A reaction *scheme* only gives us an *overview* of what is happening during a chemical reaction at the molecular level; a *reaction mechanism* reveals the detail of what's actually happening as valence electrons are redistributed, bonds are broken, and new bonds are formed.

For example, this reaction scheme:

$$CH_3CH_2Br + OH^- \rightarrow CH_3CH_2OH + Br^-$$

tells us that one mole of bromoethane reacts with one mole of hydroxide ions to form one mole of ethanol and one mole of bromide ions. However, it tells us nothing about the movement of electrons that constitutes this reaction, or the bonds that must be broken or formed to effect the change from reactants to products.

By contrast, the reaction mechanism for this reaction:

$$H{-}O^- \quad H{-}\underset{CH_3}{\overset{H}{C}}{-}Br \longrightarrow OH{-}\underset{CH_3}{\overset{H}{C}}{-}H + Br^-$$

shows us exactly what's happening at the level of the electrons. The curly arrows show us how a pair of valence electrons (in this case, non-bonding electrons) associated with the oxygen of the hydroxyl ion are attracted by the slight positive charge on the carbon atom, to form a new C–O covalent bond. At the same time, the pair of valence electrons in the C–Br bond are moving away from the carbon to the bromine, breaking the existing C–Br bond.

Stable compounds must have a full valence shell, with each component atom possessing four pairs of valence electrons. (The exception is hydrogen, whose full valence shell contains just one pair of valence electrons.) Look at the reaction mechanism again, and notice how the atoms comprising bromoethane and ethanol all have full valence shells. While electrons may have been redistributed during the reaction, redistribution happens in such a way that full valence shells are maintained.

- ***A reaction scheme gives us a molecular overview of how two or more reactants interact to form new products.***
- ***The reaction mechanism fills in the detail, telling us how electrons are redistributed during the change from reactants to products.***

Let's now move on to explore chemical reactions that are characterized by the heterolytic and homolytic movement of electrons. First, we consider heterolytic reactions.

17.3 Heterolytic reactions

A heterolytic chemical reaction is the process of bond cleavage and formation–the change from reactants to products–that is driven by valence electrons moving together, in pairs.

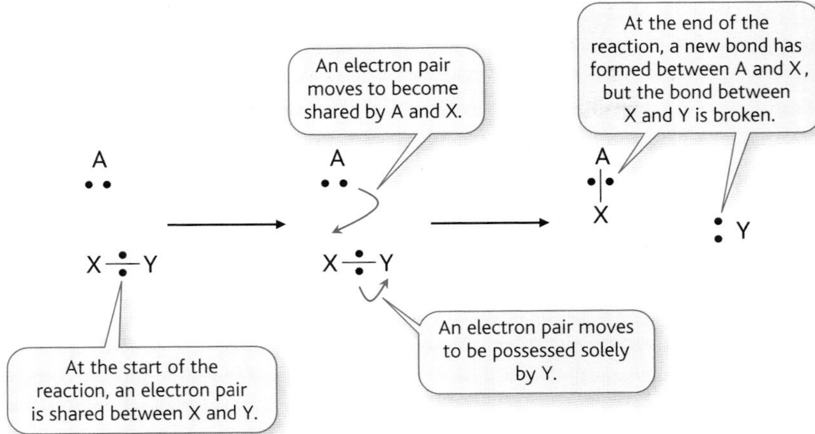

Figure 17.3 A simplified view of what is happening during the reaction $A + XY \rightarrow AX + Y$. During the course of the reaction–and as a result of the movement of electrons–the A-X bond is formed, but the X-Y bond is broken.

Let's consider the hypothetical chemical reaction:

$$A + X–Y \rightarrow A–X + Y$$

Look at Figure 17.3, which gives a simplified view of what is happening during this reaction. Notice how two electron pairs move such that one bond is broken and a new bond is formed; the electrons comprising each pair move together as a discrete unit.

Look again at Figure 17.3, and notice how heterolytic bond cleavage and formation requires two particular entities:

1. A source of an electron pair
2. An atom that can accept, and subsequently share, the electron pair.

In our hypothetical reaction, the source of an electron pair is a pair of non-bonding electrons associated with atom A[5]; the atom that can accept, and then share, the electron pair is atom X.

These two entities are what we call a **nucleophile** and an **electrophile**.

A **nucleophile** is an electron-rich species, which has an electron pair that can be donated to form a covalent bond. The name 'nucleophile' comes from the Greek 'nucleo-' = nucleus and 'phile' = 'loving', such that 'nucleophile' = 'nucleus-loving'. A nucleus is an area of low electron density, so its name tells us that a nucleophile is attracted to regions of low electron density–that is, a positive centre. To be attracted to a positive centre, the nucleophile itself carries a full or partial negative charge (or can easily become polarized to exhibit such a charge).

There are two key groups of nucleophile: molecules or ions with non-bonding pairs of electrons, and molecules with a double (or triple) covalent bond. For example, the molecules and ions displayed in Figure 17.4 all have at least one pair of non-bonding electrons.

Notice how some of these nucleophiles are anions (negatively charged ions). However, some nucleophiles are uncharged molecules that merely possess one or

---

5 We know this valence electron pair must be a non-bonding pair because A has no chemical bond, unlike X–Y.

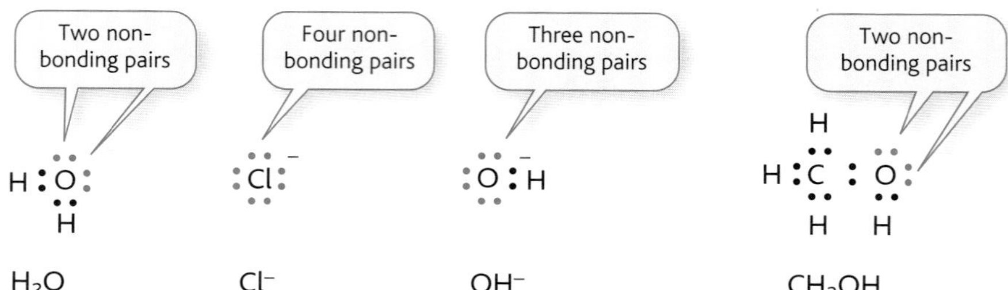

Figure 17.4 Examples of some nucleophilic species. Notice how all of these species have at least one pair of non-bonding electrons.

more pairs of non-bonding electrons, and can be readily polarized to exhibit an area of negative charge, as noted above. Either way, such species are attracted towards positive centres in molecules.

The double bond is a rich source of electrons: it contains one more pair of electrons than the joined atoms actually need to remain bonded. Figure 17.5 shows how two atoms joined by a double bond can donate the electron pair that comprises the π bond, and still remain joined by a σ bond. Consequently, a double bond may also act as a nucleophile.

Why does the π bond donate an electron pair rather than the σ bond? The electrons occupying a π bond are of higher energy than those in the σ bond, and so are less strongly held in the bond. Consequently, they are more readily available (that is, they can be redistributed more easily) than those occupying a σ bond, which are held much more tightly. Remember our essential concept that stability is preferable to instability; higher energy equates to greater instability, so the higher-energy (more unstable) π electrons are donated in preference to the lower-energy (more stable) σ electrons; the π electrons are donated because it lowers their energy.

An **electrophile** is an electron-poor (or 'electron-deficient') species, which can accept a complete electron pair, and share it with the nucleophile as a covalent bond. The name 'electrophile' comes from the Greek 'electro-' = electron or electronic charge; and 'phile' = 'loving'. Therefore, 'electrophile' = 'electron-loving'. The name 'electrophile' tells us that an electrophile is attracted to regions of high electron density (that is, areas of full or partial negative charge); to be attracted to an area of negative charge, an electrophile must carry a full or partial positive charge (or must readily be able to carry such a charge).

Remember: valence electrons must move in such a way that full valence shells are always maintained by the end of a chemical reaction.

For the electrophile to be able to accept a complete electron pair, and for a covalent bond to be formed, the electrophile must possess (or be able to possess) an **empty** orbital.

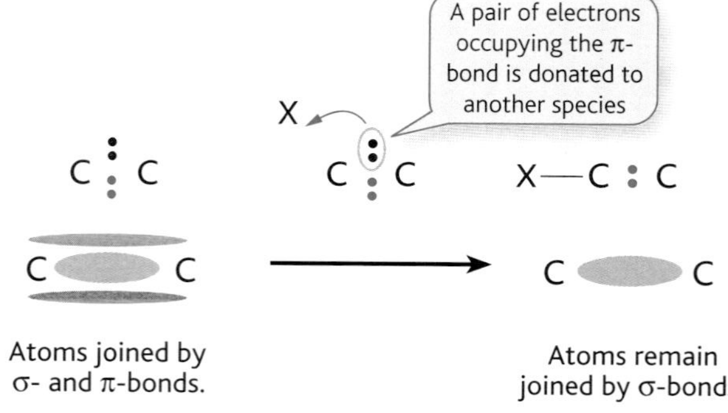

Figure 17.5 Two atoms joined by a double bond can donate the electron pair that comprises the π bond and still remain joined by a σ bond.

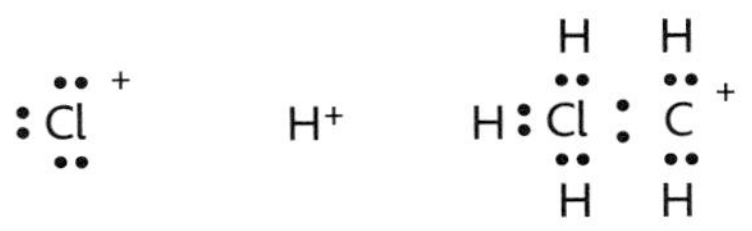

$:\ddot{\underset{..}{Cl}}^{+}$ H^+ $H:\ddot{\underset{..}{Cl}}:\overset{H}{\underset{H}{C}}^{+}$

Figure 17.6 Examples of some electrophilic species.

The species depicted in Figure 17.6 all possess an empty orbital. Notice how these electrophiles are cations (positively charged ions), in contrast to some nucleophiles, which are anions.

Self-check 17.1 Identify which of the following species are electrophiles and which are nucleophiles:

(a) CN^-; (b) NO_2^+; (c) CH_3^+; (d) CH_3NH_2

We often use Nu^- or $N^{\delta-}$ as shorthand to denote a nucleophile, and E^+ or $E^{\delta+}$ to denote an electrophile.

Nucleophiles and electrophiles must operate as complementary pairs: a nucleophile must have an electrophile to which it can donate a pair of electrons. Similarly, an electrophile must have a nucleophile from which to accept a pair of electrons.

So, at heart, heterolytic reactions comprise nucleophiles seeking out electrophiles and donating electrons to them. We see in Chapter 18 how this concept underpins a range of elegant and important chemical reactions.

- *A* ***nucleophile****, denoted Nu^- or $Nu^{\delta-}$, is an electron-rich species, which has a valence electron pair (which may be a non-bonding pair) that can be donated to form a covalent bond.*
- *An* ***electrophile****, denoted E^+ or $E^{\delta+}$, is an electron-poor species, which can accept a complete electron pair and share it with the nucleophile as a covalent bond.*

Heterolytic reactions and the polarization of bonds

Not all electrophiles have an empty orbital at the start of a reaction. However, they must be able to make an empty orbital available readily. An electrophile must have an empty orbital to accept a pair of electrons from a nucleophile to form a new covalent bond.

For example, hydrogen bromide (HBr) is an electrophile. Its Lewis structure is shown in the margin (1).

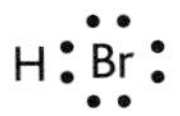

$H:\ddot{\underset{..}{Br}}:$

(1)

Notice how neither the hydrogen atom nor the bromine atom in HBr possesses an empty orbital with which to accept a new pair of electrons from a nucleophile. So how can HBr act as an electrophile?

The answer lies in the ability of HBr to readily make available an empty orbital. To do this, either the hydrogen or the bromine atom must empty one of its currently occupied orbitals, as illustrated in Figure 17.7.

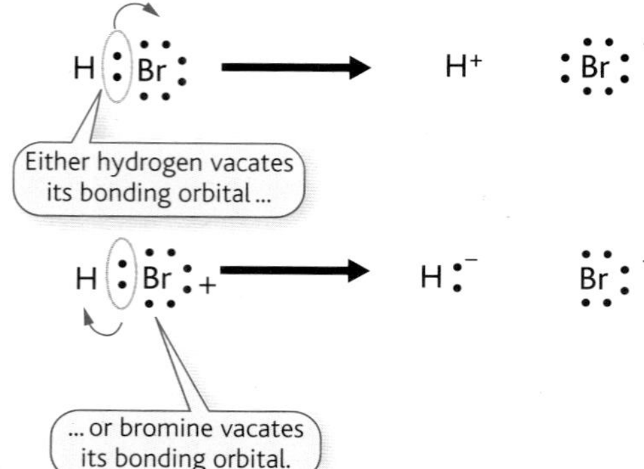

Figure 17.7 For HBr to readily make available an empty orbital, either the hydrogen or bromine atom must empty one of its currently occupied orbitals.

But which of the two situations depicted in Figure 17.7 happens in reality? A pair of electrons is donated to the *most electronegative* of the two joined atoms. We see in Chapter 3 how atoms of electronegative elements draw electrons towards themselves. As a result, covalent bonds become polarized: electrons are not shared equally between two bonded atoms; instead, the distribution of electrons is skewed towards the most electronegative atom.

Bromine is very much more electronegative than hydrogen. Consequently, the H–Br bond is highly polarized. Look at Figure 17.8, which depicts how electrons in the H–Br bond are more heavily distributed towards the Br atom than the H atom.

Because of its greater electronegativity, bromine can accept the complete electron pair (the one that it shares with hydrogen) much more readily than can hydrogen. As a result, hydrogen is the atom that can most readily change to have an empty orbital to make HBr a true electrophile.

So, an electrophile can be a molecule possessing a polarized bond between two atoms, of which the most electronegative atom can readily withdraw the shared electron pair. The weakly electronegative atom changes to have an empty orbital, and can accept an electron pair from a nucleophile.

- ***An electrophile must have an empty orbital with which to accept a pair of electrons from a nucleophile.***
- ***The orbital may become empty when an electronegative atom within the electrophile withdraws electrons from an orbital of another atom.***

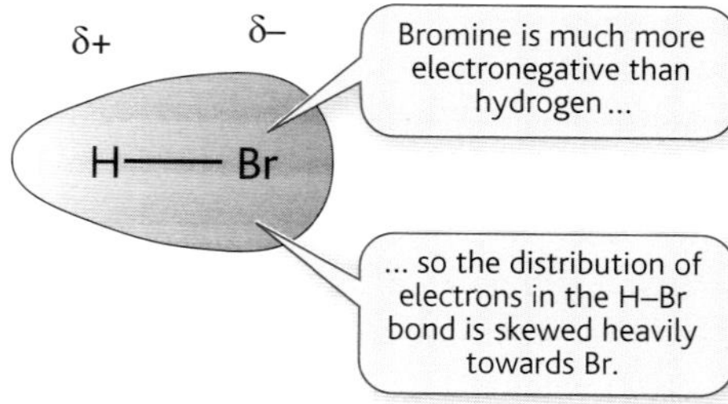

Figure 17.8 The polarization of the H–Br bond. Electrons in the H–Br bond are more heavily distributed towards the Br atom than the H atom because Br is more electronegative than H.

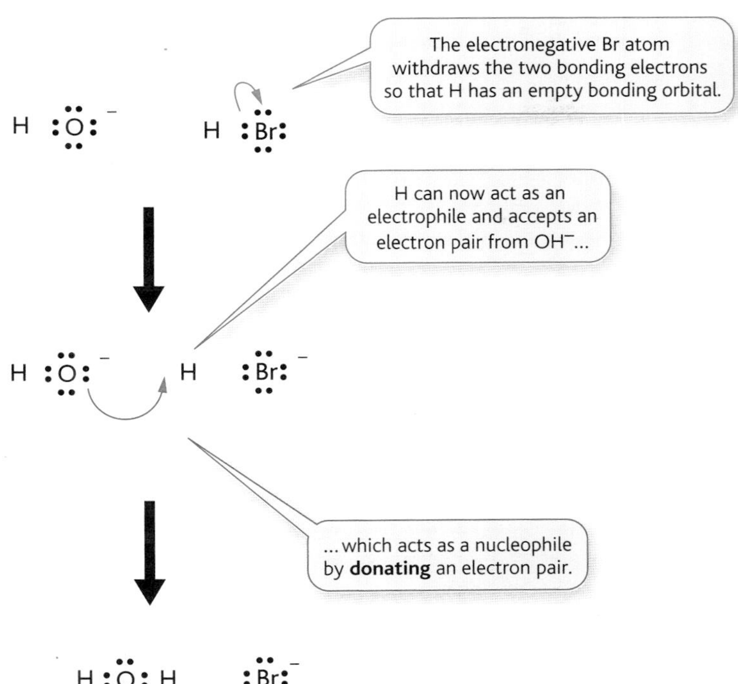

The electronegative Br atom withdraws the two bonding electrons so that H has an empty bonding orbital.
H can now act as an electrophile and accepts an electron pair from OH⁻...
... which acts as a nucleophile by **donating** an electron pair.

Figure 17.9 HBr in action as an electrophile. During the course of this reaction, the electronegative Br withdraws electrons from the H-Br bond, making the H nucleus vulnerable to attack by the nucleophile, OH^-.

Figure 17.9 shows HBr in action as an electrophile, in association with the nucleophile OH^-. Look at this figure, and notice how, at the end of the reaction both products now have full valence shells (including the Br^- anion).

In reality, the transfer of electrons from the H–Br bond to the bromine atom, and from the OH^- ion to the hydrogen atom, does not happen as two completely separate steps. Instead, the movement of both pairs of electrons happens at the same time, initiated by the attraction of the electrons on the OH^- ion for the positive end of the H–Br molecule, to give one smooth process.

So we can write both occurrences of electron pair movement as a single reaction mechanism:

Reaction mechanism H:Ö:⁻ H:Br: ⟶ H:Ö:H + :Br:⁻

Reaction scheme $OH^- + HBr \longrightarrow H_2O + Br^-$

Now look back at Figure 17.3: why did A donate an electron pair to X and not Y? We can now conclude that Y must represent a more electronegative element than X. Consequently, the electron pair shared between X and Y (to form the X–Y bond) is most readily transferred to Y, leaving X with an empty orbital. X is now a true electrophile; X fills its empty orbital by accepting an electron pair from the nucleophile, A.

17.4 Oxidation and reduction

During many chemical reactions, a nucleophile will donate one or more electrons to an electrophile such that the nucleophile actually *loses* electrons, while the electrophile simultaneously *gains* electrons.

When a chemical species loses (or donates) electrons, we say that it undergoes **oxidation**–it is **oxidized**.

When a chemical species gains (or accepts) electrons, we say that it undergoes **reduction**–it is **reduced**.

The processes of oxidation and reduction merely tell us how the relative number of electrons that a particular chemical species possesses changes during the course of a chemical reaction: if it loses electrons, we say it has been oxidized; if it gains electrons we say it has been reduced.

We see in Section 17.3 how the donation of electrons by a nucleophile is coupled to the acceptance of electrons by an electrophile. Consequently, the processes of oxidation and reduction are usually coupled too: as one species is oxidized and loses electrons, another species must be reduced, and accept the electrons that the first molecule has lost. We call a reaction that involves the simultaneous oxidation of one species and reduction of another species a **redox** reaction (a contraction of **red**uction and **ox**idation).

For example, the oxidation of an alcohol in biological systems is often coupled to the reduction of the coenzyme NAD^+ (about which we learn more below), as illustrated in Figure 17.10. Look at this figure, and notice how the alcohol is oxidized to give a carbonyl group during the course of the reaction, as a result of which it loses electrons. By contrast, NAD^+ is reduced to NADH, and gains two electrons. Also notice how these two processes–of oxidation and reduction–are coupled: the species being oxidized (the alcohol) loses two electrons to the species being reduced (NAD^+).

- ***Oxidation** occurs when a chemical species loses electrons.*
- ***Reduction** occurs when a chemical species gains electrons.*
- *A **redox reaction** is one in which the oxidation of one chemical species is coupled to the reduction of a second chemical species.*

We should also note that the processes of oxidation and reduction are sometimes considered in terms of the loss or gain of hydrogen atoms–with oxidation being the loss of hydrogen, and reduction being the gain of hydrogen. What has the loss and gain of hydrogen to do with the loss and gain of electrons? In biological systems, the movement of electrons is often tightly associated with the movement of protons: when an electron is taken up by a molecule (such that the molecule becomes reduced), the electron may be joined by a proton (H^+) from solution. Alternatively, two electrons may be transferred in the form of a hydride ion, H^-, which comprises a proton and two electrons.

Therefore, it is not unusual to see the inclusion of hydrogen-based species (protons, hydride ions, or hydrogen atoms) when we see redox reactions written down. Indeed, if we look again at Figure 17.10 we see how a hydrogen with two electrons moves from the alcohol to NAD^+ during the course of the reaction. We see further examples of the movement of both protons and hydride ions as part of redox reactions below.

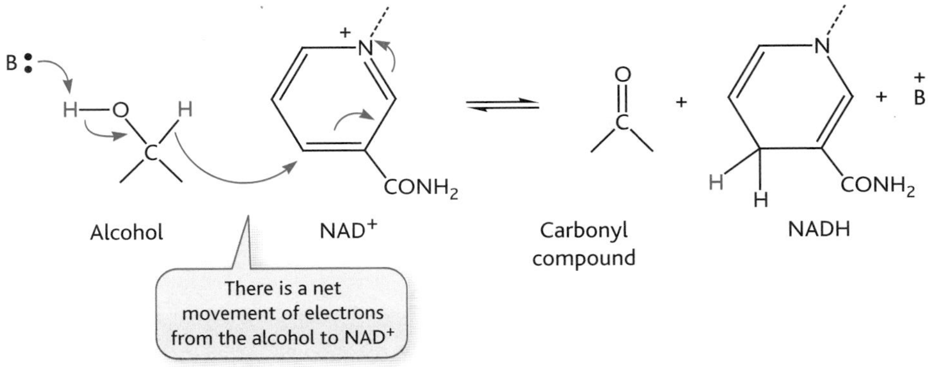

Figure 17.10 The oxidation of an alcohol coupled to the reduction of NAD^+. (Adapted from McMurry and Begley, 2005.)

Self-check 17.2 Do the following depict oxidation, reduction, or redox reactions?

(a) $2Br^- \rightarrow Br_2$

(b) $Zn + Cl_2 \rightarrow ZnCl_2$

(c) $Ca \rightarrow Ca^{2+}$

(d) $Cu^{2+} \rightarrow Cu$

(e) $Cl_2 + 2NaBr \rightarrow 2NaCl + Br_2$

Oxidation and reduction in biological systems

Just like metals and oxygen, the organic molecules found in biological systems are able to accept and donate electrons, and many important reactions in the cell involve the transfer of electrons between molecules. Indeed, the metabolism of glucose:

$$C_6H_{12}O_6 + 6\,O_2 \rightarrow 6\,CO_2 + 6\,H_2O$$

is a redox reaction, in which glucose is oxidized (and oxygen is reduced). The oxidation of glucose depicted by this overall scheme happens during the course of a remarkable series of coupled reaction pathways, which, collectively, drive the synthesis of ATP. (ATP, in turn, is then used as a source of energy to drive a range of other biological processes and reactions.)

We learn more about the reaction mechanisms that feature in the oxidation of glucose in Chapter 18.

A particular molecule may accept one or two electrons from one species and pass them on to a molecule with similar properties. Molecules have different affinities for accepting and donating electrons: some are strongly oxidizing or strongly reducing, and will readily remove electrons from another molecule or readily donate electrons to another molecule, respectively. By contrast, others are only weakly oxidizing or reducing. Certain molecules that are particularly effective at accepting or donating electrons operate as electron transfer molecules in a wide range of biochemical reactions. We call these molecules **electron carriers**.

In biological systems, the ultimate electron acceptor is oxygen itself, which is reduced to water, in this case by accepting four electrons:

$$O_2 + 4\,e^- + 4\,H^+ \rightarrow 2\,H_2O$$

Notice how the transfer of electrons to oxygen is associated with the take-up of protons from solution, as noted above. Within the cell, electrons are transferred to water via a series of electron carriers, which work in pairs called redox couples: one carrier undergoes oxidation, while the other undergoes reduction.

Remember: the processes of oxidation and reduction are coupled: when an electron is lost from one molecule–such that the molecule becomes oxidized–the electron must be transferred to another molecule, such that the second molecule becomes reduced.

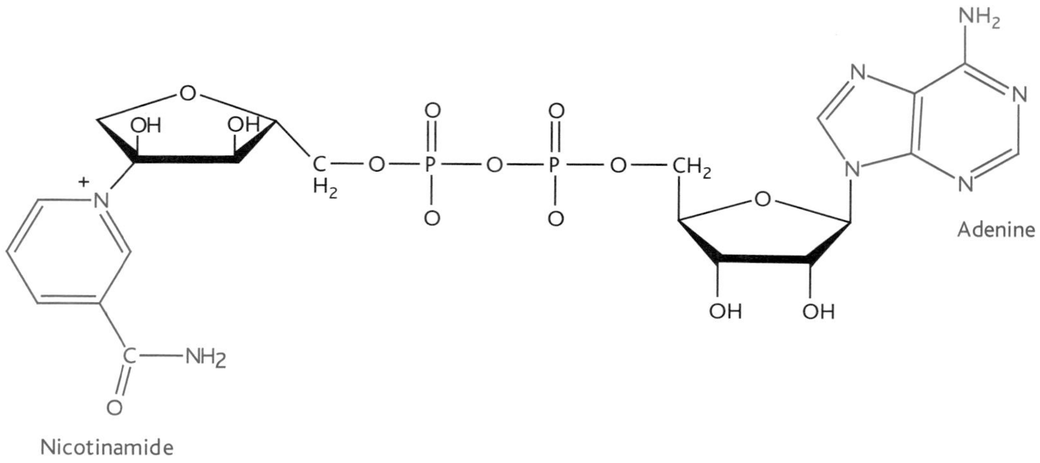

Figure 17.11 The structure of nicotinamide adenine dinucleotide (NAD^+).

One of the most important electron acceptors in biological systems is nicotinamide adenine dinucleotide (NAD^+), whose structure is illustrated in Figure 17.11. NAD^+ can accept two electrons to yield reduced nicotinamide adenine dinucleotide (NADH). The two electrons that are transferred don't move as 'free' electrons (that is, electrons that are completely independent from some kind of atomic structure). Instead, the electrons are part of a hydride ion, H^-, which comprises two electrons and one proton. Look at Box 17.1 to learn how NAD^+ is involved in the metabolism of glucose, during the process of glycolysis.

- *The transfer of electrons associated with oxidation and reduction in biological systems is often mediated by specific electron carriers.*
- *Electron carriers work together as redox couples.*

Oxidation and reduction during enzyme catalysis

Redox reactions lie at the heart of many important enzyme-catalyzed reactions within biological systems. However, enzymes themselves don't always possess the chemical entities required to participate in reduction or oxidation–that is, the amino acid residues from which enzymes are formed may not possess the chemical properties required to support efficient redox reactions. Instead, the active site of an enzyme can often feature a **cofactor**, a non-amino acid entity, which is closely associated with the active site. If a cofactor associates reversibly with an enzyme, we call it a **coenzyme**; if it associates irreversibly (such that it becomes an intrinsic part of the enzyme itself) we call it a **prosthetic group**.

For example, NAD^+ and NADH act as coenzymes, and reversibly associate with many different enzymes that catalyze oxidation–reduction reactions. Figure 17.14(a) shows how the enzyme malate dehydrogenase catalyzes the oxidation of malate to give oxaloacetate (an example of an alcohol being oxidized to a ketone); this is one of the steps in the citric acid cycle. The oxidation of malate (a loss of electrons) takes the form of the loss of a hydride ion from the –OH group of malate. It is this hydride ion that reduces NAD^+ to NADH during the course of the reaction.

BOX 17.1 Reductive reasoning: the metabolism of glucose

Some of the energy contained in the covalent bonds of a glucose molecule is harnessed by the cell in a series of reactions called glycolysis. During glycolysis, glucose is broken into two molecules of pyruvic acid; Figure 17.12 shows how one of the steps in this process involves the transfer of two electrons, in the form of a hydride ion, to NAD^+. As a result of this transfer of electrons, NAD^+ becomes reduced to NADH.

The reduction of NAD^+ is not a one-way process; if it was, the levels of NAD^+ in the cell would quickly become depleted. Instead, to maintain a source of NAD^+, the cell requires a way of transferring the electrons from NADH to another molecule in order to regenerate NAD^+. One way the cell achieves this regeneration is to reduce pyruvate to lactate (lactic acid), a process that is coupled to the oxidation of NADH, as illustrated in Figure 17.13. Specifically, pyruvate accepts two electrons from NADH and is reduced to lactate, whereas NADH is oxidized back to NAD^+. And so the cycle can continue.

$$NAD^+ + H^+ + 2e^- \rightleftharpoons NADH$$

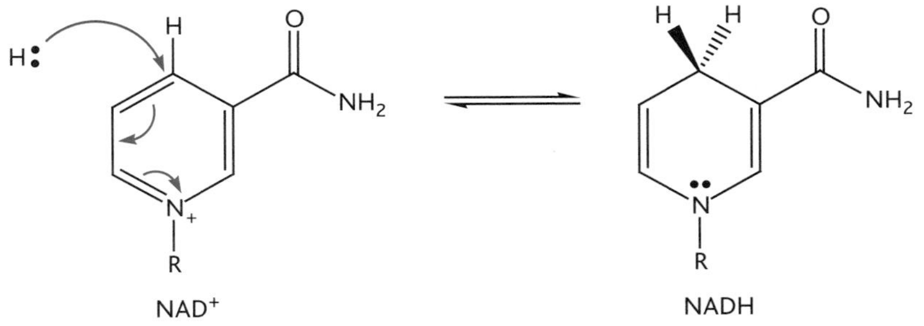

H: NAD⁺ NH₂ R NADH

Figure 17.12 The transfer of two electrons, in the form of a hydride ion, to NAD^+. As a result of this transfer of electrons, NAD^+ becomes reduced to NADH.

The reduction of pyruvate to lactate is particularly prevalent in muscle cells during periods of strenuous activity. During these times, there is insufficient oxygen available for pyruvic acid to participate in aerobic respiration. Instead, it is metabolized by anaerobic respiration, generating the lactate we mention above. This process has the benefit of regenerating NAD^+ at just the time that cells need NAD^+ to be maintained–namely, during periods of high metabolic activity that we associate with strenuous exercise.

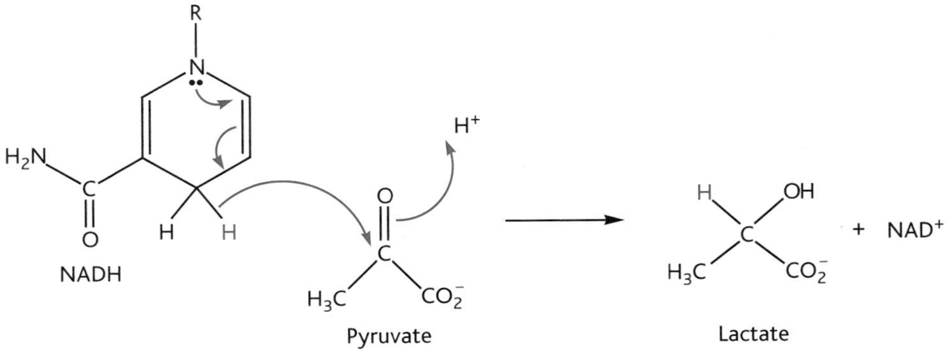

NADH + Pyruvate → Lactate + NAD^+

Figure 17.13 The reduction of pyruvate to lactate, coupled to the oxidation of NADH. Notice how the reaction involves the transfer of two electrons in the form of a hydride ion from NADH to pyruvate. Also notice how the reaction requires an acid catalyst (denoted here by the free proton, H^+). In the cell, this proton is donated by the side chain of histidine, which is present in the active site of the enzyme lactate dehydrogenase. (Adapted from McMurry and Begley, 2005.)

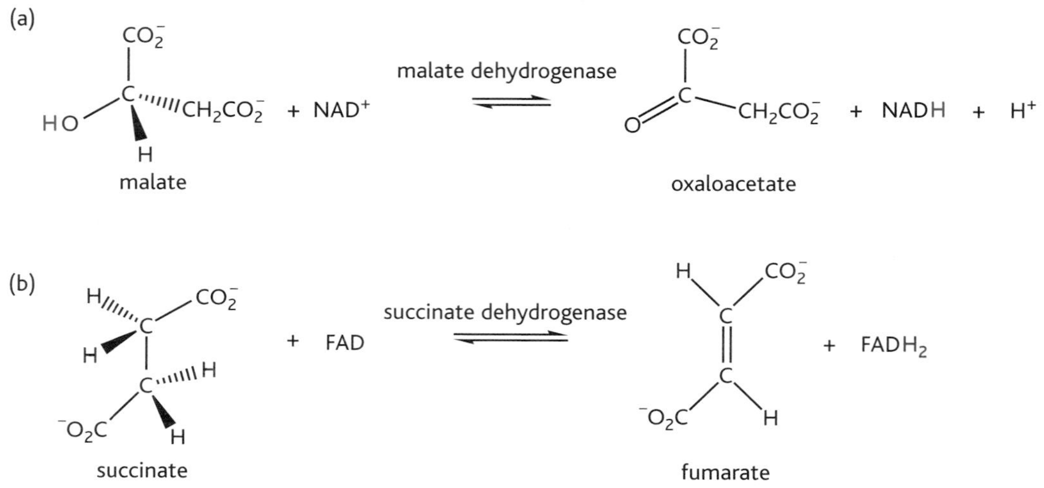

(a)

$HO-C(H)(CO_2^-)-CH_2CO_2^-$ + NAD^+ $\xrightleftharpoons{\text{malate dehydrogenase}}$ $O=C(CO_2^-)-CH_2CO_2^-$ + NADH + H^+

malate — oxaloacetate

(b)

$^-O_2C-CH_2-CH_2-CO_2^-$ + FAD $\xrightleftharpoons{\text{succinate dehydrogenase}}$ $^-O_2C-CH=CH-CO_2^-$ + $FADH_2$

succinate — fumarate

Figure 17.14 (a) The oxidation of malate to oxaloacetate by NAD. Notice how the oxidation of malate takes the form of a loss of a hydride ion (shown here in blue), which is used to reduce NAD^+ to NADH. (b) The oxidation of succinate to fumarate, involving FAD. Notice how FAD is reduced to $FADH_2$ during the course of the reaction. (Reproduced from Petsko and Ringe, 2004.)

By contrast, the cofactor flavin adenine dinucleotide (FAD) behaves as a prosthetic group; it binds irreversibly at the active site of an enzyme. For example, another step in the citric acid cycle involves the oxidation of succinate to fumarate in a reaction catalyzed by the enzyme succinate dehydrogenase, as illustrated in Figure 17.14(b). Look at this figure, and notice how FAD is reduced to $FADH_2$ during the course of the reaction. (The gain of two hydrogen atoms by FAD is an indirect way of it gaining electrons, because each hydrogen atom brings with it one electron.) FAD and its reduced counterpart $FADH_2$ are important because they accept and donate one electron at a time (unlike NAD^+ and NADH, which accept or donate two electrons).

Molecules which accept and donate electrons during enzyme-catalyzed reactions often appear to have complicated structures. However, the important part of the molecule, where electrons are accepted and donated, always contains ring structures with extensive delocalized double bonds, which can act as electron "sinks". (An electron sink is a group of atoms that can draw electrons towards themselves.) Look back at Figure 17.10, and notice the ring structure within NAD^+/NADH. Now look at Figure 17.15, which shows the structure of FAD/$FADH_2$. Again, notice the ring structure that is also present in these two molecules.

Self-check 17.3 During an electron transfer process, the following change occurs: NADH $\rightarrow$ NAD^+. Is NADH being oxidized or reduced here?

- ***Many enzymes feature cofactors—electron carriers that are either reversibly or irreversibly associated with the enzyme's active site.***

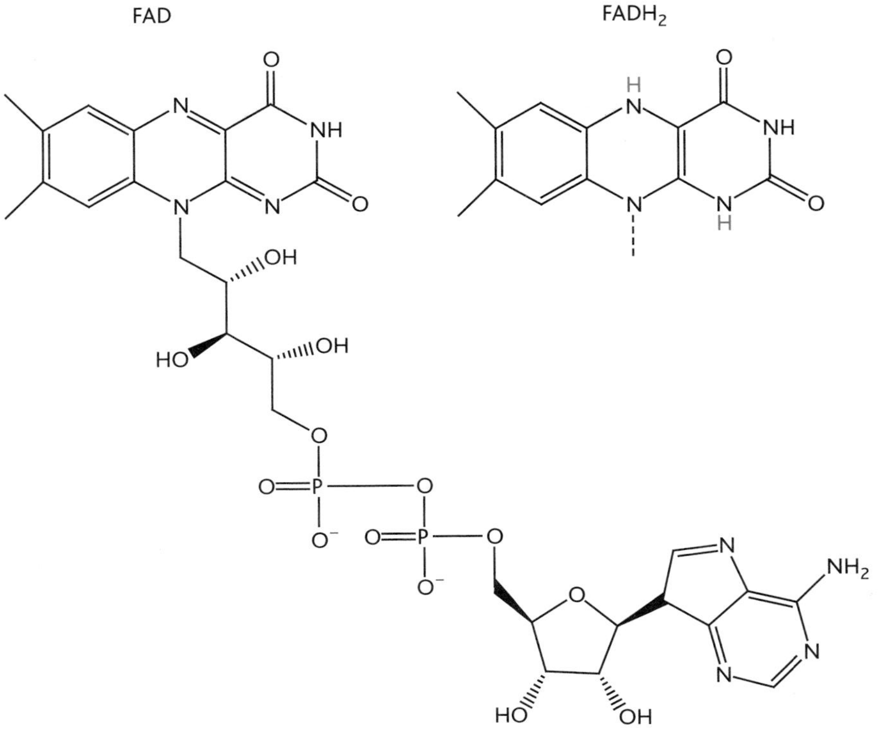

Figure 17.15 The structure of FAD and $FADH_2$. The bulk of the two structures is the same, varying only in the addition of two hydrogens to the ring structure of $FADH_2$, as shown here in blue.

17.5 Homolytic reactions

Let's now turn to those chemical reactions that involve the *equal* splitting of a pair of valence electrons between two previously joined atoms. We call these reactions **homolytic** reactions.

Look at Figure 17.16, which depicts a general homolytic reaction. Notice how we use half-arrows instead of full curly arrows; each arrow depicts the movement of a *single* valence electron, rather than a pair moving together.

Remember: 'homo' = the same; 'lytic' = splitting.

The product of homolytic bond cleavage is a special chemical species called a **free radical**, as indicated in Figure 17.16. A free radical is characterized by having one unpaired electron in its valence shell. We indicate a free radical by writing a dot against the atom bearing the unpaired electron:

X

- *A free radical is a chemical species that possesses an unpaired valence electron.*

The unpaired electron makes a free radical extremely reactive and also gives it great instability[6]. It is this reactivity that is central to the behaviour of free radicals; it also makes them extremely dangerous, as we see below.

6 Remember one of our essential concepts: low energy equates to stability; conversely, high energy equates to instability.

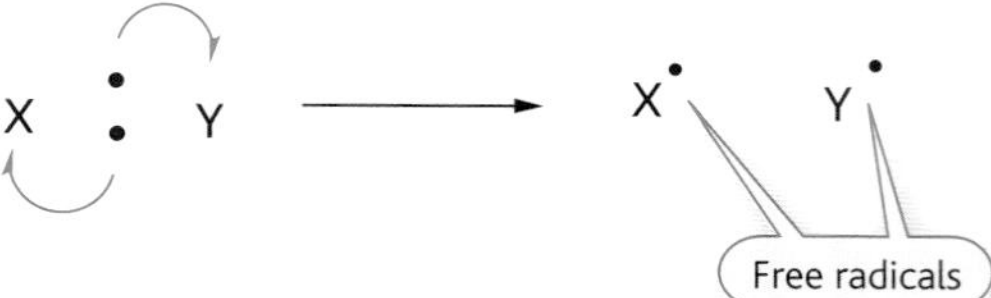

Figure 17.16 A generalized homolytic reaction.

Homolytic versus heterolytic cleavage

What determines whether a particular bond cleaves homolytically or heterolytically? The answer generally lies in the **polarity** of the bond.

A diatomic chlorine molecule, Cl_2, may undergo homolytic cleavage when exposed to sunlight (as explained on p. 601). By contrast, hydrogen chloride, HCl, undergoes **heterolytic** cleavage, during addition across the double bond of ethene, for example. What do we know about the polarity of the Cl–Cl and H–Cl bonds to help us explain this behaviour?

A homonuclear bond–a bond joining atoms of the same element–is non-polar: the two atoms have equal electronegativity, and so exert the same 'pull' on the electrons shared in their connecting bond (Section 3.12). When it comes to *breaking* the Cl–Cl bond, the two shared electrons are attracted to each atom equally, such that the most likely outcome of bond cleavage is that *both* atoms will receive one atom each: we see **homolytic cleavage** to form two $Cl^•$ species, as illustrated in Figure 17.17(a).

By contrast, chlorine is very much more electronegative than hydrogen. Consequently the H–Cl bond is highly polarized: the electronegative chlorine atoms pulls the shared electrons strongly towards itself, and so the distribution of electrons in HCl is uneven. During bond cleavage, *both* shared electrons are attracted most strongly to the highly electronegative chlorine atom, such that the most likely outcome of bond cleavage is that only the chlorine atom will receive the electrons: we see **heterolytic cleavage**, as illustrated in Figure 17.17(b).

- ***A non-polar bond is most likely to break homolytically.***
- ***A polar bond is most likely to break heterolytically.***

Self-check 17.4 Are the following bonds likely to break homo- or heterolytically?

(a) Br–Br; (b) H–Br; (c) C–H; (d) C–Cl

Homolytic reactions are referred to as **free radical reactions** and feature three distinct stages:

- Initiation
- Propagation
- Termination

Let's now consider each of these stages in turn.

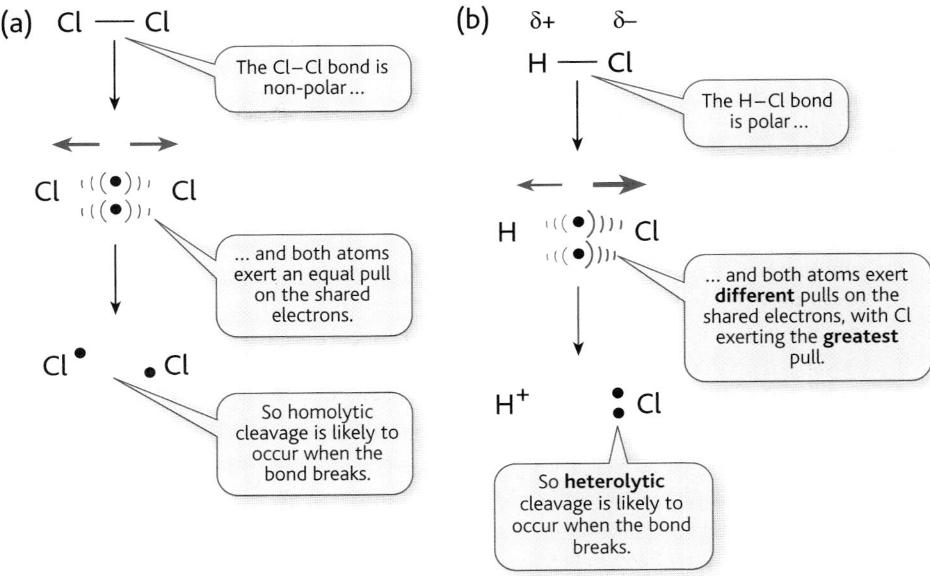

Figure 17.17 (a) When cleavage of a homonuclear bond (joining two atoms of the same element) occurs, it is most likely to be homolytic, such that each nucleus retains a single electron after cleavage. (b) When cleavage of a heteronuclear bond (joining atoms of two different elements) occurs, it is most likely to be heterolytic, with the nucleus of the most electronegative element receiving a pair of electrons, and the other nucleus receiving none.

Initiation

Initiation is the stage in a homolytic reaction during which free radical species are first formed. As we see above, free radicals are formed from the homolytic cleavage of a covalent bond:

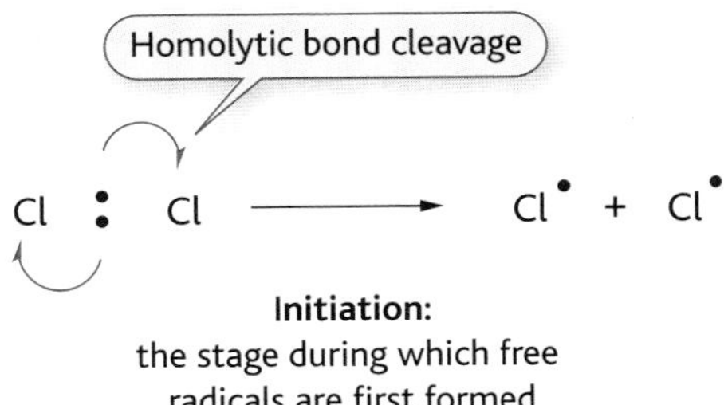

Initiation:
the stage during which free radicals are first formed

What triggers homolytic bond cleavage during an initiation step? One of the best ways of forming free radicals is by irradiating with electromagnetic radiation in the ultraviolet and visible regions of the electromagnetic spectrum. We see in Section 11.7 that many molecules with extensive double bonds can absorb UV–visible light such that their electrons become excited and move to antibonding orbitals. In other situations, light with similar wavelengths in the UV–visible region of the electromagnetic spectrum has sufficient energy to break certain covalent bonds. For example, the blue–green region of the spectrum can cause homolytic cleavage of the Cl–Cl bond in a Cl_2 molecule[7].

7 The sunlight-powered splitting of a covalent bond is called **photolysis**–Greek for 'light-splitting'.

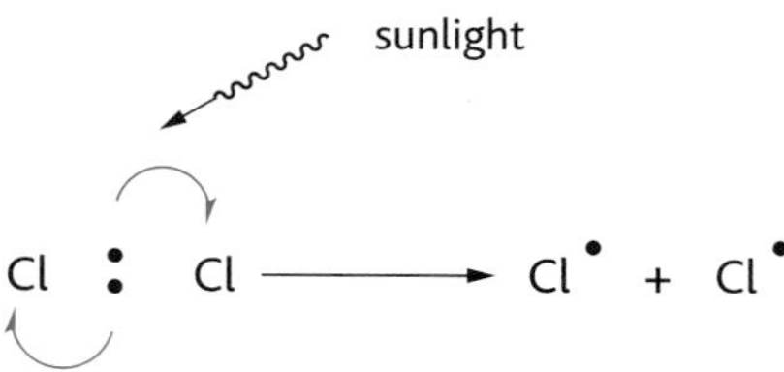

Figure 17.18 The wavelengths of sunlight corresponding to the blue region of the UV–visible spectrum drive the homolytic cleavage of the Cl–Cl bond to generate two Cl• free radicals.

Ultraviolet light is the component of sunlight that causes our skin to tan. (This is why sun beds expose the user to a source of UV light.) UV light is also the component of sunlight that is blocked by some suntan lotions: if our skin is exposed to UV light, of certain wavelengths, the UV light causes free radical damage to our skin cells, which can speed up the ageing process and contribute to an increased risk of skin cancer.

We see a potent example of the ability of UV light to drive free radical formation–to **initiate** a free radical reaction–when we consider the reaction of chlorine with methane:

$$CH_4(g) + Cl_2(g) \rightarrow CH_3Cl(g) + HCl(g)$$

When methane and chlorine are mixed in the dark, nothing happens: the two compounds fail to react. However, when the two compounds are exposed to sunlight, an explosive reaction occurs. Why do we see this difference between reactivity in the dark and in the sunlight?

Chlorine and methane react via a free radical chain reaction which can *only* occur when there is sunlight available to **initiate** free radical formation. Specifically, the wavelengths of sunlight corresponding to the blue region of the UV–visible spectrum drive the homolytic cleavage of the Cl–Cl bond to generate two $Cl^{\bullet}$ free radicals, as illustrated in Figure 17.18.

It is this ability to drive the formation of free radicals that makes ultraviolet light particularly dangerous to us. The unleashing of free radicals in the cells of our body can have devastating effects, as explained in Box 17.2.

Propagation

A free radical is a highly reactive, highly unstable species. The only way that a free radical can become stable is for it to *gain* a further electron to pair up with its existing unpaired electron. It does this by abstracting or ferociously obtaining ('scavenging') an electron, usually in the form of a hydrogen atom, from almost any molecule in its vicinity. However, this scavenging process generates a new free radical–it **propagates** free radical formation.

Let's return to our example of the reaction of chlorine and methane in bright sunlight. During initiation, Cl_2 undergoes homolytic cleavage to yield two $Cl^{\bullet}$ free radicals. A $Cl^{\bullet}$ radical then propagates the free radical reaction by scavenging an electron from a neighbouring methane molecule to form a molecule of hydrogen chloride, HCl, as illustrated in Figure 17.20, step 1.

Look at this figure and notice the following:

- The chlorine free radical causes homolytic cleavage of another covalent bond.
- The chlorine free radical acquires a second unpaired electron, which can pair with its existing unpaired electron. It is no longer a free radical, but now forms a stable molecule.
- Homolytic cleavage results in the formation of a *new* free radical, a methyl radical, $CH_3^{\bullet}$.

BOX 17.2 Dangerous liaisons: single electron transfer to oxygen

We see in Section 17.4 that some biological reducing agents such as FAD are able to transfer single electrons in redox reactions. And we also see in Section 17.4 that oxygen can accept four electrons as it is reduced to water. The transfer of electrons to oxygen can occur one electron at a time, a requirement that has important implications for us all.

Usually the electron transfer reactions occur while the oxygen is tightly bound to an enzyme with the electron donor close by. Consequently, the four electrons transfer successfully from the donor to oxygen and water is formed. But nothing works perfectly every time. Occasionally, an oxygen atom which has accepted only one electron can become disassociated from the enzyme, with potentially dire consequences for the cell in which the oxygen is present. So why is the outlook so gloomy?

When oxygen accepts just one electron, the highly reactive superoxide radical, $O_2^{\bullet}$, is formed. Superoxide is an example of a *reactive oxygen species* (ROS). Look at Figure 17.19, which depicts the formation of three key intermediates when an oxygen molecule undergoes single electron transfer. Notice how oxygen is ultimately reduced to form two molecules of water, (Remember: reduction denotes the *gain* of electrons from a reacting species.)

Why is the formation of the superoxide free radical such a problem? As we see above, free radicals carry a single unpaired electron. This makes them highly reactive: to complete the pair (and, hence, become more stable) they aggressively scavenge additional electrons from other molecules. In the case of the superoxide radical, in seeking to satisfy its requirement for another electron, superoxide will form even more dangerous molecules unless it is destroyed.

When superoxide gains a second electron (and two protons) hydrogen peroxide is formed. The addition of a third electron gives rise to a hydroxyl radical, probably the most dangerous molecular species a cell can encounter.

The following quote describes what happens next, and explains why free radicals in biological systems are so dangerous:

"When a hydroxyl radical reacts with a protein, lipid, or DNA molecule, it snatches an electron to itself and then sinks back into the sublime chemical stability of water. But of course the act of snatching an electron leaves the reactant short of an electron, just as a mugger leaves the victim short of a handbag. In the case of the hydroxyl radical, another radical is formed, this time part of the protein, lipid, or DNA. It is as if having your handbag snatched deranges your mind and turns you into a mugger yourself, restless until you have snatched someone else's bag." (Nick Lane, *Oxygen: the Molecule That Made the World*, Oxford University Press, 2002)

Free radical damage to DNA is one of the causes of genetic mutation, which itself leads to an increased susceptibility to cancer, inflammation, ageing problems and autoimmune diseases. For example, hydroxyl-radical attack on deoxyguanosine (G), one of the four nucleotides from which DNA is formed, is believed to generate 8-hydroxydeoxyguanosine (8-OHdG):

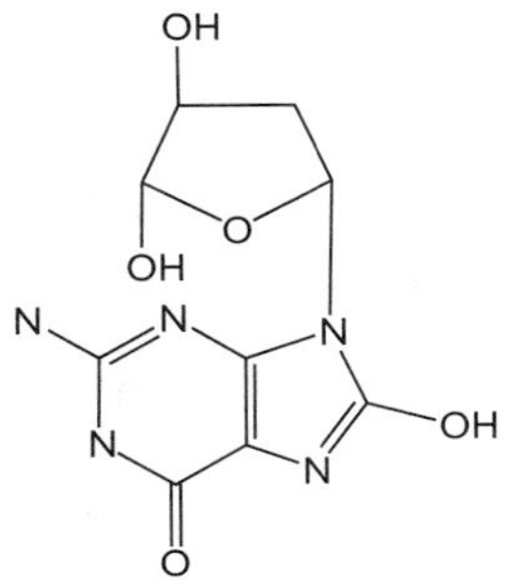

This chemically modified nucleotide is mispaired with the base adenine (A) instead of with its regular partner cytosine (C) during DNA replication, leading to a point mutation (mutation at a specific location in the DNA sequence). The accumulation of mutations such as these can cause the disruption of the regulation of cell activity that is associated with cancer.

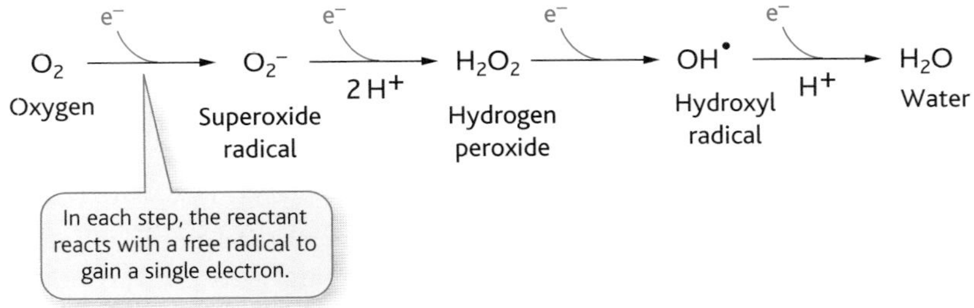

Figure 17.19 During the four-electron reduction of oxygen to water by single electron transfer, a sequence of three radical species is formed: the superoxide, hydrogen peroxide, and the hydroxyl radical.

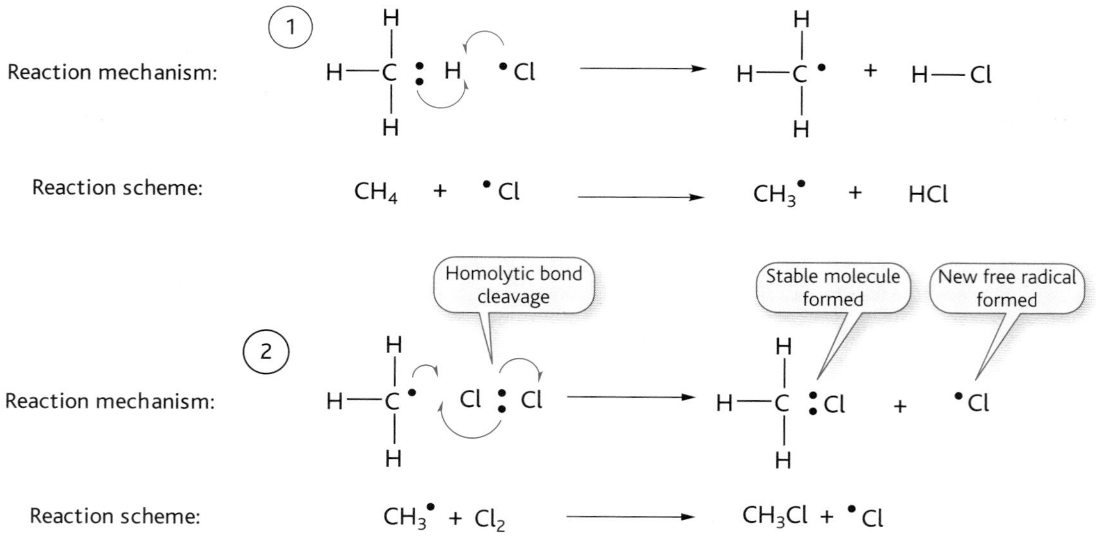

Reaction mechanism: (1) H—C(H)(H) : H •Cl ⟶ H—C(H)(H)• + H—Cl

Reaction scheme: CH_4 + •Cl ⟶ $CH_3^•$ + HCl

Reaction mechanism: (2) H—C(H)(H)• Cl : Cl ⟶ H—C(H)(H) : Cl + •Cl

Homolytic bond cleavage

Stable molecule formed

New free radical formed

Reaction scheme: $CH_3^•$ + Cl_2 ⟶ CH_3Cl + •Cl

Figure 17.20 The propagation step of a free-radical reaction. Notice how bond cleavage in each step (a C–H bond in step 1, and a Cl–Cl bond in step 2) is homolytic, generating one single covalent bond (which stabilizes the free radical present at the start of the reaction), and a new free radical species.

It is this last point which is so critical: in scavenging for an electron to become more stable, a free radical drives the formation of a *new* free radical.

Next, Figure 17.20, step 2 shows how the newly formed $CH_3^•$ free radical scavenges an electron from a neighbouring uncharged species in the gas mixture, usually a Cl_2 molecule. The scavenging of an electron from Cl_2 yields CH_3Cl and a new Cl• radical.

Notice how we've now gone full circle: the second propagation reaction drives the formation of a new Cl• free radical, which can then re-enter the first propagation reaction. We have a **chain reaction**, in which the two reactions shown in Figure 17.20 form a continual loop:

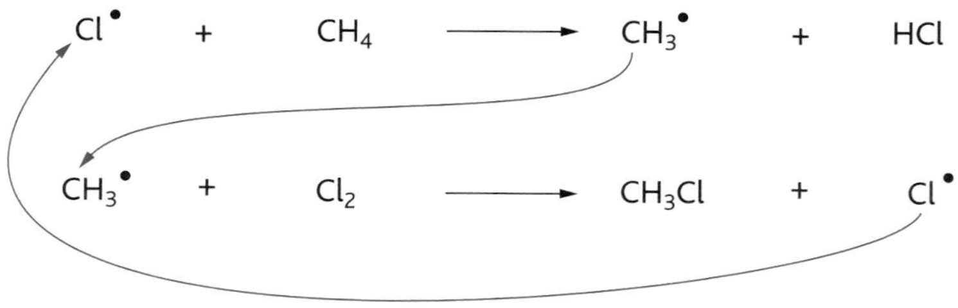

$$Cl^• + CH_4 \longrightarrow CH_3^• + HCl$$

$$CH_3^• + Cl_2 \longrightarrow CH_3Cl + Cl^•$$

It is the chain reaction that is created during the propagation stage of some free radical reactions that is potentially so damaging: the reacting species enter what is quite literally a vicious circle, with free radicals being constantly consumed and liberated as the reaction continues.

Free radical chain reactions drive the destruction of ozone by chlorofluorocarbons (CFCs), the refrigerant and propellant compounds released into the atmosphere from fridges, freezers, and aerosols. The formation of just one chlorine free radical (derived from a single CFC molecule) is enough to destroy up to 100 000 ozone molecules. We find out more about this particular free radical reaction on p. 182.

Termination

Free radical reactions do not continue indefinitely. **Termination** of a free radical reaction occurs when two free radicals collide, and combine to form a stable molecule. In the case of our example of the free radical reaction of chlorine and methane, there are three possible termination steps:

1. $Cl\cdot + \cdot Cl \longrightarrow Cl{:}Cl$ (Cl_2)

2. $H_3C\cdot + \cdot Cl \longrightarrow H_3C{:}Cl$ (CH_3Cl)

3. $H_3C\cdot + \cdot CH_3 \longrightarrow H_3C{-}CH_3$ (CH_3CH_3)

Notice how each of these termination steps yields a single, stable molecule, in which all valence electrons are paired up.

Radical termination provides a way of controlling free radical reactions: if we increase the proportion of free radicals involved in a termination reaction, instead of a propagation reaction, the 'pool' of free radicals decreases, and the reaction slows down. When we forcibly increase the proportion of free radicals entering a termination step we are said to be **quenching** the reaction.

The quenching of free radical reactions is a vital part of minimizing the damage caused by free radicals in the cells of our body, as explained in Box 17.3.

Effective antioxidants include vitamin C and vitamin E, both of which form unreactive free radical species. Because they quench cellular free radical damage, antioxidants such as vitamins C and E are an important natural weapon against the onset of cancer, the risk of which is increased by cellular free radical damage. Other natural antioxidants include lycopene (2), the pigment that gives tomatoes their red colour.

(2)

Free radicals in biological systems

Free radicals and homolytic reactions are an unavoidable feature of biological systems. We spend much of our time exposed to sunlight which, as discussed above, can yield free radicals in the cell. The activity of such free radicals can cause cellular damage, which can be the precursor to skin cancer. We can take steps to limit the

BOX 17.3 Antioxidants and the quenching of free radical reactions

We see in Box 17.2 how free radicals results in the multi-step reduction of oxygen to water with the associated production of reactive oxygen species (ROS), including the hydroxyl radical, $OH^{\bullet}$. Such free radical reactions are an inevitable by-product of many enzyme-catalyzed reactions, including those involved in respiration. But how can their effects be minimized?

When a radical is formed as an intermediate during an enzyme reaction it is tightly bound at the active site of an enzyme and is in close proximity to the molecule with which it will react. Such radicals are not 'free' but are enzyme-bound; they are of no danger to the cell as they quickly react. However, when a radical is formed in the cytosol or dissociates from an enzyme to become 'free' (as superoxide might) it can cause a great deal of damage if it is not rapidly destroyed.

The propagation of free radical reactions in the cell can be inhibited by a group of molecules called **antioxidants**. As their name suggests, antioxidants prevent the oxidation associated with electron transfer and the activity of free radicals. Specifically, we say that antioxidants **quench** free radical reactions.

The quenching properties of antioxidants stem from their unusual lack of reactivity when they encounter a free radical. The free radical scavenges a single electron from the antioxidant; however, despite becoming a free radical itself as a result of this activity, the antioxidant remains surprisingly unreactive, and does not act to propagate further free radical reactions. As a result, the chain reaction–of sequential free radical attack–simply fizzles out.

formation of free radicals: we can minimize our exposure to high-intensity sunlight; we can avoid exposure to those chemical carcinogens that are believed to induce free radical formation; and we can coat our skin with an antioxidant, e.g. sunscreen. (We have to be careful here, however: some sunscreens contain chemicals which produce free radicals in the presence of sunlight.)

Paradoxically, however, a large proportion of the damage caused by the free radicals we face comes from oxygen, a molecule that we depend on for our survival.

We require a constant supply of oxygen to drive respiration, the process by which our body generates energy from the metabolism of glucose:

$$\underset{\text{Glucose}}{C_6H_{12}O_6} + \underset{\text{Oxygen}}{6\,O_2} \longrightarrow \underset{\text{Carbon dioxide}}{6\,CO_2} + \underset{\text{Water}}{6\,H_2O} + \text{energy}$$

However, some oxygen molecules in our cells scavenge single electrons from other molecules to yield the same highly reactive free radicals–the reactive oxygen species–described in Box 17.2.

This phenomenon–of molecular oxygen propagating free radical damage–is known as **oxygen toxicity**, and has been likened to the damage caused by radiation poisoning. Over our lifetimes the cells of our bodies succumb to more and more damage as the result of free radical activity; this damage is part of what we experience as **ageing**[8].

So, our relationship with oxygen is something of a love–hate affair. We cannot live without it, yet it is damaging to us if we live *with* it.

8 To learn more about oxygen and its positive and negative impacts on our life, read the excellent book *Oxygen: The Molecule That Made the World*, by Nick Lane (OUP, 2002).

Check your understanding

To check that you've mastered the key concepts presented in this chapter, review the checklist of key concepts below, and attempt the multiple-choice questions available in the book's Online Resource Centre at **http://www.oxfordtextbooks.co.uk/orc/crowe2e/**.

Checklist of key concepts

The nature of chemical reactions

- A **chemical reaction** is a chemical change occurring within one compound, or an interaction between two or more starting compounds, to form one or more end compounds, which are different from the compound(s) we started with
- **Reactants** are the substances that interact at the start of a chemical reaction
- **Products** are the substances that are formed as a result of the chemical reaction
- A **reaction scheme** tells us the reactants and products that are involved in a particular reaction
- A reaction scheme also tells us the **stoichiometry** of the reaction: the ratios in which the reactants and products are present

The movement of electrons during chemical reactions

- Chemical reactions involve the **redistribution of valence electrons** as existing chemical bonds are broken, and new chemical bonds are formed
- Bonds cleave in a **heterolytic** manner when the electrons that represent the bond redistribute **unequally**, with two electrons going to the atom at one end of the bond, but none to the other atom
- Bonds cleave in a **homolytic** manner when the electrons that represent the bond redistribute **equally**, with one electron going to the atom at one end of the bond, and one to the other atom
- A **reaction mechanism** denotes the movement of valence electrons during the course of a chemical reaction
- We depict the movement of valence electrons with **curly arrows**
- **Full** arrows denote the movement of a valence electron pair
- **'Fish hook'** arrows denote the movement of a single electron

Heterolytic reactions

- A **nucleophile** is an electron-rich species, which has a valence electron pair that it can donate to another atom to form a covalent bond
- An **electrophile** is an electron-poor species, which possesses–or can readily possess–an empty orbital, and can accept a valence electron pair from a nucleophile
- A heterolytic reaction must involve a nucleophile and an electrophile: the nucleophile must have an electrophile to which it can donate an electron pair

Oxidation and reduction

- When a species loses electrons, we say it is **oxidized**: it undergoes oxidation
- When a species gains electrons, we say it is **reduced**: it undergoes reduction
- A **redox reaction** is one in which two species are simultaneously oxidized and reduced: one species loses electrons to the other
- Oxidation and reduction can also be considered in terms of the loss or gain of hydrogen
- **Electron carriers** are compounds that are particularly effective at accepting or donating electrons
- Many enzyme-catalyzed reactions involve redox processes, whereby the redox reaction itself is mediated by a **cofactor**
- A cofactor is a non-amino acid entity, which is closely associated with the enzyme's active site

Homolytic reactions

- Homolytic reactions usually involve **free radicals**
- A free radical is a chemical species possessing an unpaired valence electron
- Non-polar bonds favour homolytic cleavage
- Polar bonds favour heterolytic cleavage

- Homolytic reactions occur in three stages: initiation, propagation, and termination
- **Initiation** is the stage in which free radicals are first formed
- Free-radical formation can be driven by the photolytic (light-driven) cleavage of a covalent bond
- **Propagation** is the continued generation of free radicals in a sequential manner
- Propagation may lead to the setting up of a **chain reaction**
- **Termination** is the ending of a chain reaction, which occurs when two free radicals react with one another

Chemical reactions 2: reaction mechanisms driving the chemistry of life

INTRODUCTION

The food we eat supplies the cells of our body with the molecules they needs to stay alive. The molecules absorbed by our bodies take part in a remarkable series of chemical reactions which we collectively call **metabolism**. Large molecules such as carbohydrates are broken down into smaller molecules by a series of reactions know as **catabolism**. The breakdown of large molecules is accompanied by the release of chemical energy, which is used by the body for its growth and development.

The other key aspect of metabolism is the synthesis from small molecules of new, complex molecules such as DNA, proteins, and lipids. The synthesis of new molecules is called **anabolism;** these reactions often require energy to be supplied–with the source often being the energy released by catabolic reactions.

In this chapter we look first at the nature of reaction mechanisms to discover, at the molecular level, the chemical changes that ultimately underpin both anabolic and catabolic processes. We then use our understanding of reaction mechanisms to explore some of the key reactions involved in the initial stages of the catabolism of carbohydrates–the process by which our bodies obtain energy from food.

18.1 An introduction to reaction mechanisms

Do all chemical reactions show the same pattern of bond cleavage and formation as reactants change into products? If we look back at the preceding chapter we see that the answer is 'no'. There is a range of different reaction mechanisms, each with its own distinct pattern of bond cleavage and bond formation. Each mechanism involves the movement of electrons. However, it is the *way* that the reactants interact, and the nature of the products that form, that differs from reaction to reaction.

There are four key reaction mechanisms that we explore in this chapter:

- Substitution
- Addition

- Elimination
- Condensation

These reactions are named for the change in components of reactants and products during the course of the reaction. Different chemical reactions may share the same *general* pattern of bond cleavage and formation and so be assigned the same reaction mechanism name, just as different organic compounds may share the same functional group and so be assigned the same family name (for example, 'alcohol' or 'amine').

Before we look at the different types of reaction mechanism, however, we need to find out about transition states and intermediates, two species that are central to chemical reactions and the mechanisms they adopt. We use some simple chemical examples here while we explore some fundamental concepts. Exactly the same principles apply in biological systems, however; we see plenty of examples of biological reactions in which these principles apply during the course of the chapter.

Transition states and intermediates

We see in Section 17.1 how a chemical reaction can proceed in just one step, or during the course of two or more steps.

The substitution reaction of bromoethane and hydroxide ion occurs in a single step:

Remember: a full arrow denotes the movement of a pair of valence electrons.

OH^- H–C(H)(CH$_3$)–Br → HO–C(H)(CH$_3$)–H + Br^-

overall: $CH_3CH_2Br + OH^- \longrightarrow CH_3CH_2OH + Br^-$

However, the reactants do not suddenly 'switch' to being products. Instead, there is a progressive movement of electrons–from the nucleophilic OH^- ion to the electrophilic carbon atom as a new C–O bond is formed, and from the C–Br bond to the Br^- ion as the C–Br bond is broken[1].

It is during this progressive movement of electrons that the reactants make the **transition** from being reactants into products. As they make this transition, the reactants pass through what we call a **transition state**[2]. A transition state is the point of **highest energy** of a chemical reaction. It represents the nature of the reactants and products part way through a reaction: when some bonds are part way to being broken, and some bonds are part way to being formed.

Look at Figure 18.1, which shows how the reactants pass through a transition state during the one-step reaction of bromoethane and hydroxide ion. Notice how the transition state represents the point at which the C–O bond is partly formed, and the C–Br bond is partly broken.

1 We look at the mechanism of this reaction in more detail in the next section.

2 A **transition state** is a high-energy, unstable species. As the name implies, it exists only transiently. One way in which enzymes are able to speed up chemical reactions is to stabilize the transition state. We learn more about transition states, and energy changes associated with chemical reactions, in Chapter 13.

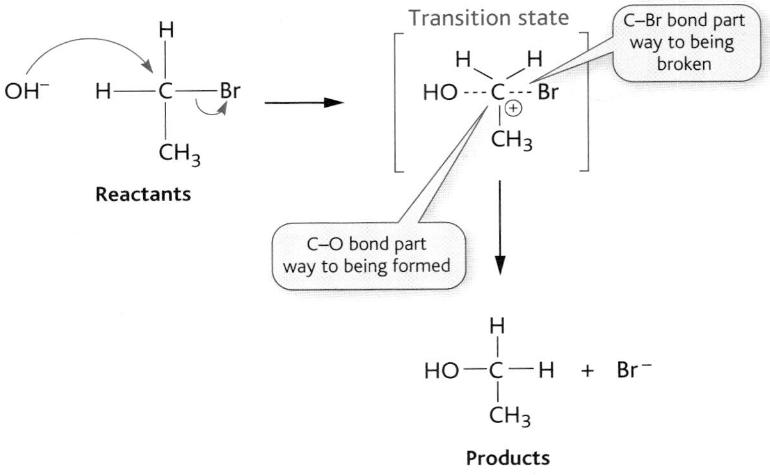

Figure 18.1 The one-step substitution reaction of bromoethane and a hydroxide ion. The reactants pass through a **transition state** as they form products.

So what happens during a *multi-step* reaction? During a multi-step reaction the reacting species form an **intermediate** compound. This intermediate compound undergoes further chemical change to form the final products of the reaction.

For example, let's consider the reaction involving hydroxide ion with 2-chloro-2-methylpropane, $(CH_3)_3CCl$. This reaction occurs via the two-step mechanism depicted in Figure 18.2. Notice how step 1 yields an intermediate compound, in this case a **carbocation**[3].

The intermediate, once formed, then reacts with the hydroxide ion during step 2 to form the products and complete the reaction.

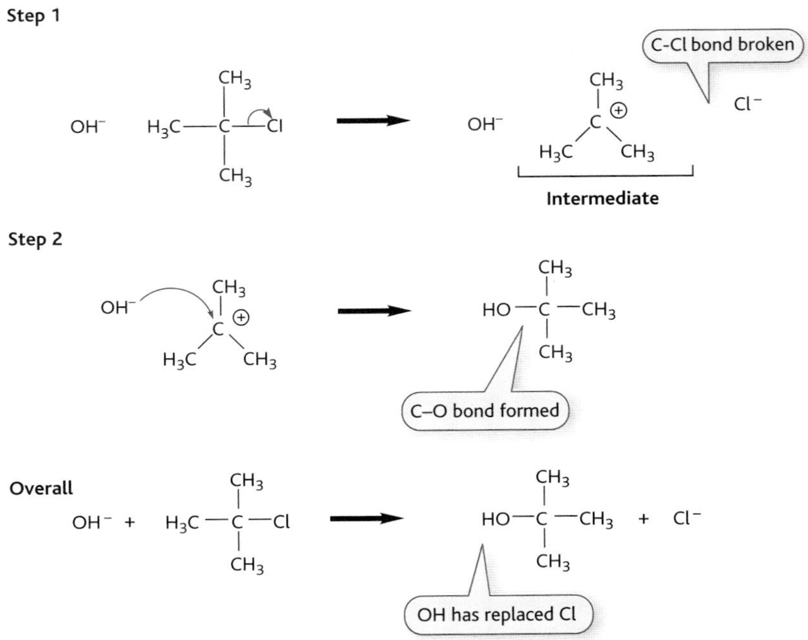

Figure 18.2 An example of a two-step reaction, in which an intermediate species is formed after the first step.

3 A carbocation is the general name for any positively charged ion in which the positive charge resides directly on a carbon atom.

Transition states are not only a feature of one-step reactions. *Every* step in a chemical reaction involves a transition state. Look at Figure 18.3, which shows how reactants pass through a transition state as they form products in a one-step reaction. Notice how reactants also pass through a transition state as they form an intermediate during the first step of a two-step reaction; the intermediate then passes through a *second* transition state as it changes into products during the *second* step of a two-step reaction.

Look again at Figure 18.3 and notice the following:

- An intermediate is of higher energy than the reactants or products.
- A transition state is of higher energy than reactants, products, or intermediates.

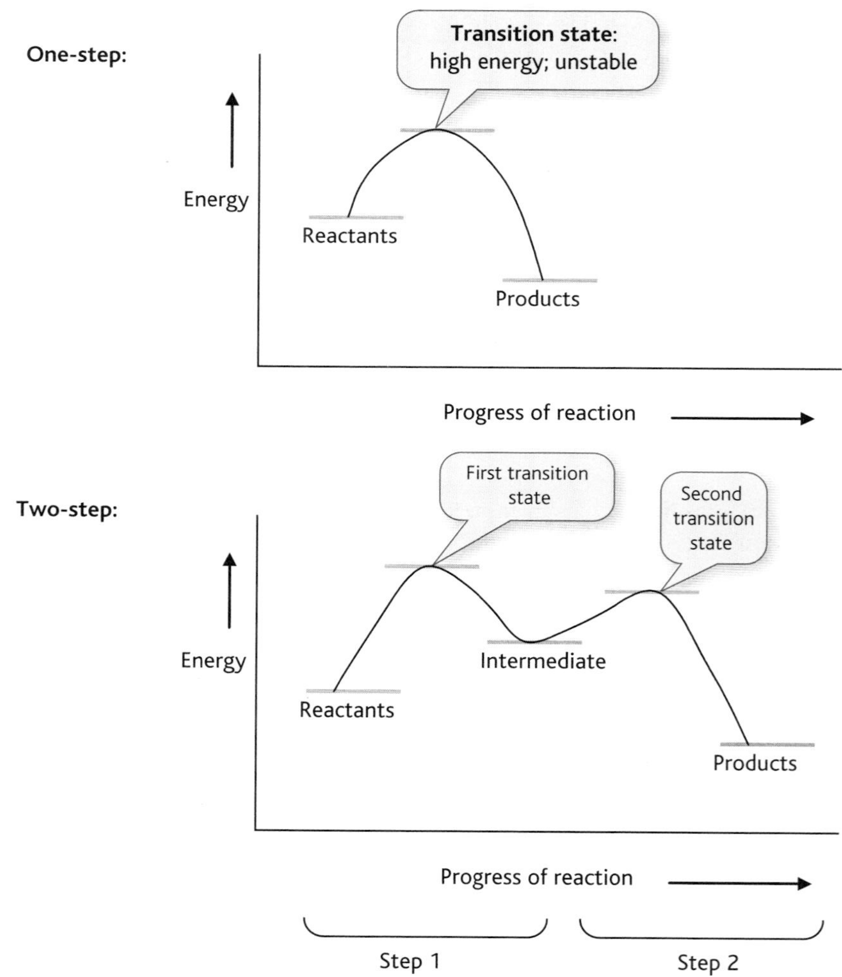

Figure 18.3 Each step in a multi-step reaction features a transition state. Reactants pass through a transition state as they form products in a one-step reaction. Reactants also pass through a transition state as they form an intermediate during the first step of a two-step reaction; the intermediate then passes through a *second* transition state as it changes into products during the *second* step of a two-step reaction.

Intermediates are different from transition states in several important ways: they have a much longer life time and may undergo different, independent reactions giving several products.

- *Each step in a chemical reaction involves a high-energy, unstable transition state. A one-step reaction passes through one transition state. A two-step reaction passes through two transition states.*

We find out more about the energy profile of a reaction–and the influence of transition states and intermediates–in Chapter 13.

Let's now consider the first type of reaction mechanism: substitution.

18.2 Substitution

During a substitution reaction an atom (or group of atoms) on the reactant molecule is replaced by a different atom or group of atoms:

$$A + B\text{—}X \longrightarrow B + A\text{—}X$$

A is **substituted** for B

A substitution reaction is described as nucleophilic or electrophilic, depending on the nature of the species that is interacting with the molecule, which changes as a result of the substitution reaction. In this section we consider both nucleophilic and electrophilic substitution in turn.

Nucleophilic substitution

Look at the substitution of a bromide ion (Br^-) on bromoethane by a hydroxide ion (OH^-), as shown in Figure 18.4. In this reaction, the molecule being changed by the reaction is bromoethane, and the species 'attacking' this molecule is the nucleophile, OH^-. So this is a nucleophilic substitution reaction. Notice the following:

- The OH^- ion acts as a nucleophile, donating an electron pair to the central carbon atom.
- The bond in bromoethane that breaks (to accommodate the newly formed C–OH bond) is the highly polarized C–Br bond.

The Br^- group, which is lost from bromoethane during the course of the substitution reaction, is called a **leaving group**.

Why was it the Br^- group that was lost from bromoethane? Why did we not see

$$OH^- \quad H\text{—}\underset{CH_3}{\overset{H}{C}}\text{—}Br \longrightarrow HO\text{—}\underset{CH_3}{\overset{H}{C}}\text{—}Br + H^-$$

for example, with H^- as the leaving group?

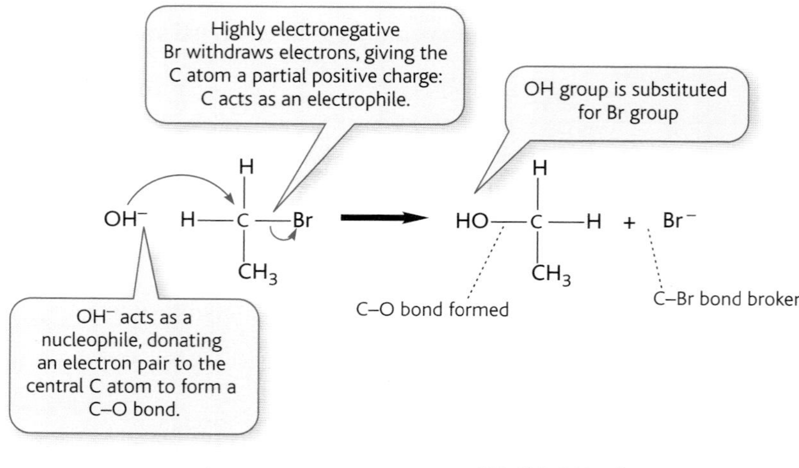

Figure 18.4 The substitution of a bromide ion on bromoethane by a hydroxide ion. Notice how the OH^- ion acts as a nucleophile and donates an electron pair to the central carbon atom, which acts as an electrophile.

Br^- is the best *leaving group* because it is the most **electronegative** group: it is good at withdrawing the electrons shared with the central carbon atom at which substitution occurs. With the electrons it shares with bromine withdrawn, the carbon atom carries a partial positive charge. In this state, the carbon atom can act as an electrophile and accept a pair of electrons from the nucleophile, OH^-.

- ***The best leaving group during nucleophilic substitution is the most electronegative species attached to the atom attacked by the nucleophile.***

Self-check 18.1 Which is the most likely leaving group when a nucelophile attacks each of the following?

(a) H_3C—C(H)(OH)—H

(b) H_3C—C(Cl)(NH_2)—H

One step versus two step substitution

Nucleophilic substitution reactions can occur in one or two steps, as illustrated in Figure 18.5. During a two-step reaction, the end of the first step is marked by the formation of an intermediate cation, as seen in Figure 18.5(b). The incoming

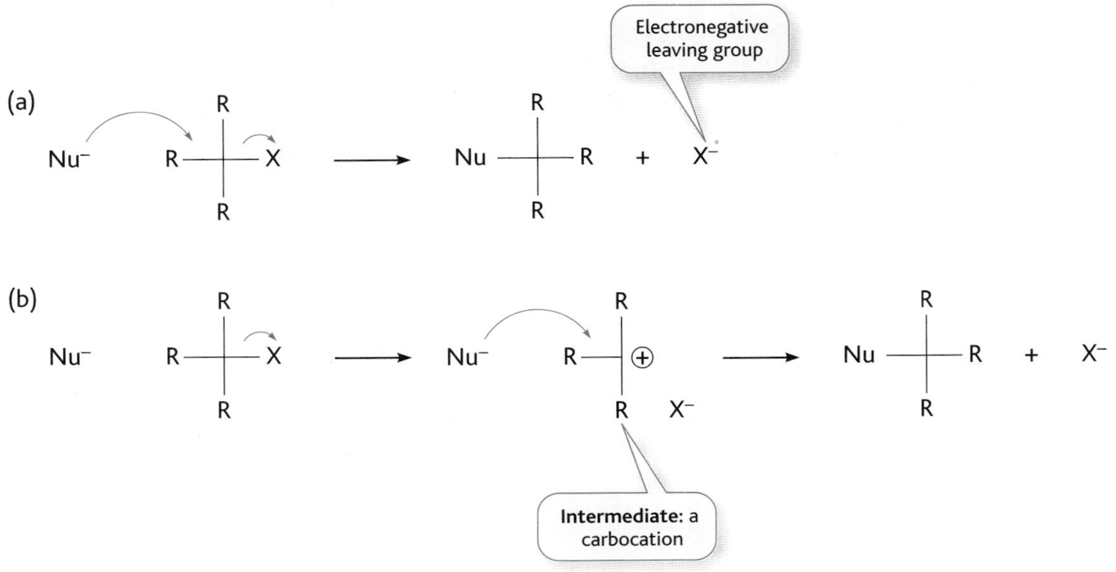

Figure 18.5 One-step versus two-step substitution reactions. (a) A generalized S_N2 (one-step) reaction, in which a nucleophile (Nu^-) attacks the central carbon atom, and a leaving group (X) leaves, all within a single step. (b) A generalized S_N1 (two-step) reaction. During the first step, a carbocation is formed when the electrophilic leaving group, X, withdraws electrons from the C–X bond. During the second step, the nucleophile attacks the carbocation to form the end product.

nucleophile attacks this cation during the second step of the reaction. By contrast, Figure 18.5(a) shows how a one-step substitution reaction does not involve an intermediate. Instead, the nucleophile simply attacks the target molecule, which passes through a transition state–but *not* a stable intermediate species–before forming the product of the reaction:

Nu⁻ R R R X → [Nu R R R X]⊖ → Nu R R R + X⁻

Transition state

Remember: a transition state is not a stable species, so the formation of a transition state does not count as one discrete step in a reaction; it is merely the point of highest energy as the reacting species move in a single step towards forming products.

A one-step nucleophilic substitution reaction is called an **S_N2 reaction**, and a two-step nucleophilic substitution reaction is called an **S_N1 reaction**, where S tells us that it is a substitution reaction; $_N$ tells us that the reaction is promoted by nucleophilic attack; and the numbers 1 and 2 tell us how many species are involved in the rate-limiting step[4].

Because a one-step reaction has just the one step, this must be the rate-limiting one; look again at Figure 18.5(a) and notice how this reaction involves two species–the nucleophile and the molecule that it is attacking. This is the S_N2 mechanism.

By contrast, the rate-limiting step of a two-step reaction is the first step–the formation of the intermediate cation. Look at Figure 18.5(b) and notice how this step involves just one species, the molecule that forms the intermediate. Also notice how the nucleophile doesn't become involved until the second step (which *isn't* the rate-limiting step). This is the S_N1 mechanism[5].

- ***An S_N2 reaction is a one-step nucleophilic substitution reaction in which the rate-limiting step involves two species.***
- ***An S_N1 reaction is a two-step nucleophilic substitution reaction in which the reactants pass through a stable intermediate form before the reaction is completed. The rate-limiting step involves one species.***

So what determines whether a substitution reaction proceeds via an S_N1 or S_N2 mechanism? The mechanism is dictated by a number of factors, including the nature of the leaving group (whether or not it is a 'good' leaving group); the nature of the nucleophile; and the nature of the groups attached to the central carbon atom (the atom that is attacked by the incoming nucleophile). For our purposes, however, we don't need to worry too much about whether a nucleophilic substitution reaction happens in one or two steps; the end product is the same regardless.

Nucleophilic attack–such as that seen during both one-step and two-step nucleophilic substitution reactions–is a common feature of a great number of reactions occurring within biological systems. For example, Figure 18.6 shows how nucleophilic attack lies at the heart of both the hydrolysis of peptide bonds by the enzyme chymotrypsin, and the synthesis of DNA from an RNA primer.

Substitution reactions are not always desirable; read Box 18.1 to find out how substitution reactions involving DNA can lead to the onset of cancer.

Electrophilic substitution

In our examples of substitution so far, the groups of atoms acting as the substituent (replacing other atoms on the 'parent' molecule) are nucleophiles. However, in a number of important substitution reactions, including some that occur in biological systems, the substitute atoms are **electrophiles**, not nucleophiles. We call this type of process an electrophilic aromatic substitution reaction.

Look at Figure 18.8, which shows the general mechanism for the substitution of an electrophile (represented by the symbol 'E^{+}') for a hydrogen atom on benzene.

4 The rate-limiting step is the step in a multi-step reaction that determines the rate of the reaction overall. We can think of the rate-limiting step as the bottleneck in the reaction.

5 Take care here: don't be tricked into thinking that the '1' and '2' denote the number of steps in a substitution reaction, such that an S_N2 reaction is a two-step reaction, and an S_N1 reaction is one-step. The opposite is true!

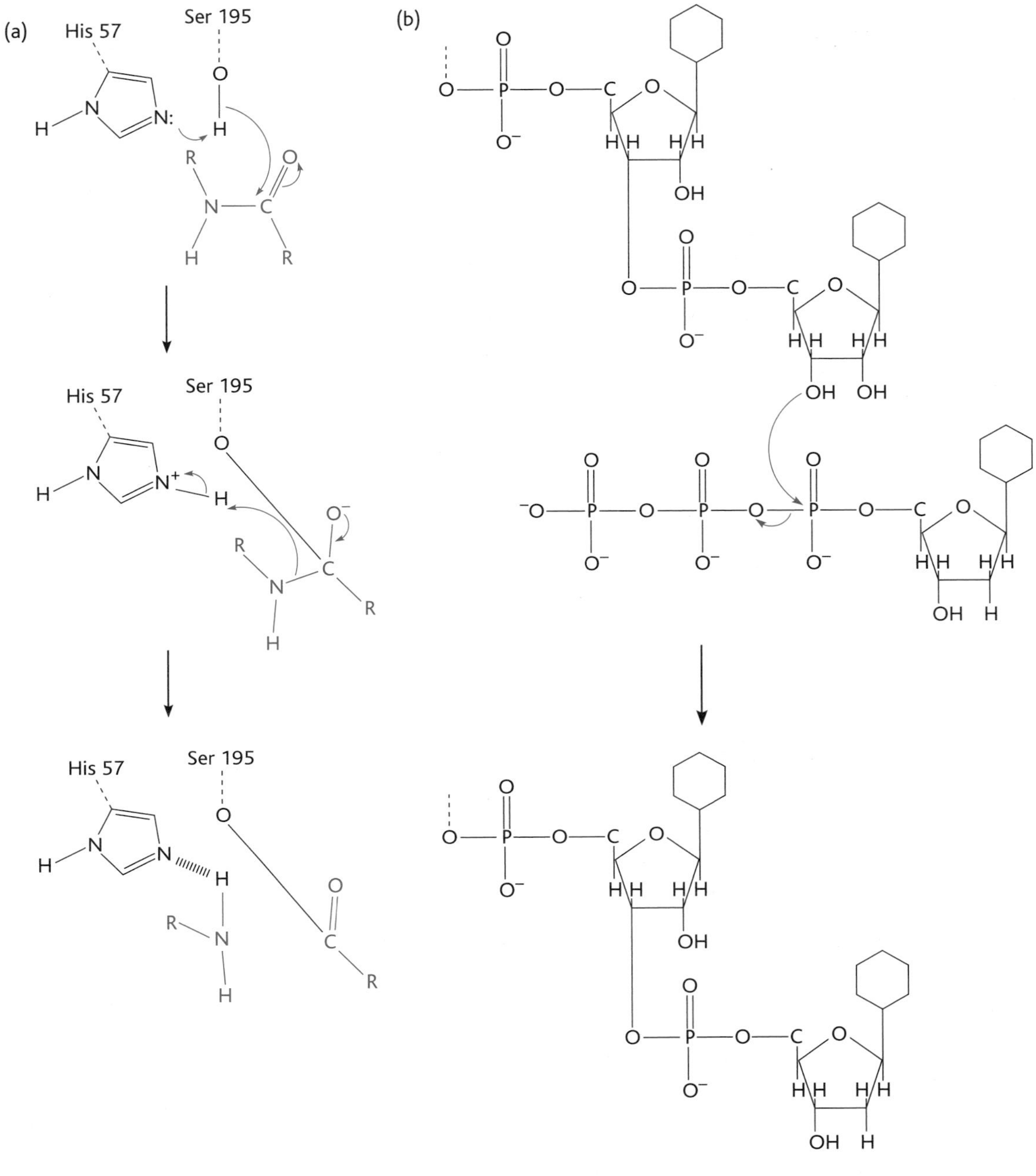

Figure 18.6 Examples of nucleophilic attack in biological systems. (a) The cleavage of a peptide bond by the enzyme chymotrypsin. Chymotrypsin features His and Ser residues in its active site. Notice how the Ser mounts a nucleophilic attack on the carbonyl carbon of a peptide bond, which precipitates cleavage of a C–N bond. Cleavage of the C–N bond results in the formation of two peptide fragments. (b) The synthesis of DNA from an RNA primer. Notice how the –OH attached to the ribose group of the RNA primer mounts a nucleophilic attack upon a deoxyribonucleotide molecule, forming an O–P bond; this bond subsequently links the deoxyribonucleotide (the first nucleotide of the DNA strand) to the RNA primer.

BOX 18.1 Substitution and chemical carcinogenesis

Proper cell growth and division requires the accurate replication and transmission of DNA from a parent cell to its daughters. Errors in the DNA sequence can jeopardize the proper functioning of the cell, leading to a possible breakdown in the regulation of a cell and its division–a prerequisite for many types of cancer.

But how can errors in DNA sequences arise? Some chemical substances are called **carcinogens**–they are known to directly increase the risk of cancer. Some carcinogens have their effect by participating in substitution reactions with components of DNA.

For example, look at Figure 18.7(a), which shows how the carcinogen *N*-methylnitrosourea drives a substitution reaction involving the nucleotide base guanine. This alkyation reaction (that is, the addition of an alkyl group such as $-CH_3$) follows an S_N1 reaction mechanism.

The insertion of inappropriate groups of atoms during a substitution reaction causes inappropriate base-pairing, as illustrated in Figure 18.7(b), which results in the distortion of the three-dimensional double-helical structure of the DNA. Such distortion impairs the accurate replication of DNA, leading to possible mutations during DNA replication. It is these DNA mutations that can lead to changes in gene function or gene regulation, which can lead to the onset of cancer.

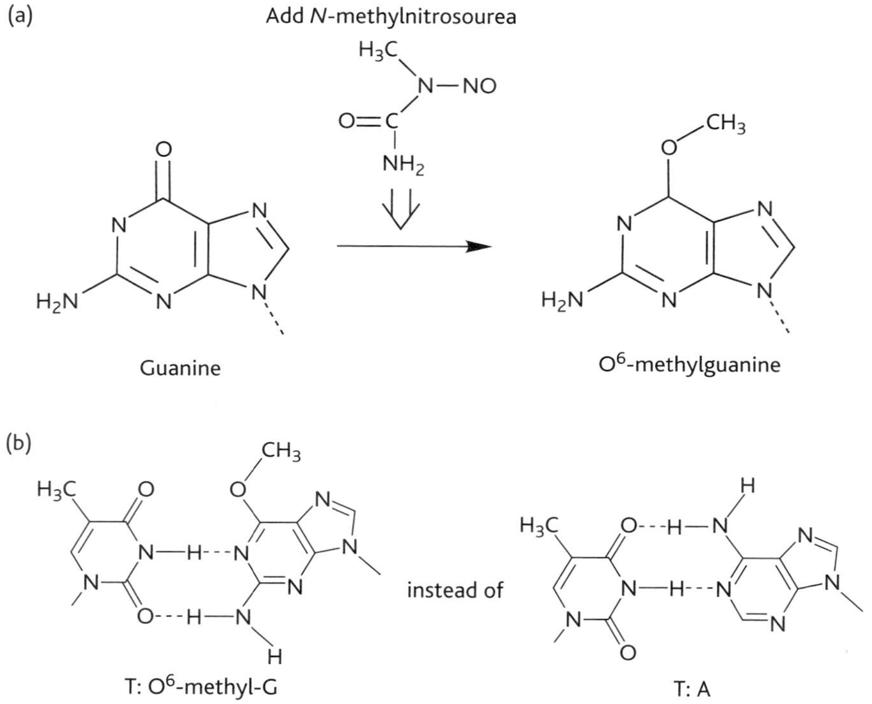

Figure 18.7 (a) The carcinogen *N*-methylnitrosourea drives a substitution reaction involving the nucleotide base guanine, forming the modified base O^6-methylguanine. (b) Such substitution reactions impair the accurate replication of DNA, leading to possible mutations arising during DNA replication. For example, thymine (T) mispairs with O^6-methylguanine instead of adenine (A).

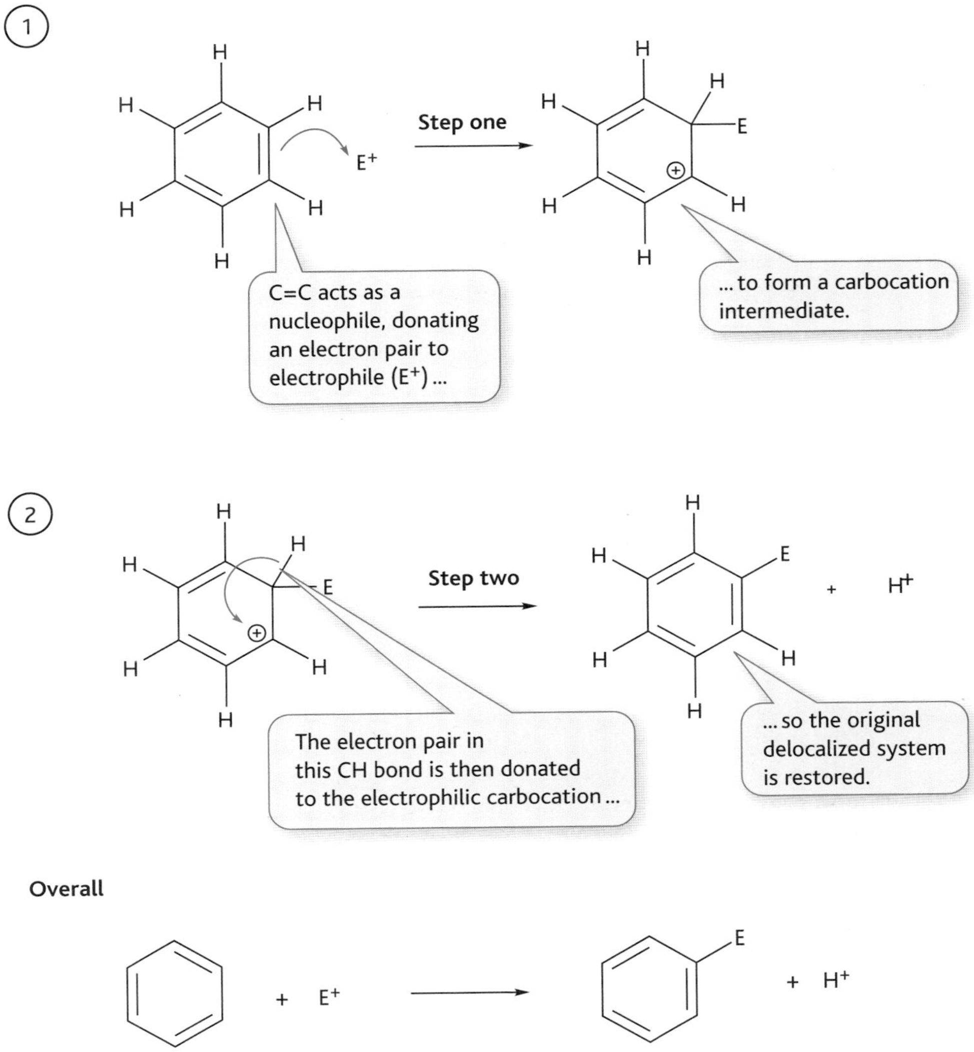

1

Step one

C=C acts as a nucleophile, donating an electron pair to electrophile (E^+) ...

... to form a carbocation intermediate.

2

Step two

The electron pair in this CH bond is then donated to the electrophilic carbocation ...

... so the original delocalized system is restored.

Overall

$$C_6H_6 + E^+ \longrightarrow C_6H_5E + H^+$$

Figure 18.8 The general mechanism for the substitution of an electrophile (represented by the symbol 'E^+') for a hydrogen atom on benzene.

Notice how this reaction is a two-step process, involving a stable carbocation intermediate.

Electrophiles that can undergo substitution reactions with benzene-based compounds include highly reactive (and highly unstable) groups including Br^+, Cl^+, and NO_2^+ (the nitronium ion)[6].

- ***During electrophilic aromatic substitution, two electrons in the pi system of an aromatic ring structure attack an electrophile.***

Electrophilic substitution reactions can be a useful tool in modifying biological molecules for the purposes of characterizing events that are happening at the

6 We usually encounter Br and Cl as negatively charged Br^- and Cl^- ions, which are relatively stable. By contrast, the extremely rare Br^+ and Cl^+ are far less stable, and react readily.

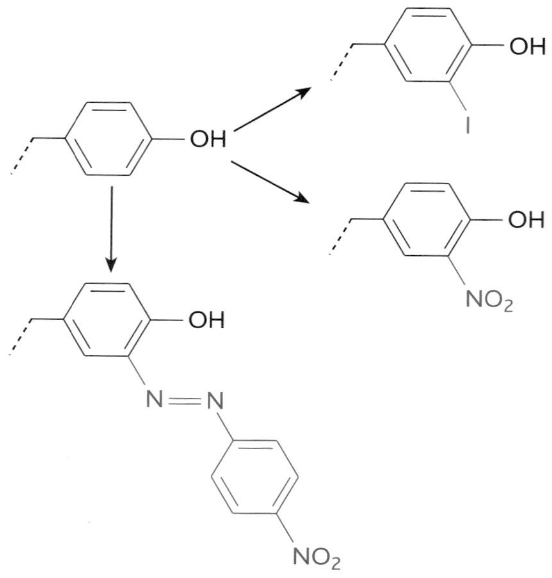

Figure 18.9 Examples of modified tyrosine side-chains, as can be achieved through electrophilic substitution.

molecular or cellular level in biological systems. For example, the amino acid tyrosine can be modified through electrophilic substitution to give the kinds of products illustrated in Figure 18.9. Such modification of tyrosine might lead to the alteration of the properties of a protein of which it is part, allowing us to study different aspects of the behaviour of that protein–for example, the location in the cell to which it is transported after synthesis (which may, then, give a clue to its function). The modification of biological molecules in this way is called **bioconjugation**.

Let's now consider addition, a second class of reaction mechanism by which organic compounds react.

18.3 Addition

During an addition reaction, two molecules combine with one another such that the product contains *all* the atoms of both molecules; the component atoms of one molecule are **added** to the atoms of another:

$$A + B \rightarrow AB$$

The most common and widespread addition reactions involve the addition of small molecules to the double bonds of **alkenes**.

We see in Chapter 5 how alkenes are unsaturated hydrocarbons: compounds comprising atoms of carbon and hydrogen, which contain at least one **double bond**.

We see in Section 17.3 how a double bond behaves as a nucleophile: the electron pair occupying the π orbital of the double bond can be donated to an electrophile, while the σ bond remains intact, so the two atoms originally connected by the double bond remain joined. (Look again at Figure 18.8 and notice how the double bond acts as a nucleophile during the electrophilic substitution of benzene.)

Let's see the addition process that occurs when the alkene ethene (1)

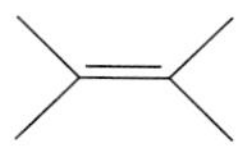

(1)

reacts with hydrogen bromide, HBr, to yield bromoethane, CH_3CH_2Br, according to the following reaction scheme:

$$CH_2CH_2 + HBr \rightarrow CH_3CH_2Br$$

The reaction is a two-step process. We begin with step 1:

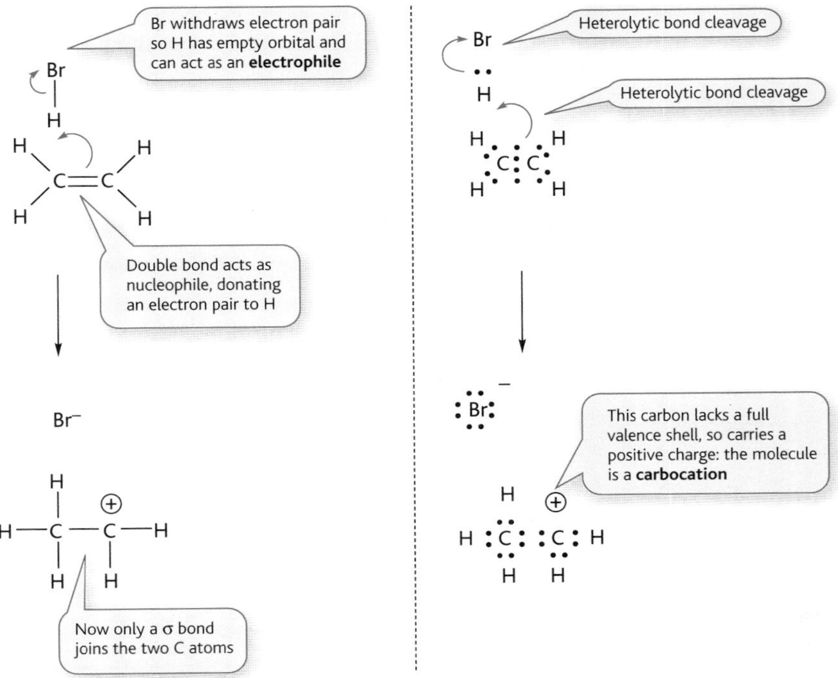

Notice how, at the end of step 1, the carbon atom to which the hydrogen atom has *not* added now carries a positive charge: it no longer has a full valence shell because the electron pair it was originally sharing in the π bond is now incorporated in a new bond between the other carbon and the hydrogen atom of HBr.

Remember: we call the intermediate formed at the end of step 1 a **carbocation**–a carbon-based cation.

Let's now consider step 2:

bromide ion acts as a **nucleophile** ...

... and carbocation acts as an **electrophile** – it has an empty orbital.

Notice how the bromide ion formed by the heterolytic cleavage of the H–Br bond now acts as a nucleophile: it donates a pair of electrons to the carbocation. (Look carefully and notice that the right-hand carbon has an empty orbital; the electrons previously in this orbital vacated when the π bond was broken during step 1.) The carbocation simultaneously acts as an electrophile: it has an empty orbital and can accept the pair of electrons being donated by the bromide ion.

What do you notice about the relative locations in the product (bromoethane) of the H and Br atoms, which started out as the HBr molecule? During the addition reaction, the H and Br atoms attach to **opposite ends** of the double bond: one atom joins to the carbon atom at one end of the double bond, and the other atom joins to the carbon atom at the other end of the double bond. We say that hydrogen bromide is added **across** the double bond.

- *During an addition reaction, the components of a molecule become added across a double bond of another molecule.*
- *During the course of the reaction, one component of the molecule being added acts as an electrophile, and accepts a pair of electrons from the double bond.*

Addition across a carbonyl double bond

A double bond does not always behave as the nucleophile (the initial electron donor) during an addition reaction. Let's consider what can happen during addition across the carbonyl double bond in aldehydes and ketones.

The carbonyl group is highly polarized; the electronegative oxygen atom pulls electrons towards it, leaving the carbonyl carbon relatively electron-deficient:

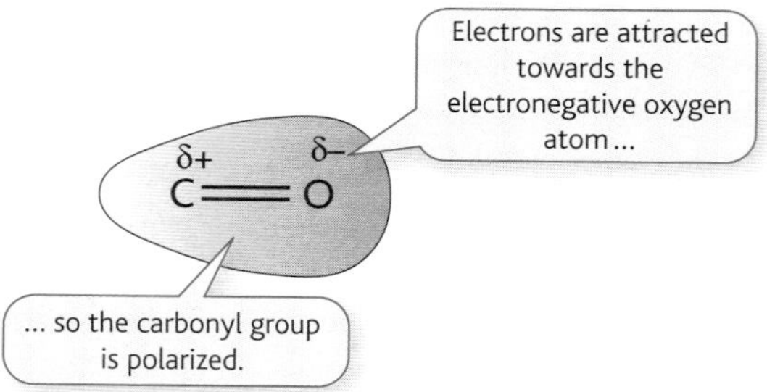

Figure 18.10 shows how this polarization enables the carbonyl carbon to act as an electrophile; it loses its share of the electron pair that comprises the π bond between C and O, and accepts an electron pair from a nucleophile instead. As you look at this reaction, notice the following:

- The reaction proceeds via two steps; the first step forms a tetrahedral carbon intermediate
- In the tetrahedral carbon intermediate the oxygen now has a full negative charge and is able to attract a proton
- The addition of this proton stabilizes the intermediate, forming an addition product.

Remember: we use 'Nu^-' to denote a generic nucleophile, just as we use 'R' to denote a generic alkyl group.

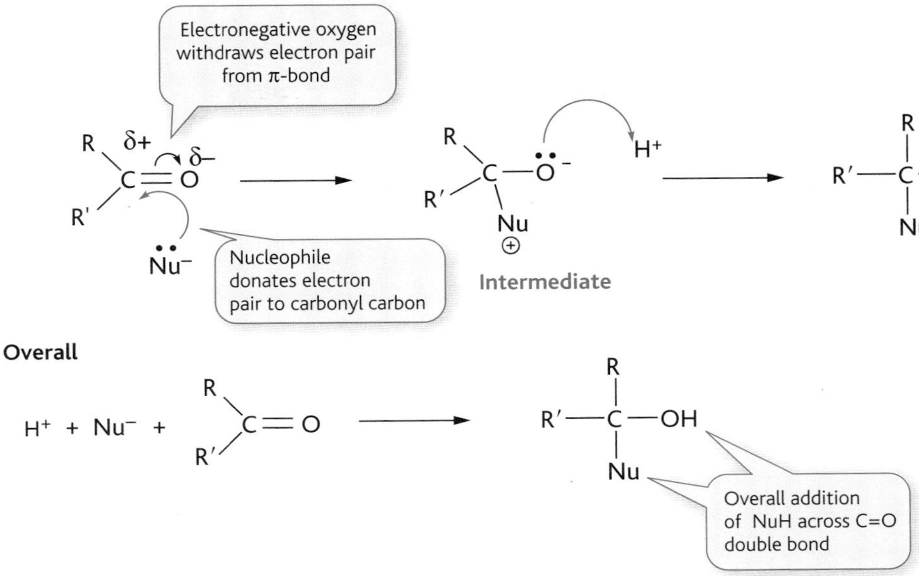

Figure 18.10 The polarization of the C=O bond in the carbonyl group enables the carbonyl carbon atom to act as an electrophile. The electronegative O withdraws electrons, giving the carbonyl carbon a partial positive charge. This positive charge is then a target for attack by a nucleophile, as illustrated here. Notice how the reaction results in the addition of two species–Nu and H–across the double bond.

Overall, we see the addition of Nu^-H^+ across the C=O bond.

Let's take a real example, using the nucleophile CN^-:

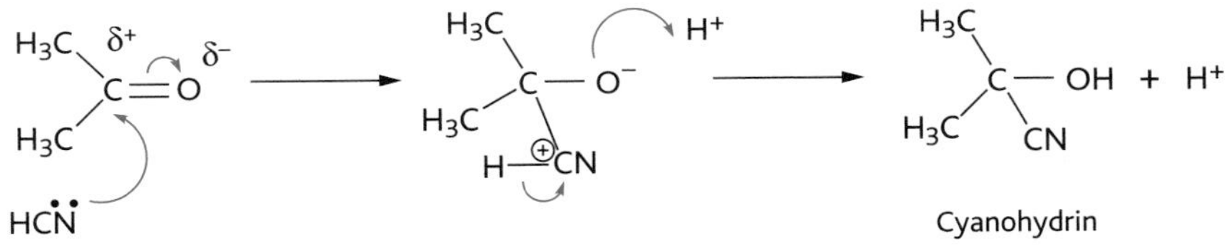

This addition reaction yields a cyanohydrin, which contains the cyano functional group, –C≡N.

Cyanohydrins are used as natural defence weapons in plants, as described in Box 18.2.

Addition processes that give the net addition of the components of a **water molecule** across a double bond are called **hydration** reactions:

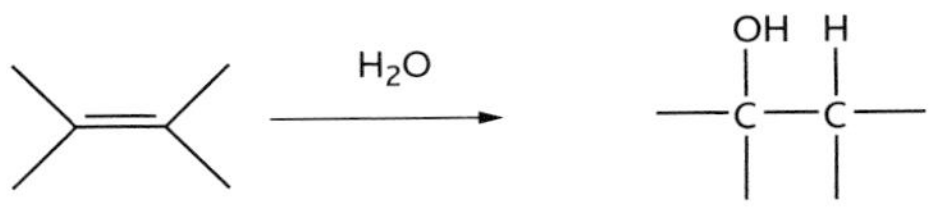

Look out for an example of a hydration reaction on p. 636.

The addition of non-polar molecules across a double bond: inducing polarization

In the examples above, we see how addition across a double bond involves partially or fully charged electrophiles or nucleophiles. Does this mean that non-polar

BOX 18.2 In the name of self-defence: cyanide production by plants

A number of plants have been found to use members of the cyanohydrin family as a unique method of self-defence. Some plants contain enzymes that catalyze the process of cyanogenesis; Figure 18.11 shows how the breakdown of cyanohydrins generates hydrogen cyanide, which is toxic to many insects and animals (and humans!).

It is thought that injured plants initiate cyanogenesis to prevent further damage being inflicted by microbes or herbivores (plant-eating animals).

Cyanohydrins are produced in plants from the degradation of a group of compounds called cyanoglucosides: a family of sugar molecules which contain the cyano (C≡N) functional group.

For example, cassava is a root crop found widely in South America. Cassava contains the cyanoglucoside linamarin. When its self-defence mechanism is activated, cassava hydrolyzes linamarin into an unstable cyanohydrin, which undergoes cyanogenesis to liberate the toxic hydrogen cyanide, as illustrated in Figure 18.12. Notice that this reaction is the reverse of an addition to a carbonyl group.

Cassava must be cooked before consumption to degrade the linamarin it contains; eaten raw, the root of cassava is poisonous and potentially fatal: our digestive system metabolizes linamarin, producing toxic hydrogen cyanide in our body.

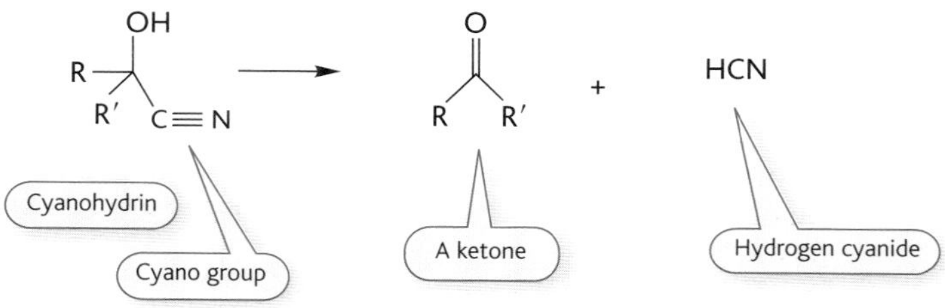

Figure 18.11 The breakdown of cyanohydrins generates hydrogen cyanide, which is highly toxic to many insects and animals.

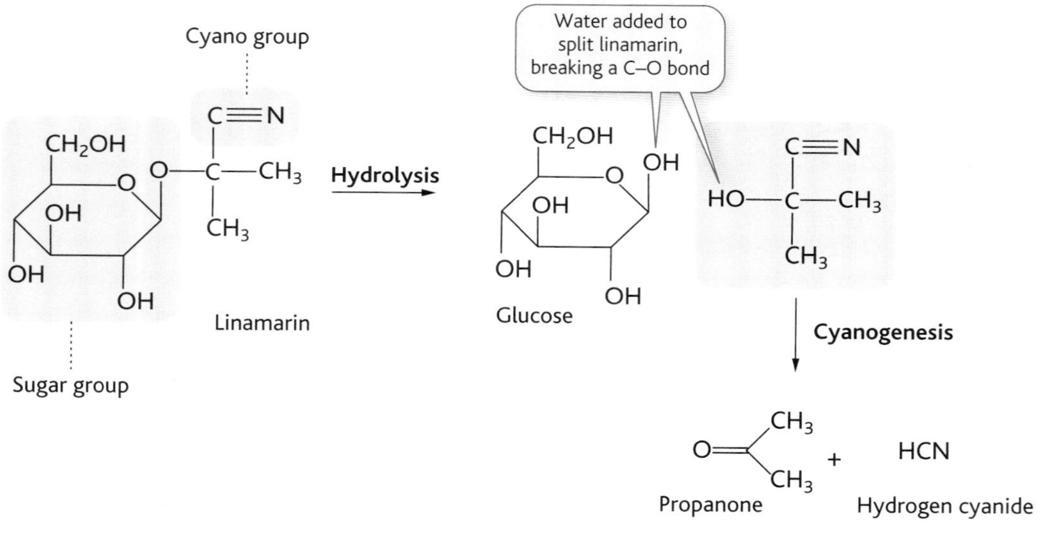

Figure 18.12 Linamarin undergoes hydrolysis to form an unstable cyanohydrin, which undergoes cyanogenesis to liberate the toxic hydrogen cyanide.

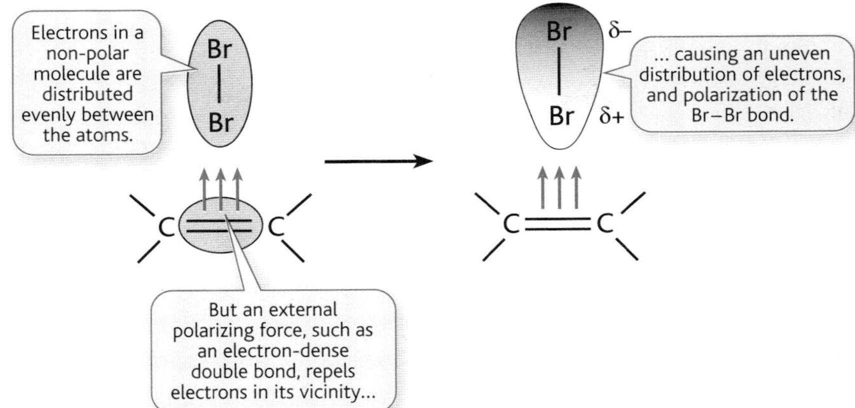

Figure 18.13 A double bond can effect polarization of a non-polar molecule such as Br_2. A double bond acts as a polarizing force because it is electron rich.

species, which carry no overall charge, cannot be involved in an addition reaction across a double bond? The answer is 'no'. So why is this the case?

Electron-rich entities such as double bonds are able to **induce** polarity in otherwise non-polar molecules such that they can then participate in an addition reaction. Areas of high electron density (such as a double bond) repel electrons in their vicinity, so that the area from which electrons are repelled acquires a relatively positive charge[7]. We say that the bond has been **polarized**.

Consider a molecule of bromine, Br–Br. We see in Section 3.12 how molecules that comprise two identical atoms are **non-polar** because both atoms have the same electronegativity, and so exert the same 'pull' on the electrons shared between them. However, if we expose a non-polar molecule such as Br_2 to an 'external' polarizing force–such as a double bond–the distribution of electrons is no longer even. Look at Figure 18.13 and notice how a double bond can effect polarization of a non-polar molecule such as Br_2 as the Br_2 molecule approaches the double bond.

As a result of induced polarization, the Br_2 molecule behaves in just the same way as the permanent dipole, HBr: the bromine atom carrying the partial negative charge withdraws the shared electron pair, so that the bromine atom carrying the partial positive charge can readily obtain an empty orbital, and can behave as an electrophile.

Consequently, most Br_2 adds across the double bond of a molecule such as ethene in the way HBr does, as illustrated in Figure 18.14.

- ***A double bond can induce polarization in a previously non-polar molecule, after which the molecule may become involved in an addition reaction across the double bond.***

Self-check 18.2 Write down the two-step reaction mechanism for the addition of HCl to ethene, whose structure is shown on p. 620.

7 Remember: areas of high electron density *repel* electrons in their vicinity, just as the south poles of two bar magnets repel one another if in close proximity. We explore polarization in more detail in Section 3.12.

Step 1

Step 2

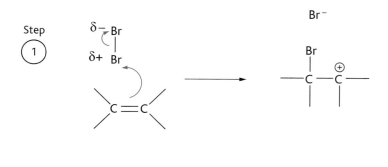

Figure 18.14 The addition of Br_2 across the double bond of ethene.

18.4 Elimination

In an elimination reaction, a molecule *loses* some of its component atoms to form a compound with a new double bond, while the components that are eliminated come together to form one or more new, separate molecules:

$$\text{A-X—X-B} \longrightarrow \text{X=X} + \text{A—B}$$

We learn more about unsaturated and saturated compounds in Chapter 5.

Notice how elimination provides a means of generating an unsaturated compound (containing a double bond) from a previously saturated compound.

In essence, an elimination reaction is the reverse of an addition reaction: in an addition reaction, the components of one molecule are *added* to another molecule; in an elimination reaction, the components of one molecule are *eliminated* from another molecule.

For example, compounds containing an ester functional group, COOR, undergo hydrolysis in the presence of an acid to form a carboxylic acid and an alcohol, where the alcohol is the compound eliminated from the ester during the course of the reaction. The final steps of this hydrolysis reaction are those in which the elimination reaction occurs, as illustrated in Figure 18.15.

So what movement of electrons do we see during an elimination reaction? Let's consider the reaction of a nucleophile with bromoethane. Once again, polarization

Figure 18.15 During the final steps in the hydrolysis of an ester an elimination reaction occurs, as depicted here. Notice how this reaction eliminates an alcohol (ROH), and results in the formation of a carboxylic acid, which features a new double C=O bond.

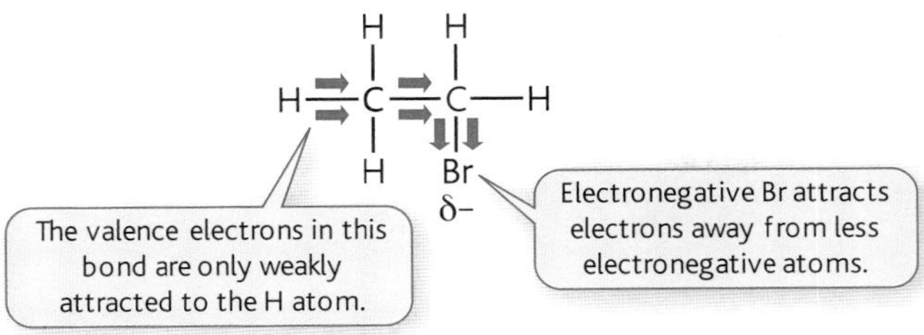

The valence electrons in this bond are only weakly attracted to the H atom.

Electronegative Br attracts electrons away from less electronegative atoms.

Figure 18.16 The highly electronegative Br pulls electrons through the molecule towards itself, leading to an uneven distribution of electrons throughout the molecule as a whole. Consequently, the molecule is polar.

has a big part to play in the reaction. Look at the structural formula of bromoethane shown in Figure 18.16. Notice how the highly electronegative bromine atom pulls towards itself electrons from the rest of the molecule; these electrons are pulled through the bonds in the molecule rather than simply being pulled through space. As a result of this movement of electrons, the molecule becomes polarized. The other, less electronegative atoms in bromoethane attract electrons only weakly.

Now look at Figure 18.17 to see what happens when a nucleophile such as an OH^- ion, approaches. Notice how a double bond is formed as a pair of electrons move from being shared in a C–H bond to being shared in a C–C π bond. Also, notice how the nucleophile is acting as a base–it is extracting a proton from the compound it is attacking. (We see the significance of the nucleophile acting as a base shortly.)

The overall reaction is:

$$CH_3CH_2Br \rightarrow CH_2CH_2 + HBr$$

As a result of the reaction, a molecule of HBr is **eliminated** from a molecule of CH_3CH_2Br. This elimination reaction only happens in the presence of a **catalyst**–something that makes the reaction happen more quickly. In this instance, the catalyst is the nucleophile OH^-. The reacting species must also be heated. We find out more about the action of catalysts–and why they are required–in Chapter 14.

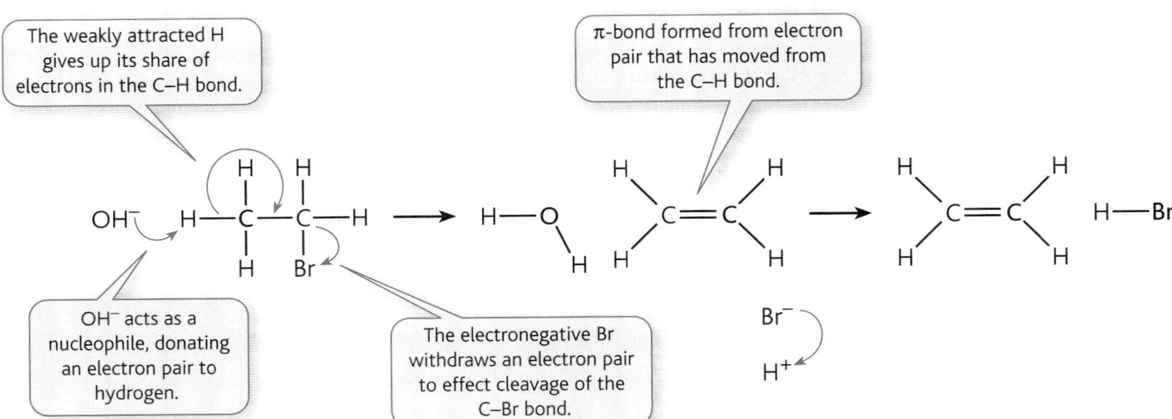

Figure 18.17 Nucleophilic attack on bromoethane by OH^-. The nucleophilic attack leads to the formation of a C=C bond and the elimination of HBr.

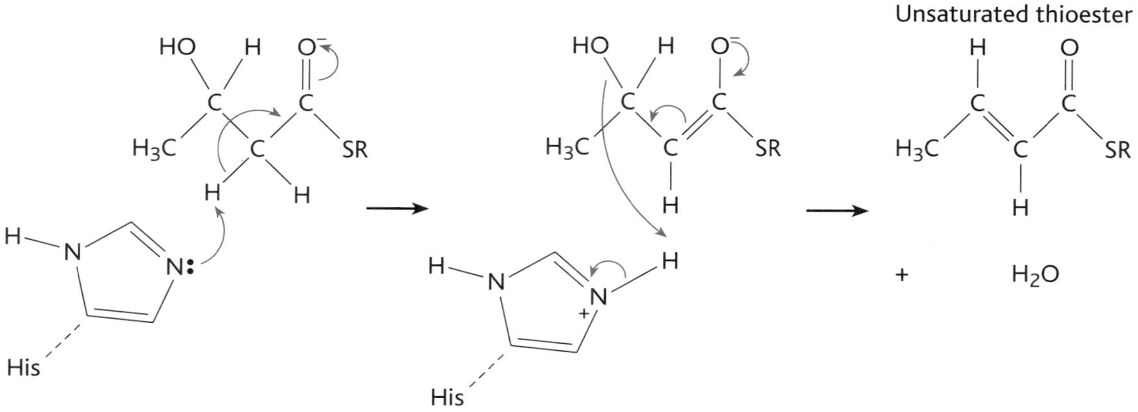

Figure 18.18 An example of a biological elimination reaction, which occurs during the synthesis of fatty acids. Notice how the reaction is catalyzed by a His side chain (which would be part of an enzyme active site), which acts as the nucleophile. Adapted from McMurry and Begley, 2005.

- ***Elimination reactions result in the introduction of a double bond in the molecule from which some components have been eliminated.***

Elimination reactions are also prevalent in biological systems. For example, Figure 18.18 illustrates how elimination occurs during the biological synthesis of fatty acids, which, as we see in Chapter 7, are a key component of lipids. Figure 18.18 shows how such reactions are enzyme-catalyzed; in this instance, the side chain of a histidine residue (which forms part of the enzyme itself) acts as the nucleophile. Also, notice how this elimination reaction results in the formation of a new carbon–carbon double bond, and how a water molecule is eliminated.

One-step versus two-step elimination

We see in Section 18.2 how nucleophilic substitution reactions can proceed via a one- or two-step reaction mechanism. Elimination reactions can occur in an entirely analogous way–either proceeding via an intermediate to give a two-step reaction (denoted the E1 mechanism), or directly (without an intermediate) through a one-step reaction (denoted the E2 mechanism)[8]. These two mechanisms are illustrated in Figure 18.19.

Elimination versus substitution

The commonalities between substitution and elimination reactions don't end with them both exhibiting two possible reaction mechanisms. Look at Figure 18.20,

8 Once again, the '1' and '2' denote the number of species involved in the rate-limiting step, in an equivalent way to the naming of nucleophilic substitution reactions. The E2 reaction is a one-step reaction in which two species–the nucleophile and target compound–are involved; the E1 reaction is a two-step reaction in which only one species–the target compound–is involved in the rate-limiting step.

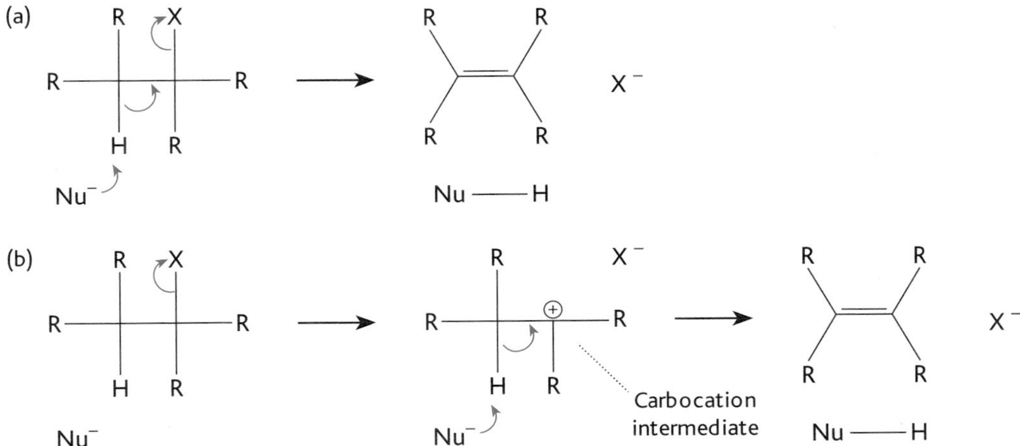

Figure 18.19 One-step versus two-step elimination reactions. (a) A generalized E2 (one-step) reaction, in which a nucleophile (Nu^-) attacks a hydrogen atom, and a leaving group (X) leaves, all within a single step. (b) A generalized E1 (two-step) reaction. During the first step, a carbocation is formed following the departure of the electrophilic leaving group, X. During the second step, the nucleophile attacks a hydrogen attached to the carbocation, and electrons in the C–H bond move to form a π bond between the two carbon atoms.

which compares S_N1 and E1, and S_N2 and E2 mechanisms. What do you notice? Both types of reaction–substitution and elimination–involve the attack by a nucleophile on a target molecule. The key difference, however, is where on the molecule the attack occurs.

Look again at Figure 18.20, and see where the nucleophile is attacking the target molecule in each case. Notice how, in a substitution reaction, the nucleophile attacks a carbon atom; the nucleophile becomes joined to the carbon atom, while a leaving group that had previously been attached to the carbon atom is lost. By contrast, during an elimination reaction, the nucleophile attacks a proton and *not* a carbon atom: it acts as a base to extract the proton from the molecule, but does not join to the molecule itself. However, the same leaving group is lost from the molecule.

- ***If a nucleophile attacks a carbon atom, a substitution reaction occurs, and the nucleophile becomes joined to the target molecule.***
- ***If a nucleophile attacks a hydrogen atom, an elimination reaction occurs, and the nucleophile does not become joined to the target molecule.***

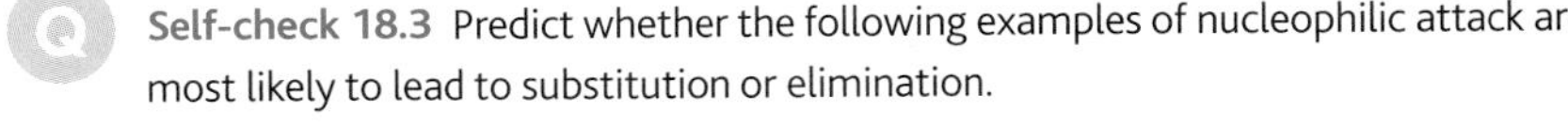

Self-check 18.3 Predict whether the following examples of nucleophilic attack are most likely to lead to substitution or elimination.

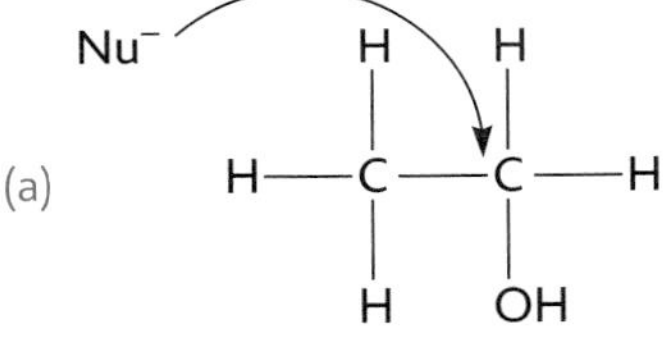

(b)

So, what determines whether a nucleophile attacks a carbon atom or a hydrogen atom? There are several factors that influence the behaviour of the nucleophile, but the primary factor is how effective it is as a base. If the nucleophile is a strong base, it will be very effective at extracting a proton from the molecule, and so it is most likely that elimination will occur. By contrast, if the nucleophile is a relatively weak base, it is more likely that the nucleophile will attack the carbon atom and substitution will occur.

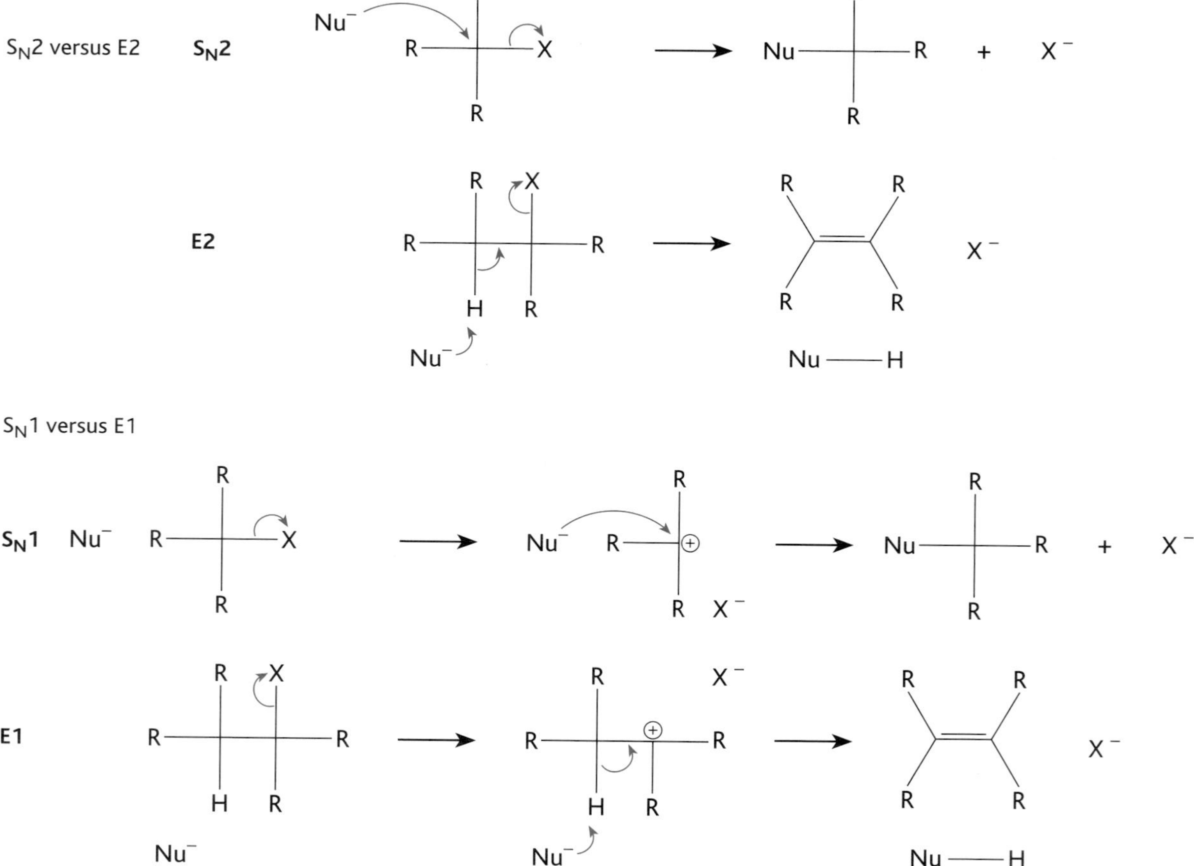

Figure 18.20 A comparison of the S_N1 and S_N2 substitution mechanisms with the E1 and E2 elimination mechanisms. Notice how the nucleophile attacks a carbon atom during substitution, but a hydrogen atom during elimination.

18.5 Condensation

We now move on to consider our final type of reaction, the condensation reaction. A condensation reaction is a combination of both an addition reaction *and* an elimination reaction; two molecules combine into a single product, *and* a further molecule is eliminated during the reaction:

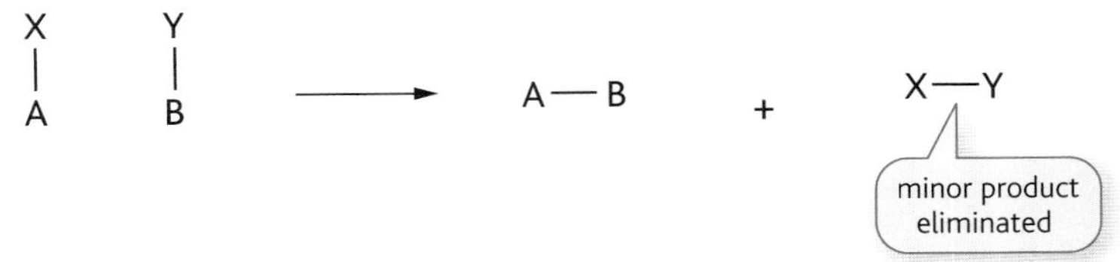

$$X{-}A + Y{-}B \longrightarrow A{-}B + X{-}Y$$

minor product eliminated

During a condensation reaction, the eliminated molecule is always small.

Unlike addition reactions, condensation reactions leave the majority of the two initial reactants intact once the reaction has occurred. What condensation reactions provide, however, is a way of neatly **joining** two molecules together, to form a larger product.

One of our essential concepts is that small, repetitive units join together to form larger, more complicated structures. Condensation reactions are one way of joining small units together; they play a vital part in the polymerization of amino acids to form long polypeptide chains, and the joining of nucleotides to form nucleic acids, such as DNA and RNA.

Look at Figure 18.21, which shows how the polymerization of amino acids to form polypeptides is an example of a condensation reaction. The reaction is initiated by the non-bonding electrons on a nitrogen atom of one amino acid, which are attracted to the slightly positive carbon atom in the carboxylic acid of a second amino acid. The unstable tetrahedral intermediate is stabilized by the elimination of water and the reforming of the carbonyl group. Thus:

- The two amino acid residues join to one another–they undergo **addition**.
- A water molecule is **eliminated** during the course of the reaction[9].

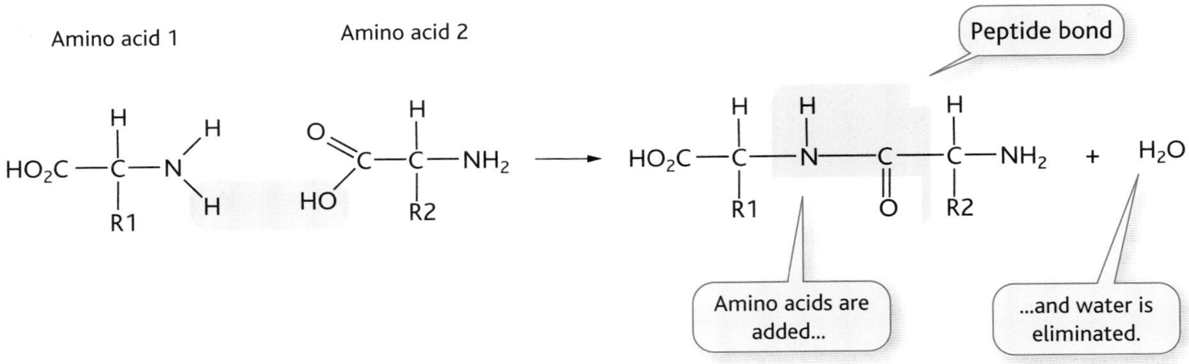

Amino acid 1

Amino acid 2

Peptide bond

$$HO_2C{-}CH(R1){-}NH_2 + HO{-}C(=O){-}CH(R2){-}NH_2 \longrightarrow HO_2C{-}CH(R1){-}NH{-}C(=O){-}CH(R2){-}NH_2 + H_2O$$

Amino acids are added...

...and water is eliminated.

Figure 18.21 The polymerization of amino acids to form polypeptides is an example of a condensation reaction. Two amino acid residues join together and a molecule of water is eliminated.

9 Condensation processes that result in the elimination of water are also called **dehydration** reactions.

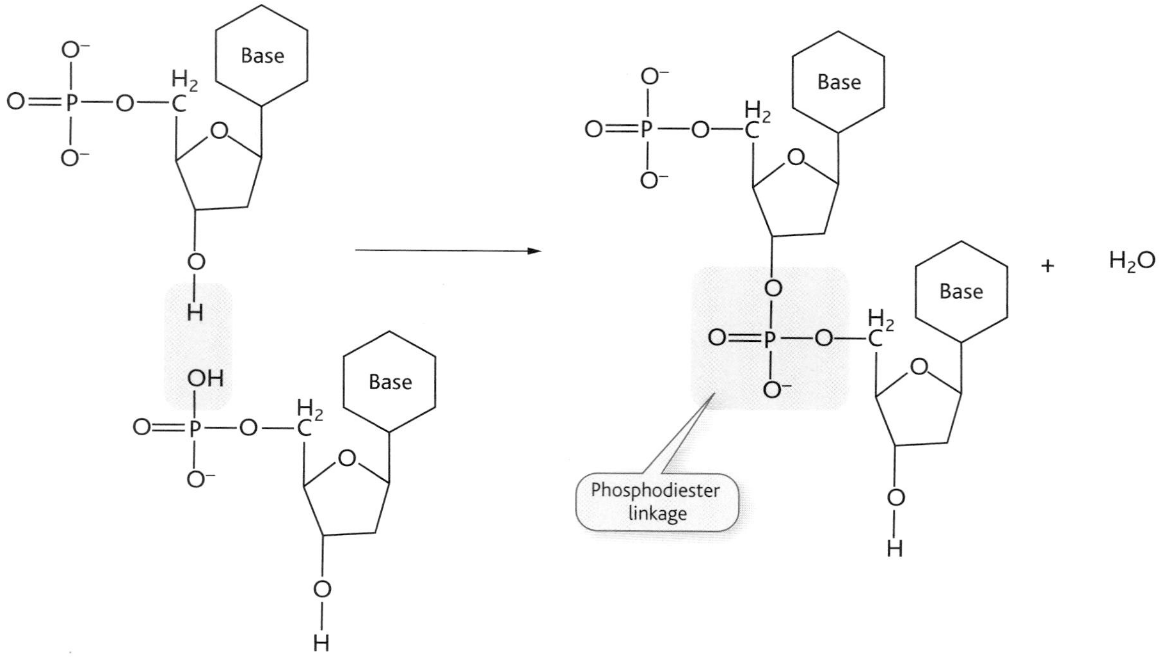

Figure 18.22 The joining of nucleotides to form nucleic acids such as DNA and RNA is also a condensation reaction.

Contrast this reaction with the addition reaction to the carbonyl group of aldehydes and ketones described in Section 18.3 (p. 622) where the tetrahedral intermediate is stabilized by addition of a proton to the carboxyl oxygen to form an addition product, rather than by the elimination of water to give an elimination product.

The joining of a nitrogen atom of one amino acid residue to the carbonyl group of another amino acid residue is called a **peptide bond**.

Now look at Figure 18.22, which shows how the joining of nucleotides to form nucleic acids such as DNA and RNA is also a condensation reaction. Notice that this reaction is also a dehydration reaction because a water molecule is eliminated during the joining of the two nucleotides.

We also see an example of a condensation reaction when we consider **esterification**, the condensation reaction of a hydroxyl group with a carboxyl group, which occurs when we react an alcohol with a carboxylic acid. We see in Chapter 6 how esters comprise a modified carboxyl group, in which the hydrogen atom of the hydroxyl group is replaced by an alkyl group:

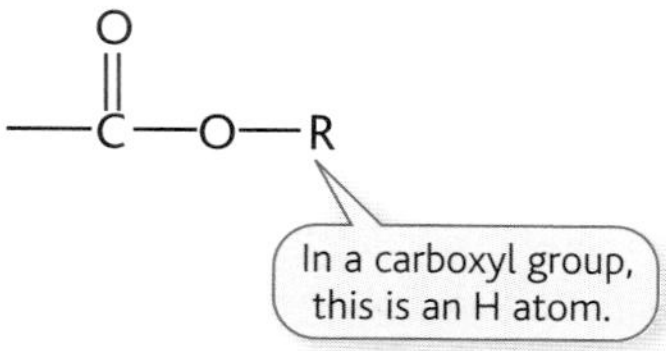

$$-\overset{\overset{\large O}{\|}}{C}-O-R$$

In a carboxyl group, this is an H atom.

The process of esterification, the formation of an ester from the condensation of an alcohol with a carboxylic acid, is illustrated in Figure 18.23. Many of us find ourselves thankful for the process of esterification; it was the condensation of the

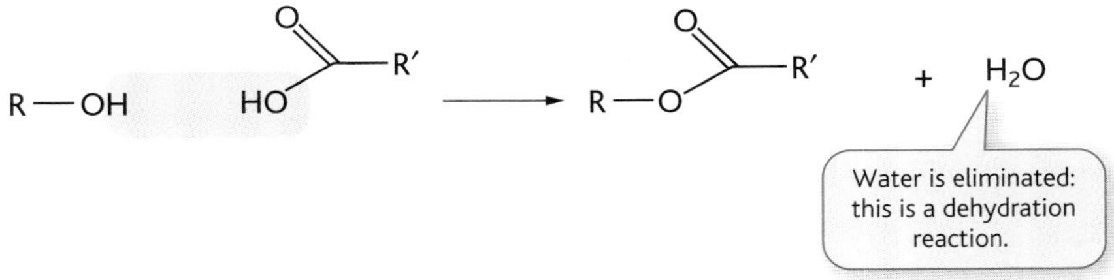

$$R{-}OH \quad HO{-}C(=O){-}R' \longrightarrow R{-}O{-}C(=O){-}R' + H_2O$$

Water is eliminated: this is a dehydration reaction.

Figure 18.23 Esterification, the condensation reaction of a hydroxyl group with a carbonyl group. Notice how water is eliminated, making this a dehydration reaction.

carboxylic acid ethanoic acid with the hydroxyl group of salicylic acid that first led to the synthesis of the painkiller aspirin:

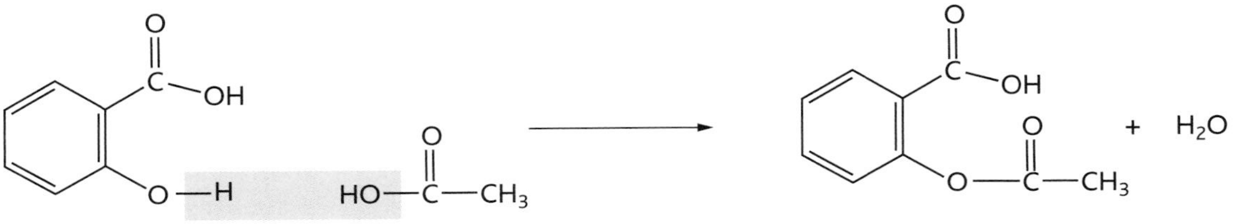

Salicylic acid + Ethanoic acid ($HO{-}C(=O){-}CH_3$) ⟶ Acetylsalicylic acid **(aspirin)** + H_2O

Aspirin is far from being the most important ester in our lives. Box 18.3 explains how esters of phosphoric acid are vital to the biological processes occurring in all living organisms.

BOX 18.3 Phosphoric acid esters

Carboxylic acids aren't the only acids to form esters; other organic acids can do likewise. Phosphoric acid (2)

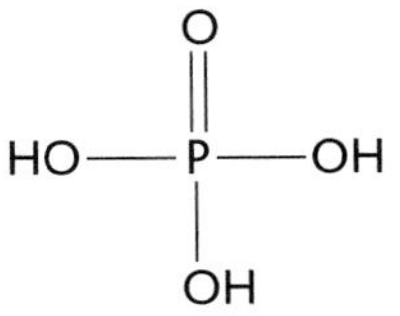

$$HO{-}P(=O)(OH){-}OH$$

(2)

is a major industrial chemical that is widely used in the production of fertilizers, detergents, and insecticides. Its use as a fertilizer underlines its importance to all life processes.

In aqueous solution, phosphoric acid has three ionizable hydroxyl groups, which are readily stabilized by resonance[10], as illustrated in Figure 18.24. (The presence of the three ionizable groups means that phosphoric acid behaves as what we call a triprotic acid.) When ionized, we typically call these groups phosphate groups.

Each of the OH groups can form esters with alcohols such as methanol and ethanol. Importantly, however, the OH groups of two phosphoric acid molecules can also react with each other to form larger molecules linked by ester groups, as depicted in Figure 18.25.

Such molecules are called phosphoric acid anhydrides, species that have two atoms joined by a bridging oxygen atom, and in which each of the two atoms is also joined to a second oxygen atom by a double bond:

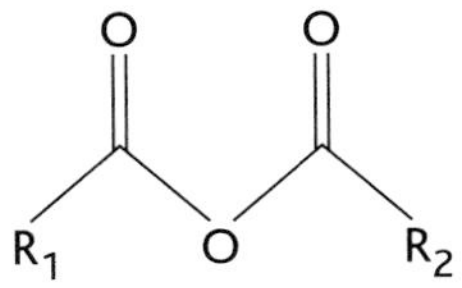

$$R_1{-}C(=O){-}O{-}C(=O){-}R_2$$

10 The more resonance structures you can draw, the more stable the compound.

BOX 18.3 **Continued**

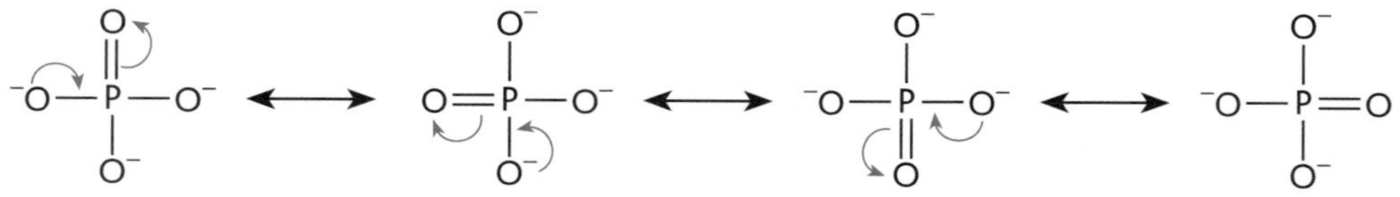

Figure 18.24 In aqueous solution, the three hydroxyl groups of phosphoric acid become ionized, and the overall structure is stabilized by resonance, as illustrated here.

Acid anhydrides are particularly susceptible to hydrolysis, a characteristic that is exploited by biological systems, as we see below.

But what have phosphoric acids to do with biological systems? In the cellular environment, phosphoric acid esters don't exist in the form shown in Figure 18.25; instead, they join to molecules such as sugars and ribonucleic acids to form an array of different compounds.

For example, glucose 6-phosphate, one of the intermediates in the metabolism of glucose during glycolysis, is a phosphoric acid (or phosphate) ester, formed from the condensation of an –OH group of glucose and a phosphate group from ATP, as illustrated in Figure 18.26.

Indeed, ATP is also a phosphoric acid ester. Together with the closely-related adenosine diphosphate (ADP) and adenosine monophosphate (AMP). ATP plays a central role in the transfer of energy within the cell: it is the formation and hydrolysis of the phosphate ester groups of molecules such as ATP and ADP that drives the metabolic reactions of the cell.

We learn more about how molecules such as ATP provide energy for life in Box 18.5, and in more detail in Chapter 13.

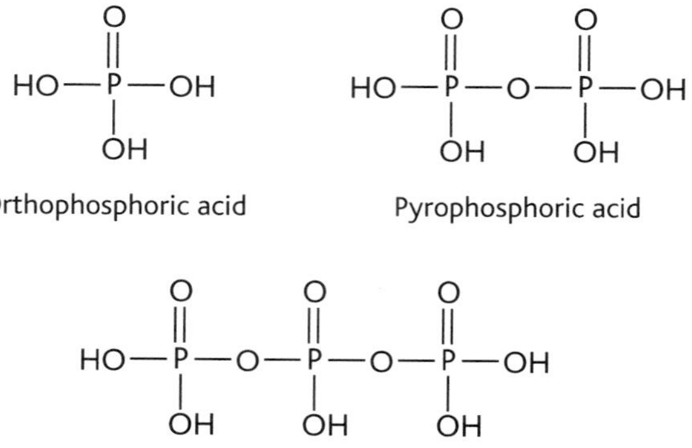

Orthophosphoric acid

Pyrophosphoric acid

Tripolyphosphoric acid

Figure 18.25 Multiple phosphoric acid molecules can react together to form larger molecules, called phosphoric acid anhydrides.

Hydrolysis: reversing condensation

Condensation reactions are a vital part of life: they allow the assembly of complicated biological molecules from smaller, simpler molecules–proteins from amino acids; polysaccharides from simple sugars; nucleic acids from simple nucleotides.

BOX 18.3 **Continued**

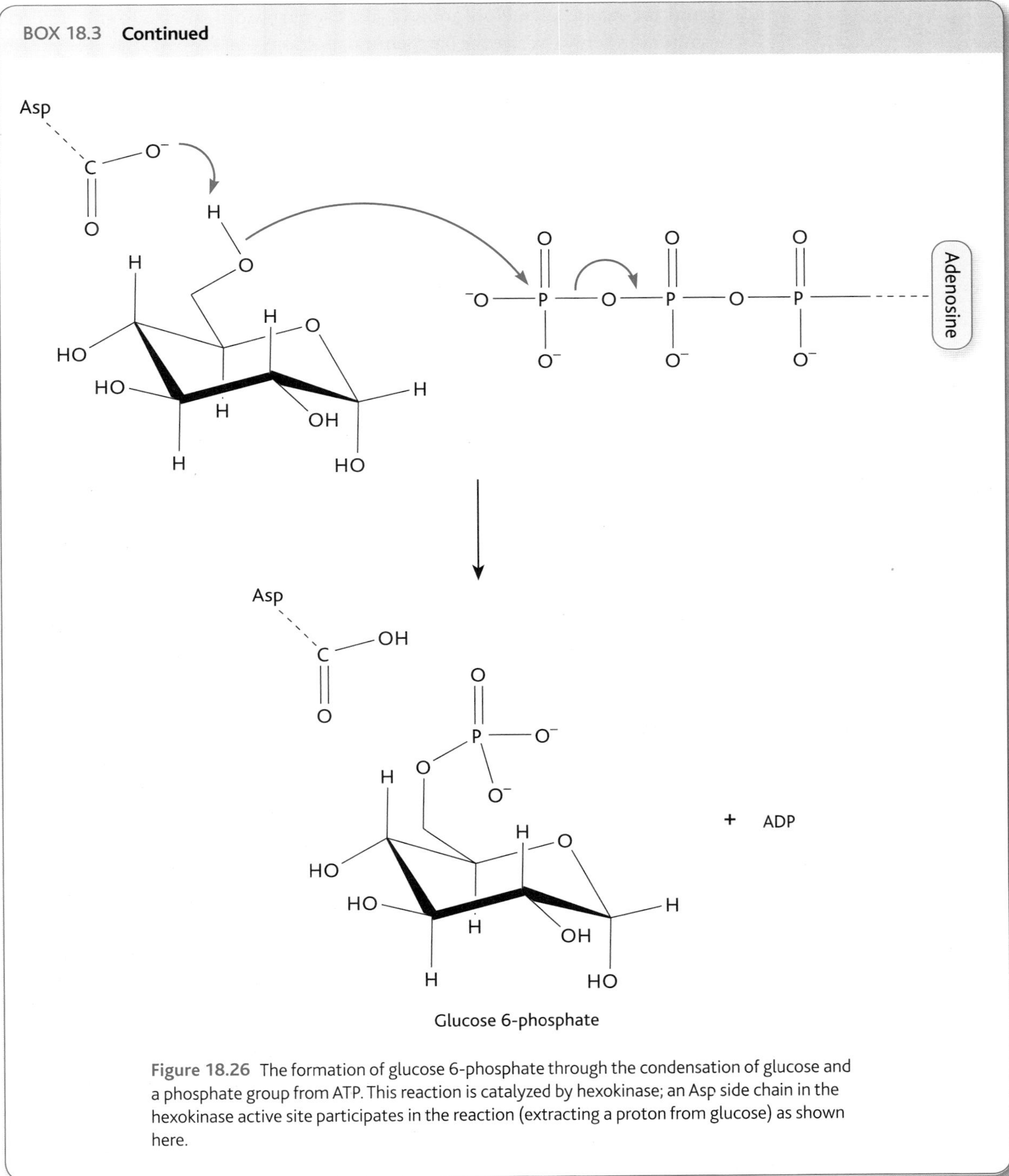

Figure 18.26 The formation of glucose 6-phosphate through the condensation of glucose and a phosphate group from ATP. This reaction is catalyzed by hexokinase; an Asp side chain in the hexokinase active site participates in the reaction (extracting a proton from glucose) as shown here.

However, of equal importance to the assembly of large, complicated molecules is the *breakdown* of large molecules into their component parts.

Every time we eat something we ingest foods that comprise large, complicated molecules: carbohydrate-rich foods such as pasta and potatoes contain polymers comprising small sugar molecules; protein-rich foods such as meat and dairy products contain polymers of amino acids. The purpose of eating is to give our body a source of energy to drive the biochemical reactions that keep us alive. To generate energy from the foods we eat, however, these foods must be broken down into their small component molecules.

We see above how the assembly of small subunits into large molecules is achieved through condensation reactions. Most biological condensation reactions are dehydration reactions, during which water is eliminated. The breakdown of large molecules is the exact *opposite* of condensation: large molecules are split into their smaller subunits. The splitting of large molecules is effected not by the elimination of water, but by it being *added*.

'Hydro' = 'water'; 'lysis' = 'splitting'. 'Hydrolysis' = 'splitting with water'.

The process of breaking down large molecules through the addition of water is called **hydrolysis**. For example, we see in Chapter 7 how the fats and oils found in foods are triacylglycerols, which are formed from the joining of fatty acids with the three-carbon alcohol, glycerol:

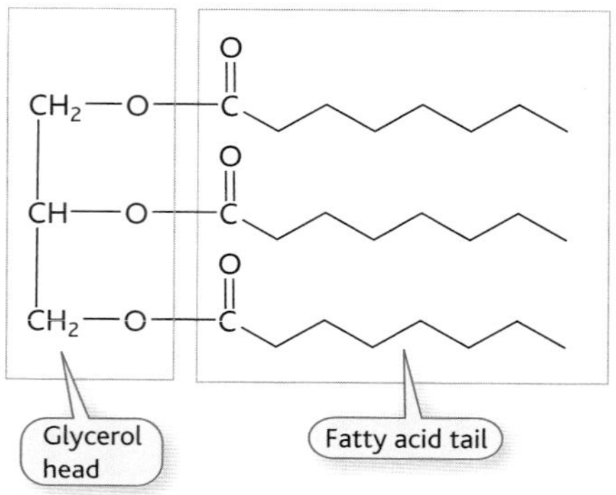

CH_2—O—C(=O)~~~
CH—O—C(=O)~~~
CH_2—O—C(=O)~~~

Glycerol head

Fatty acid tail

When we consume fats and oils, our bodies hydrolyze them into their glycerol and fatty acid components:

BOX 18.4 Hydration versus hydrolysis

Don't get hydration and hydrolysis confused. Hydration involves the addition of water to a larger molecule. By contrast, hydrolysis involves the cleavage of a larger molecule, the splitting process involving water. Look at Figure 18.27, which depicts this difference.

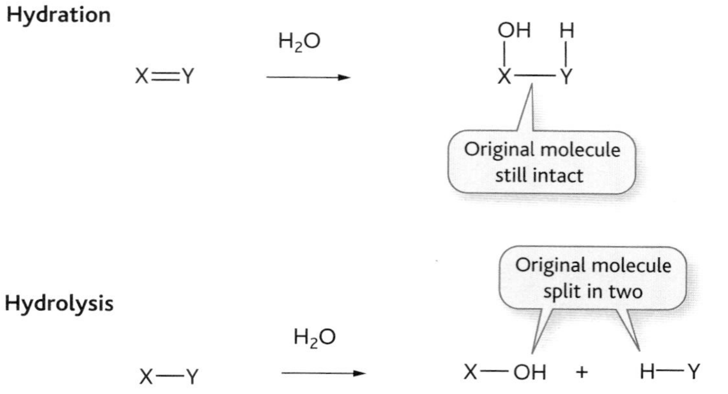

Hydration

$$X{=}Y \xrightarrow{H_2O} X(OH){-}Y(H)$$

Original molecule still intact

Hydrolysis

$$X{-}Y \xrightarrow{H_2O} X{-}OH + H{-}Y$$

Original molecule split in two

Figure 18.27 The difference between hydration and hydrolysis. Hydration involves the addition of water to a larger molecule. By contrast, hydrolysis involves the cleavage of a larger molecule, the splitting process involving water.

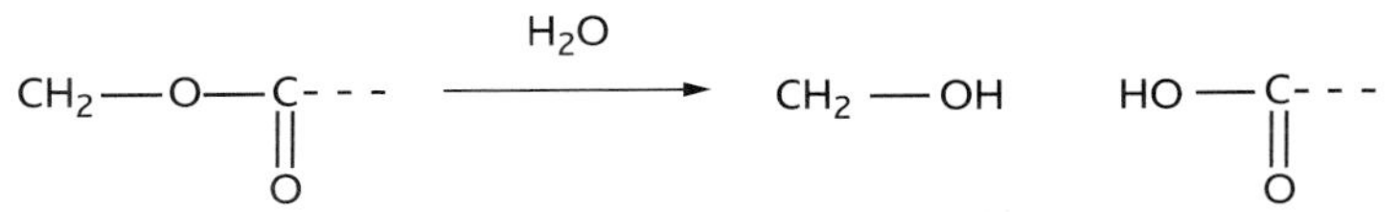

$$CH_2{-}O{-}\underset{\displaystyle \|\atop O}{C}{-}{-}{-} \xrightarrow{H_2O} CH_2{-}OH \quad HO{-}\underset{\displaystyle \|\atop O}{C}{-}{-}{-}$$

Notice how, after hydrolysis, a C–O bond has been broken, and replaced with the components of a water molecule (H and OH): the bond has been split with water.

The mechanism of hydrolysis involves an addition reaction followed by an elimination reaction. Look at Figure 18.28, which depicts the hydrolysis of ethanoyl ethanoate (whose common name is acetic anhydride). The reaction requires either an acid or a base catalyst.

During this reaction, the acid proton, which acts as a catalyst, is attracted to the non-bonding electrons on a carbonyl group. This makes the carbonyl group susceptible to nucleophilic attack, resulting in an addition reaction across the carbonyl C=O bond. This addition reaction generates a tetrahedral intermediate, as depicted in Figure 18.28. This intermediate then undergoes further rearrangement, giving the elimination of an ethanoic acid molecule.

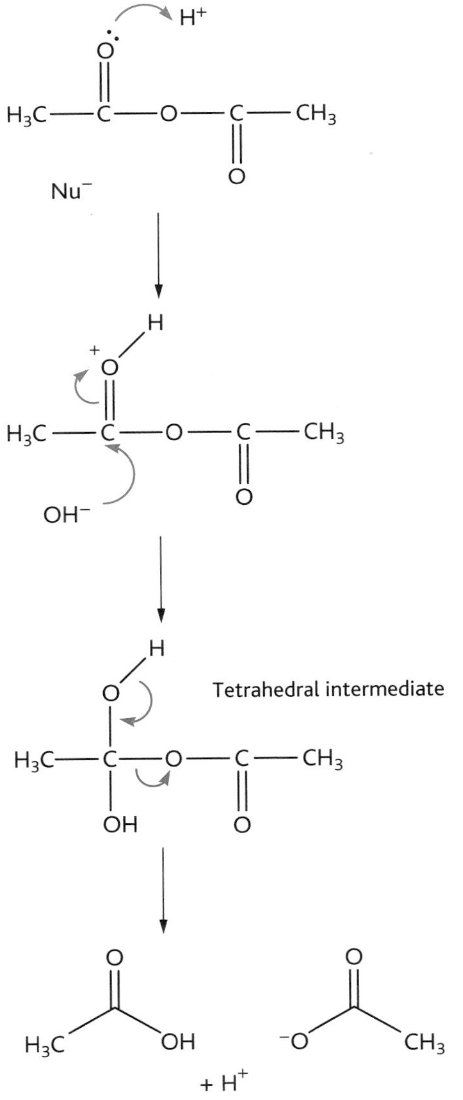

Figure 18.28 The mechanism of hydrolysis of ethanoyl ethanoate. Notice how the overall process of hydrolysis includes an addition reaction (which generates a tetrahedral intermediate) and an elimination reaction, whereby a molecule of ethanoic acid is released.

- ***During condensation, two molecules become joined to form a single product.***
- ***During hydrolysis, a single molecule is cleaved into two separate product molecules.***

One of the most important hydrolytic reactions in biological systems is the hydrolysis of ATP, as described in Box 18.5.

Self-check 18.4 Identify the reaction mechanisms that are represented by the following:

(a) $H_3C\text{—}CH_2\text{—}Br + OH^- \longrightarrow H_3C\text{—}CH_2\text{—}OH + Br^-$

(b) $H_2N\text{—}CH_2\text{—}CO_2H + H_2N\text{—}CH(CH_3)\text{—}CO_2H \longrightarrow H_2N\text{—}CH_2\text{—}C(=O)\text{—}NH\text{—}CH(CH_3)\text{—}CO_2H + H_2O$

(c) $H_3C\text{—}CH{=}CH_2 + HBr \longrightarrow H_3C\text{—}CH_2\text{—}CH_2\text{—}Br$

BOX 18.5 Powering the body: the hydrolysis of ATP

Our daily food consumption is used in part to power the synthesis of the nucleotide adenosine triphosphate, ATP. ATP acts as the cell's source of energy: ATP is broken down to yield energy to drive chemical reactions in the cell, just as power stations burn fuels, such as coal, to ultimately produce energy in the form of electricity. (Unlike coal in power stations, however, ATP isn't stored in large quantities in the cell. Instead, a cell only has enough ATP to last it for a few minutes at any one time.)

Specifically, the 'breaking down' of ATP involves the hydrolysis of a phosphodiester bond that joins a phosphate group with its neighbour[11]–either the bond between the terminal phosphate group and the middle phosphate group, or the middle group with the phosphate group attached to the sugar, as illustrated in Figure 18.29. Hydrolysis of either or both of these P–O bonds is highly exergonic: it is associated with a large negative Gibbs free energy change (see p. 455). It is the energy from this highly exergonic reaction that is utilized to drive other endergonic reactions in the cell. We learn more about the energy associated with ATP hydrolysis in Box 13.9.

11 When representing these reactions on paper, rather than drawing out the full chemical structure of phosphoric acid every time, we use a convention in which the symbol P*i* represents a single phosphate ion, PO_4^{-3} (which is called orthophosphate), and PP*i* represents $P_2O_7^{-4}$ (which is called pyrophosphate).

BOX 18.5 **Continued**

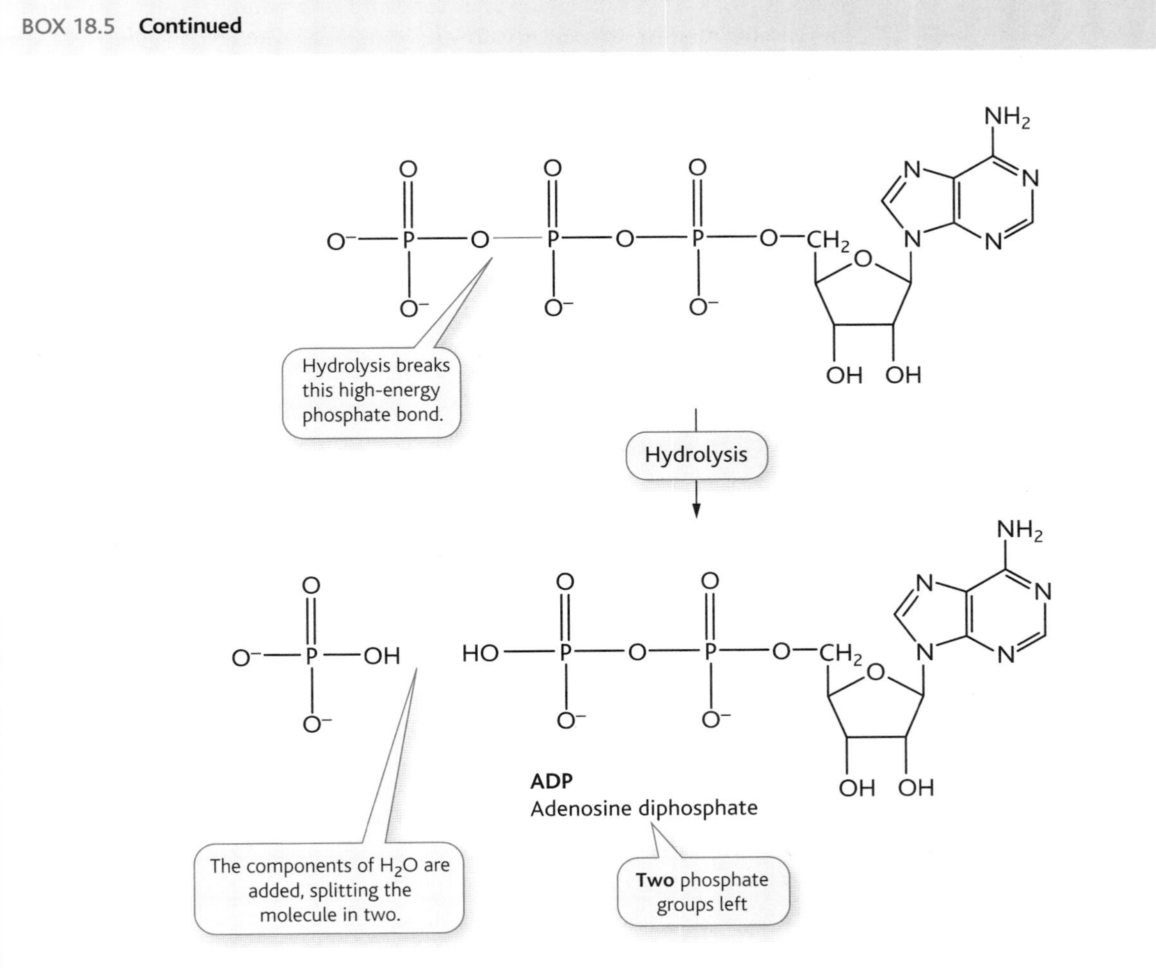
Hydrolysis breaks this high-energy phosphate bond.
Hydrolysis
ADP
Adenosine diphosphate
The components of H2O are added, splitting the molecule in two.
Two phosphate groups left

Figure 18.29 The hydrolysis of a phosphodiester bond in ATP. Both the phosphodiester bonds between the terminal and middle phosphate groups, and the middle phosphate group and the phosphate group joined to the sugar, are considered 'high energy'. This figure denotes the hydrolysis of the first of these.

18.6 Biochemical reactions: from food to energy

At the start of Chapter 1, we are introduced to the molecular machine, ATP synthase. ATP sythase, and the synthesis of ATP that it mediates, represents the end-point of one of the most fundamental processes of life: the process through which our bodies generate energy from the food we consume. At the start of the process is glucose; at the end of the process is a molecular machine whose rotation generates ATP. And between the two lies an interconnected network of biochemical pathways, which operate in concert to link one to the other.

Respiration–the process by which we generate energy from food–is summarized by the following reaction scheme:

$$C_6H_{12}O_6 + 6\,O_2 \rightarrow 6\,CO_2 + 6\,H_2O + \text{energy}$$

(where $C_6H_{12}O_6$ is glucose). However, beneath this seemingly simple scheme is a complicated multi-step reaction pathway, comprising three key stages:

- **Glycolysis**–the metabolism of glucose to form pyruvate
- The **citric acid cycle**–the metabolism of pyruvate to form CO_2 and H_2O
- The **electron transport pathway**

These reaction pathways, on which we depend to stay alive, provide numerous examples of the reaction mechanisms featured earlier in this chapter in action. We now look at a couple of these processes in detail to see how the reaction mechanisms we encounter earlier in the chapter come into play and how our bodies achieve the remarkable feat of producing energy, in the form of the energy 'carrier' ATP, from glucose.

The reaction mechanisms underpinning glycolysis

Glycolysis is a series of coupled reaction which converts glucose, a six-carbon-containing molecule, into two molecules of pyruvate, each of which contains three carbon atoms.

Step 1

The first reaction in glycolysis is the reaction of glucose and ATP to form glucose 6-phosphate and ADP, as illustrated in Figure 18.30. The reaction is catalyzed by the enzyme hexokinase. This reaction is an example of a hydrolysis reaction, and results in the transfer of a phosphoryl group from ATP to glucose.

Look at Figure 18.30 to see what is happening at the molecular level during this reaction. At the active site of the enzyme, a carboxylic acid side chain on an aspartic acid residue acts as a base to remove the proton from the C6 hydroxyl group of the glucose molecule. It is through this step that the enzyme acts to catalyze the reaction in a chemical sense: the removal of the proton from the hydroxyl group activates the oxygen atom to which it was attached. The oxygen atom then acts as a nucleophile, and attacks the phosphate atom of the terminal phosphoryl group on ATP.

Once linked to a negatively charged phosphate group in the form of glucose 6-phosphate, the glucose is now trapped within the cell and is on the way to being broken down into two molecules of pyruvate[12].

Remember: isomers are compounds that are assembled from the same atoms, but in which the atoms are connected to each other in different ways.

Step 2

In the next step of the process, glucose 6-phosphate is isomerized to fructose 6-phosphate in a reaction catalyzed by phosphohexose isomerase, as illustrated in Figure 18.31(a).

The conversion of glucose to fructose involves the opening of the ring structure of glucose. Figure 18.31(b) shows how, initially, opening of the ring gives a compound with an aldehyde group; this group then rearranges to form an enol group–an alkene group with a hydroxyl group attached to one end of the double bond. Finally, this enol structure rearranges to the open chain fructose structure with a

We learn more about isomers in Chapter 10.

12 Don't get hung up on terminology: the terms pyruvic acid and pyruvate are often used interchangeably. The carboxylic acid form ($CH_3COCOOH$) is referred to as pyruvic acid. In the ionized state (CH_3COCOO^-) the carboxylate anion is referred to as pyruvate. The form taken by the molecule at any time depends upon the pH of the solution.

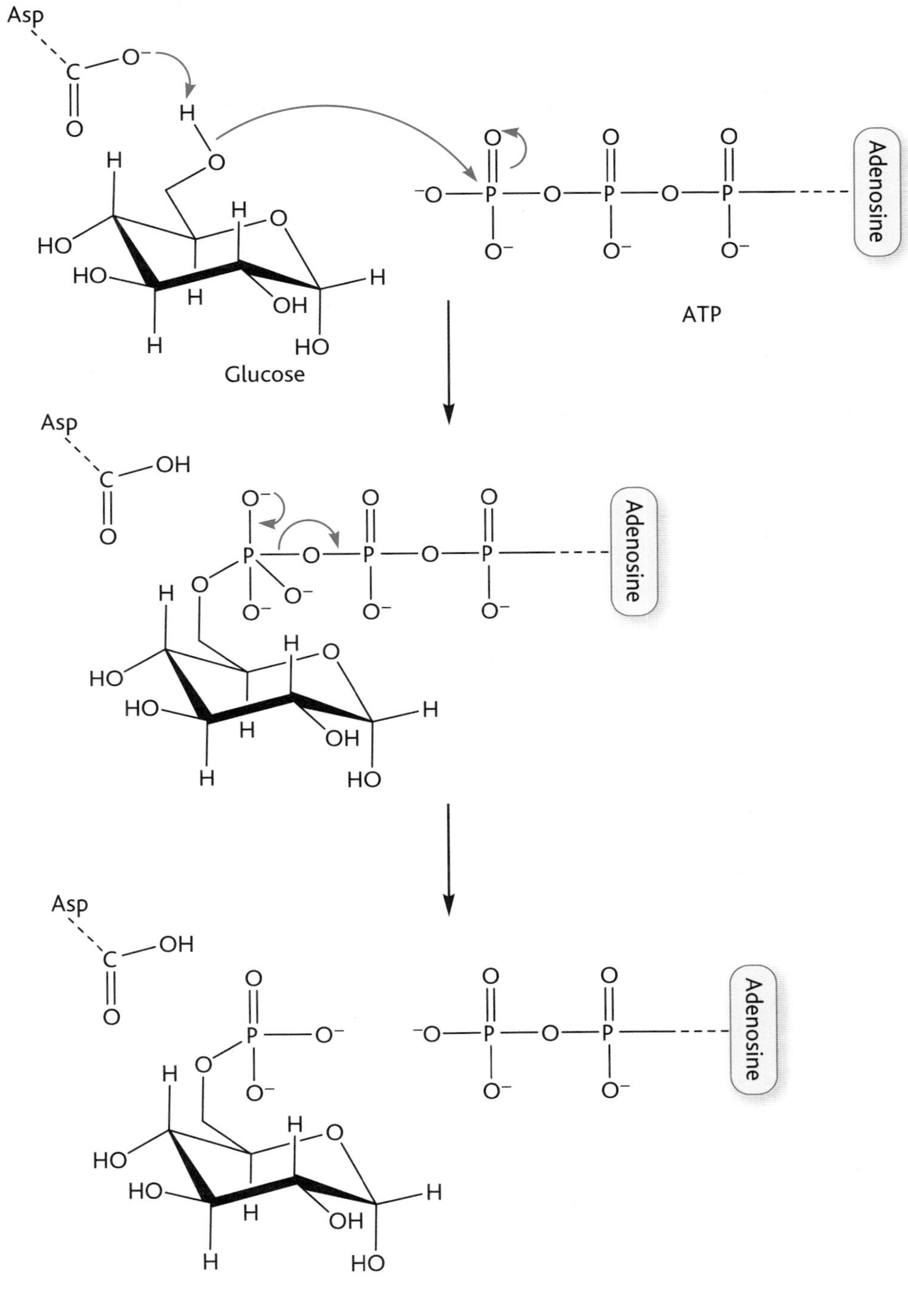
Asp
Glucose
ATP
Adenosine
Asp
Adenosine
Asp
Adenosine
Glucose 6-phosphate

Figure 18.30 The first step in glycolysis is the hexokinase-catalyzed phosphorylation of glucose to yield glucose 6-phosphate.

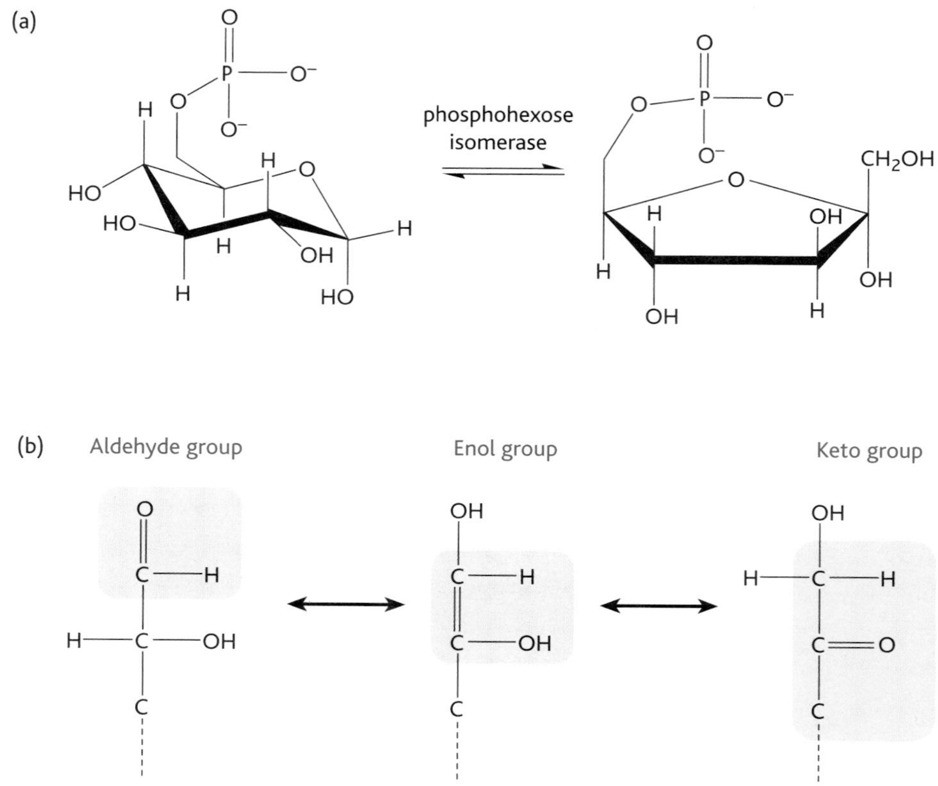

(a)

phosphohexose isomerase

(b) Aldehyde group Enol group Keto group

Figure 18.31 (a) Isomerisation of glucose 6-phosphate to fructose 6-phosphate. (b) This process involves keto–enol tautomerism in which the aldehyde group of open-chain glucose first rearranges to give an enol group, before rearranging once again to generate a keto group.

keto group (RCOR'). The open chain form of fructose can now cyclize to generate the more common representation of fructose; it is this form of fructose that we see as part of the structure of fructose 6-phosphate in Figure 18.31(a). Notice how this rearrangement has the overall effect of moving the carbonyl group one carbon down from the terminal carbon. When the opposite end of the open-chain sugar now loops round to react with the carbonyl group during cyclization, the resulting ring contains just five carbons, rather than the six we would see with the open-chain glucose structure (where cyclization would involve the terminal carbon).

Step 3

In the next step of glycolysis, fructose 6-phosphate is converted to fructose 1,6-bisphosphate by reaction with ATP. The mechanism of this reaction is similar to the one we discussed in the first step of glycolysis, and is catalyzed by phosphofructokinase. Look at Figure 18.32, and notice how this reaction leads to the addition of a second phosphoryl group to fructose 6-phosphate[13].

13 The numbers between the two parts of the compound name identify the carbon atoms to which the two phosphate groups are attached–in this case, carbons 1 and 6 of the fructose group.

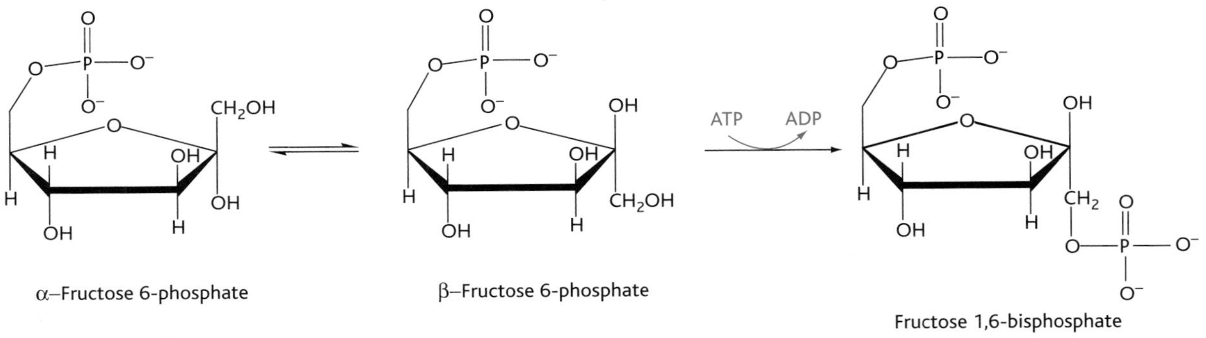

α–Fructose 6-phosphate

β–Fructose 6-phosphate

ATP ADP

Fructose 1,6-bisphosphate

Figure 18.32 Formation of fructose 1,6-bisphosphate involves conversion of α-fructose 6-phosphate to β-fructose 6-phosphate, and the subsequent reaction of β-fructose 6-phosphate with ATP, resulting in phophorylation at C1 of fructose. Once again the reaction is catalyzed by a kinase enzyme.

Step 4

With the formation of the fructose 1,6-bisphosphate, the stage is set for the splitting of the six-carbon sugar group into two three-carbon molecules. The reaction is catalyzed by the enzyme aldolase and involves breaking the bond between C3 and C4 in fructose, as illustrated in Figure 18.33(a). The breaking of this bond is catalyzed

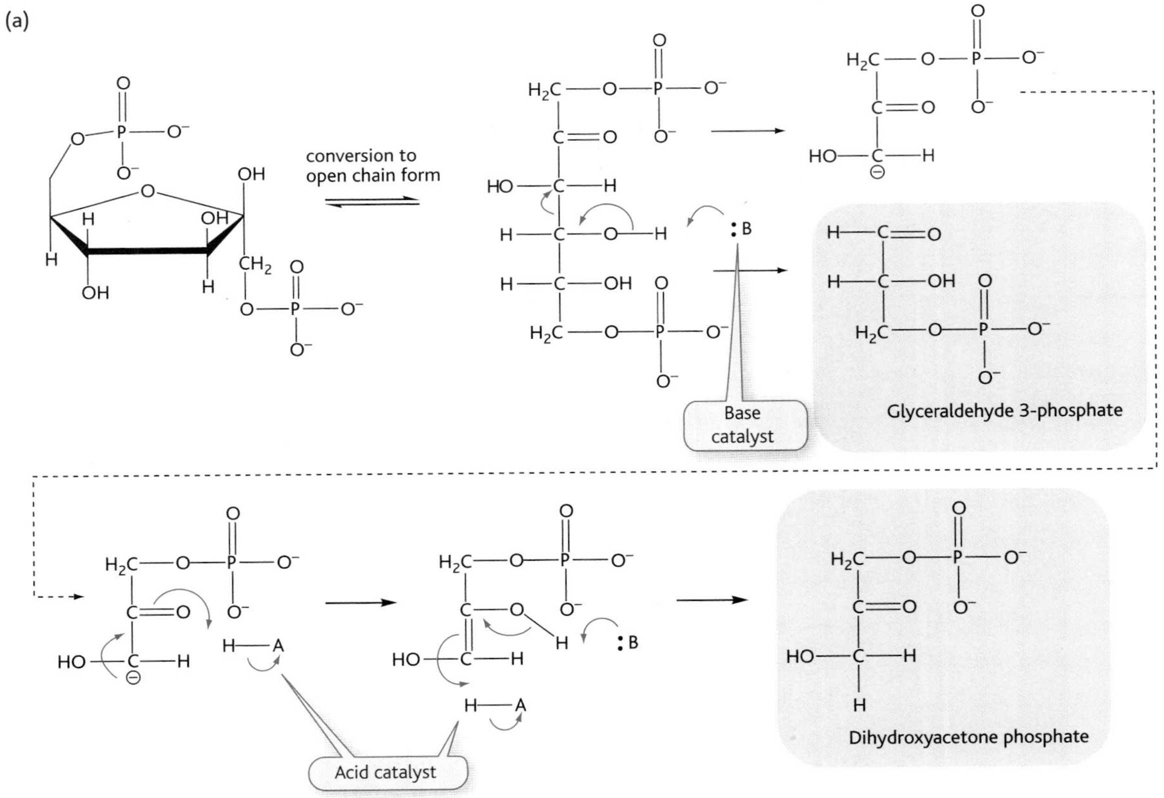

(a)

conversion to open chain form

Base catalyst

Glyceraldehyde 3-phosphate

Acid catalyst

Dihydroxyacetone phosphate

Figure 18.33 (a) Fragmentation of fructose 1,6-bisphosphate into glyceraldehyde 3-phosphate and dihydroxyacetone phosphate. Notice how this reaction requires both acid and base catalysts at different stages.

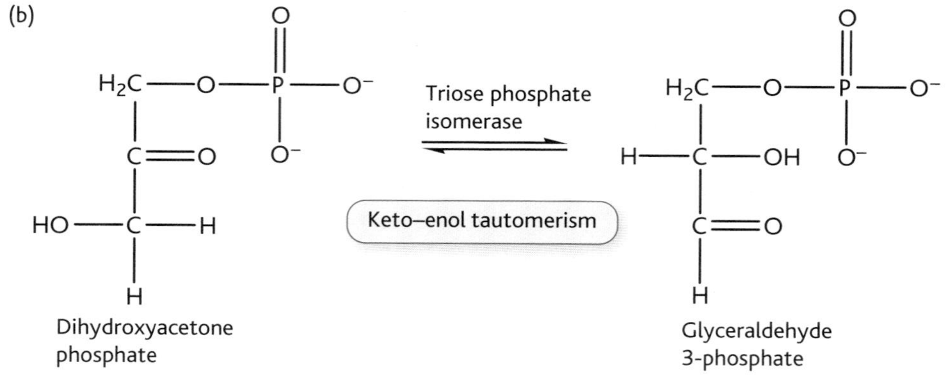

(b)

Dihydroxyacetone phosphate ⇌ (Triose phosphate isomerase) Glyceraldehyde 3-phosphate

Keto–enol tautomerism

Figure 18.33 (*Continued*) (b) The dihydroxyacetone phosphate produced during the reaction shown in (a) undergoes isomerization to form more glyceraldehyde 3-phosphate. Therefore, the net end product of the breakdown of one molecule of fructose 1,6-bisphosphate is two molecules of glyceraldehyde 3-phosphate.

by acid and base side chains in the enzyme, which form a ketone group at C4. Look at Figure 18.33(a), and notice how this is an example of an elimination reaction: it results in the formation of a double bond.

Figure 18.33(a) shows how the reaction yields two different three-carbon molecules: one half of the fructose molecule gives glyceraldehyde 3-phosphate, while the other gives dihydroxyacetone phosphate. If we look closely at Figure 18.33, we notice how glyceraldehyde 3-phosphate and dihydroxyacetone phosphate are isomers. The next step of glycolysis uses only glyceraldehyde 3-phosphate; however, the dihydroxyacetone phosphate formed here isn't wasted: a subsidiary reaction, catalyzed by triose phosphate isomerase, isomerizes dihydroxyacetone phosphate into more glyceraldehyde 3-phosphate, as illustrated in Figure 18.33(b). So there are now two molecules of glyceraldehyde 3-phosphate to undergo all subsequent reactions.

Step 5

We learn more about the involvement of NAD^+ in electron transfer processes in Chapter 17.

In the next step of glycolysis, the aldehyde group of glyceraldehyde 3-phosphate is oxidized and phosphorylated. Both modifications occur in the active site of the enzyme glyceraldehyde 3-phosphate dehydrogenase and involve electron transfer to NAD^+.

Figure 18.34 shows how a thiol group in the enzyme (featuring the –SH group from a cysteine side chain) helps to bring about the reaction by attacking the electrophilic aldehyde carbon atom to form a thiohemiacetal[14]. (Notice how this is an

14 A hemiacetal has the general formula

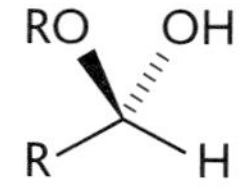

RO OH / R H

A thiohemiacetal replaces the O of the OR group with a sulfur, giving

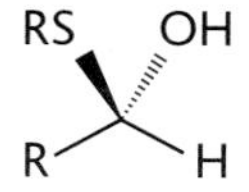

RS OH / R H

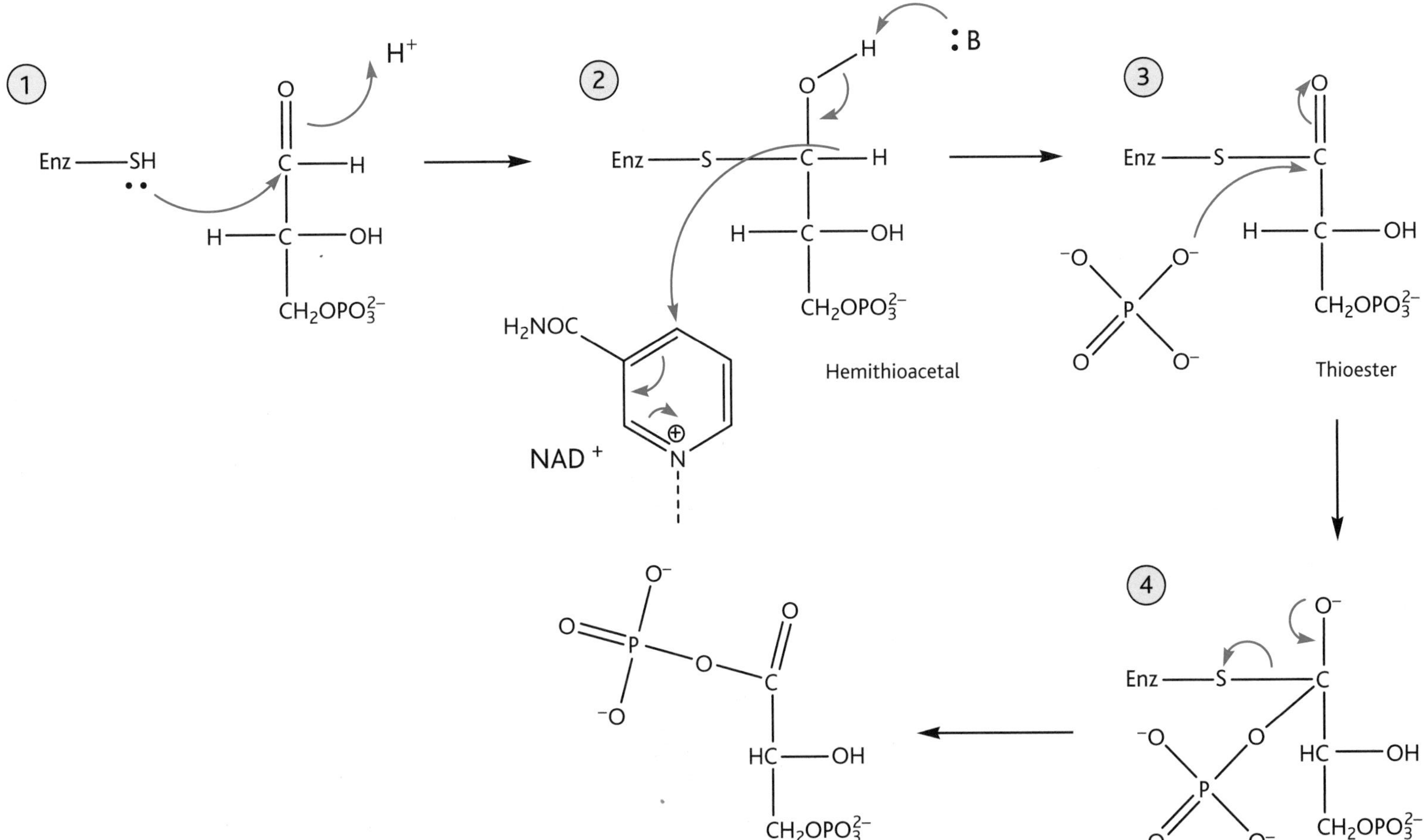

Figure 18.34 The conversion of glyceraldehde 3-phosphate into 3-phosphoglycerate. At first glance (if we compare the structure at the start of (a) with that at the end of (b)) the reaction looks like a simple isomerization reaction. However, the reaction actually involves two different phosphate groups at distinct stages. (a) The conversion of glyceraldehyde 3-phosphate to 1,3-bisphosphoglycerate, a high-energy phosphate compound carrying two phosphate groups. Notice how this process requires the electron carrier NAD^+, which is reduced to NADH during step 2. Throughout this process, the reacting species is associated with the enzyme glyceraldehyde 3-phosphate dehydrogenase.

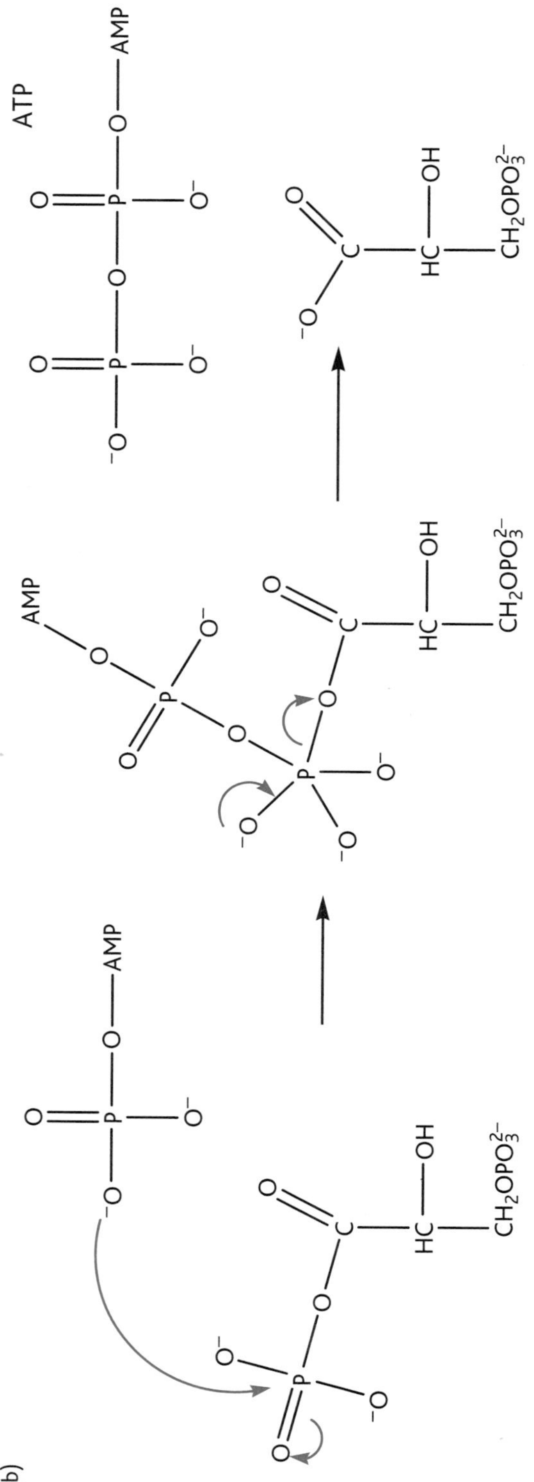

Figure 18.34 *(Continued)* (b) The hydrolysis of 1,3-bisphosphoglycerate to the final product, 3-phosphoglycerate. Notice how this process generates one molecule of ATP from ADP.

example of a condensation reaction, whereby two species–in this case, the –SH group of the enzyme and glyceraldehyde 3-phosphate–join together.)

This species is then readily oxidized by the enzyme to form a thioester. We see in Chapter 17 how oxidation is the loss of electrons (or loss of hydrogen species, H^+ and H^-); look at Figure 18.34a and notice how the oxidation of the thiohemiacetal during step 2 involves the loss of two hydrogen species: one is accepted by NAD^+ (which becomes reduced to NADH), and one is accepted by a base in the enzyme's active site (represented by 'B' in the figure).

At this stage, the oxidized species is still bound to the enzyme. During step 3 it is attacked by a phosphate group to form 1,3-bisphosphoglycerate. 1,3-bisphosphoglycerate is a high-energy phosphate compound (in parallel with compounds such as ADP and ATP). If we look carefully at the structure of 1,3-bisphosphoglycerate, we notice that it is a mixed anhydride of phosphoric acid and a carboxylic acid.

We note in Box 18.3 how acid anhydrides are readily hydrolyzed. (It is this hydrolysis that releases the energy 'stored' in the high-energy bonds of such a compound.) In line with this general characteristic of acid anhydrides, Figure 18.34(b) shows how 1,3-bisphosphoglycerate undergoes hydrolysis through the attack of ADP to form ATP and 3-phosphoglycerate in a reaction catalyzed by phosphoglycerate kinase. Reactions in which ATP is synthesized by transfer of phosphate to ADP are referred to as **substrate-level phosphorylation**.

Step 6

The next step in glycolysis converts 3-phosphoglycerate to 2-phosphoglycerate. This is an isomerization reaction, the net effect of which is to move the phosphoryl group from C3 to C2 of the glycerate molecule. This step involves two reactions, as illustrated in Figure 18.35. First, a condensation reaction occurs between the C2 hydroxyl group of the 3-phosphoglycerate and a phosphoryl group present in the active site of the enzyme that catalyzes this reaction, phosphoglycerate mutase. This condensation reaction is followed by a hydrolysis reaction during which the C3 phosphate group is removed.

Look at Figure 18.35 and notice how the net effect of this reaction is the movement of the phosphoryl group from C3 to C2. Also notice how the enzyme is regenerated at the end of the reaction: the phosphoryl group attached to C3 moves to the enzyme, such that the enzyme regains the structure it possessed at the start of the reaction.

BOX 18.6 Regenerating NADH

During the conversion of glyceraldehyde 3-phosphate to 1,3-diphosphoglycerate, the coenzyme NAD^+ is reduced to NADH. In order for the cell to maintain the pool of NAD^+ required to catalyze such reactions, it must necessarily find a way to oxidize the NADH back to NAD^+.

In aerobic organisms, such as humans, this is achieved when NADH enters the process of oxidative phosphorylation, during which most of the cell's ATP is generated as the electrons are transferred eventually to oxygen. However, anaerobic organisms need to find alternative pathways to regenerate NAD^+. We see in Section 17.4 how, in the absence of oxygen (in exercising muscles, for example) pyruvate is reduced to lactate by lactate dehydrogenase; this enzyme uses NADH as a coenzyme, regenerating NAD^+ as a consequence.

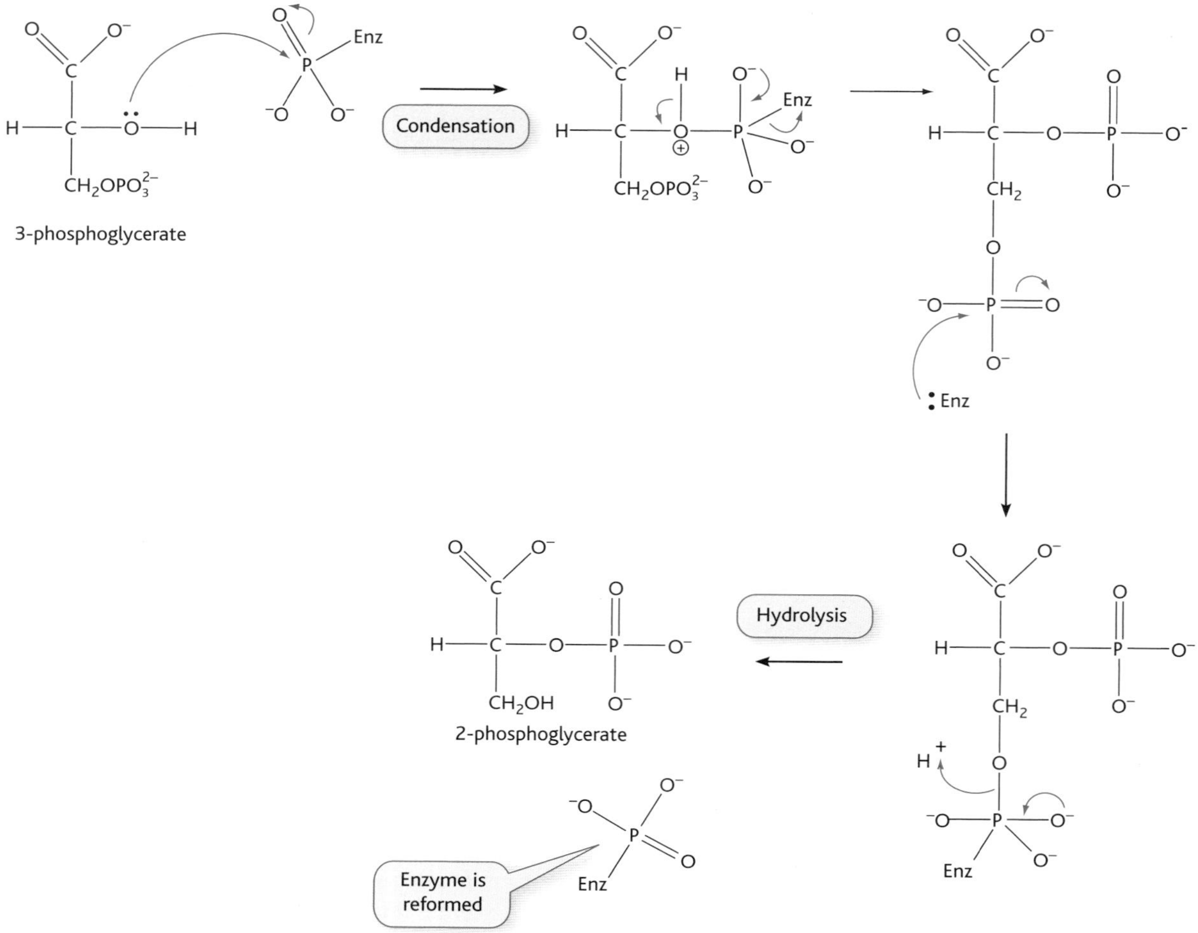

Figure 18.35 The isomerization of 3-phosphoglycerate to 2-phosphoglycerate. Notice how this reaction comprises a condensation reaction (involving a phosphate group attached to the enzyme, phosphoglycerate mutase), followed by a hydrolysis reaction.

Step 7

The penultimate step in glycolysis involves an elimination reaction. Look carefully at the structure of 2-phosphoglycerate depicted in Figure 18.36, which shows how two possible leaving groups are present during the course of elimination: water (H–OH) and phosphoric acid (H–OPO_3^{2-}). Elimination reactions favour the most stable leaving group–in this case, water.

Therefore, the penultimate step in glycolysis is the dehydration of 2-phosphoglycerate to generate phosphoenolpyruvate (PEP) as depicted in Figure 18.36(a), a reaction catalyzed by enolase.

Step 8

The final step in glycolysis has two main phases. Firstly, pyruvate kinase catalyzes the transfer of the phosphoryl group from phosphoenolpyruvate to ADP, forming enolpyruvate and ATP. The mechanism of this reaction involves a condensation reaction between PEP and ADP, followed by an elimination reaction to form

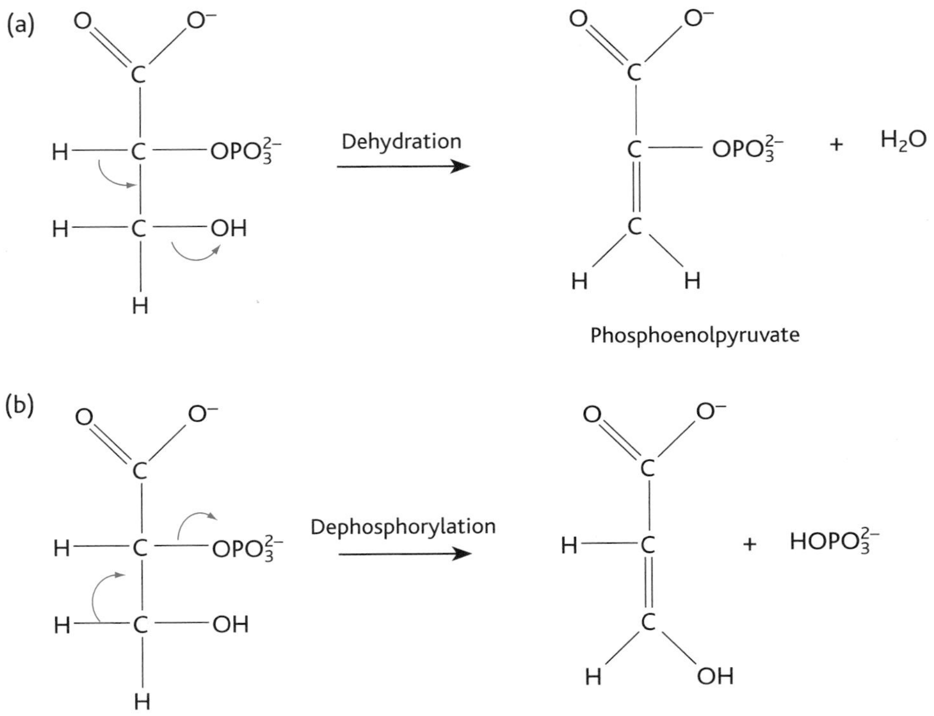

(a)

Dehydration

Phosphoenolpyruvate

+ H_2O

(b)

Dephosphorylation

+ $HOPO_3^{2-}$

Figure 18.36 Two elimination reactions involving 2-phosphoglycerate are possible. (a) The leaving group is H_2O, making it a dehydration reaction. This reaction would yield phosphoenolpyruvate (PEP). (b) The leaving group is $HOPO_3^{2-}$, making it a dephosphorylation reaction. In practice, the most stable leaving group is favoured; this is water, as seen in (a).

enolpyruvate, as illustrated in Figure 18.37. Notice how the condensation reaction once again involves nucleophilic attack by the terminal phosphoryl group of ADP on the P of the phosphoryl group of PEP.

The second phase of the final step in glycolysis is a spontaneous step in which enolpyruvate rapidly converts to its structural isomer, pyruvate. This reaction is an example of tautomerization, a particular type of chemical reaction in which a proton switches atoms, and a double bond switches positions within a molecule, as illustrated in the final step of Figure 18.37.

Beyond glycolysis: how the oxidation of glucose ultimately powers the cell

One of the main purposes of the oxidation of glucose in our cells is to generate molecules of ATP, which can be harnessed to drive forward energetically unfavourable reactions. During glycolysis, however, ATP is both generated *and* consumed as one molecule of glucose is converted to two molecules of pyruvate. Figure 18.38 gives a summary of glycolysis as a whole; look at this figure and notice that, while ATP is generated during two steps, it is also consumed during two other steps. Because two molecules of glyceraldehyde 3-phosphate are produced from one molecule of glucose, however, the second half of the pathway happens at twice the frequency of the first half, leading to a net gain of two molecules of ATP overall.

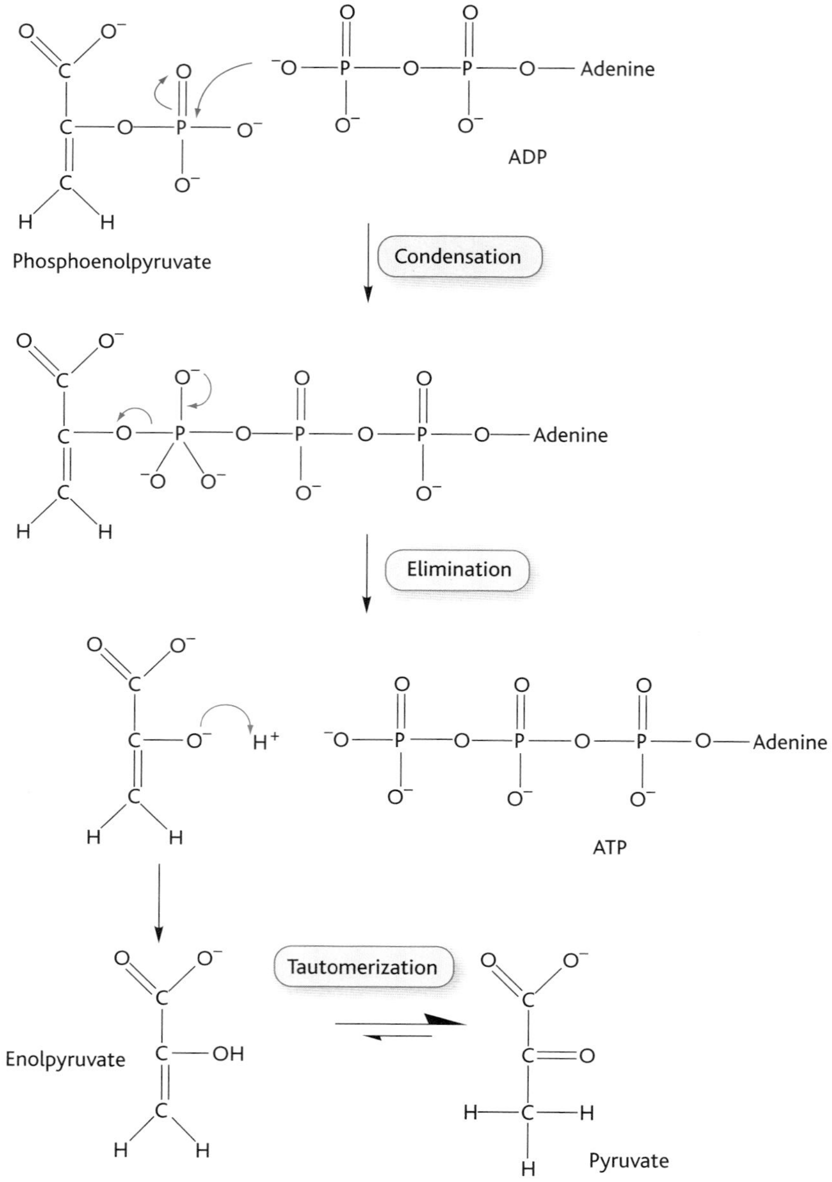

Phosphoenolpyruvate

ADP

Condensation

Elimination

ATP

Enolpyruvate

Tautomerization

Pyruvate

Figure 18.37 The final step in glycolysis–the formation of pyruvate from phosphoenolpyruvate. This reaction has two main stages: a condensation reaction involving ADP, and an elimination reaction. This elimination reaction generates enolpyruvate and a molecule of ATP. Enolpyruvate then undergoes tautomerization to give the final product, pyruvate. Notice that, although the tautomerization is an equilibrium reaction, it lies heavily to the right, favouring the formation of pyruvate.

A net yield of two molecules can sometimes be a pretty trivial amount–and certainly not enough to supply all cells with the quantities of ATP needed for survival. So, how do such cells achieve their goal of generating a sufficient number of ATP molecules through the oxidation of glucose? In humans, pyruvate is oxidized further by a series of chemical reactions involving pyruvate dehydrogenase; the product of this reaction enters the citric acid cycle to be further oxidized. (This oxidation

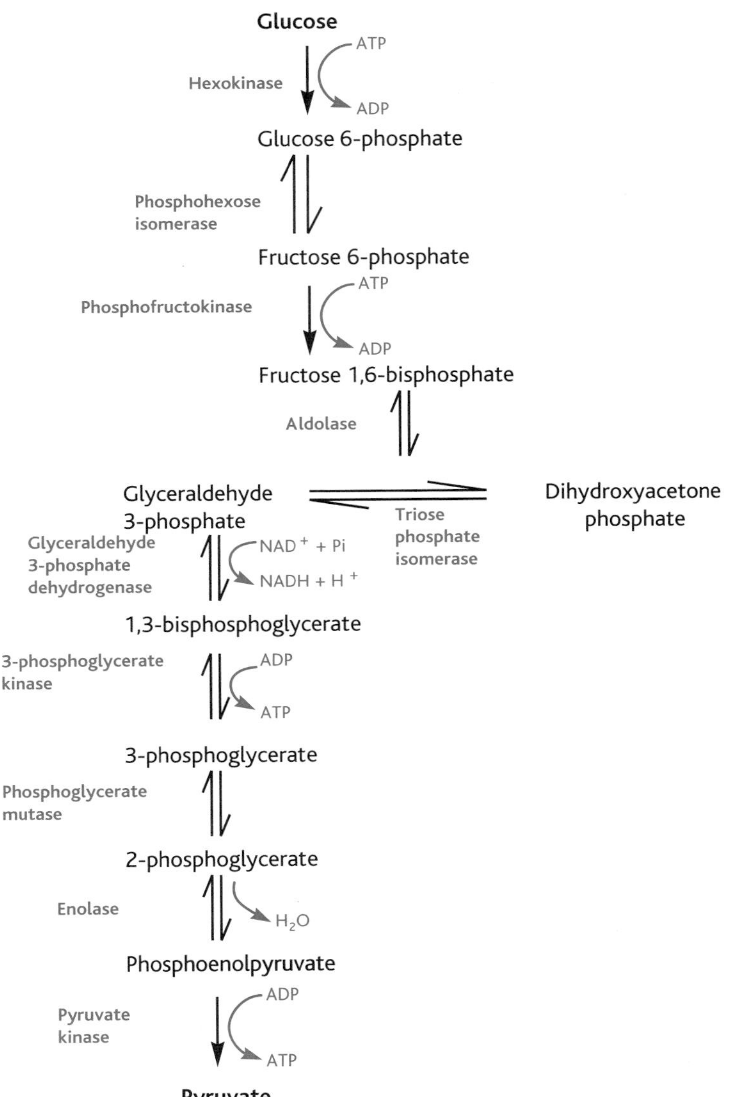

Figure 18.38 The overall process of glycolysis. Note that, for every one molecule of glucose entering the pathway, two molecules of pyruvate are formed. (This is because fructose 1, 6-bisphosphate ultimately generates two molecules of glyceraldehyde 3-phosphate, as described in the text.) There is also the net synthesis of two molecules of ATP for each glucose molecule entering the pathway.

is coupled to the reduction of two electron carriers, as we see below.) These reduced products are then consumed during the process of oxidative phosphorylation. Ultimately, during the course of these processes, the energy released by the oxidation of pyruvate is transformed by oxidative phosphorylation to yield ATP.

If we consider just a part of one of these subsequent stages, the citric acid cycle, we see in action again several of the reactions we explore in the previous few pages. Look at Figure 18.39, which depicts the first four steps in the citric acid cycle, and one pre-cycle step (the conversion of pyruvate to acetyl coenzyme A). Notice how,

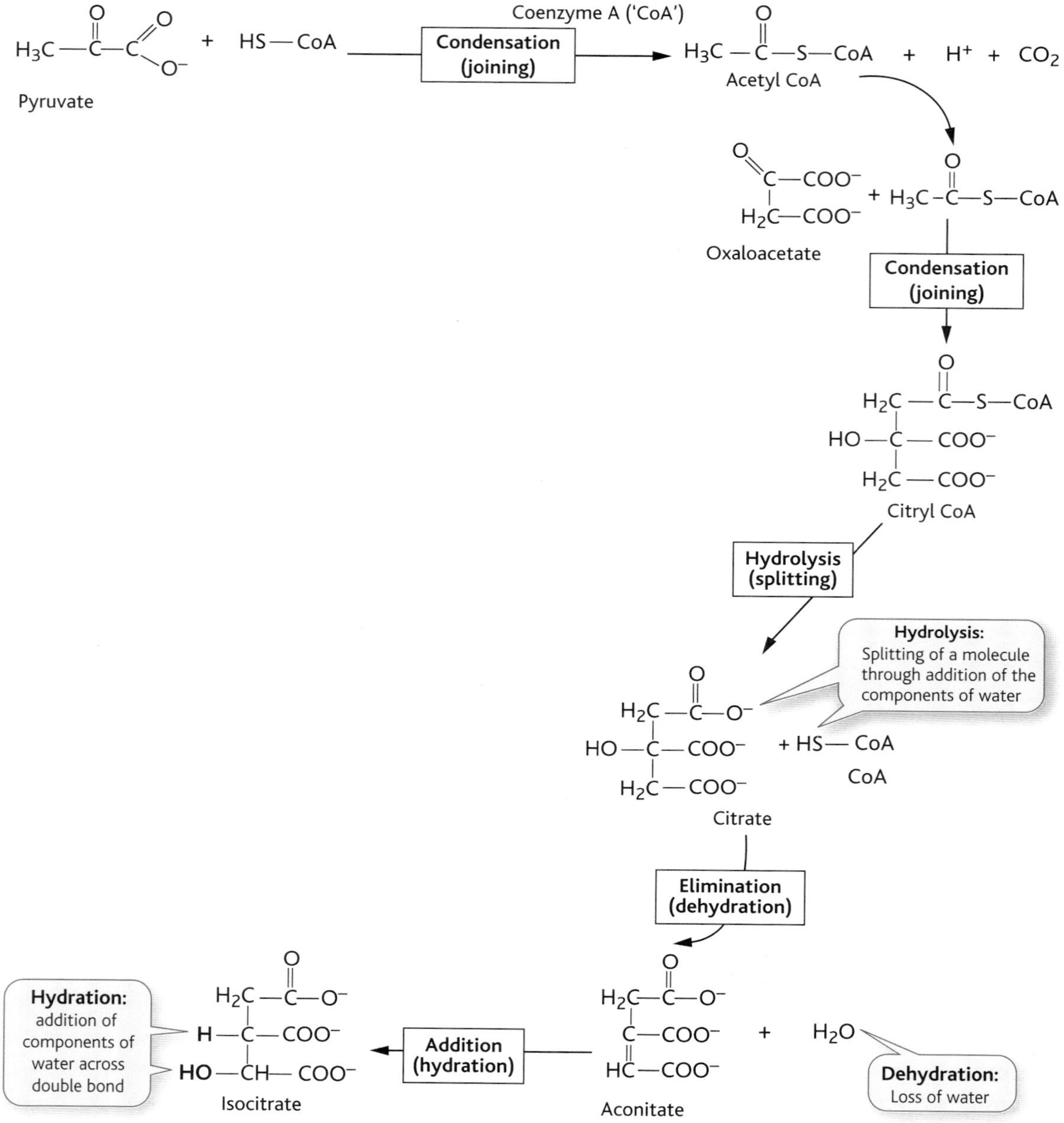

Figure 18.39 The first four steps in the citric acid cycle, and one pre-cycle step (the conversion of pyruvate to form acetyl coenzyme A).

once again, we are seeing the exploitation of a number of different reaction mechanisms, including condensation, elimination, and addition.

The citric acid cycle still only yields the equivalent of two more ATP molecules per glucose molecule oxidized. However, it also has the essential role of reducing the electron carriers NAD^+ and CoQ to NADH and $CoQH_2$ respectively. It is the subsequent oxidation of these two electron carriers that really pays dividends in terms of generating ATP.

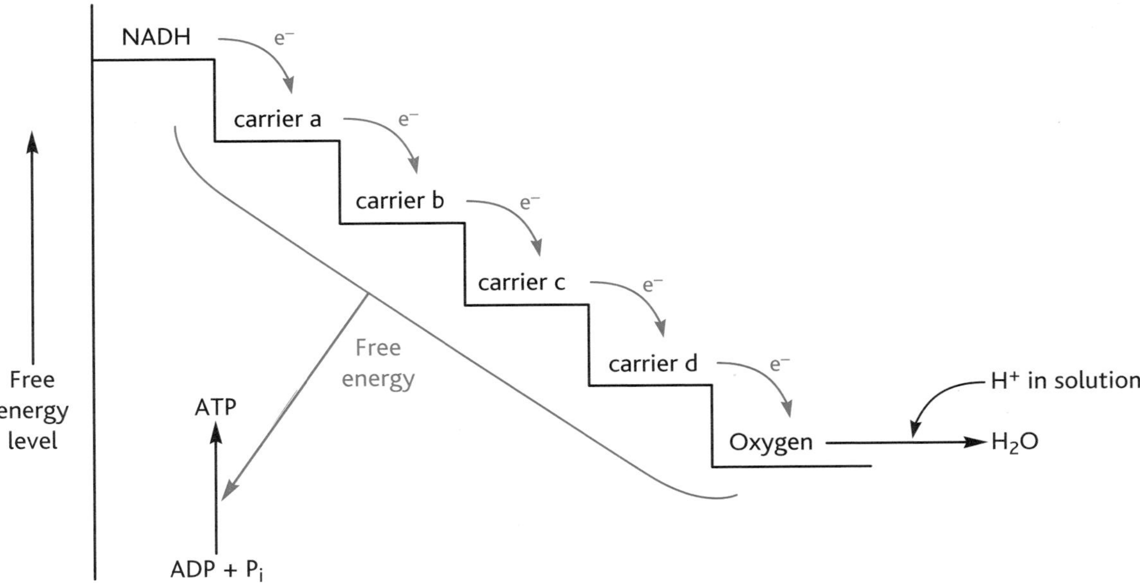

Figure 18.40 A simplified overview of the electron transport chain through which electrons are passed from the electron carrier NADH to oxygen, which ultimately becomes reduced to water. Notice how this electron transfer process yields free energy, which is coupled to the synthesis of ATP.

We see above how some ATP is produced through substrate-level phosphorylation–that is, the generation of ATP from the condensation of an ADP molecule and a phosphoryl group as part of particular glycolytic chemical reactions. During the last phase of the oxidation of glucose, however, ATP is not generated through substrate-level phosphorylation, but through what we call **oxidative phosphorylation.**

Central to this process is the electron transport chain, illustrated in Figure 18.40, through which the electron carrier NADH is oxidized (that is, it releases electrons). These electrons are taken up by a series of electron carrier molecules that are embedded in the inner mitochondrial membrane. Notice how the last step in the electron transport chain sees the electrons being accepted by oxygen, which becomes reduced to water.

We learn more about the activity of oxygen as an electron carrier in Box 17.2.

Importantly, the transfer of electrons along the chain results in the pumping of protons across the mitochondrial inner membrane. This movement of protons forms a gradient, whereby the concentration of protons on one side of the membrane is higher than on the other. It is this gradient that is used to achieve ATP synthesis, using a remarkable cellular machine called the ATP synthase. ATP synthase, whose structure is depicted in simplified form in Figure 18.41(a), is a multi-subunit protein; part of the protein is anchored in the mitochondrial membrane, while the rest protrudes into the mitochondrial matrix. The part of the protein embedded in the membrane– the F_0 subunit–acts as a channel for protons, allowing them to flow across the mitochondrial membrane. By contrast, the F_1 subunit, which is attached to the F_0 subunit through a stalk region, possesses the catalytic activity that drives the synthesis of ATP from ADP and a phosphoryl group. The three-dimensional structure of the F_1 subunit is depicted in Figure 18.41(b).

The channelling of protons through the F_0 subunit is coupled to rotation of the F_0 subunit, which, in turn, is coupled to the ATP-synthesizing activity of the F_1 subunit. As such–and though this brief description doesn't do the elegance of the operation of ATP synthase justice–this multi-subunit enzyme really does behave like a molecular machine, where physical movement of part of the machine is used to drive a chemical reaction–the synthesis of ATP–elsewhere in the machine[15].

The ultimate yield of ATP from the activity of ATP synthase is significant: while the previous steps of glucose oxidation–glycolysis and the citric acid cycle–yield just four ATP molecules between them from one molecule of glucose, oxidative phosphorylation yields either 26 or 28 ATP molecules[16]–giving a total yield of either 30 or 32 molecules of ATP from just one molecule of glucose.

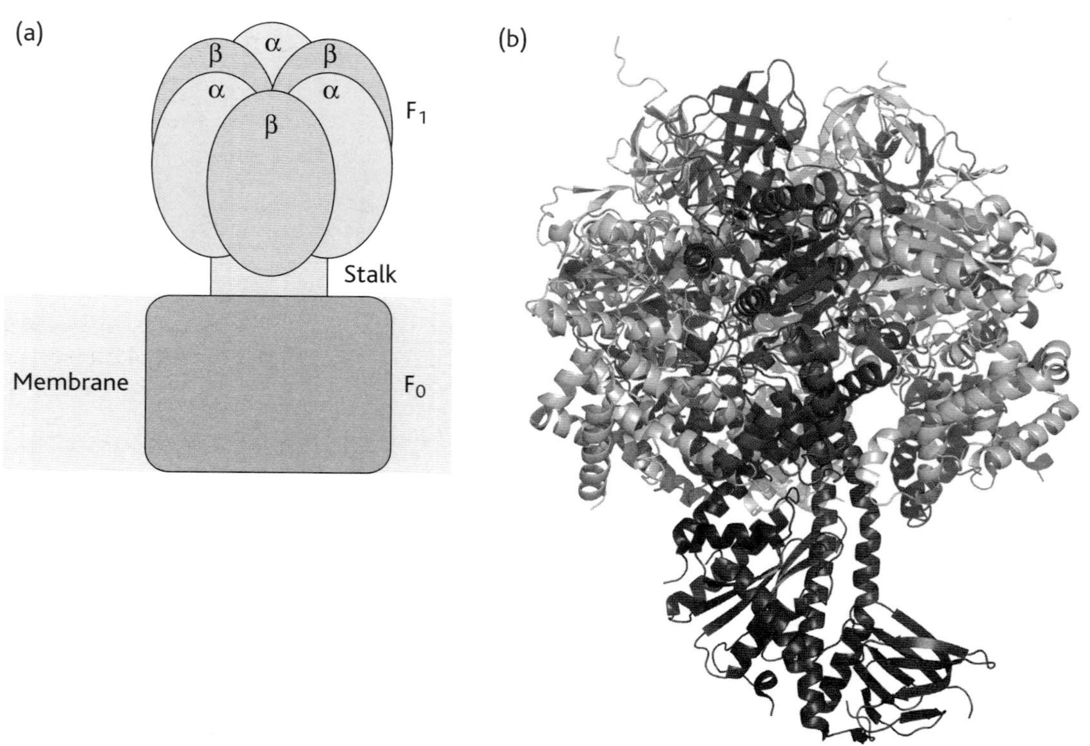

Figure 18.41 The ATP synthase. (a) Simplified diagram showing the major components of ATP synthase. The F_0 has multiple subunits and embeds within the lipid bilayer membrane. The F_1 is composed of a six-subunit ring of alternating α and β subunits enclosing two central subunits, γ and ε, which project downwards and contact the F_0 unit. (Reproduced from Elliott and Elliott, 2009.) (b) The three-dimensional structure of the F_1 subunit. The alternating α and β subunits are shown in blue and light grey, with the two central γ and ε subunits in dark grey.

15 For a more detailed explanation of how the ATP synthase operates, read Elliott and Elliott, *Biochemistry & Molecular Biology* (Oxford University Press).

16 The actual number depends on the precise source of the NADH.

(c)

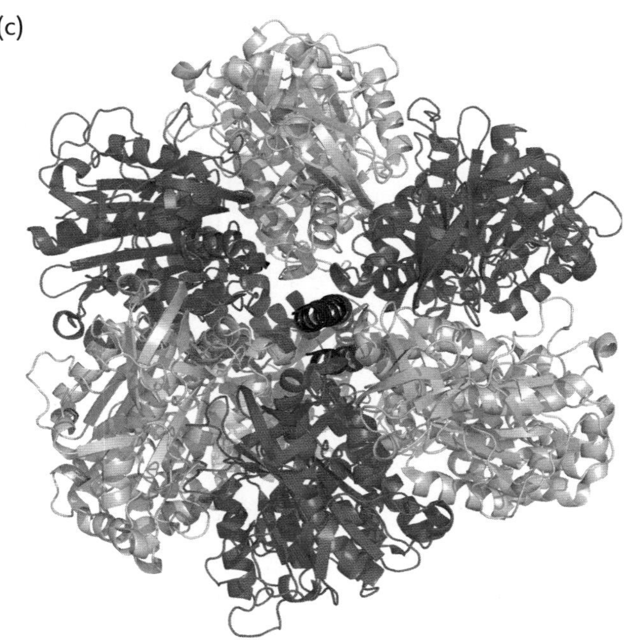

Figure 18.41 (*Continued*) The alternating arrangement of the α and β subunits is more obvious when the complex is viewed from above, as shown in (c).

Check your understanding

To check that you've mastered the key concepts presented in this chapter, review the checklist of key concepts below, and attempt the multiple-choice questions available in the book's Online Resource Centre at **http://www.oxfordtextbooks.co.uk/orc/crowe2e/**.

Checklist of key concepts

Reaction mechanisms

- A heterolytic reaction proceeds via one of four key mechanisms: substitution, addition, elimination, or condensation
- Chemical reactions may proceed in a single step, or via multiple steps
- When reactants react in a single step they pass through a high-energy, unstable transition state
- When reactants react via more than one step, an intermediate is formed at the end of each interim step
- A carbocation is an ion containing a positively charged carbon

Substitution

- During substitution, an atom or group of atoms from one reactant is substituted for an atom or group of atoms on a second reactant molecule
- Substitution may be electrophilic or nucleophilic, depending on the nature of the molecule that changes as a result of the reaction
- An S_N2 reaction is a one-step nucleophilic substitution reaction in which the rate-limiting step involves two species
- An S_N1 reaction is a two-step nucleophilic substitution reaction in which the reactants pass through a stable

intermediate form before the reaction is completed. The rate-limiting step involves two species

Addition

- During an addition reaction, two molecules combine in such a way that the product contains all the atoms of both reactants
- Addition usually occurs across a double bond, resulting in the breaking of a π bond
- An addition reaction that gives the net addition of the components of a water molecule across a double bond is what we call a hydration reaction
- Nucleophiles induce polarization in electrophiles, often facilitating bond cleavage

Elimination

- During an elimination reaction, a reactant loses some of its component atoms; these atoms form one or more new compounds
- Elimination provides a means of generating an unsaturated compound from a previously saturated compound
- Elimination may proceed via a two-step E1 mechanism, or a one-step E2 mechanism
- During an elimination reaction, the nucleophile attacks a hydrogen atom; during a substitution reaction the nucleophile attacks a carbon atom

Condensation

- During a condensation reaction two reactants combine to form a single product; a further molecule is eliminated during the reaction, which forms an additional product
- Condensation reactions feature in the polymerization of some monomers to form polymers
- Hydrolysis reactions reverse the effect of condensation: they use the components of water to split large molecules into smaller constituent molecules

Biochemical reaction pathways

- Glycolysis is a series of reactions which converts glucose into two molecules of pyruvate
- The oxidation product of pyruvate enters the citric acid cycle, which serves to reduce two electron carriers
- The reduced electron carriers are consumed during oxidative phosphorylation, through which ATP is ultimately generated

Epilogue

Whatever else it may be, at the level of chemistry life is fantastically mundane: carbon, hydrogen, oxygen and nitrogen, a little calcium, a dash of sulfur, a light dusting of other very ordinary elements–nothing you wouldn't find in any ordinary pharmacy–and that's all you need. The only thing special about the atoms that make you is that they make you. That is, of course, the miracle of life.

Bill Bryson, *A Short History of Nearly Everything*, 2003

Bibliography

Books

Atkins, P. and de Paula, J. (2005). *Elements of Physical Chemistry* (4th edn, Oxford: Oxford University Press).

Atkins, P. and de Paula, J. (2006). *Physical Chemistry for the Life Sciences.* Oxford: Oxford University Press.

Atkins, P. and Jones, L. (2005). *Chemical Principles: The Quest for Insight* (3rd edn, New York: W. H. Freeman and Company).

Atkins, P., Overton, T., Rourke, J., Weller, M., and Armstrong, F. (2006). *Inorganic Chemistry* (4th edn, Oxford: Oxford University Press).

Bloomfield, M. and Stephens, L. (1966). *Chemistry and the Living Organism* (6th edn, USA: John Wiley).

Brown, T., LeMay, H., and Bursten, B. (2000). *Chemistry: The Central Science* (8th edn, New Jersey: Prentice Hall).

Bryson, B. (2003). *A Short History of Nearly Everything* (London: Doubleday).

Buckberry, L. and Teesdale, P. (2001). *Essentials of Biological Chemistry* (Chichester: John Wiley & Sons).

Chang, R. (1994). *Chemistry* (5th edn, USA: McGraw Hill).

Clayden, J., Greeves, N., Warren, S., and Wothers, P. (2000). *Organic Chemistry* (Oxford: Oxford University Press).

Dobson, C., Gerrard, J., and Pratt, A. (2001). *Foundations of Chemical Biology* (Oxford: Oxford University Press).

Duckett, S. and Gilbert, B. (2000). *Foundations of Spectroscopy* (Oxford: Oxford University Press).

Elliott, W. and Elliott, D. (2009). *Biochemistry and Molecular Biology* (4th edn, Oxford: Oxford University Press).

Emsley, J. (2001). *Nature's Building Blocks: An A–Z Guide to the Elements* (Oxford: Oxford University Press).

Frausto da Silva, J. and Williams, R. (2001). *The Biological Chemistry of the Elements: The Inorganic Chemistry of Life* (Oxford: Oxford University Press).

Frohne, Pfänder, and Alford (2005). *Poisonous Plants: A Handbook for Doctors, Pharmacists, Toxicologists, Biologists, and Veterinarians* (Chichester: Wiley-Blackwell).

Gibson, G. and Muse, S. (2004). *A Primer of Genome Science* (2nd edn, Sunderland, MA: Sinauer Associates).

Glusker, J. P. and Trueblood, K. N. (1972). *Crystal Structure Analysis: A Primer* (Oxford: Oxford University Press).

Hammes, G. (2007). *Physical Chemistry for the Biological Sciences* (New Jersey: Wiley-Blackwell).

Harris, D. (2005). *Exploring Chemical Analysis* (3rd edn, New York: W. H. Freeman and Company).

Higson, S. (2004). *Analytical Chemistry* (Oxford: Oxford University Press).

Hill, G. and Holman, J. (2000). *Chemistry in Context* (5th edn, Cheltenham: Nelson Thornes).

Hill, J., Baum, S., and Feigl D. (1997). *Chemistry and Life* (New Jersey: Prentice Hall).

Hill, J., Petrucci, R., McCreary, T., and Perry, S. (2005). *General Chemistry* (New Jersey: Prentice Hall).

Hornby, M. and Peach, J. (1993). *Foundations of Organic Chemistry* (Oxford: Oxford University Press).

Jackson, A. and Jackson, J. (2004). *Forensic Science* (Harlow: Pearson Education Limited).

Keeler, J. and Wothers, P. (2003). *Why Chemical Reactions Happen* (Oxford: Oxford University Press).

Keeler, J. and Wothers, P. (2008). *Chemical Structure and Reactivity* (Oxford: Oxford University Press).

King, R. (2000). *Cancer Biology* (2nd edn, Harlow: Pearson Education Limited).

Lane, N. (2002). *Oxygen: The Molecule that Made the World* (Oxford: Oxford University Press).

McMurry, J. and Begley, T. (2005). *The Organic Chemistry of Biological Pathways* (Greenwood Village: Roberts and Company).

McMurry, J. and Castellion, M. (2006). *Fundamentals of General, Organic and Biological Chemistry* (4th edn, New Jersey: Prentice Hall).

Monk, P. (2004). *Physical Chemistry: Understanding our Chemical World* (Chichester: John Wiley & Sons).

Moore, J., Stanitski, C., and Jurs, P. (2002). *Chemistry: The Molecular Science* (Florence, KY: Thomson Learning Inc.).

Patrick, G. (2005). *An Introduction to Medicinal Chemistry* (3rd edn, Oxford: Oxford University Press).

Petsko, G. and Ringe, D. (2004). *Protein Structure and Function* (Oxford: Oxford University Press).

Pocock, G. and Richards, C. D. (2009). *The Human Body* (Oxford: Oxford University Press).

Price, N. C., Dwek, R. A., Ratcliffe, R. G., and Wormald, M. R. (2001). *Principles and Problems in Physical Chemistry for Biochemists* (3rd edn, Oxford: Oxford University Press).

Price, N. C. and Nairn, J. (2009). *Exploring Proteins* (Oxford: Oxford University Press).

Schreiber, S., Kapoor, T., and Wess, G. (2007) *Chemical Biology: From Small Molecules to Systems Biology and Drug Design* (Weinheim: Wiley-VCH).

Timberlake, K. (2002). *General, Organic, and Biological Chemistry: Structures of Life* (San Francisco: Benjamin Cummings Publishing Company).

Weinhold, F., and Landis, C., (2005). *Valency and Bonding: a Natural Bond Orbital Donor–Acceptor Perspective* (Cambridge: Cambridge University Press).

Winter, M. (1994). *Chemical Bonding* (Oxford: Oxford University Press).

Websites

(Date last accessed indicated in square brackets, where known)

http://agambiae.vectorbase.org/SequenceData/Genome/ [11/04/09]

http://en.wikipedia.org/wiki/Thermodynamics

http://news.bbc.co.uk/1/hi/sci/tech/7827148.stm [07/03/09]

http://www.absw.org.uk/Briefings/Nitric%20oxide.htm

http://www.bact.wisc.edu/Bact303/ [16/10/05]

http://news.bbc.co.uk/1/hi/sci/tech/7784531.stm [25/01/09]

http://www.bbc.co.uk/climate/ [17/01/09]

http://www.biologie.uni-hamburg.de/b-online [29/08/04]

http://www.diabetes.org.uk/Guide-to-diabetes/Introduction-to-diabetes/What_is_diabetes/ [28/02/09]

http://www.diabetes.org.uk/Professionals/Information_resources/Reports/Diabetes-prevalence-2008/ [28/02/09]

http://www.drlam.com/ [09/08/04]

http://www.faqs.org/faqs/ozone-depletion/ [05/02/05]

http://www.fda.gov/FDAC/features/2003/103_msg.html [25/01/09]

http://www.genoscope.cns.fr/spip/Anopheles-gambiae-vector-for-a.html [11/04/09]

http://www.geo-pie.cornell.edu/gmo.html [09/08/04]

http://www.microbe.org/microbes/ [16/10/05]

http://www.nhs.uk/conditions/Diabetes/Pages/Introduction.aspx?url=Pages/What-is-it.aspx [28/02/09]

http://www.nhs.uk/Conditions/Hyperglycaemia/Pages/Symptoms.aspx?url=Pages/What-is-it.aspx [28/02/09]

http://www.nlm.nih.gov/medlineplus/encyclopedia.html [16/08/04]

http://www.nmsea.org/Curriculum/Primer/energy_physics_primer.htm

http://www.plantphys.net/ [29/8/04]

http://www.secondlaw.com/ [01/11/05]

http://www.teachmetuition.co.uk/Chemistry/

http://www.umich.edu/~gs265/society/greenhouse.htm [17/01/09]

Mills, C. E. 1999-present. Bioluminescence of *Aequorea*, a hydromedusa. Electronic internet document available at http://faculty.washington.edu/cemills/Aequorea.html. Published by the author, web page established June 1999. [16/05/09]

Journal and periodical articles

A taste of umami, *Nature* (2009) volume 457, page 160.

Ball, P. (2006). Quantum sketches, *New Scientist*, 8 July 2006.

Berger, S. (2007). The complex language of chromatin regulation during transcription, *Nature*, **447**: 407–412.

Berman, H. M., Westbrook, Z. F., Gilliland, G., Bhat, T. N., Weissig, H., Shindyalov, I. N., and Bourne, P. E. (2000). The Protein Data Bank, *Nucleic Acids Research*, **28**: 235–242.

Bird, A. (2007). Perceptions of epigenetics, *Nature*, **447**: 396–398.

Cereda, M. and Mattos, M. (1996). Linamarin–the toxic compound of cassava, *J. Venom. Anim. Toxins*, **2**(1): 6–12.

Doherty, E. (2001). Ribozyme structures and mechanisms, *Annual Review of Biophysics and Biomolecular Structure*, **30**: 457–475.

Gruber, K. and Kratky, C. (2004). Biopolymers for biocatalysis: structure and catalytic mechanism of hydroxynitrile lyases, *Journal of Polymer Science*, Part A, **42**: 479–486.

Dário E. Kalume, Suraj Peri, Raghunath Reddy, Jun Zhong, Mobolaji Okulate, Nirbhay Kumar, and Akhilesh Pandey (2005) Genome annotation of *Anopheles gambiae* using mass spectrometry-derived data, *BMC Genomics*, **6**: 128.

Kiernan, J. (2000). Formaldehyde, formalin, paraformaldehyde and glutaraldehyde: what they are and what they do, *Microscopy Today*, **00–1**: 8–12.

Knight, J. (2003). Drugs in sport: no dope, *Nature*, **426**: 114–115.

Kushner, D. J., Baker, A., and Dunstall, T. G. (1999) Pharmacological uses and perspectives of heavy water and deuterated compounds, *Can. J. Physiol. Pharmacol.*, **77**: 79–88.

Lambert, F. (2002). Disorder–a cracked crutch for supporting entropy discussions, *J. Chem. Ed.*, **79**(2): 187–192.

Nature (2008) volume 456, Issue 7222, page xiii.

Pan, K., Baldwin, M., Nguyen, J., Gasset, M., Serban, A., Groth, D., Mehlhorn, I., Huang, Z., Fletterick, R., Cohen, F., and Prusiner, S. (1993). Conversion of α-helices into β-sheets features in the formation of the scrapie prion proteins, *Proceedings of the National Academy of Sciences USA*, **90** (23): 10962–10966.

Prusiner, S. (1998). Prions, *Proceedings of the National Academy of Sciences USA*, **95**(23): 13363–13383.

Ritter *et al.* (2005). Correlation of structural elements and infectivity of the HET-s prion, *Nature*, **435**: 844–848.

Sanderson, K. (2009). Emissions control, *Nature*, **459**, 500–503.

Answers to self-check questions

This section shows the **final** answer for each self-check question. For some numerical questions, the final answer is obtained by carrying out a calculation comprising several steps. For these questions (which are marked here with a*), the *full* answer (showing every step in the calculation) is available in the book's Online Resource Centre. Visit **http://www.oxfordtextbooks.co.uk/orc/crowe2e/**.

Chapter 2

2.1 Period 3. Group 13.

2.2 16 protons, 16 electrons, 16 neutrons

2.3 An oxygen ion carries two negative charges (its overall charge is −2). It is represented by O^{2-}.

2.4* 35.5

2.5 18 electrons

2.6 Two electrons occupy the 1s orbital. Two electrons occupy the 2s orbital. Two electrons occupy 2p orbitals.

2.7 $1s^2\ 2s^2\ 2p^2$

2.8

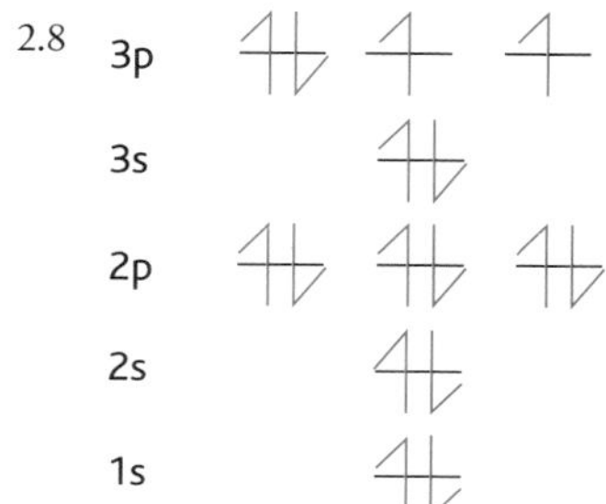

3p
3s
2p
2s
1s

2.9 Ultraviolet light has more energy than infrared light because ultraviolet light has a shorter wavelength.

2.10 Shell 4. $4s^2$

Chapter 3

3.1 Sodium: the valence electron occupies shell 3.
Oxygen: the valence electrons occupy shell 2.

3.2 (a) Mg (b) Br

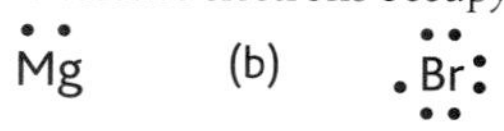

3.3 O

Oxygen has two non-bonding pairs of electrons.

3.4 (a) Ionic bonding
(b) Covalent bonding

3.5 Bromine must gain one electron.

3.6 Sodium ion carries a charge of +1. Oxide ion carries a charge of −2. **Two** sodium atoms must react with **one** oxygen atom to give a neutral compound.

3.7 $CaBr_2$. (Calcium forms the Ca^{2+} ion; bromine forms the Br^- ion.)

3.8 (a) A molecule of propane 1-thiol comprises three atoms of carbon, eight atoms of hydrogen, and one atom of sulfur. (b) $C_{10}H_{20}O$

3.9

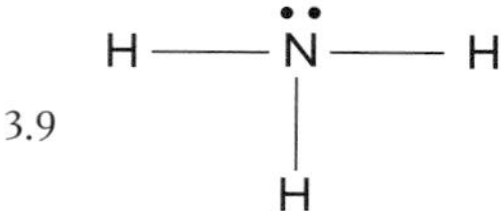

H
N
H
H

3.10 3

3.11 (b) C, O; (d) C, C

3.12 (a) C, N; (d) C, C

3.13 (c) is drawn correctly. The errors are as follows:

(a)

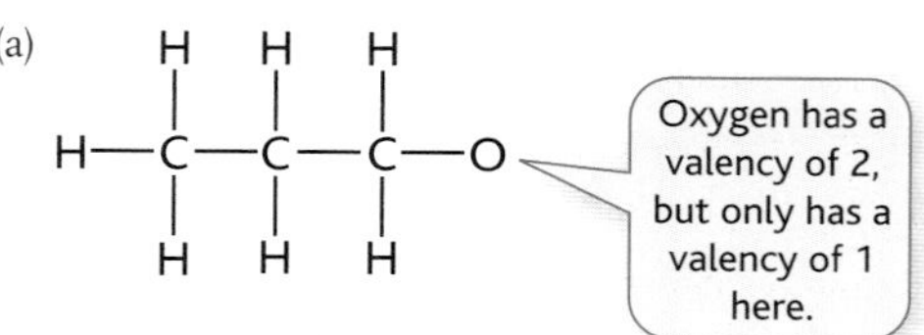

Oxygen has a valency of 2, but only has a valency of 1 here.

(b)

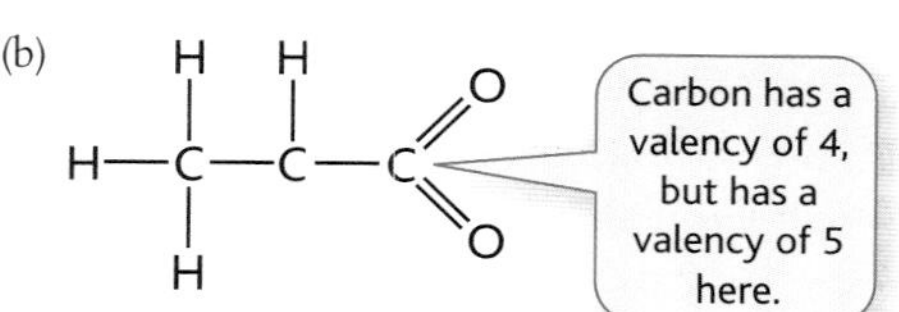

Carbon has a valency of 4, but has a valency of 5 here.

(c)

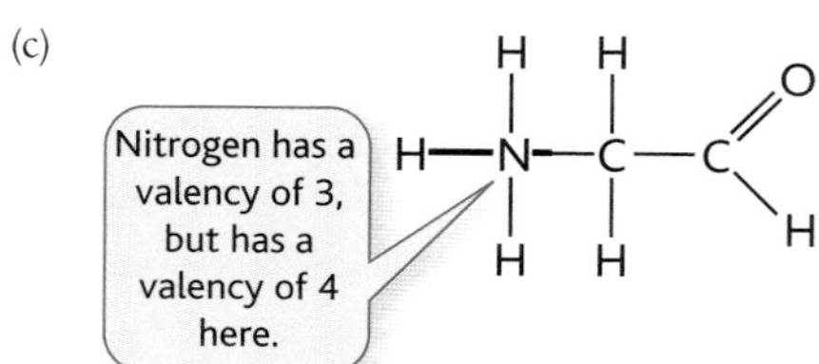

Nitrogen has a valency of 3, but has a valency of 4 here.

3.14 (c) is correct.

3.15 4

3.16 Li_2CO_3

3.17 There are two anions (OH^-) per cation (Ca^{2+}). The chemical formula is $Ca(OH)_2$.

3.18 (a) Difference in electronegativity values = 1.0 ⇒ **covalent** bonding is most likely.

(b) Difference in electronegativity values = 2.0 ⇒ **ionic** bonding is most likely.

3.19 (a), (c), (b), (d)

Chapter 4

4.1 Non-polar. Even though the C–Cl bond is polar, the four Cl nuclei are arranged symmetrically around the central carbon atom, making the molecule non-polar overall.

4.2 Molecule (b) experiences the greatest dispersion forces, because it is a larger molecule.

4.3 9: H must be bonded to O, F, or N (three different arrangements). In each instance, the H can form a hydrogen bond with O, F, or N (three further different arrangements). 3 × 3 = 9.

4.4 Hydrogen bonds are short-range interactions. When a polypeptide chain is stretched out, the parts of the chain between which hydrogen bonds may form are too far apart for interactions to occur.

4.5 Threonine and asparagine

4.6 (a) (ii) is a salt bridge–it shows an interaction between two oppositely charged amino acid side chains.

(b) (i) Hydrogen bond

(ii) Dispersion force

(iii) Permanent dipolar interaction

(iv) Ionic force

4.7 (a) Hydrophilic. The relatively electronegative $-NH_2$ group makes this molecule polar. (b) Hydrophilic. The highly electronegative C=O group makes this a polar molecule. (c) Hydrophobic. This is a non-polar molecule.

4.8 (a) (i), (iii), (ii)

Molecules of (i) are held together only by dispersion forces.

Molecules of (iii) are held together by permanent dipolar forces **and** dispersion forces.

Molecules of (ii) are held together by relatively strong hydrogen bonds.

(b) Ethanol. Ethanol can form hydrogen bonds, whereas ethane cannot.

Chapter 5

5.1 Many naturally occurring compounds can be synthesized in the lab. Examples include:

Penicillin: this was first isolated when produced naturally by bacteria. Now it is manufactured synthetically on a commercial scale.

Methane: this is produced naturally as, for example, a by-product of the digestive system of the cow, and through the decomposition of carbon deposits (e.g. fossilized bacteria). It can also be synthesized artificially.

Ethanol: this is produced naturally by the metabolism of sugars by yeast, but it can also be synthesized in the lab.

5.2 (a)  (b)

(c)  OH

5.3 Octane

5.4 Ethene has a lower boiling point (–104 °C). It experiences fewer intermolecular forces because it is a smaller molecule than pentane, contains fewer electrons, and so experiences fewer dispersion forces.

5.5 (a) –42°C; (b) –6°C

5.6 (a) Six carbon atoms; alkene

(b) Two carbon atoms; alkyne

(c) Ten carbon atoms; alkane

5.7 Alkyl group name is 'heptyl' group: $CH_3CH_2CH_2CH_2CH_2CH_2CH_2-$

Chapter 6

6.1 Butanol

6.2 Propoxypropane

6.3 (a) A ketone: hexanone (hexan-2-one)

(b) An aldehyde: pentanal

(c) A ketone: butanone

6.4 A ketone is defined as having two carbons attached to the carbonyl group. Therefore the smallest possible ketone must have **three** carbon atoms in total (the carbonyl carbon plus two others). 'Ethanone' would have only two carbon atoms, and so it couldn't be a ketone.

6.5 Hexanoic acid

6.6 We would predict ethanoic acid to be more soluble in water than ethane. Ethanoic acid can form hydrogen bonds with water molecules, which enhances its water solubility. By contrast, ethane cannot form hydrogen bonds with water molecules. Consequently, it is insoluble in water.

6.7 Propyl butanoate. Its structural formula is:

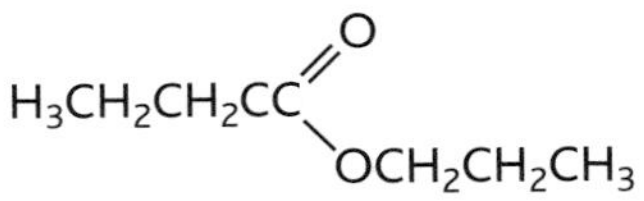

O

$H_3CH_2CH_2CC$

$OCH_2CH_2CH_3$

6.8 (a) Butanamine

(b) N-dimethyl ethanamine

(c) N-ethyl propanamine

6.9 (b), (a), (c), (d)

6.10 Butanamide

6.11 (a) Bromomethane

(b) 2-fluoropentane

(c) 1-chlorobutane

Chapter 7

7.1 Cysteine

7.2

7.3 An aldose

7.4 Glucose and fructose

7.5 (a) As we move from lauric acid to palmitic acid to arachidic acid, the molecules increase in size. Larger molecules experience greater dispersion forces than smaller molecules (they have more electrons) so more energy is needed to change the compounds from solid to liquid. Consequently, we see an increase in melting point. (b) Arachidonic acid: polyunsaturated; Linoleic acid: polyunsaturated; Oleic acid: monounsaturated; Lauric acid: saturated; Palmitic acid: saturated; Arachidic acid: saturated.

7.6 Serine

7.7 Thymine has an additional methyl group (–CH_3) attached to the ring structure.

Chapter 8

8.1 (a) 180 pm; (b) 110 pm; (c) 151 pm

8.2 (a) (i) H–H; (ii) C=O; (iii) N≡N

(b) C–O bond = 150 pm. Therefore C=O bond is longer than the C–O bond.

8.3* Linear geometry

8.4 (a) Two non-bonding pairs

(b) sp3 hybridization

8.5 The valence orbitals on a carbon atom undergo sp3 hybridization when the carbon atom reacts to form ethane. The C–C bond in ethane is formed from the overlap of two sp3 hybrid orbitals.

8.6 (a) only. The bond in (b) is part of a cyclic structure so cannot rotate. The bond in (c) sits between two double bonds, so will form part of a delocalized system, preventing bond rotation.

Chapter 9

9.1 A beta sheet.

9.2 Adenine and uracil don't possess the groups of atoms necessary to form a third hydrogen bond. By contrast, guanine and cytosine possess the correct atoms, which are arranged in the correct orientation relative to each other, for a third hydrogen bond to form.

9.3 Yes. A coiled coil may form from the association of alpha helices from within a single polypeptide chain, or from separate polypeptide chains. If the alpha helices are from the same polypeptide chain, the coiled coil would be a tertiary structure; if the alpha helices are from different polypeptide chains, it would be more accurate to consider the coiled coil a quaternary structure.

Chapter 10

10.1 False. It is a structural isomer because it comprises the same atoms as the compounds shown in Figure 10.1: 6 carbon atoms and 14 hydrogen atoms.

10.3 Four.

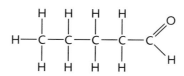

10.4 (a) Hex-2-ene; (b) Butan-2-ol; (c) Hexan-1-ol

10.5 (a) Penta-1,4-diene; (b) Hexane 2,3,5-triol

10.6 (a) 5-methylhexan-2-ol; (b) 2-methylhepta- 1, 4-diene

10.7 (a) (a) and (c); (b) and (d); (b)

10.8 One pair: (a) and (f)

10.9 (a) = *trans* isomer; (b) = *cis* isomer

10.10 *E*-isomer = (b) ; *Z*-isomer = (a)

10.11 (a), (c), (d)

10.12 (a) (b) (c) (d)

10.13 (d)

10.14 One.

10.15 The *S*-enantiomer

Chapter 11

11.1 (a) is a fatty acid with a long alkyl tail, which makes it non-polar. It is therefore most likely to partition into the organic phase. (b) is a small alcohol, in which the relatively short alkyl group doesn't offset the polarity of the –OH group in terms of water solubility. Therefore it is most likely to partition into the aqueous phase.

11.2 32 amu: ((C: 1 × 12 amu) + (H: 4 × 1 amu) + (O: 1 × 16 amu))

11.3 (b) m/z value of propane = 44.0953. m/z value of ethanal = 44.0524.

11.4 (c). The carbons in the CH_3 groups share the same chemical environment, so generate a single double-height peak; the carbons in the CH_2 groups also share the same chemical environment, so generate another double-height peak. The carbon to which the –OH group is attached is in a unique chemical environment, so generates a third peak. This is half the height of the other two peaks, because it is generated by just one carbon.

11.5 The O–H peak is at a frequency consistent with –OH groups that are participating in hydrogen bonds (3230 to 3550 cm^{-1}). This tells us that hydrogen bonds **do** occur in ethanol.

11.6 Phenylalanine. A peak at 200 nm is characteristic of an aromatic group. Out of the three amino acids listed, only phenylalanine possesses an aromatic group: a benzene ring.

Chapter 12

12.1 (a) 14 g; (b) 19 g

12.2* (a) 0.64 mol; (b) 1.1 mol; (c) 3.6 mol

12.3 (a) 44 g mol^{-1}; (b) 32 g mol^{-1}; (c) 60 g mol^{-1}

12.4* (a) 0.68 mol; (b) 0.50 mol; (c) 0.56 mol

12.5* (a) 0.025 mol (0.03 mol to 1 s.f.)
(b) 0.1 mol; (c) 0.2 mol

12.6* (a) 1 mol; (b) 2 mol; (c) 0.5 mol

12.7* $\frac{\text{mass of sample}}{\text{number of moles}} = \text{molar mass}$

12.8* (a) 90 g; (b) 11.7 g; (c) 0.373 g

12.9* (a) 2 mol L^{-1}; (b) 3 mol L^{-1}; (c) 0.4 mol L^{-1}; (d) 2.5 L

12.10* (a) 0.2 mol L^{-1}; (b) 0.13 mol L^{-1}; (c) 0.1 mol L^{-1}

12.11* (a) 0.8 L; (b) 1.57 L; (c) 0.32 L

12.12* 120 bacteria/mL

12.13 Ultraviolet light and visible light have shorter wavelengths than infrared light. Light with short wavelengths has more energy than light with long wavelengths. So spectrophotometry uses light of higher energy than infrared spectroscopy.

12.14* 1.22×10^{-4} mol L^{-1}

12.15 0.5 mol (The stoichiometry 1:2 tells us that, for every one mole of A that is consumed by the reaction, two moles of B are consumed–that is, for each mole of B consumed, half as many moles of A are consumed. If one mole of B is needed to convert all of the A into C, then half as much A must be present in the solution initially. So there must be 0.5 mol of A present initially.)

Chapter 13

13.1* 10,000 J (or 10 kJ)

13.2* 640 J

13.3 (a) Endothermic; (b) Exothermic; (c) Endothermic

13.4* (a) $\Delta H = -1018$ kJ mol^{-1}. The reaction is exothermic.
(b) $\Delta H = +1985$ kJ mol^{-1}. The reaction is endothermic.

13.5 (a) Increased. There is a net increase in the number of entities in the system.
(b) Increased. The unfolding of a protein represents an increase in disorder and, hence, an increase in entropy.
(c) Decreased. There is a net decrease in the number of entities in the system (and the system is also becoming more organized as the free amino acids are organized into a polypeptide chain).
(d) Decreased. There is a net decrease in the number of entities in the system as compounds come together to form the gallstone. Also, the solid gallstone will represent a more organized entity than the compounds from which it was formed.

13.6* +16.28 J K^{-1}

13.7* (a) 183.15 J K^{-1}; (b) 134.05 J K^{-1}

13.8* −131.2 kJ mol^{-1}

13.9 (a) Spontaneous, at the level of the thermodynamic system
(b) Non-spontaneous

Chapter 14

14.1 (a) No; (b) Yes; (c) No; (d) Yes

14.2* 9.5×10^{-4} mol L^{-1}/min

14.3 The graph is in the form of an approximately linear curve (i.e. a straight line). This tells us that the rate of the reaction remains constant at all time points. Therefore we can predict this to be a zero-order reaction.

14.4 (a) As the hydrolysis of sucrose is a zero-order reaction, we can predict that the half-life of sucrose will decrease as the concentration in the reaction mixture decreases. (b) 157.5 min. Figure 14.10(a) shows how the half-life for a zero-order reaction halves for each half-life that passes. A reduction in concentration from 8 mmol L^{-1} to 1 mmol L^{-1} represents three half-lives: 8 mmol L^{-1} to 4 mmol L^{-1}; 4 mmol L^{-1} to 2 mmol L^{-1}; and 2 mmol L^{-1} to 1 mmol L^{-1}.

If the first half-life lasts 90 min, the second half-life lasts 45 min, the third 22.5 min. The total time that has elapsed is therefore 157.5 min.

14.5* 5.53 mmol mg^{-1} min^{-1}

14.6 (a) Lineweaver–Burk plot: K_M = 3.58 µM; V_{max} = 1.14 nmol min^{-1}. Hanes plot: K_M = 3.29 µM; V_{max} = 1.12 nmol min^{-1}.
(b) A Hanes plot gives K_M = 3.1 mM and V_{max} = 0.94 µmol min^{-1} mg^{-1}

Chapter 15

15.1* (a) $K = \frac{[NH_3]^2}{[N_2][H_2]^3}$
(b) K = 664

15.2 (a) Lies to left; (b) Lies to right
(c) Lies in the middle; (d) Lies to right

15.3 The reactants ($CH_3COOH + H_2O$) predominate at equilibrium.

15.4* 0.003

15.5* Q = 0.099, so the reaction must proceed from left to right for equilibrium to be reached.

15.6 K_d for adrenaline = 500 nM; K_d for alprenolol = 3.1 nM. A large value for K_d equates to relatively weak binding, therefore the larger value of K_d here must belong to the more weakly binding ligand, which is adrenaline.

15.7 (a) Shifts to the right
(b) Remains unchanged
(c) Shifts to the left

15.8 (a) The reaction will shift to the left. (b) The reaction will shift to the right. (Remember here that an enthalpy change > 0 = an endothermic reaction; an enthalpy change < 0 = an exothermic reaction.)

15.9* (a) (i) 36.4 kJ mol^{-1}; (ii) −70.4 kJ mol^{-1};
(iii) 0.153 kJ mol^{-1}; (iv) −465 kJ mol^{-1}
(b) (i) Non-spontaneous
(ii) Spontaneous
(iii) Non-spontaneous
(iv) Spontaneous

15.10* $K = 5.22 \times 10^7$

Chapter 16

16.1 (a) Acid; (b) Base; (c) Acid and base; (d) Base

16.2 (a)

Conjugate acid–base pair 2
Base 2
Acid 2
HCl + H2O ⇌ H3O+ + Cl−
Acid 1
Base 1
Conjugate acid–base pair 1

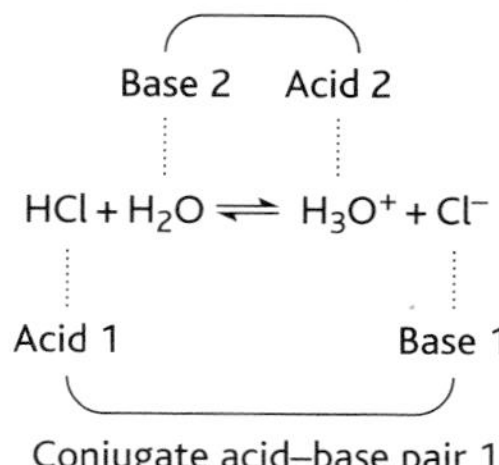

(b) Acid = H_2SO_4; base = H_2O

(c)

Conjugate acid–base pair 2
Acid 2
Base 2
CO3 2− + H2O ⇌ HCO3− + OH−
Base 1
Acid 1
Conjugate acid–base pair 1

16.3 (a) Sulfurous acid; formic acid; benzoic acid; ethanoic acid; boric acid
(b) Ethanoate ion; sulfite ion; ammonia; cyanide ion; phosphate ion

16.4* (a) 4.7; (b) 4.7

16.5* 10^{-7} M

16.6* (a) $K_a = 4.2 \times 10^{-7}$ M; (b) $K_b = 5 \times 10^{-11}$ M

16.7* (a) 4.3; (b) 6.5

16.8* (a) pH = 2.5; (b) pH = 1.55

16.9* pH = 5.75

16.10* (a) pOH = 0.3; (b) pOH = 2.5

16.11 (a) More than 50% ionized
(b) Less than 50% ionized
(c) Roughly 50% ionized

16.12* pH = 3.80

16.13* (a) 0.71:1 (b) 83 mL ammonia and 117 mL of ammonium ion.

Chapter 17

17.1 (a) Nucleophile; (b) Electrophile
(c) Electrophile; (d) Nucleophile

17.2 (a) Oxidation; b) Redox; (c) Oxidation
(d) Reduction; (e) Redox

17.3 Oxidized.

17.4 (a) Homolytically; (b) Heterolytically
(c) Heterolytically; (d) Heterolytically

Chapter 18

18.1 (a) OH^-; (b) Cl^-

18.2

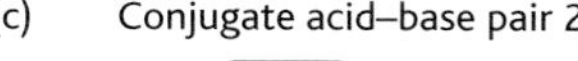

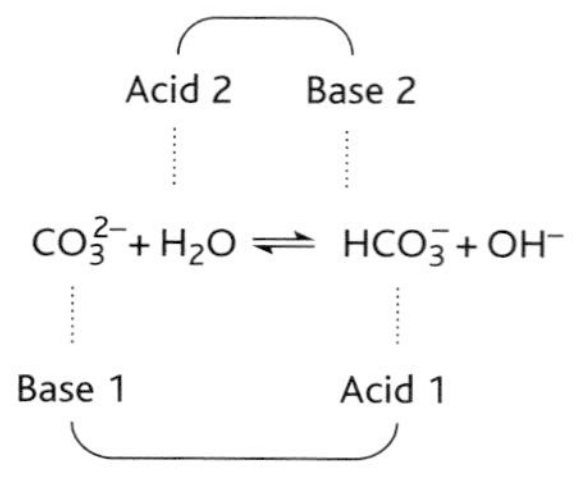

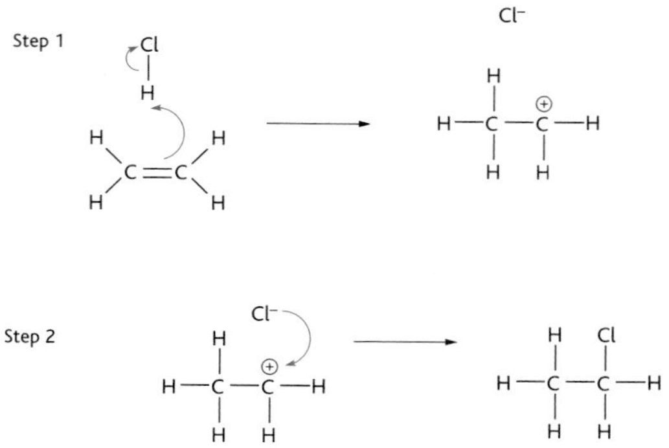

Step 1
Step 2

18.3 (a) Substitution; (b) Elimination

18.4 (a) Substitution; (b) Condensation; (c) Addition

Index

B

D

E

F

I

J

K

L

M

N

O

P

T

U

V

W

X

Z